PROCEEDINGS OF SPIE

Optical Microlithography XXIII

Mircea V. Dusa
Will Conley
Editors

23–25 February 2010
San Jose, California, United States

Sponsored by
SPIE

Cooperating Organization
SEMATECH Inc. (United States)

Published by
SPIE

Volume 7640
Part Two of Two Parts

Proceedings of SPIE, 0277-786X, v. 7640

SPIE is an international society advancing an interdisciplinary approach to the science and application of light.

Please use the following format to cite material from this book:
   Author(s), "Title of Paper," in *Optical Microlithography XXIII*, edited by Mircea V. Dusa, Will Conley, Proceedings of SPIE Vol. 7640 (SPIE, Bellingham, WA, 2010) Article CID Number.

ISSN 0277-786X
ISBN 9780819480545

Published by
SPIE
P.O. Box 10, Bellingham, Washington 98227-0010 USA
Telephone +1 360 676 3290 (Pacific Time) · Fax +1 360 647 1445
SPIE.org

Printed in the United States of America.

Publication of record for individual papers is online in the SPIE Digital Library.

SPIE
Digital Library

SPIEDigitalLibrary.org

Paper Numbering: Proceedings of SPIE follow an e-First publication model, with papers published first online and then in print and on CD-ROM. Papers are published as they are submitted and meet publication criteria. A unique, consistent, permanent citation identifier (CID) number is assigned to each article at the time of the first publication. Utilization of CIDs allows articles to be fully citable as soon they are published online, and connects the same identifier to all online, print, and electronic versions of the publication. SPIE uses a six-digit CID article numbering system in which:
  - The first four digits correspond to the SPIE volume number.
  - The last two digits indicate publication order within the volume using a Base 36 numbering system employing both numerals and letters. These two-number sets start with 00, 01, 02, 03, 04, 05, 06, 07, 08, 09, 0A, 0B ... 0Z, followed by 10-1Z, 20-2Z, etc.
The CID number appears on each page of the manuscript. The complete citation is used on the first page, and an abbreviated version on subsequent pages. Numbers in the index correspond to the last two digits of the six-digit CID number.

Contents

Part One

SESSION 1   INVITED SESSION

SESSION 2   FREEFORM AND SMO

SESSION 3   DOUBLE PATTERNING I

POSTER SESSION: DOUBLE PATTERNING

POSTER SESSION: FREEFORM AND SMO

Author Index

Towards Ultimate Optical Lithography with NXT:1950i Dual Stage Immersion Platform.

Tom Castenmiller, Frank van de Mast, Toine de Kort, Coen van de Vin, Marten de Wit, Raf Stegen, Stefan van Cleef, ASML Netherlands B.V. De Run 6501, 5504 DR Veldhoven, The Netherlands

ABSTRACT

Optical lithography, currently being used for 45-nm semiconductor devices, is expected to be extended further towards the 32-nm and 22-nm node. A further increase of lens NA will not be possible but fortunately the shrink can be enabled with new resolution enhancement methods like source mask optimization (SMO) and double patterning techniques (DPT). These new applications lower the k1 dramatically and require very tight overlay control and CD control to be successful. In addition, overall cost per wafer needs to be lowered to make the production of semiconductor devices acceptable. For this ultimate era of optical lithography we have developed the next generation dual stage NXT:1950i immersion platform. This system delivers wafer throughput of 175 wafers per hour together with an overlay of 2.5nm. Several extensions are offered enabling 200 wafers per hour and improved imaging and on product overlay.

The high productivity is achieved using a dual wafer stage with planar motor that enables a high acceleration and high scan speed. With the dual stage concept wafer metrology is performed in parallel with the wafer exposure. The free moving planar stage has reduced overhead during chuck exchange which also improves litho tool productivity.

In general, overlay contributors are coming from the lithography system, the mask and the processing. Main contributors for the scanner system are thermal wafer and stage control, lens aberration control, stage positioning and alignment. The back-bone of the NXT:1950i enhanced overlay performance is the novel short beam fixed length encoder grid-plate positioning system. By eliminating the variable length interferometer system used in the previous generation scanners the sensitivity to thermal and flow disturbances are largely reduced. The alignment accuracy and the alignment sensitivity for process layers are improved with the SMASH alignment sensor. A high number of alignment marker pairs can be used without throughput loss, and furthermore the GridMapper functionality which is using the inter-die and intra-die scanner capability can reduce overlay errors coming from mask and process without productivity impact.

In this paper we will present the main design features and discuss the system performance of the NXT:1950i system, focusing on the improvements made in overlay and productivity. We will show data on imaging, overlay, focus and productivity supporting the 3X-nm node and we will discuss next improvement steps towards the 2X-nm node.

Keywords: immersion lithography, Exposure System, Double patterning, 32-nm, 22-nm, Overlay, Throughput

1. INTRODUCTION

Resolution shrink still drives the semiconductor industry. However the technical options to do so have changed. Historical methods as increasing the lens NA and/or lowering the optical wavelength have been shown not feasible for volume production in 2011. Fortunately new resolution enhancement techniques like source mask optimization (SMO) and double patterning techniques (DPT) can bridge the gap towards EUV. These new applications enable a further lowering of the k1 and require a very tight overlay control and CD control. Nowadays several solutions are available to print line-space patterns beyond the single expose resolution limit of 193 nm hyper NA systems, e.g. Litho-Etch-Litho-Etch (LELE), Spacer technology, Litho-Freeze-Litho (LFL) and Litho-Process-Litho (LPL). [5, 6, 7]. For both the Spacer and the LELE technology the wafer has to leave the litho-cluster (track-scanner combination) in order to achieve the intended dense line pattern. This results in higher complexity of the process and lower throughput. Dual line processes (LFL and LPL) are processed completely inside the litho-cluster and therefore result in simpler processing and improved throughput, which makes these processes attractive for our customer since the costs of a simpler process will be lower. [8]

Optical Microlithography XXIII, edited by Mircea V. Dusa, Will Conley, Proc. of SPIE Vol. 7640,
76401N · © 2010 SPIE · CCC code: 0277-786X/10/$18 · doi: 10.1117/12.847025

All these processes have in common that they have strict requirements on key performance parameters such as on-product overlay and critical dimension uniformity as is shown in the Figure 1. In addition, overall cost per wafer needs to be lowered to make the production of semiconductor devices acceptable. For this ultimate era of optical lithography we have developed the next generation dual stage NXT:1950i immersion platform. This system delivers 175 wafers per hour throughput and layer to layer overlay error of 2.5 nm.

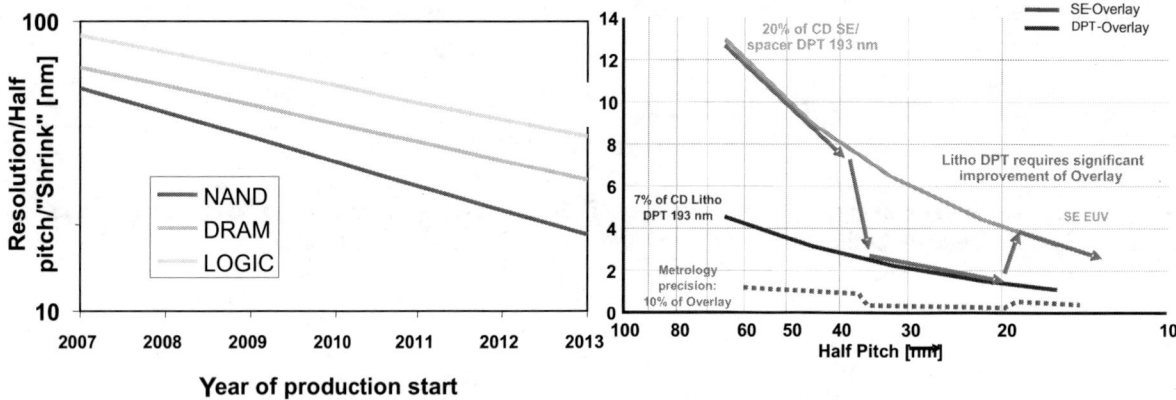

Figure 1 The resolution shrink roadmap (left) drives the on product overlay roadmap down (right).

On-product overlay can be separated into tool-contributions and process contributions. To deal with future overlay requirements both tool and process performance need to be addressed. In the new system presented in this paper ASML has improved tool overlay of 2.5 nm with a capability down to 1.5 nm. This is achieved using a novel stage position measurement concept based on short fixed beam-length encoder grid-plate system. This concept is nearly insensitive to refractive index changes of air, eliminating one of the biggest errors sources of a conventional long variable beam length interferometers measurement system.

Process overlay improves through the use of multiple align markers which allows local process-induced wafer deformation to be measured. A zone-alignment scheme divides the wafer in multiple zones, of which each can be described by a set of alignment parameters. This enables compensation for coping with local wafer deformation. There is no throughput loss due to the introduction of SMASH XY & GridAlign: a faster alignment scheme which uses a new (bi-directional) mark-type combined with faster alignment speeds. The left picture of Figure 2 shows with SMASH XY & GridAlign[TM], one can align many more marks, thus improving on product overlay, without throughput loss. Also for high dose applications an extension can be made on the NXT:1950i. The right picture of Figure 2 the PEP high dose extension ensures maximum throughput up to 50mJ dose usage.

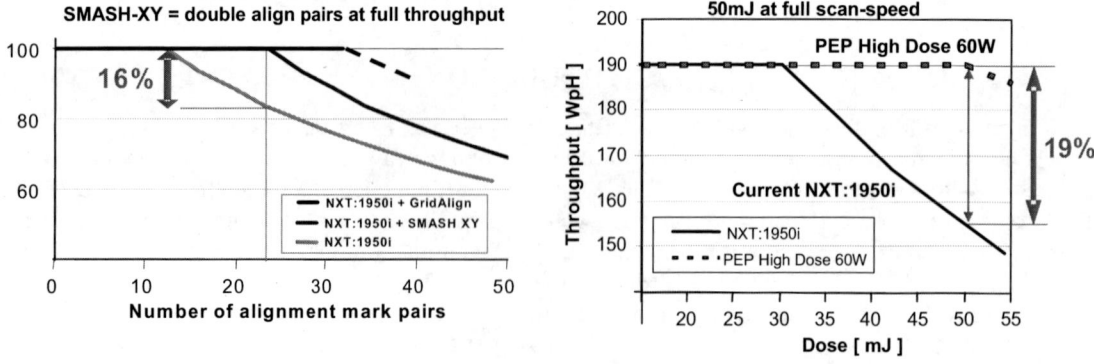

Figure 2 Left: SMASH XY & GridAlign[TM] enables zone alignment without throughput loss to improve on-product overlay; Right: the PEP high dose option 60W maintains productivity at higher doses.

The platform on which the system presented in this paper is based, delivers 175 wafers per hour and can be extended to a throughput of more than 200 wafers per hour. This is achieved by an increase of speed and acceleration of the expose chuck as well as a reduction of non-exposure time. This constitutes a performance increase of 30%, enabling a productivity of well over 4000 wafers per day for volume manufacturing.

Summarizing, we present a novel immersion tool, incorporating extremely tight overlay and very high throughput, which is highly suitable for advanced single-patterning techniques on a resolution of 38 nm. It also enables an economic solution for double-patterning on 32 and 22 nm imaging node. In the remainder of this paper we will discuss the technical details and show results.

2. TOOL OVERLAY IMPROVEMENTS

The overlay of the tool is determined by the performance of several contributors. Stage Positioning is the accuracy of how well the wafer & reticle stages can be positioned with respect to each other. Besides servo control of the stages, also the accuracy of the measurement system plays an important role. The optical column performance is determined by the design and control of the aberration level of the lens. The performance of the reticle and wafer measurement is determined by the sensor contributions and, again the accuracy of the measurement system. The clamping of wafer and reticle contributes marginally; reticle heating is the main part of this error post. Finally the stability / reproducibility of thermal wafer & reticle deformations complete the budget. In Figure 3 the tool overlay budget breakdown for dedicated chuck usage is shown for the NXT (right) and his predecessor the XT:1900Gi (left). In the white blocks the specific improvement areas are denoted.

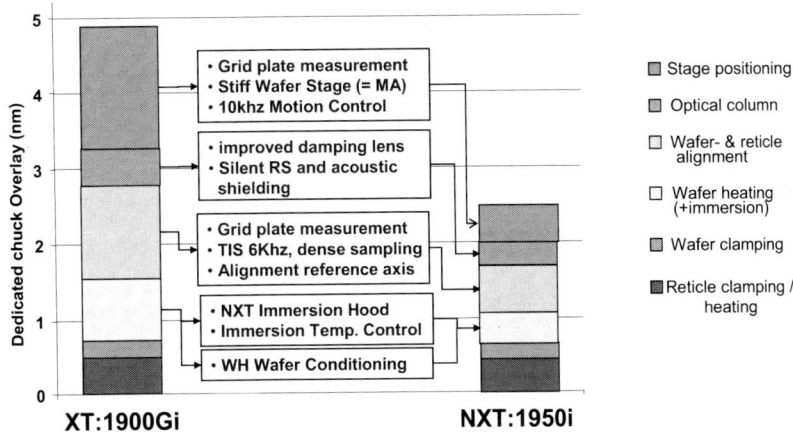

Figure 3 NXT:1950i Overlay Budget (Dedicated Chuck), NXT platform improvements activities relative to the XT:1900Gi

Parts of these improvements are described in earlier publications and also applied in the XT:1950Hi. Details on the Airdrag Immersion technology can be found in [1]. For the improved damping of the lens and the silent reticle stage and acoustic shielding see [1][4]. In this paper we will focus on the specific NXT improvements: the novel stage position measurement system and the novel light weight wafer-stage enabling high throughput.

3. DESIGN FOR ACCURACY & STABILITY

The new NXT platform introduces a novel measurement positioning system. The drivers for the design choices were short term accuracy & long term stability.

3.1 Design for accuracy

The short term accuracy is mainly improved by the choice of a short fixed beam-length encoder system in stead of a long variable beam-length interferometer system. Figure 4 shows schematically the differences between these 2 systems. Interferometers have been used in the industry as a stage-position measurement system for more than 20 years. The basic principle of an interferometer is the measurement of an optical path difference of a beam reflected on a stage mirror with an internally reflected beam. These two beams give an interference pattern of which the fringes can be counted. From this the change in length of the external beam can be calculated in terms of parts of the wavelength in air. To transfer this into a position measurement one needs to know the wavelength of the light in the air through which the beam has traveled. Uncompensated changes in the refractive index of air in the beam lead to measurement noise and errors, especially in the region 0.5-50 Hz. Even in very well conditioned systems these can be in the order of 1 nm and higher. Large sources of these disturbances are the movements of the stage itself: this causes the beam to be disturbed with local pressure- and temperature-changes as well as changes in air-composition. These change the refractive index and thus lead to positioning errors.

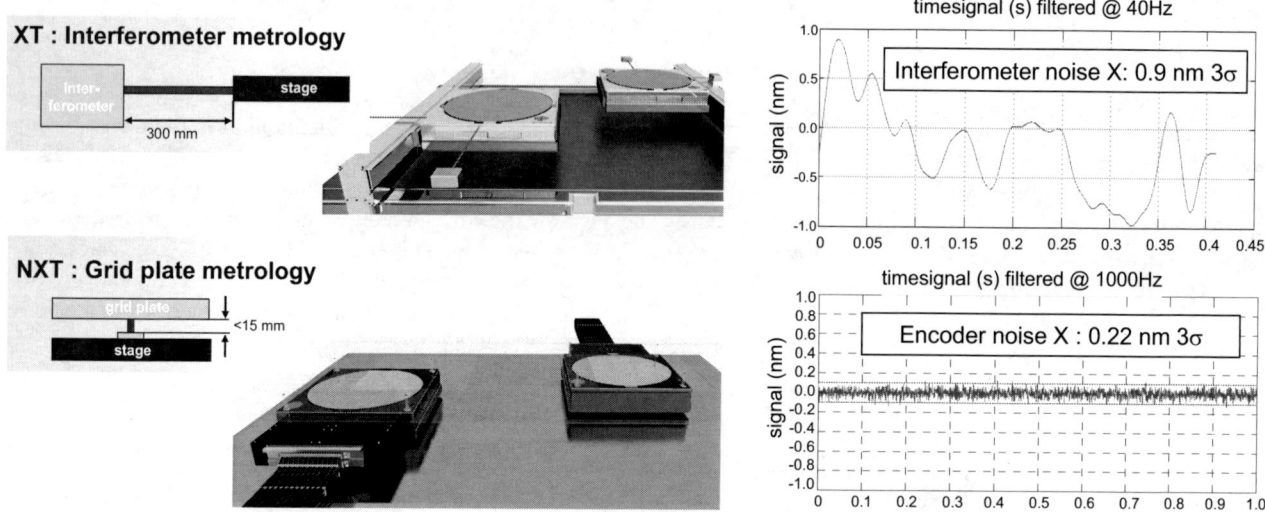

Figure 4   Schematic views of a conventional interferometer system with long variable beams (top) and the new encoder system with short fixed beam interferometers and grid-plates (bottom). On the left the noise levels of both systems is plotted.

A dramatically reduced sensitivity to refractive index changes can be achieved by using encoders. These are able to have short, vertical, fixed beam-path lengths (as short as 15 mm) while executing a horizontal travel range in the operational plane of hundreds of millimeters. The type used can be considered as short fixed beam interferometers, signal generation and interpolation is done on basis of interference of two combined light-beams. These are very close to each other so that both beams will travel through air with very similar refractive index. This reduces the sensitivity to refractive index variations even further. In the design the encoder sensing part is connected to the chuck and measures relative to a grating, which is supported by a grid-frame mounted on a vibration isolated metrology frame. The measured noise levels of both systems are shown on the right part of Figure 4.

3.2 Design for Stability

Essential in the design of the NXT is the choice to put the encoder grid-frame on the vibration isolated metrology frame, the so called quiet stable world, see Figure 5. This metrology frame provides a very stable support; it also holds the exposure lens and measuring equipment like alignment- and focus-sensor at the measure area. The grid-frame consists of a stiff 'zero'-expansion material carrier and four mounted grid-plates together behaving as one monolithic body. The grid-plate itself is kinematic mounted onto the metrology frame. Although the sensitivity for thermal effects is small even aerodynamic precautions are taken to control the air around the grid-plate to the level of 10 mK. The result is a very stable grid on sub-nm level with thermal time-constants of several hours which only influence lower order deformations. These lower order deformations are for each wafer corrected by the stage & wafer align and an additional inline encoder consistency calibration.

The encoder heads are mounted on the corners of the chuck and measure in two degrees of freedom: a translation in the plane of the grating and the distance between sensor and grid-frame grating.. The particular measurement orientation chosen, tangential with respect to the center of the chuck, makes the measurement insensitive for the lowest order of deformation of the chuck which is homogeneous expansion. Any other encoder position change is corrected by stage & wafer align and a new inline encoder consistency calibration. In this way the long term stability is secured in the stable quite world and the encoder stability on the wafer stage has only to be secured over one wafer cycle which does not last for more than typically 20 seconds. Below we further explain how encoder position drift is corrected by the alignment measurements.

Figure 5 Left the quiet stable Metroframe world versus the dynamic waferstage world. Right two cinematically mounted low expansions Grid-frames with Gridplates and encoder heads mounted under 45° on the chuck to secure a stable grid on sub-nm level.

Measuring directions @ 45° with XY perpendicular to homogeneous chuck expansion

Kinematic mounted

"zero" expansion material
0.8*0.8 m

Exp. Coeff ~ 10^{-8} m/K
→ 0.16 nm

Gridplates behave like monolithic.

Waferstage compartment temperature controlled air

Thermal time constant:
Multiple hours

Given this system layout any encoder position drift can be described in 4 orthogonal vectors: Translation y (Ty);

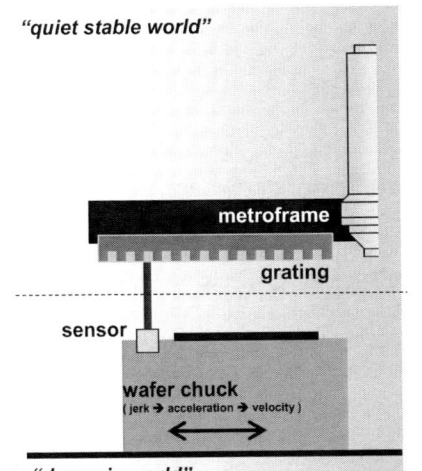

"quiet stable world"

metroframe

grating

sensor

wafer chuck
(jerk → acceleration → velocity)

"dynamic world"

Translation x (Tx); Symmetrical wafer Rotation (RwS) and Encoder Measurement

$[+ - - +]$ Ty

$[+ + + +]$ RwS

$[- - + +]$ Tx

$[+ - + -]$ 'EMSI'

Sensor Integrity (EMSI). Figure 6 explains these vectors.

Figure 6 Orthogonal drifts can create intrinsic positioning errors in an encoder system. Each arbitrary change in the 4 encoder positions is a linear super-position of above 4 "orthogonal drifts".

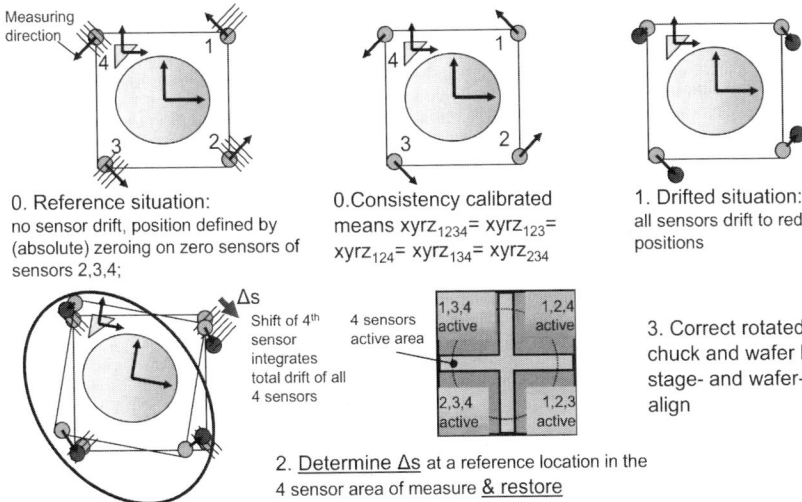

Measuring direction

0. Reference situation:
no sensor drift, position defined by (absolute) zeroing on zero sensors of sensors 2,3,4;

0. Consistency calibrated means $xyrz_{1234} = xyrz_{123} = xyrz_{124} = xyrz_{134} = xyrz_{234}$

1. Drifted situation:
all sensors drift to red positions

3. Correct rotated chuck and wafer by stage- and wafer-align

Δs
Shift of 4th sensor integrates total drift of all 4 sensors

1,3,4 active   1,2,4 active
2,3,4 active   1,2,3 active
4 sensors active area

=> error in consistency and rotated chuck and wafer

2. Determine Δs at a reference location in the 4 sensor area of measure & restore consistency by correcting position of 4th sensor (1) by Δs

The EMSI measurement makes uses of overlapping areas in the grid plate layout. In those areas all encoders signals are valid. Together with the stage and wafer-align measurements the inconsistency denoted as Δs in the figure below can be calculated with respect to the other three sensors. Figure 7 describes the EMSI sequence in detail.

Figure 7 EMSI: Encoder Measurement Sensor Inconsistency calibration sequence. All sensor drifts are corrected by the consistency correction and stage & wafer alignment.

4. TOOL THROUGHPUT IMPROVEMENTS

Higher throughput is achieved by speeding up the exposures and minimizing the non-exposure time. The graph below shows the breakdown for the throughput improvements with respect to the XT:1900Gi.

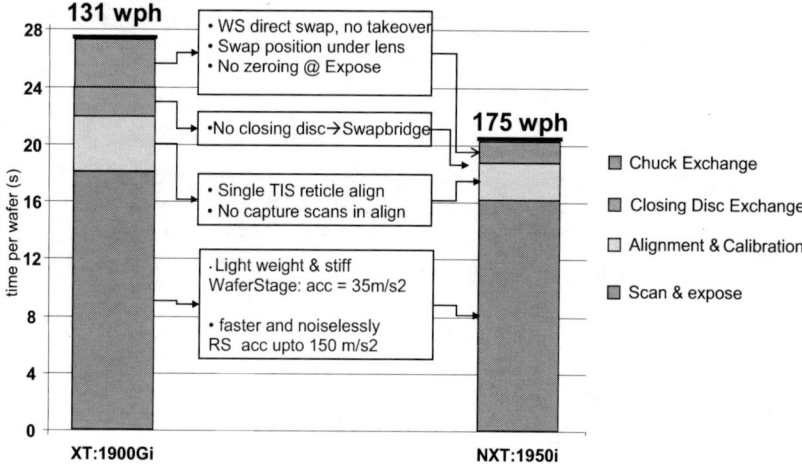

Figure 8 NXT:1950i Throughput breakdown , NXT improvement activities relative to an actual XT:1900Gi (16x32mm, 125 exposures).

The non-exposure time consists of three parts: chuck exchange; closing disc exchange and alignment & calibration. In NXT, chuck exchange has been significantly reduced by a much faster stage swap enabled by a new stage concept allowing a direct swap (so no chuck takeover) immediately after the last exposure right under the lens. Because in the new grid-plate measurement system the chuck remains in encoder control during the swap from measure to expose, the zeroing of the chuck @ expose position has been eliminated from the sequence. Another skipped sequence part is the closing disc. The NXT uses a swap-bridge which is docked during a swap between the two stages and enables a cross-over without taken extra time consuming actions for the immersion hood. The alignment time is shortened due to function absolute zeroing. Absolute zeroing takes place during the wafer-load sequence and secures that the stage is positioned in 6 degrees of freedom (x,y,z, rx, ry & rz) under encoder control on sub-nm accuracy. Due to this accuracy a second reticle align on another stage position is not needed and the remaining reticle align scan be shortened because capture scans can be eliminated. The exposure time is decreased by applying higher acceleration and speed. This is made possible by reducing the total moving mass in the stage design, up to almost one-third of the original moving mass. With equal force this increases the possible accelerations by the same ratio.

The combination of these efforts results in a platform with an introduction throughput oif 175 wafers per hour and a capability more than 200 wafers per hour when applying the PEP-200 package, see chapter extensions

5. A NEXT WAFERSTAGE GENERATION

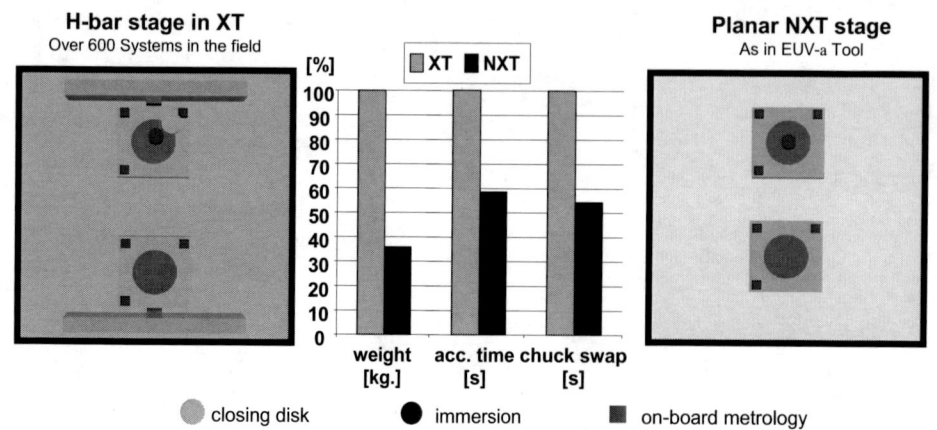

The novel wafer-stage is based upon a planar stage design with magnetic levitation. Figure 9 shows schematically the difference between the XT stage concept and the NXT stage concept.

Figure 9 The planar stage using magnetic levitation (right) has a weight of almost 1/3 of the conventional H-bar design (left).

Where possible high stiffness/low weight materials is used to reduce the weight of the chuck. No heavy H-bars are needed to drive the rough positioning of the stage and last but not least due to the use of encoder-heads the bulky stiff mirrors are removed from the chuck. The support frame for the encoder heights is significant smaller. The result is an almost 3x reduced weight while even improving on the dynamics by creating a stiffer wafer stage in xy.

In the graph below the open loop dynamics of the XT wafer-stage and the NXT wafer-stage are compared as measured on real chucks. As can be seen is that the NXT has a 40% higher first resonance frequency and also parasitic forces shift in this amount to higher frequencies. This enables a higher Bandwidth servo control which is the main driver for servo positioning improvements.

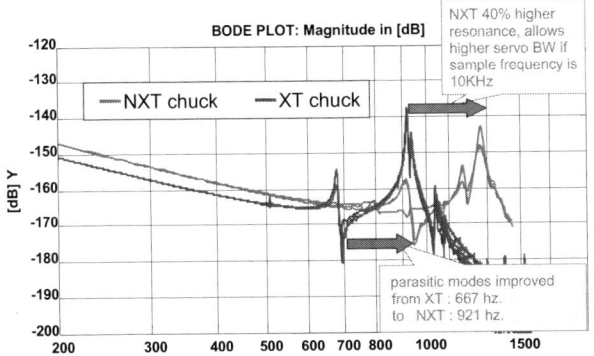 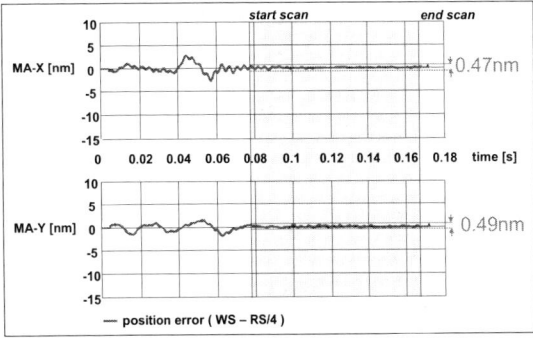

Figure 10 On the left the open loop mechanics showing a much stiffer NXT due to the use of new materials and design. This results in higher bandwidth control and consequently improved servo control which is demonstrated on the right were time traces of the x and y |RS/4 – WS| MA servo errors are plotted @ 175 wafers per hour throughput settings.

6. DATA

In this chapter an overview of NXT:1950i system performance data is presented on the area's Overlay, Stability, Imaging, Defectivity and Productivity.

6.1 Overlay

Figure 11 shows the dedicated machine, dedicated chuck and the matched overlay performance. The experiment and reporting was done conform the standard ASML ATP protocol using a scan speed of 0.6m/s. The dedicated chuck overlay results contain about 0.5 nm alignment contribution. The other part is determined by the stage positioning accuracy, thermal control, clamping residues and optics. This system performance result confirms the subsystem accuracy data on the grid plate measurement system as shown in Figure 4. The matched overlay result contains the differences in positioning grid and lens distortion between the XT:1900Gi and the NXT:1950i. Taking out the lens fingerprint shows that the gridplate contribution after calibration is in the order of 2 nm.

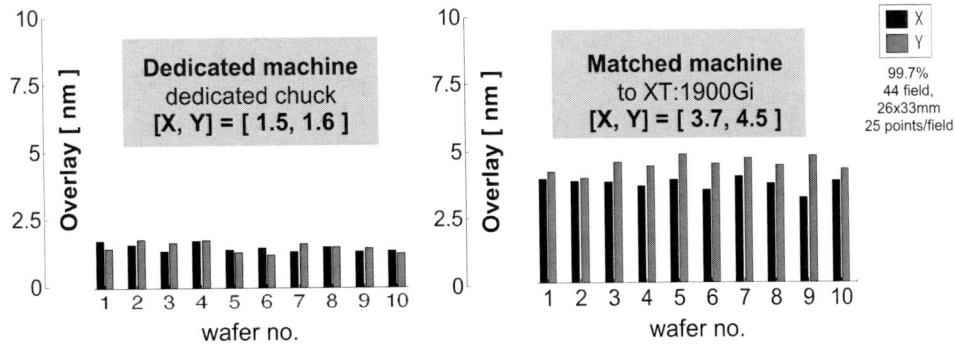

Figure 11 Dedicated Chuck Overlay and Matched Overlay to a XT:1900Gi measured on a NXT:1950i.

In Figure 12 the overlay results of a full coverage wafer is almost equal to the inner-fields wafer layout as shown in Figure 11.

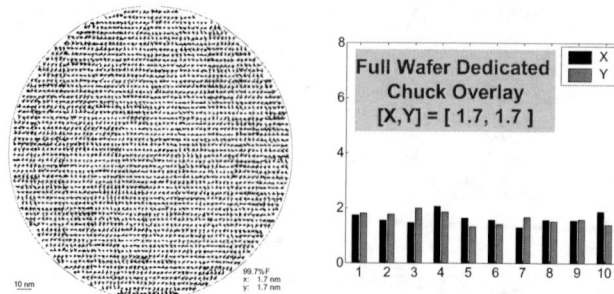

Figure 12 Full coverage wafer dedicated chuck overlay including the edges.

6.2 Overlay and Focus stability

In the left side of Figure 13 The matched machine overlay over 6 consecutive days is plotted. The layout used is the standard ASML overlay layout and contains 44 full field exposures. The reference wafers used are made on a XT:1450. During the 6-day experiment, no intermediate calibrations were performed. The performance remains stable on a level of 5.0 nm - 5.6 nm. In the right side of Figure 13 the focus stability over 7 days is printed.

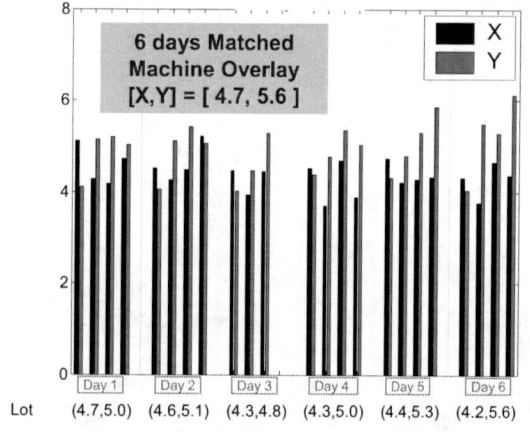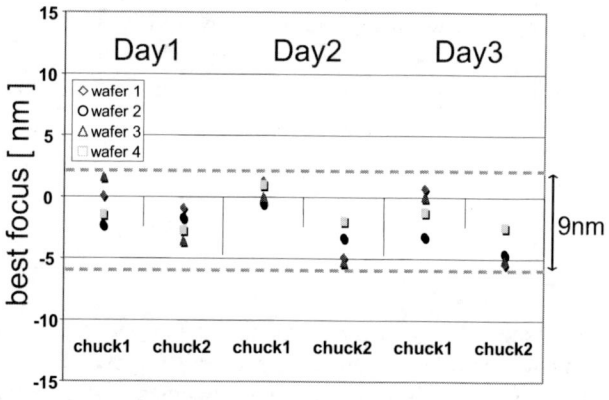

Figure 13 left: 6 days of Matched machine Overlay to a XT:1450 and right: 3 days of Focus Stability data measured on a NXT1950i using a scan speed of 0.6m/s.

6.3 Focus uniformity data

Focus performance measured with the Leveling Verification method (LVT) is printed below in Figure 14.

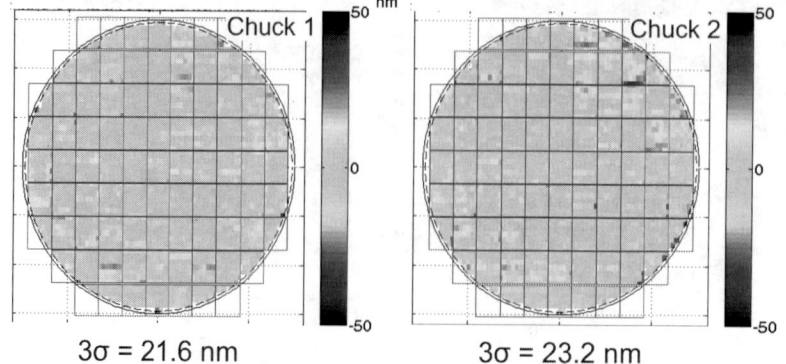

Figure 14 Focus uniformity data from both chucks using a scan speed of 0.6m/s.

6.4 Imaging data

40nm lines critical dimension uniformity (CDU) data of isolated and dense patterns are on the level on 1.0nm for dense patterns and 1.3nm for isolated patterns as is shown in Figure 15

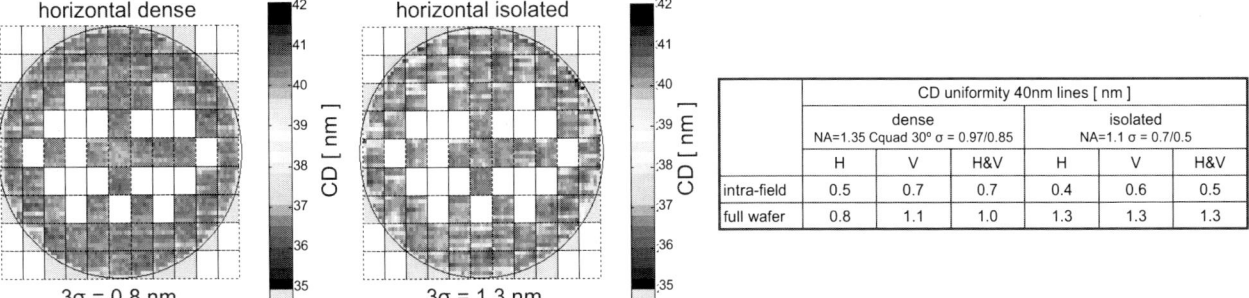

	CD uniformity 40nm lines [nm]					
	dense NA=1.35 Cquad 30° σ = 0.97/0.85			isolated NA=1.1 σ = 0.7/0.5		
	H	V	H&V	H	V	H&V
intra-field	0.5	0.7	0.7	0.4	0.6	0.5
full wafer	0.8	1.1	1.0	1.3	1.3	1.3

Figure 15 Isolated line and dense line Imaging CDU data measured on a NXT1950i with a scanspeed of 0.6m/s. The CDU is measured with a scatterometer (Yieldstar)

Double expose

Two relevant imaging applications for the 32nm imaging node are double expose imaging and double patterning imaging using a freeze process (LFL). A very important parameter to meet required CD uniformity levels is the overlay between two images within such a layer. In Figure 16 the overlay results for a typical double exposure application is shown.

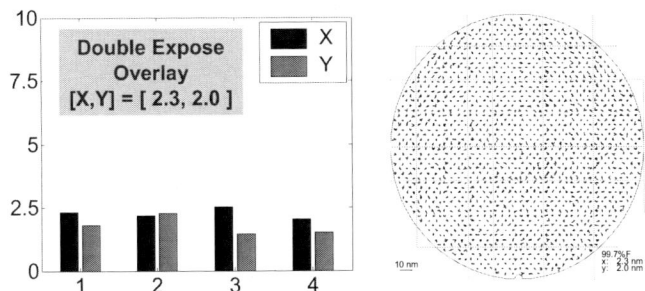

Figure 16 Double Expose Overlay data using an ABAB wafer sequence. 1st layer: Annular, 1.35 NA, σo-σi 0.94-0.79, XY pol and 2nd layer: CQuad30, 1.35 NA, σo-σi 0.94-0.79, XY pol. In between the exposures the reticle is realigned. The overlay is measured on a full wafer coverage.

Litho Freeze Litho

The second relevant application is a Litho Freeze Litho (LFL) application using a thermal cure process as described in [2]. The CDU of the spaces is determined by the overlay between the 2 exposures and the CDU of the lines following equation 1[10].

$$Space\_CDU = \sqrt{\left(\frac{CDU_{L1}}{2}\right)^2 + \left(\frac{CDU_{L2}}{2}\right)^2 + \left(OVL_{3\sigma}\right)^2 + \left(3*OVL_{mean}\right)^2} \qquad \text{Equation 1}$$

The experiment uses as setting: CQuad30 NA: 1.15 Sigma: 0.75/0.55, XY-pol and the exposures are optimized with DOSEMAPPER. The structures are measured on a SEM. The overlay on resolution is calculated using the position measurement of S1 and S2 according the equation 2:

$$Overlay\_on\_resolution = \frac{(S1 - S2)}{2} \qquad \text{Equation 2}$$

CDU's of the first exposure L1 are around 1.2-1.4nm, the second exposure L2 is around 1.9nm-2.2nm. The spaces CDU result is due to the on resolution overlay around the 3.0nm.

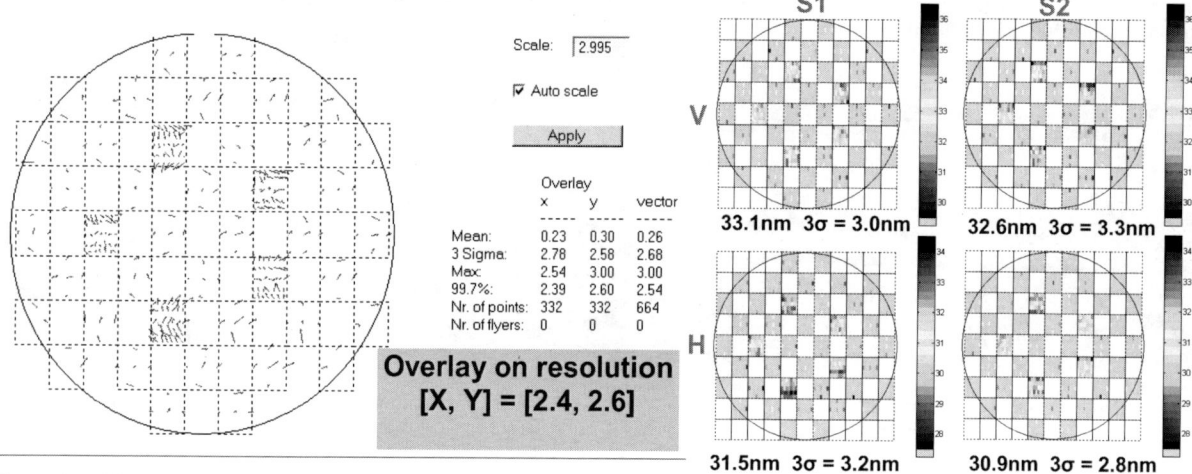

Figure 17 Litho Freeze Litho overlay on resolution results and spaces CDU using a process as described in [2].

6.5 Defectivity

In Figure 18 defectivity results are plotted for 2 different types of resist. A topcat (low contact angle) resist TCX041 and a topcoatless (high contact angle) resist AIM5484. The scan-speed used in this experiment is 0.61 m/s. Clearly the topcoatless resist outperforms the topcoat resist showing almost no defects. On 0.7 m/sec the performance degrades somwwahat but is well within acceptable limits.

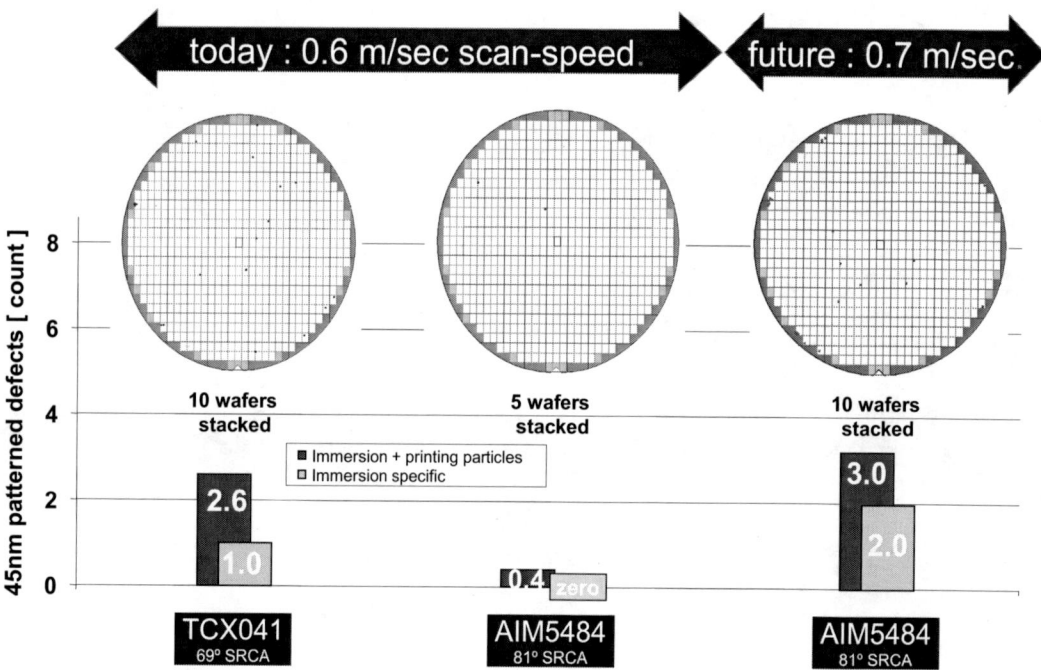

Figure 18 Defectivity results of the NXT:1950i using 0.61m/s scan speed on a low contact angle resist TCX041 and a high contact angle resist AIM5484.

6.6 Productivity

The 175wafer per hour productivity of NXT:1950i enables 4000+ wafer per day scenario's. A 6 hrs productivity experiment was performed to demonstrate this capability. The job used contained full coverage, full field size 26mmx33mm, 96 shots , 30 mJ shows the result.

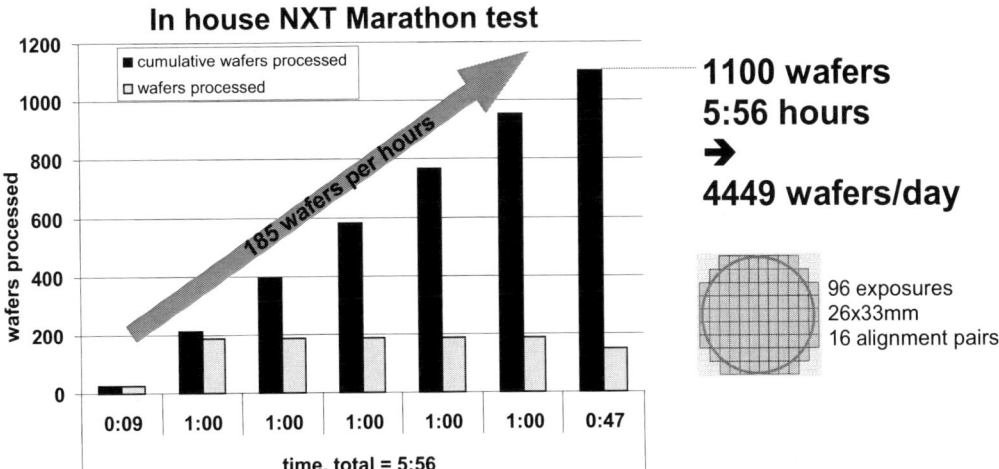

Figure 19 A 6hr marathon test on the NXT body demonstrated 185 wafer per hour throughput and 1100 wafers which is equivalent to 4449 wafers per day peak productivity.

System Extensions

Several extensions are already discussed in the beginning of this paper: SMASH-XY and GridAlign to improve process overlay; the PEP200 package to bring the system to 200wph and the High Dose extension to enable full throughput up to 45mJ dose usage. Two relevant extensions are to be added to this list: Programmable Illumination FlexRay and programmable wavefront FlexWave.

FlexRay is meant to improve the process windows on high-NA immersion systems by offering a fully programmable illumination system for generation of freeform sources [9].

Flexwave makes it possible to make corrections on the wave-front in the pupil of the projection lens on local scale. This can be applied for improving wave-front stability due to for instance lensheating effects or as a process window improvement by applying chip imaging structure specific corrections. For more details see [1][11].

programmable illumination

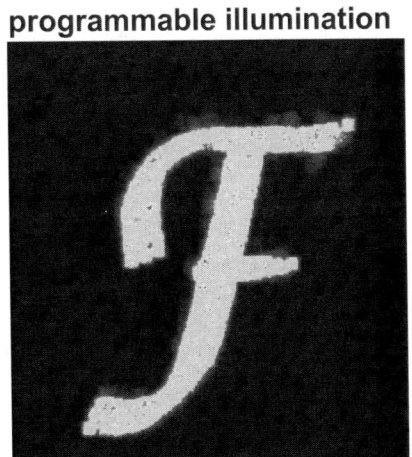

intensity in the pupil plane of the illuminator
measured at the center of the exposure slit

programmable wave front

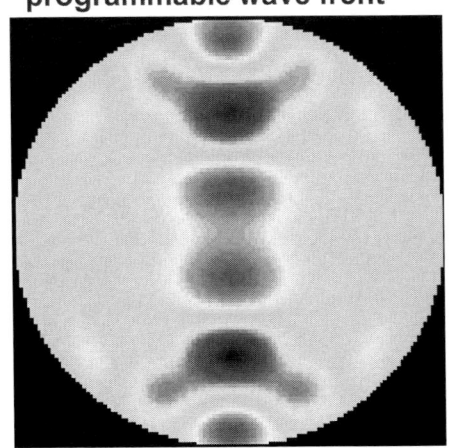

wave front in the pupil plane of the projection lens
measured at the center of the exposure slit

Figure 20 A FlexRay generated pupil and a Flexwave generated wave front as measured with the Ilias sensor demonstrating the programmable functionality of both imaging enhancements extensions.

7. CONCLUSIONS

We presented a lithography tool with a new stage design combined with a positioning system based on encoders. This system boost the productivity and improves the overlay enabling new lithographic processes which are needed for further resolution shrink which cannot be provided otherwise by higher NA or smaller wavelengths in the near future. Results on all key parameters are presented and show a complete system raising the overlay and productivity standards to new levels. Overlay capability is demonstrated to the level of 1.5nm in a dedicated chuck usage application. Matched overlay can be as good as 4.5nm. Productivity data showing the capability of more than 4000 wafers per day is shown. Together with several Overlay, Imaging and Productivity extensions which are coming available, the performance levels of the NXT:1950i provides an economically viable solution for future nodes on the lithographic roadmap.

ACKNOWLEDGEMENT

The people contributing to this paper are too numerous to mention individually. Still, the authors do want to express their gratitude to the integration and performance teams who optimized the NXT:1950i in a short time and provided the necessary material to generate this paper.

REFERENCES

[1] André Engelen, Igor Bouchoms, Andre Engelen, Jan Mulkens, Herman Boom, Richard Moerman, Paul Liebregts, Roelof de Graaf, Marieke van Veen, Patrick Thomassen, Wolfgang Emera and Frank Sperling, *Extending Single-Exposure Patterning Towards 38-nm Half-Pitch using 1.35 NA Immersion*, SPIE 2009 proceedings 7274_56.

[2] Chiew-seng Koay, Steven Holmes, Karen Petrillo, Kuang-Jung Chen, Matthew Colburn (IBM), Youri von Dommelen, Michael Crouse, Aiqin Jiang, Michael Many, Robert Routh (ASML Albany), Jason Cantoneb, David Hetzer, Kenichi Uedab, Andrew Metzc ,Shannon Dunn (TEL), Cherry Tang, Mark Slezak (JSR), Michael Reilly, Vaishali Vohra (DOW) Sumanth Kini, Tony Dibiase (KLA), *Evaluation of Double Exposure Processes for Advanced Logic Nodes*. To be published SPIE 2010 proceedings

[3] Fred de Jong, Bert van der Pasch, Tom Castenmiller, Bert Vleeming, Richard Droste, Frank van de Mast, *Enabling the lithography roadmap: an immersion tool based on a Novel Stage Positioning System*, SPIE 2009 proceedings 7274-64.

[4] Igor Bouchoms, Jan Mulkens, Sander de Putter, Pieter Gunter, Roelof de Graaf, Marcel Beems, Erwin Verdurmen, Hans Jasper, Frank Bornebroek (ASML), Nils Dieckmann (CARL ZEISS), *Advanced imaging with 1.35 NA immersion systems for volume production*, to be published SPIE 2010 proceedings

[5] Jo Finders, Mircea Dusa, Bert Vleeming, Henry Megens, Birgitt Hepp, Mireille Maenhoudt, Shaunee Cheng, and Tom Vandeweyer, "*Double patterning for 32nm and below: an update*", Proc. SPIE 6924, 692408 (2008), DOI:10.1117/12.772780

[6] H.Tanaka, K. Hoshiko, T. Shimokawa, H.F. Hoefnagels, D.E Keller, S. Wang, O. Tanriseven, R. Maas and J. Mallmann, K. Shigemori, C. Rosslee, *Simplified – "Litho-Cluster-Only" – Solution for Double Patterning*, Proc. SPIE 7639-81, to be published (2010).

[7] Mircea Dusa, Bill Arnold, Jo Finders, Hans Meiling, Koen van Ingen Schenau, and Alek C. Chen, "*The lithography technology for the 32 nm HP and beyond*", Proc. SPIE 7028, 702810 (2008), DOI:10.1117/12.796016

[8] Goji Wakamatsu, Yusuke Anno, Masafumi Hori, Tomohiro Kakizawa, Michihiro Mita, Kenji Hoshiko, Takeo Shioya, Koichi Fujiwara, Shiro Kusumoto, Yoshikazu Yamaguchi, and Tsutomu Shimokawa, "*Double patterning process with freezing technique*", Proc. SPIE 7273, 72730B (2009), DOI:10.1117/12.814073

[9] Melchior Mulder et all, *Performance of FlexRay:a fully programmable illumination system for generation of freeform sources on high-NA immersion systems*, SPIE proceedings 7640-59, 2010, to be published.

[10] Gerald Dicker: GID Double Patterning and Double Exposure CDU Models, internal ASML publication.

[11] Jo Finders et all, Litho & patterning challenges for memory and logic applications at the 22-nm node, SPIE 2010 proceedings 7640-11.

Latest performance of immersion scanner S620D with the *Streamlign* platform for the double patterning generation

Hirotaka Kohno, Yuichi Shibazaki, Jun Ishikawa, Junichi Kosugi
Yasuhiro Iriuchijima and Masato Hamatani
Nikon Corporation, Miizugahara, 201-9, Kumagaya, Saitama 360-8559 Japan

ABSTRACT

Currently, it is considered that one of the most favorable options for the 32 nm HP node is pitch-splitting double patterning, which requires the lithography tool to achieve high productivity and high overlay accuracy simultaneously. In the previous work [1], we described the concepts and the technical features of Nikon's immersion scanner based on our newly developed platform, *Streamlign*, designed for 2nm overlay, 200wph throughput, and short setup time. In this paper, we present the latest actual performance of S620D with the *Streamlign* platform.

Owing to the high repeatability of our new encoder metrology system, Bird's Eye Control, and Stream Alignment, S620D achieves less than 2 nm overlay accuracy, less than 15nm focus accuracy, and successful 32 and 22 nm L/S pitch-splitting double patterning exposures. Furthermore, the results at high scanning speed up to 700 mm/s are fine and we have successfully demonstrated over 4,000 wpd throughput, which confirms the potential for high productivity. Nikon has developed this *Streamlign* as an optimized long life platform based on the upgradable Modular^2 structure for upcoming generations. The performance of S620D indicates the possibility of immersion extension down through the 22 nm HP node and beyond.

Keywords: exposure tool, scanner, double patterning, overlay, throughput, modular structure, encoder, fluctuation

1. INTRODUCTION

The technology roadmap suggests continuous device shrinkage in the coming generations [2] as shown in Figure 1. It is widely recognized that the 32nm HP node will be the age of double patterning [3]. One of the most favorable options is pitch-splitting double patterning, as shown in the inset of Figure 1. Pitch-splitting double patterning requires double process steps and severe overlay accuracy because the overlay errors cause space CD errors in the final pattern. So the mission of the lithography tool supplier is to achieve high productivity and high accuracy simultaneously. Moreover, in the case that EUV will be delayed due to the remaining technical obstacles, it should be a challenge to extend the double patterning by immersion down through the 22 nm HP node and beyond.

To meet these requirements, Nikon has developed NSR-S620D based on what we call the *Streamlign* platform [1]. Figure 2 represents an overview of S620D, which is equipped with a newly developed projection lens of 1.35 NA. This platform has three main features: First, Bird's Eye Control for 2 nm overlay; second, Stream Alignment for 200 wph throughput; and third, Modular^2 (Modular Squared) structure for short setup time. The target productivity is 4,000 wafer outs per day. In the present paper, the latest performance of S620D with high speed scanning up to 700 mm/s will be described.

Optical Microlithography XXIII, edited by Mircea V. Dusa, Will Conley, Proc. of SPIE Vol. 7640,
76401O · © 2010 SPIE · CCC code: 0277-786X/10/$18 · doi: 10.1117/12.846485

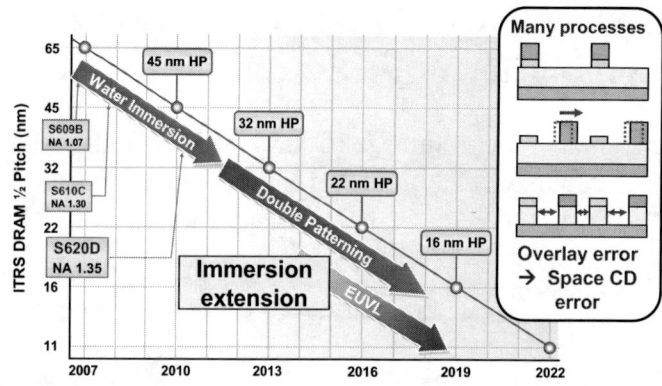

Figure 1. ITRS technology and lithography roadmap. If EUV will be delayed by the remaining obstacles, it should be a challenge to extend double patterning by immersion down through the 22 nm HP node and beyond. Inset: Schematic explanation of overlay influence on the space CD uniformity in pitch-splitting double patterning.

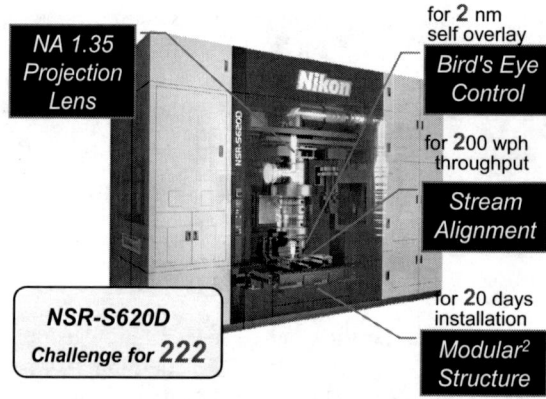

Figure 2. Overview of NSR-S620D with 1.35 NA projection lens and three main features of the *Streamlign* Platform.

2. CONCEPT

Let us begin with an explanation of the main concepts for S620D. The first innovation is a new encoder metrology system called Bird's Eye Control. The left side of Figure 3 shows a schematic configuration of Bird's Eye Control. The encoders measure a grating scale on the top of the wafer stage from above, like a bird looking for its prey, which leads to negligible Abbe errors. The working distance is only 2mm, thus air fluctuations are also negligibly small. It should be noted that these encoders are used together with interferometers. The right side of Figure 3 is a table which compares the merits and demerits of interferometer servo, encoder servo and our new hybrid system, Bird's Eye Control. The remarkable merit in the hybrid system is mutual compensation. In principal, the linearity of the encoder grating cannot be guaranteed, however, the interferometers, which have perfect linearity, can be used to calibrate the grating scale. Moreover, the hybrid system provides stability. When the encoder servo is interrupted, e.g., by a particle or water droplet on the scale, the interferometers will take over immediately to maintain the accuracy for continuous exposure. Then after the interrupting object moves away, servo control will speedily be switched back to the encoders again.

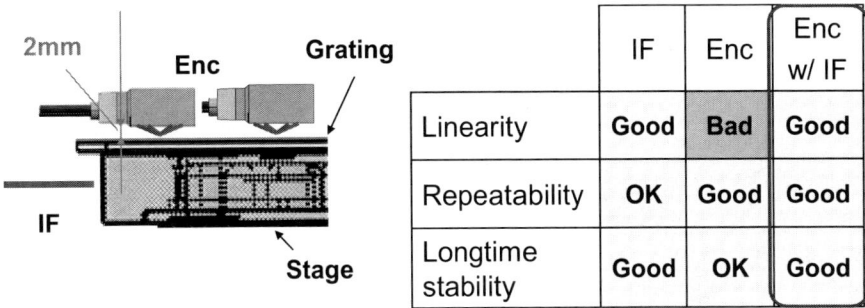

	IF	Enc	Enc w/ IF
Linearity	Good	Bad	Good
Repeatability	OK	Good	Good
Longtime stability	Good	OK	Good

Figure 3. Left: Schematic configuration of the Bird's Eye Control system. The working distance is only 2 mm. Right: The merits and demerits of interferometer servo, encoder servo and our hybrid system. In the hybrid system, encoder and interferometer compensate each other.

The second feature is a new metrology sequence: Stream Alignment. Figure 4 shows the schematic configuration of the Stream Alignment system, which consists of 5 alignment microscopes called Five-Eye FIA and a wide area auto focus sensor array named Straight Line AF. The multipoint alignment and AF mapping are carried out simultaneously during one short, straight trajectory of the wafer stage. Wafer alignment begins in the upper area of the wafer when the first alignment site comes exactly under the FIA, and is repeated while the wafer stage moves straight along the scanning direction to the exposure area. Halfway thorough the wafer alignment, simultaneous AF mapping begins when the wafer edge first comes under the Straight Line AF and continues until the whole wafer surface has been measured. This optimized configuration enables dramatic reductions in the alignment time and achieves high throughput even with a large number of alignment sites.

The third concept, Modular^2 structure, will be taken up later in chapter 4. From the next chapter, the latest performance of S620D with the *Streamlign* platform will be presented.

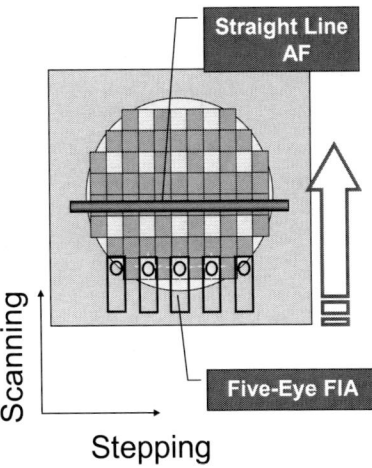

Figure 4. Schematic configuration of Stream Alignment system. The five circles correspond to the measurement points of Five-Eye FIA. The red line denotes the area of Straight Line AF.

3. LATEST PERFORMANCE

3.1 Imaging

The numerical aperture of the projection lens for S620D is 1.35, and it has sufficiently low wavefront aberration. Figure 2(a) shows the total RMS of wavefront aberration and each component of the first S620D lens plotted with three S610C lenses. The total RMS of the first S620D is 3.9mλ. The fine performance has been confirmed compared to the recent lenses of S610C. Another important point is its improved polarized aberration. Figure 2(b) shows the difference of polarized wavefront, V-H, of the Zernike coefficient Z5 and Z12. Compared to the recent S610C, it is improved remarkably. Of course these improvements lead to higher image contrast and lower OPE impact.

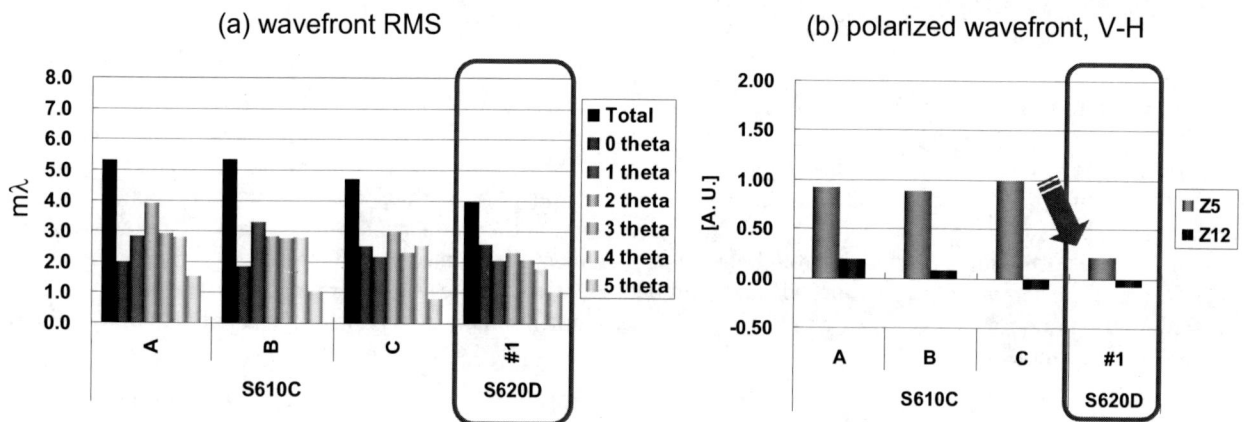

Figure 5. (a) The wavefront RMS across the slit of the first S620D plotted with that of S610C. (b) The difference in polarized wavefront, V-H, of Z5 and Z12 of S620D and S610C.

The lens of S620D has a new function useful for the severe requirements of double pattering. The schematic drawing of this function, which is the Adaptive 2θ Compensator, is shown in the left of Figure 6. A deformable mirror with piezo drive is introduced in the new catadioptric projection lens. By pushing and pulling the mirror mechanically, Zernike 2θ components Z5 and Z6 can be independently adjusted as shown in the right of figure 6. The deformable mirror controls thermal aberrations like astigmatism in dipole illumination, as represented in Figure 7. The merit of the adaptive optics is that its response is very quick, within one second. So not only static adjustment but also dynamic adjustment is possible.

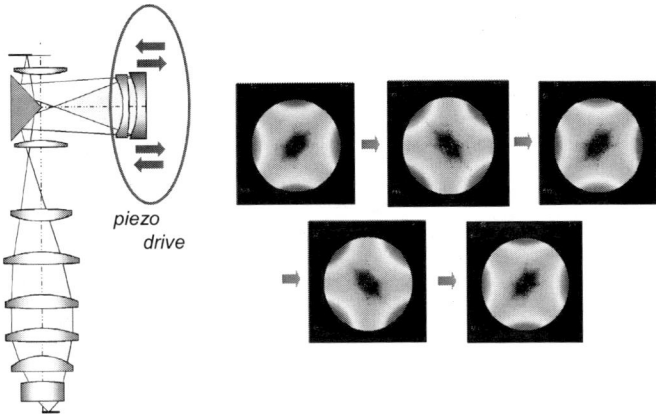

Figure 6. Left: Schematic drawing of the Adaptive 2θ Compensator with piezo drive. Right: The example of independent adjustment of Zernike 2θ components Z5 and Z6.

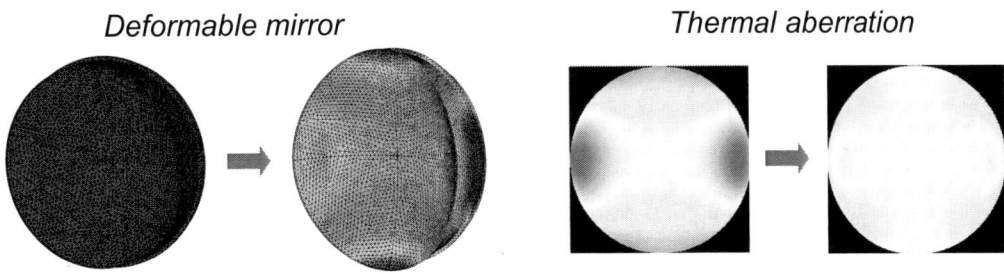

Figure 7. The example of the thermal aberration control. The deformable mirror enables the compensation of the astigmatism in dipole illumination.

3.2 Accuracy

In this section, we would like to focus on the most important point, the accuracy performance. As a beginning, let us discuss briefly the optimum concept for high accuracy. From the viewpoint of overlay accuracy, the single wafer stage has a principal advantage, since there is no grid matching error between the different stages. In addition, Nikon's current Tandem Stage system, which consists of a single wafer stage and a calibration stage, is improving based on experience from prior machines. These considerations lead to Nikon's decision to continue the Tandem Stage system in the new *Streamlign* platform.

Confirmation of Nikon's decision is clearly proved by S620D's overlay results. Figure 8 shows the overlay champion data. S620D has actually achieved excellent overlay accuracy less than 2nm. The best result is 1.22 nm and 1.05 nm for X and Y, respectively. Of course this result is enabled by the high repeatability of the new encoder system, Bird's Eye Control. This result also shows that alignment of every exposure site is successfully measured by Stream Alignment. Furthermore, it should be emphasized here that even at a high scanning speed of 700 mm/s excellent overlay accuracy better than 2.4nm is achieved. It follows that S620D also has the potential for high productivity. Of course, the final overlay target at 700 mm/s is also 2 nm or better, and we plan to achieve this improvement soon, using software upgrades only.

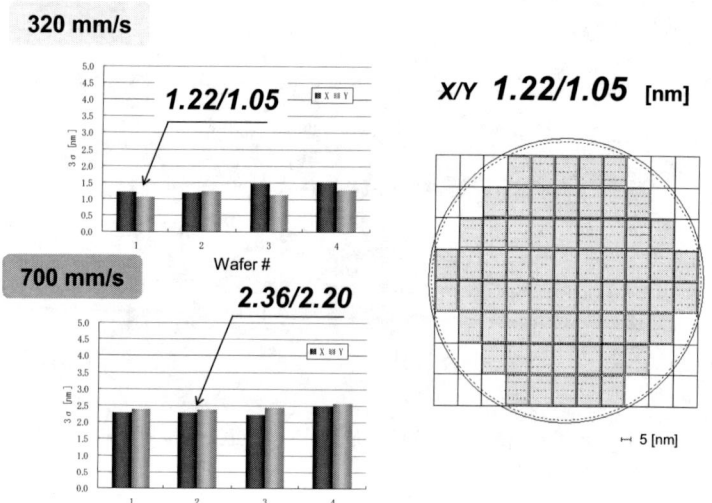

Figure 8. The single machine overlay accuracy of S620D. The top left and the top right show results for 320 mm/s; the bottom left shows results for 700 mm/s.

Figure 9 shows the overlay results through one lot, continuous exposure of 20 wafers. The total 3σ within the lot is 1.94 nm and 1.80 nm for X and Y, respectively, at 320 mm/s. 2 nm overlay is achieved through the lot. In this evaluation, the corrected components are common within the lot, so it can be said that this is the practical overlay. Furthermore the result at 700 mm/s is also quite good, with through-the-lot overlay better than 3nm.

The next interest must be overlay stability. Figure 10 shows the overlay results through 3 lots, 60 wafers total, over 3 days. The 3σ in 3 lots are 2.20 nm and 1.92 nm for X and Y, respectively. S620D achieves overlay stability of approximately 2 nm. It proves that the stability of the Stream Alignment is good even through several days.

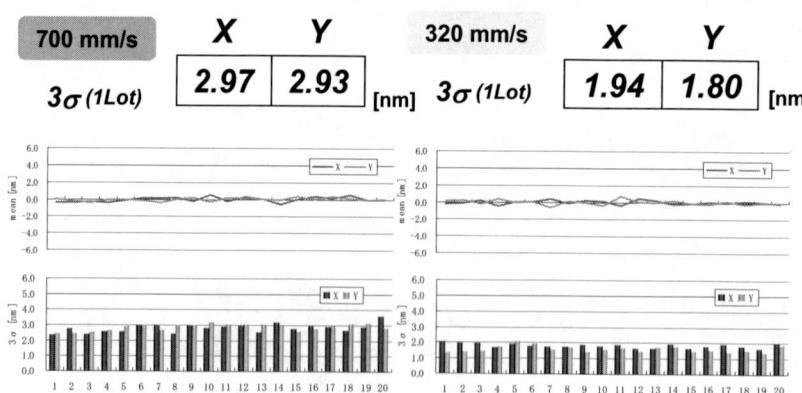

Figure 9. The overlay results through a lot. The left and the right represent the data at 700 mm/s and 320 mm/s, respectively.

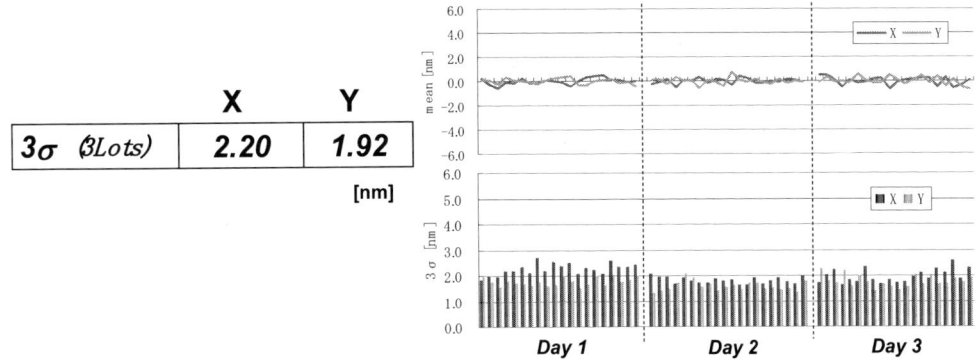

	X	Y
3σ (3Lots)	2.20	1.92
		[nm]

Figure 10. The overlay stability results through 3 lots, 60 wafers during 3 days.

Next, let us turn our attention to the performance of auto focus, which consists of the straight line focus sensor to measure wafer surface topology and Z sensors to control the stage height. Figure 11 shows the focus control uniformity data measured by the Phase Shift Focus Monitor (PSFM) method. S620D has achieved excellent accuracy of 14.3 nm (3σ) including edge shots (All shot) and 11.8 nm for full field shots, which definitely meets the budget for the next semiconductor generation. Figure 12 shows the focus control uniformity results in different three wafers. Owing to the high repeatability of the Z sensors, less than 15 nm is constantly achieved. The important point to be noted here is that these exposures were made with the maximum scanning speed, 700 mm/s. The straight Line AF of S620D is accurate and stable even at high scan velocities.

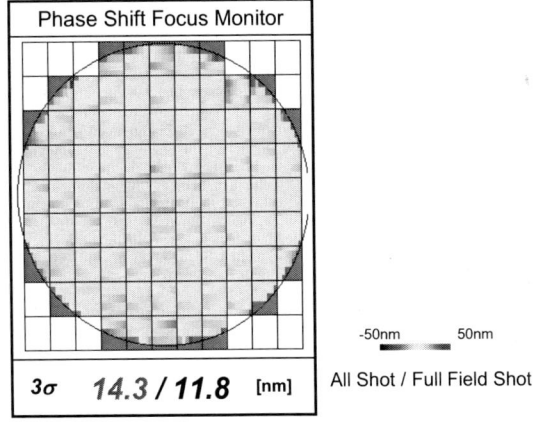

Figure 11. The auto focus accuracy result at 700 mm/s measured by PSFM method.

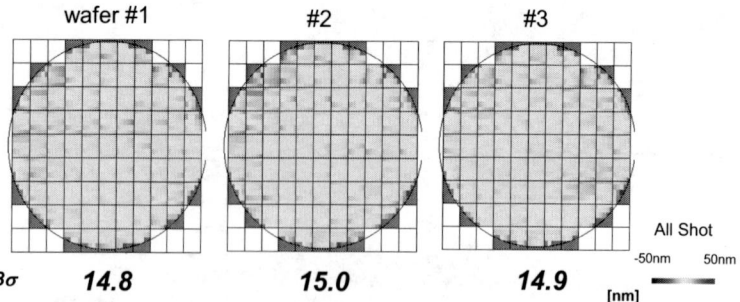

wafer #1   #2   #3

All Shot

-50nm   50nm

3σ   **14.8**   **15.0**   **14.9**   [nm]

Figure 12. The repeatability of auto focus accuracy at 700 mm/s measured by PSFM method, which is the result in different three wafers.

Taking advantage of these high accuracies of S620D, the following imaging result has been obtained. Figure 13 shows SEM images of a 32nm L/S pattern produced by pitch-splitting double patterning using S620D. The exposure conditions are as follows: Dipole-Y, σ0.85 / Ratio0.77, NA 1.00, and Self Freezing Process. The fine patterns are successfully exposed whole over the wafer. The CDU of pooled space across the wafer is 3.3 nm and that of pooled line is 2.5 nm. These sufficiently meet the budget of the 32 nm HP node in the ITRS roadmap[2]. This achievement of difficult space CDU specification is the evidence of high overlay accuracy, as mentioned in the beginning of this paper. The detailed analysis of double patterning results are described elsewhere [4].

CDU : Line 2.5   Space 3.3 *(3σ [nm])*

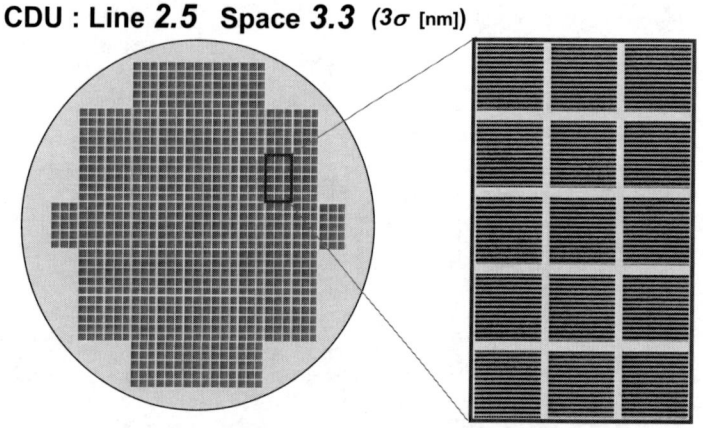

Figure 13. SEM images of 32 nm HP by pitch-splitting double patterning using S620D. The exposure conditions: Dipole-Y, σ0.85 / Ratio0.77, NA 1.00, and Self Freezing Process.

Let us now look at some key technologies of S620D, which is the base of total accuracy. In the *Streamlign* platform, the reticle stage, wafer stage, and the projection lens are protected from vibrations by an effective isolation structure, which includes reticle stage isolation and Sky Hook isolation[1]. Regarding motion control, we introduced a new method called Multi-shot Iterative Learning Control (MILC)[5], which can reduce errors in the stage trajectory by iterative motion learning. Moreover, to improve CDU, we introduced a new system called CDU Master[6], which is the dose and focus optimization from CDU scatterometry results.

One of the good measures of the basic body performance is the scanning synchronization accuracy. Figure 14 shows the scanning synchronization accuracy measured at the maximum scan speed of 700 mm/s. Moving Average error and MSD are shown in the left and right, respectively. X, Y, and Z values all meet the specifications. It should be noted that settling time in this measurement is zero. The exposure fields are represented by the areas between the two vertical green lines. Even at the beginning of the exposure, the synchronization is also remarkably accurate. It is obvious that the excellent body dynamics and motion control system are a great help for these results. Combining these enhanced dynamics with the up-to-date encoder measurement, S620D has achieved excellent stage repeatability. Figure 15 shows the stage repeatability results measured by the overlay difference between two sequential exposures on a single wafer (the wafer is not unloaded between exposures), which corresponds to the stage positioning component of the actual overlay exposure. Stage repeatability of approximately 1 nm is achieved. The bar chart in the figure shows stage repeatability results at three different scanning speeds. It looks as if there is no significant dependence on the scanning speed up to 700mm/s.

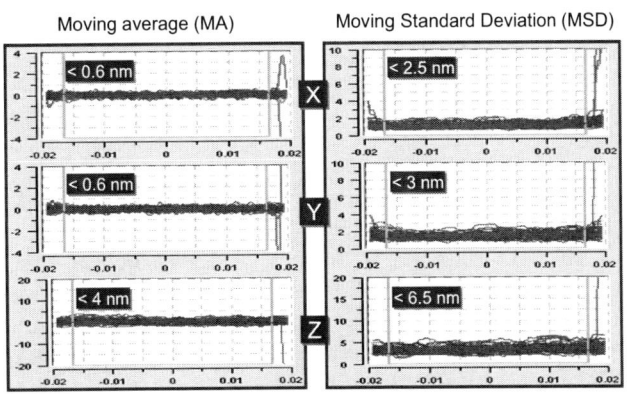

Figure 14. The scanning synchronization accuracy measured at the maximum speed of 700 mm/s with zero settling time.

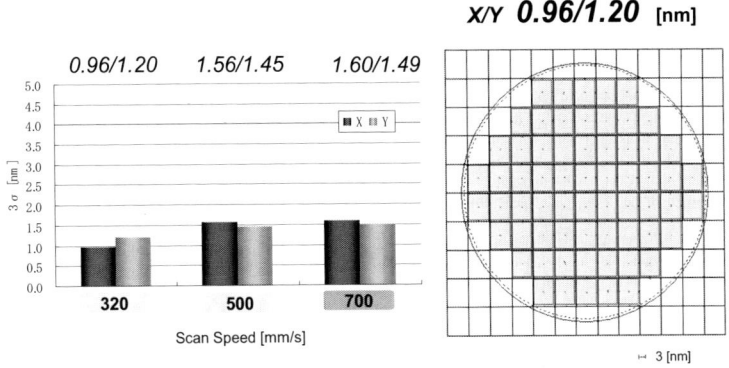

Figure 15. The stage repeatability results at three different scanning speeds.

As the last data in this accuracy section, results from the CDU control system, called CDU Master, is shown in Figure 16. With CDU Master, in addition to the existing optimal dose control, the focus parameters can also be extracted from the metrology data, which are CD and pattern height information measured by scatterometry, and the focus across the wafer can be corrected. This system is the result of joint work with KLA-Tencor, and the details is described elsewhere[6]. In the example shown in Figure 16, the CDU 3σ is remarkably improved after correction, from 3.36 nm to 1.28 nm.

A more detailed data analysis of overlay, focus, and dose control in S620D is presented in another paper [7].

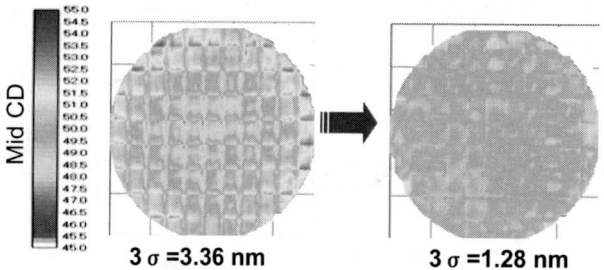

$3\sigma = 3.36$ nm $3\sigma = 1.28$ nm

Figure 16. Example of CDU control by CDU Master. 3σ is remarkably improved from 3.36 nm to 1.28 nm.

3.3 Productivity

The next question for S620D must be to see what the actual productivity in mass-production will be. To answer the question, we conducted a test running the machine for 24 hours. The wafers in this test are not production wafers, but the conditions are similar. The results were excellent and S620D achieved over 4,000 wafers per day using a single wafer stage, as shown in Figure 17(a). Compared to the successful wafer output of S610C, this result is a great advance. Figure 17(b) shows the bar chart of wafer out versus running time in the test#3 of Figure 17(a). It should be emphasized here that no error or interrupt occurred during 24 hours. In addition, the raw throughput in this test was nearly 200wph. These results prove the capability for 4,000 wpd production with S620D.

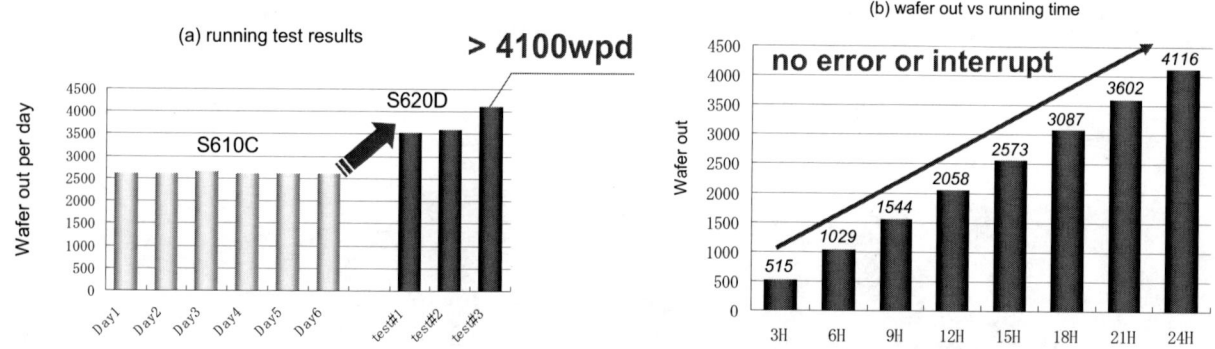

Figure 17. (a) 24H continuous running test results. S620D achieved over 4,000 wafers per day demonstration by single wafer stage. (b) Wafer out versus running time in test#3. No error or interrupt occurred during 24H.

Another important point for actual mass-production is defectivity. Figure 18 shows a result for printing particles and defects originating in the Top-Coat at 700 mm/s. The total count is less than 10, and the immersion specific defect count is nearly zero. Results for both Top-Coat-less and Top-Coat resists meet the specification. Especially the results with Top-Coat-less resist are excellent. These results prove that, from the viewpoint of defectivity, the new immersion nozzle of S620D has the capability required for high scanning speed.

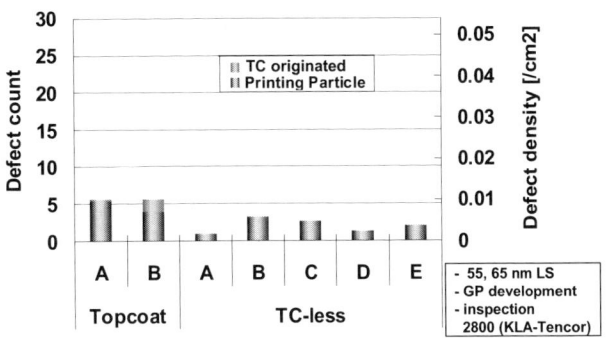

Figure 18. The defectivity result for Top-Coat originated defects and printing particles at 700 mm/s.

4. FUTURE PROSPECTS

Lastly, we would like to discuss briefly the future prospects of our *Streamlign* platform. As mentioned above, the *Streamlign* platform has a Modular^2 (Modular Squared) structure. The main body consists of several independent modules, some of which can be further divided into sub modules, creating a hierarchical module structure [1]. This structure provides us the benefits of easy maintenance and short installation time. The important point to note here is that this structure also enables easy upgrades. The upgrade of an individual module is possible, depending on the various requirements of upcoming generations. Nikon has developed *Streamlign* as the optimized long life platform.

Finally, another double patterning result of S620D is shown in Figure 19. These are the SEM images of 22 and 25nm L/S exposed by pitch-splitting double patterning using S620D. Both the first and the second patterns are exposed successfully. This result definitely indicates the possibility of immersion extension down through the 22 nm HP node and beyond. We believe that our *Streamlign* platform makes it possible, and Nikon will drive the immersion extension.

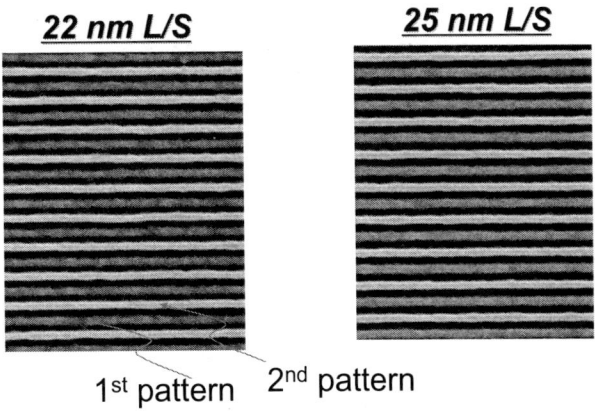

Figure 19. SEM images of 22 and 25 nm L/S by pitch-splitting double patterning using S620D. This result indicates the possibility of immersion extension down through the 22 nm HP node and beyond.

5. SUMMARY AND CONCLUSION

S620D with *Streamlign* platform achieves the following results:

High Accuracy

    32 and 22 nm L/S pitch-splitting double patterning

    < 2 nm overlay accuracy

    < 15 nm focus accuracy

High Productivity

    High scanning speed up to 700 mm/s

    > 4000 wpd throughput demonstration

Long Life Platform

    Immersion extension is possible

6. ACKNOWLEDGMENTS

The authors are grateful to S. Owa, A. J. Hazelton, S. Wakamoto, and T. Kikuchi for valuable discussions and helpful advices. We also thank Nikon data acquisition project team for SPIE and all the Nikon people who contributed to the development and evaluation of S620D.

REFERENCES

1. Y. Shibazaki, *et al.*, "An innovative platform for high-throughput, high-accuracy lithography using a single wafer stage" Proc. of SPIE Vol.7274, 72741I (2009)
2. International Technology Roadmap for Semiconductors, 2009 edition, http://www. itrs.net/
3. A. J. Hazelton *et al.*, "Exposure Tool for 32 nm Lithography: Requirements and Development Progress" Proc. of SPIE Vol. 7140, 714028 (2008).
4. T. Kikuchi, *et al.*, "Double-patterning lithography study with high-overlay accuracy," Proc. of SPIE Vol.7640, 7640-16 (2010)
5. K. Hirano, *et al.*, "Latest Results from the Nikon NSR-S620 Double Patterning Immersion Scanner" Proc. of SPIE Vol. 7520, 75200Z (2009)
6. T. Toki, *et al.*, "Simultaneous optimization of dose and focus controls in advanced ArF immersion scanners," Proc. of SPIE Vol.7640, 7640-40 (2010)
7. S. Wakamoto *et al.*, "Actual performance data analysis of overlay, focus and dose control of an immersion scanner for double patterning" Proc. of SPIE Vol. 7640, 7640-09 (2010).

Performance of FlexRay, a fully programmable Illumination system for generation of Freeform Sources on high NA immersion systems

Melchior Mulder, André Engelen, Oscar Noordman, Gert Streutker, Bert van Drieenhuizen, Cas van Nuenen, Wilfred Endendijk, Jef Verbeeck, Wim Bouman, Anita Bouma, Robert Kazinczi, Robert Socha[a],

Dirk Jürgens, Joerg Zimmermann, Bastian Trauter[b],

Joost Bekaert, Bart Laenens [c]

Daniel Corliss, Greg McIntyre [d]

ASML, De Run 6501, 5504 DR, Veldhoven, The Netherlands

a ASML Brion Technologies, Inc., 4211 Burton Dr., Santa Clara, CA, 95054

b Carl Zeiss SMT AG, D-73446 Oberkochen, Germany

c IMEC vzw, Kapeldreef 75, B-3001 Leuven, Belgium

d IBM Microelectronics, Hopewell Junction, NY

ABSTRACT

This paper describes the principle and performance of FlexRay, a fully programmable illuminator for high NA immersion systems. Sources can be generated on demand, by manipulating an array of mirrors instead of the traditional way of inserting optical elements and changing lens positions. On demand (freeform) source availability allows for reduction in R&D cycle time and shrink in k1. Unlimited tuning allows for better machine to machine matching.
FlexRay has been integrated in a 1.35NA TWINSCAN exposure system. We will present data of FlexRay using measured traditional and freeform illumination sources. In addition system performance qualification data on stability, reproducibility and imaging will be shown. The benefit of FlexRay for SMO enabling shrink is demonstrated using an SRAM example.

Keywords: Programmable Illuminator, FlexRay, Freeform sources, Source Mask Optimization

1   INTRODUCTION

In this paper the principle and performance of FlexRay, the world's first fully programmable illuminator for high NA immersion systems, will be described. Arbitrary sources can be generated on demand, by manipulating an array of mirrors instead of the traditional way of inserting diffractive optical elements (DOE) and changing lens positions.
This next generation illuminator is developed to give lithographers easy access to the enhanced process windows and lower MEEF as predicted by state of the art simultaneous Source and Mask Optimization software (SMO) and to give virtually unlimited source tuning capability that can be used to improve matching or correct for example mask bias errors. Numerous papers have been published on the benefit of freeform sources as method to extend the limits of single exposure [2,4,5,6,8].
Although it is possible to produce freeform sources with diffractive optical elements there are some limitations that do not exist for a fully programmable illuminator[7]. The first limitation that is taken away is that the lead-time for a new source is now eliminated since no new hardware has to be ordered and installed to produce the required source. Furthermore the amount of different sources that can be handled is not limited anymore by the amount of DOE's that can be stored in the DOE exchanger. Finally a fully programmable illuminator allows for virtually unlimited tuning and does not rely on compatibility of existing source manipulators with new freeform sources. FlexRay therefore offers a powerful pupil knob for ASML's computational litho product family. Instant access and unlimited tuning make the use of freeform source shapes in production both convenient and practical.

Optical Microlithography XXIII, edited by Mircea V. Dusa, Will Conley, Proc. of SPIE Vol. 7640,
76401P · © 2010 SPIE · CCC code: 0277-786X/10/$18 · doi: 10.1117/12.845984

In section 2 of this paper the principle of the FlexRay illuminator is explained.
To evaluate the performance of FlexRay, the illuminator has been integrated in a 1.35NA TWINSCAN XT:1950Hi exposure system.
In section 3 we will present data of the key characteristics of FlexRay using measured traditional and freeform illumination sources: Pupil stability and repeatability, system reliability, matching capability towards traditional Aerial illuminators and polarization performance. In addition basic system performance data on imaging and overlay will be shown. The benefit of FlexRay in combination with SMO is demonstrated for the case of downscaling an SRAM cell. Conclusions and acknowledgements can be found in sections 4 and 5 respectively.

2   PRINCIPLE OF THE FLEXRAY ILLUMINATION SYSTEM

In the FlexRay illumination system the angular distribution of the light required to make a desired source is controlled by thousands of micro mirrors. Each micro mirror reflects a spot of 193nm light in the pupil plane. For reference the spot reflected by a single mirror in the pupil is shown in Figure 1.

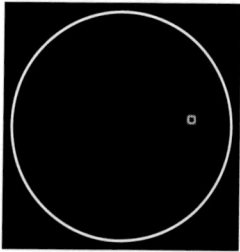

Figure 1: Spot created by a single mirror in the pupil. The white circle refers to sigma 1 at maximum NA (simulation).

Changing the angles of the mirrors will change the positions of the spots in the pupil and thus change the source shape as is illustrated in Figure 2. On the left hand side all mirrors are in a neutral position resulting in a small conventional source. In the picture on the right hand side half of the mirrors is rotated clockwise and the other half counter clockwise resulting in a dipole source. It is important to note that all mirrors are always used to create the desired source. This means that there is no light loss when switching from one source shape to another. To allow creation of any source shape it is required to have full freedom to place any of the spots anywhere in the pupil. Therefore all of the mirrors can reflect their spot to all locations in the pupil. For more information on the FlexRay design the reader is referred to [1]

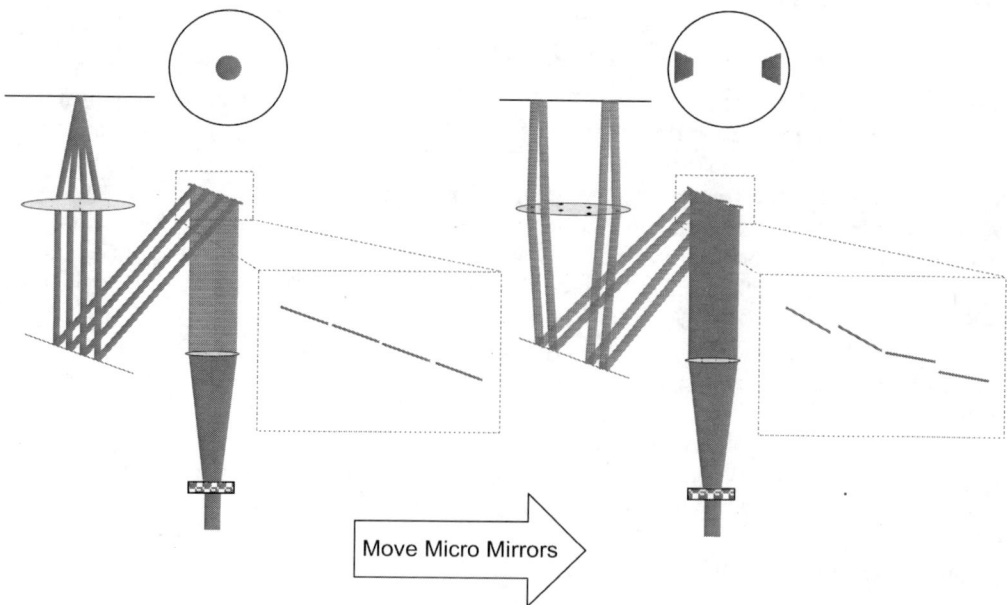

Figure 2 : Changing source shape with FlexRay illustrated for four mirrors instead of the available thousands

2.1 Distribution of the mirrors

There are thousands of mirrors that need to be distributed such that the target pupil is reproduced accurately. The algorithm that calculates the optimum mirror angle is described in this section. The mirrors are allocated by first randomly distributing the mirrors in the pupil and then keeping the mirrors that are on good positions. This process is repeated until the target source is reproduced. An illustration of the distribution process of the mirrors can be seen in Figure 3. Note that this is a graphical representation of the calculation process. The mirrors are only set after the calculation is finished and the set-points for all mirrors are known. This method of distributing the mirrors allows for accurate reproduction of freeform and traditional source shapes. Since there are thousands of mirrors, the end result is not effected if (thousands minus a few) mirrors are used. Should there be non functional mirrors, measures are in place to make sure these mirrors do not reflect light into the pupil plane so optimal source quality is maintained.

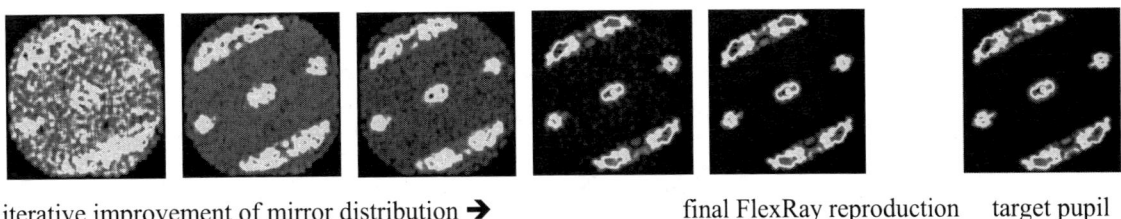

iterative improvement of mirror distribution ➜   final FlexRay reproduction   target pupil

Figure 3 : Graphical representation of the computational mirror distribution process

2.2 Closed loop mirror control

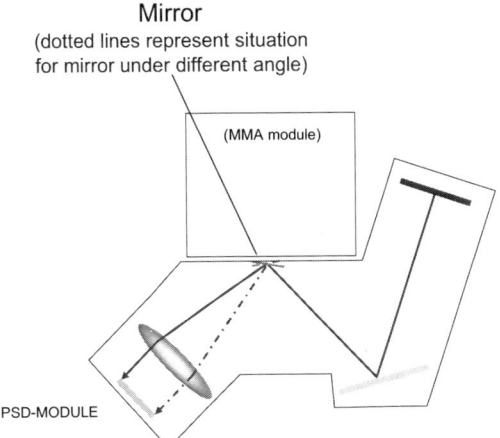

Figure 4 : Measuring mirror angle also during scan

The angles of the individual mirrors together determine the locations of all the spots that form the complete pupil. Pupil quality and stability depend on how well these mirror angles are controlled. In the FlexRay illumination system all mirror-angles are under closed loop control to ensure they are on the right position during exposure of an image. A metrology laser-beam illuminates the mirror. The reflected light is passed through a Fourier lens and hits a position sensitive device (PSD). The centre of gravity of the incoming metrology light on the PSD represents the angle of the mirror as is depicted in Figure 4. This angle information is fed to the servo electronics to allow closed loop mirror control. The mirror servo performance will be logged in the lot reports.

3   MEASURED FLEXRAY PERFORMANCE

The results presented in this section are obtained on two fully operational FlexRay Proto illuminators that are integrated in a TWINSCAN XT:1950Hi system and an FlexRay Test stand with the complete optical path up to reticle level.

3.1   Regression testing

FlexRay is designed to improve the illumination capabilities without any detrimental effect on other key performance areas. Below a summary of the functional regression tests that were performed.

Table 1: results of regression tests

What was tested	Result
Existing Calibration and performance tests for non illumination subsystems	Complete setup of system performed and OK.
Delta Polarization performance	Delta polarization performance as measured with PMM reticle before and after upgrade is well within boundaries of normal system to system variation.
Delta Overlay performance	No detectable impact of FlexRay on overlay.
Existing recipes run on FlexRay	FlexRay systems reads "old" recipes and reproduces the source shape a DOE based system would make automatically without changing look and feel for the operator….
Reliability	On track to be on par with existing illumination system early 2010.

3.2   FlexRay three day Reproducibility and Accuracy

On an illumination test stand (complete optical path up to reticle level) a three day marathon test was performed to check stability of the FlexRay illumination system. 28 different sources were measured over three days. In total ~1600 source measurements were performed. The CD performance of the measured sources was simulated for vertical and horizontal line-space and line-end features. For each source shape the minimum line-space pitch that still has acceptable process window is determined. The target CD is then 0.5x minimum pitch. Then OPC is determined for pitches up to 2x minimum pitch. All pitches that fulfill process window requirements are taken into consideration. For line-ends the isolated line-end and the minimum line end pitch that still has acceptable process window (5% exposure latitude, 80nm DoF) is taken into account. Reproducibility and delta to target numbers are reported in % of the target CD. Since the absolute value of the target CD is varying per source this allows for a better comparison for the different source shapes. In Figure 5a and b the reproducibility is presented. Maximum delta to target is shown in Figure 6a and b.

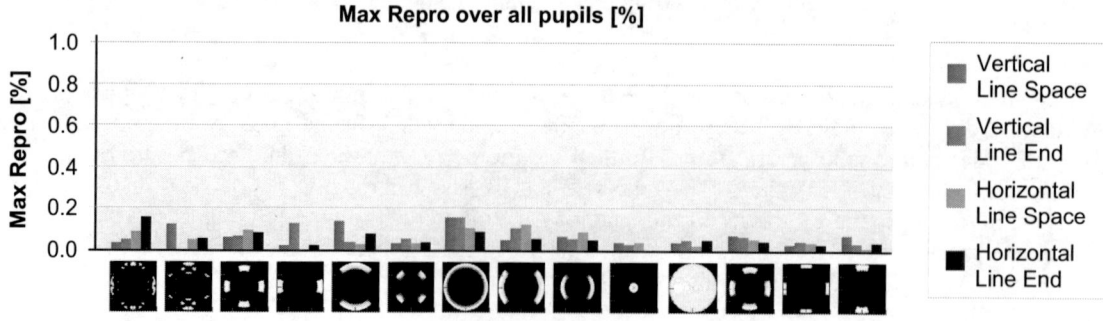

Figure 5a: overview of reproducibility performance of the first 14 source shapes

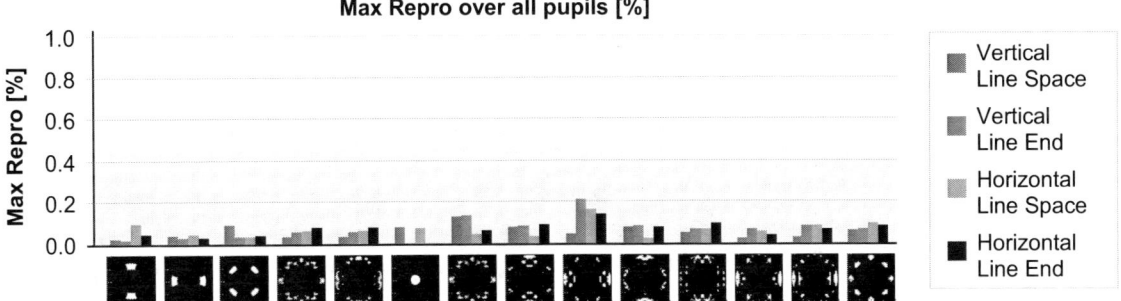

Figure 5b: overview of reproducibility performance of source shapes 15 to 28

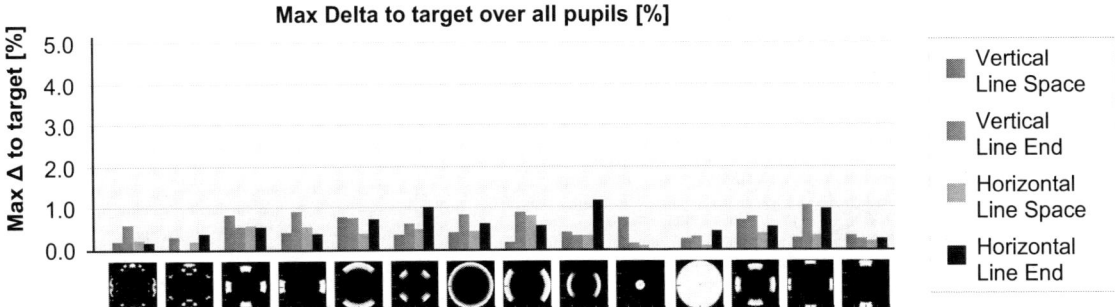

Figure 6a: overview of simulated delta to target difference for the first 14 source shapes

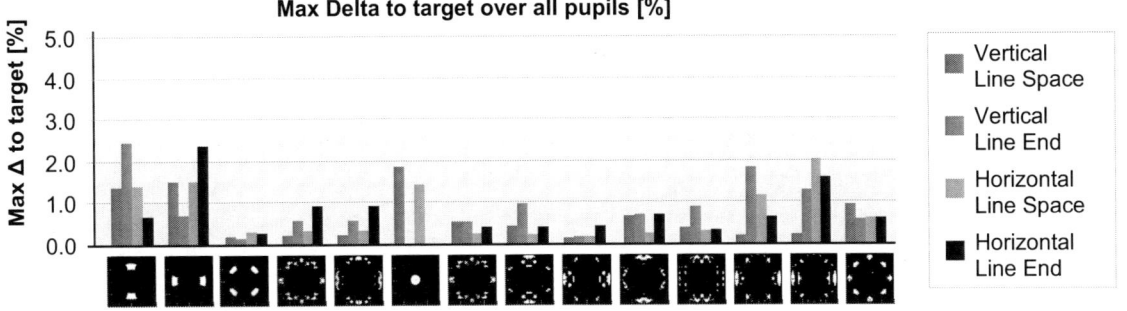

Figure 6b: overview of simulated delta to target difference for source shapes 15 to 28

The simulated imaging impact of reproducibility of the measured sources is less than 0.2% of the target CD for all targets (~0.17nm). Maximum delta to target stays below 2.5% (~1.3nm for Line Space, 1.8nm for Line end). This demonstrates FlexRay's capability to both accurately and reproducibly create a wide variety of sources.

Qualifying a source by simulating the imaging performance takes all aspects of the pupil into account. This is also referred to as CD based qualification which is described in further detail in [9].

3.3 Tuning a freeform source with the FlexRay illuminator

One of the advantages of having a programmable illuminator is that there is full freedom to adjust sources if required. There is freedom in both the nature and the amplitude of the correction since there is no limitation imposed by mechanics or optics of existing pupil correction mechanisms. Reasons for adjusting could be matching to other litho clusters or compensation of effects outside the scanner like the mask induced effects. The tuning capability is demonstrated using the freeform fit-model described in ref [2,7]. For the freeform source shown in Figure 7, ellipticity was introduced using freeform fit parameter CA2. The measured sources and the definition of the CA2 parameter are shown in Figure 7. For this rotated source the CA2 parameter especially shows up in the upper and lower part of the source. The measured sources are analyzed with the freeform fit-model to check if the induced ellipticity was as intended. Results of this analysis are shown in Figure 8a. For the measured source with CA2 input value = 0 more detailed analysis results are shown in Figure 8b.

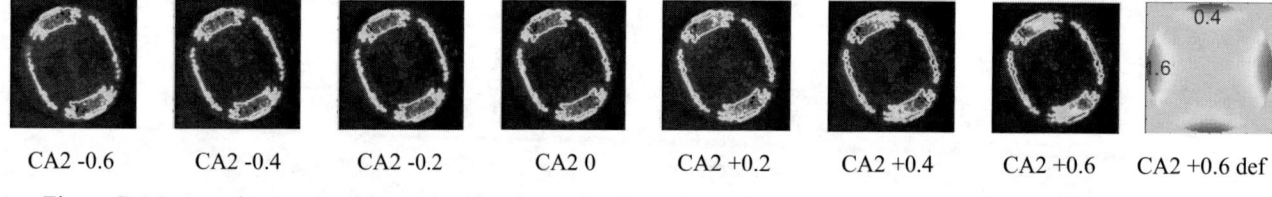

| CA2 -0.6 | CA2 -0.4 | CA2 -0.2 | CA2 0 | CA2 +0.2 | CA2 +0.4 | CA2 +0.6 | CA2 +0.6 def |

Figure 7: Measured sources with varying freeform ellipticity parameter CA2 demonstrating FlexRay's pupil tuning capability

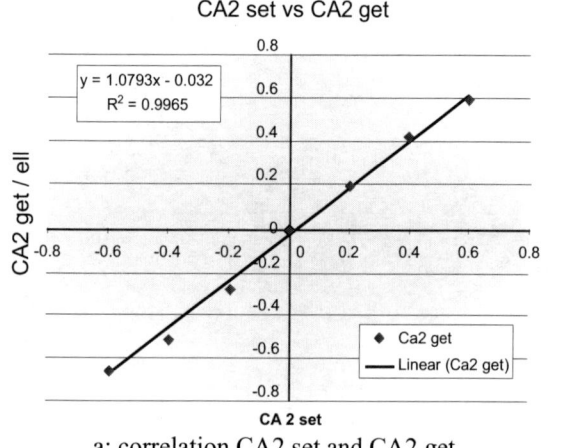

a: correlation CA2 set and CA2 get

Geometric sigma inner set	0.6865	[σ]
Geometric sigma inner get	0.6854	[σ]
Delta sigma inner	-0.0011	[σ]
Geometric sigma outer set	0.9157	[σ]
Geometric sigma outer get	0.9144	[σ]
Delta sigma outer	-0.0012	[σ]
Intensity ellipticity HV	-1.00%	[%]
Intensity ellipticity ST	-1.66	[%]

b: further source analysis for CA2 set = 0

Figure 8: Quantitative analysis of sources shown in Figure 7

For test purposes the CA2 parameter was varied over an extremely large range. A CA2 of 0.6 corresponds to an ellipticity of ~50%! The current ellipticity manipulator has a range of ~10%. The graph in Figure 8a demonstrates the capability of FlexRay to correct any source over a large range. The analysis of the nominal source shows a geometric sigma set-get errors less than 1.2 mσ compared to the requested target.

3.4 Comparison of a FlexRay exposure and a DOE based exposure

In Figure 9 a and c a DOE created source is shown and a single exposure image of a 22nm SRAM metal layer created using that source. Smallest pitch is 90nm, smallest CD is 40nm. This highly aggressive exposure is matched with a FlexRay illumination system. The measured FlexRay source and resulting resist image are shown in Figure 9 b and d respectively.

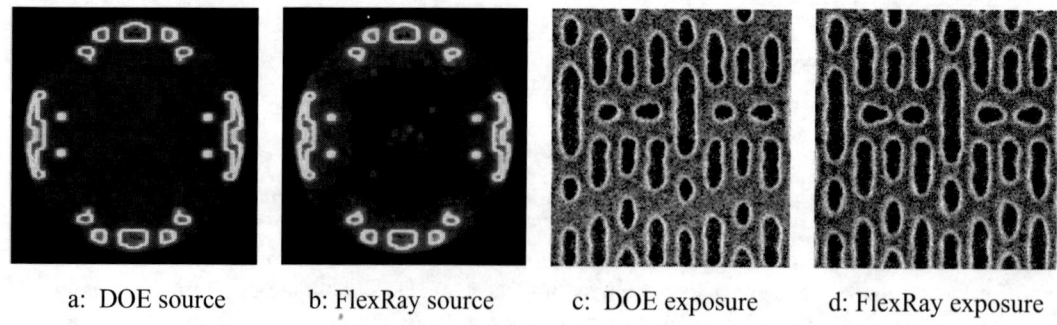

| a: DOE source | b: FlexRay source | c: DOE exposure | d: FlexRay exposure |

Figure 9: comparison FlexRay and DOE based exposure.

3.5 Comparison of a process window before and after FlexRay upgrade

Also process windows derived from resist measurements before and after FlexRay upgrade compared. The results obtained for a 40nm rotated brick-wall structure are shown in Figure 10. Within the noise of determining the process window, no difference can be observed which is to be expected since the sources match so closely.

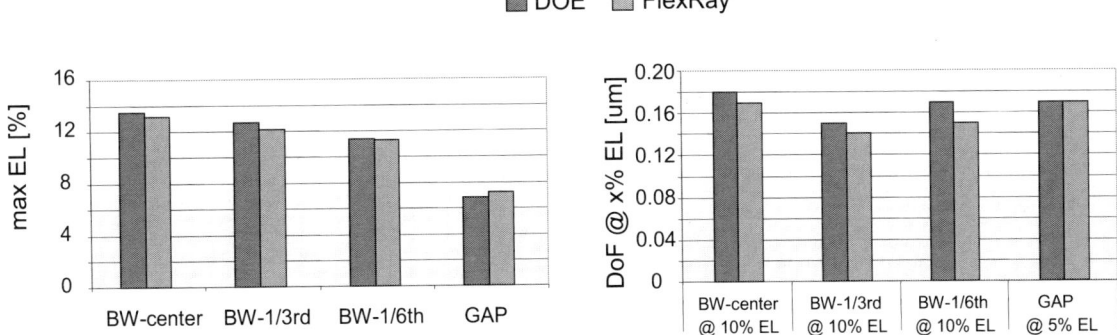

Figure 10: comparison of measured process windows of a FlexRay and DOE based exposure for 40nm rotated brick-wall structure

3.6 Comparison of CD uniformity of a FlexRay exposure and a DOE based exposure

For various features full wafer CD uniformity is compared before and after FlexRay upgrade. For pre and post upgrade the same source shapes were used. For a properly matched FlexRay system it is to be expected that the CDU before and after upgrade are equivalent. In Table 2 an overview is given of the measurements.

Table 2 Comparison of CD uniformity of a FlexRay exposure and a DOE based exposure

Feature	Illumination condition	System	Across Wafer Nominal Focus[nm]	Across Field Best Focus[nm]	Across Field 50nm of Focus[nm]	HV [nm]
40nm Isolated	NA 1.1 σ 0.70/0.50 Annular XY pol	Aerial	0.9	0.5	1.5	0.5
		FlexRay	1.5	0.7	1.7	0.8
40nm dense	NA 135 σ 0.97/0.85 C-quad XY polarized	Aerial	1.1	0.7	1.2	0.6
		FlexRay	1.0	0.6	1.0	0.6
40nm Brickwall	NA 1.35 Freeform XY polarized	Aerial Centre/1/3rd/1/6th	2.7/3.0/3.6	2.0/2.4/2.8	2.5/2.8/3.1	
		FlexRay Centre/1/3rd/1/6th	2.7/2.9/3.6	2.0/2.3/3.0	2.5/2.8/3.1	
38nm Flashgate	NA 1.35 σ 0.98/0.90 Dipole X20 Y polarized	Aerial L1/ S1	1.3/1.6	0.82/1.0		
		FlexRay L1/ S1	1.0/1.2	0.60/0.86		

From the results listed in Table 2 it can be concluded that for a wide range of features across field and across wafer CD uniformity performance of FlexRay matches the CD uniformity performance before the upgrade.

3.7 Matching of FlexRay to different DOE systems.

Measured source-maps of four different Aerial systems were used as input for FlexRay. Both the input source-maps and the measured FlexRay reproduction of the input source were analyzed for CD through pitch. In the figure below the various CD through pitch curves as a result of only changing the source are compared. The spread between the four Aerial systems is ~0.6nm. FlexRay's error in matching any of the pitch-curves is less than 0.25nm. This illustrates that FlexRay can accurately match to any other system by offering a measured source-map as target.

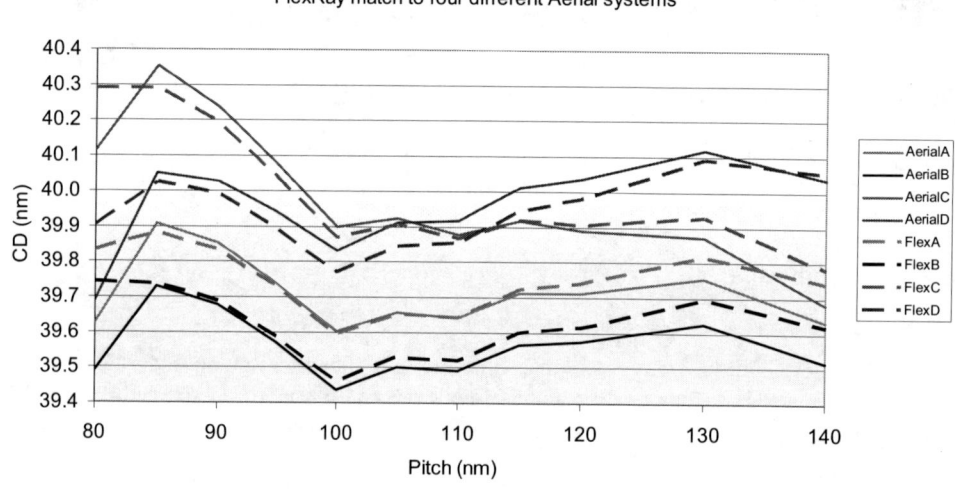

Figure 11: Matching a single FlexRay system to different Aerial systems for C-quad NA 1.35 s 0.97/0.82 XY polarization. Simulation condition: source impact only.

3.8 FlexRay and Source Mask optimization

In this paragraph a classical shrink R&D use case is described. FlexRay's flexibility was used to quickly verify the performance of a source calculated with SMO in resist. Source and mask where derived using Tachyon SMO where mask and source are co-optimized as is illustrated in Figure 12. Depending on the restrictions on the source the output is either a choice from the available DOE based settings or a true freeform. The SMO flow is illustrated in Figure 12.

In Figure 13 the process windows for the best standard source and freeform source as found by Tachyon SMO are compared illustrating the benefit of freeform sources. For this application the process window is increased and MEEF is reduced compared to the best traditional source solution.

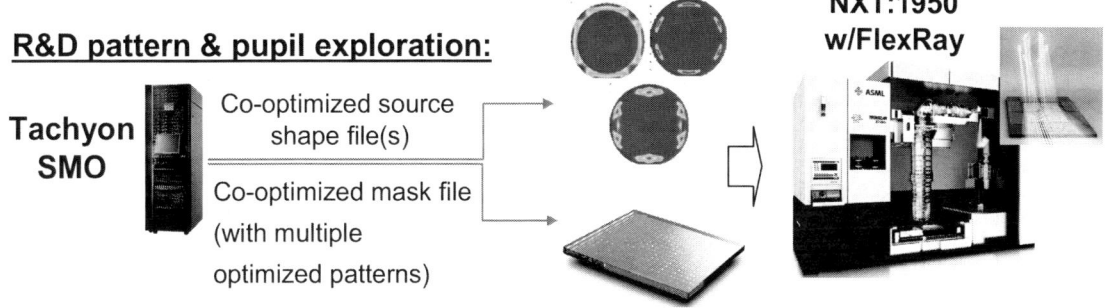

Figure 12 : Tachyon Source mask optimization flow

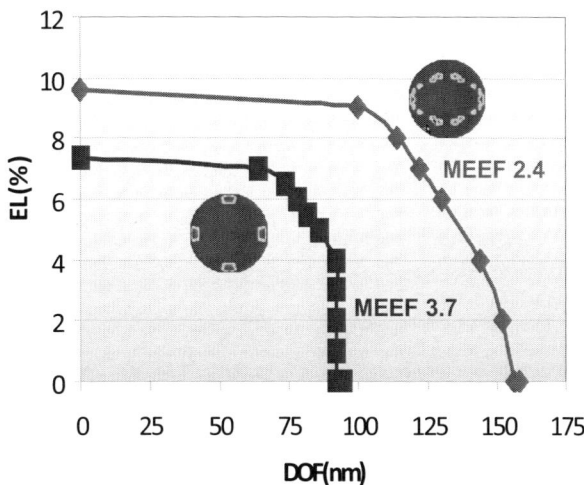

Figure 13: Benefit of freeform illumination for SRAM : increase process window and reduced MEEF. Cut-line used for process window simulation is indicated with "PW" in Figure 14c.

The successfully printed image is shown in Figure 14d. This is a first illustration how the flexibility of a programmable illuminator can help in reducing research and development cycle-times. It was already possible to place many clips on a singe R&D mask. With FlexRay it is also possible to test each of these clips with the optimum source.

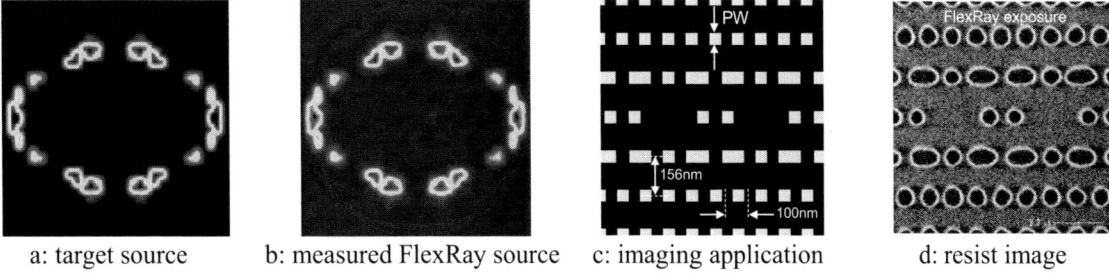

a: target source   b: measured FlexRay source   c: imaging application   d: resist image

Figure 14 : Optimum source for 20% downscaled clip, calculated using Tachyon SMO and resulting imaging

4   CONCLUSION

In this paper system performance of a TWINSCAN XT: 1950Hi with FlexRay illumination system is described.
Unique fully programmable illumination capability is offered allowing unlimited tuning of the pupil and a practical way of introducing freeform illumination into a production environment. Source stability, reproducibility and excellent print to target accuracy have been demonstrated for a wide variety of freeform and traditional sources. Process window comparison between FlexRay and DOE based sources demonstrated that, as expected, a DOE exposure can be matched in both CD and process window. CDU data for various features further confirmed the capability of FlexRay to deliver full field imaging performance well within specifications. The advantage of combining source mask optimization and on demand availability of any source is demonstrated for an SRAM case.
The capability to generate both "traditional" sources and fully freeform sources on demand, both with virtually unlimited tuning capability makes the FlexRay illuminator compatible with the existing nodes and ready for the coming nodes.

5   ACKNOWLEDGEMENTS

A large group of people worked hard to build the FlexRay illuminator and obtain the results presented in this paper. The author would like to thank members of the MMA and MASU team for their hard work to get all the measurements done as well as the ZEISS team for their work on optics and the mirror distribution algorithm. The software and electronics teams are acknowledged for their contribution to a speedy integration and the imaging team and IMEC for the wafer data. The IBM Albany team is acknowledged for sharing data on their FlexRay illumination system. Furthermore I would like to thank Frank Coppelmans, Anthony Ngai and Joep Bonthond for technical discussions and Jeannot Driedonkx for valuable advice on the presentation.

REFERENCES

[1]   **Melchior Mulder et al.**, *"Performance of a programmable illuminator for generation of freeform sources on high NA immersion systems,"* Proc. SPIE 7520, (2009)

[2]   **Andre Engelen et al.,** *"Imaging solutions for the 22nm node using 1.35NA"*, Proc. SPIE 7274, (2009)

[3]   **Igor Bouchoms et al.,** *"Extending Single-Exposure Patterning Towards 38-nm Half-Pitch Using 1.35 NA Immersion"*, Proc. SPIE 7274, (2009)

[4]   **Kafai Lai et al.,** *"Experimental result and simulation analysis for the use of pixelated illumination from source mask optimization for 22nm logic lithography process"*, Proc. SPIE 7274, (2009)

[5]   **Alan E. Rosenbluth.,** *"Intensive Optimization of Masks and Sources for 22nm Lithography"*, Proc. SPIE 7274, (2009)

[6]   **David O. S. Melville et al.,** *"Demonstrating the benefits of source-mask optimization and enabling technologies through experiment and simulation"*, Proc. SPIE 7640, (2010)

[7]   **Joerg Zimmermann et al.,** *"Generation of arbitrary freeform source shapes using advanced illumination systems in high-NA immersion scanners"*, Proc. SPIE 7640, (2010)

[8]   **Joost Bekaert et al.,** *"Freeform illumination sources: an experimental study of source-mask optimization for 22-nm SRAM cells"*, Proc. SPIE 7640, (2010)

[9]   **Jin-hyuck Jeon et al.**, *"Analysis of the impact of pupil shape variation by pupil fit modeling"*, Proc. SPIE 7640, (2010)

High Reliability ArF Light Source for Double Patterning Immersion Lithography

Rostislav Rokitski; Toshi Ishihara; Rajeskar Rao; Rui Jiang;
Mary Haviland; Theodore Cacouris; Daniel Brown, Cymer Inc.

ABSTRACT

Double patterning lithography places significant demands not only on the optical performance of the light source (higher power, improved parametric stability), but also on high uptime in order to meet the higher throughput requirements of the litho cell. In this paper, we will describe the challenges faced in delivering improved performance while achieving better reliability and resultant uptime as embodied in the XLR 600ix light source from Cymer, announced one year ago. Data from extended life testing at 90W operation will be shown to illustrate these improvements.

KEYWORDS: immersion lithography, double patterning, excimer laser, deep ultraviolet

1. INTRODUCTION

Double patterning (DP) lithography is gaining widespread use in 32 and sub-32nm technology nodes as an extension to immersion lithography. While many resolution enhancement technologies (RET) have been developed recently, including source-mask optimization (SMO)[1], and pixilated illumination[2] schemes for the scanner, the overarching requirement for the light source in double patterning has been a need for improved optical performance stability and higher power. With the introduction of the XLR 600ix light source from Cymer last year, these requirements have been met and integrated on the most advanced DP immersion scanners on the market. Key areas of improvement include higher power with flexibility to address a wide range (60 to 90W), improved energy stability, improved bandwidth stability and improved wavelength stability. The details of these improvements were reported on a previous paper[3]. These characteristics have enabled improved CD uniformity along with higher throughput operation for the litho cell to counteract the impact of the higher cost of DP lithography.

In addition to providing improved performance, the light source needs to have higher reliability and uptime in a DP environment, as the impact of down time is magnified further when the litho cell throughput is increased dramatically. In this area, the XLR 600ix was designed to address this need by building on a proven platform and introducing features that further enhance reliability and uptime. In this paper, we will describe the challenges faced in delivering improved performance while achieving better reliability and uptime on the XLR 600ix. Areas of improvement include development of advanced optics materials and coatings to provide stable performance over a wide power range (60 to 90W), an improved control system delivering faster closed-loop feedback for optical stability over extended periods, and 'smart' on-board diagnostics with predictive capability to prevent unscheduled downtime. Data from extended life testing as well as field performance data will be presented to illustrate these improvements.

2. TECHNOLOGY ADVANCEMENTS

Several technologies have been introduced in this light source to enable not only high power operation, but sustained performance stability under continuous operation at high power. The development of optical materials and coatings that can endure fluences in excess of 20mJ internal to the light source (in order to deliver an output of 15mJ) while staying impervious to thermal effects has been a key enabling technology. Similarly, advanced control algorithms that further reduce parametric variability in wavelength, bandwidth and energy have enabled the use of fewer pulses to achieve a desired on-wafer dose stability, which in turn leverages the use of higher energy to improve wafer throughput at the scanner. An example of such performance improvements is shown for four different light sources tested under varying repetition (rep) rates from 1.5 to 6kHz in Figure 1, where energy stability is measured.

Optical Microlithography XXIII, edited by Mircea V. Dusa, Will Conley, Proc. of SPIE Vol. 7640,
76401Q · © 2010 SPIE · CCC code: 0277-786X/10/$18 · doi: 10.1117/12.849065

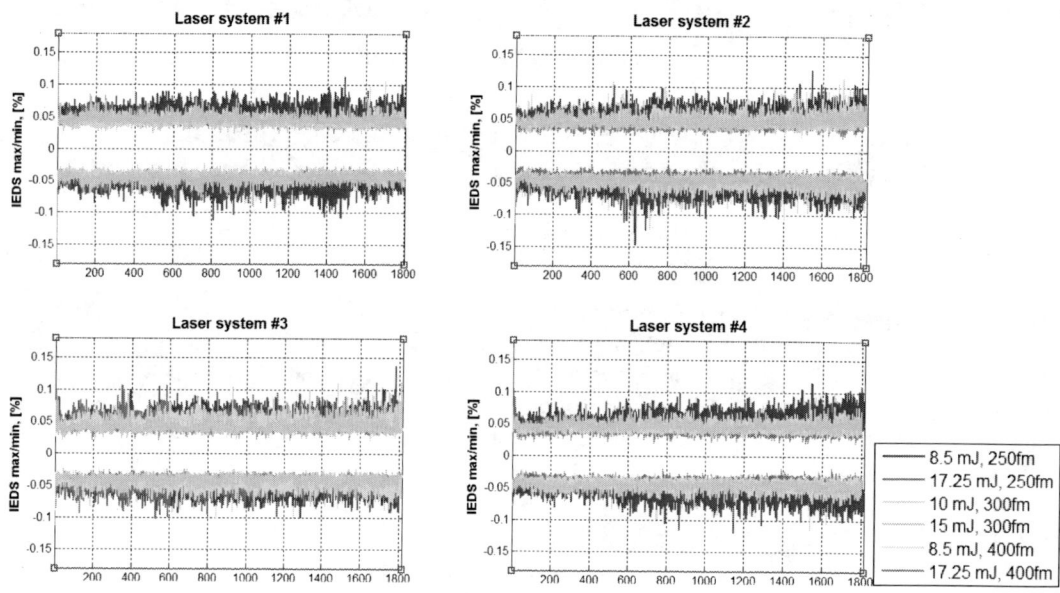

Figure 1 – Dose stability of 4 different light sources measured across various energy and bandwidth settings varying from 8.5mJ (51W) to 17.25mJ (103.5W), which covers the nominal 60 – 90W operation. The horizontal axis represents a sequence of 'bursts' or firing patterns of the test suite that include variations in rep rate, starting with 6kHz (leftmost data) and decreasing to 1.5kHz.

More recently, a new advancement in excimer laser discharge chamber technology was introduced in the XLR 600ix to further improve performance stability over life. Excimer discharge chambers have a finite operating life that is modulated by several key aging mechanisms: (a) discharge electrode erosion, (b) chamber window damage due to DUV exposure, and (c) accumulation of particulate debris, or 'dust' that are byproducts of gas interaction with chamber materials. While advances in chamber design and material selection have progressively extended the life of discharge chambers by minimizing or mitigating the aging mechanisms, they have not fundamentally addressed the observed trending of operating parameters that result from aging. For example, while electrode erosion rates can be reduced with the appropriate selection of materials and design, the change in electrode gap due to erosion is still present and can result in beam property changes over time. The new technology recently introduced with the XLR 600ix includes a new discharge chamber design that automatically compensates for electrode erosion by simply moving the electrodes physically to maintain a constant gap (Figure 2). While this concept is not new, the ability to realize it in a production-worthy light source has been elusive until now. The benefit of maintaining a constant gap between electrodes in the discharge chamber is a more stable beam characteristic in physical dimensions, and the secondary effects that can contribute to parametric stability. This translates to better stability in pupil fill, less variation in bandwidth and a resulting improved focus, overlay and CD control on the wafer.

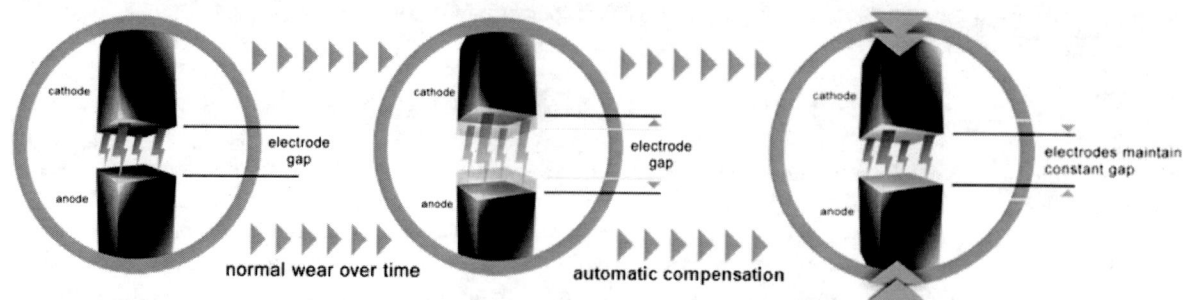

Figure 2 – Discharge chamber electrode aging results in a change in the electrode gap over time; new chamber technology recently introduced automatically compensates for this effect to maintain a constant gap as illustrated here, resulting in more stable performance over the life of this chamber.

3. EXTENDED RELIABILITY TESTING

While performance testing and reliability have been extensively demonstrated for light sources operating at 10mJ, 6kHz (60W), this is the first instance where extensive testing has been performed and demonstrated on a light source running 15mJ, 6kHz (90W). Previous reports on the XLR 600ix have centered on the challenges of maintaining improved light source stability while achieving higher power (90W). We report here, for the first time, extensive testing that represents ~ 1year of continuous operation at a high utilization memory fab.

In the aforementioned extended testing, a XLR 600ix light source operating at 90W was subjected to continuous operation similar to a high utilization fab environment, simulating about 1 year of operation. This testing was performed in an accelerated manner to enable completion of this test within 30 weeks, accumulating 30 billion pulses of DUV light. Interspersed with the continuous operation were periodic test suites that collected detailed parametric data to monitor light source performance. Figure 3 illustrates an example of this data set where wavelength stability is analyzed in terms of average wavelength error around the central wavelength, where the data is clustered mostly within ±5fm compared to a performance requirement for the scanner of ±12fm. Improved wavelength stability directly affects on-wafer contrast and focus, that results in improved CD uniformity.

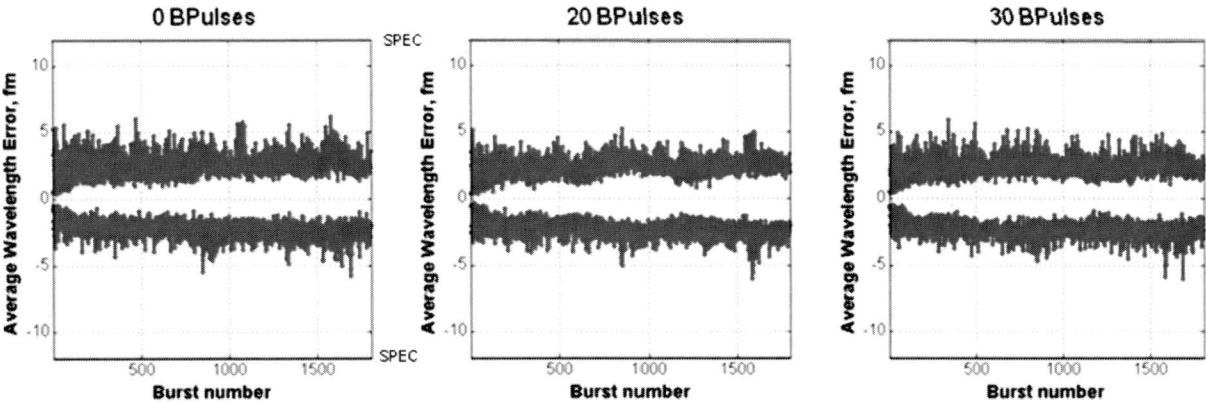

Figure 3 – Average wavelength error measured at the beginning of the extended test (0 Bpulses), at 20Bpulses and at the end of the test (30Bpulses). Wavelength error remains unchanged and mostly clustered within ±5fm around the center wavelength. The horizontal axis represents a sequence of 'bursts' or firing patterns of the test suite that include variations in rep rate, duty cycle and energy to test the effects of wavelength stability across all operating conditions.

The test suites used to evaluate light source performance periodically include subjecting the light source to variations in rep rate, duty cycle and energy to capture the performance across all operating space. Figure 4 further explores wavelength stability through wavelength sigma, indicating that most of the data across all operating conditions is clustered below 30fm, with a scanner requirement of <50fm for this technology generation.

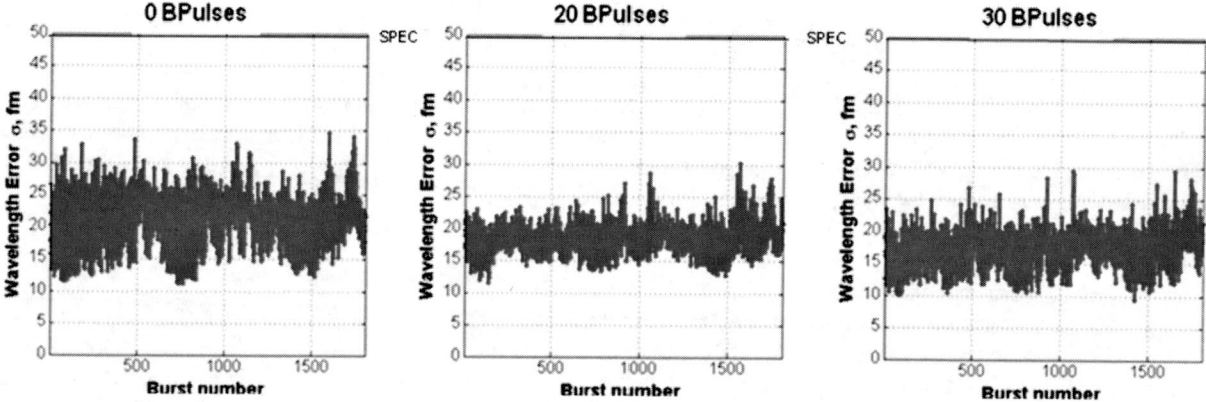

Figure 4 – Wavelength sigma measured at the beginning of the extended test (0 Bpulses), at 20Bpulses and at the end of the test (30Bpulses). Wavelength sigma remains unchanged and mostly clustered below 30fm. The horizontal axis represents a sequence of 'bursts' or firing patterns of the test suite that include variations in rep rate, duty cycle and energy to test the effects of wavelength stability across all operating conditions.

In addition to wavelength stability measurements, bandwidth stability was similarly evaluated, as shown in Figure 5. Here, the nominal bandwidth target is 300fm and is maintained through the use of active controls to support optical proximity correction (OPC) design features in the mask set with minimal variation. This in turn results in the high contrast necessary on-wafer to achieve the desired CD uniformity. The data shown in Figure 5 shows that throughout the extensive testing, the nominal bandwidth stays centered at 300fm and the variation observed across varying operating conditions (rep rate, duty cycle and energy) are mostly within 25fm.

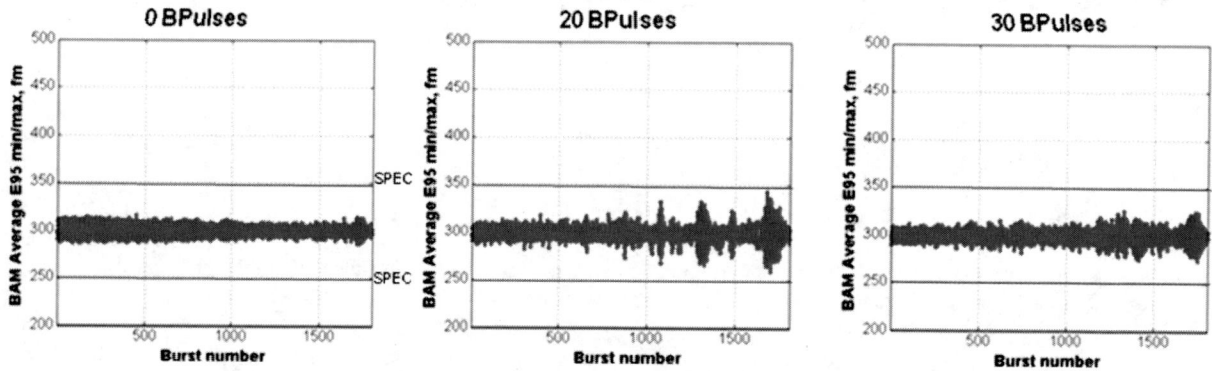

Figure 5 – Bandwidth stability measured at the beginning of the extended test (0 Bpulses), at 20Bpulses and at the end of the test (30Bpulses). Nominal bandwidth target is remains unchanged at 300fm and bandwidth variation is mostly within 25fm. The horizontal axis represents a sequence of 'bursts' or firing patterns of the test suite that include variations in rep rate, duty cycle and energy to test the effects of bandwidth stability across all operating conditions.

Energy stability was also monitored throughout the extended testing and the calculated dose stability based on a 35-pulse window was well below ±0.1%, especially at the high rep rates (Figure 6). Raw energy sigma also showed low values, mostly below 3%, with some deviations to 4% at the lower rep rates and near the end of the test period (Figure 7).

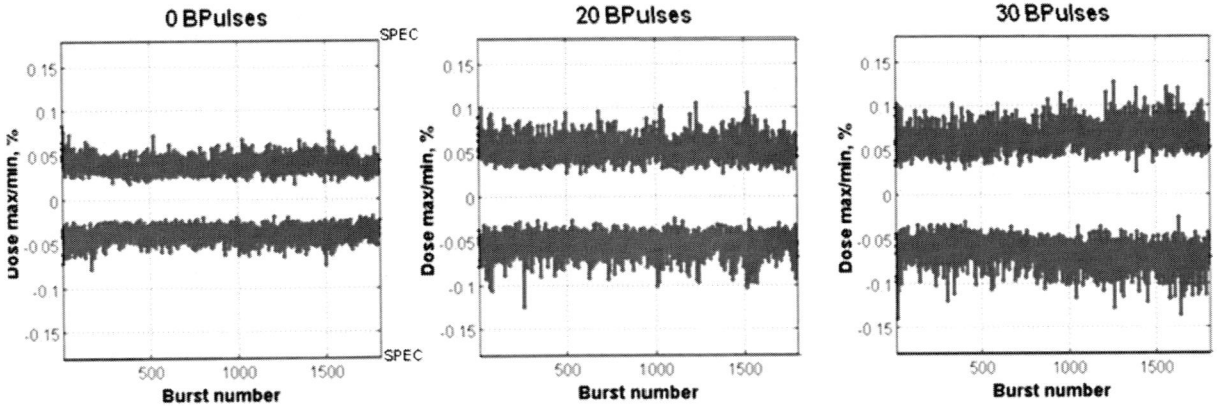

Figure 6 – Dose stability measured at the beginning of the extended test (0 Bpulses), at 20Bpulses and at the end of the test (30Bpulses). Dose variation is mostly below ±0.1%, especially at the high rep rates (6kHz) that are captured at the leftmost set of data. The horizontal axis represents a sequence of 'bursts' or firing patterns of the test suite that include variations in rep rate, duty cycle and energy to test the effects of bandwidth stability across all operating conditions.

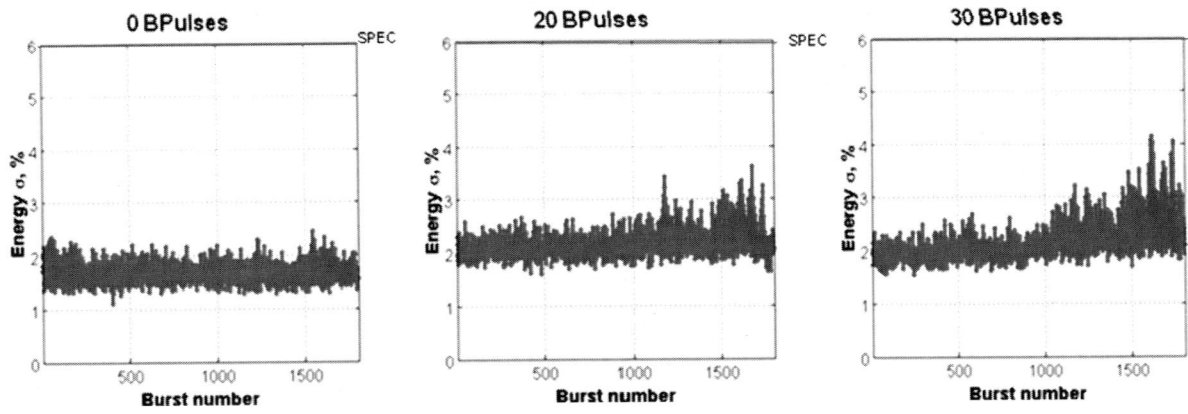

Figure 7 – Raw energy stability (sigma) measured at the beginning of the extended test (0 Bpulses), at 20Bpulses and at the end of the test (30Bpulses). Energy sigma is mostly below 3%, especially at the high rep rates (6kHz) that are captured at the leftmost set of data. The horizontal axis represents a sequence of 'bursts' or firing patterns of the test suite that include variations in rep rate, duty cycle and energy to test the effects of bandwidth stability across all operating conditions.

4. PERFORMANCE MONITORING

While fundamental improvements in performance and reliability have been introduced with this light source, a complementary, an operational infrastructure is key in maximizing uptime and utilization. In particular, serviceability and maintainability of the light source can dramatically enhance the litho cell performance with better performance and productivity. Cymer light sources have historically been 'connected' to provide near-real-time performance data to a centralized monitoring station. Initially, this connectivity enabled remote monitoring by experts who could better direct field service personnel on a particular maintenance activity, based on symptoms and their experience. Statistical-process-control (SPC) type charting augmented this capability by flagging light sources that were showing signs of less than ideal performance, helping proactively schedule maintenance before an unscheduled event occurred. More recently, we introduced algorithms that mine this data to extract unique performance signatures, essentially automating the analysis and providing specific guidance to the field service engineer. The on-board light source control system continuously monitors the system state for anomalous behavior. When an anomaly is encountered, a high-data-rate log of key signals is automatically sent to the central monitoring station. Sophisticated fault-signature detection (FSD) algorithms that reside on the central station that employ pattern recognition functions analyze this data to identify known patterns associated with a particular subsystem. As this data is amassed, a picture of the health of the system is developed and a maintenance action can be scheduled to proactively correct the issue. The net result of this approach is that fewer unscheduled events occur, and when a maintenance action is required, it can be scheduled to ensure the proper resources and parts are in place to minimize down time (Figure 8).

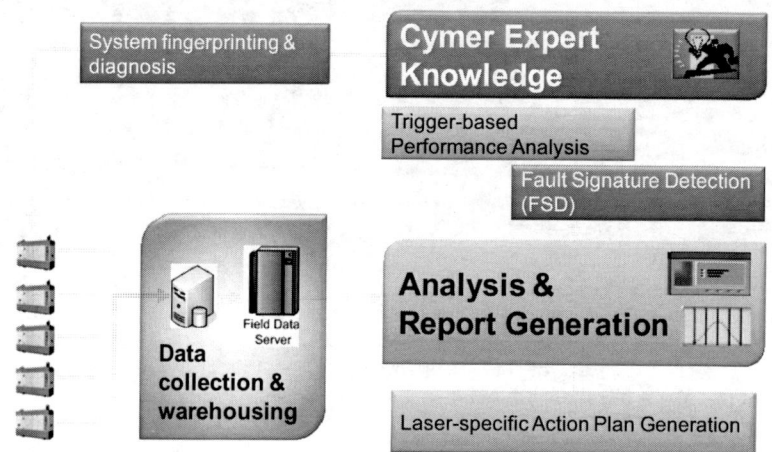

Figure 8 – Down time reduction through automation of data analysis of light source performance data. Centralized data collection and warehousing enables the use of fault signature detection (FSD) to automate knowledge derived from laser experts, which can trigger a scheduled maintenance action. Continuous diagnostics generation shortens or eliminates the time required to troubleshoot a fault.

In parallel, a routine that runs periodically on the light source collects a 'fingerprint' or performance state of the system which can be used during a service event to quickly identify the source of degraded performance, thereby minimizing the time required for troubleshooting. This function tracks key performance indicators so that once a service event is concluded, the light source can be restored back to its 'healthy' state.

5. SUMMARY

As 193nm immersion lithography is further challenged with double patterning applications, stringent demands are placed on the light source to enable improved CD uniformity, overlay and OPC performance. The XLR 600ix light source described here exceeds these requirements and further enhances the lithographer's toolbox by providing flexibility in power output. Demonstrated extended performance at 90W supports the requirements of high-throughput double patterning applications, where high uptime and utilization are expected. Such high uptime has been achieved through the use of new optics and chamber technology, as well as a data infrastructure and analysis capability.

[1]Matsuyama, T., Nakashima, T., Noda, T., "A study of source and mask optimization for ArF scanners," Proc. SPIE 7274 (2009)

[2]Lai, K., Rosenbluth, A. E., Bagheri, S., Hoffnagle, J. A., Tian, K., Melville, D. O., Tirapu-Azpiroz, J., Fakhry, M., Kim, Y., Halle, S. D., McIntyre, G., Burr, G. W., Burkhardt, M., Corliss, D. A., Flagello, D. G., Zimmermann, J., Kneer, B., Rohmund, F., Hartung, F., Russ, C., Maul, M., Kazinczi, C. R., Engelen, A., Mulder, M., "Experimental result and simulation analysis for the use of pixelated illumination from source mask optimization for 22-nm logic lithography process", Proc. SPIE 7274 (2009).

[3] Rokitski, R., Ishihara, T., Rao, R., Jiang, R., Riggs, D., Haviland, M., Cacouris, T., Brown, D., "Flexible 60-90W ArF light source for double patterning immersion lithography in high volume manufacturing", Proc. SPIE Lithography Asia 7520-37 (2009).

Advanced imaging with 1.35 NA immersion systems for volume production

Igor Bouchoms, Jan Mulkens, Sander de Putter, Pieter Gunter, Roelof de Graaf, Marcel Beems, Erwin Verdurmen, Hans Jasper, Nils Dieckmann[a], Frank Bornebroek

ASML Netherlands B.V.
De Run 6501, 5504 DR Veldhoven,
The Netherlands

(a) Carl Zeiss SMT AG
Rudolf-Eber-Strasse 2
73447 Oberkochen, Germany

ABSTRACT

The semiconductor industry has adopted water-based immersion technology as the mainstream high-end litho enabler for 5x-nm and 4x-nm devices. Exposure systems with a maximum lens NA of 1.35 have been used in volume production since 2007, and today achieve production levels of more than 3400 exposed wafers per day. Meanwhile production of memory devices is moving to 3x-nm and to enable 38-nm printing with single exposure, a 2nd generation 1.35-NA immersion system (XT:1950Hi) is being used. Further optical extensions towards 32-nm and below are supported by a 3rd generation immersion tool (NXT:1950i).

This paper reviews the maturity of immersion technology by analyzing productivity, robust control of imaging, overlay and defectivity performance using the mainstream ArF immersion production systems. We will present the latest results and improvements on robust CD control of mainstream 4x-nm memory applications. Overlay performance, including on-product overlay control is discussed. Immersion defect performance is optimized for several resist processes and further reduced to ensure high yield chip production even when exposing more than 15 immersion layers.

Key words: Immersion lithography, Exposure systems, Defects, Overlay, CD control, low k1

1. INTRODUCTION

When evaluating the lithography resolution shrink over the years for DRAM, logic and NAND type of applications, it becomes clear from figure 1a that the shrink is driven by NAND memory applications. For a long time the enabling technology for shrink is reduction of wavelength and increase of the NA. The maximum NA the industry is using today is 1.35 NA at a wavelength of 193-nm. Immersion scanners with this wavelength are currently being used in volume manufacturing of 5x-nm and 4x-nm devices. In the near future the wavelength is expected to move to 13.5nm (EUV), however until than immersion with 1.35 NA at 193-nm wavelength is the main technology. In practice this means that low k1 optical enhancements are being practiced to support resolutions between 40-nm and 20-nm. These low k1 enhancements are based on two main pillars: 1) double patterning technology and 2) source-mask optimization (figure 1b).

In this paper we discuss the state-of-art of immersion technology in volume manufacturing by analyzing litho tool productivity, robust overlay and CD control and latest defectivity performance. This is done using the mainstream ArF immersion production systems XT:1900Gi (first generation 1.35 NA) and XT:1950Hi (2nd generation 1.35 NA). These tools are currently being used for mainstream 4x-nm and early 3x-nm applications. In section 2 we discuss the maturity of immersion scanners in volume manufacturing. Imaging performance and latest developments are discussed in section 3. For upcoming 2x-nm applications we developed additional optical enhancements which enable a higher resolution wave-front manipulation. For double patterning we developed a new immersion scanner platform with improved overlay and productivity. This new platform (NXT:1950i) is discussed in a separate paper [2]. In section 4 we discuss

Optical Microlithography XXIII, edited by Mircea V. Dusa, Will Conley, Proc. of SPIE Vol. 7640,
76401R · © 2010 SPIE · CCC code: 0277-786X/10/$18 · doi: 10.1117/12.845597

overlay performance including on-product overlay budget analysis. It is shown that new scanner improvements are supporting the on-product overlay roadmap. Immersion defect performance is analyzed in section 5. For memory and logic applications the defects requirements are different and we show that latest immersion related defect results for several resist processes are in the stable single digit level through lot.

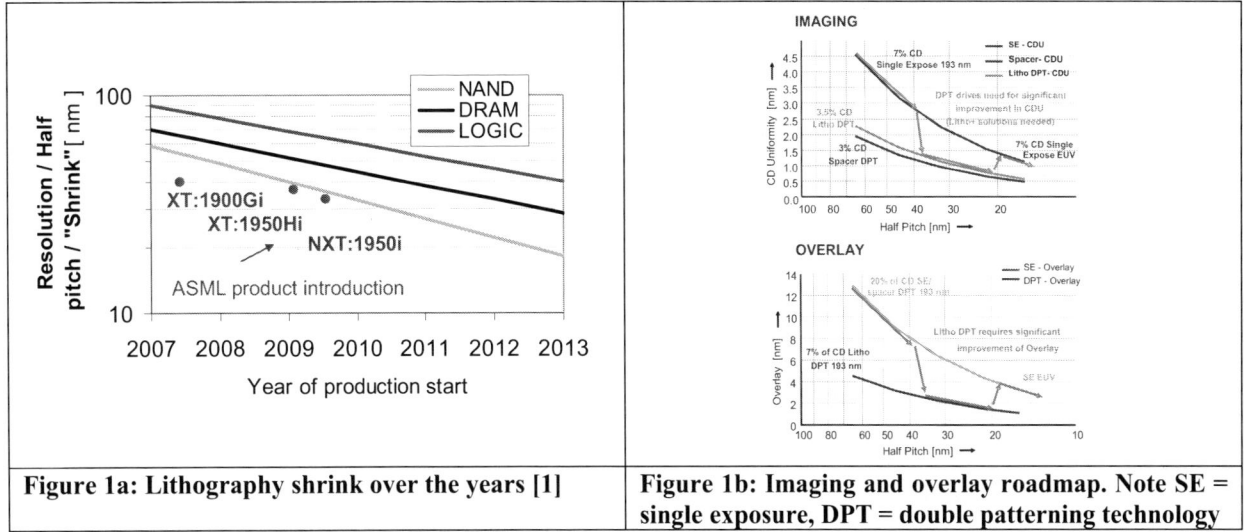

Figure 1a: Lithography shrink over the years [1]	Figure 1b: Imaging and overlay roadmap. Note SE = single exposure, DPT = double patterning technology

2. MATURITY OF IMMERSION SCANNERS IN VOLUME MANUFACTURING

We start analyzing the maturity of immersion litho systems with analyzing litho tool productivity. Productivity can be measured in the amount of wafers output per day. We analysed the best wafer per day output of 1.35 NA immersion systems starting from 2007 when the first tools came on the market. Figure 2 shows that the maximum wafer output increased to above 3400 wafers per day. This immersion wafer output gets close to the mainstream output levels of dry ArF as measured today (figure 3).

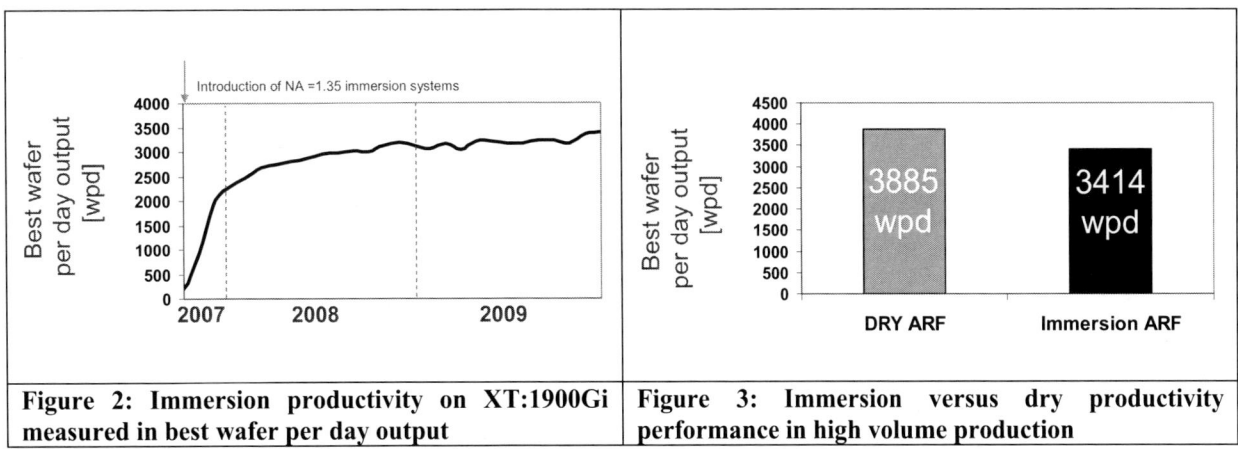

Figure 2: Immersion productivity on XT:1900Gi measured in best wafer per day output	Figure 3: Immersion versus dry productivity performance in high volume production

Besides tool productivity, mature volume manufacturing also requires robust product yield. First enabling step for this is stable long term overlay, imaging and defectivity performance. We have determined the stability of imaging CD uniformity performance of the XT:1900Gi tool by monitoring the intra field CD uniformity of 40 nm dense lines and spaces for a 10 week interval. Result is a stable CD uniformity variation within 0.4 nm (see figure 4).

Robustness of long term overlay performance was investigated by measuring dedicated chuck overlay for a 10 week interval. Multiple wafers were exposed every week and in figure 21 < 3.5-nm full wafer dedicated chuck overlay is shown on the XT:1950Hi.

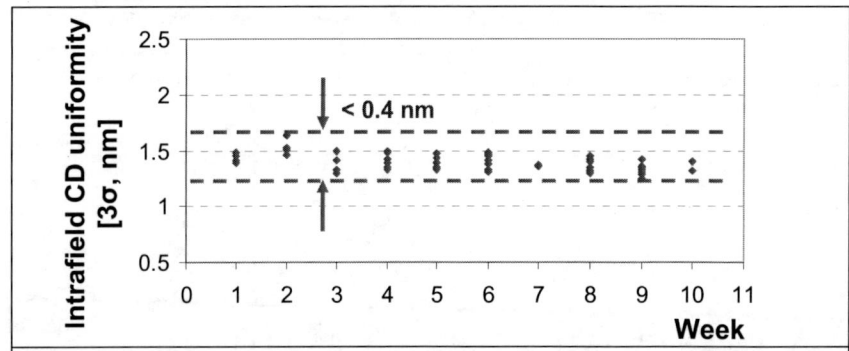

Figure 4: Monitoring of XT:1900Gi CD uniformity (40 nm Dense lines spaces, exposed with annular illumination). Reticle error correction is applied.

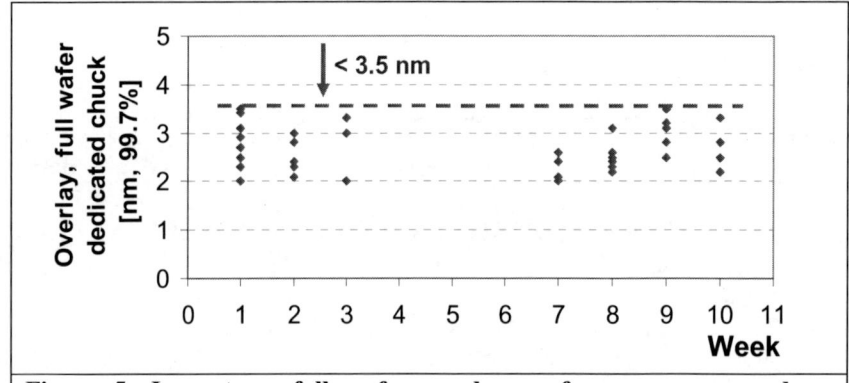

Figure 5: Long term full wafer overlay performance measured on XT:1950Hi. Zerolayer wafer alignment marks are etched in the wafers.

The stability of long term defect performance is tested during a 2.5 months period on a XT:1900Gi that is equipped with an improved (air drag) immersion hood. Less than 4 immersion defects including printing particles are measured on a high contact angle (81° SRCA) topcoat less resist (see figure 6).

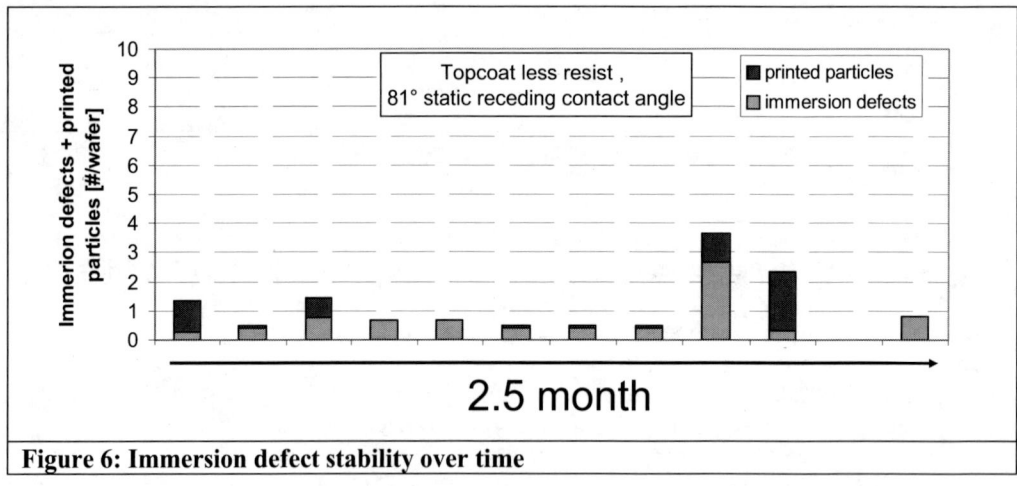

Figure 6: Immersion defect stability over time

The next three chapters investigate in more detail the performance areas that highly influence the product yield being imaging, overlay and immersion related defect performance.

3. IMAGING PERFORMANCE AND FURTHER IMPROVEMENTS

3.1 Imaging critical contributors for high volume Memory

Current mainstream high volume is especially on 4X-nm memory applications. In this chapter we will analyze the main budget contributors in the CD uniformity budget We start with the investigation of the most critical contributors in CD uniformity for a 45-nm NAND application. The CD uniformity analysis was done for the various litho-tool contributors being NA, sigma, dynamics, focus and dose performance based on current mainstream immersion scanner XT:1900Gi tool performance. Also the processing (resist) and reticle contribution were taken into account. Based on Monte Carlo simulation analysis we determined the most critical parameters in the CD uniformity budget. For each individual contributor to CD uniformity, a transfer function, or sensitivity, has been simulated by varying that contributor over its effective range, while keeping the other contributors fixed at their mean value. All transfer functions between the various contributors and the resulting wafer CD were treated as linear except for focus and dynamics (MSD), for which a quadratic formulation was used.

The results are depicted in figure 8. The total full wafer CD uniformity of 3 nm is supporting the 45 nm FLASH node. Especially the reticle CD control and process induced CD variation appear to be main critical contributors to CD uniformity. On the scanner part the aberration control and dynamic image position control (MSD-xy) of the litho tool appear to be main critical contributors.

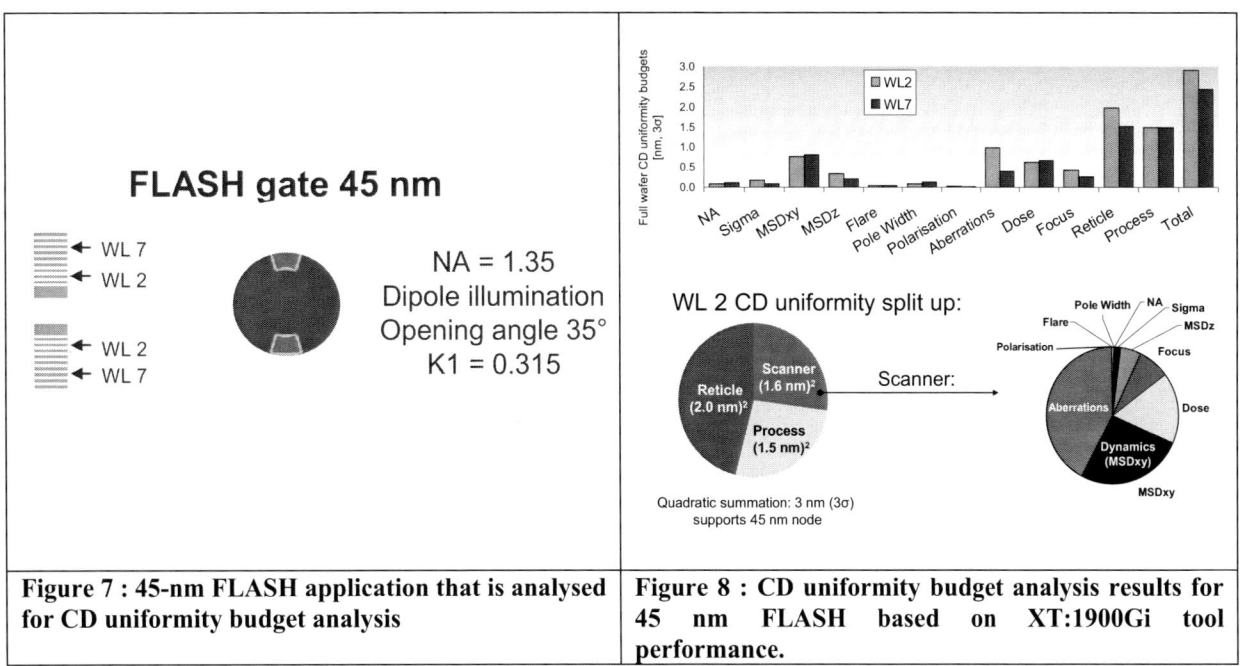

Figure 7 : 45-nm FLASH application that is analysed for CD uniformity budget analysis	Figure 8 : CD uniformity budget analysis results for 45 nm FLASH based on XT:1900Gi tool performance.

3.2 Dynamic control

Dynamic image control is the first main CD uniformity contributor that we evaluate in more detail. Dynamic image control can be expressed in terms of how well the aerial image is kept stable in position during exposure onto the wafer. The aerial image is produced by the illuminated reticle and is transferred by the projection lens onto the wafer. The position of the aerial image at wafer level is influenced by the servo control of the reticle stage and wafer stage and also the position stability of the projection lens. We investigated experimentally the dynamic position performance of both the stages and the lens during exposures. The dynamic performance of the reticle and wafer stage position control is measured with exposing both stages at maximum scan speed. The position of the reticle stage is measured with encoders and the position of the wafer stage with interferometers. The synchronised position error between the reticle and wafer stage is recorded during the exposures and the mean standard deviation of this synchronization error is translated into aerial image contrast loss for 40 nm dense lines with annular illumination as explained in reference [3]. The dynamic position performance of the lens is measured in x-direction by means of a reticle containing a single chrome line that is exposed with full reticle stage speed onto an optical sensor in the wafer stage. Position errors are translated in contrast variations. In order to get an indication of the mechanical stability of the lens, accelerometers on the top and bottom of exposure lens measure the accelerations during these exposures.

We determined a dynamics induced contrast loss reduction of more than 25% for the XT:1950Hi with respect to the XT:1900Gi for the stage position control and more than 60% reduction for the lens dynamics (see figure 9a). These improvements are mainly originated from a new design of the reticle stage on this system which causes less vibration coupling with the exposure lens. Furthermore the stage servo control is improved resulting in less synchronization errors between the exposure chucks. The quadratic transfer function of dynamic performance to CD control results in up to 25% improvement of dynamics MSD-xy induced CD uniformity (see figure 9b).

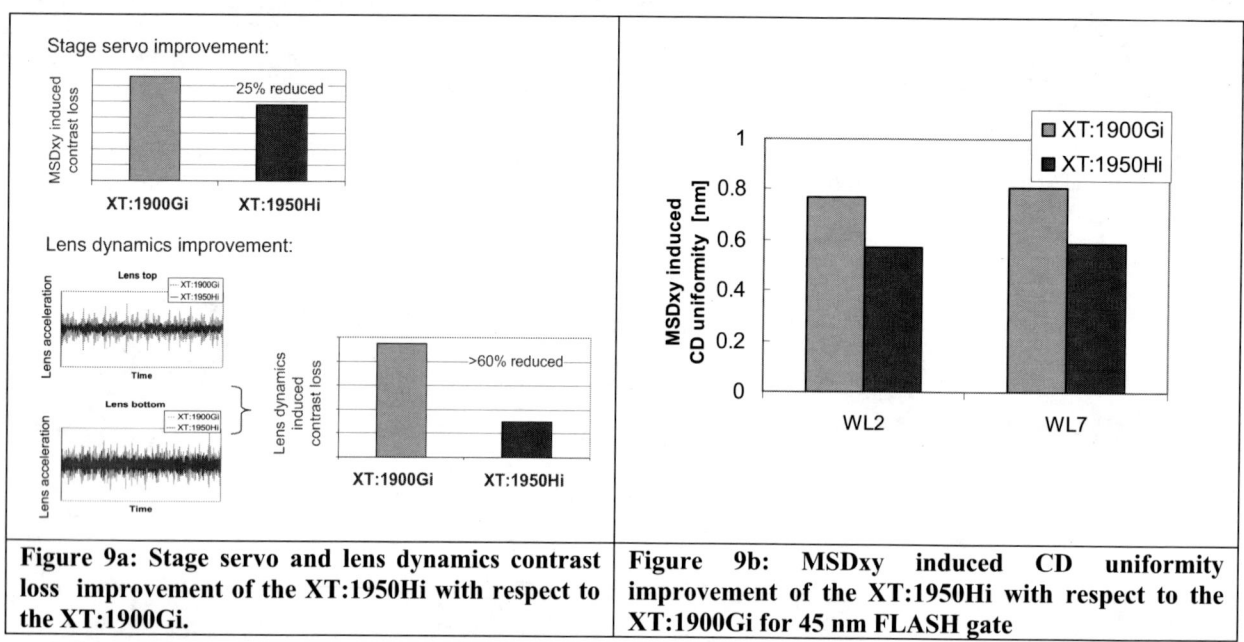

Figure 9a: Stage servo and lens dynamics contrast loss improvement of the XT:1950Hi with respect to the XT:1900Gi.

Figure 9b: MSDxy induced CD uniformity improvement of the XT:1950Hi with respect to the XT:1900Gi for 45 nm FLASH gate

3.3 Wave front control

The second main CD uniformity contributor in today's high immersion volume production application (see section 3.1) that is evaluated in more detail is wave front aberration control on the litho tool. During wafer exposures the wave front is not perfectly stable due to the (minimal) absorbed amount of DUV light by the exposure lens. In order to control this wave front error or aberration drift, there are several actuators in the lens that can generate counterbalancing wave front corrections. A feed forward model is used to predict the time dependent drift in aberrations. The results of the feed forward model are translated by the lensmodel into necessary amount of lens manipulators adjustments needed to compensate the aberrations. After the manipulators are tuned, the exposures are performed. The resulting wave front is measured with image sensors and if the measured wave front differs from the predicted one then the feed forward model is adjusted (feedback (FB) loop).

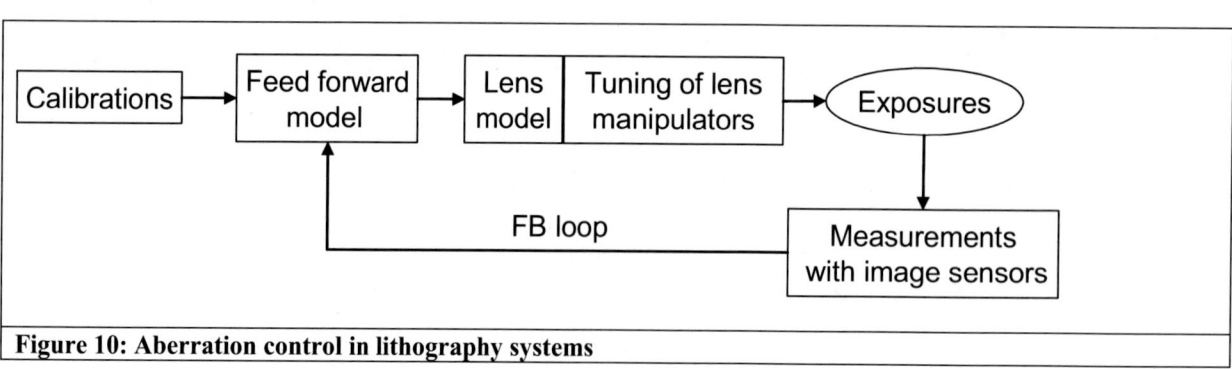

Figure 10: Aberration control in lithography systems

Last year we introduced [4] wave front control that is able to correct independently for wave front Zernikes Z6, Z10 and Z17 aberrations drifts in the system, especially designed for 4x and 3x nm DRAM applications using rotated structures and rotated illumination types in the $6F^2$ cell layout. We expand the application range for 4x and 3x applications to higher order astigmatism and 4 foil wave front correction specifically designed for FLASH type of structures oriented in xy orientation using non rotated dipole illumination modes (see figure 11). We increase the flexibility in wave font control even further for generic application use and high throughput by introducing a new high spatial manipulator that is able to control Zernike aberrations up to Zernike Z64 independent of illumination mode. The improvement in aberration control for this new manipulator is shown in figure 12.

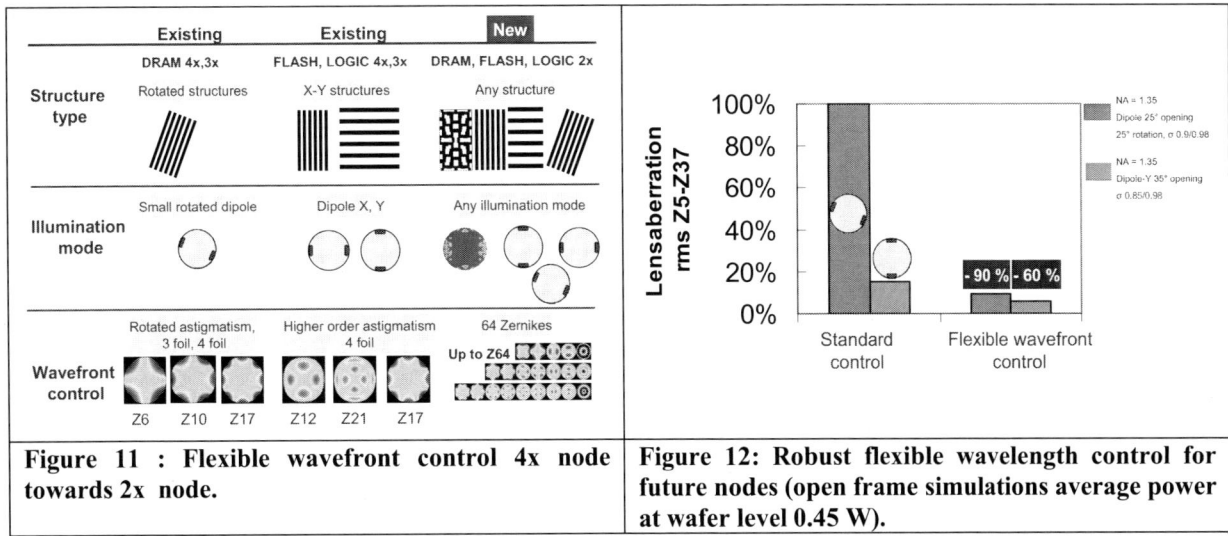

Figure 11 : Flexible wavefront control 4x node towards 2x node.

Figure 12: Robust flexible wavelength control for future nodes (open frame simulations average power at wafer level 0.45 W).

3.4 Process control

The CD uniformity on wafer level is not only determined by the litho-tool performance itself, but also by the resist processing and etch steps that are required. We have shown in section 3.1 process control is one of the major contributors in CD uniformity in today's used high volume immersion applications. Examples of process induced CD variations are temperature variations on the post exposure bake plate or the applied etch process. Elimination of these effects, for instance by optimizing temperature uniformity of the bake plates is possible, but there will always remain residual process induced intra wafer CD fingerprints. Since the structures to be exposed on the wafer are sensitive to dose variations it is possible to introduce a process counter balancing corrective dose fingerprint across the wafer. We performed an experiment by exposing a full wafer with 45-nm FLASH type of structures and corrected the intra wafer CD fingerprint using dose scanner corrections. We were able to reduce the intra wafer CD fingerprint by 60% (see figure 13). The correction mechanism of applying the dose correction is applicable in volume production environment since the corrections are loaded automatically onto the scanner.

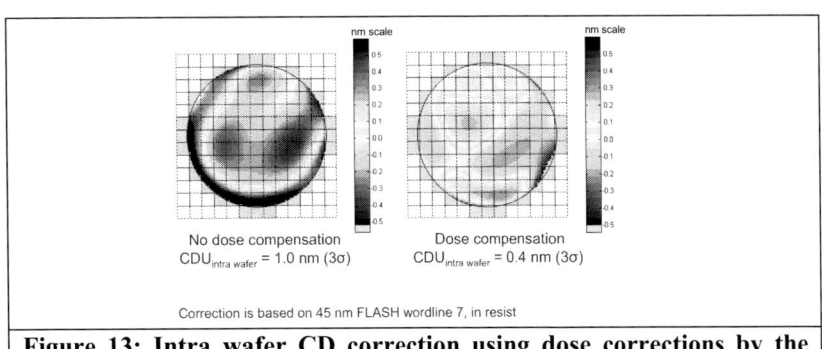

Figure 13: Intra wafer CD correction using dose corrections by the litho scanner on 45-nm FLASH structures using XT:1900Gi scanner

3.5 Reticle and litho cluster matching

As explained in section 3.1, CD variations on reticle itself also contribute to a large extent to the CD uniformity on wafer level in today's 4x memory application. Compensation for the intra field reticle profile can be done with intra field dose corrections in 2 dimensions.

High volume production applications run on multiple litho clusters (scanner and track combinations). Especially in logic application volume production, more than 1 reticle is used for the same product. Reticle copies will not all look exactly the same in terms of CD on nanometer level and will not yield exactly the same on all litho clusters in the fab. Reticle matching and litho cluster matching is therefore key for optimal yield. We did 2 experiments to match 2 contact hole reticles and 2 litho clusters (also using a contact hole reticle) to each other (see figure 14).

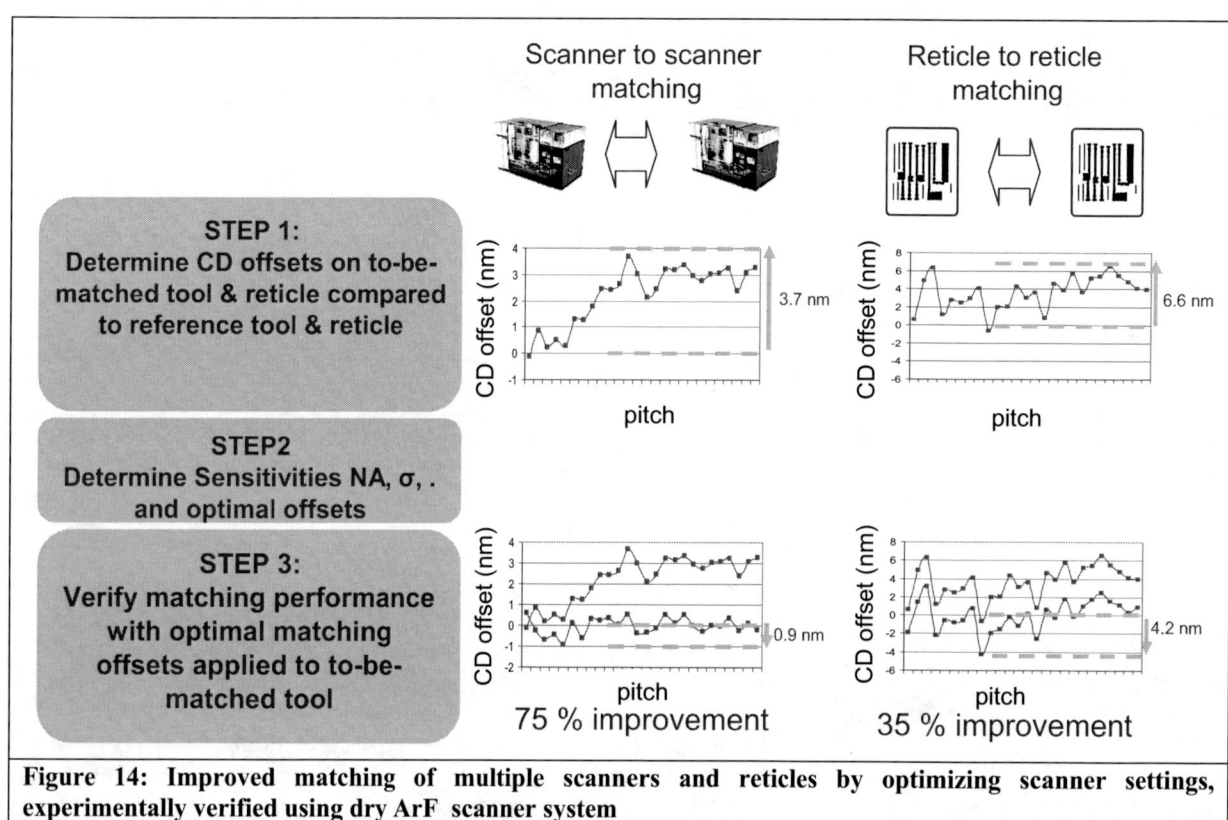

Figure 14: Improved matching of multiple scanners and reticles by optimizing scanner settings, experimentally verified using dry ArF scanner system

First we determined on the reference litho cluster and with the reference reticle the CD versus pitch behaviour. We repeated this on the to-be- matched litho cluster and with the to-be-matched reticle. The CD matching offset across pitch was determined and then we calculated optimal dose, focus and sigma offset such that the matching offset was reduced. We verified the resulting optimized CD offset across pitch and were able to improve the reticle matching CD offset by 35% and the litho cluster CD matching error by 75%. Also this technique can be fully automated and is useful in volume production today.

4. OVERLAY CONTROL

4.1 On product overlay

The yield of a product in volume production is not only determined in terms of imaging CD uniformity control but also in terms of how well the critical imaging structures are aligned with respect to each other. When analyzing this so called on product overlay performance we can make a division in three categories. The first category is the overlay capability of the litho scanner itself. How well the exposure stages can be aligned to each other is a quality of the litho tool, but beside this intrinsic capability also the specific application that is exposed can cause specific overlay errors. The reticle itself contains writing errors, the density of the structures on the reticle in combination with the specific illumination mode can result in heating of the reticle or projection lens resulting in overlay errors. Finally the exposure field grid is calibrated which might also not be perfect, resulting in field matching errors.

The second category for on product overlay contains the errors introduced by the use of the specific litho process itself. For instance chemical mechanical polishing (CMP) steps can degrade alignment marks on the wafer affecting alignment quality and also etching steps might deform the exposure grid.

The last category for on product overlay contains the stability control of the litho cluster in the fab. Automatic process correction (APC) strategies are used in fab environment to maintain stable performance, but the quality of these corrections is determined by the accuracy of the metrology that is used to determine the corrections and the effectiveness of the control algorithms themselves.

In order to determine the maturity of immersion systems in volume production concerning overlay, we investigate the intrinsic immersion high volume overlay capability and latest improvements that have been made. Furthermore we investigate a realistic on product overlay example.

		Contributors
System	System intrinsic	Intrinsic scanner
System	Application dependent	Lens heating, reticle heating, ill. setting dependency, routing dependency, reticle writing errors
System	Matching	Interfield and intrafield matching
Process	Grid deformation	Grid deformation due to processing
Process	Mark optimization	Alignment mark degradation due to processing
Control	Metrology	Metrology accuracy
Control	Stability	Stability control

Figure 15: From tool overlay to on product overlay

4.2 Intrinsic scanner overlay

The first category of on product overlay, the intrinsic tool overlay capability, is investigated in this section. The immersion scanners have a dual stage concept, and the stage to stage (single machine) overlay budget for the three generations of 1.35 NA scanners is depicted in figure 16a. Last year we introduced the XT:1950Hi (see ref [4]) and explained improved chuck deformation control (1) and thermal control of the wafer table (3).We have improved the overlay further on this tool to achieve sub 4 nm overlay performance by improving the alignment accuracy on the tool (2). Before an exposure takes place the reticle position has to be aligned to the wafer and this is done by exposing a reticle alignment mark to its equivalent at wafer level (factor 4 reduction) that is placed on a transmission image sensor (TIS) in the exposure chuck. The aerial image that is produced at the TIS sensor is sampled by a photo sensor under the TIS grating by scanning in x,y and z direction, and the optimum position is found as highest intensity being the best aligned position. We optimized the aerial image sampling scheme resulting in an overall repro improvement of the aerial image position of 20% (see figure 16b). The improved dynamics of the reticle stage as explained in section 3.1 results in more stable lens positioning and thus reduced overlay errors (4). We also have improved the conditioning of the interferometer beams that are used to position the stages (5). The overlay on the NXT:1950i is further improved by using encoders for stage positioning instead of interferometers (6) and also the wafer stage has improved servo control (7). These improvements are further explained in ref [2].

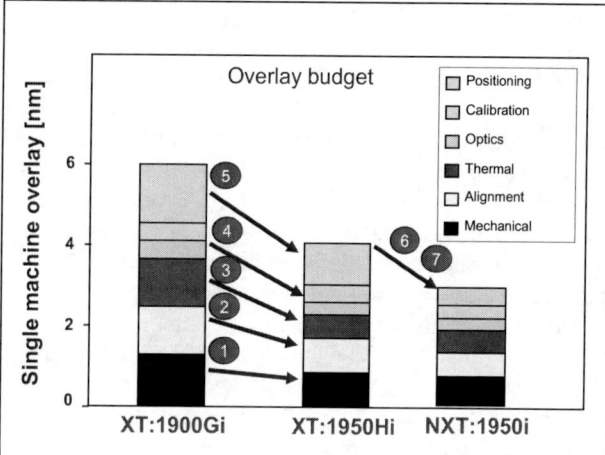

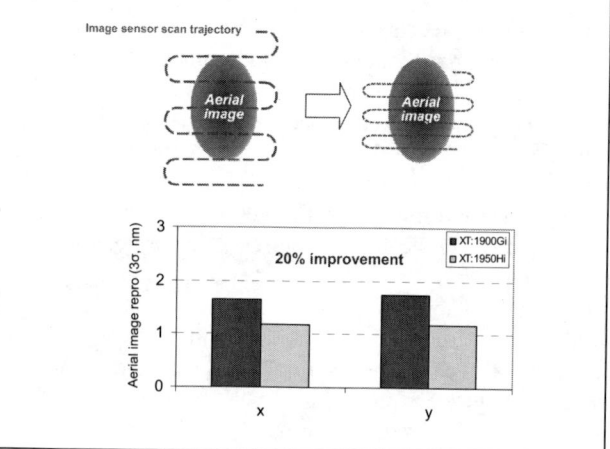

Figure 16a Single machine overlay budget for three generations of 1.35 NA immersion scanners	Figure 16b: Improved aerial image capturing by optimizing the sampling scheme on XT:1950Hi

We tested the single machine overlay performance in a three day period on the XT:1900Gi and XT:1950Hi. In figure 17 we show that the improvements as described above result in a sub 4 nm overlay performance on the XT:1950Hi on multiple systems. In dedicated dual chuck mode these improvements resulted in sub 3 nm overlay through the lot. This is a full throughput mode where the first and the second layer are exposed on the same chuck.

Thermal control of the wafer table is shown in figure 18. In this experiment we exposed 10 wafers in single chuck mode with the second layer in reversed wafer sequence onto the first layer to enhance thermal effects. We achieved full wafer overlay performance similar to a reduced layout.

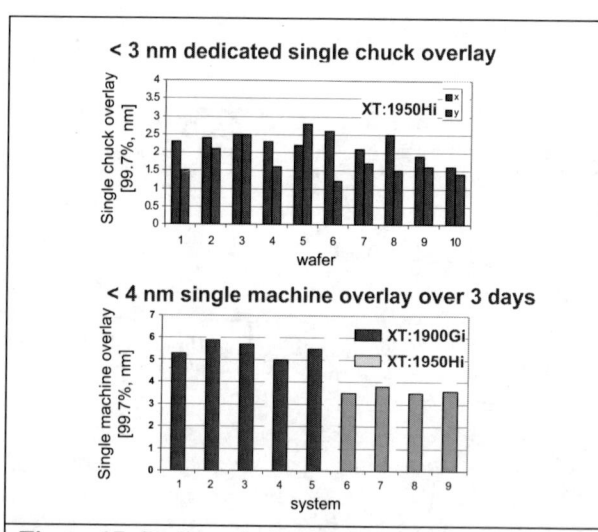

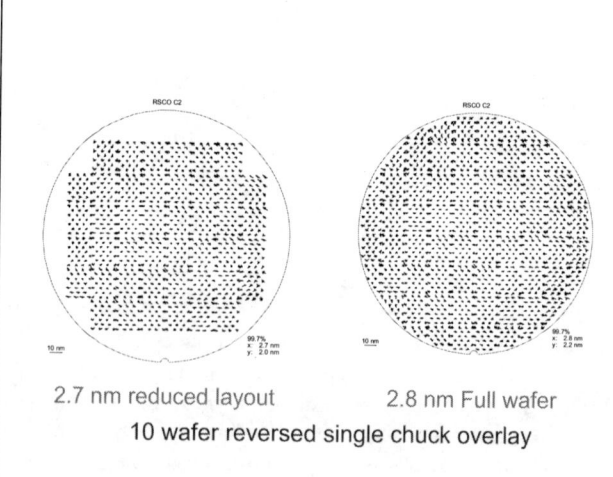

Figure 17: Dedicated chuck overlay and 3 day chuck to chuck overlay on multiple systems.	Figure 18: Single chuck overlay with the second layer in reversed wafer sequence, full wafer versus reduced layout, exposed on XT:1950Hi

4.3 Product overlay verification: Contact-to-gate overlay

Immersion overlay in volume production is not only influenced by the intrinsic scanner overlay performance but also highly depends on the specific application that is used (see section 4.1). In this section we verify the on product overlay performance experimentally with a generic application used in IC industry: placement of contacts on top of transistor gates. We applied realistic production circumstances as close as possible.

The first layer mimics a gate layer exposure with a reticle transmission of 17%. The gate layer reticle contains 45-nm dense lines and spaces illuminated with dipole x illumination at max NA with a dose of 31 mJ/cm2. The machine is conditioned to start with a cold lens to show a worst case lens heating transient effect at maximum scan speeds.

The second layer mimics a contact layer with a full chromed reticle with low transmission (1.5%) including a pellicle exposed with conventional illumination at reduced NA of 1.2 with a dose of 50 mJ/cm2 which is appropriate for contact hole printing. Settings are chosen to show the worse case effect of lot timing variations starting with a hot lens that is cooling down during the lot, combined with the transient effect of reticle heating. To keep the lens in its hot state, the machine exposes at least two dummy lots of 25 wafers with the layer 1 reticle and exposure settings before the second layer exposure is started. For the second layer exposure, the 23 wafer lot is split up in 5 separate lots. This is done to create a mix of chuck to chuck overlay and dedicated chuck overlay. Furthermore, the 5 lots are defined such that they include the occurrence of an occasional single wafer lot for process verification and simulation of 1 rework wafer between the layers, which actually means that there is a delay in time to the next wafer. Reticle writing errors are part of on product overlay and are not corrected for in this experiment that uses 2 reticles with a full field exposure size 26 x 33 mm.

No separate alignment marks (zero layer) are exposed before layer 1, which means that the first layer is exposed without wafer alignment. For exposure of the second layer, wafer alignment is performed on scribelane marks that are printed in the first layer exposure.

In order to correct for the inherent overlay errors introduced by the direct alignment scheme, and to imitate production environment, process corrections are applied. For reasons of timing, these corrections are not determined by means of send-ahead wafers and physically applied during the exposures. Instead, the combined data set of the first and last wafer of the first lot of 4 wafers will act as send-ahead lot for all 5 second layer lots: a batch correction will be modeled on wafer 1 and 4 of this first lot, and the resulting 10 parameter correction set is applied to all 5 lots during the data analysis to simulate the use of process corrections.

The overlay is measured using scatterometry, with box-in-box targets or on the exposure tool itself. In order to include full wafer coverage overlay, detailed intra field overlay effects (lens aberration drift effects, reticle heating and writing errors), and detailed inter field overlay effects (exposure grid) we applied a sampling schema as depicted in figure 20.

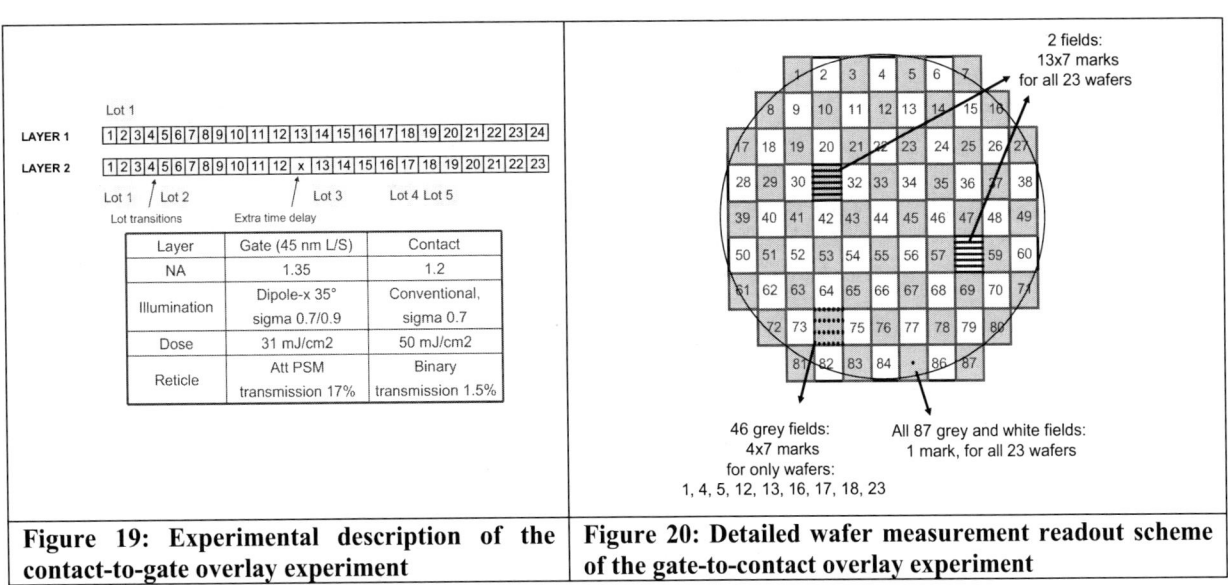

Figure 19: Experimental description of the contact-to-gate overlay experiment	**Figure 20: Detailed wafer measurement readout scheme of the gate-to-contact overlay experiment**

Results of this extensive gate versus contact overlay experiment are depicted in figure 21. Sub 7-nm overlay was achieved on the XT:1950Hi which meets the overlay requirement for the 45 nm node and even for the 32-nm single patterning node. Clear improvement is seen with respect to the XT:1900Gi which is the result of the system overlay improvements as described in the previous section. This experiment shows the robustness of these high NA immersion systems for overlay critical applications in a production like environment.

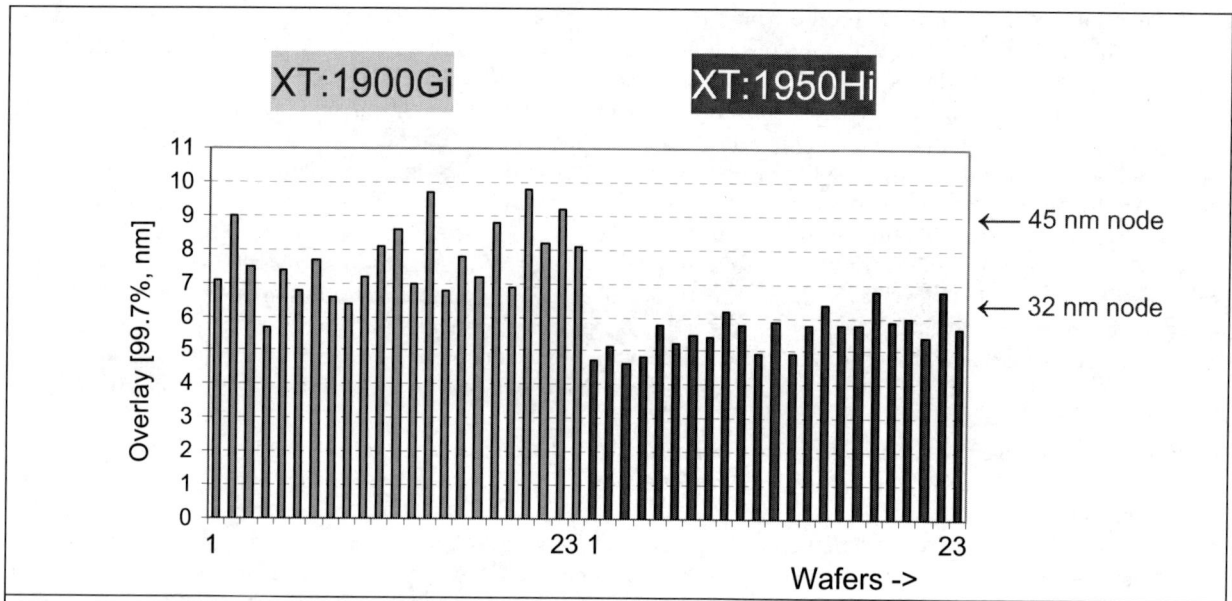

Figure 21: Experimental overlay results of the gate versus contact layer application on XT:1900Gi and XT:1950Hi; Overlay number is determined by removal of edge points < 3mm outer radius, 5 sigma threshold for spot outliers, 10 parameter APC correction based on wafer 1 and 4. A 99.7% criterium is used to present the final overlay numbers.

5.   IMMERSION DEFECT CONTROL

5.1 Immersion defect control for DRAM, FLASH and logic for 45 and 32-nm node

Volume production with high NA immersion systems requires tight control of immersion related defects due to their impact on yield. These printed defects are caused by the interaction of water, resist, particles and the exposure system itself. The immersion defect requirements for volume production are application dependent since they are highly dependent on the amount of immersion layers that are used in production and the die size. The tightest immersion defect control demands are posed by logic type of applications, which can be illustrated with a simple model describing the impact of immersion defects on yield (see figure 22). Extrapolating the logic requirements from the 45-nm node to the 32-nm node we show that at least a factor of 1.8 better defect performance is required to reach the same yield level. This is caused by the increased number of immersion layers that will be used for this node (estimated from 10 to 18 layers).

5.2 Air drag Immersion hood defect control

A very important factor in immersion defect control on immersion litho scanners is the so-called immersion hood that controls the water based optical element between the lens and the wafer. When the wet exposure takes place and the wafer moves, the meniscus of the water based lens is affected by viscous forces that work to pull water from the lens bath onto the wafer. The water containment is influenced by the specific design of the immersion hood and also by the surface properties of the resist stack, most importantly, the contact angle.

Last year, we introduced a new immersion hood design and showed the first results [4]. In figure 23 we compare the previous and new design by means of defect population data of the XT:1900Gi [5] and the XT:1950Hi. The immersion defect control is measured with a patterned defect test that uses a 45-nm dense lines and spaces reticle to expose the wafers. Defect inspection of patterned wafers is done with optical inspection tooling, followed by a SEM review of the found defects. The measured defects are classified into types which relate to their root causes. In [6] we discuss the classes we use in our immersion tool qualification process. The defect qualification for 45-nm patterns is done with KLA 2800 and UVision inspection tools.

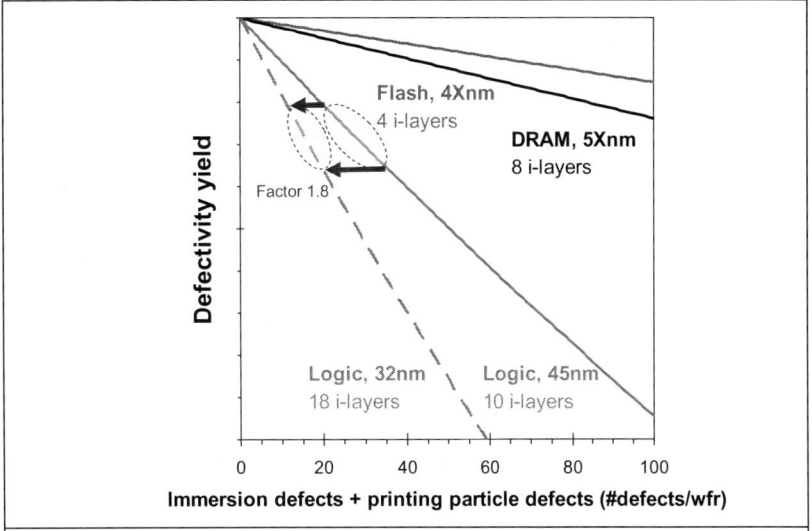

Figure 22: Immersion defect versus yield for DRAM, FLASH and logic applications.

Model Assumptions:

Defectivity yield = e -(#layers x (die size/wafer area) x (defects/wafer))

Die size FLASH: 150 mm^2, die size DRAM 50 mm^2; die size logic: 250 mm^2

Comparing the two populations in figure 23, we can see more than a factor of 3 improvement in immersion defect control, making the new immersion system well equipped for extension to 32-nm node volume production.

As stated above an important parameter in immersion defect control is the contact angle of the resist stack that is used. We tested the various commonly used resists in volume production environment on defect control. In figure 24, we show robust defect control with single digit immersion defects (including printing particles) for topcoat less resist stacks with a contact angle (SRCA) between 59° and 81°.

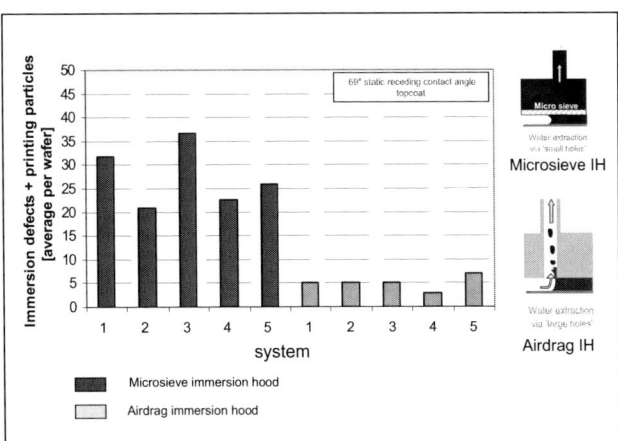

Figure 23: Comparison of population immersion defect performance of micro sieve immersion hood and improved air drag immersion hood

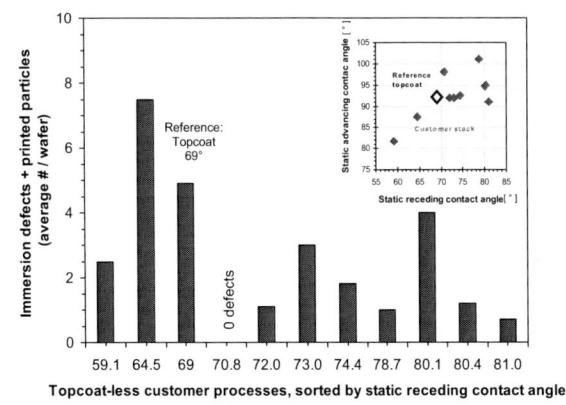

Figure 24: Immersion defect control for various topcoat less resists ranging from low to high contact angles using air drag immersion hood

6. SUMMARY AND CONCLUSIONS

The semiconductor industry has adopted water-based immersion technology as the mainstream high-end litho enabler for 5x-nm and 4x-nm devices. Exposure systems with a maximum lens NA of 1.35 have been used in volume production since 2007, and today achieve production levels of more than 3400 exposed wafers per day.
In 2009 ASML introduced a new improved immersion exposure system, the XT:1950Hi, that enables the production of 38-nm memory devices [4].
This paper analyses the maturity of immersion technology by analyzing robust control of tool productivity, imaging, overlay and defectivity performance using the mainstream ArF immersion production systems XT:1900Gi and XT:1950Hi.

The first step of maturity of the ArF immersion tools is shown in wafer output performance which approaches the level of the main stream dry ArF systems. We have shown stable system performance required for high volume manufacturing: 40 nm dense lines CD uniformity stability of < 0.4-nm, < 3.5-nm dedicated chuck overlay control and single digit immersion defects per wafer.

Besides having a stable system, the imaging control is improved on the critical areas required for imaging 4x-nm memory applications. We have analysed the CD uniformity budget of a 45-nm FLASH application and have shown that actual tool performance supports the 45-nm node requirements and beyond.

Dynamic performance of the litho tool appears to be one of the critical parameters and we have shown with CD budget analysis an improvement of dynamics induced CD uniformity of 25 % on the XT:1950Hi with respect to its predecessor. We have shown flexible aberration control optimized for rotated (DRAM) and non-rotated (FLASH) dipole applications. Further flexibility towards future imaging nodes is shown with a new flexible high spatial wave front manipulator for control of up to 64 Zernikes.
Another major CD contributor is the litho process itself. We have shown experimental results of exposure dose compensation on FLASH application reducing the intra wafer CD uniformity fingerprint by more than 60%.
Especially for logic applications, which rely on using multiple reticles for the same application on multiple tools in production, we have shown experimental results of litho tool reticle matching and litho cluster matching improvement of 75%.

On the overlay part we have shown the latest improvements in overlay on the XT:1950Hi resulting in < 4-nm single machine overlay control in dual chuck mode on multiple systems. Reticle align control is improved by better aerial image capturing. Also we have shown that the new wafer table heating control concept in the XT:1950Hi is able to maintain overlay performance for all dies on the wafer including the edge dies. Based on experimental data we have shown application robustness with contact-to-gate overlay < 7-nm, supporting the 45-nm node down to 32-nm node layer to layer overlay requirements for single patterning.

We have indicated that logic applications have the strongest defectivity requirements. We demonstrated more than a factor of 3 immersion defect control improvement with the air drag immersion hood that was introduced on the XT:1950Hi, making it extendable for the 32-nm node logic applications. Further, long term defect robustness as well as single digit immersion defect control for various topcoat less resists used in the industry were shown.

All in all we have shown excellent maturity of 1.35 NA immersion litho scanners not only in basic scanner tool performance but also in the production like environment for today and in the future.

ACKNOWLEDGEMENTS

We would like to thank the XT:1900Gi and XT:1950Hi project teams for making this work possible. Special gratitude is expressed to Youri van Dommelen, Bart Lemmen, Antoine Loock, Joost Beke, Paul Derks, Marinus Jochemsen, Frank Coppelmans, Bert Vleeming, Martijn Leenders, Jos Benschop, Alex van der Hoff, Alexander Padiy, Hans van der Laan,

Hoite Tolsma, Jos Maas, Marco Pieters, Mark van de Kerkhof, Alexander Germeau, Bert Vleeming, Clemens Beijers, Alberto Colina, Orion Mouraille, Joost Bekaert, Ton van Gaans, Paul van Adrichem, Peter Lee, Bill Arnold.

REFERENCES

[1] Customer data as collected by ASML (2010)

[2] Tom Castenmiller et. al, ' Towards ultimate optical lithography with NXT:1950i dual stage immersion platform', Proc SPIE , to be published (2010)

[3] Alexander Straaijer, " Formulas for lithographic parameters when printing isolated and dense features", J. Microlith., Microfab., Microsyst. Okt-Dec 2005, vol 4

[4] Igor Bouchoms et. al., "Extending single-exposure patterning towards 38-nm half pitch using 1.35 NA immersion", Proc SPIE 7274 (2009)

[5] Jos de Klerk et. al., "Performance of a 1.35 NA ArF immersion lithography system for 40-nm applications XT:1900i System Performance", Proc SPIE 6520 (2007)

[6] Jan Mulkens et. al. "Latest Developments on Immersion Exposure Systems", Proc. SPIE 6924 (2008)

The impact of resist model on mask 3D simulation accuracy beyond 40nm node memory patterns

Kao-Tun Chen[a], Shin-Shing Yeh[a], Ya-Hsuan Hsieh[a], Jun-Cheng Nelson Lai [a],
Stewart A. Robertson[b], John J. Biafore [b], Sanjay Kapasi [b] , Arthur Lin [b]

[a]Powerchip Semiconductor Corp., No. 12 Li-Hsin RD, Hsinchu, Taiwan
[b]KLA-Tencor Corp., 8834 N. Capitol of Texas Highway, Austin, TX 78759, USA

ABSTRACT

Beyond 40nm lithography node, mask topograpy is important in litho process. The rigorous EMF simulation should be applied but cost huge time. In this work, we compared experiment data with aerial images of thin and thick mask models to find patterns which are sensitive to mask topological effects and need rigorous EMF simulations. Furthur more, full physical and simplified lumped (LPM) resist models were calibrated for both 2D and 3D mask models. The accuracy of CD prediction and run-time are listed to gauge the most efficient simulation. Although a full physical resist model mimics the behavior of a resist material with rigor, the required iterative calculations can result in an excessive execution time penalty, even when simulating a simple pattern. Simplified resist models provide a compromise between computational speed and accuracy.

The most efficient simulation approach (i.e. accurate prediction of wafer results with minimum execution time) will have an important position in mask 3D simulation.

Keywords: EMF, 3D mask, Mask topography, Lumped parameter resist model, Full physical calibrated resist model

1. INTRODUCTION

In ArF immersion process where mask pattern pitch (3X and 2X nodes) is many times smaller than exposure wavelength where strong RET and high NA are required. The light diffraction can not be correctly predicted by the Kirchhoff approximation mask model (or thin mask model) – as most frequently used in many imaging simulations today. Precise and accurate forecasting of the wafer pattern requires rigorous electromagnetic field analysis (EMF or 3D mask mode) which fully considered mask topography effects [1]. Many studies have indicated significant differences in patterning prediction between Kirchhoff approximations and 3D mask models[2][3][4].

Besides the difference of aerial image, ArF resist kinetics also play an important role which impacts the real image

Optical Microlithography XXIII, edited by Mircea V. Dusa, Will Conley, Proc. of SPIE Vol. 7640,
76401S · © 2010 SPIE · CCC code: 0277-786X/10/$18 · doi: 10.1117/12.846010

obtained on the wafer. However, 3D mask model is well-known for longer execution time and also consumes large computing resources [4], even for small mask areas of several microns. In product development stage, it's much predicted accuracy and time-concerned to cost and market. There is another choice, simplified resist model, which is generally called LPM model (Lumped-Parameters Model) and may provide a compromise between computational speed and accuracy.

In this paper, we quantify the CD bias between experiments and aerial image simulations across a range of 40nm node flash memory patterns. The difference will be used to gauge the importance of using rigorous EMF model. We also calibrate both the full physical and simplified resist models for 2D and 3D mask simulations. The comparison of CD accuracy and running time will be used to gauge the importance of using any resist model.

2. 2D & 3D MASK AERIAL IMAGE SIMULATION

First, we focus on some well-known typical features of flash product beyond 40nm node to figure out what kind of feature is sensitive to mask topological effects. For 1D patterns, select-gate features are studied, and for 2D patterns, landing pad and cut line features are studied (Fig. 1).

Figure 1 : Typical features of memory, (a)1D features : Select-gate area, (b)2D features : landing pad area, (c) 2D features : cut line.

In first part, we observe pure optical simulation accuracy of select-gate area design 1. WL1 to WL12 are chosen to monitor the accuracy because these features are close to SG features which may suffer much proximity effect caused by strong variation of pattern deployment. △CD is defined with ADI CD difference between simulation and experimental data.

In fig. 2(a), pure optical aerial image simulation is done with Kirchhoff approximation (2D mask) and EMF mask (3D mask) compared to experimental data with commercial PR A. It's obviously that ADI CD bias of EMF mask is much better than of Kirchhoff approximation, especially WL1 to WL6 CD bias which are closer to SG feature with much optical proximity effect suffered. The maximum. ADI CD bias is only about 1.2nm of pure optical behavior of EMF mask.

However, it is quite process related. In fig. 2(b), experimental ADI CDs of two different PR are compared to simulation results of EMF mask. The maximum. ADI CD bias is increased to 5.6nm with commercial PR B. It implies that litho process change may induce worse accuracy by different PR parameters. Only aerial image with EMF mask simulation is surely enough. Different calibrated PR models of commercial PR A will be discussed in following sections.

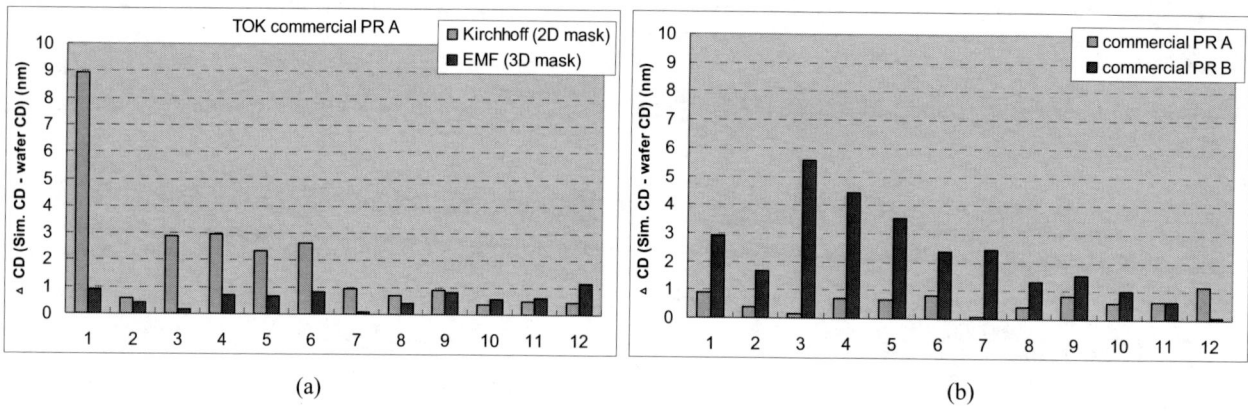

Figure 2: △CD of aerial image simulation result benchmark with experimental data of SG features : (a)PR A ADI CD errors with 2D mask Kirchhoff approximation and with 3D EMF mask, (b) PR A and PR B CD errors with 3D EMF mask.

In Fig. 3, L1, L2, and L3 of landing pad features through different focus with 2D mask Kirchhoff approximation and with 3D EMF mask are observed. In Fig. 3(a), it's also obvious that with 3D EMF mask the simulation compensation is better than with 2D Kirchhoff approximation, and the pure optical model simulation difference of two mask approaches is shown in Fig. 3(b). The difference range is from 1.4nm to 2.6nm, which is compensated by 3D EMF mask simulation. Compared to select-gate features, which the simulation difference of 2D and 3D mask is large as 8nm in WL1, the simulation difference of landing pad seems acceptable due to dense structure. Though it indeed a good way to adopt 3D EMF mask for landing pad feature to get more accuracy, it also takes a huge time difference to gain.

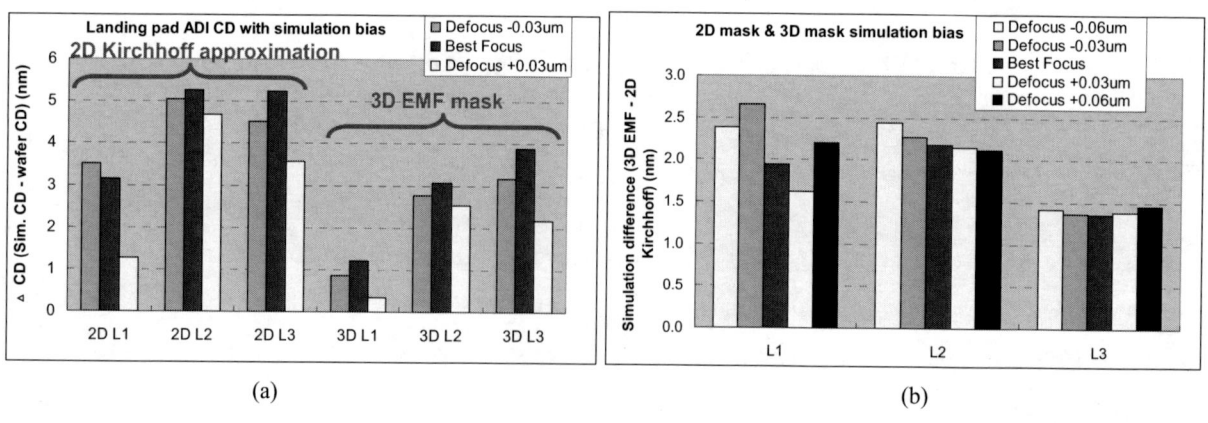

Figure 3: △CD of aerial image simulation result benchmark with experimental data of landing pad features : (a)PR A ADI CD errors with 2D mask Kirchhoff approximation and with 3D EMF mask through focus, (b) pure optical model simulation difference of 2D mask Kirchhoff approximation and 3D EMF mask through focus

In Fig. 4, L1, L2, and L3 of cut-line features through different focus with 2D mask Kirchhoff approximation and with 3D EMF mask are observed. In Fig. 4(a), There is no obvious difference between with 3D EMF mask and 2D Kirchhoff approximation, and the pure optical model simulation difference of two mask approaches is shown in Fig. 4(b). The difference range is from 0.4nm to 2.4nm, which is compensated by 3D EMF mask simulation. Compared to select-gate features, which the simulation difference of 2D and 3D mask is large as 8nm in WL1, the simulation difference of cut-line feature seems acceptable due to dense structure.

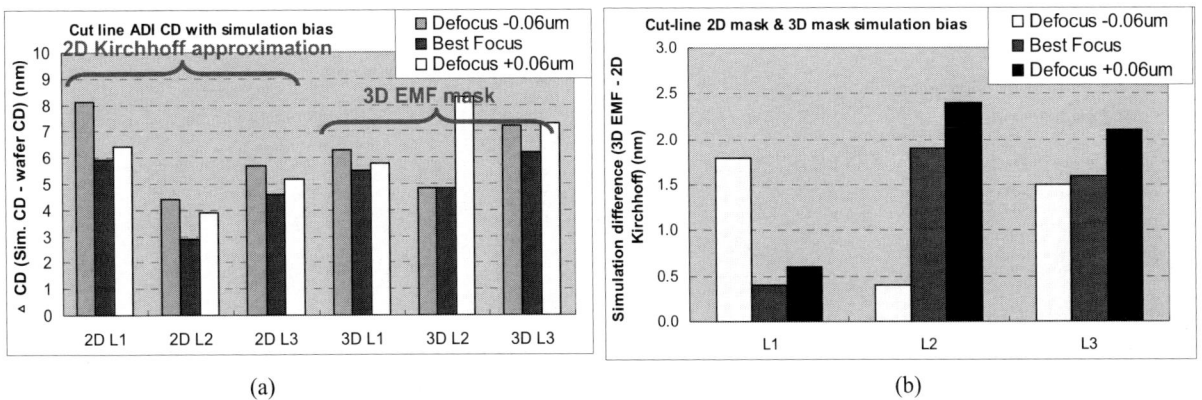

(a)   (b)

Figure 4: △CD of aerial image simulation result benchmark with experimental data of cut-line features : (a)PR A ADI CD errors with 2D mask Kirchhoff approximation and with 3D EMF mask through focus, (b) pure optical model simulation difference of 2D mask Kirchhoff approximation and 3D EMF mask through focus

3.   DATA COLLECTION AND PR CALIBRATION

3.1 Evaluating goodness of fit with the root mean square error

Model calibration is performed by minimizing the RMS error between simulation and actual data, yielding a set of kinetic parameters for each modeled resist. The goodness of model fit is estimated by the standard deviation of the error between the data and predictions [5]:

$$rmse = \left((N-1)^{-1} \sum_N w_i \left(\langle x_i \rangle - f(x) \right)^2 \right)^{1/2} \qquad (1)$$

$$w_i = \sigma_i^{-2} N^{-1} \sum_i \sigma_i^{-2} \qquad (2)$$

RMSE is the weighted root mean square of the error associated with the second moment about the mean. The weighting function *wi* is calculated from the statistics of repeat trials.

3.2 Feature selection and data collection

The physically-rigorous resist model and LPM model are calibrated using CDSEM S9380 metrology measurements from wafers processed above. The dataset collected for calibration consisted of 5 focus exposure matrices, including

140nm pitch with line width 70nm and space width 70nm; 180nm pitch with line width 80nm; and 600nm pitch with line width 100nm and space 100nm. The line sizes on the mask were 67 *nm*, 70 *nm* and 73*nm*. Each *F-E* matrix was measured on 4 duplicate wafers, so that an estimation of process variability could be made.

The process conditions for the calibration data:

- Wavelength: 193 *nm*
- Topcoat: 30nm OC-301
- Resist: 110 *nm* commercial *ArF* PR A
- Process: 110°/60*s* PRE, 110°/60*s* PEB, 72*s* development
- BARC: NISSAN NCA41074
- Mask: 6% attenuated PSM
- Exposure: 1.30 NA, Annular, 0.9/0.6

3.3 Calibration of rigorous resist model and LPM model

In this work, we calibrated four models by accounting for two different mask characterization – one with 2D mask (Kirchhoff approximation) and one with 3D EMF mask (FDTD – Finite Difference Time Domain). These models are calibrated by minimizing the RMS error. After calibration, the quality of the match can be evaluated by inspection of the RMS error.

	Total RMS (nm)	Max. RMS (nm)	Max. RMS features
Rigorous Resist Model (Kirchhoff)	2.2	3.8	100nm space / 600nm pitch
LPM Model (Kirchhoff)	2.2	3.8	100nm space / 600nm pitch
Rigorous Resist Model (EMF mask)	2.6	4.9	80nm line / 180nm pitch
LPM Model (EMF mask)	2.6	5.4	100nm space / 600nm pitch

Table 1: Calibration fitting result : Summary table of total RMS and Max. RMS features.

Table 1 summarizes the comparison amongst four different models and Table 2 summarizes calibration results for all the features and shows calibration fit results for all the features used for calibration. Fig. 5 shows max. RMS error matching features of each model with experimental and simulation data. With calibrated PR qualification, cross-section images of experiment and simulation show good compatible result in fig. 6. Simulated PR profiles of rigorous resist models show good matching to empirical wafer profiles, especially profiles with 3D EMF mask get compatible top rounding and PR footing. With LPM model, there is much difference from empirical wafer profiles, which means insufficient physical parameters to describe the detail reactions and development phenomenon during wafer process. In this study, we focused on ADI CD matching, which is mainly determined by top view CDSEM image, and the longitudinal PR profile is not our first concern. LPM model with simplified physical parameters can be accepted.

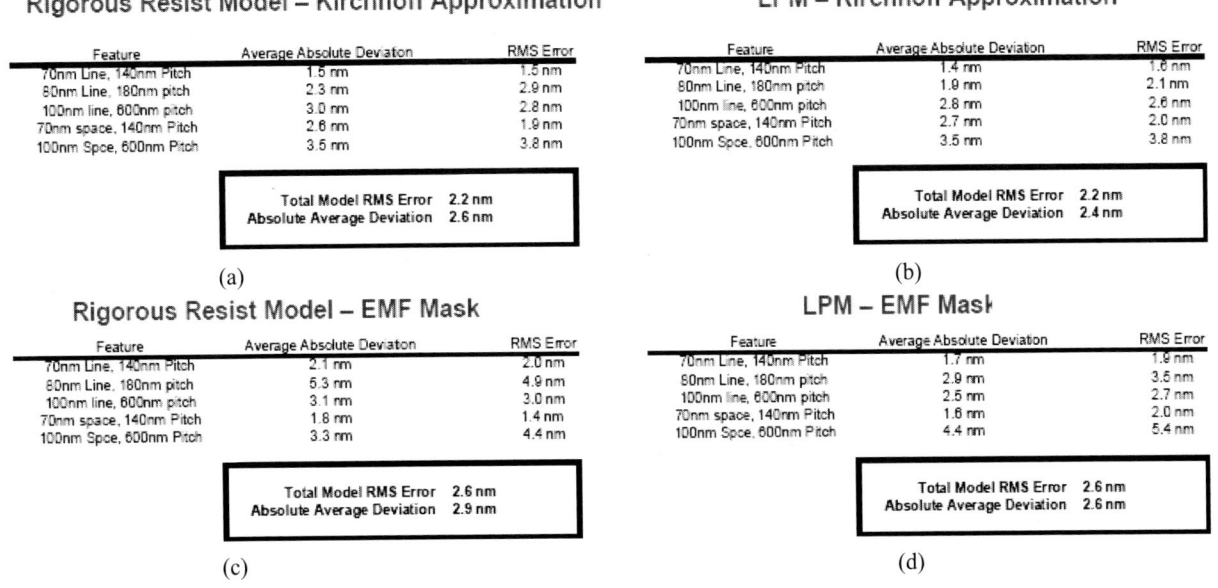

Rigorous Resist Model – Kirchhoff Approximation

Feature	Average Absolute Deviation	RMS Error
70nm Line, 140nm Pitch	1.5 nm	1.5 nm
80nm Line, 180nm pitch	2.3 nm	2.9 nm
100nm line, 600nm pitch	3.0 nm	2.8 nm
70nm space, 140nm Pitch	2.6 nm	1.9 nm
100nm Spce, 600nm Pitch	3.5 nm	3.8 nm

Total Model RMS Error   2.2 nm
Absolute Average Deviation   2.6 nm

(a)

LPM – Kirchhoff Approximation

Feature	Average Absolute Deviation	RMS Error
70nm Line, 140nm Pitch	1.4 nm	1.6 nm
80nm Line, 180nm pitch	1.9 nm	2.1 nm
100nm line, 600nm pitch	2.8 nm	2.6 nm
70nm space, 140nm Pitch	2.7 nm	2.0 nm
100nm Spce, 600nm Pitch	3.5 nm	3.8 nm

Total Model RMS Error   2.2 nm
Absolute Average Deviation   2.4 nm

(b)

Rigorous Resist Model – EMF Mask

Feature	Average Absolute Deviation	RMS Error
70nm Line, 140nm Pitch	2.1 nm	2.0 nm
80nm Line, 180nm pitch	5.3 nm	4.9 nm
100nm line, 600nm pitch	3.1 nm	3.0 nm
70nm space, 140nm Pitch	1.8 nm	1.4 nm
100nm Spce, 600nm Pitch	3.3 nm	4.4 nm

Total Model RMS Error   2.6 nm
Absolute Average Deviation   2.9 nm

(c)

LPM – EMF Mask

Feature	Average Absolute Deviation	RMS Error
70nm Line, 140nm Pitch	1.7 nm	1.9 nm
80nm Line, 180nm pitch	2.9 nm	3.5 nm
100nm line, 600nm pitch	2.5 nm	2.7 nm
70nm space, 140nm Pitch	1.6 nm	2.0 nm
100nm Spce, 600nm Pitch	4.4 nm	5.4 nm

Total Model RMS Error   2.6 nm
Absolute Average Deviation   2.6 nm

(d)

Table 2 : Calibration fitting result among test features, rigorous resist model with Kirchhoff approximation(a), LPM model with Kirchhoff approximation(b), rigorous resist model with EMF mask(c), LPM model with EMF mask(d).

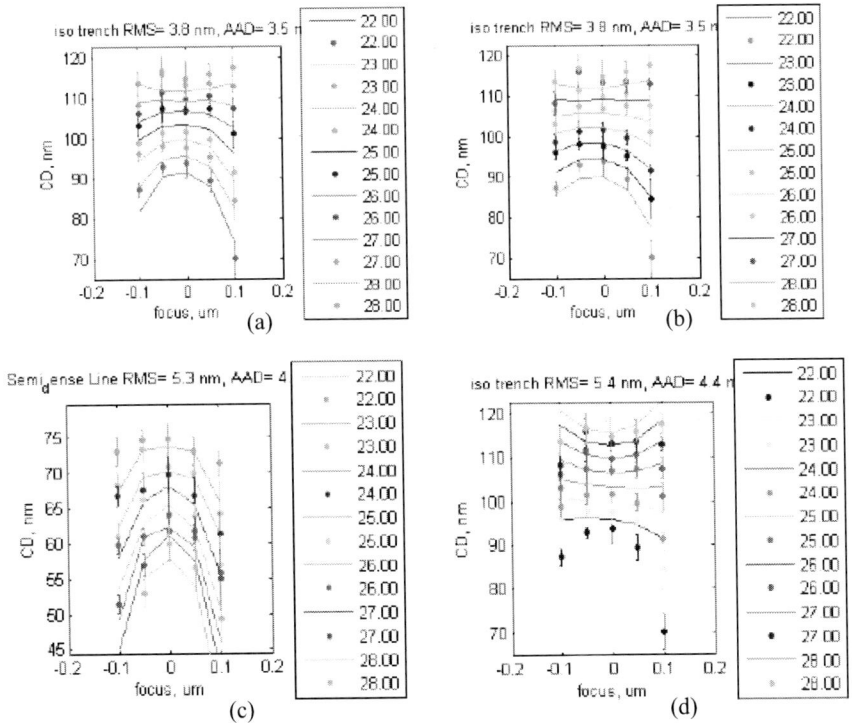

Figure 5 : Max. RMS error matching features of each model with experimental(points) and simulation(lines) data. Rigorous resist model (Kirchhoff) with 100nm space / 600nm pitch(a), LPM model (Kirchhoff) with 100nm space / 600nm pitch (b), rigorous resist model (EMF mask) with 80nm space / 180nm pitch (c), LPM model (EMF mask) with 100nm space / 600nm pitch (d).

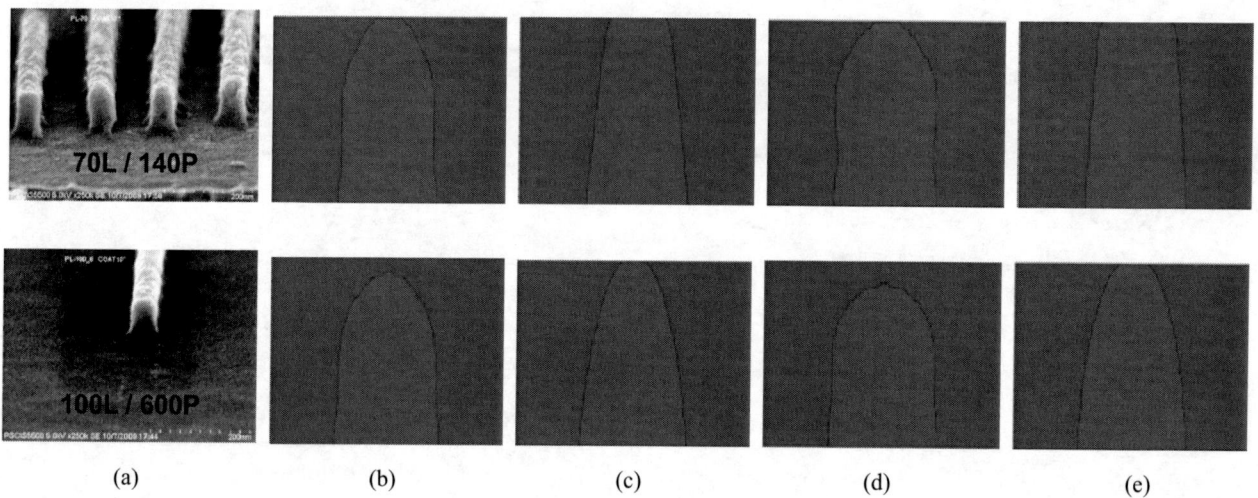

| (a) | (b) | (c) | (d) | (e) |

Figure 6 : Comparison of experimental cross-section image and simulation cross-section image for different test features. Experimental wafer cross-section profile(a), rigorous resist model with Kirchhoff approximation(b), LPM model with Kirchhoff approximation(c), rigorous resist model with EMF mask(d), LPM model with EMF mask(e).

4.   COMPARISON WITH REAL DEVICE EMPIRICAL RESULT

4.1 **2D Kirchhoff approximation mask with PR full physical model and lumped model simulation**

We generate full physical model and simplified lumped model to compensate the model accuracy.

Fig. 7 shows 42nm to 100nm proximity features fitting results of LPM model and full physical model. For LPM model, RMS is 4.61nm, and for FPM model, RMS is 3.42nm. FPM shows better fitting result than LPM does.

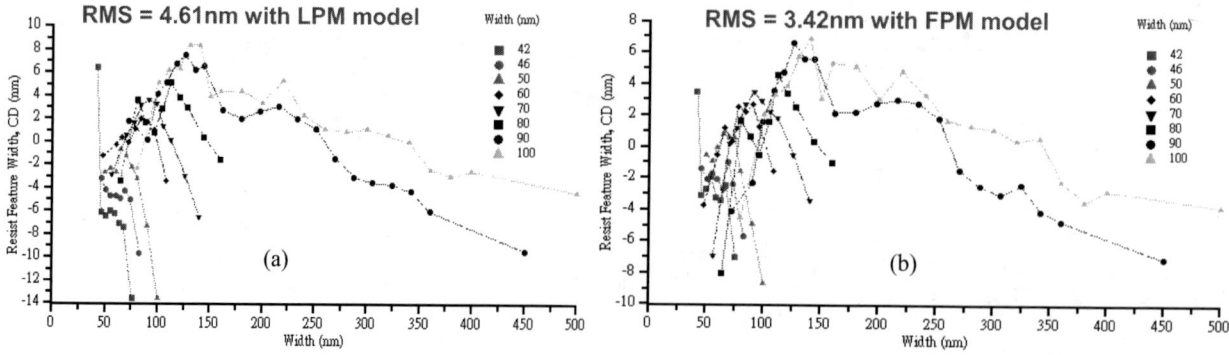

Figure 7 : Proximity features fitting result by 2D Kirchhoff approximation simulation (a) LPM model (b) Full physical model

In fig. 8, we still choose select-gate design 1&2 WL1 to WL12 to monitor model accuracy. In design 1, the total △CD is well compensated by full physical PR model, which including more details of PR parameters to illustrate litho behavior. Especially △CDs of WL1 to WL6, which are closer to SG feature with much optical proximity effect suffered, are also well compensated.

By simplified LPM model, in design1, even the most critical pattern WL1 is well compensated, but △CDs of WL2 to WL9 become worse than full physical model. From Table 1 shown, Kirchhoff LPM model is well generated with total RMS only 2.2nm, which is same as full physical model dose. But in-sufficient PR descriptions of parameters will cause wrong result with total ADI CD fitting, even in some critical patterns the fitting result is good. In design 2, full physical model is also better compensated than simplified LPM model, but the tendency is not so obvious as design 1 showed. It may be caused by measurement error or more test patterns and parameters need to be considered during PR calibration progress.

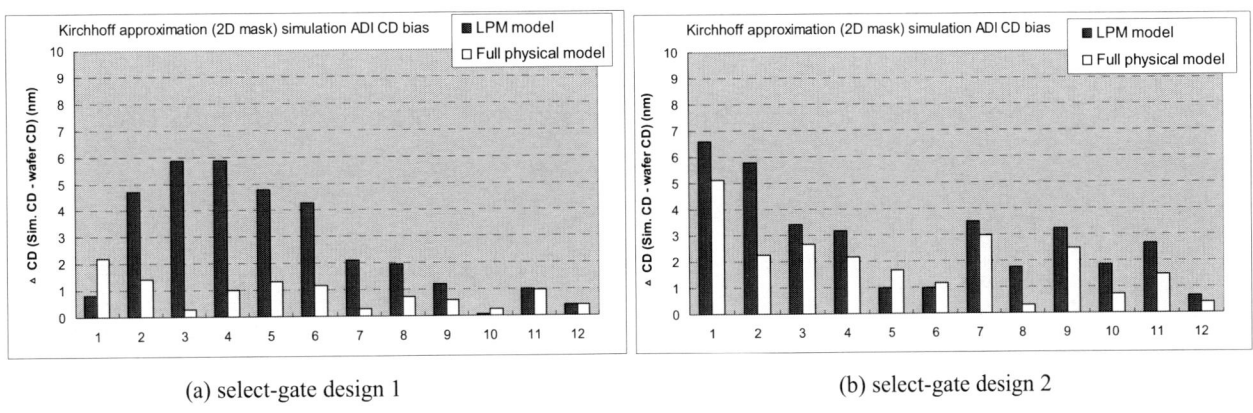

(a) select-gate design 1   (b) select-gate design 2

Figure 8 : △CD of Kirchhoff approximation simulation result benchmark with experimental data of SG features (LPM model and full physical model) : (a) select-gate design 1, (b)select-gate design 2.

4.2 3D EMF mask with PR full physical model and lumped model simulation

In this section, in order to verify 3D EMF mask distribution to simulation accuracy, we also generate full physical model and simplified lumped model with 3D EMF mask calibration. Select-gate design 2 is applied to verify the difference between these two models.

Fig. 9 shows 42nm to 100nm proximity features fitting results of LPM model and full physical model. For LPM model, RMS is 4.92nm, and for FPM model, RMS is 4.17nm. FPM shows better fitting result than LPM does.

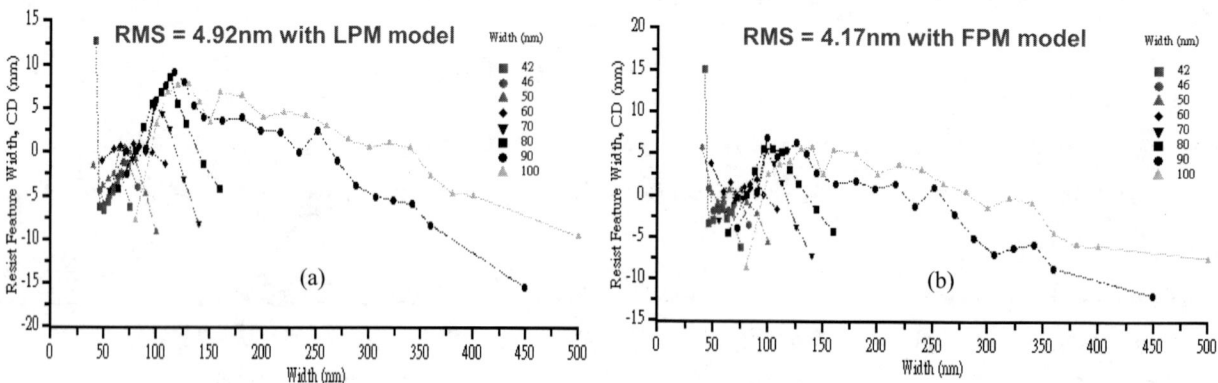

Figure 9 : Proximity features fitting result by 3D EMF mask simulation (a) LPM model (b) Full physical model

In Fig. 10(a), it shows that with full physical model gets better compensation than with simplified LPM model. But notice that both are time consuming compared to 2D Kirchhoff approximation simulation. In device development early stage, accuracy is the most important, but also time to market efficiency is needed to be concerned. Depends on user's host equipment, Either 3D EMF mask with LPM model or with full physical model is chosen to provide a better compensation in simulation accuracy and running time.

In Fig. 10(b), full physical models with 2D Kirchhoff approximation and 3D EMF mask are compared. Result shows better fitting result with 3D EMF mask, but also it's time-consuming both in calibration and simulation stage.

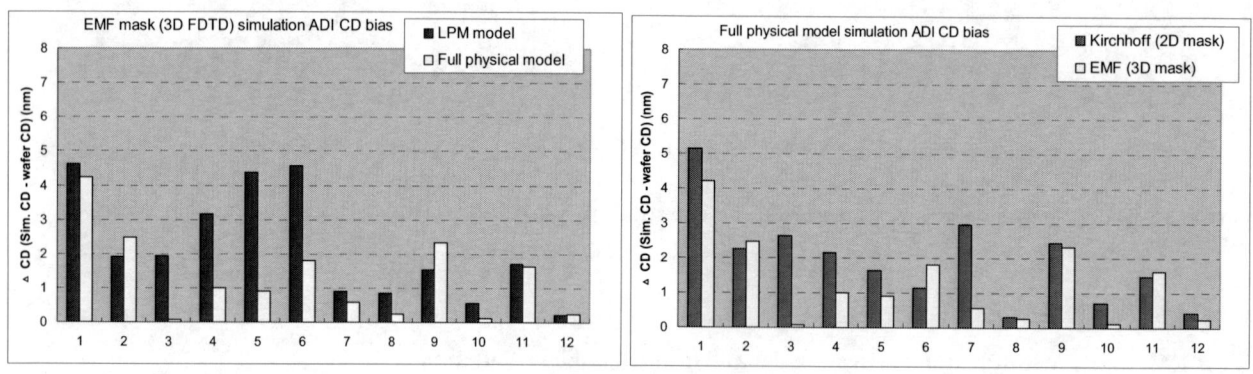

Figure 10 : (a) △CD of 3D EMF mask simulation result benchmark with experimental data of SG features (LPM model and full physical model), (b) △CD of full physical model simulation result benchmark with experimental data of SG features (2D Kirchhoff approximation and 3D EMF mask).

4.3 Time consuming benchmark

In Table 3, we compute the running time of single resist simulation of 2D Kirchhoff approximation and 3D EMF mask on 1D and 2D patterns. Even for a standard line-end pattern, single aerial image (AI) simulation of 3D EMF mask model costs half hour which is tens of run-time of 2D Kirchhoff mask model.

In 2D Kirchhoff approximation model, LPM model saves 80%~90% run-time compared to FPM model; In 3D EMF mask model, LPM model saves 6%~10% run-time compared to FPM model. Combining with the CD accuracy, LPM can be a candidate if 3D EMF mask model is needed.

	2D_AI	2D_LPM	2D_FPM	3D_AI	3D_LPM	3D_FPM
Dense L/S (1D)	2 s	2 s	2 s	12 s	12 s	12 s
Select-gate (1D)	4 s	6 s	16s	92s	96s	106 s
Dense Lineend (2D)	16s	56s	6m46s	38m8s	38m34s	44m56s
Landing pad (2D)	30s	1m58s	4m26s	>2hours	>2hours	>2hours

Table 3: Running time summary table of different models with 1D and 2D test patterns

5. CONCLUSION

Three typical features of flash memory are examined by 2D Kirchhoff approximation and 3D EMF mask optical models. Select-gate patterns is the most sensitive to mask topological effects due to it's various pitch environment.

Four different PR models with LPM and full physical by Kirchhoff approximation and 3D EMF mask calibrated are observed. All these models are calibrated well by test features and total RMS is from 2.2nm to 2.6nm.

3D EMF mask application significantly improves the simulation accuracy of select-gate pattern beyond 40nm, especially critical patterns which suffer much optical proximity effect. For LPM model and full physical model application, definitely full physical model provides more accurate simulation result than LPM model due to more parameters are involved during calibration progress, but full physical model is time-consuming both in calibration and simulation stage. More precise parameters in LPM model calibration will provide compatible simulation result to full physical model, and it would be a compensation solution between simulation accuracy and simulation time.

ACKNOWLEDGEMENT

The authors would like to thanks KLA-Tencor lithography teams for their contributions to this paper. PROLITH is a trademark of KLA-Tencor.

REFERENCES

1. A. Erdmann, "Mask modeling in the low k1 and ultrahigh NA regime: phase and polarization effects" Proc. of SPIE Vol. 5835, pp.69-81(2005).

2. M. Saied, F. Foussadier, et. al., "Three-dimensional mask effects and source polarization impact on OPC model

accuracy and process window" Proc. of SPIE Vol. 6520 (2007).

3. K. Sato, M. Itoh, T. Sato, "Mask 3D effect on 45-nm imaging using attenuated PSM" Proc. of SPIE Vol. 6520 (2007).

4. Peter D. Bisschop, T. Muelders, et. al., "Impact of mask three-dimensional effects on resist-model calibration", JM3 letters Vol. 8(3), Jul.-Sep. 2009.

5. P. Bevington, et. al., "Data Reduction and Error Analysis for the Physical Sciences" McGraw-Hill, 2003.

Comparison of OPC models with and without 3D-mask effect

Jung-Hoon Ser*[a], Tae-Hoon Park[a], Moon-Gyu Jeong[a], Eun-Mi Lee[a], Sung-Woo Lee[a], Chun-Suk Suh[a], Seong-Woon Choi[a], Chan-Hoon Park[b], and Joo-Tae Moon[a]

[a]Process Development Team, NRD Center, Semiconductor Business, Samsung Electronics Co., LTD.; [b]Photo Technology Team, Memory Division, Semiconductor Business, Samsung Electronics Co., LTD., Hwasung-City, Gyeonggi-Do, Korea

ABSTRACT

OPC models with and without thick mask effect (3D-mask effect) are compared in their prediction capabilities of actual 2D patterns. We give some examples in which thin-mask models fail to compensate the 3D-mask effect. The models without 3D-mask effect show good model residual error, but fail to predict some critical CD tendencies. Rigorous simulation predicts the observed CD tendencies, which confirms that the discrepancy really comes from 3D-mask effect.

Keywords: OPC model, 3D mask, thick mask

1. INTRODUCTION

In OPC modeling, it has long been assumed that photomask is infinitely thin and that mask function is a simple constant function of positions. As feature size becomes smaller, mask topography dimension becomes significant compared to the scanner exposure wavelength. In lithography textbooks, it is told that 3D-mask effect should be taken care of when mask dimension is smaller than the scanner wavelength. Then, the near-field image after a mask is quite different from Kirchhoff approximation. In a year or so, there have been many discussions on whether 3D-mask effect already appears or not in current lithography process. [1,2] In reality, current critical CDs are out of the thin-mask regime. But the inherent limitation of OPC model makes the 3D mask effect unseen. Since Hopkins formulation can not easily take 3D-mask effect into account, still, OPC models do not consider the 3D-mask effect in common. In large amount, this is because the resist part of OPC model erroneously fits the optical or mask properties like the 3D-mask effect.

2. OPC MODEL AND 3D-MASK EFFECT

2.1 Resist model and model fitting

Constructing an ordinary OPC model requires a fitting process of parameters and coefficients based on multitudes of wafer CD measurements. It is largely accepted that optical part of current OPC models are fairly accurate, but that mask part and resist part are open for further improvement. In this sense, it might be expected that 3D-mask effect is compensated in the fitting process of resist model by proper choice of test patterns. (Fig. 1) In some cases, it can be true since there is some flexibility in the resist model to compensate the mask effect. But in other cases, it is far from reality, as shown in this study.

Optical Microlithography XXIII, edited by Mircea V. Dusa, Will Conley, Proc. of SPIE Vol. 7640, 76401T · © 2010 SPIE · CCC code: 0277-786X/10/$18 · doi: 10.1117/12.848317

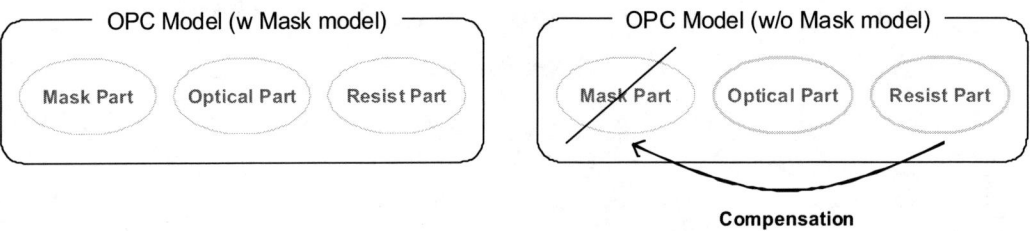

Figure 1. OPC model and its substructure. When mask part is absent, resist part tries to compensate the thick-mask effect.

One of the reason that an engineer believes that 3D-mask effect is well accounted for without the mask part is that the model residual error can be reduced by trial and error for a given test pattern set. But even in that case, the model can give faulty results in real 2D patterns.

General test patterns for OPC modeling consists of various 1D and 2D patterns. For 1D patterns, through-pitch L/S (line and space), ISO line, and ISO space are included. In general, for ease of SEM measurement and for high image quality of SEM image, the CDs are measured at the center of the test patterns by SEM. (Fig. 2 (a)) So, even when a model shows small residual error in 3D-mask region, it only guarantees good prediction for that periodic 1D pattern. The prediction capability for real 2D pattern is still open. One of the simple example is the prediction of CDs at the block-end of the same 1D test pattern. (Fig. 2 (b))

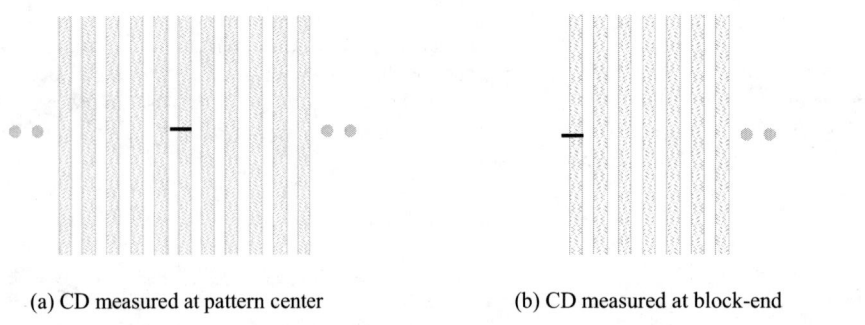

(a) CD measured at pattern center   (b) CD measured at block-end

Figure 2. Position of SEM CD measurement in the test patterns for OPC modeling. Case (a) is a common practice.

2.2 Mask model for OPC

There have been many attempts to take 3D-mask effect into OPC model. [1,2] But, current OPC tools adopt somewhat different approach. [3,4] We used 3D-mask model of ordinary OPC tools for this study. Besides the accuracy study, runtime issue is also to be addressed in the future.

3. COMPARISON OF OPC MODELS

3.1 Model generation

With a set of test patterns, we generated two kinds of OPC models – one with 3D-mask effect (Model B-1), one without it (Model A-1). The test patterns consist of 1D L/S ranging from 90-nm pitch to ISO. A few small-space patterns, whose space CDs are less than 35nm, are intentionally included in the set. As mentioned in section 2.1, CD is measured at the center of a test pattern. And mask manufacturing CD error is corrected to the test patterns for the modeling.

For each model, RMS minimization is tried a few times for best result.

Table 1. Models generated for the tests.

Test patterns for OPC modeling	Without 3D-mask model	With 3D-mask model
1D L/S patterns (CD measured at center)	Model A-1	Model B-1
Block-end CDs also included	Model A-2	Model B-2

3.2 Model comparison

For the OPC models with and without 3D-mask effect, residual fitting errors are compared in Fig. 3. First, total model fitting error RMS' for both models are almost the same, which means that both have similar amount of overall errors for given test patterns. As expected, Model A-1 does not cover the very small-space patterns, whereas, Model B-1 with 3D-mask effect nicely fits them within 3nm.

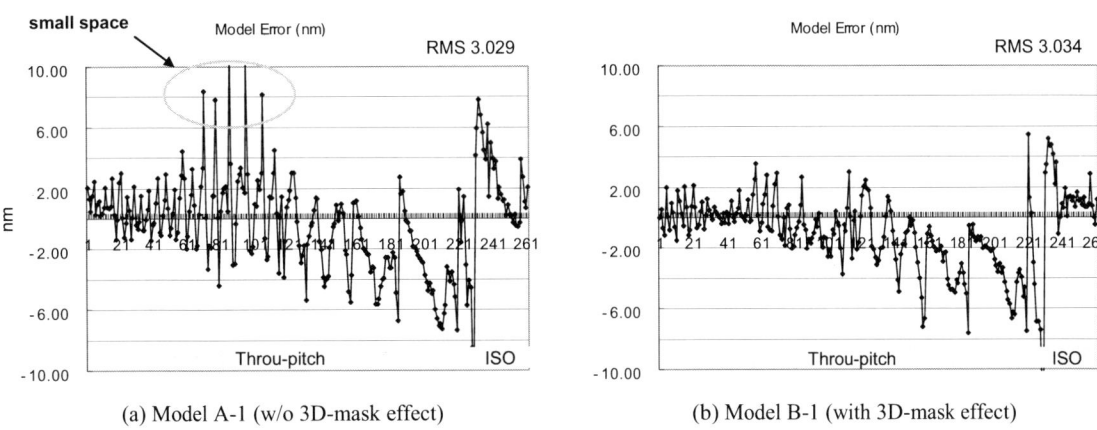

(a) Model A-1 (w/o 3D-mask effect)   (b) Model B-1 (with 3D-mask effect)

Figure 3. Residual fitting errors of the OPC models.

To further compare the model capabilities, block-end CDs' for a semi-1D patterns are predicted by both models. Since pattern fidelity of block-end bars (Fig. 2(b)) are too poor for pitch near 90 nm, we tried to compare CDs of OPCed L/S pattern. With Model A-1, some of test patterns are OPCed and incorporated in a test mask. If Model A-1 is accurate enough, every bar in the block-end must show nearly constant CD throughout.

In Fig. 4, CDs of the OPCed block-end pattern are compared. Contrary to expectation, wafer CD of 2nd bar drops by 2nm compared to inner bar CDs. Thus, Model A-1 (without 3D-mask effect) is not correct for the pattern, which is also seen by the almost constant CDs predicted by Model A-1. Whereas, Model B-1 (with 3D-mask effect) predicts all four block-end CDs to the wafer CDs.

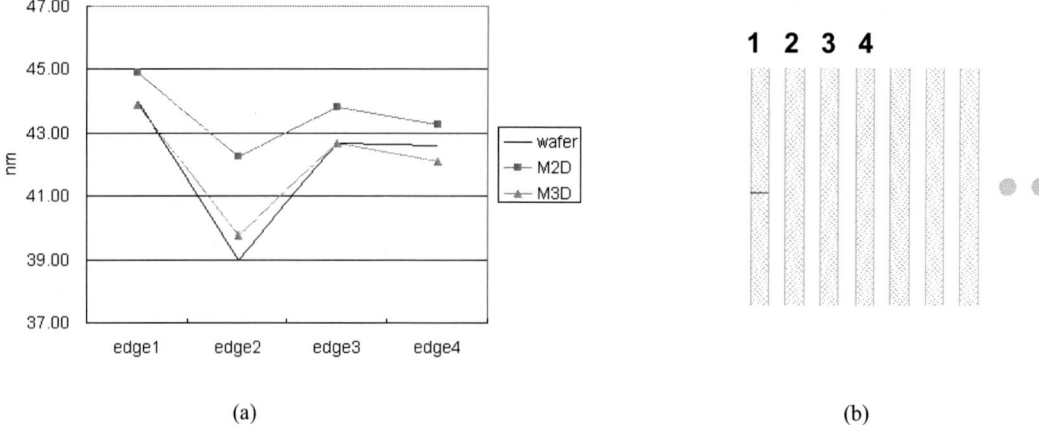

(a)   (b)

Figure 4. CDs for an OPCed block-end pattern. (Wafer CD, Model A-1, Model B-1) Figure (b) shows the numbering
of each bar.

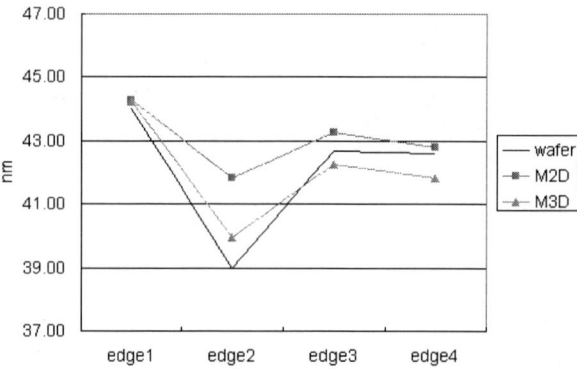

Figure 5. CD predictions for an OPCed block-end pattern. (Wafer CD, Model A-2, Model B-2) Block-end CDs are
included in the OPC modeling.

To reduce model errors for block-ends, block-end CDs are measured and included in OPC modeling with weighting
factor of 10. But the model trends shown in Fig. 5 are almost the same as in Fig. 4.

Finally, the same test pattern is inspected with rigorous litho simulator. Simulations are conducted both with and without
3D-mask effect. Exact values between CDs by simulation and by OPC models are different up to 1.3nm, but the trend of
Fig. 5 is same as in Fig. 4. This is another evidence that 3D-mask OPC model is accurate for this block-end pattern.

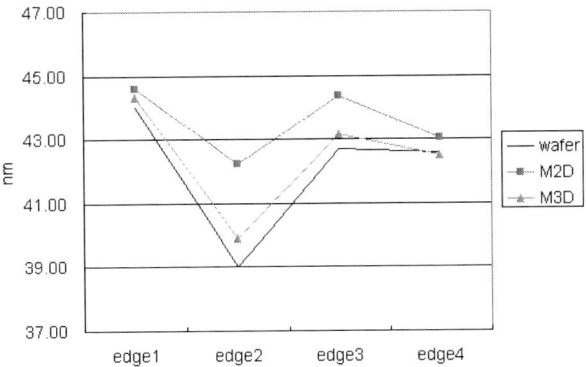

Figure 6. CD predictions by a litho simulator. Both with and without mask effect show similar tendency as Fig. 4.

4. MASK MODEL AND OVERFIT

We checked stability of the models with and without 3D-mask effect. As a measure of stability, amount of overfit is quantified as below (Eqn. 1), where Overfit Ratio (O.R.) is defined. It is the sum of absolute value of a coefficient of each resist term divided by coefficient of the optical term of an OPC model.

$$\text{O.R.} = \frac{\sum_i Abs(c_i)}{Abs(C_{OPT})} \tag{1}$$

To observe general tendency, models are generated several times which result in somewhat different resist term coefficients, and their O.R. are calculated and compared as in Fig. 7. Average O.R. for thin-mask OPC models (M2D) is 0.464, whereas, that for 3D-mask OPC models is 0.263. Thus, thin-mask models are more susceptible to errors in their prediction capability.

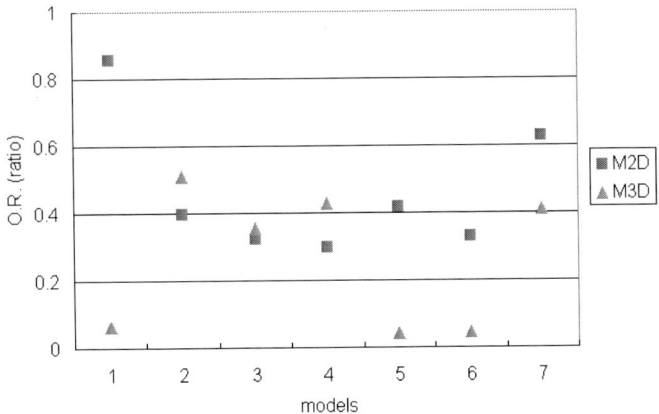

Figure 7. Overfit Ratio of several OPC models with and without 3D-mask effect.

Table 2. Average Overfit Ratios for OPC models with and without 3D-mask effect.

Item	Average O.R.
Model without 3D-mask effect	0.464
Model with 3D-mask effect	0.263

5. SUMMARY

OPC models with and without 3D-mask effect are compared. It is shown that, with normal procedure of OPC modeling, one can not perceive whether the OPC model can correctly predict real patterns. It is true even for semi-1D patterns.

Lack of predictability can be also verified by large Overfit Ratio of the M2D models. With 3D-mask effect, Overfit Ratio is much smaller, and the model is optical-term-dominated, which guarantees more reliable CD prediction capability.

REFERENCES

1. Kim, Y., Kim, I., Park, J., Kim, S., Suh, S., Cheon, Y., Lee, S., Lee, J., Kang, C., Moon, J., Cobb, J., and Lee, S., "OPC in memory-device patterns using boundary layer model for 3 dimensional mask topographic effect," SPIE 6520, T-1, T-10 (2007)
2. Suh, S., Lee, S., Back, K., Lee, S., Kim, Y., Kim, S., and Chun, Y., "Three-dimensional mask effect approximate modeling for sub-50 nm node device OPC," SPIE 6521, 03-1-03-8 (2007)
3. Melvin III, L. S., Schmoeller, T., and Li, J., "32nm Half pitch node OPC process model development for Three dimensional Mask effects using rigorous simulation," SPIE 6730, 41-48 (2007).
4. Liu, P., Cao, Y., Chen, L., Chen, G., Feng, Mu., Jiang, J., Liu, H., Suh, S., Lee, S., and Lee, S. , "Fast and accurate 3D mask model for full-chip OPC and verification," SPIE 6520, R1-R12 (2007).
5. Fukuda, H., Hattori, K., and Hagiwara, T., "Impact of Acid/Quencher behavior on lithography performance," SPIE 4346, 319-330 (2001).

Virtual Fab Flow for Wafer Topography Aware OPC

Hans-Jürgen Stock[a], Lars Bomholt[b], Dietmar Krüger[a], James Shiely[c],
Hua Song[c], Nikolay Voznesenskiy[d]

a Synopsys GmbH, Karl-Hammerschmidt-Straße 34, Aschheim/Dornach, Germany
b Synopsys Switzerland LLC, Thurgauerstrasse 40, Zurich, Switzerland
c Synopsys Inc., Cornelius Pass Rd, Hillsboro, Or, US
d Riia 185A, Tartu, Estonia

ABSTRACT

Small feature sizes down to the current 45 nm node and precision requirements of patterning in 193 nm lithography as well as layers where the wafer stack does not allow any BARC require - not only correction of optical proximity (OPC) effects originating from mask topography and imaging system, but also correction of wafer topography proximity (WTPC) effects as well. In spite of wafer planarization process steps, wafer topography (proximity) effects induced by different optical properties of the patterned materials start playing a significant role, and correction techniques need to be applied in order to minimize the impact.

In this paper, we study a methodology to create fast models intended for effective use in OPC and WTPC procedures. In order to be short we use the terms "OPC\WTPC modeling" and "OPC\WTPC models" through the paper although it would be more correctly to take the terms "mask synthesis modeling" and "mask synthesis models".

A comprehensive data set is required to build a reliable OPC model. We present a "virtual fab" concept using extensive test pattern sets with both 1D and 2D structures to capture optical proximity effects as well as wafer topography effects.

A rigorous lithography simulator taking into account exposure tool source maps, topographic mask effects as well as wafer topography is used to generate virtual measurement data, which are used for model calibration as well as for model validation.

For model building, we use a two step approach: in a first step, an OPC model is built using test patterns on a planar, homogenous substrate; in a second step a WTPC model is calibrated, using results from simulated test patterns on shallow trench isolation (STI) layer. This approach allows building models from experimental data, including hybrid approaches where only experimental data from planar substrates is available and a corresponding OPC model for the planar case can be retrofitted with capabilities for correcting wafer topography effects.

We analyze the relevant effects and requirements for model building and validation as well as the performance of fast WTPC models.

Keywords: optical lithography simulation, optical proximity correction, wafer topography

Introduction

In this study, we describe a flow which generates "virtual data" by rigorous lithography simulation in order to build a fast model for optical proximity corrrection[2,3] (OPC model). In a next step we compare this model with an OPC model calibrated against real measurements data obtained on the same pattern set. In the subsequent section we study the influence of the wafer topography for an OPC model. We compare a model which takes the underlying wafer topography into account to a standard OPC model which is based only on the actual layer layout, and explain the theory of this approach.

Virtual Fab flow

The first step is the generation of the test pattern mask and the measurement points containing a broad set of 1D and 2D test structures for bright and dark field conditions. For this study test patterns with underlying

Optical Microlithography XXIII, edited by Mircea V. Dusa, Will Conley, Proc. of SPIE Vol. 7640,
76401U · © 2010 SPIE · CCC code: 0277-786X/10/$18 · doi: 10.1117/12.848440

wafer topography were also created. For this work we have used a test pattern generator[4] (TPG), where the patterns can be parameterized and will be drawn into one GDSII file.

This GDSII file was used as input for the rigorous lithography simulator[1] as well as for the OPC model calibrator[2]. Instead of real, measured CD values, the simulated CD values (virtual data) were used as input for the OPC model calibrator[2], as shown in the scheme of Figure 1a. We distinguish two cases of test patterns – planar and topographic ones – in the extended scheme of Figure 1b where we show the combined two cases with rigorous simulation and experimental (real measured) data flows.

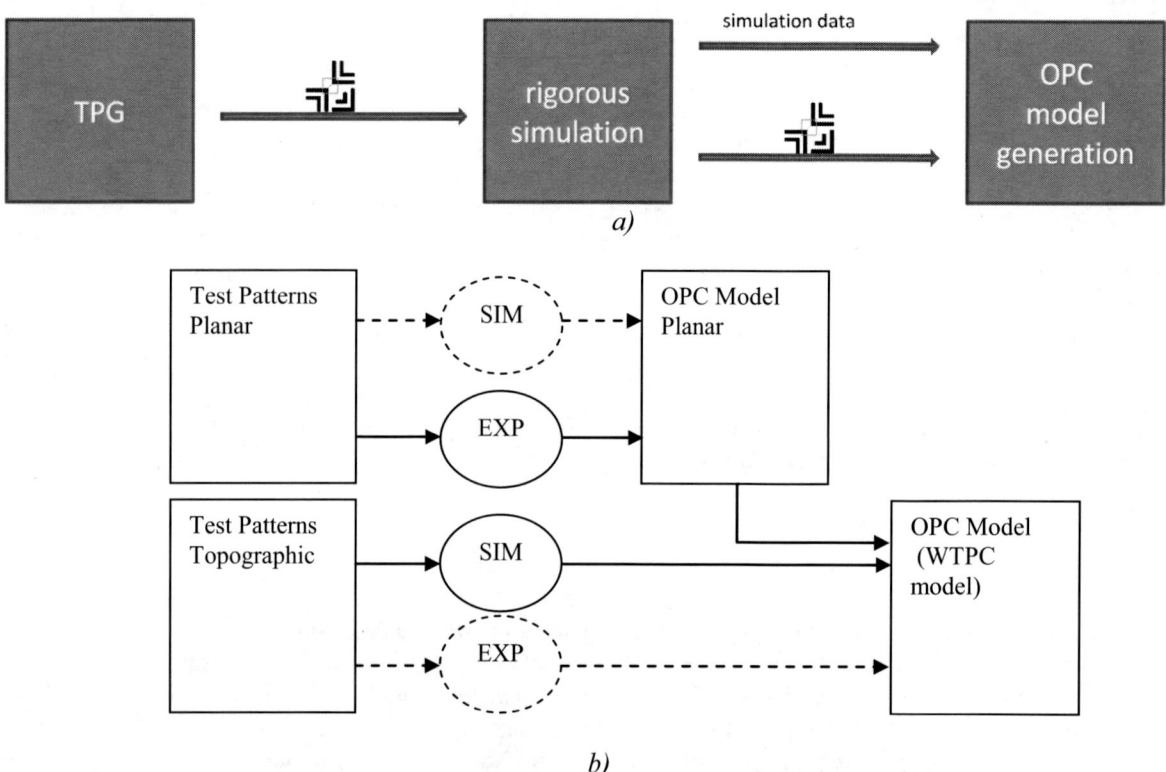

Fig. 1 Virtual fab flow: a) test pattern (planar case) is coming from TPG, two data flows for the OPC model: measurement results (upper arrow) and rigorous simulation results (bottom arrow); b) extended scheme with planar and topography test patterns, the experimental data come from metrology tool measurements, simulated data come from rigorous calculations.

Comparison of experimental data model versus virtual data model

In order to demonstrate the performance of a virtual fab flow, we compare two planar OPC models based on the same test pattern set, a subset (about 750 patterns) of the IMEC OPC6[5] data set. The subset contains Lines & Spaces pattern of different line widths / space widths, pitches, isolated lines/spaces and 2D line end structures. Verification was performed against the whole data set of about 5000 patterns.

One model was built on the measured CD data. For the second OPC model a rigorous lithography simulator produced the input data from the same test pattern set. The simulator uses a physical resist model which is calibrated against a different, non-overlapping subset of OPC6 data. The RMS value for this resist model is 1.9 nm. In both cases the verification was done against the measurement data. The results for the OPC models are shown in Table 1.

The comparison shows that the performance of an OPC model built from a calibrated rigorous simulation model is comparable to that of an OPC model directly built on experimental data. We see a minor deterioration which can be expected from effects not captured in the physical model as well as the fitting error of the physical model.

Table 1 Comparison of the experimental data model versus the virtual data model. The RMS values for the calibrated 1D and 2D pattern set are nearly the same. The virtual model has a little bit worse values for the verification against the experimental data set, which is to be expected since the use of a limited set of physical models in the rigorous simulation should have an effect of accuracy. But the difference is inside the expected range for a resist model with a RMS value of 1.9 nm.

Model	RMS calibration 1D [nm]	RMS calibration 2D [nm]	RMS verification [nm]
experimental	1.6	3.0	3.7
virtual	1.9	2.8	4.7

Principles of OPC and WTPC large scale modeling

In order to develop a fast model for wafer topography proximity correction (WTPC model), we start from the planar wafer case which is equivalent to OPC modeling conditions. In a next step we extend this model to the wafer topography case so that new effects are to be described independently of optical proximity effects and without necessity to modify parameters of the planar model. This means that the OPC model which is already fitted to the planar case can be incorporated into the WTPC model as its component – the concept of a "retrofitted" OPC model.

In the planar case there are mainly two effects which must be captured by the model: 1) mask optical projection resulting in bulk intensity distribution – aerial image, and 2) effects of resist exposure and post exposure processing. Such model is actually a linear system operator applied to the pattern data. This model can be schematically illustrated in Fig. 2.

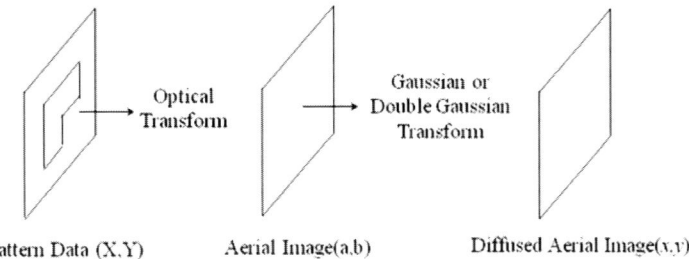

Fig.2 Planar case (OPC model) as projection of the mask pattern data to the image space.

No doubt, the most important effect which needs to be captured by the OPC model – planar model – is optical imaging. But evaluation of capturing ability only by the aerial image is not sufficient, resist effects need to be considered as well in order to obtain the final image contour. In this study we used a rather simple resist model – Gaussian blurring. While model is sufficient for our investigations, and it does not limit us using more advanced resist model at a later stage.

For OPC modeling, we use ProGen\Proteus[2], which offers a number of object *types* for creating OPC models. The object types incorporate mask data, imaging tool settings including light source description and a set of free parameters used for calibration of virtual models. Through regression of the free parameters, OPC effects can be adequately captured for a great variety of practical test patterns.

According to the above mentioned concept we extend the planar model to the wafer topography case by incorporating the OPC model as a whole and extending it with wafer topography effects. This concept is in good agreement with physics because effects of non-planar wafer arise independently as an extension to mask optical projection. Consequently there is no need to change those parameters, which are fixed after OPC calibration, while fitting the WTPC model. This limits a degree of freedom for a user but on the other side has the advantage that OPC will not deteriorate when capturing wafer topography effects.

The approach has, however, also practical benefits, since the OPC model is unchanged in structure, quality and accuracy for regions free of wafer topography.

In the course of linear system OPC modeling illustrated in Fig. 2 construction of a WTPC model can be understood from Fig. 3 which shows a similar linear system simulation of wafer topography effects. This approach refers to the "virtual mask" idea[6].

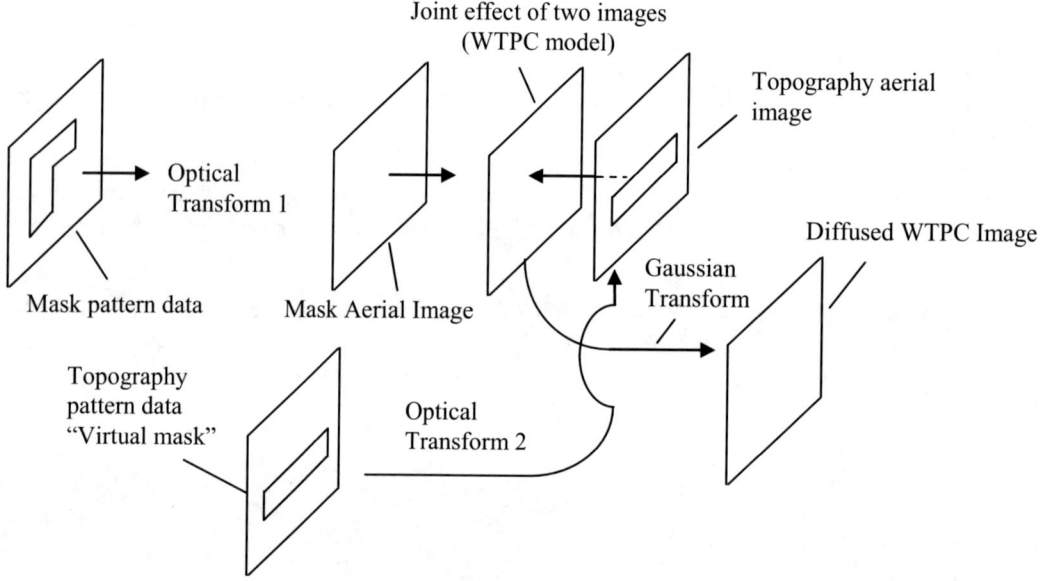

Fig. 3 WTPC linear system model is considered as formation and interaction of two images: 1) image of the mask (mask pattern data) and 2) image of the "virtual mask" – wafer topography (topography pattern data).

The equations below describe principles of OPC and WTPC models in a simplified operator form.

Optical transform:

$$K(x_1, y_1, x_2, y_2) = Pupil(x_1, y_1) \cdot Pupil(x_2, y_2) \cdot Source(x_1 - x_2, y_1 - y_2), \tag{1}$$

where $Pupil(x_1, y_1)$ is the pupil function of the optical system, $Source(x_1 - x_2, y_1 - y_2)$ is description of the light source, $K(x_1, y_1, x_2, y_2)$ is the large scale *optical kernel* describing aerial image formation.

Gaussian transform with a given standard deviation:

$$K_{opt}(x_1, y_1, x_2, y_2) = Gaussian(diff\_stddev) \otimes K(x_1, y_1, x_2, y_2), \tag{2}$$

where $Gaussian(diff\_stddev)$ is Gaussian filtering transform imitating exposure and post exposure processing, $K_{opt}(x_1, y_1, x_2, y_2)$ is the optical diffused kernel describing the resultant image.

Components of the WTPC model in operator form:

$$P_{opc}(x,y) = conval\left(K_{opt}(x_1, y_1, x_2, y_2), M(x,y)\right), \qquad (3)$$

$$P_{tpc}(x,y) = conval\left(K_{opt}(x_1, y_1, x_2, y_2), T(x,y)\right), \qquad (4)$$

where $M(x,y)$ is the *mask pattern data*, $T(x,y)$ is the STI *pattern data*, $conval(...)$ is a linear system operator which describes application of the *optical kernel* to patterns data, $P_{opc}(x,y)$ and $P_{tpc}(x,y)$ are mask and wafer topography images.

The first variant of a WTPC model can be constructed from components (3) and (4) in the form of a linear combination as follows:

$$P(x,y) = \left(P_{opc}(x,y) + r1 \cdot P_{tpc}(x,y)\right)/(1 + r1), \qquad (5)$$

where $r1$ is a free parameter.

Equation (5) describes summing intensities and is a first approximation of the WTPC effects. This model can capture only small topography effects in case of BARC placed on the topography layer. We checked capturing ability of such a modeling approach without BARC on the non-planar wafer layer. The results are summarized below.

In case without BARC simulations according to Equation (5) become inaccurate because on the wafer there occurs a near-field interference between the incident field carrying mask image and the field reflected from the wafer topography features, as it is schematically illustrated in Fig. 4.

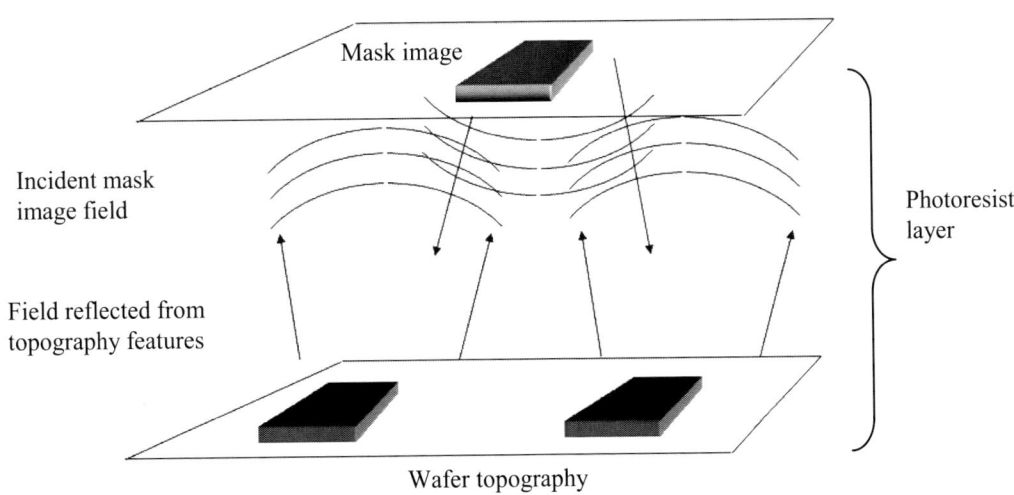

Fig. 4 Near-field interference of the fields in the wafer (scheme).

As drawn in Fig. 4 the incident and reflected fields interfere in the thick photo resist layer. The resultant mask image is formed by the system of standing waves which is non-trivial modified because of the presence of topographic features. Such understanding of the process helps to obtain a better approximation to reality rather than simple adding intensities. This cannot be described in terms of the model (5), where interference effects are completely lost.

In order to capture interference effects a new term can be introduced in the WTPC model:

$$K_{tpc}(x_1, y_1, x_2, y_2) = T(x, y) \otimes K(x_1, y_1, x_2, y_2),$$

$$P_{int}(x, y) = Re\left[conval\left(K_{tpc}(x_1, y_1, x_2, y_2), M(x, y)\right)\right],$$

where the STI layout $T(x, y)$ plays the role of a filter depending on wafer geometry, $K_{tpc}(x_1, y_1, x_2, y_2)$ is a *mixed optical kernel*, $P_{int}(x, y)$ is the real part of the linear operator performed over the mask pattern and mixed optical kernel. We take the real part of this linear operation because it is a complex function, in contrary to $P_{opc}(x, y)$ and $P_{tpc}(x, y)$ which are intensities. This is an interference component. Herewith the WTPC model which takes into account interference will look like:

$$P(x, y) = \left(P_{opc}(x, y) + r1 \cdot P_{tpc}(x, y) + r2 \cdot P_{int}(x, y)\right)/(1 + r1 + r2), \tag{6}$$

where both $r1$ and $r2$ are free parameters.

Test pattern generation

In this paper we give results of calibration and validation of the OPC model and WTPC models corresponding to Equations (5) and (6). The same OPC model is included in both equations (5) and (6). OPC modeling with planar wafer substrate is performed with both $r1$ and $r2$ set to zeros. If $r1$ is non-zero we have two variants of WTPC modeling – with zero $r2$ and non-zero $r2$, respectively. We have chosen some critical cases of wafer topography in order to check capturing ability of both variants of the WTPC model. All the considered test patterns contain straight rectangular features resembling the so-called "Manhattan" geometry. However, this is not a generic restriction; in principle, features of arbitrary shape and orientation could be included in the model building.

For calibration and validation of such a model a set of test patterns needs to be arranged, and all of them to be placed in one large test chip. This test chip with the mask layout and stack data is an input for the "virtual fab" flow. In Fig. 5 a view of the used stack with parameters is given.

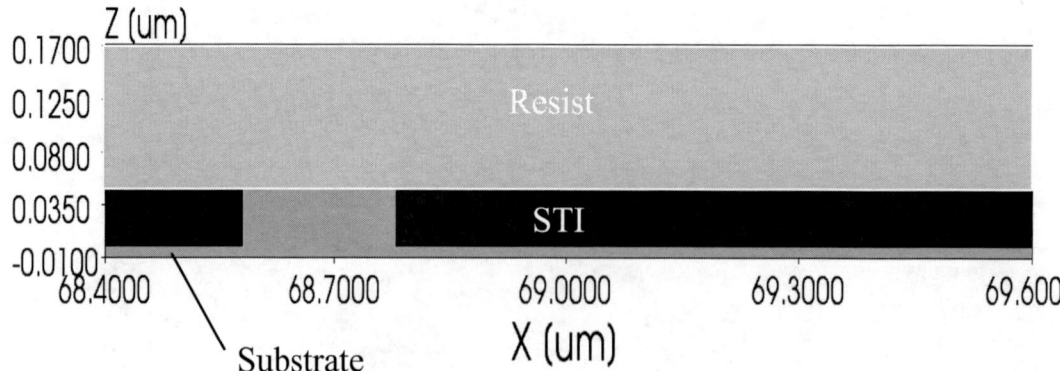

Fig. 5 The stack structure with a topography feature; resist material: TOK TArF-Pi6-00- me; STI material: Silicon-Dioxide; substrate and topography material: Silicon.

In order to construct a comprehensive test chip we used the above mentioned test pattern generator[4] to arrange a number of typical mask layouts – *isolated lines, L&S* and *line ends* with *simple* and *hammer-like* shapes of different sizes, with and without *sub resolution assist features (SRAFs)*. Some of these test cases are used for calibration and the others for validation of the OPC and WTPC models. Actually, the processes of calibration and validation are performed simultaneously. The total number of test patterns in this test chip is over 150, each occupying the area of 1.2 μm × 1.2 μm (wafer scale). Some test chip fragments are shown in Fig. 6.

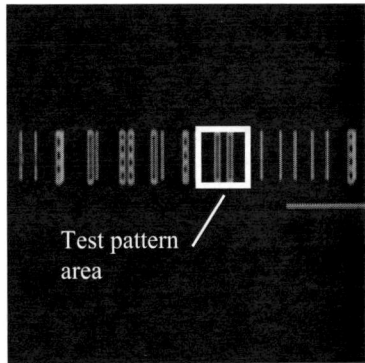 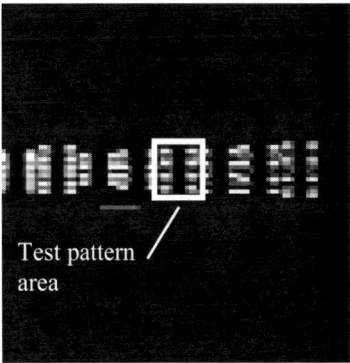

Fig. 6 Small fragments of the test chip with some test pattern areas highlighted.

Firstly the images of test patterns are calculated by rigorous simulation[1], thereafter the test chip and rigorous imaging results – virtual CD data – are transferred to OPC\WTPC calibration.

The image contour obtained by the rigorous simulator can be easily compared to the image contour obtained by OPC and WTPC models. In this paper some critical features are taken from the test chip in order to illustrate the result of capturing. Capturing is evaluated graphically by the overlay of rigorous and modeled contours for these patterns.

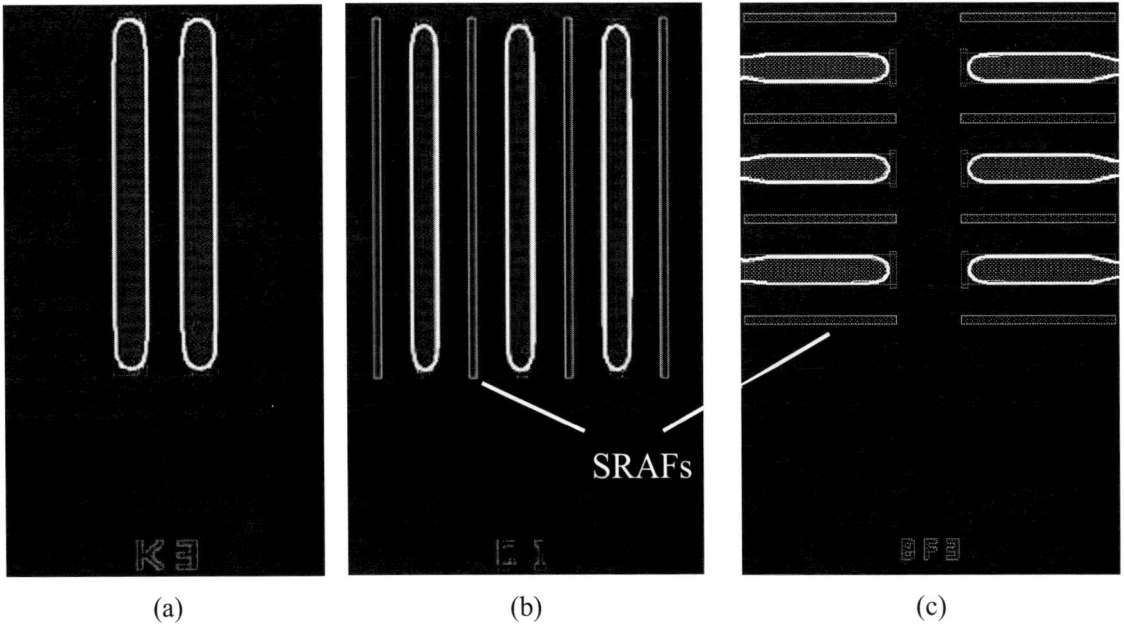

(a)   (b)   (c)

Fig. 7 OPC modeling for a) two isolated lines, b) L&S with SRAFs, c) hammer-like ends with SRAFs.

OPC model assessment

The variant of the OPC model presented in the paper (see Equation (3)) is rather simple but demonstrates a sufficient ability to capture OPC effects for all used test patterns. In Figs. 7(a, b, c) there are given some typical examples of image contours overlays: the contours of the OPC model (solid white curves) well coincide with and entirely cover those from rigorous simulation.

WTPC model assessment

We compare the performance of WTPC models according to Equations (5) and (6) with respect to quality and run-time, for two different stacks. This will give us an estimate to what level of accuracy we need to model the effects and what the expected performance impact is.

Capturing wafer topography effects is a much more challenging task especially for the BARC-less cases. With BARC, effects of non-planar wafer are negligible[7]. However, when BARC is removed, the field reflected from the silicon features is rather strong to damage or even "*erase*" the implant image (see Figs. 7, 8, 10). The large test pattern set used for WTPC model calibration and verification allows us to select and demonstrate some critical test cases where wafer topography induces effects which a simple WTPC model, built without taking *near-field interference* (NFI) into account, cannot capture. In this paper it is shown (see Figs. 10, 12) that a WTPC model, built with taking NFI into account, performs much better for the entire test pattern set, in particular for the critical cases.

A first critical case is "line disintegration" when the nominal line goes in-between two rectangular topography features in the STI layer. In Fig. 8 there shows a rigorous 3D simulation of the image.

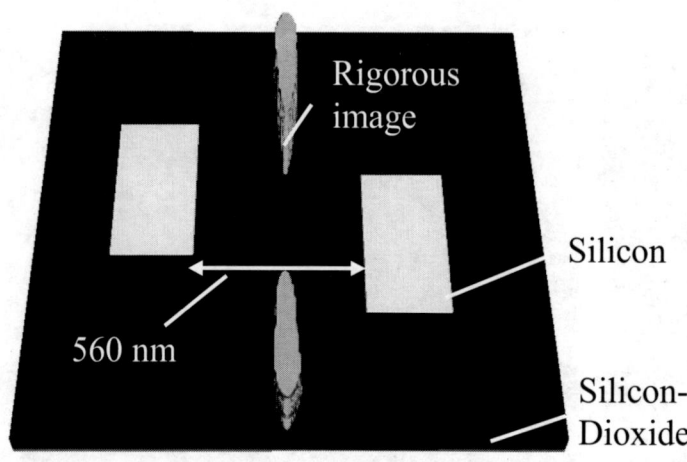

Fig. 8 Rigorous 3D simulation: 100 nm line image without BARC is broken ("line disintegration") when the topography features are at some distance from the line image.

For small gaps between Silicon features less than 560 nm the line goes through. Once the distance increases beyond 560 nm, the line is broken into pieces and is restored again for larger distances (see Fig. 9).

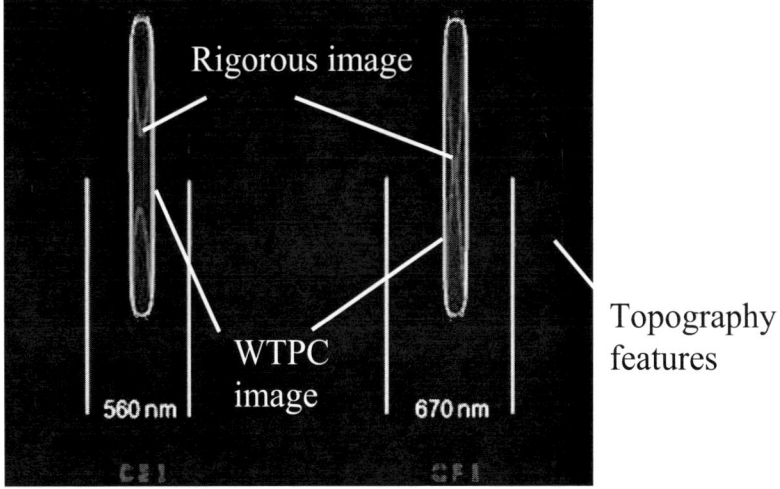

Fig. 9 Changes of the rigorous image when the distance between topography features increases from 560 nm to 670 nm: rigorous image comes from rigorous calculations, WTPC model results come from Equation (5).

In order to capture this extreme topography effect the WTPC model must contain the interference term describing NFI (see Equation (6)). In Fig. 10 there are given images modeled without/with taking NFI into account. From these results it follows that considering NFI in modeling is critical for correct capturing WTPC effects.

A second critical case for examination contains two lines running over the cross-like STI topography (see Fig. 11). In this case the field reflected by the vertical bar of the cross partially disintegrates the line images except those parts which are placed directly on the horizontal bar. This is exaggerated "footing" effect when only the thickest parts of lines remain.

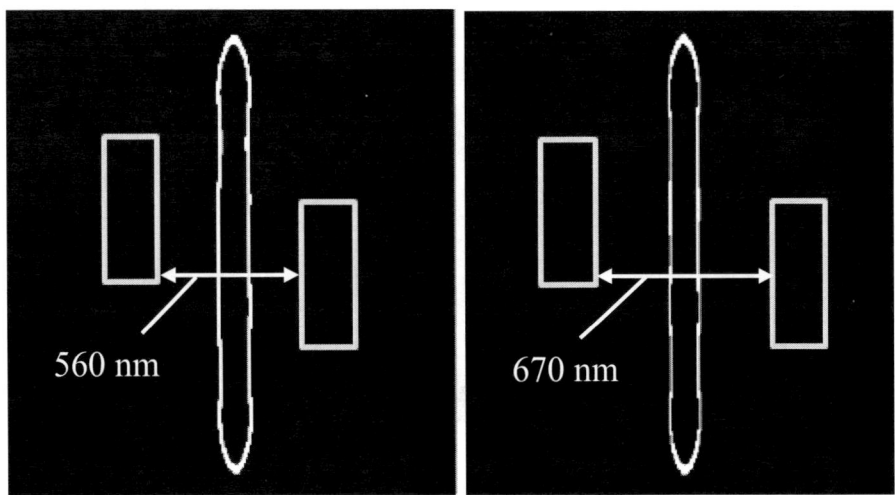

(a) WTPC model results without taking NFI into account;

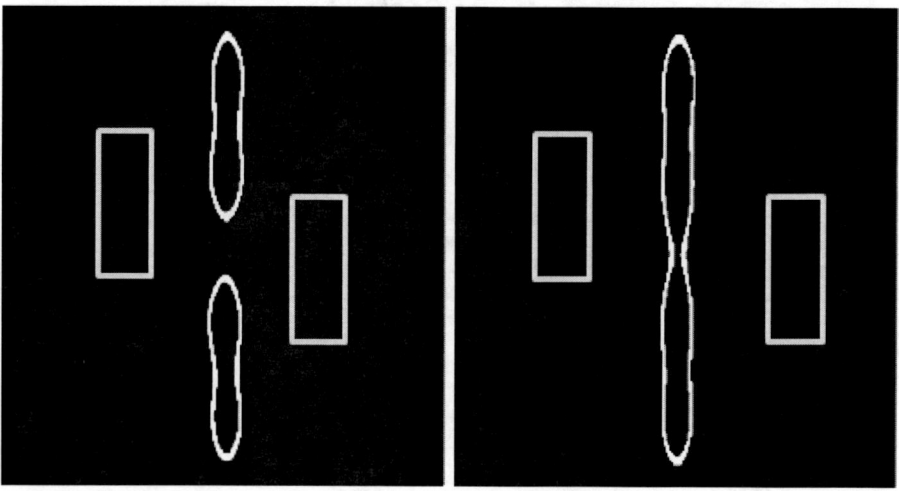

(b) WTPC model results with taking NFI into account.

Fig. 10 WTPC modeling with/without NFI for the case of Fig. 8: a) topography effect is not captured without taking NFI into account; b) topography effect is captured with taking NFI into account.

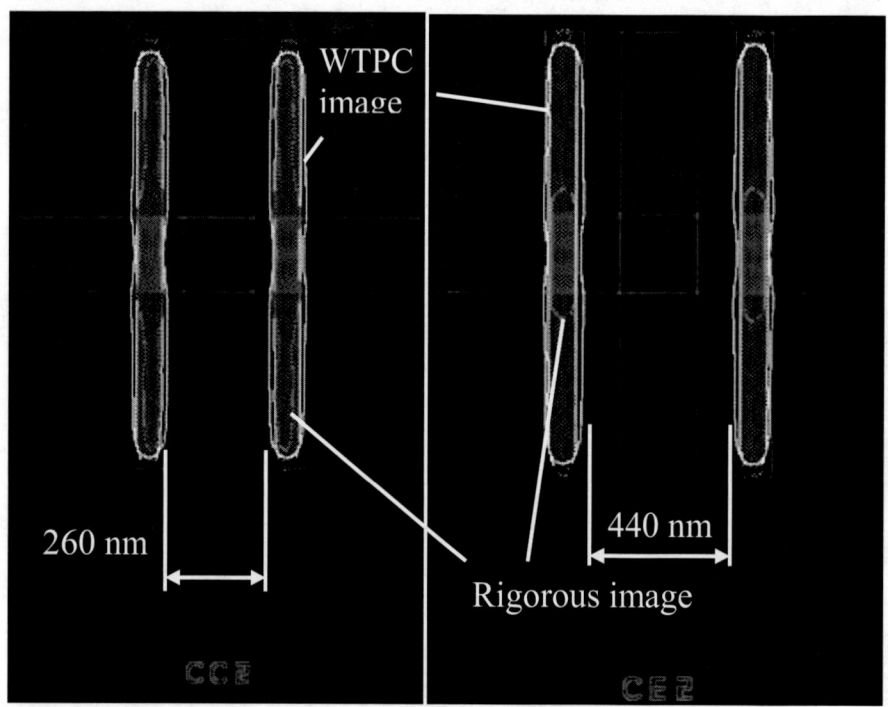

Fig. 11 Rigorous image of two lines over the cross-shaped topography: effect of the field reflected by the vertical bar of the cross is stronger along with increasing distance between the implant lines from 260 nm to 440 nm; rigorous image comes from rigorous calculations, WTPC model results come from Equation (5).

It is shown in Fig. 12 that taking NFI into account in modeling allows capturing such phenomenon, whereas the absence of NFI description in the model leads to its failure.

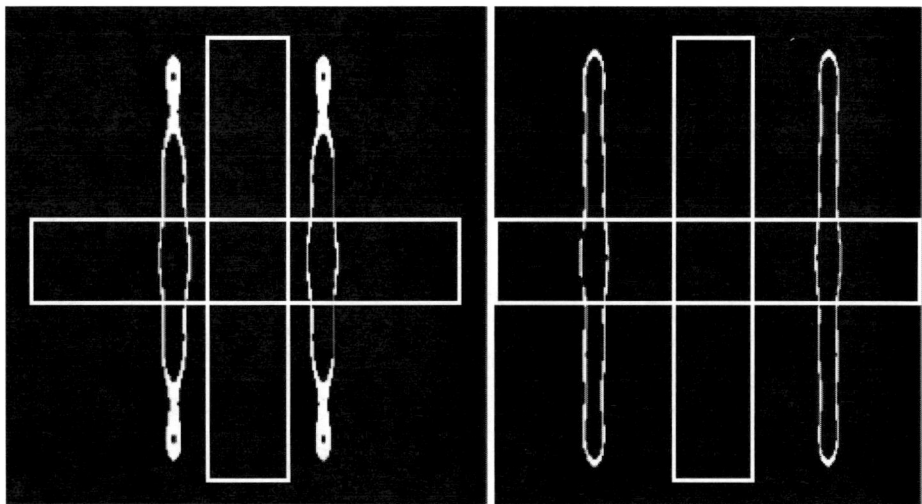

a) WTPC model results without taking NFI into account;

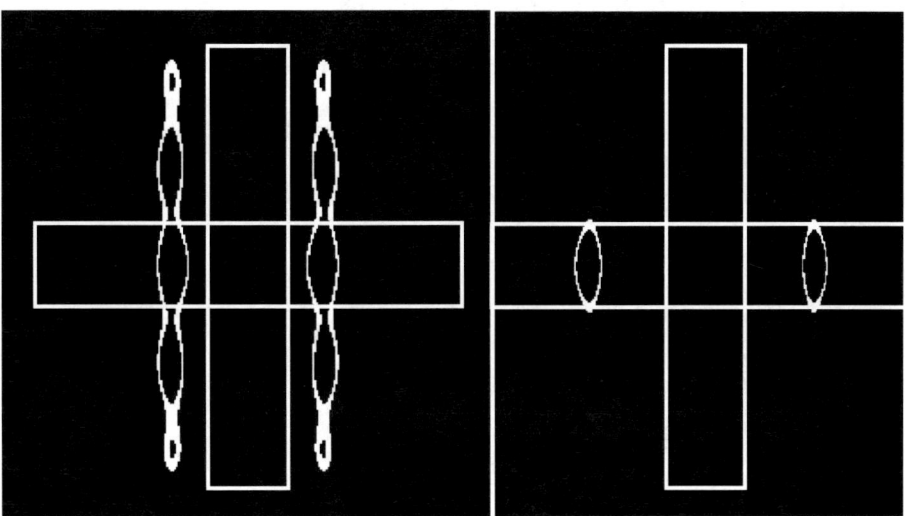

b) WTPC results with taking NFI into account.

Fig. 12 WTPC modeling with/without NFI for the case of Fig. 10: a) topography effect is not captured without taking NFI into account; b) topography effect is captured with taking NFI into account.

In this study we approximately evaluated time to results (TTR) for the standard OPC model and the OPC incorporating the WTPC model. Performance of both variants – without/with taking NFI into account – is evaluated. Run-time experiments have shown that the wafer topography model is able to capture the main effects so that the quality of results (QOR) looks satisfactory. The results are obtained within reasonable time.

In the Table 2 there are given run-times normalized to a *single* CPU performance of 50 Mflops and the test chip area of 200 µm × 200 µm in mask scale (50 µm × 50 µm in wafer scale).

The run-times for the WTPC models show that we can build fast approximate models for full-chip applications that compare reasonably to run-times of OPC models for the planar case.

Table 2 The run-times (TTR) of OPC and WTPC models

Model	TTR
OPC (planar) (3)	2'
OPC + WTPC (5)	4'
OPC + WTPC (6)	5'

Summary and Outlook

We presented a "virtual fab" concept that allows us to quickly generate test patterns and simulation data for creating fast, chip-scale models that can be used for mask synthesis. This flow allows us to efficiently test different model forms for chip-scale models and assess their performance. We then apply the "virtual fab" flow to the case of wafer topography, a particularly challenging case both due to the difficulty of obtaining experimental data as well as due to the computational complexity involved in getting rigorous data. Our approach for chip-scale modeling is based on the idea of a two staged modeling with a planar wafer OPC model and additional terms for capturing wafer topography effects. The reason for this approach is that it lends itself to retrofitting a highly accurate conventional OPC model for the planar case with a simulation-based or experimentally-based second stage for areas with wafer topography. We propose two model forms for chip scale modeling and compare their performance with respect to ability to capture critical patterns as well as run-time. We show that simple model forms as they are available today may be sufficient for cases where wafer topography effects are moderate, as for instance when a BARC layer is present. Where wafer topography effects are stronger, such as the case of a resist layer over STI's without BARC, we need to take into account near-field interference. Both models are suitable for mask synthesis, from a run-time point of view.

Going forward, the developed "virtual fab" flow will help us to improve modeling for the chip scale by providing a test ground for developing, assessing and perfecting model forms to accurately and efficiently capture specific effects depending on the situation at hand. The approach chosen for modeling wafer topography effects, which will be further improved, also shows a novel combination of experimental and simulation methods to minimize the model building effort and maximize the value of experimental data.

REFERENCES

1. Synopsys Sentaurus Lithography C-2009.06

2. Synopsys Progen 2008.12

3. R. Zimmermann, M. Schulz, W. Hoppe, H.-J. Stock, W. Demmerle, A. Zepka, A. Isoyan, L. Bomholt, S. Manakli, and L. Pain, Predictive Modeling for EBPC in EBDW, Proc. SPIE 7488, 74883J-13 (2009).

4. Synopsys Proteus MetroKit 2008.12

5. P. De Bisschop, T. Muelders, U.Klostermann, T. Schmoeller, J. J. Biafore, S.A. Robertson, M. Smith, Impact of mask three-dimensional effects on resist-model calibration, JM3 Letters Vol. 8(3) , 030501-F

6. H. Song, J. Shiely, I. Su, L. Zhang, W-K. Lei, Wafer topography proximity effect modeling and correction for implant layer patterning, Proc. SPIE, Vol. 7488, 74883F (2009).

7. I. Guilmeau, A. F. Guerrero, V. Blain, S. Kremer, V. Vachellerie, D. Lenoble, P. Nougueira, S. Mougel, J.-D. Chapon, Evaluation of wet-developable KrF organic BARC to improve CD uniformity for implant application, Proc. SPIE (2004), vol. 5376, pp. 461-470.

Inter-layer Self-Aligning Process for 22nm Logic

Michael C. Smayling*[a], Stewart Robertson[b], Damian Lacey[c], Sanjay Kapasi[b]

[a]Tela Innovations, Inc., 655 Technology Pkwy, Suite 150, Campbell, CA, USA 95008
[b]KLA-Tencor, 8834 N. Capital of Texas HW, Suite 301, Austin, TX USA 78759
[c]Cavendish Kinetics, Inc., 3833 N. First St., San Jose, CA USA 95134

ABSTRACT

Line/space dimensions for 22nm logic are expected to be ~35nm at ~70nm pitch for metal 1. However, the contacted gate pitch will be ~90nm because of contact-to-gate spacing limited by alignment. A process for self-aligning contact to gates and diffusions could reduce the gate pitch and hence directly reduce logic and memory cells sizes.

Self-aligned processes have been in use for many years. DRAMs have had bit-line and storage-node contacts defined in the critical direction by the row-lines. More recently, intra-layer self-alignment has been introduced with spacer double patterning, in which pitch division is accomplished using sidewall spacers defined by a removable core.[1] This approach has been extended with pitch division by 4 to the 7nm node.[2]

The introduction of logic design styles which use strictly one-directional lines for the critical levels gives the opportunity for extending self-alignment to inter-layer applications in logic and SRAMs. Although Gridded Design Rules have been demonstrated to give area-competitive layouts at existing 90, 65, and 45nm logic nodes while reducing CD variability[3], process extensions are required at advanced nodes like 22nm to take full advantage of the regular layouts.

An inter-layer self-aligning process has been demonstrated with both simulations and short-loop wafers. An extension of the critical illumination step for active and gate contacts will be described.

Keywords: Gridded design rules, restricted design rules, self-aligned structures

1. INTRODUCTION

The era of continual improvement in patterning equipment resolution has ended, with no improvement in optical performance since 1.35NA immersion scanners [4] were introduced in 2007. With λ/NA = 143nm, the Rayleigh equation CD = $k_1 \lambda$/NA implies that further improvements in resolution depend on k_1.

k_1 has been decreasing for recent logic technology node as shown in Figure 1. To maintain pattern fidelity at k_1 values below ~0.6, resolution enhancement techniques (RET) such as optical proximity correction (OPC), off-axis illumination (OAI), and phase shift masks (PSM) have been introduced. Note that the "22S" point is not realizable with $k_1 < 0.25$.

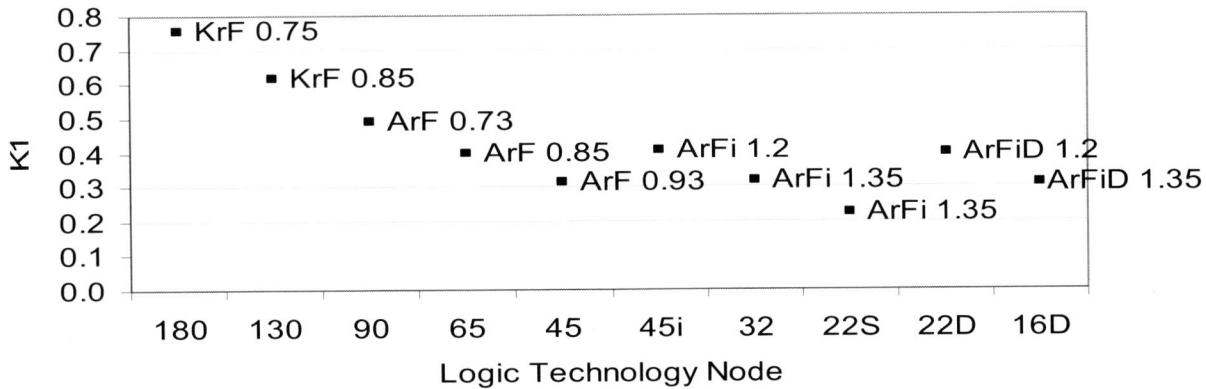

Fig. 1. k_1 trend for sub-200nm logic technology nodes.

*mike@tela-inc.com; phone +1 408 558-6321; fax +1 408 559-4600

Optical Microlithography XXIII, edited by Mircea V. Dusa, Will Conley, Proc. of SPIE Vol. 7640,
76401V · © 2010 SPIE · CCC code: 0277-786X/10/$18 · doi: 10.1117/12.846562

As k_1 decreases, "practical limits" are imposed by the design style.[5] 2D layouts with bent polygons are limited to ~0.35. A 1D layout style with parallel straight lines looking much like a grating pattern, and has a limit of ~0.28. Extensive efforts are being made to define "restricted design rules" which allow bends but with constraints on widths or spaces.[6] A 1D layout style with further requirements for keeping lines on a regular grid permits using a simplified set of design rules described as "gridded design rules" (GDR).[7,8]

2. SELF-ALIGNING PROCESS

2.1 Importance of contacted-gate pitch

Even with gridded design rules, logic elements like standard cells are limited by the grid pitch in the X and Y directions. As shown in Figure 2, with gate electrodes in a vertical direction, and metal-1 wires in the horizontal direction, the logic cell area is set by the number of lines in each direction multiplied by the pitch for those lines. The metal-1 pitch is set purely by lithography resolution, and can be extended for many nodes by using pitch division.[2]

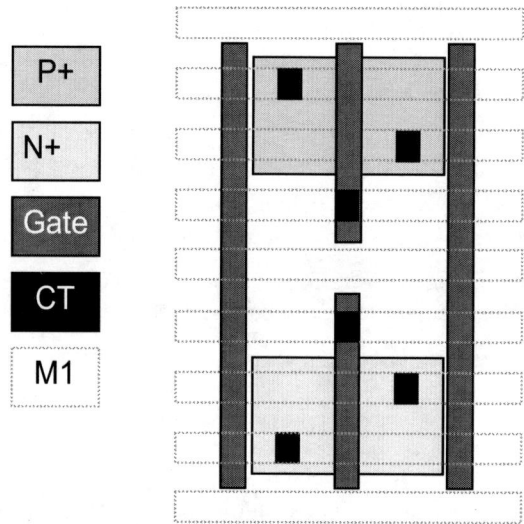

Figure 2. Logic inverter layout showing transistor layers.

However, the gate pitch is limited by both intra-layer resolution and inter-layer alignment between the gates and adjacent active contacts as shown in Figure 3. Anything done to reduce the contact-to-gate-space will directly improve cell area.

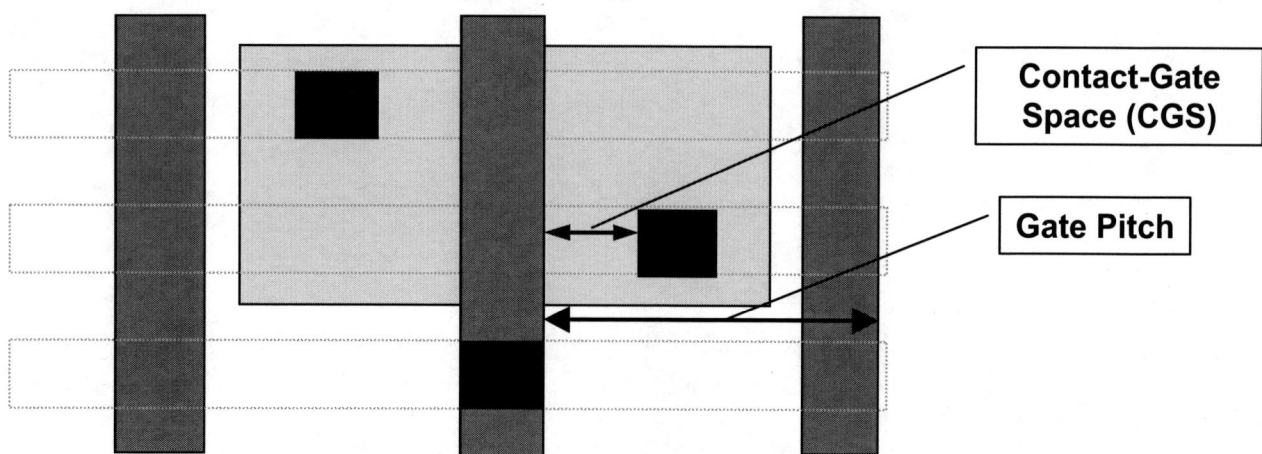

Figure 3. Detail of contacted-gate pitch.

Since both the active and gate contacts are on a grid set by the gates in the horizontal direction, this 1D gridded style has the potential to use the gate location to align contacts either to the gate or centered between the gates. This is not the case for conventional 2D random logic layout.

2.2 Structure to be evaluated

The self-aligning process involves multiple layers in the device structure, so additional structure information is needed to describe the steps. Figure 4 shows the cross-section of the structure to be tested in simulation and wafer form.

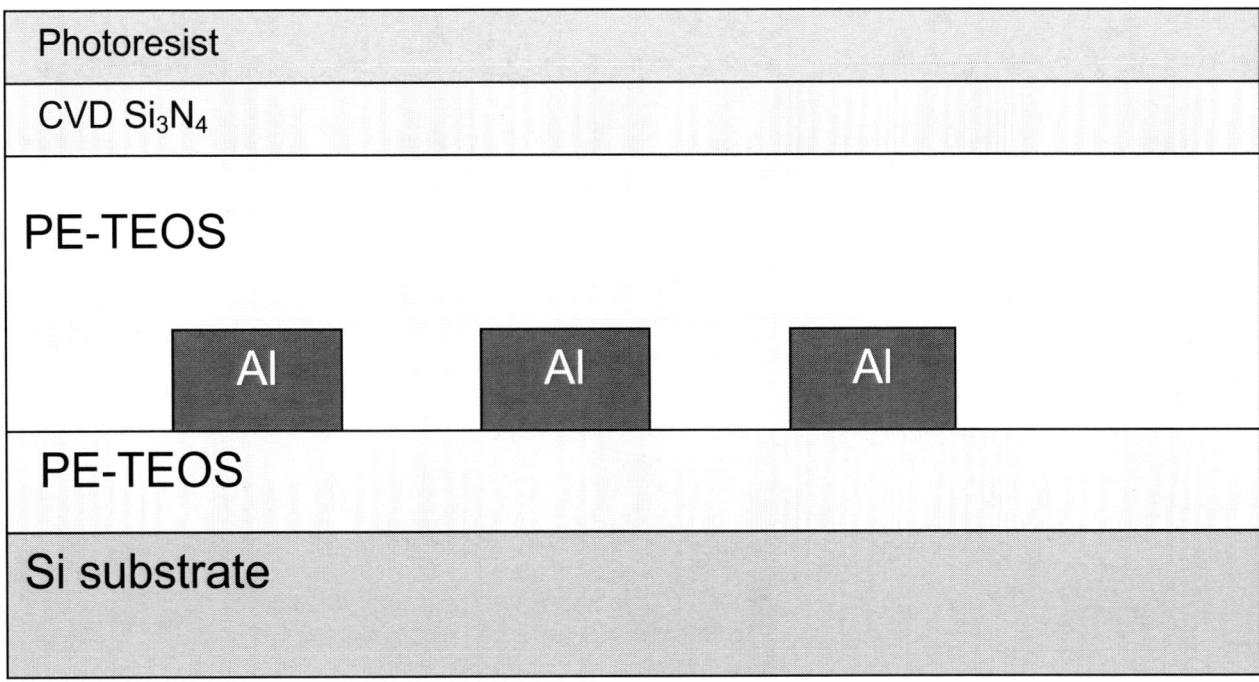

Figure 4. Cross-section of the structure under evaluation.

The structure in Figure 4 was fabricated at SVTC with standard process equipment and recipes. The gates were built with aluminum instead of silicided polysilicon for the first experiments to have better control of their optical properties. The upper plasma-enhanced TEOS layer was processed in several steps, with an intermediate CMP step to get a flat surface over the gates. The post-polish oxide thickness was measured, and then additional TEOS was deposited to reach the final target thickness over the gates. A silicon nitride layer was deposited to serve as a hardmask for subsequent self-alignment steps. A thin layer of photoresist was spun-on over the nitride to complete the structure preparation.

The critical step in the self-aligning sequence is the blanket illumination of the top layer of photoresist. Using the "wafer as a mask" allows a blanket illumination to be used. The exposure dose is an important factor; the dose is chosen such that regions of the resist which receive both incoming light as well as light reflected from the gates are above the develop threshold, while regions with only the incoming light are below the threshold. The resist can be positive or negative, and the develop process can be positive or negative, so either lines or trenches can be formed above the gates.

The experimental processing was completed up to the blanket illumination step. Future experiments are anticipated to complete a series of blanket deposition and etching steps to allow forming contacts which are self-aligned in the horizontal direction to either the gates or centered between the gates.

3. SIMULATIONS AND WAFER RESULTS

3.1 Simulations

Although many photolithography simulation tools can produce aerial images from source and mask data, many have an implicit assumption that there is no light reflected from the substrate. This is often not a good assumption, and in the case of the self-aligning process, reflected light is of paramount importance.

The recently introduced PROLITH X3 is ideally suited for simulations in which wafer topography has an impact on the patterning results. It can read in a GDS file with the underlying pattern, and use this in the preparation of the simulation substrate.

For simplicity, a cross-section was constructed with a single line having spaces on each side. Figure 5 shows the structure and the expected intensity pattern during the blanket illumination.

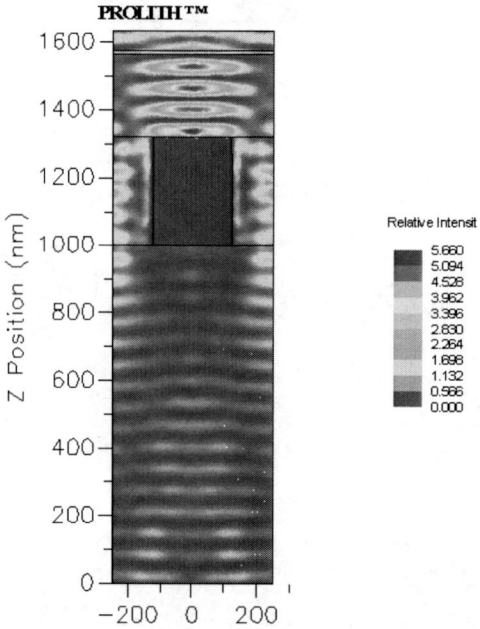

Figure 5. Prolith X3 simulation cross-section with a central line having gaps on each side.

As figure 5 shows, a pattern of standing waves is created above the central gate line. The gate is clearly reflecting light, while the gaps on either side are permitting light to pass deeper into the structure. There are several high intensity regions directly above the gate; control of the TEOS thickness is required to put a peak of intensity within the photoresist layer.

In addition to looking at the cross-sections to tune the stack thickness, aerial images were also created. For example, figure 6 shows a segment of a gate line and the resulting image based on the blanket illumination of the gate. The aerial image shows some bowing and widening compared to the original line width, and some line-end shortening. However, having this pattern in the resist above the hardmask can be useful for the self-aligning sequence since the aerial image is still congruent with the original pattern in the horizontal direction.

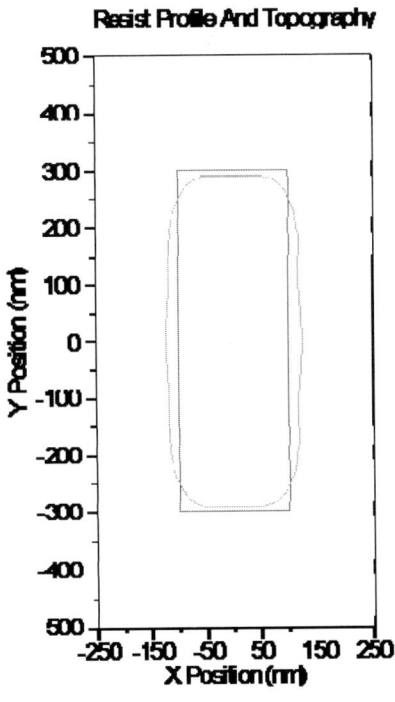

Figure 6. Aerial image (light brown) resulting from the original gate pattern (blue rectangle).

In addition to looking at just nominal conditions, simulations were done to look at the resist gap sensitivity to the underlying pattern line width as shown in Figure 7a. Further work is planned in the subwavelength regime. The gap width was also studied as a function of the film stack. Figure 7b shows the relationship with the nitride hardmask thickness; there is a 27.4nm change in linewidth for each 1nm change in nitride thickness, so process control is important.

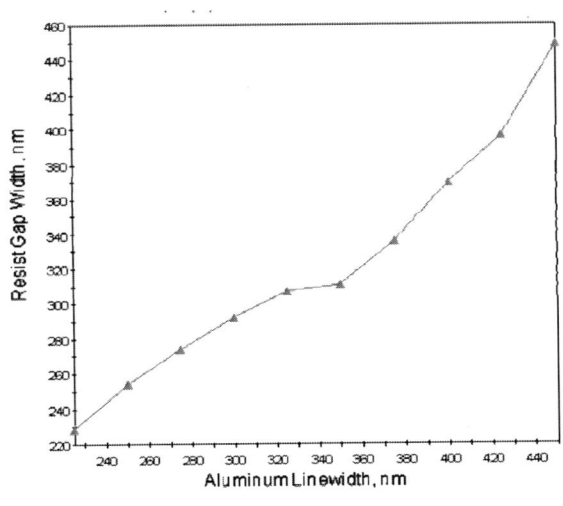

Figure 7a. Resist gap width versus mask width.   Figure 7b. Resist gap width versus nitride cap thickness.

The effect of a masked exposure was also studied. A slot wider than a contact is formed in the normal mask, then the wafer with an underlying pattern is exposed. Figure 8a-c shows the resulting final pattern in the top resist. Offset of the mask pattern was introduced to represent misalignment. The resulting pattern in the top resist had an offset reduced to a third of the original offset in the mask. The final CD was changed by less than 0.5% as a result of the offsets.

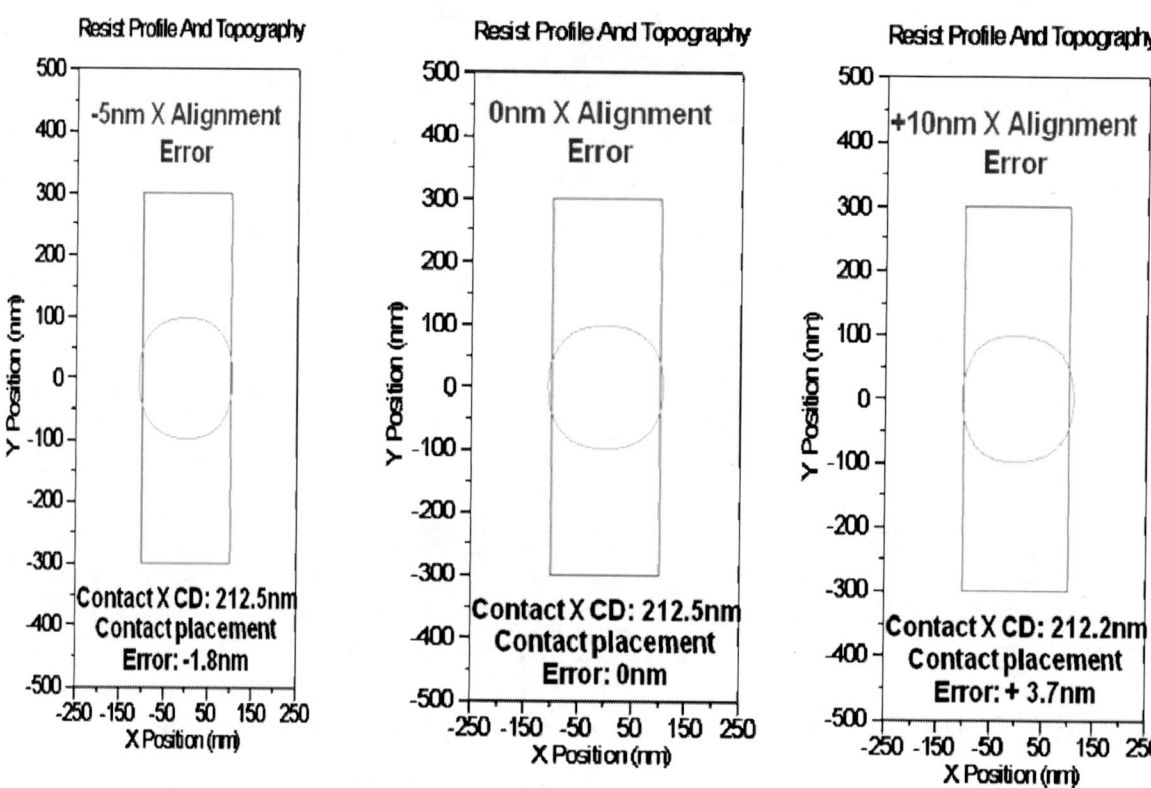

Fig. 8a. Slot exposure with -5nm alignment offset.

Fig. 8b. Slot exposure with 0nm alignment offset.

Fig. 8a. Slot exposure with +10nm alignment offset.

3.2 Wafer results

The critical step in the process sequence is the blanket illumination step. Test wafers were prepared with the film stack shown in figure 4. The oxide thickness above the aluminum lines was controlled by a measurement after CMP planarization, then an adjustment to the deposition time to achieve the desired final thickness. The thicknesses were initially simulated with Prolith and adjusted to get a reasonable intensity profile in the resist.

The photoresist type and deposition were adjusted to get the thin layer required. Shin-Etsu SXM-1754 was hand-dispensed with a thickness of 55.0nm with a sigma of 1%. Exposure was done with an ASML 1250XT (with no reticle pattern) to permit precise exposure control. An exposure matrix was run, with a dose of 3.2mJ/cm^2 giving the best results.

SEM photos are shown in Figures 9a and 9b. The top-down image in Figure 9a shows that the pattern of lines in the underlying aluminum gate layer was replicated into the resist lying on top of the hardmask. The line/space pattern as well as the line ends were resolved, as predicted b the simulations. The tilted SEM in Figure 9b shows that the resist was removed in the double-exposed regions as expected. However, work remains on the develop process to reduce edge roughness and eliminate any remaining resist in the exposed regions.

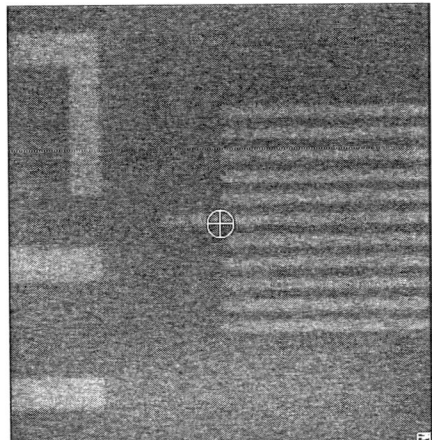

Figure 9a. Top-down SEM of the top resist pattern.

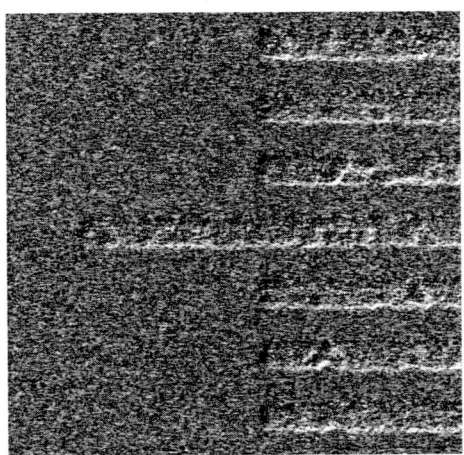

Figure 9b. Tilted SEM of the top resist pattern.

4. CONCLUSIONS

A self-aligning process, using the wafer-as-a-mask, has been simulated. This kind of process can be very useful to reduce logic cell size at the 22nm node where gate pitch is dependent on contact alignment. Wafer data have been presented down to 250nm linewidths. The new capabilities of PROLITH X3 have allowed the study of realistic logic cell layouts. The approach appears feasible for both gate and active contacts.

We would like to thank the staff at Tela Innovations for their help in creating layouts, the team at KLA for the simulations, and the team at Cavendish for the wafer processing. We appreciate the continued support of our executive management.

REFERENCES

[1] C. Bencher, H. Dai, Y. Chen, "Gridded Design Rule Scaling: Taking the CPU toward the 16nm node," Proc. of SPIE, vol. 7274 (2009).
[2] Y. Borodovsky, "Lithography 2009: Overview of Opportunities," SemiCon West (2009).
[3] M. C. Smayling, H. Y. Liu, L. Cai, "Low k_1 logic design using gridded design rules," Proc. of SPIE, vol. 6925 (2008).
[4] B.J. Lin, "Immersion lithography and its impact on semiconductor manufacturing," Proc. of SPIE, vol. 5377 (2004).
[5] W. Arnold, "Lithography for the 32nm Technology Node," IEDM 32nm Technology Short Course (2006).
[6] L. Capodieci, "From Optical Proximity Correction to Lithography-Driven Physical Design (1996-2006): 10 years of Resolution Enhancement Technology and the roadmap enablers for the next decade," Proc. of SPIE vol. 6154 (2006).
[7] M. C. Smayling, "Gridded Design Rules – 1-D Design Enables Scaling of CMOS Logic," Nanochip Technology Journal, vol. 6(2), (2008).
[8] M. Fritze, B. Tyrrell, T. H. Fedynyshyn, M. Rothschild, P. Brooker, "High-Throughput Hybrid Optical Maskless Lithography: All-Optical 32-nm Node Imaging," Proc. of SPIE, vol. 5751, (2005).

Process Window and Integration Results for Full-Chip Model-Based Assist-Feature Placement at the 32 nm Node And Below

Ji Li[*1], Gerry Luk-Pat[2], Amyn Poonawala[2], Kevin Lucas[3], Ben Painter[4]

[1]Synopsys Inc., 1027 Changning Road, Shanghai 200051, China PRC
[2]Synopsys Inc., 700 East Middlefield Road, Mountain View, CA 94043, USA
[3]Synopsys Inc., 1301 S. Mopac Expressway, Austin, TX 78746 USA
[4]Synopsys Inc., 2025 NW Cornelius Pass Road, Hillsboro, OR 97124 USA

ABSTRACT

Model-based assist-feature (MBAF) placement has been shown to have considerable lithographic benefits vs. rule-based assist-feature (RBAF) placement for advanced technology-node requirements. For very strong off-axis illumination, MBAF-placement methods offer improved process window, especially for so-called forbidden pitch regions, and greatly simplified tuning of AF-placement parameters. Historically, however, MBAF-placement methods had difficulties with full-chip runtime, friendliness to mask manufacturing (e.g., mask rule checks or MRCs), and methods to ensure that placed AFs do not print on-wafer. Therefore, despite their known limitations, RBAF-placement methods were still the industry *de facto* solution through the 45 nm technology node. In this paper, we highlight recent manufacturability advances for MBAFs by a detailed comparison of MBAF and RBAF methods. The MBAF method employed uses Inverse Mask Technology (IMT) to optimize AF placement, size, shape, and software runtime, to meet the production requirements of the 28 nm technology node and below. MBAF vs. RBAF results are presented for process window performance, and MBAF vs. OPC results are presented for full-chip runtimes. The final results show that MBAF methods have process-window advantages for technology nodes below 45 nm, with runtimes that are comparable to OPC.

Keywords: model-based assist feature (MBAF), rule-based assist feature (RBAF), assist-feature printing, process window, mask-rule check (MRC)

1. INTRODUCTION

For placing assist features, model-based methods have demonstrated benefits over rule-based methods [1, 2], both in terms of larger process window for strong off-axis illumination, and greater ease of use in adapting to changes in illumination and layout pattern. Despite these MBAF advantages, RBAF methods have remained the tool of choice up to the 45 nm node, and even beyond. Hesitation over MBAF has stemmed from a gap in full-chip runtimes, much like the gap between rule-based and model-based OPC; increases in shot count or mask complexity, exemplified by recent efforts towards pixellated masks; and concerns over printing assist features, particularly when assists are counter-intuitively larger than the main features themselves.

However, for advanced nodes, the time-consuming effort to extract a rule table is magnified by the increasing complexity of layouts. It is especially challenging to construct rules for two-dimensional (2D) semi-dense random patterns. Often such patterns do not receive RBAFs, but are readily covered by MBAFs, with a resulting boost in process-window size.

Our MBAF solution is centered on Inverse Mask Technology (IMT) [3]. Inverse mask design is an image-synthesis problem, which consists of finding a mask image that produces the desired wafer image. Since the entire mask field is synthesized, the primary, secondary, and shared AFs are computed simultaneously, allowing complex AF configurations.

*jili@synopsys.com; phone +86.212.307.2495;

Optical Microlithography XXIII, edited by Mircea V. Dusa, Will Conley, Proc. of SPIE Vol. 7640,
76401W · © 2010 SPIE · CCC code: 0277-786X/10/$18 · doi: 10.1117/12.848443

In this work, we show the suitability of MBAF for the 32 nm technology node and below. Comparisons of MBAF vs. RBAF are presented for process window, and comparisons of MBAF vs. OPC are shown for full-chip runtimes. The rest of this paper is organized as follows. Section 2 characterizes the different data flows for RBAF and MBAF, pausing to illustrate the differences between our IMT-AF flow, and a conventional inverse-lithography flow. Section 3 describes the data flow that we used to evaluate the process window for RBAF and MBAF. In Sections 4 and 5, we present process-window results for contact-hole layouts and line-space layouts, respectively. Section 6 compares full-chip runtimes for MBAF and OPC, and Section 7 concludes.

2. ASSIST-FEATURE FLOWS

In creating a mask with assist features, four tasks are required: (i) AF placement, (ii) AF-print fixing, or the prevention of AF printing, (iii) Main-Feature Correction, and (iv) MRC enforcement or MRC clean-up. What follows are different data flows for carrying out these four tasks.

2.1 RBAF Flow

In the rule-based flow shown in Fig. 1, AF Placement and AF-Print Fixing are carried out simultaneously in Step 1. Both tasks are limited in that rule-based AF Placement explores a limited AF space, and AF-Print Fixing relies on severely restrictive rules to ensure non-printing of AFs. In Step 2, traditional OPC is performed, consisting of iteration between Main-Feature Correction and MRC Clean-Up.

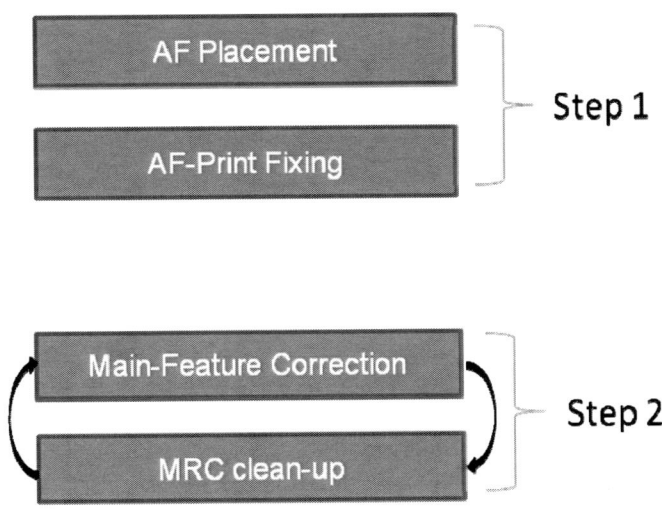

Figure 1: RBAF flow.

2.2 Traditional Inverse Flow

For the traditional inverse flow shown in Fig. 2, inverse methods are applied simultaneously for both AF Placement and for OPC, which can require very long runtimes. Also, not all models are invertible. For example, modeling of 3D-mask effects and mask-corner rounding may not be amenable to inverse lithography. Furthermore, since inverse-mask images can be quite complex, the MRC Clean-Up task is a major one; it must extract manufacturable shapes from the inverse-mask image. However, this MRC Clean-Up is not easily separated from the previous three tasks. The resulting compromises on contour positions will likely require iteration between Steps 1 and 2, further increasing the runtime.

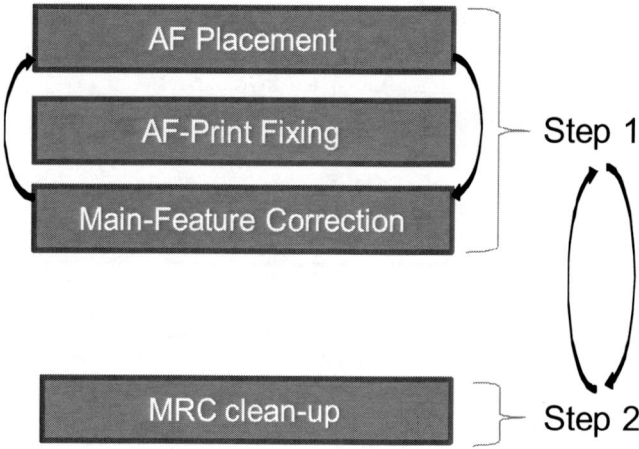

Figure 2: Traditional inverse flow.

2.3 Improved Inverse Flow for MBAF

For our MBAF, we use the inverse flow show in Fig. 3. In Step 1 of this flow, IMT is used for AF Placement. In Step 2, traditional OPC is performed but with an AF-Print Fixing step added to the usual iteration of Main-Feature Correction and MRC Clean-Up. Compared to the traditional inverse flow, runtimes are much lower because the inverse-mask image is only computed once – for AF Placement, and IMT is not used for OPC. However, accuracy is maintained because the AF-Print Fixing can use a fully complex model, including modeling of 3D-mask effects and mask-corner rounding.

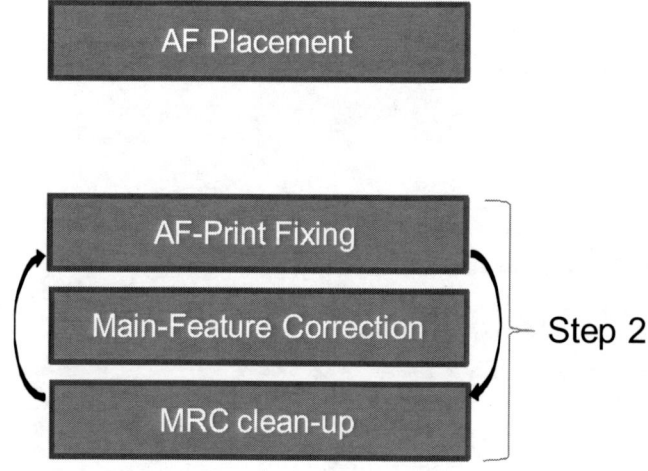

Figure 3: Improved inverse flow for MBAF.

3. EVALUATION FLOW

Our experimental flow for MBAF and RBAF evaluation and process window (PW) comparison is shown in Fig. 4. We selected 32nm and 28nm contact-hole and line-space layouts as the test cases. Rules are applied for RBAF placement, and IMT-MBAF is used for MBAF placement. The OPC is the same for MBAF and RBAF except that, for MBAF results, AF-Print Fix is included in the OPC recipe to prevent unwanted AF printing. Process-window analysis is performed by simulating contours under process variation, and measuring the CDs. Finally, we compare the process-window results of IMT-MBAF and RBAF.

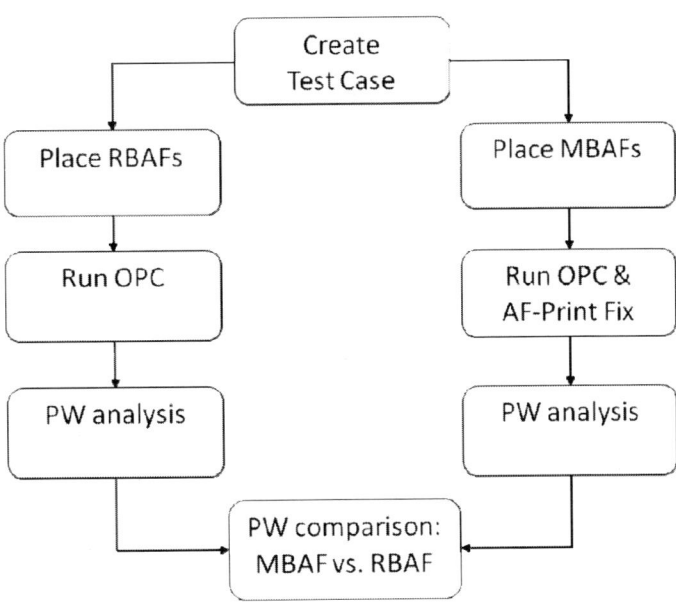

Figure 4: Evaluation flow.

4. IMT-MBAF COMPARISON WITH RBAF FOR CONTACT-HOLE LAYOUTS

4.1 Test Case

For the comparison of IMT-MBAF vs. RBAF for contact-hole layouts, we built the IMT model and OPC model with a strong OAI (off-axis illumination) source and a 1.35 NA optical illumination system. A defocus variation of 100nm and a dose variation of +/-2.5% were used for process-model generation. The test layout consisted of symmetrically arranged through-pitch contact arrays. The target CD of the contact pattern was 65nm. The acceptable range of CDs for process window analysis was between 90% of target CD and target CD.

4.2 Results

Figure 5 shows the contact-hole CD distribution plots for both IMT-MBAF and RBAF at -50nm defocus with nominal-dose or under-dose conditions. In these plots, the range of acceptable CD values is between the target CD, the solid red line, and 90% of the target CD, the dashed red line. IMT-MBAF has better or essentially equal process window compared to RBAF for all pitches, especially for the pitches where the number of AFs has a transition, for example from 1 AF per edge to 2 AFs per edge. Examples of these transition pitches are 300nm and 360nm. Due to the 1D spacing geometric constraint, RBAF cannot insert more AFs to improve the process window at the transition pitches. However, at these pitches, IMT-MBAF not only covers the 1D spaces but also makes use of the diagonal spaces between the contacts for AF placement.

In addition, for the semi-dense pitches, IMT-MBAF also shows good CD improvement over RBAF. As shown in Fig. 6, RBAFs only cover the 1D spaces because it is difficult to extract AF rules for the 2D spaces. However, since IMT-MBAF has good AF coverage in the diagonal spaces, the contact patterns show CD improvement.

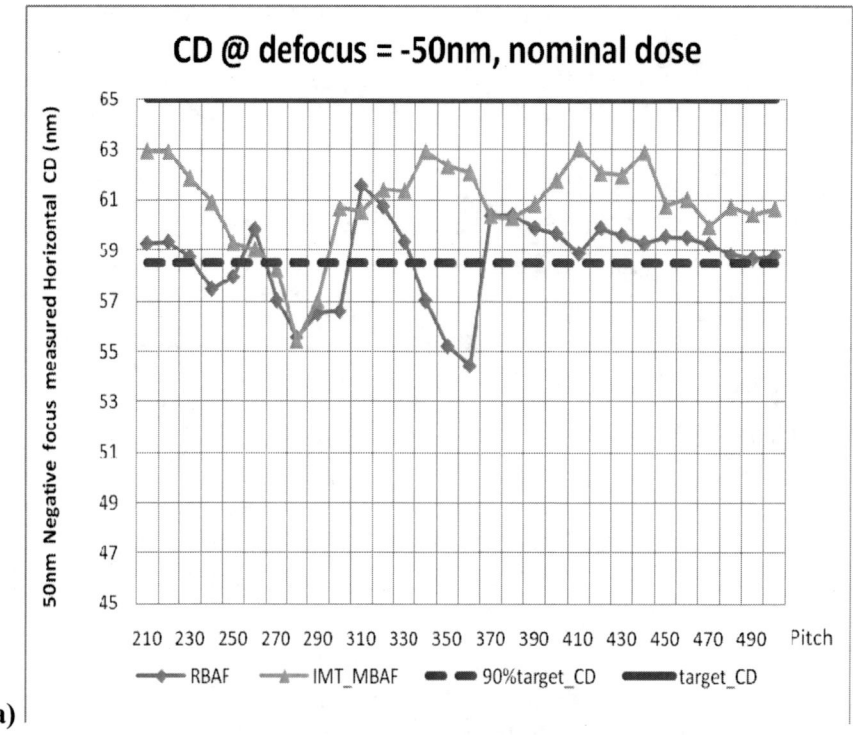

(a)

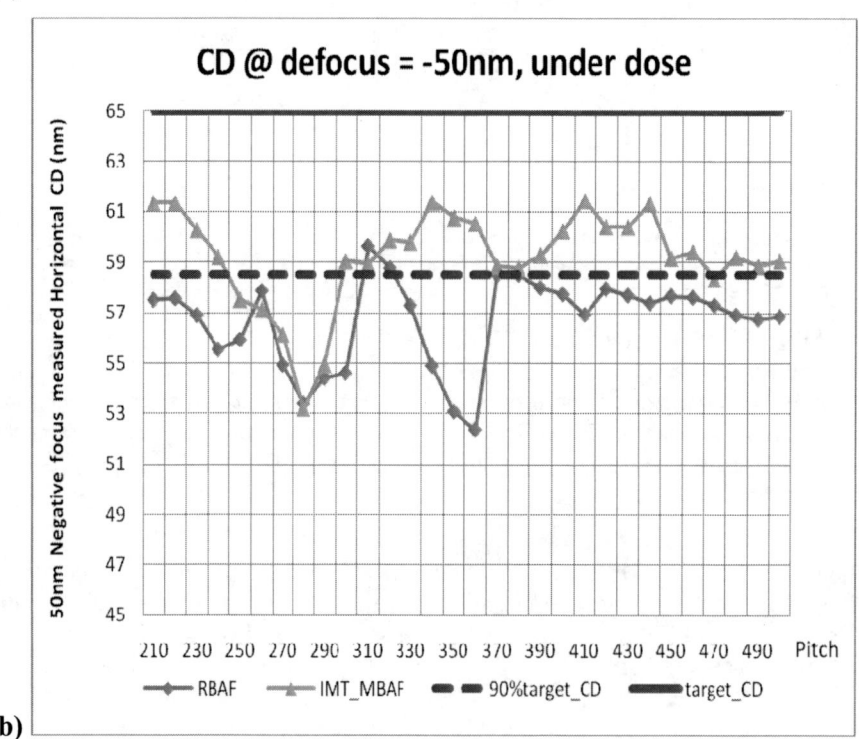

(b)

Figure 5: Contact-hole CD distributions at -50 nm defocus. The range of acceptable CD values is between the target CD, the solid red line, and 90% of the target CD, the dashed red line. IMT-MBAF has better process control than RBAF because it allows more pitches to achieve acceptable CD values. (a) Nominal dose. (b) Under-dosed by 2.5%.

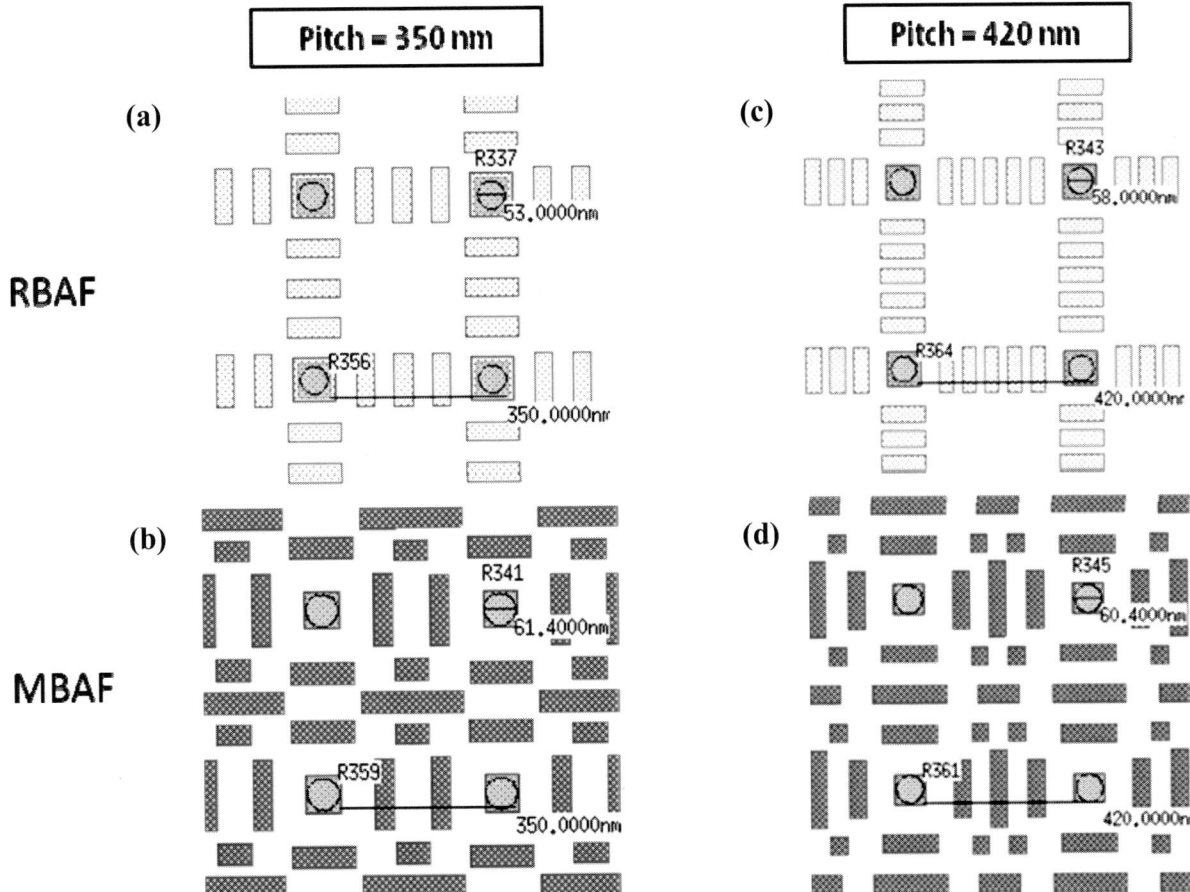

Figure 6: Contact-hole contours measured at -50 nm defocus and 2.5% under-dose. IMT-MBAF makes better use of the diagonal spaces between the contacts. (a) RBAF at a pitch of 350 nm. (b) IMT-MBAF at a pitch of 350 nm. (c) RBAF at a pitch of 420 nm. (d) IMT-MBAF at a pitch of 420 nm.

5. IMT-MBAF COMPARISON WITH RBAF FOR LINE-SPACE LAYOUTS

5.1 Test Case

For the comparison of IMT-MBAF vs. RBAF for line-space layouts, we again built the IMT model and OPC model with a strong OAI source and a 1.35 NA optical illumination system. As before, a defocus range of 100nm and a dose variation of +/-2.5% were used for process-model generation. The test case layout includes one-dimensional (1D) through-pitch line-space patterns, and through-pitch shared-contact patterns of different aspect ratios. The CD acceptance range for process window analysis was between 90% of target CD and target CD.

5.2 Results

The line-space CD distribution plots for different aspect ratio patterns are shown in Fig. 7. The CDs are measured at -50nm defocus and nominal-dose condition. Again, in these plots, the range of acceptable CD values is between the target CD, the solid red line, and 90% of target CD, the dashed red line. At isolated pitches, IMT-MBAF particularly has larger CDs than RBAF for all the aspect ratio patterns. At the semi-dense and dense pitches, IMT-MBAF also shows process-window advantage over RBAF for small aspect ratios, as it has better AF coverage for 2D areas. For example, in the RBAF plot for shared contacts, there is a noticeable valley from pitch 110nm to 150nm. However, for these pitches, IMT-MBAF makes use of the diagonal spaces between the contacts to greatly improve the CDs.

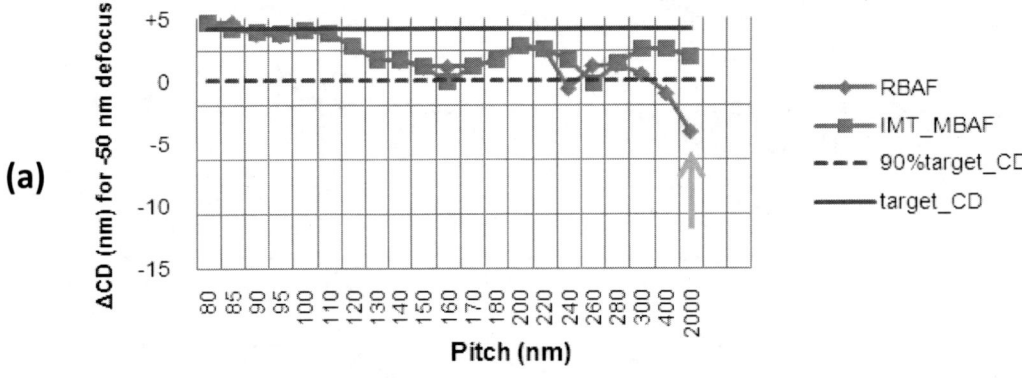

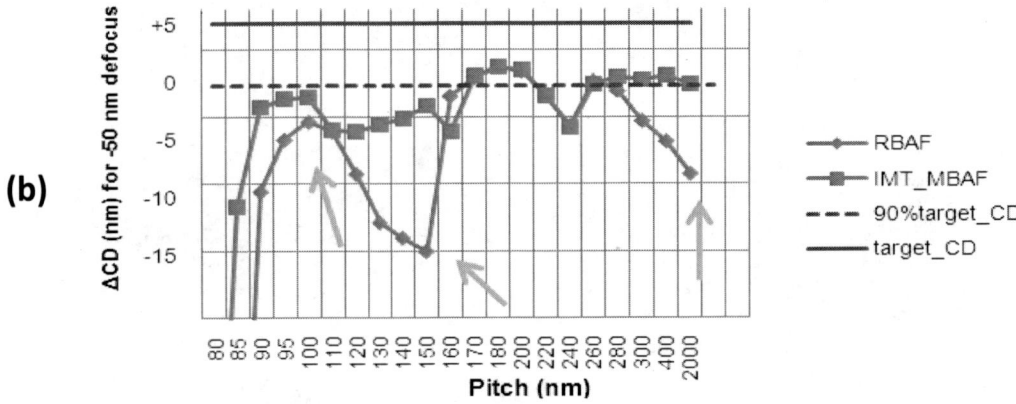

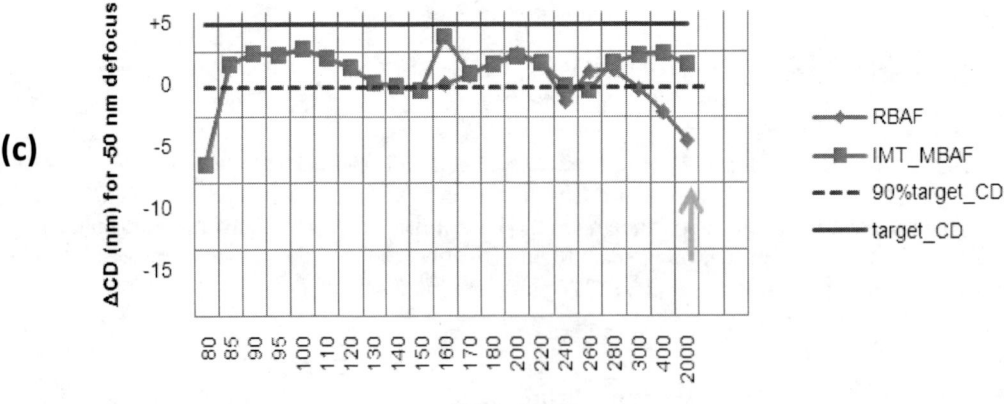

Figure 7: Line-space CD distributions at -50 nm defocus and nominal dose, for different aspect ratios. The range of acceptable CD values is between the target CD, the solid red line, and 90% of the target CD, the dashed red line. IMT-MBAF has larger process window than RBAF for isolated and semi-isolated features, and for small aspect ratios where it can exploit 2D spaces. The blue arrows show where RBAF CDs are particularly worse than IMT-MBAF CDs. (a) 1D features. (b) Shared contacts. (c) Tall shared contacts.

RBAF and IMT-MBAF placements for a semi-isolated pitch of 400 nm are shown in Fig. 8. The line-space contours are measured at nominal dose and -50nm defocus condition. Although RBAF has closer AFs and more AFs, it does not allow the main-feature contour to meet the target CD. While IMT-MBAF only places three AFs between the main features, its simulated contour CD exceeds the target CD because IMT-MBAF optimizes AF distance as a function of pitch.

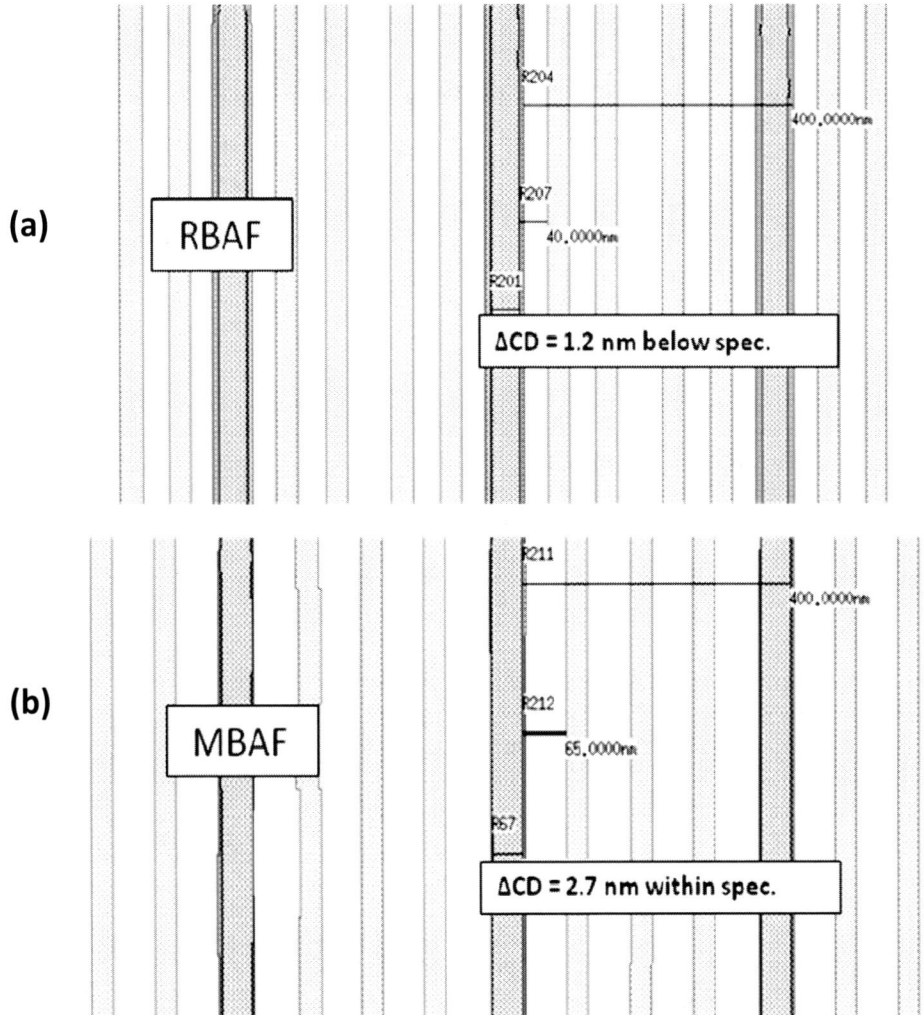

Figure 8: Line-space contours measured at -50 nm defocus and nominal dose, for a semi-isolated pitch of 400 nm. (a) RBAF. (b) IMT-MBAF.

6. FULL-CHIP RUNTIME

Full-chip turn-around time results for IMT-MBAF, and for OPC with AF-Print Fix are shown in Table 1. The areas of the full-chip layouts in the table are approximate. In addition, the runtimes for both IMT-MBAF and for OPC with AF-Print Fix are reported relative to a fixed quantity, 'T'. The results show that IMT-MBAF has comparable runtime to OPC operation. On average, the IMT-MBAF runtime is 0.5 times the OPC with AF-Print Fix runtime.

Table 1. Full-chip runtimes for IMT-MBAF and OPC with AF-Print Fix. On average, MBAF runtime is half of the OPC runtime.

Full-chip layout	Area (mm^2)	TAT (CPU hrs)	
		IMT-MBAF	OPC with AF-Print Fix
Layout 1	100	6T	9T
Layout 2	5	T	2T
Layout 3	10	T	3T

7. CONCLUSIONS

IMT-MBAF has been shown to have advantages over RBAF for advanced technology nodes. IMT-MBAF offers better process window than RBAF, especially for complex 2D semi-dense random patterns where it is challenging to construct RBAF rules. The MBAF solution greatly shortens the setup time compared to RBAF, since it is very time-consuming to extract a suitable rule table for the complex layouts of advanced nodes. Finally, we found that full-chip runtimes for MBAF are comparable to OPC runtimes.

ACKNOWLEDGEMENTS

The authors are grateful for technical support from Shanhu Shen and Charlie Zhang and for helpful discussions with them.

REFERENCES

[1] Levi D. Barnes, Benjamin D. Painter, and Lawrence S. Melvin III, "Model-based placement and optimization of subresolution assist features", Proc. SPIE, Vol. 6154, 61542C (2006);

[2] Chi-Yuan Hung, Qingwei Liu, Kyohei Sakajiri, Shumay D. Shang, and Yuri Granik, "Model based insertion of assist features using pixel inversion method: implementation in 65nm node", Proc. SPIE, Vol. 6283, 62832Y (2006);

[3] Amyn Poonawala, Benjamin Painter, and Jeffrey Mayhew, "Model-based assist feature placement: an inverse imaging approach", Proc. SPIE, Vol. 7122, 71220U (2008);

The role of mask topography effects in the optimization of pixelated sources

Vitaliy Domnenko[a], Bernd Küchler[b], Thomas Mülders[b],
Thomas Schmöller[b], Hans-Jürgen Stock[b], Georg Viehöver[b]

Synopsys, Inc.: [a]VO 7th line 76-A #402, St.-Petersburg, Russia;
[b]Karl-Hammerschmidt-Strasse 34, Aschheim/Dornach, Germany.

ABSTRACT

Ongoing technology node shrinkage requires the lithographic k_1 factor to be pushed closer to its theoretical limit. The application of customized illumination with multi-pole or pixelated sources has become necessary for improving the process window. For standardized exploitation of this technique it is crucial that the optimum source shape and the corresponding intensity distributions can be found in a robust and automated way. In this paper we present a pixelated source optimization procedure and its results. A number of application cases are considered with the following optimization goals: i) enhancement of the depth of focus, ii) improvement of through-pitch behavior, and iii) error sensitivity reduction. The optimization procedure is performed with fixed mask patterns, but at multiple locations. To reduce optical proximity errors, mask biasing is introduced. The optimization results are obtained for the pixelated source shapes, analyzed and compared with the corresponding results for multi-pole shaped sources. Starting with the 45 nm node mask topography effects as well as light polarization conditions have significant impact on imaging performance. Therefore including these effects into the optimization procedure has become necessary for advanced process nodes. To investigate these effects, the advanced topographical mask illumination concept (AToMIC) for rigorous and fast electromagnetic field simulation under partially coherent illumination is applied. We demonstrate the impact of mask topography effects on the results of the source optimization procedure by comparison to corresponding Kirchhoff simulations. The effects of polarized illumination sources are taken into account.

Keywords: optical lithography simulation, mask topography effects, polarized illumination, customized illumination, pixelated source, source mask optimization

1. INTRODUCTION

Ongoing technology node shrinkage requires the lithographic k_1 factor to be pushed closer to its theoretical limit, utilizing high-NA immersion lithography and a number of resolution enhancement techniques. One of these techniques is the application of customized illumination with multi-pole or pixelated structure sources [1]. The design of customized illumination is done by optimization of the source shape and intensity distribution with the following goals:

- improvement of process window (enhancement of the depth of focus (DOF) and exposure latitude(EL));
- improvement of through-pitch behavior (enhancement of CD uniformity and minimization of "forbidden pitch" effect);
- improvement of image fidelity for critical layer patterning (reduction of optical proximity effects);
- improvement of process robustness (reduction of sensitivity to mask errors, illumination and other process errors).

In the exposure tools itself, illumination are controlled by approaches like ASML's FlexRay [2] or Nikon's sPure [3] as well as by diffractive optical elements [4].

Starting with the 45 nm node, the electromagnetic field (EMF) induced in the mask by oblique incident and polarized light is significantly affected by the mask topography. In high-NA projection it has significant impact on imaging performance, and consideration for EMF effects in process optimization procedures [5] and in source mask optimization techniques [6] have become a must for advanced process nodes. The impact of mask topography effects on the results of the source optimization procedure is investigated in this paper.

Section 2 gives an overview of optimization procedure, including image intensity calculation details and the form of merit function. A combinatorial optimization algorithm is used to find the optimum source shape and the corresponding intensity distributions in a robust and automated way. In section 3 the process setup, mask and other parameter settings

Optical Microlithography XXIII, edited by Mircea V. Dusa, Will Conley, Proc. of SPIE Vol. 7640,
76401X · © 2010 SPIE · CCC code: 0277-786X/10/$18 · doi: 10.1117/12.845759

are defined in details. We study the imaging of 45 nm dense line/space patterns with 193 nm immersion lithography process in the following optimization test cases:

- through-pitch behavior improvement;
- absorber thickness error sensitivity reduction;
- mask error enhancement factor (MEEF) improvement; and
- polarization error sensitivity reduction.

In section 4 results obtained in the listed cases are discussed. The optimization is performed with rigorous EMF simulations as well as with Kirchhoff simulations. Corresponding imaging results of both solutions are compared, and the role of mask topography effects in the optimization of pixelated sources is investigated. In addition, the optimization results obtained for the pixelated source shapes are analyzed in comparison with the corresponding results for multi-pole sources.

2. OPTIMIZATION PROCEDURE DESCRIPTION

The procedure for finding a desired source shape in automated way requires an optimization strategy, which of choosing:

- optimization variables and constraints (the ranges of value for changed parameters);
- merit function (a fitness of solution metrics);
- optimization algorithm (defining the sequence of actions to find a desired solution).

In our optimization procedure the main variables are source point intensities, which can be in the range from 0 to 1. In order to reduce optical proximity errors, mask biasing can be applied as well, and the best biases for different pitches are chosen automatically on each optimization iteration. The optimization is performed for several pitches on periodic patterns or multiple metrology locations on arbitrary pattern in several focus positions. Additional parameters (absorber thickness, polarization degree, etc.) can be included into the optimization routine as noise factors to reduce the error sensitivity.

The simulation is based on the assumption [7] that the aerial image intensity $I'(x', y')$ is a weighted and normalized sum of coherent images $I'(x_s, y_s; x', y')$ corresponding to each separate source point:

$$I'(x', y') = \frac{\iint_{-\infty}^{+\infty} S(x_s, y_s) \cdot I'(x_s, y_s; x', y') dx_s dy_s}{\iint_{-\infty}^{+\infty} S(x_s, y_s) dx_s dy_s}$$

where $S(x_s, y_s)$ is the source point intensity. The integral over the source intensity distribution in the denominator is used for result normalization. In this study, simulations are performed on a rectangular pupil mesh with 0.05σ step (in terms of coherence factor) and source symmetry is taken into account. That corresponds to 1185 points over the pupil. All coherent images are precomputed with a commercially available lithography simulation tool [8]. It provides a rigorous and fast EMF computation capability [9] and an advanced topographical mask illumination concept (AToMIC) for accurate simulation of the mask topography under partially coherent illumination according to the light incidence direction and polarization conditions. Although rigorous precomputation can be resource consuming, computational cost of optimization with the topographical mask model is the same as with the thin model. For comparison all optimization sessions are also repeated with simplified Kirchhoff (or thin mask) simulation method. In both cases images are calculated in the immersion medium with a scaled defocus approach [10], although any wafer stack can be easily taken into account. The constant threshold resist model [11] is used.

The pixelated source optimization procedure is started without any particular initial source intensity distribution. Different combinations of source point intensities provide different imaging performance, and the optimal combination is found by applying combinatorial optimization algorithm. The harmony search algorithm [12] works well for the pixelated source optimization task, and provides high speed of convergence. The numerical investigations in different cases show that the optimization converges to solution in less than 1500 iteration, faster and more reliably than some other optimization algorithms also tested. In each iteration, the evaluation of image intensity and metrics for the current source intensity combination can be performed relatively fast.

The source optimization procedure is aimed at finding compromise solutions which allow obtaining optimal results under certain conditions. The optimality is determined by merit function, which can include different image metrics like

CD errors, image log-slopes, image contrasts as well as DOF or MEEF. The results for this paper are obtained with a merit function in the form of the following weighted sum:

$$M = \frac{1}{P \cdot F} \sum_{p}^{P} \sum_{f}^{F} w_{p,f} \left(NILS_{p,f}^{-1} + |CD - CD_{p,f}| \right)$$

where P is the number of pitches, p is the corresponding index, F is the number of focus positions, f is the corresponding index, and $w_{p,f}$ is the weight which controls the importance of the corresponding pitch and focus position, $NILS_{p,f}^{-1}$ is the inverse of normalized image log-slope, CD is the target line width, and $CD_{p,f}$ is a result line width. This form of merit function focuses the optimization on minimizing a CD error (i.e. to improve CD uniformity in a broader sense). At the same time, minimization of NILS should provide for an increase in exposure latitude, since exposure latitude is linearly related to NILS [13]. Results for several focus positions are also included into the optimization, thus increasing the DOF.

Thus, the optimization procedure consists of the following steps:

- precomputation of coherent images corresponding each separate source point;
- generation of source intensity distribution, which is performed with a combinatorial optimization algorithm;
- evaluation of merit function for the current source intensity distribution, and
- iteration of the second and third step, until a stable solution is obtained.

3. APPLICATION TEST CASE DESCRIPTION

In this study, the source optimization procedure is applied to improve the process of 193 nm immersion lithography for the imaging of dense line/space patterns with 45 nm width at various pitches. In this case, a "forbidden pitch" phenomenon occurs, and the purpose of the source optimization is to minimize this effect. A number of papers [14, 15] have analyzed this phenomenon in details and have shown results for conventional source optimization approaches.

Mathematical analysis of diffractive imaging [13] shows that a maximum contrast in an image of dense line/space pattern with relatively small pitch p can be achieved with dipole illumination. While 0th and 1st diffraction orders are used to produce an image, the optimal location of the source pole centers σ can be easily determined as:

$$\sigma = \pm \frac{\lambda}{2 \cdot p \cdot NA}$$

where λ is the wavelength, and NA is the numerical aperture. When a range of pitches need to be imaged, the optimal position of source poles is not so obvious. Moreover, when the pitch becomes larger than

$$p > \frac{n \cdot \lambda}{2 \cdot NA}$$

the projection lens starts to collect the n diffraction orders and the optimal illumination conditions for different pitches can vary significantly. Moreover, the analysis of pupil filling shows [15], that for better image contrast and DOF it is necessary to increase the overlap area between 0th and higher diffraction orders on the pupil. Thus, the improvement of some image metric for one pitch can be accompanied by its degradation for others.

The optimization procedure is performed with the following process setup:

- Wavelength: 193nm;
- NA: 1.35;
- Refractive index of the immersion liquid (water): 1.437;
- Polarization: Y-polarized.

The optimization of through-pitch behavior is performed for the pitch range 110-220 nm, where the EMF effects are most prominent. The investigations show that optimization can be performed with reduced selection of specific pitches. The imaging properties are dependent on the number of diffraction orders which are collected by the projection lens. We start with 110 nm pitch, while two diffraction orders are inside the pupil. This will occur up to ~140 nm pitch, and three-order diffraction range begins at ~150 nm. The pitch 180 nm is in the center of a three diffraction-order range, the end of which is at ~210 nm pitch. Thus, although the optimization is performed only for the key pitches 110 nm, 140 nm, 150 nm, 180nm, and 210 nm, the imaging behavior is improved for the continuous pitch range as well. The optical proximity correction is performed by mask biasing, which is explored in the range of 60-70 nm with 1 nm step. The threshold is determined by an anchor structure (220 nm pitch and 70 nm line width) for target in 45 nm. The problems of large pitches (> 220 nm) can be solved with subresolution assist features, which also can produce significant EMF effects, but they are not considered in the current paper.

In this study, an attenuated phase shifting mask is used (6% AttPSM) with the following materials:

- Substrate: SiO_2 (n=1.563);
- Absorber: MoSi (n=2.343; k=0.586) with thickness 72 nm.

During Kirchhoff simulations a thin mask model is used, in which absorber area has 6% transmission and produces 180 degree phase delay.

4. RESULTS AND DISCUSSION

4.1 CD uniformity improvement

The first optimization session is performed for the set of pitches (110 nm, 140 nm, 150 nm, 180nm, and 210 nm) in two defocus positions (0 and 80 nm). The exploration of mask biases is performed in the range 60-70 nm line widths with 1 nm step. The optimization using the thin mask model leads to source intensity distribution presented in figure 1a and to set of line widths #1.1 presented in table 1. The line widths for pitches which are not involved into optimization directly are determined in postoptimality exploration. Figures 1c-e show that optimization procedure provides a solution with a DOF about 200 nm, NILS > 2 and MEEF close to 1 except for the pitches 110 and 120 nm. In the case of small pitches, NILS has lower (down to 1.25) and MEEF has higher values (up to 2.5).

The verification of thin mask model optimization results with the topographic mask model performed with the same mask biasing shows that Kirchhoff simulations predict NILS and MEEF well for pitches > 120 nm and DOF slightly higher (Fig. 1c-e). The rigorous simulations show that in the case of 110 nm pitch the most of CDs are out of 10% tolerance and DOF is significantly lower.

The results can be improved, if a mask biasing exploration is repeated with the same source, but using rigorous simulations. New optimal line widths set #1.2 is presented in table 1. Apparently, different mask biases (mainly 1-2 nm larger) are required. In the figure 2c-e results for this case is compared with the topographic mask model optimization results. The corresponding line widths set #1.3 is also presented in table 1. In this case the optimal mask bias is mainly constant for all pitches < 180 nm. The source (Fig. 2a) rigorously optimized with topographic mask model provides better DOF (mainly 30 nm higher) for medium pitches, slightly higher NILS (by 0.5) and lower MEEF (by 0.2) for small pitches. For the rest the imaging performance of the source optimized with thin mask model and the source optimized with topographic mask model is approximately the same.

Thus, thin mask model and topographic mask model during pixelated sources optimization provides different image metrics, and optimizer is directed to slightly different optimal solutions. After optimization which is performed with thin mask model it is necessary to verify results with topographic mask model and refine mask biasing to get comparable results.

The comparison for source intensity distributions (Fig. 1a and 2a) shows that optimization with thin mask model as well as with topographical mask converges to similar solutions. Although intensity distributions are different, but it is possible to find a conventional multi-pole source which is closely matching both source shapes. The optimized sources can be transformed to discrete source segments, which are presented in figures 1b and 2b. To get desired image metrics with target in 45 nm it is necessary to use another range for mask biasing (40-55 nm) and anchor structure (220 nm pitch and 55 nm line width). Table 2 contains optimal line width sets after exploration with thin mask model (set #2.1) and topographic mask model (set #2.2). The difference is again no more than 2 nm. Corresponding image metrics are compared in figure 3. In both cases DOF is less than 200 nm, but MEEF is lower than 2 for all pitches. The NILS values produced with multi-pole source is slightly lower in comparison with pixelated source results, therefore lower exposure latitude are expected.

Thus, the optimization results in comparison with conventional source shape simulations shows that pixelated source can provide better CD uniformity and a process window enlargement owing to improvements in DOF. The pixelated source solutions, which are obtained in described way, can be used as a starting point for further parametric optimization of conventional source shape.

Table 1. Sets of mask line widths (in wafer scale) for the test cases with pixelated sources. The line width values typed in a bold style are determined during optimization, whereas others are in postoptimality exploration. Set #1.1 is the result of optimization with thin mask model. Set #1.2 is refining result after simulation with topographic mask model. Set #1.3 is the result of optimization with topographic mask model.

Set	Pitch, nm											
	110	120	130	140	150	160	170	180	190	200	210	220
#1.1	**60**	65	65	**61**	**60**	59	59	**61**	64	67	**70**	74
#1.2	59	66	66	63	61	61	61	62	64	67	70	74
#1.3	**60**	61	61	**60**	**60**	60	60	**62**	64	67	**70**	74

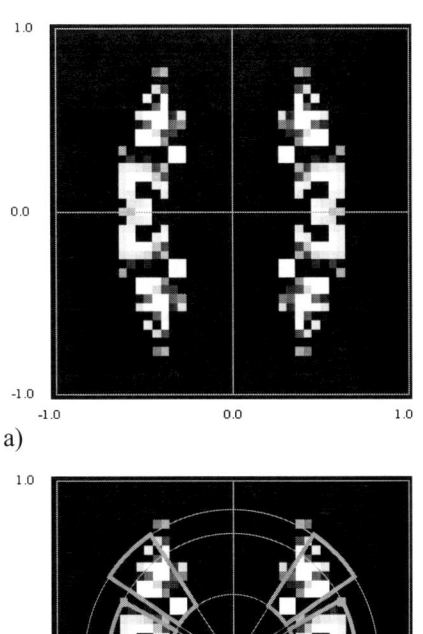

a)

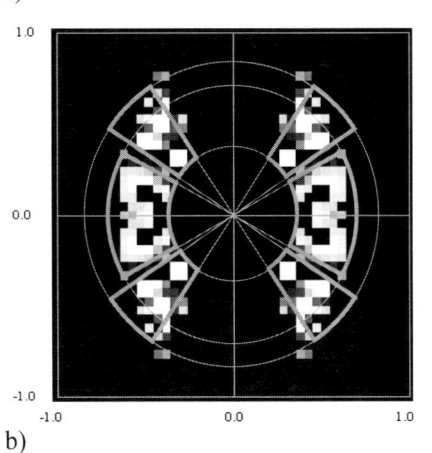

b)

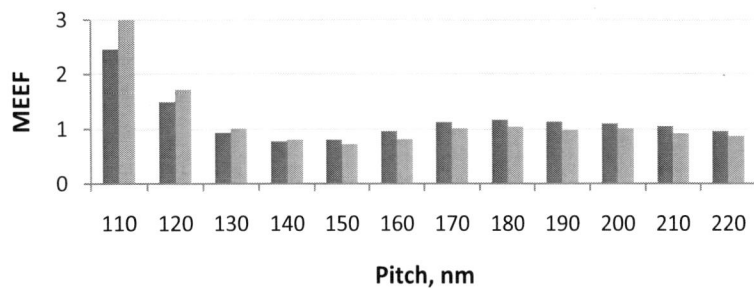

c)

d)

e)

Fig. 1 Results of the pixelated source optimization with thin mask model:

a) source intensity distribution;
b) corresponding multi-pole source;
c) DOF vs. pitch histogram (CD tolerance is ±10%);
d) NILS vs. pitch plot at 80 nm defocus;
e) MEEF vs. pitch histogram at 80 nm defocus.

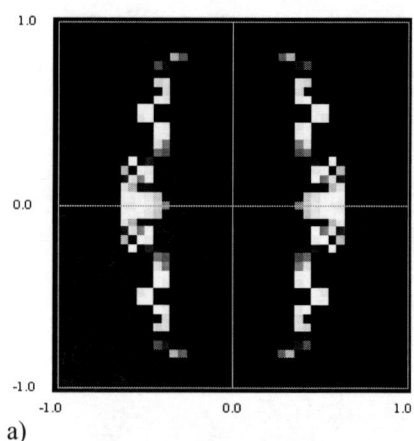

a)

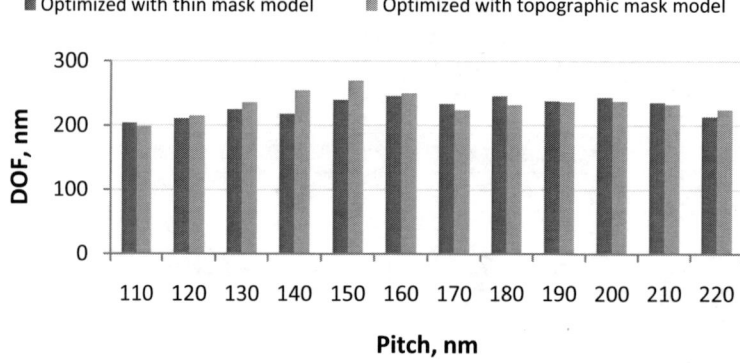

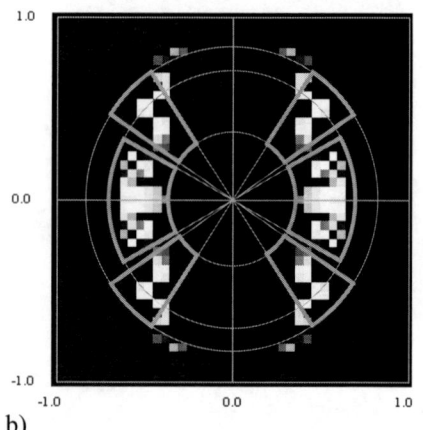

b)

c)

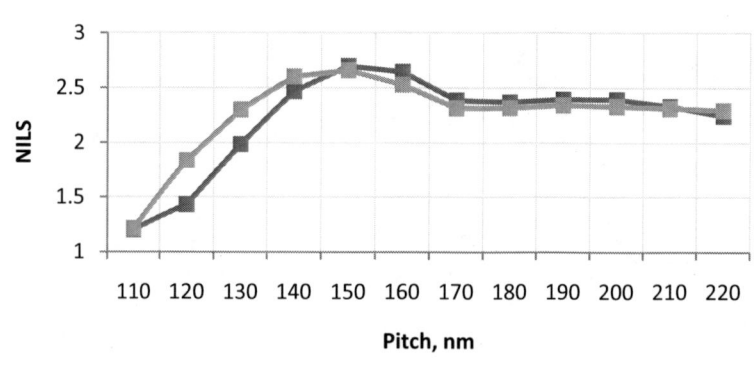

d)

Fig. 2 Results of the pixelated source
    optimization with topographic
    mask model:

a) source intensity distribution;
b) corresponding multi-pole source;
c) DOF vs. pitch histogram (CD
tolerance is ±10%);
d) NILS vs. pitch plot at 80 nm
defocus;
e) MEEF vs. pitch histogram at
80 nm defocus.

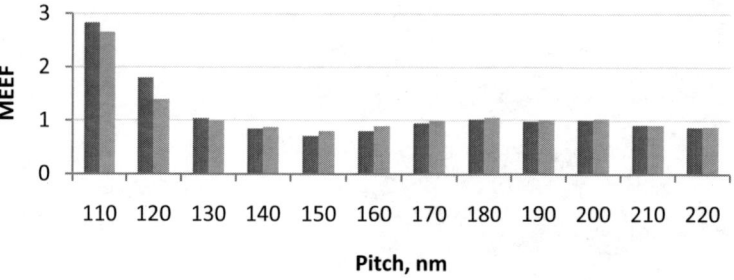

e)

Table 2. Sets of mask line widths (in wafer scale) for test cases with multi-pole sources. Set #2.1 is the result of exploration with thin mask model. Set #2.1 is the result of exploration with topographic mask model.

Set	Pitch, nm											
	110	120	130	140	150	160	170	180	190	200	210	220
#2.1	52	54	47	44	44	44	45	47	50	53	55	55
#2.2	51	54	48	43	42	43	44	45	48	52	55	55

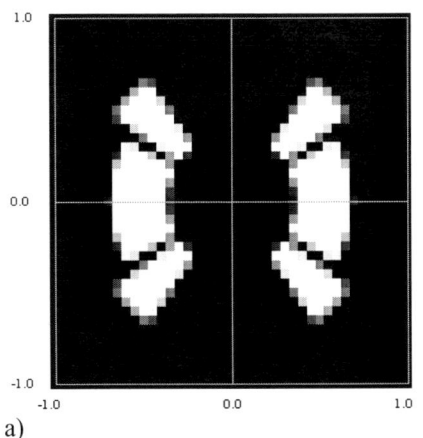

a)

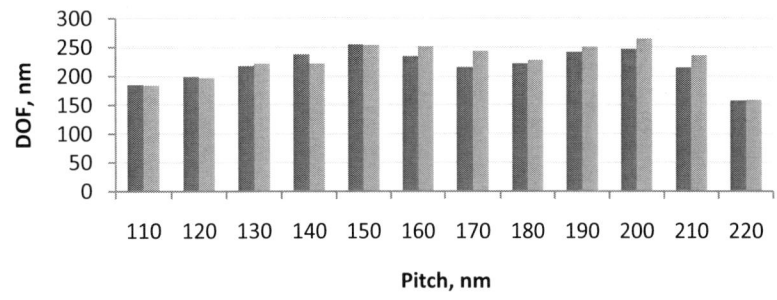

b)

Fig. 3 Results of the multi-pole source simulation:

a) multi-pole source (dipole with $\sigma_i = 0.36$, $\sigma_o = 0.69$, and opening angle = 60 deg; quasar with $\sigma_i = 0.36$, $\sigma_o = 0.82$, and opening angle = 20 deg);
b) DOF vs. pitch histogram (CD tolerance is ±10%);
c) NILS vs. pitch plot at 80 nm defocus;
d) MEEF vs. pitch histogram at 80 nm defocus.

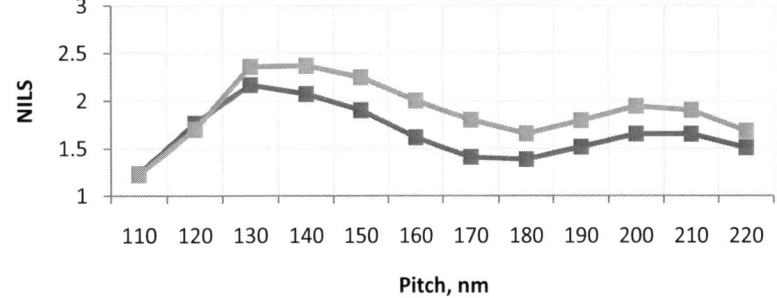

c)

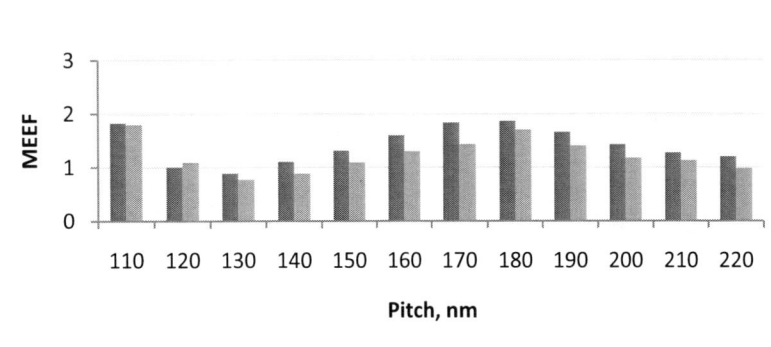

d)

4.2 Absorber thickness error sensitivity reduction

The second optimization session is aimed at reducing the sensitivity of the solutions from section 4.1 to absorber thickness error. In order to demonstrate the effect a hypothetical variation of the absorber film thickness by 5 % can be introduced through simulation. It results in 4 nm line width error (dotted lines in Fig. 4b, 4c, and 4e).

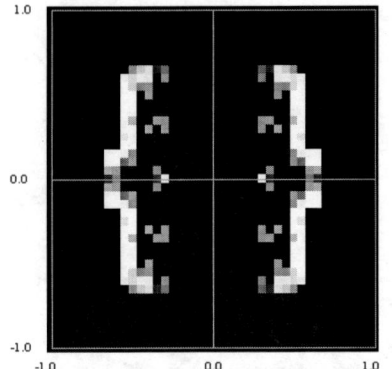

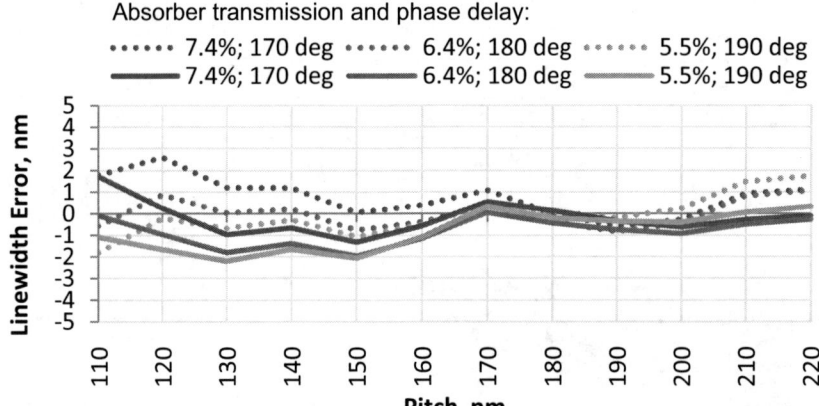

a) source intensity distribution after optimization with thin mask model;

b) linewidth error vs. pitch plots at 80 nm defocus corresponding to source a) and simulated with thin mask model;

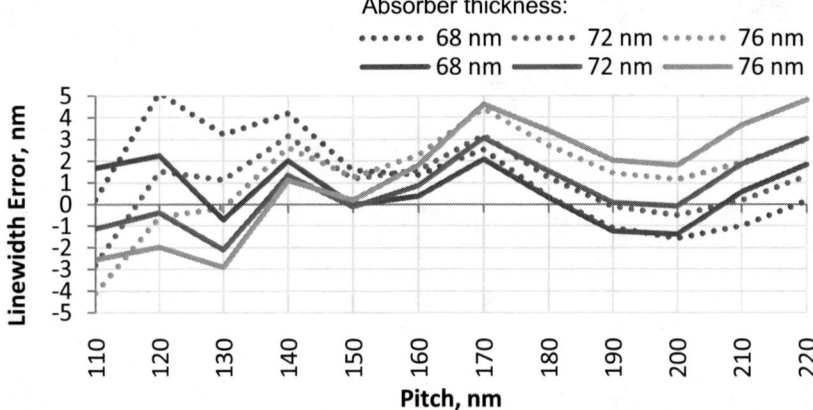

c) linewidth error vs. pitch plots at 80 nm defocus corresponding to source a) and verified with topographic mask model;

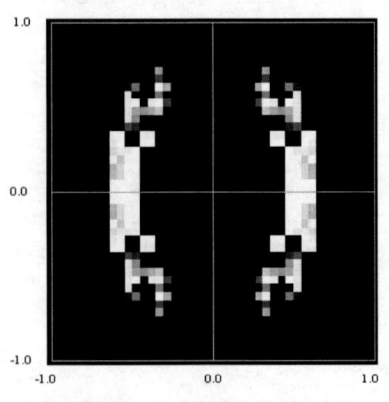

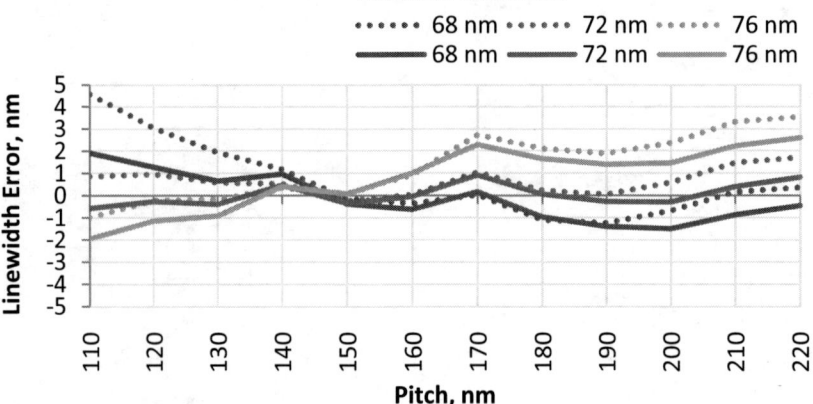

d) source intensity distribution after optimization with topographic mask model;

e) linewidth error vs. pitch plots at 80 nm defocus corresponding to source d) and simulated with topographic mask model.

Fig. 4 Results of the pixelated source optimization to reduce sensitivity to absorber thickness (dotted lines are results before optimization; solid lines are results after optimization).

Absorber thickness can be used during optimization as a noise factor, but it should be done differently for thick and thin mask models. For thick mask model the real thickness is specified directly, but for thin mask model absorber transmission and phase delay need to be defined. The absorber thickness values 68, 72, and 76 nm corresponds to absorber transmissions of 7.4%, 6.4%, and 5.5% and phase delays of 170, 180, and 190 degrees. For pixelated source optimization the set of pitches are the same as in previous section. The correspondent line widths are taken from table 1 (the set #1.1 is for Kirchhoff simulations and the set #1.3 is for topographical mask simulations). Since, the changes in absorber thickness especially have highest impact on smaller pitches; a higher weight factor in the merit function is empirically defined for the case of 110 nm pitch.

In the case of optimization based on thin mask model (Fig. 4a and 4b) the line width error induced by the absorber thickness error is reduced, but not significantly. Moreover, if we verify the solution with rigorous simulations (Fig. 4c), the line width error is still large (up to 4 nm), and the through-pitch behavior is highly non-uniform. The optimization with topographic mask model (Fig. 4d and 4e) provides a reduction in line width down to 2 nm for 110 nm pitch.

The comparison of pixelated sources produced by optimization with thin and topographic mask model shows that optimization leads to absorber thickness error sensitivity reduction. At the same time, the verification of optimization results with topographic model shows that optimization with thin mask model does not provide any improvements. Kirchhoff's approximation does not operate with absorber thickness directly and cannot provide accurate image metrics in this case. These results show that the rigorous simulation of mask topography effects would be preferred to perform the reduction of sensitivity to absorber thickness error.

4.3 MEEF improvement

The third optimization session is focused on MEEF improvement. This is an important application, because MEEF is not only a measure for mask error sensitivity, but also used as an overall process quality metric. The optimization is performed with the same set of pitches, but with three line widths for each pitch: with nominal mask bias (Table 1) and also relative ±2 nm (in wafer scale) mask biases. Higher empirical weights in the merit function for smaller pitches are used to provide better MEEF minimization.

The source intensity distribution after optimization with thin mask model is presented in figure 5a and corresponding results are plotted in figure 5b. These results are shown that optimization leads to MEEF reduction by 0.5 for the 110 nm pitch. If we verify the solution again with topographical model simulations, the benefit of MEEF optimization is also presented, but absolute values are higher by about 0.5 (Fig. 5c). The source intensity distribution after optimization with thin mask model is presented in figure 5d. The optimization with topographic mask model also shows MEEF improvement by 0.5 for small pitches (Fig. 5e).

All these results show that both optimizations procedures produce consentient MEEF reduction with marginal differences in MEEF values (< 0.5).

4.4 Polarization error sensitivity reduction

Finally, the results on polarization error sensitivity reduction are presented. In reality it is not possible to create fully polarized light, and it is always a mix of polarized and unpolarized illumination, i.e. the degree of polarization is always lower than 1. Thus, the degree of polarization can be included within the optimization as a noise factor to reduce the sensitivity to polarization condition changes. An analysis of previous solutions (dotted lines in Fig. 6b, 6c, and 6e) shows that changes in line width induced by degree of polarization variation do not almost depend on pitch, thus it is not necessary to use any special weights in the merit function.

In the case of optimization using the thin mask model (Fig. 6a and 6b) the CD error is reduced by 2-3 nm, but the verification with rigorous simulations shows an error increase up to 7 nm (Fig. 6c). It shows that error related to polarization effects cannot be reduced with Kirchhoff-based optimization. At the same time, the optimization with topographic mask model (Fig. 6d and 6e) is successful and shows reduction in line width errors and improvement in CD uniformity.

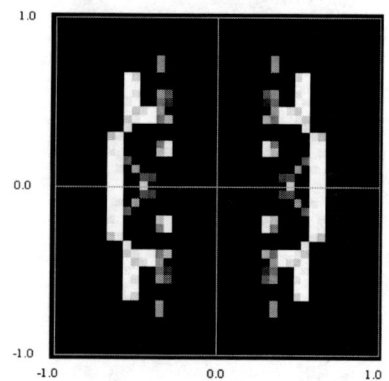

a) source intensity distribution after optimization with thin mask model

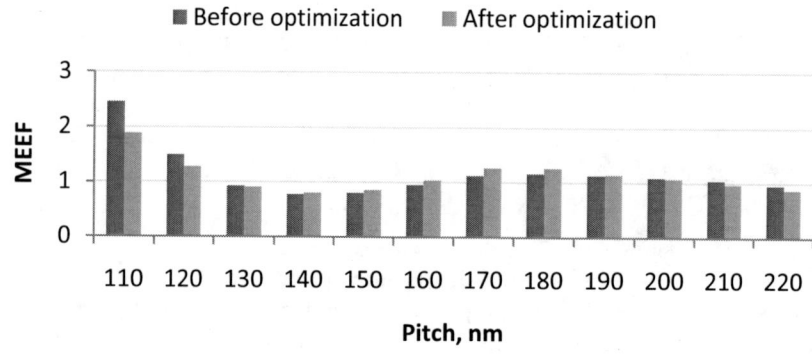

b) MEEF vs. pitch histogram at 80 nm defocus corresponding to source a) and simulated with thin mask model;

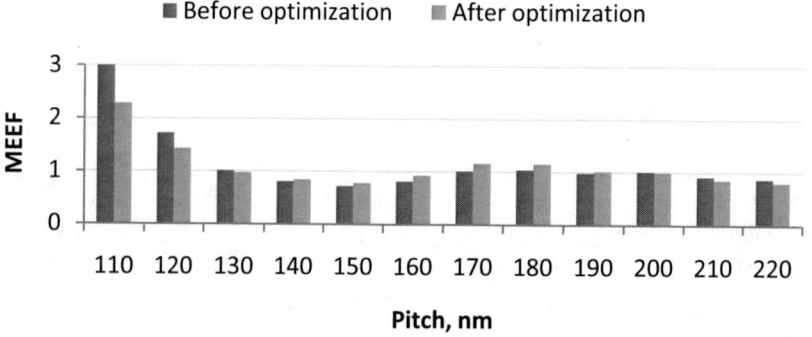

c) MEEF vs. pitch histogram at 80 nm defocus corresponding to source a) and verified with topographic mask model;

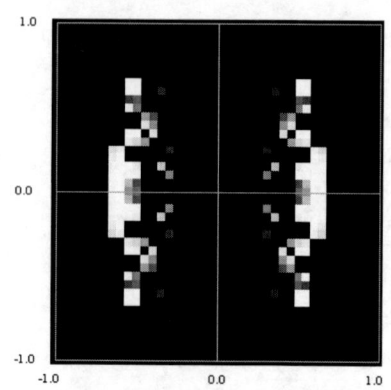

d) source intensity distribution after optimization with topographic mask model;

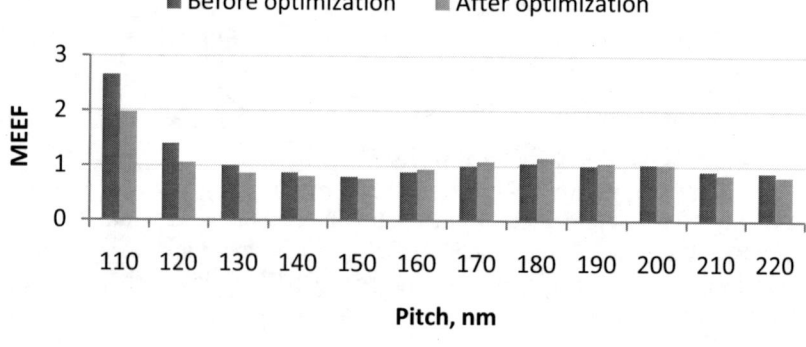

e) MEEF vs. pitch histogram at 80 nm defocus corresponding to source d) and simulated with topographic mask model.

Fig. 5 Results of the pixelated source optimization to improve MEEF.

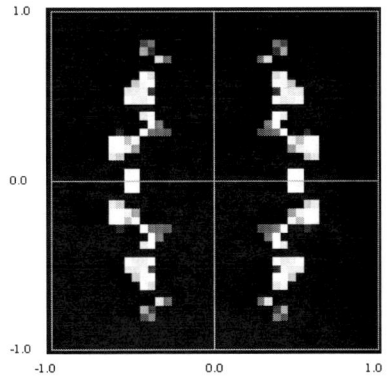

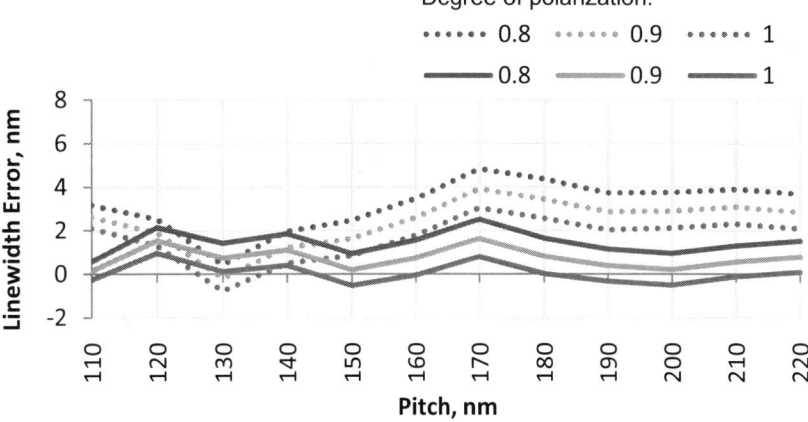

a) source intensity distribution after optimization with thin mask model;

b) linewidth error vs. pitch plots at 80 nm defocus corresponding to source a) and simulated with thin mask model;

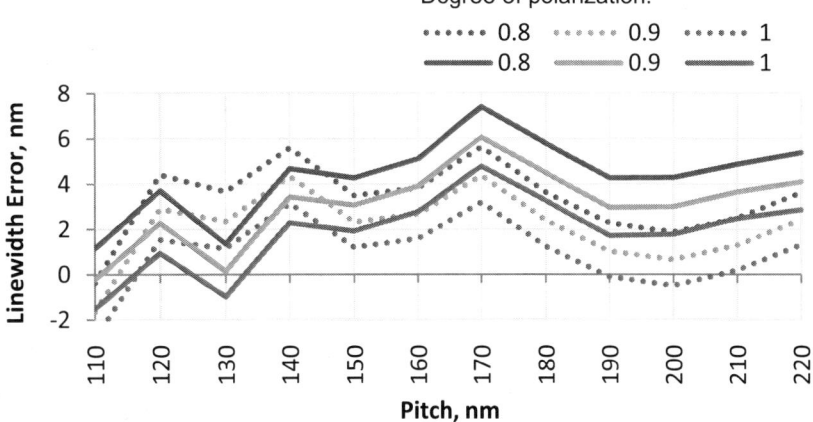

c) linewidth error vs. pitch plots at 80 nm defocus corresponding to source a) and verified with topographic mask model;

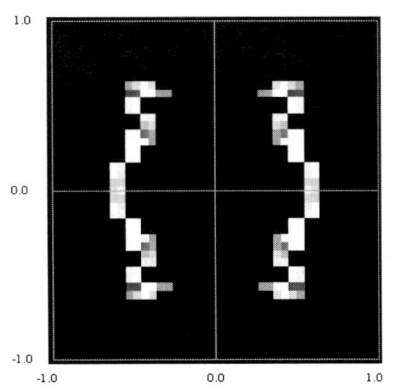

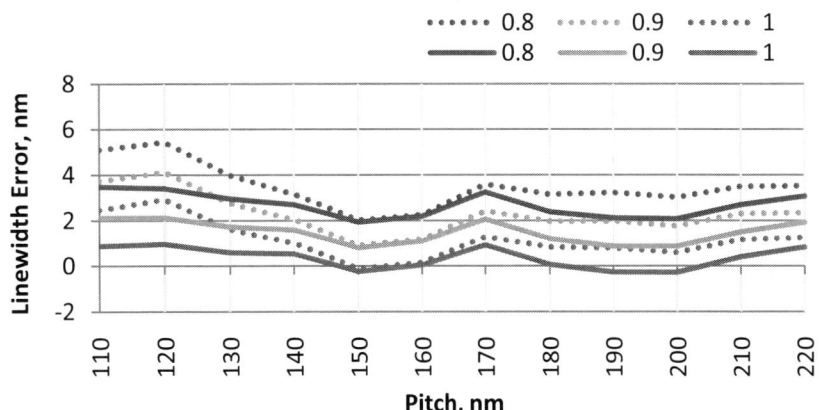

d) source intensity distribution after optimization with topographic mask model;

e) linewidth error vs. pitch plots at 80 nm defocus corresponding to source d) and simulated with topographic mask model.

Fig. 6 Results of the pixelated source optimization to reduce sensitivity to light polarization degree (dotted lines are results before optimization; solid lines are results after optimization).

CONCLUSIONS

The analysis of the presented results leads to the following conclusions regarding the role of mask topography effects and in pixelated source optimization. In the studied case the thin mask and topographic mask model optimizations both can provide improvements in CD uniformity and enlargement of process window with pixelated source. The optimization procedures converge to slightly different optimal solution, because of difference in image metrics provided to optimizer by different models. After optimization with thin mask model it is necessary to verify results with topographic mask model and refine mask biasing to get comparable results.

The comparison of results produced by optimization with thin mask model against verification results produced with topographic mask model shows mainly marginal difference in MEEF (< 0.5), NILS (< 0.5), and DOF (< 30 nm). Although it is not the case when the optimization cases are aimed to reduction of sensitivity to errors which are related to EMF effects (absorber thickness error and polarization degree variations). The verification of thin mask model optimization results with topographical mask model simulation does not show desired improvements (line width error is up to 7 nm). Thus, it can be reached only under consideration of mask topography and the use of rigorous simulation is highly preferred.

Although intensity distributions after optimization with thin mask and topographic mask models are different, but it is possible to find a conventional multi-pole source which is closely matching both source shapes. The pixelated source solutions, which are obtained in described way, can be also used as a starting point for further parametric optimization of conventional source shape.

REFERENCES

1. Granik Y. "Source optimization for image fidelity and throughput," *J. Microlith. Microfab. Microsyst.* Vol. 3 No. 4, 509-522, (2004).
2. Bekaert J., et al. "Freeform illumination sources: Source mask optimization for 22 nm node SRAM," on 6th International Symposium on Immersion Lithography Extensions (Prague, 2009).
3. Matsuyama T., et al. "A study of source mask optimization for future optical lithography," on 22nd International Microprocesses and Nanotechnology Conference (Sapporo, 2009).
4. Leonard J., et al. "An improved process for manufacturing diffractive optical elements (DOEs) for off-axis illumination systems," *Proc. SPIE* 6924, 69242O, (2008).
5. Sato K., et al. "Mask 3D effect on 45-nm imaging using attenuated PSM," *Proc. SPIE* 6520, 65201J, (2007).
6. Lai K., et al. "Experimental result and simulation analysis for the use of pixelated illumination from source mask optimization for 22nm logic lithography process," *Proc. SPIE* 7274, 72740A, (2009).
7. Hopkins H. H. "Image formation with coherent and partially coherent light," *Photographic science and engineering* Vol. 21 No. 3, 114-123, (1977).
8. Synopsys TCAD Sentaurus Lithography (http://www.synopsys.com/Tools/TCAD/ProcessSimulation/Pages/SentaurusLithography.aspx)
9. Lucas K. D., et al. "Efficient and rigorous three-dimensional model for optical lithography simulation," *J. Opt. Soc. Am. A* 13 (11), 2187-2199 (1996).
10. Bernard D. A. "Simulation of Focus Effects in Photolithography," *IEEE Transactions on Semiconductor Manufacturing* Vol. 1 No. 3, 85–91, (1988).
11. Brunner T. A. and Ferguson R. A., "Approximate Models for Resist Process Effects," *Proc. SPIE* 2726, 198-207, (1996).
12. Z. W. Geem, Music-Inspired Harmony Search Algorithm: Theory and Applications, Studies in Computational Intelligence, Volume 191, Springer (2009).
13. C. Mack, Fundamental Principles of Optical Lithography: The Science of Microfabrication, Wiley (2008).
14. Ling M. L., et al. "Customized illumination shapes for 193nm immersion lithography," *Proc. SPIE* 6924, 692435, (2008).
15. Ling M. L., et al. "Forbidden pitch improvement using modified illumination in lithography," *J. Vac. Sci. Technol. B* 27 (1), 85-91, (2009).

Experimental Study of Effect of Pellicle on optical Proximity Fingerprint for 1.35 NA immersion ArF Lithography

Lieve Van Look *[a], Joost Bekaert[a], Bart Laenens[a], Geert Vandenberghe[a],
Jan Richter[b], Karsten Bubke[*b], Jan Hendrik Peters[b], Koen Schreel[c], Mircea Dusa[c]
[a] imec, Kapeldreef 75, B-3001 Leuven, Belgium
[b] Advanced Mask Technology Center GmbH, Raehnitzer Allee 9, D-01109 Dresden, Germany
[c] ASML BV, De Run 6501, 5504 DR Veldhoven, The Netherlands

ABSTRACT

Pellicles are mounted on the masks used in ArF lithography for IC manufacturing to ensure defect-free printing. The pellicle, a thin transparent polymer film, protects the reticle from dust. But, as the light transmittance through the pellicle has an angular dependency, the pellicle also acts as an apodization filter.

In the current work, we present both experimental and simulation results at 1.35 NA immersion ArF lithography showing the influence of two types of pellicles on proximity and intra-die Critical Dimension Uniformity (CDU). To do so, we mounted and dismounted the different pellicle types on one and the same mask. The considered structures on wafer are compatible with the 32 nm logic node for poly and metal. For the standard ArF pellicle (thickness 830 nm), we experimentally observe a distinct effect of several nm's due to the pellicle presence on both the proximity and the intra-die CDU. For the more advanced pellicle (thickness 280 nm) no signature of the pellicle on proximity or CDU could be found.

By modeling the pellicle's optical properties as a Jones Pupil, we are able to simulate the pellicle effects with good accuracy. These results indicate that for the 32 nm node, it is recommended to take the pellicle properties into account in the OPC calculation when using a standard pellicle. In addition, simulations also indicate that a local dose correction can compensate to a large extent for the intra-die pellicle effect. When using the more advanced thin pellicle (280 nm), no such corrections are needed.

Keywords: Pellicle, optical proximity, Jones Pupil, CDU, pellicle thickness uniformity

1. INTRODUCTION

In current optical lithography for IC manufacturing, it is very common to use high Numerical Apertures (NA's) (1 < NA ≤ 1.35) for critical layer patterning. These high NA's are needed to address the resolution requirements of the 32 nm node. The standard pellicles that are used to keep the reticles dust-free are optimized for maximum light transmittance under normal incidence of light. However, in these high NA immersion systems, the incidence of light on the pellicle is oblique, leading to transmission and phase errors for the light passing through it. At the same time, the CD and CD uniformity specifications for high NA applications become ever tighter, leading to increasing concern of the effects that pellicles may have on proximity and CDU performance.

In literature, a number of studies can be found where this concern is quantified by simulations, and clear guidelines are provided on how to model the pellicle effects[1, 2]. As far as experimental wafer data is concerned, there are only few examples and, to our knowledge, none of them are at 1.35 NA. The paper[3] by Luo *et al.* contains experimental and modeling data at 1.2 NA showing the pellicle effect of several nm's for lines through pitch with varying CDs ranging from 55 nm to 65 nm. For the effect of a pellicle on the CDU, we refer to an early evidence[4] of CDU increase due to the pellicle presence for 180 nm wide resist lines. In the study[5] by Morikawa *et al.* an increased CDU of 45 nm features (NA 1.4) is measured by AIMS in the presence of a pellicle. However, in none of these studies, a correlation to the pellicle thickness uniformity is made. There is also no data available on the advanced thin 280 nm pellicle.

*lieve.vanlook@imec.be; *currently working at GLOBALFOUNDRIES

Optical Microlithography XXIII, edited by Mircea V. Dusa, Will Conley, Proc. of SPIE Vol. 7640,
76401Y · © 2010 SPIE · CCC code: 0277-786X/10/$18 · doi: 10.1117/12.848219

With this work, we aim to provide experimental demonstration of the effect of a standard pellicle and a more advanced thin pellicle on both the CD through pitch (proximity) curve and the intra-die CDU fingerprint at 1.35 NA. By verifying the predictive power of the simulations based on measured pellicle characteristics, we also want to show that the pellicle effects can be accurately modeled, and can thus be taken into account during process optimization by pellicle-aware OPC[3,4] and/or (variable) local dose fingerprint application. As the lifetime of a pellicle is much shorter than that of the reticle, replacing the pellicle may lead to different intra-field properties. Adjustment of the dose map to the measured pellicle thickness map could provide a solution in this case.

In the remainder of this paper, we will first explain the setup and flow of the experiments we performed. Section 3 summarizes how the pellicle effect can be modeled and then simulated using commercially available lithographic simulation software. Section 4 describes the experimental and simulation results for proximity and CD uniformity.

2. SETUP OF THE EXPERIMENT

2.1 Experimental flow

The purpose of these experiments is to show the influence of the 830 nm thick and the 280 nm thin pellicle on the proximity and CDU printing result on wafer at NA 1.35. We therefore subsequently placed and removed a thick and a thin pellicle on one and the same mask. We exposed a set of wafers in the three conditions of the reticle (without pellicle, with thick pellicle, with thin pellicle) at several illumination conditions on an ASML Twinscan XT:1900Gi scanner, interfaced to a Sokudo RF³ˢ track. The stack used on all wafers is identical: 95 nm of ARC29SR BARC (Brewer/Nissan) + 105 nm TArF-Pi-6001 resist (TOK).

After removal of the thick pellicle, a mask clean was performed. Reticle SEM measurements proved that the reticle CDU fingerprint remained identical under such action.

2.2 Reticle design

The reticle used for this work was a 6% attenuated phase shifting MoSi mask with a unit cell that is repeated 13 times in the slit and 9 times in the scan direction (Figure 1, right panel). The unit cell contains large pitch-CD matrices of line and trench structures intended for measurement with CD-SEM. This allows the user to perform a manual bias picking that ensures a constant wafer CD through pitch. The range of available reticle CDs is very broad, such that the reticle can be used to print 45 nm as well as 90 nm features through pitch. A small number of scatterometry marks are also available which were used for YieldStar measurements. There are no assist features present on the mask.

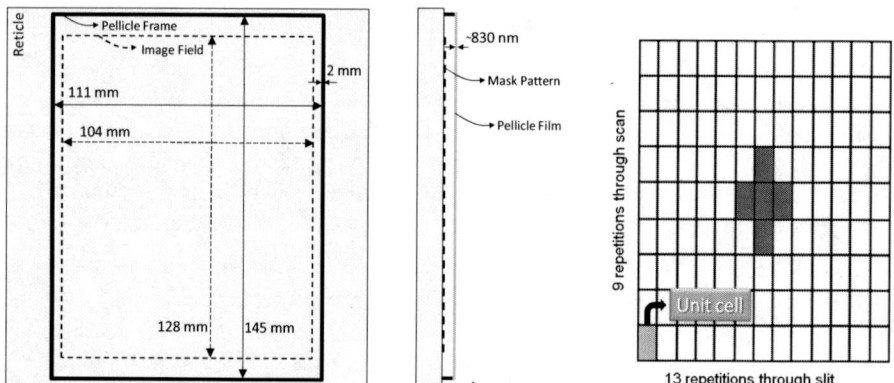

Figure 1. *Left*: Schematic representation of the reticle with a mounted pellicle (top down and side view). Indicated sizes are typical of the standard ArF thick pellicle. *Right*: Layout of the test reticle used for this work. Blue subfields are the ones used for the proximity measurements.

2.3 Pellicle thickness measurements

The thickness map of the 830 nm (thick) and 280 nm (thin) pellicle were measured using an Analyzer 5700-CDRT ellipsometer (N&K Technology, Inc.). The static measurement repeatability of this tool on identical locations was verified to be below 0.2 nm 3σ. The pellicle area that was measured covers the image field completely. The frame height was 4 mm for the thick pellicle and 2 mm for the thin pellicle.

The measured thickness map of the *thick pellicle* has a range of 3.1 nm around an average value of 824 nm. The average pellicle thickness therefore lies at the edge of the manufacturer's specification. The thickness range on this particular pellicle is found to be typical for pellicles of this type. To verify that the thickness fingerprint was stable, we measured the pellicle thickness map before mounting the pellicle frame onto the reticle, and after removal of the pellicle frame. Both measurements yielded an almost identical thickness map (3σ of the difference is 0.3 nm). For the *thin pellicle*, the measured thickness map has a range of 1.3 nm around an average value of 278 nm. Also this is within the manufacturer's specification of 281 ± 2.8 nm. Typical thickness ranges found for this type of pellicles are 1 to 1.2 nm, so the specimen used for this work can also be considered as typical.

2.4 Wafer CD measurements

For the *proximity experiments*, we selected an Annular illumination condition: 1.35 NA Annular σ0.64-0.84, XY Polarized. At the start of the experiment, a bias picking was done in order to print lines and trenches to a target of 45 and 90 nm on wafer through pitch. In total, approximately 80 structures were chosen which remained fixed for the remainder of the experiment. All metrology through pitch was performed using a KLA-Tencor e-CD2 top down CD-SEM. To reduce metrology noise, we averaged over a set of 5 sub dies positioned cross-wise central on the reticle (Figure 1 right). For the *CDU measurements*, we exposed wafers with three different illumination conditions, amongst which the Annular setting mentioned above. The other conditions used are DipoleY35° σ 0.85-0.98, X Polarized, and CQuad40° σ 0.70-0.90, XY Polarized. The CD measurements were mainly performed on an ASML YieldStar S-100 scatterometer. The CDU maps were obtained by averaging three dies on the wafer. The repeatability of the YieldStar was verified to be 0.07 nm 3σ on identical measurement locations. We also measured multiple wafer CDU maps with the KLA-Tencor e-CD2 top down CD-SEM for comparison.

2.5 Simulations

The simulations presented in this work are performed with KLA-Tencor Prolith v12.0 lithographic simulation software, using a calibrated resist model. We make use of the same Annular condition as used in the experiments: NA 1.35 Annular σ0.64-0.84, XY Polarized. The simulation sequence mimics the experimental sequence. First, a bias picking is performed to print 45 and 90 nm lines and trenches on target through pitch. Then, this set of structures remains fixed for all following simulations.

3. MODELING OF THE PELLICLE EFFECT ON IMAGING

3.1 Angular dependence of pellicle transmittance and induced phase shift

When discussing the possible influence of a pellicle on imaging, the main concern lies in the usage of large angles of incidence on the pellicle film that are intrinsically linked to the use of high NA's. At reticle side, the light rays that are used for imaging hit the pellicle film under a maximum angle $\theta = \sin^{-1}\left(\frac{NA}{4}\right)$, which amounts to ~20° for 1.35 NA applications. The pellicle film is typically fully transparent (we assume k=0) for 193 nm light, but has a refractive index of 1.4, causing reflections at its surfaces. The amount of light that makes it through the pellicle and the phase shift this light receives depends on the incidence angle and is given by textbook expressions for homogeneous dielectric films[6].

We calculated the transmittance as a function of incidence angle for the standard thick 830 nm and the thin 280 nm pellicle using a 193.3 nm wavelength (see Figure 2). The arrow in Figure 2 indicates the maximal incidence angle obtained at 1.35 NA. The standard pellicle is optimized for low angles of incidence (low NA applications), and shows a significant transmission loss for the 193 nm light at oblique angles of 20°. This transmittance loss at high angles is equivalent to an apodization effect, be it that the amount of apodization depends on the polarization state of the light. As seen in the right panel of Figure 2, the thin pellicle suffers much less from this apodization effect, even at 1.35 NA.

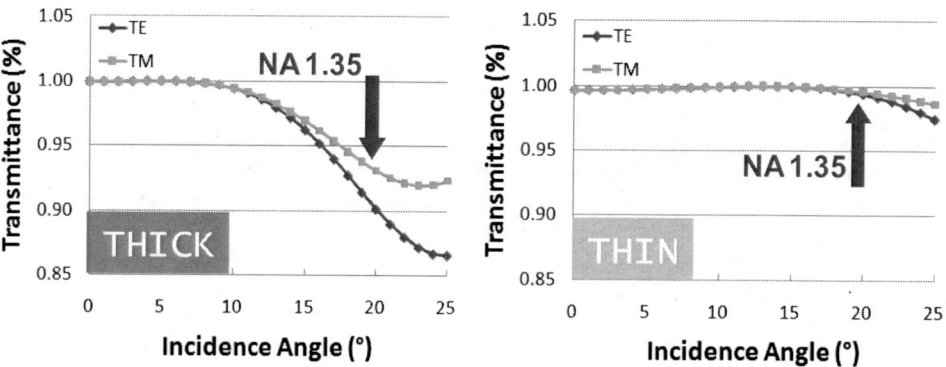

Figure 2. Pellicle transmittance as a function of incidence angle for the 830 nm thick pellicle (left) and for the 280 nm thin pellicle (right).

Because of the different path lengths through the pellicle film for light coming in at different angles, also the phase shift that is brought upon by the pellicle depends on the incidence angle. The calculated phase shift induced by the pellicle is shown in Figure 3 for both the thick and the thin pellicle. The plot shows the phases relative to the phase of the normal beam. These phase shifts are responsible for aberrations (mostly spherical Z9[1,2]) and are again clearly much smaller in the case of the thin pellicle.

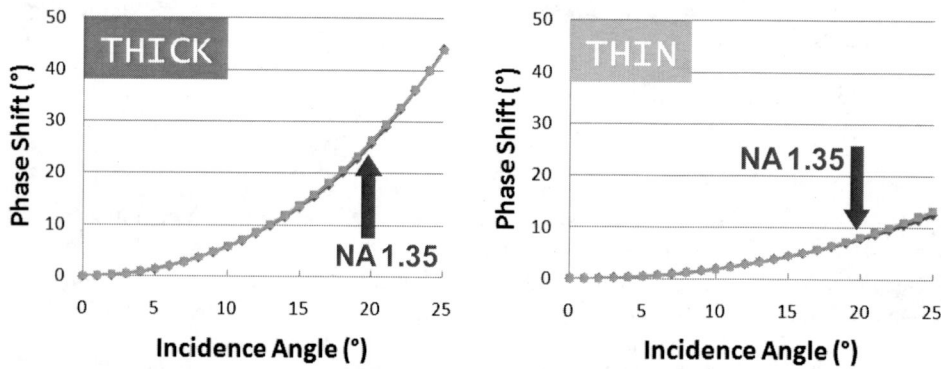

Figure 3. Phase change upon passing through the pellicle film as a function of incidence angle for the 830 nm thick pellicle (left) and for the 280 nm thin pellicle (right).

3.2 Translating pellicle effect to a Jones Pupil

A Jones Pupil describes the polarization-dependent apodization and aberration effects that are induced by the pellicle presence. As explained in literature[1,2], the above calculated transmission and phase change form the diagonal elements of the Jones Pupil in polar coordinates. After transformation to Cartesian coordinates, the cross-terms are no longer zero, implying that there is some intermixing of the two polarization states by passing through the pellicle. However, the found cross-talk terms J_{xy} and J_{yx} turn out to be very small. The Jones pupil acts on the x and y components of the incoming light vector as

$$\begin{pmatrix} \overline{E}'_x \\ \overline{E}'_y \end{pmatrix} = \begin{pmatrix} J_{xx} & J_{xy} \\ J_{yx} & J_{yy} \end{pmatrix} \begin{pmatrix} \overline{E}_x \\ \overline{E}_y \end{pmatrix}$$, with $\overline{E}_x$ and $\overline{E}'_x$ the x components of the incoming and outgoing electric field vectors

respectively. As we also take the phase changes due to the pellicle presence into account, the Jones matrix elements J_{ii}

are complex. Figure 4 depicts the amplitude of one of the Jones Pupil elements J_{xx} for both the thick (830 nm) and the thin (280 nm) pellicle. The scales of both graphs are identical for easy comparison.

The Cartesian complex Jones pupil can be imported into standard lithography simulation software to evaluate its impact on the printed CDs.

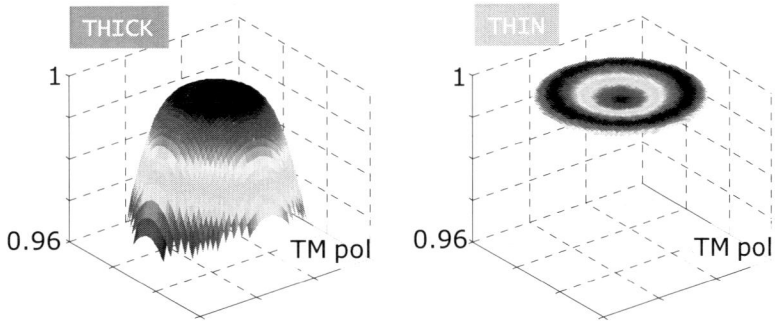

Figure 4. Amplitude of one element J_{xx} of the Jones Pupil, describing the apodization for TM polarized light for the thick (a) and the thin (b) pellicle.

4. EXPERIMENTAL AND SIMULATION RESULTS

4.1 Pellicle effect on Proximity

We exposed wafers with and without the thick and thin pellicle and measured the same set of structures (line and trench patterns that print to a CD of 45 and 90 nm through pitch) on all wafers. The dose was targeted on a pitch 90 nm line of 45 nm. The apodization effect caused by the pellicle induces a change in the dose-to size. With the thin pellicle mounted, the dose-to-size was equal to that without pellicle. When the thick pellicle was mounted, a 6.5% higher dose was needed to have the same dense structure on reticle print to target on wafer.

For the simulations, we constructed the Jones pupils for each of the pellicles using the measured pellicle thickness at the center location of the image field, i.e. at the same location as where the CD measurements are performed. These values are 822 nm for the thick and 279 nm for the thin pellicle. We then loaded the Jones Pupils for each of the pellicle types and performed a retargeting of focus and dose.

For both the experiment and the simulations, we finally subtracted the proximity curves obtained without a pellicle from those with pellicle.

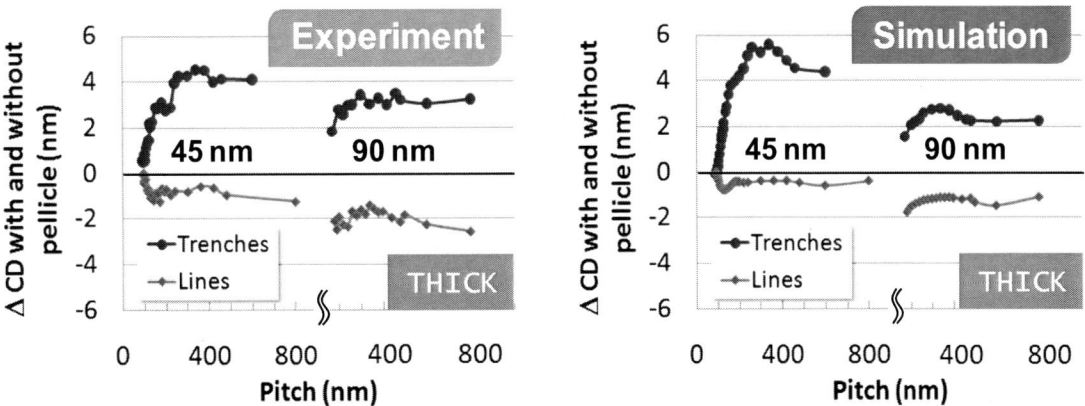

Figure 5. Measured (left) and simulated (right) CD difference through pitch between wafers exposed with and without the thick pellicle. CD differences with and without pellicle of several nm's are observed through pitch.

The obtained CD difference through pitch curves for the thick pellicle are shown in Figure 5 for the experiment (left) as well as for the simulation (right). The thick pellicle clearly changed the proximity curves for both lines and trenches. For lines of 45 nm, experimental pellicle effects up to 1.2 nm were found through pitch. The proximity curve for the larger 90 nm lines received a shift of about 2 nm over the whole pitch range. For the trenches, the influence of the pellicle on proximity is even higher (up to ~ 4 nm), mainly due to the smaller contrast of these dark field features compared to the lines. The simulations using the Jones pupils for the thick pellicle (see Figure 5, right panel) yield very similar proximity differences as the experiment with the same typical characteristics and even very similar amplitudes.

For the thin pellicle, the simulations predict a negligibly small effect. The experiments seem to confirm this trend, although small CD differences (< 1 nm) are observed.

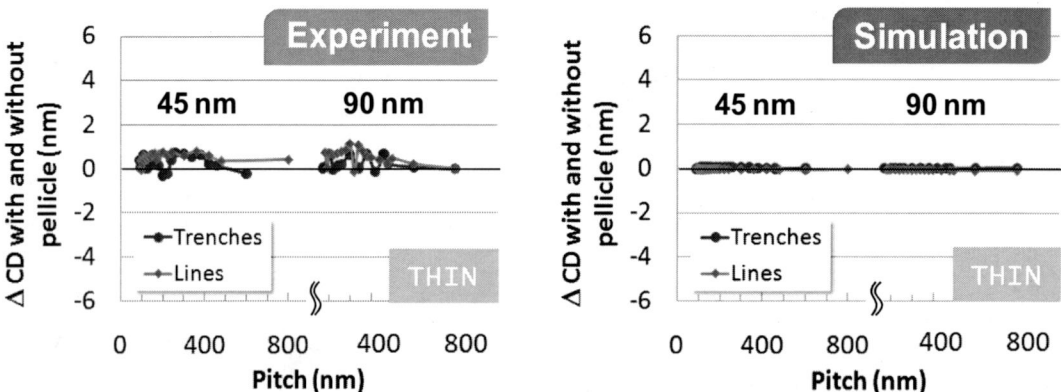

Figure 6. Measured (left) and simulated (right) CD difference through pitch between wafers exposed with and without the 280 nm thin pellicle. The observed CD differences through pitch are below 1 nm.

Since we are able to model the pellicle's effect on proximity quite accurately, it is also possible to account for the pellicle presence in optical proximity correction calculations. This would certainly be beneficial in case the thick 830 nm pellicles are used. For the thinner pellicle of 280 nm, both experiments and simulations indicate that there is no need to take the pellicle into account during optical proximity correction.

4.2 Pellicle effect on intra-field CDU

Besides the effect the pellicle has on the proximity, we now consider its contribution to the CD uniformity on wafer. The pellicle thickness is not constant over the whole image field. Due to this, the magnitude of the apodization effect also varies over the field. Consequently, identical features at different locations in the field require a different dose-to-size. The measured pellicle thickness map for the thick pellicle showed an average measured value of 824 nm and a range of 3.1 nm, being 0.4 % of the thickness, a typical value found for this type of pellicles. (upper left panel in Figure 8). The dashed rectangle indicates that part of the measured pellicle area that corresponds to the printed image on wafer. The measured thickness map of the thin pellicle is shown in the lower left panel of Figure 8. The thin pellicle has an average thickness of 278 nm with a range of 1.3 nm (0.5 % of the thickness), which seems mostly due to the corners of the pellicle, i.e. outside the printed field.

The simulated effect of the varying pellicle thickness on the pellicle transmittance is illustrated in Figure 7 for TE polarized light. The green solid lines in these plots are calculated using the nominal pellicle thickness while the solid red curves correspond to the measured average pellicle thickness. The dashed curves around the red solid curve indicate the thickness range that was measured on the used pellicles and that we know to be typical for the pellicle types used. As is immediately apparent from these plots, the variations in transmittance are much smaller for the thin pellicle than for the thick pellicle.

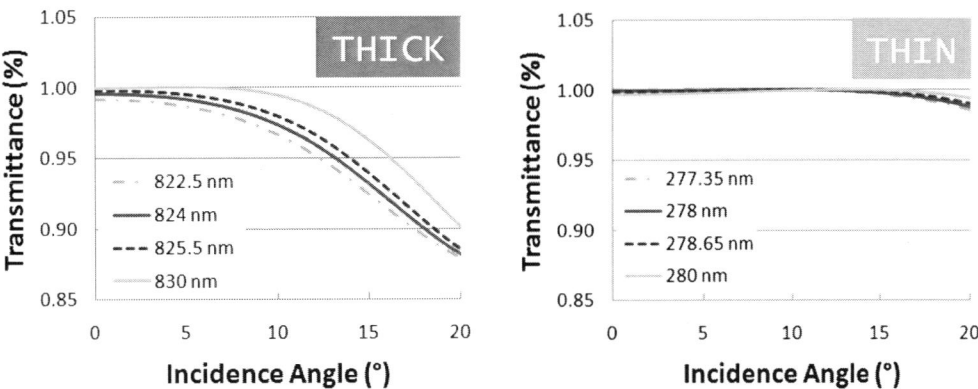

Figure 7. Calculated pellicle transmission for TE polarized light as a function of incidence angle for the nominal pellicle thickness (solid green), for the actual thickness (solid red) and for two thicknesses corresponding to the measured pellicle thickness range (dashed lines).

On top of the smaller observed thickness range for the thin pellicle (1.3 nm versus 3.1 nm), the sensitivity of the wafer CD to 1 nm pellicle thickness change is simulated to be a factor ~5 smaller for the thin pellicle. The expected effect on the intra-die CD uniformity of a thin pellicle is therefore negligible.

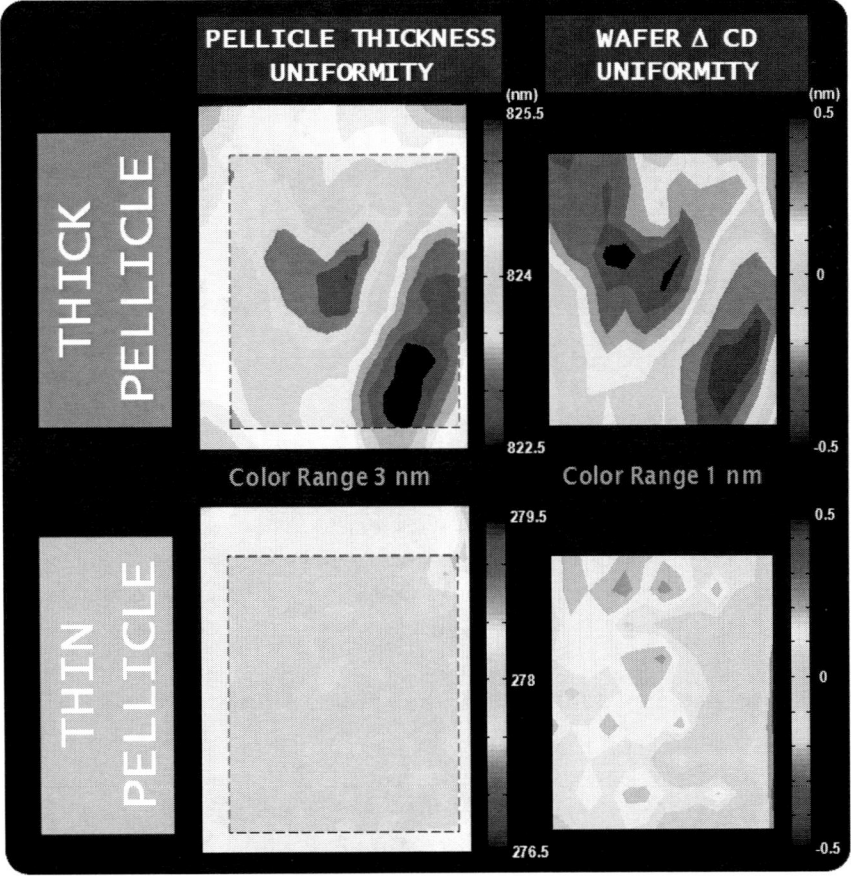

Figure 8. Measured pellicle thickness uniformity maps (left) and their corresponding experimental contribution to the CD uniformity on wafer (right) for a pitch 100 nm line printed using NA 1.35, Annular σ0.64-0.84, XY Polarized illumination.

We now investigate experimentally to what extent the observed pellicle thickness changes affect the intra-die CD uniformity on wafer. To this purpose, we measured the CDU maps on wafers exposed with and without the pellicle. Typical absolute 3σ numbers for the intra-die CD uniformity of the structures we consider here are between 0.9 and 1.4 nm. The CDU difference maps result from subtracting the CDU maps obtained without the pellicle from those obtained with the pellicle.

The right panels of Figure 8 show these CDU difference plots for a line of 45 nm printed at a pitch 100 nm. For the thick pellicle (upper right panel), even though the observed CD differences are small (3σ of this difference plot is 0.7 nm), the thickness signature of the pellicle is unmistakably reflected in to the CD uniformity. For the thin pellicle (lower right panel), no correlation between pellicle thickness and CDU difference map (3σ of 0.3 nm) could be found. This is not a surprising result, as the predicted effect of the pellicle thickness changes is negligible. For the considered pellicles, the CD sensitivity for lines to pellicle thickness changes is negative: thicker pellicle regions give rise to lower line CDs on wafer.

The CDU difference maps for a pitch 100 nm line (shown above in Figure 8) were obtained on wafers exposed using an NA 1.35, Annular 0.64-0.84 XY Polarized illumination setting, and measured with the ASML YieldStar scatterometer. Additionally, we confirmed the presence of this distinct CDU difference fingerprint for other features, using other illumination conditions and also using the CD-SEM as a metrology tool (see Figure 9).

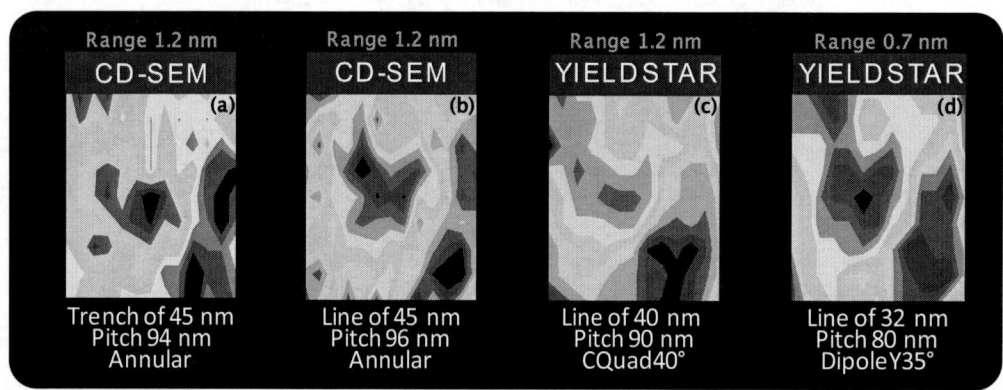

Figure 9. Confirmation of the pellicle influence on the experimental CD uniformity difference maps for multiple features, using different illumination conditions and metrology tools. The apparent shift of the maps is due to the different locations of the features on reticle.

In Figure 9(a) and (b), we used the same Annular condition as before, but now measured dense lines and trenches with CD-SEM. As can be seen in Figure 9(a), trenches react opposite to pellicle thickness changes as lines. Figure 9(c) and (d) show CDU difference maps obtained with the YieldStar scatterometer. The lines of 40 nm on a pitch 90 nm (Figure 9(c)) were obtained using a CQuad40° σ 0.70-0.90, XY Polarized illumination condition. In Figure 9(d), lines of 32 nm at a pitch 80 nm are shown that are printed using a DipoleY35° σ 0.85-0.98, X Polarized condition. The apparent shift in the maps between the different features is due to an offset (up to 2.4 mm (1x)) between their locations on the reticle.

4.3 Compensating the pellicle effect on intra-die CDU with local dose adjustment

As mentioned earlier, the pellicle thickness variations over the field lead to variations in the effective dose-to-size between different locations in the die. Moreover, it is current practice to replace the pellicle after a certain period of time, so these dose-to-size variations may change when a new pellicle is mounted. If one is able to locally adjust the dose, using a dose map that corresponds to the measured pellicle thickness map, the CDU contribution of the pellicle could be minimized. This dose map can be recalculated per pellicle.

However, when adjusting dose locally for one feature, there may be the concern what effect this dose change may have on other features through pitch. To examine the feasibility of correcting for the pellicle CDU effect of the thick pellicle using a local dose map, we first simulate the effect on CD through pitch induced by 1.6 nm thickness change at a constant dose (Figure 10, left panel). This amount of pellicle thickness variation is the one that is present on the thick pellicle we have used for our experiment and also the illumination condition is identical.

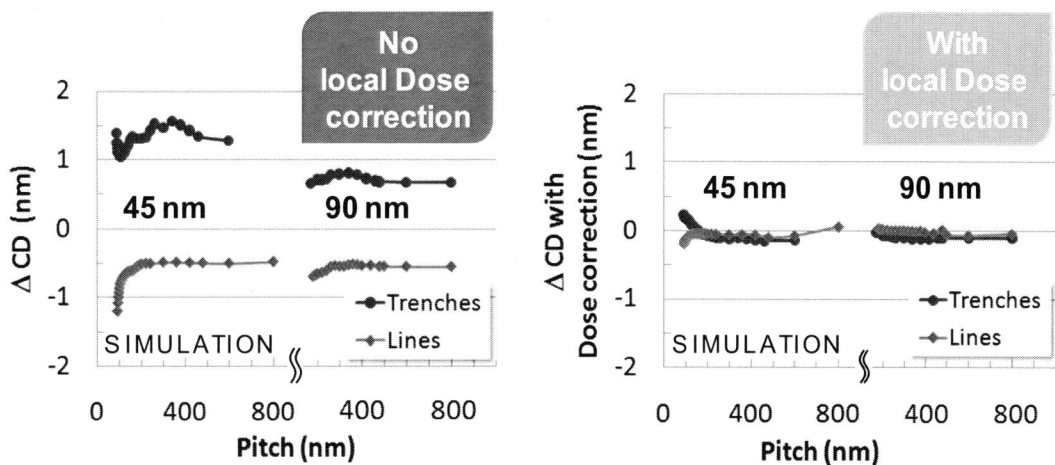

Figure 10. Through-pitch CD simulation of the effect of a local pellicle thickness change of 1.6 nm (left) and its compensation using a local Dose adjustment (right). Results are for a thick pellicle.

When we now use the dose to compensate for the effect of this pellicle thickness variation, the through-pitch 'penalty' turns out to be quite small. All features (lines and trenches, 45 nm and 90 nm) come much closer to target. There is some remaining apodization effect for the dense 45 nm features that cannot be tuned away using the dose knob, but overall, the situation is much better with than without the dose correction.

For the current example, the amplitude of the dose correction that was applied to compensate for the ± 1.6 nm pellicle thickness change is ± 0.75 % of the nominal dose.

5. CONCLUSIONS AND RECOMMENDATIONS

We performed an experimental and simulation study of the effect of a pellicle on the proximity and on the intra-die CDU fingerprint at NA 1.35. We examined two pellicle types: the standard 830 nm thick pellicle, and a more advanced 280 nm pellicle. Through their effect on transmission and phase at high incidence angles, these pellicles act as an apodization filter.

For the *thick standard pellicle* clear experimental evidence (up to ~4 nm for trenches) of the pellicle *effect on proximity* was found. The simulations, in which the effect of the pellicle is described as a Jones Pupil, are in good agreement with the experimental results. When thick (830 nm) pellicles are used, we therefore recommend taking the nominal pellicle properties into account during the Optical Proximity Correction. We also found an obvious correlation of the pellicle thickness map (range 3.1 nm) to the effect on the *intra-die CDU fingerprint*, with a pellicle contribution to the intra-die CDU on wafer of about 1 nm. As the life-time of a pellicle is much shorter than that of the reticle, the pellicle contribution to CDU will change when replacing the pellicle. In this case, one may consider a correction strategy using local dose mapping to minimize the pellicle contribution to the CDU.

For the *thinner pellicle of 280 nm*, the measured effect on proximity was much smaller (<1 nm). Moreover, the influence of the pellicle on the intra-die CDU was close to zero and it could not be correlated to the thickness map of the thin pellicle. It is therefore our recommended choice for 32 nm node applications.

ACKNOWLEDGEMENTS

The authors would like to thank Jeroen Van de Kerkhove (IMEC) for creating the gds of the reticle, Anne-Laure Charley (IMEC) for YieldStar training and help, Kevin Lucas (Synopsys), Mark Smith (KLA-Tencor) for help with the Jones Pupil calculations and Matt Lamantia (Toppan), Peter De Bisschop (IMEC) and Jo Finders (ASML) for useful discussions.

REFERENCES

[1] Bubke, K., Alles, B., Cotte, E., Sczyrba, M., Pierrat, C., "Pellicle induced aberration and apodization in hyper NA optical lithography," Proc. SPIE 6283, 628318 (2006).

[2] Sato, K., Nagai, S., Shinichiro, N., Sato, T., Itoh, M., "Influence of pellicle on hyper-NA imaging," Proc. SPIE 6924, 692451 (2008).

[3] Luo, B., Chang, C.-K., Wang, W. L., Huang, W. C., Wu, T., Lai, C. W., Liu, R.G., Lin, H.T., Chen, K.S., Ku, Y.C., "Pellicle effect on OPC modeling," Proc. SPIE 6924, 69243T (2008).

[4] Lucas, K., Gordon, J. S., Conley, W., Saied, M., Warrick, S., Pochowski, M., Smith, M. D., West, C., Kalk, F., Kuijten, J.P., "Optical issues of thin organic pellicles in 45 nm and 32 nm immersion lithography," Proc. SPIE 6349, 63490K (2006).

[5] Morikawa, Y., Sutou, T., Mesuda, K., Nagai, T., Inazuki, Y., Adachi, T., Toyama, N., Mohri, H., Hayashi, N., Stroessner, U., Birkner, R., Richter, R., Scheruebl, T., "Pupil plane analysis on AIMS^{TM} 45-193i for advanced photomasks," Proc. SPIE 6520, 65201H (2007).

[6] Born, M., Wolf, E., [Principles of Optics], Cambridge University Press, Cambridge, Section 1.6.4 (2003).

Achieving Interferometric Double Patterning through Wafer Rotation

Peng Xie, Neal V. Lafferty, Bruce W. Smith
Microsystems Engineering, Rochester Institute of Technology,
82 Lomb Memorial Drive, Rochester, NY 14623

ABSTRACT

Owing to its simplicity and ability to produce line/space gratings with the highest contrast, interferometric lithography is an ideal platform for developing novel double patterning materials and processes. However, lack of sub-10 nm alignment in most interferometric systems impedes its application. In this paper, litho-etch-litho double patterning on a two-beam interferometric system is achieved by converting Cartesian alignment into angular alignment. By concentrically rotating the wafer in the second exposure, the interleaved region between the two exposures allows for the evaluation of double patterning process and materials. Geometric analysis shows that angular alignment has greatly relaxed requirements compared to the Cartesian alignment. It is calculated that for 22 nm double patterning technologies, rotation angle larger than 0.12 degree is sufficient to produce 1 μm long frequency doubled line/space patterns with less than 10% CD variation.

Keywords: Interferometric lithography, double patterning, wafer rotation, alignment

1. INTRODUCTION

Double patterning lithography (DPL) is expected to be the technology of choice for the 32 nm and 22 nm lithographic nodes. Among many double patterning schemes, a litho-etch-litho-etch (LELE) approach and self-aligned spacer double patterning (SADP) have been intensively studied.[1,2] However, the higher cost associated with their intermediate etch or CVD step respectively has made them expensive technologies and unlikely to scale beyond the 22 nm node. It is therefore desirable to search for alternative double patterning materials and processes such as litho-freeze-litho-etch, litho-litho-etch and double exposure that meet both the technological and economical requirements for future lithographic generations.[3-5]

Interferometric imaging systems such as the Amphibian XIS microstepper serve as an ideal platform for developing novel double patterning materials and processes.[6] Two-beam interference lithography produces the highest possible contrast grating of any optical lithography, while eliminating both the mask and projection optics. High index lens materials such as sapphire and LuAG, which are hard to produce in large diameter high precision projection optics, are once again viable. Very high resolution line/space gratings down to 26 nm has been reported using interference lithography and solid immersion lithography, making it a potential candidate to be investigated as a starting template for double/multiple patterning in a research environment.[7]

Most interferometric systems lack sub-10 nm alignment required by overlaying two patterning steps. In this paper, a new approach is proposed to overcome the alignment issue in an interferometric system. Litho-etch-litho double patterning on a two-beam interferometric system is achieved by converting Cartesian alignment into angular alignment. By concentrically rotating the wafer in the second exposure, the interleaved region between the two exposures provides SEM latitude in evaluating the process and materials.

Optical Microlithography XXIII, edited by Mircea V. Dusa, Will Conley, Proc. of SPIE Vol. 7640,
76401Z · © 2010 SPIE · CCC code: 0277-786X/10/$18 · doi: 10.1117/12.848345

2. METHODOLOGY

Conventional Cartesian alignment relies on the accurate placement of the second exposure in reference to the first exposure. The tolerance for the overlay error is minimum in the X/Y direction. By rotating the wafer between the two exposures a small amount, this error tolerance is greatly enhanced depending on the actual rotation. Figure. 1 exaggerates the rotation between two groups of line/space gratings and shows the illustrative aerial image by two exposures. It is shown that periodic patterns of a frequency doubled interleave region and a single frequency overlaid region alternate along the length of the mask line. Aerial image simulation in Prolith confirms the expected frequency doubling in the region where images interleaved. The results are shown in Figure. 2.

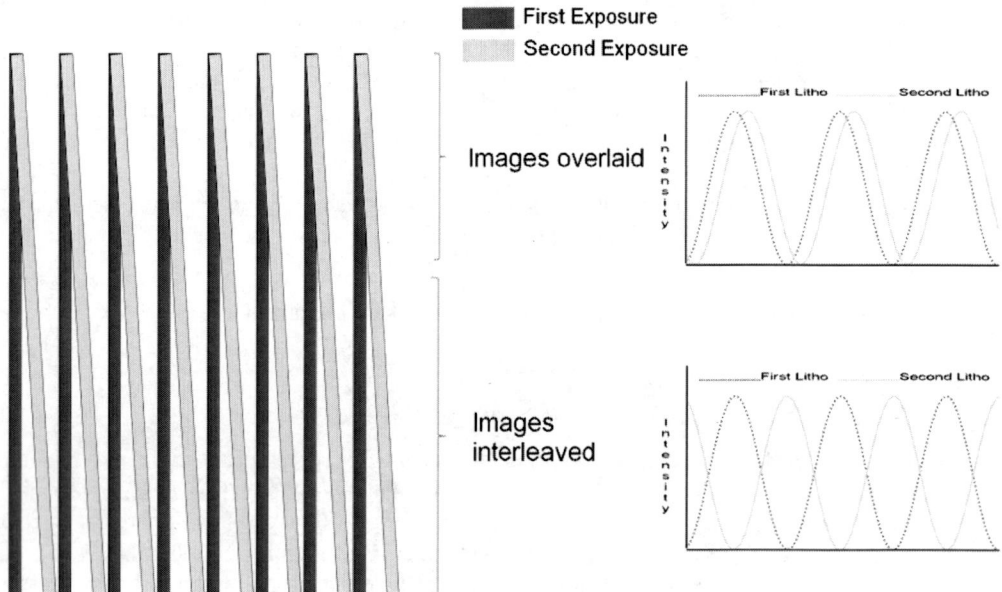

Figure 1. Illustration of rotating wafer between two exposures (effect exaggerated). Periodic patterns of overlaid region and interleaved region form along the length of the line/space grating. The respective aerial images in different regions are shown to the right. In the interleaved region, spatial frequency doubling results as a consequence of two exposures and the intermediate freeze step.

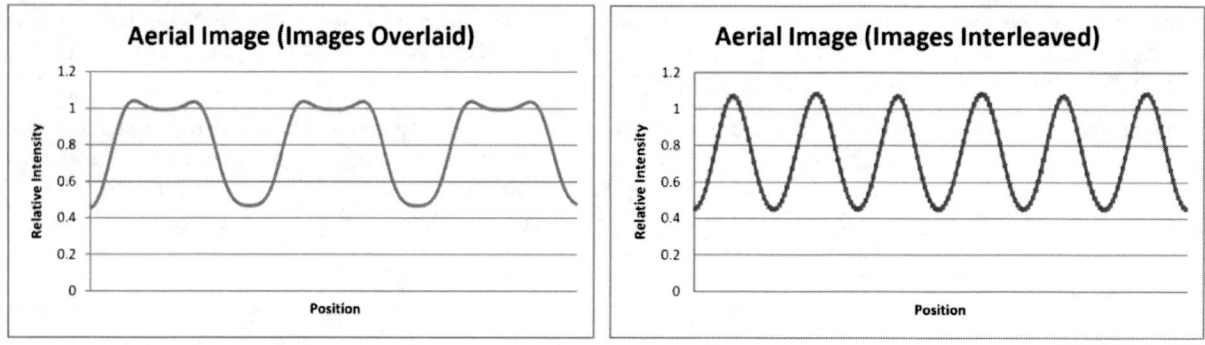

Figure 2. Aerial image simulation by Prolith showing the frequency doubling in the interleaved region compared to the 1X intensity distribution in the overlaid region. Simulation conditions include conventional source shape (σ=0.5) and two line/space mask patterns with one rotated by 0.5 degree.

To investigate how much rotation is required for one to have sufficient SEM latitude to locate and evaluate the frequency doubled region, a simple geometric analysis is performed, as shown in Figure. 3. Assuming the following parameters: rotation angle θ, lateral displacement Δ with reference to the center of the space, and interleave region d corresponding to the lateral displacement. The lateral displacement Δ can then be expressed as $d*\tan(\theta)$. If one further assumes that Δ has to be less than 10% of the line CD_L to have sufficient SEM latitude, the corresponding usable length of the interleave region within one period d_{usable} can be expressed as

$$d_{usable} \times \tan\theta < 10\%CD_L = P_{single}/40$$

where P_{single} is the single pitch by one exposures. Note here that 1:3 line/space ratio is assumed for each exposure. Table 1 summarizes the wafer rotation requirements for different lithographic nodes assuming $d_{usable} = 1$ µm. Obviously the requirements are greatly relaxed compared to the sub-10 nm Cartesian alignment strategy.

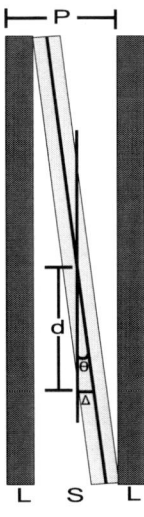

Figure 3. Illustration of geometric analysis of the usable interleaved region as a function of wafer rotation angle and mask pitch. It is assumed that the requirement for d_{usable} is such that the corresponding lateral displacement with is less than 10% of the line width.

P_{single}	128 nm	88 nm	64 nm	44 nm
P_{double}	64 nm	44 nm	32 nm	22 nm
Interleave region d	1 µm	1 µm	1 µm	1 µm
Rotation angle θ	0.183 deg	0.126 deg	0.092 deg	0.063 deg
Overlay requirement	6.4 nm	4.5 nm	3.2 nm	2.3 nm

Table 1. A summary of wafer rotation requirements for different lithographic nodes assuming $d_{usable} = 1$ µm. The Cartesian overlay requirements data are adopted from ITRS 2009. [8]

3. EXPERIMENTAL

A variable-NA optical two-beam interferometric imaging system was used to expose 4 mm by 4 mm wafer pieces. The imaging system is shown in Figure. 4. To accommodate precise rotation control, the wafer stage was mounted on a rotation stage (Newport M-481-A) with 15 arc second sensitivity. In order to simplify the process development, a 0.18 NA (~550 nm pitch) optical configuration was adopted. The first exposure dose was selected to over-expose the grating pattern to 1:3 line/space ratio. A subsequent oxygen plasma RIE etch (Trion Minilock RIE) removed the resist line and the BARC in the space, leaving a 80 nm thick BARC grating with 1:3 line/space ratio. A secondary ~30 nm BARC layer was then coated on top of the BARC grating to improve adhesion and minimize reflection. After coating the second resist layer ~130 nm, the wafer piece was mounted back on the wafer stage using a locating jig, shown on Figure. 5 on the rotation stage. The wafer stage was then rotated by 5 arc minute and exposed at the same dose as the first exposure. SEM images showing interleaved regions where spatial frequency is doubled are shown in Figure. 6.

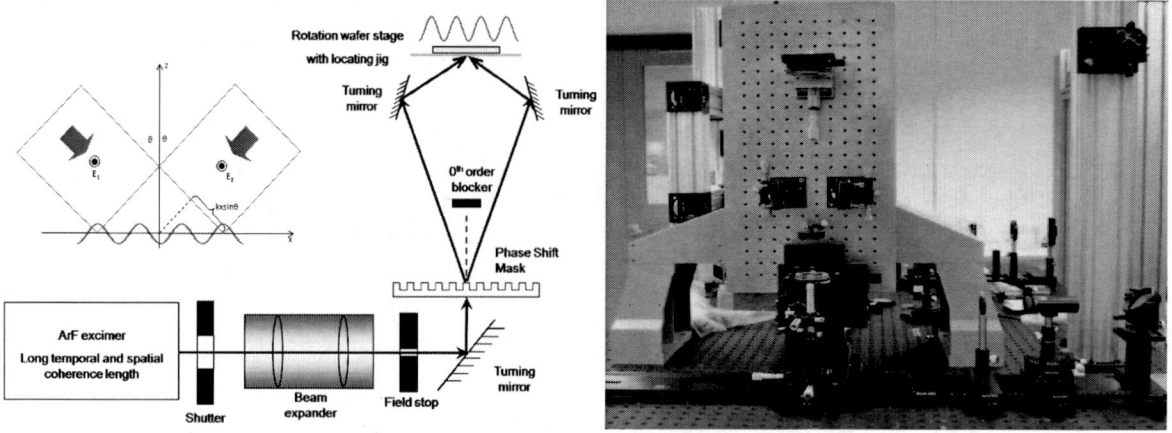

Figure 4. Schematics and snapshot of Smith-Talbot interferometric lithography

Figure 5. Rotation stage with locating jig.

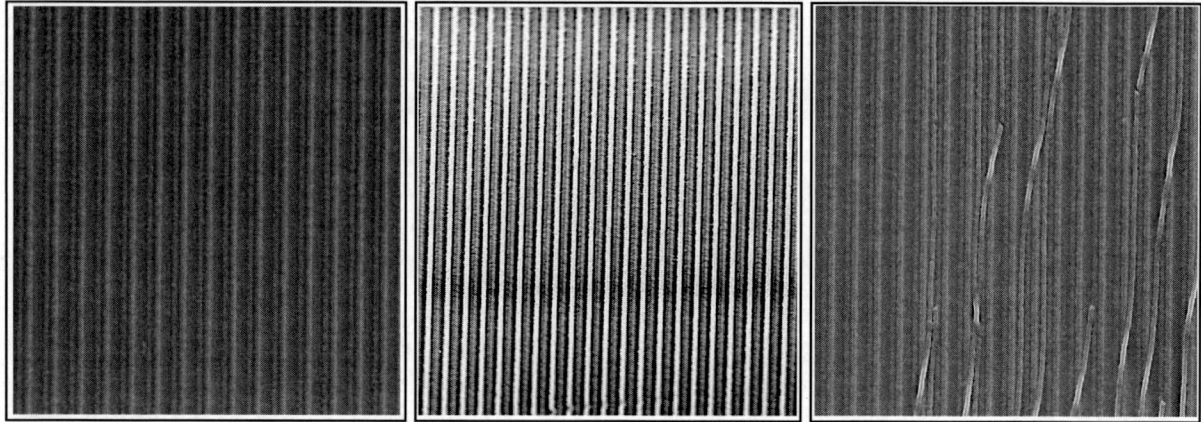

Figure 6. SEM snapshots of frequency doubled line/space patterns. One line is the BARC imaged by the first exposure while the other is the second resist line. The resist peeling in the right snapshot is due to the poor adhesion of the second resist to the wafer substrate.

4. CONCLUSION

A new approach is proposed to achieve double patterning alignment on an interferometric imaging system by rotating the wafer stage in the second exposure. The interleave region between the two exposures renders aligned frequency doubled patterns. This approach, which relies on a relaxed requirement for angular alignment, shows that angular rotation as large as ~0.1 degree can be tolerated to produce 1 μm interleave region with less than 10% CD variation. A litho-etch-litho double patterning process using the proposed alignment approach is achieved on a low-NA two-beam interferometric imaging system.

REFERENCES

[1] Maenhoudt, M.; Gronheid, R.; Stepanenko, N.; Matsuda, T.; Vangoidsenhoven, D. In *Proceedings of SPIE*; 2008; Vol. 6924, p. 69240P.

[2] Smayling, M. C.; Bencher, C.; Chen, H. D.; Dai, H.; Duane, M. P. In *Proceedings of SPIE*; 2008; Vol. 6925, p. 69251E.

[3] Lee, D. K.; Cao, Y.; Abdallah, D.; Yin, J.; Thiyagarajan, M.; Dammel, R. *Journal of Photopolymer Science and Technology* **2009**, *22*, 653–661.

[4] Ando, T.; Takeshita, M.; Takasu, R.; Yoshii, Y.; Iwashita, J.; Matsumaru, S.; Abe, S.; Iwai, T. In *Proceedings of SPIE*; 2008; Vol. 7140, p. 71402H.

[5] Bristol, R.; Shykind, D.; Kim, S.; Borodovsky, Y.; Schwartz, E.; Turner, C.; Masson, G.; Min, K.; Esswein, K.; Blackwell, J. M.; others In *Proceedings of SPIE*; 2009; Vol. 7273, p. 727307.

[6] Smith, B.; Bourov, A.; Fan, Y.; Cropanese F.; Hammond, P. In *Proceedings of SPIE*; 2005; Vol. 5754, p. 751.

[7] Smith, B.; Fan, Y.; Zhou, J.; Lafferty, N.; Estroff, A. In *Proceedings of SPIE*; 2006; Vol. 6154, p. 61540A.

[8] http://www.itrs.net/Links/2009ITRS/Home2009.htm, ITRS 2009 Edition (2009).

Novel ATHENA Mark Design to Enhance Alignment Quality in Double Patterning with Spacer Process

L.W. Chen, Mars Yang, Elvis Yang, T. H. Yang, K. C. Chen and Chih-Yuan Lu
Macronix International Co. Ltd, No. 16, Li-Hsin Rd., Science Park, Hsinchu 300, Taiwan

ABSTRACT

DPS (Double Patterning with Spacer) has been one of the most promising solutions in flash memory device manufacturing. Apart from the process complexity inherent with the DPS process, the DPS process also requires more engineering efforts on alignment technique compared to the single patterning. Since the traditional alignment marks defined by the core mask has been altered hence the alignment mark recognition could be challenging for the subsequent process layers.

This study characterizes the process influence on the traditional ASML VSPM (Versatile Scribelane Primary Marks) alignment mark, and various types of sub-segmentations within VSPM marks were carried out to enable the alignment and find out the best performing alignment marks. The design of the transverse and vertical sub-segmentations within the VSPM marks is aimed to enhance the alignment signal strength and mark detectability. Alignment indicators of WQ (Wafer Quality), MCC (Multiple Correlation Coefficient) and ROPI (Residual Overlay Performance Indicator) were used to judge the alignment performance and stability. A good correlation was established between sub-segmentations and wafer alignment signal strength.

Keyword: DPS, VSPM, alignment, WQ, MCC, ROPI, overlay

1. INTRODUCTION

Among a variety of double patterning schemes, the DPS has been one of the most promising solutions in flash memory device manufacturing. Figure 1 depicts the process sequence of a positive-type DPS to form 45nm word-lines. The process flow starts with the first exposure of core pattern at 180nm pitch, followed by the PR trimming and 1st hard mask etch to form 45nm lines. Following by the 1st hard mask etch, the CVD (Chemical Vapor Deposition) SiN spacer was formed and spacer etch was carried out for delineating the desired line width at 90nm pitch. The sacrificial core pattern was then removed followed by the 2nd exposure and etch to trim away the closed line end formed by spacer approach. The core removal can be implemented before or after core pattern removal up to the process integration concern. Before the final etch for word lines, the 3rd exposure was conducted for defining the peripheral patterns. And finally, the poly etch followed by the 2nd hard mask etch delineates the whole word line pattern. Clearly, there is high process complexity inherent with the DPS furthermore the DPS process also requires more engineering efforts on alignment technique compared to the single patterning.

As above description, the DPS was usually adopted to double the line density for only the critical line and space patterning. However, to complete the whole circuit function, two more masks are usually required to trim the closed line end and form pad and random features in peripheral area. That makes the inter-layer alignment necessary. Though the overlay requirements for trim and periphery masks are less critical, the alignment will be a challenge due to the alignment mark has been altered by DPS process. As shown in Figure 2, the traditional scribe-lane alignment marks defined by the core mask is a template structure, the pitch and cycle of alignment mark was further changed after spacer formation and core film removal hence the alignment mark recognition could be challenging for the subsequent process layers. In this paper, various types of sub-segmentations within the ASML VSPM and RVSPM marks were carried out to enable the alignment and find out the best performing alignment marks in DPS process. The design of the transverse and vertical sub-segmentations within the VSPM/RVSPM marks is aimed to enhance the alignment signal strength and mark detectability accordingly alignment and overlay performance can be enhanced. Alignment indicators of WQ, MCC and ROPI were used to judge the alignment performance and stability. In addition, process variations such as spacer height were also employed to verify the robustness of the alignment strategy.

Optical Microlithography XXIII, edited by Mircea V. Dusa, Will Conley, Proc. of SPIE Vol. 7640,
764020 · © 2010 SPIE · CCC code: 0277-786X/10/$18 · doi: 10.1117/12.846014

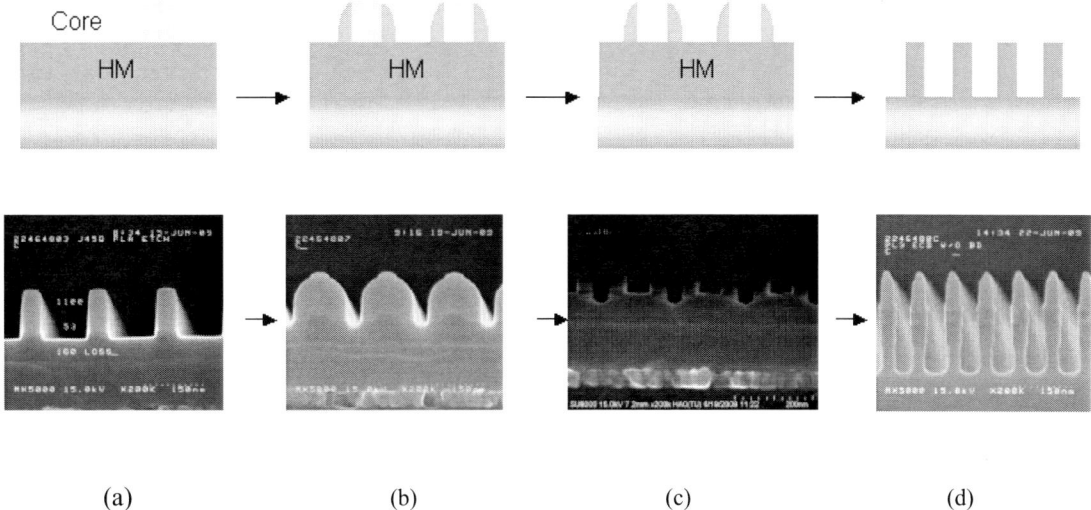

Figure1.The process sequence of DPS scheme to form 45nm line-width at 90nm pitch: (a) around 45nm core pattern was formed by photo and etch process on hard mask, (b) SiN spacer deposition and etch, (c) sacrificial core template removal, and (d) final etch step to define hard mask and line width.

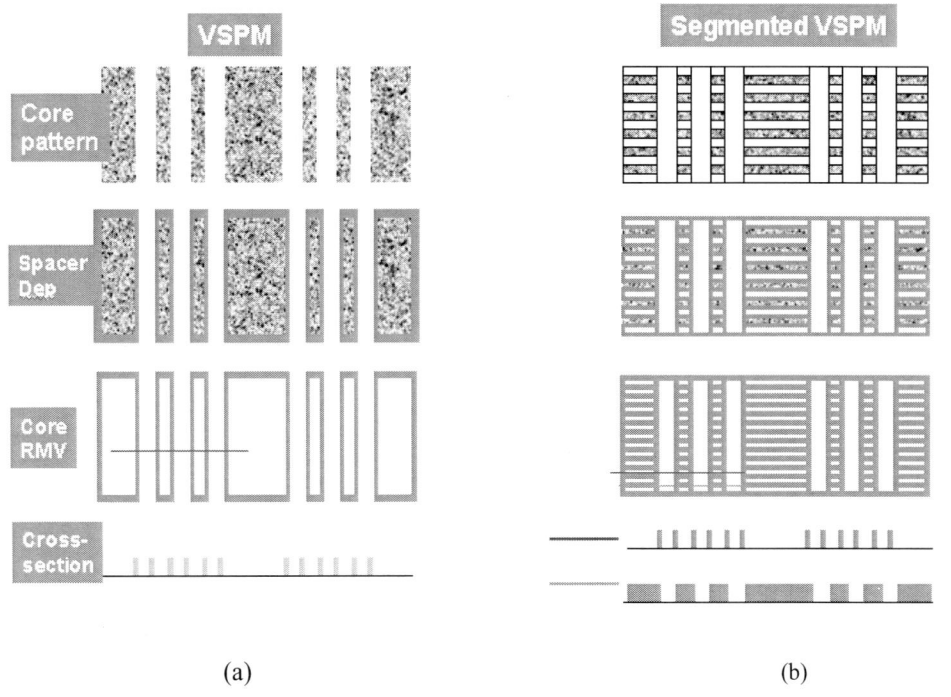

Figure 2. The evolution of (a) VSPM and, (b) segmented VSPM mark in the DPS process flow.

2. MARK DESIGN AND EXPERIMENTAL

Multiple alignment marks were usually printed in each layer for process optimization purposes. The ASML alignment mark type named VSPM has offered the feasibility within a mark for multiple mark types. To address the alignment challenge in DPS process, sub-segmentations were designed and applied in the VSPM mark. The design concept for the sub-segmentations in alignment mark is maintaining its minimum feature size to be slightly larger than the critical feature size of process layer, and extending the splits of feature size to be larger than the alignment-illuminating wavelength. Figure 3 schematically displays the examples of transverse and vertical sub-segmentations, the transverse smaller features within the mark were defined to have equal line/space with the line-width spanning from 100 to 900nm (as shown in Figure 3(a)). Likewise, the vertical sub-segmentation and reversal tonality marks were designed for evaluated in this study as well. The naming of our segmentation is defined with the acronym of ASML mark and segmenting direction ("VT" for VSPM mark with transverse sub-segmentations, "VV" for VSPM mark with vertical segmentations) plus the dimension of line width in nm. For example, VT100 stands for transverse sub-segmentation of VSPM mark with 100nm line/space and RV100 represents for vertical sub-segmentation of RVSPM mark with 100nm line/space.

The lithographic processes were performed by ASML XT1250B ArF scanner and ASML XT850 KrF scanner for the core layer and trim/peripheral patterning, respectively. The critical dimension was measured with Hitachi-9380II and overlay was measured with KLA-Tencor XP5300. The spacer height was modulated through changing the thickness of core film that ranges from 950A to 1350A in this study. Alignment indicators of WQ, MCC and ROPI were used to judge the alignment quality and stability.

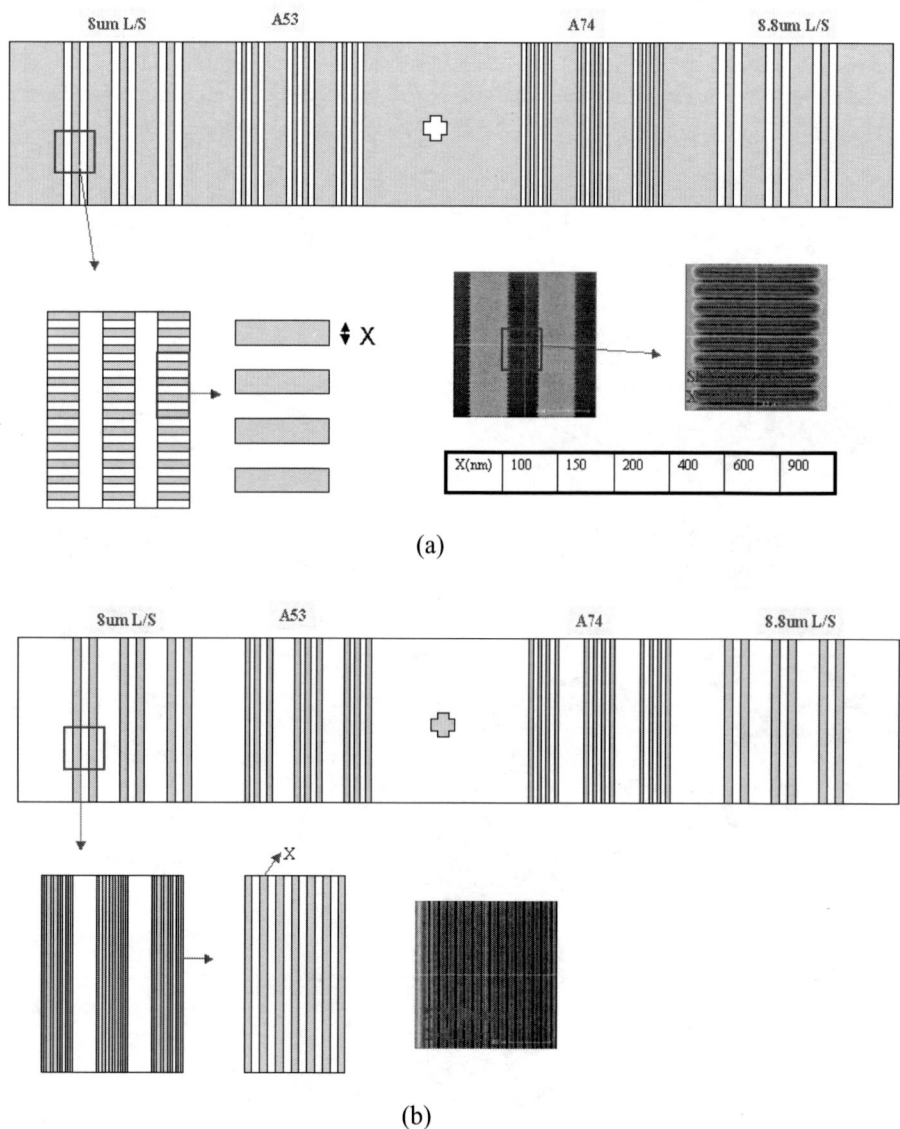

(a)

(b)

Figure 3. (a) The schematic example of transverse sub-segmentations of VSPM, the smaller features within the mark were defined to have equal line/space with the line-width spanning from 100 to 900nm. (b) The schematic example of vertical sub-segmentations of VSPM, the smaller features within the mark were defined to have equal line/space with the line-width spanning from 100 to 900nm.

3. RESULTS AND DISCUSSION

As illustrated in Figure 2 that the double patterning with spacer process alters the mark duty cycle after sacrificial core removal, the alignment mark recognition hence should be carefully verified. The alignment wafer quality was first examined among VSPM and segmented marks after core patterning, as shown in Figure 4. It is clear that the VSPM mark has the higher WQ than the segmented marks no matter the red or green light was used for alignment. The WQ decreases as the segmented line/space increasing but all segmented marks with 100 to 900nm line/space segmentations as well as traditional VSPM can be used for alignment successfully.

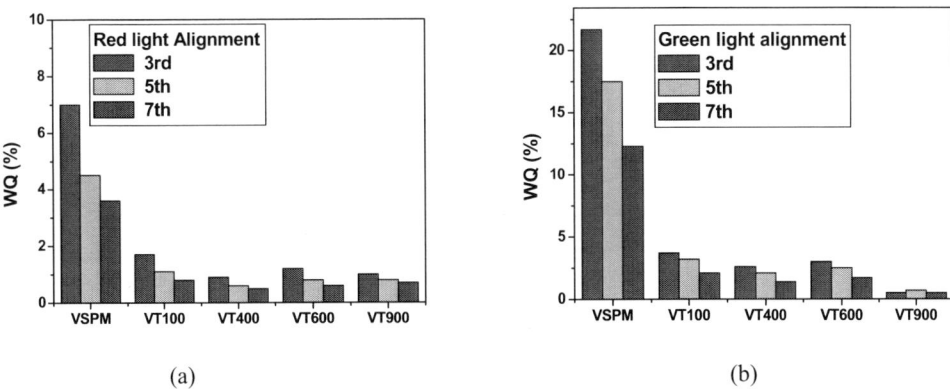

(a)   (b)

Figure 4. 3rd, 5th and 7th order WQ of ASML VSPM and segmented marks in: (a) red light alignment, and (b) green light alignment right after core patterning.

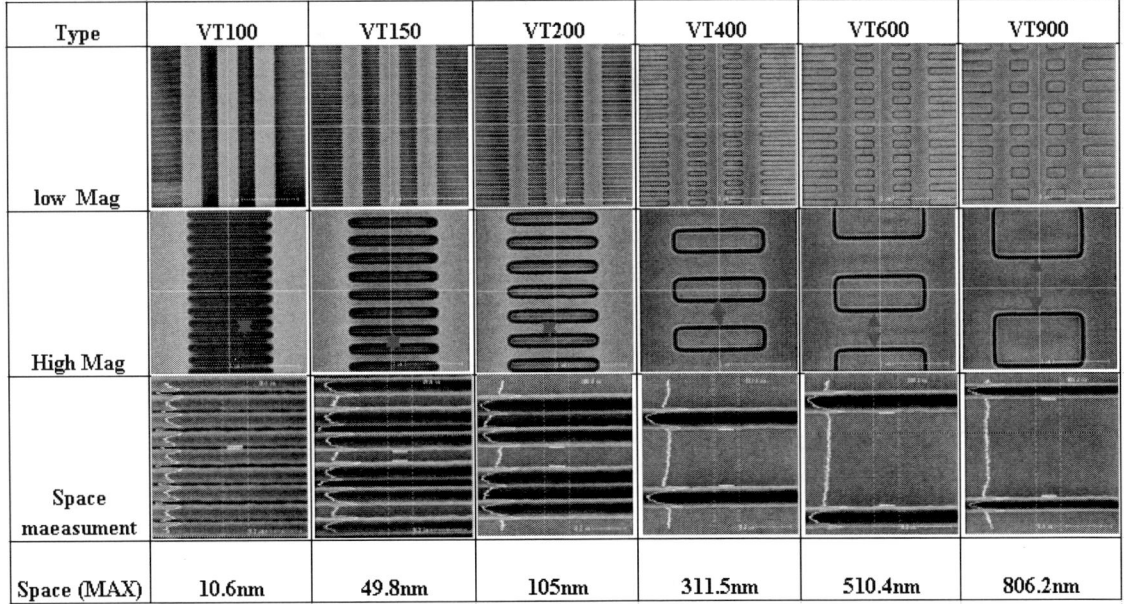

Type	VT100	VT150	VT200	VT400	VT600	VT900
low Mag						
High Mag						
Space maeasument						
Space (MAX)	10.6nm	49.8nm	105nm	311.5nm	510.4nm	806.2nm

Figure 5. The top-view images are transversely segmented Athena marks after core removal. The marks with smaller line/space segmentations create dense sidewall spacer arrays while the marks with larger line/space segmentations generate isolated sidewall spacer arrays.

Figure 5 shows the top-view images of the transversely segmented marks after sacrificial core removal. The marks eventually remained the spacer outlines. The marks with smaller line/space segmentations create denser sidewall spacer arrays while the marks with larger line/space segmentations generate more isolated sidewall spacer arrays. In the low CDSEM magnification, the VT100 mark looks very like the traditional VSPM and the maximum space between transverse segmentation is as small as ~10nm that can functionally generate diffraction patterns for the alignment wavelengths. The alignment behavior may hold true till the space is approaching or larger than the wavelengths of alignment lights, for example 510nm of VT600 for green light (532nm wavelength) and 806.2nm of VT900 for both green and red (633nm wavelength) lights. The alignment tests were carried out to compare the alignment performance between VSPM and segmented marks after core removal, as a result, the traditional VSPM failed at alignment because the WQ is lower than the threshold specification of 0.1%. Conversely, most of the segmented marks are still valid for alignment, except the VT900 for both the red and green alignments and VT600 for the green alignment. As can be seen from Figure 6(a), the VT100 to VT600 can be adopted for successful alignment by red light since the WQ is larger than the threshold. The smaller segmentation performs higher WQ, and similar behavior was observed in Figure 6(b) by green light. The reason behind the failure of VT900 for alignment could be the isolated sidewall spacer arrays generated by large line/space segmentation that can't functionally generate diffraction patterns for the alignment wavelengths accordingly the alignment signal could be affected. The MCC for all the marks by red or green alignments get comparable value of around 1. It also can be seen from the Figure 6 that the smallest segmentation mark of VT100 gets the highest WQ in both red and green alignments. However a difference is that an abrupt decrease of WQ occurs at segmentation larger than 200nm for red alignment but gradual decrease happens in green alignment. And also, the highest WQ value of VT100 in red alignment scheme is higher than that of the green alignment that may imply the relatively smaller gratings than the alignment wavelengths generated by segmentations are better for enhancing alignment signal.

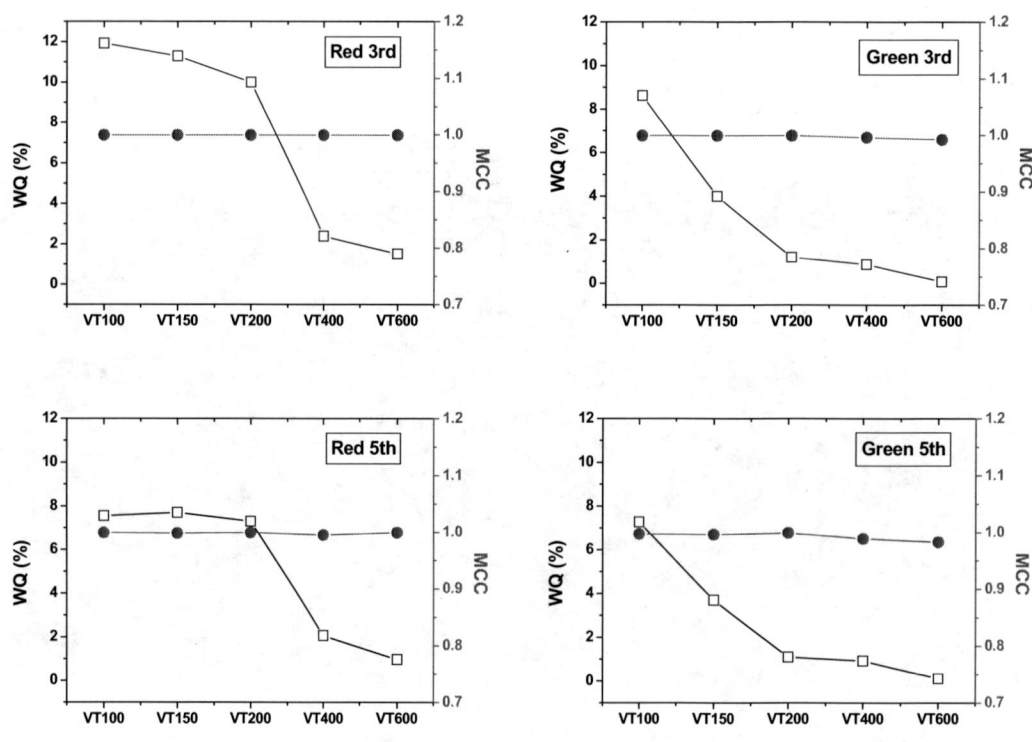

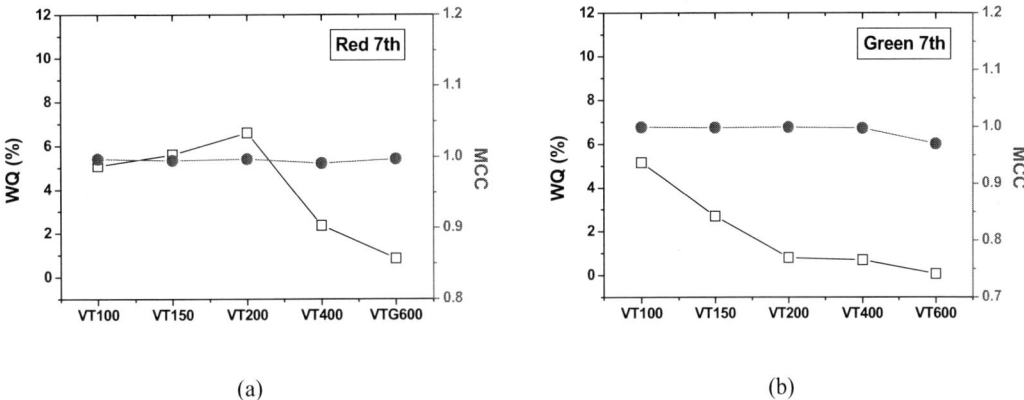

(a)   (b)

Figure 6. 3rd, 5th and 7th order WQ/MCC of segmented marks in: (a) red alignment, and (b) green alignment after sacrificial core removal.

As aforementioned results, the highest WQ can be achieved through alignment by VT100 mark. The assessment of alignment stability was hence carried out for VT100, and the results were depicted in Figure 7. The WQ and MCC for various wafers by red/green alignments are stable enough.

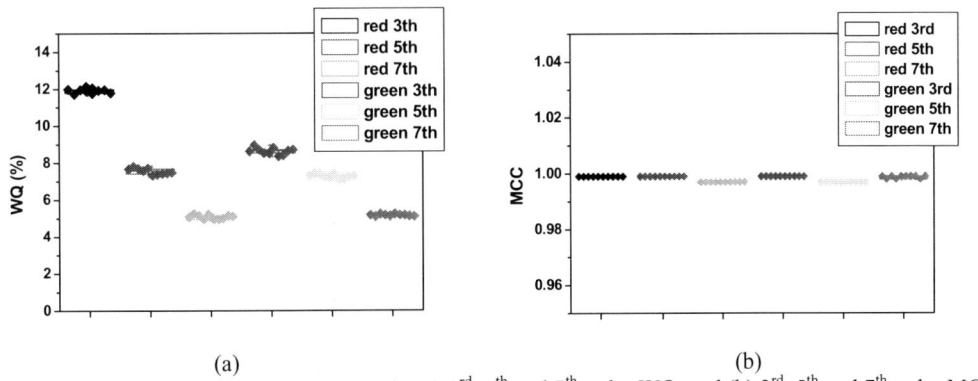

(a)   (b)
Figure 7. The alignment stability of VT100 mark: (a) 3rd, 5th and 7th order WQ, and (b) 3rd, 5th and 7th order MCC.

The ROPI is defined as 3σ of wafer residuals of the measured minus the modeled positions. The alignment strength with lowest ROPI was generally recommended for obtaining the best overlay and alignment result [2-3]. The ROPI comparison among the segmented marks was shown in Figure 8(a). The ROPI by red alignment is comparable for smaller than 400nm segmented marks but larger in the 600nm segmented mark. While the 600nm segmented mark failed at green alignment thereby an abrupt increase of ROPI is displayed in the plot. Figure 8(b) and 8(c) show the mean overlay error and 3σ value, respectively, for both red and green alignments. The 3σ values for all successfully aligned mark sets are comparable, but the red alignment behaves slightly smaller mean overlay errors than the green alignment. In short, all the successfully aligned segmented mark sets can get acceptable overlay performance with no big difference.

Signal strength directly affects the alignment repeatability and accuracy, weak signal strength results in large noise to signal ratio and introduces large alignment errors. However, the alignment signal strength varies much with the alignment mark profile, which can be affected by many process parameter settings of DPS process. Hence it is also crucial to verify how process parameter influents the alignment performance. The most import factors in DPS process affect the alignment mark profile are spacer width and spacer height, the spacer width is typical set to meet the final CD target but the spacer height remains an important factor for tailoring in etch hard mask resistance. Figure 9 illustrates the relationship between WQ and sidewall spacer height by taking VT100 mark as an example. Though the 7th order signal strength is the lowest but it is still strong enough for successful alignment no matter the spacer height ranging from 950A to 1350A. As taking a close look at the plots, the WQ of red alignment increases as the sidewall spacer height increasing, but contrary behavior is observed for the green alignment.

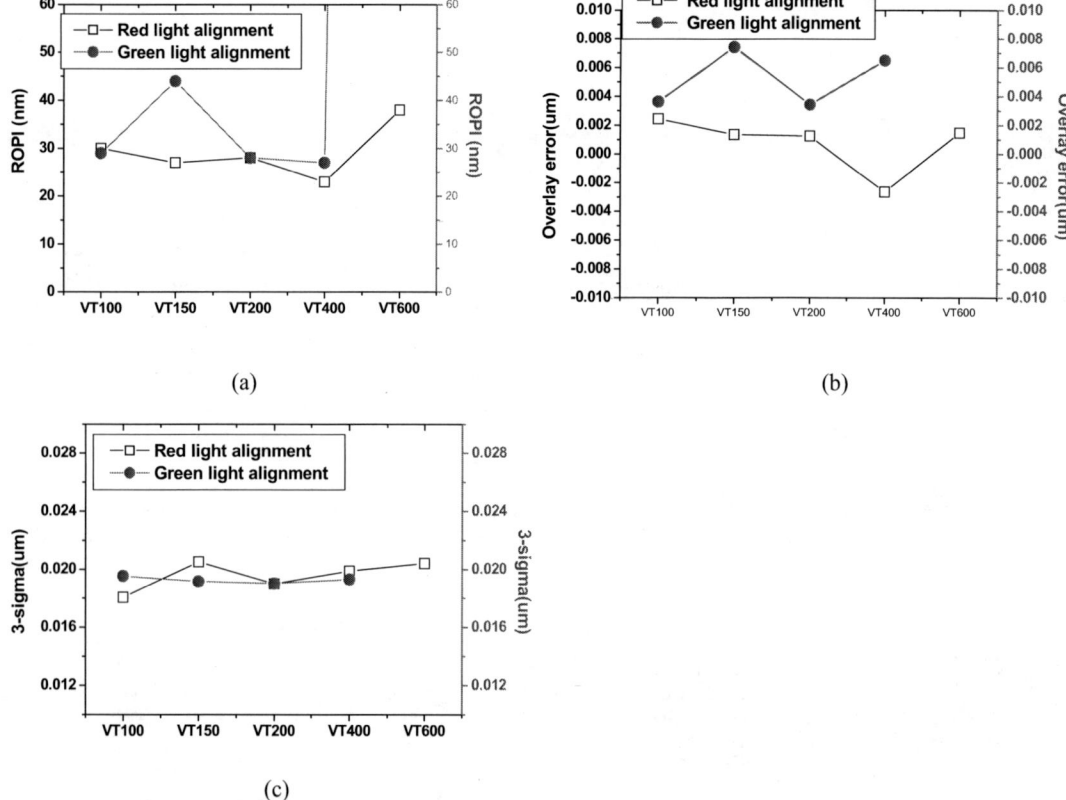

(a)

(b)

(c)

Figure 8. (a) The ROPI comparison among the segmented marks, (b) a comparison of mean overlay errors among the segmented marks, and (c) a comparison of 3σ overlay errors among the segmented marks. The alignment was performed on the KrF scanner to segmented marks generated in core layer by the ArF scanner.

Different types of segmented marks were also evaluated and compared for optimizing the mark design to facilitate robust alignment in DPS flow. Figure 10 displays the 3rd, 5th and 7th order WQ of 3 kinds of segmented marks including VT100, RVT100 and RVV100. It is obvious that the reverse tone mark, RVT100, performs comparably to VT100 in red and green alignment signals. The vertically segmented mark, RVV100, has larger alignment signal strength than VT100 and RVT100. The origin could be the vertical segmentation is parallel to the original VSPM grating (as shown in Figure 10(b)), hence a better integrity in mark edge can be anticipated.

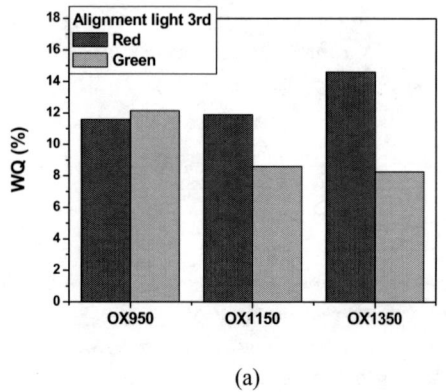

(a)

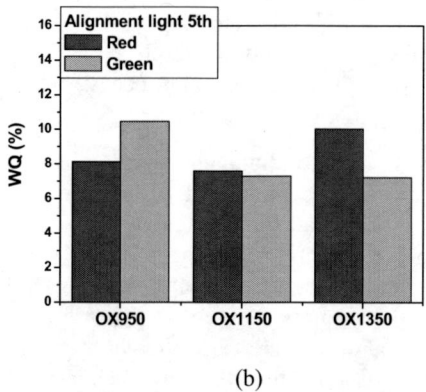

(b)

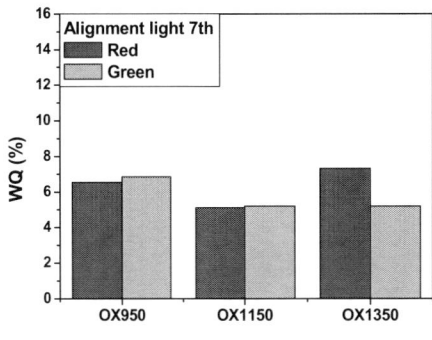

(c)

Figure 9. The relationship between WQ and spacer height of DPS after core removal: (a) 3rd order, (b) 5th order and (c) 7th order light alignment.

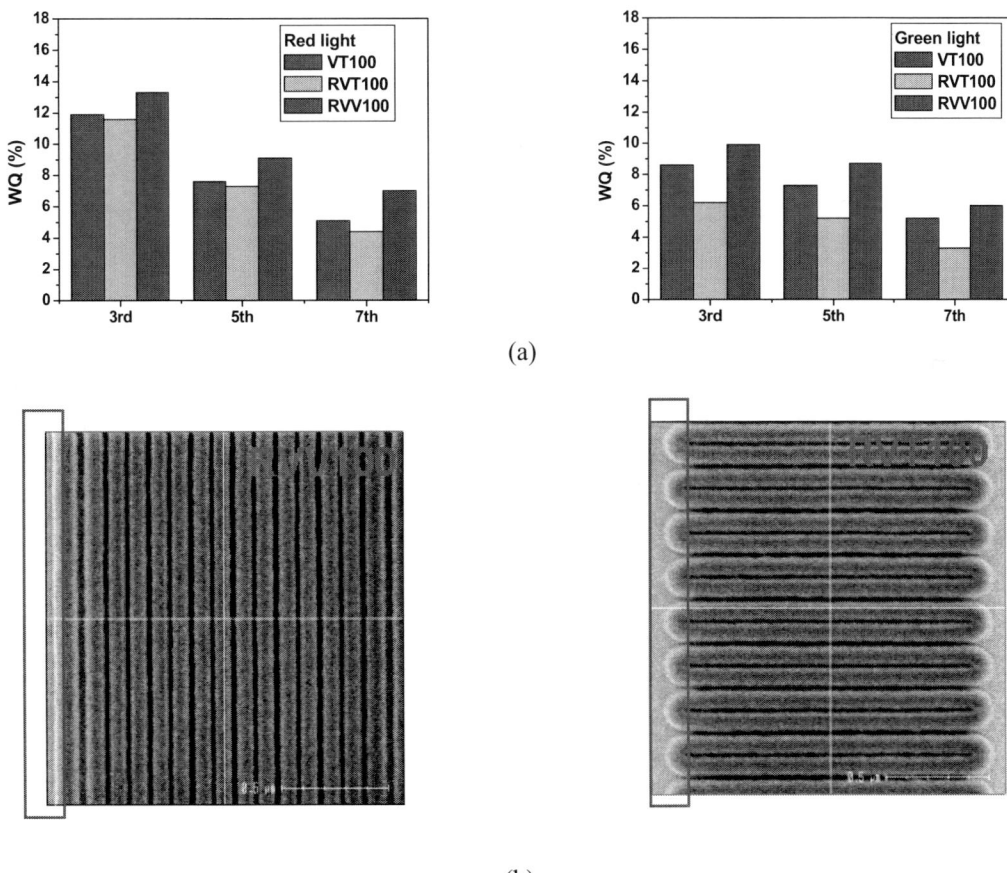

(a)

(b)

Figure 10. (a) WQ comparison among various segmented marks including VT100, RVT100 and RVV100, (b) the top-view images of RVV100 and VT100 to show the mark edge corresponds to the edge of traditional VSPM.

4. CONCLUSIONS

The traditional ASML VSPM mark is no more valid for alignment after sacrificial core removal in DPS process flow. In this paper, we have shown the alignment mark modifications that need to be done in alignment for DPS process. Various sub-segmentations in VSPM mark were carried out to enable the alignment, and alignment indicators including WQ, MCC and ROPI, together with overlay performance on wafers were used to find the best performing alignment marks.

The study has suggested that core marks with smaller line/space sub-segmentations tend to have higher alignment signal strength. In practice, the sub-segmentation should be sufficiently small but not become imaging critical, because it would make the alignment performance sensitive to imaging performance. Besides, both the transverse and vertical sub-segmentations work for alignment but the vertical sub-segmentation is slightly superior to the transverse on signal strength.

REFERENCES

1. H. Dai, C. Bencher, Y. Chen, S. Sun, X. Xu, C. Ngai, "Alignment and Overlay Improvements for 3x nm and Beyond Process with CVD Sidewall Spacer Double Patterning," *Proceedings of SPIE 7274*, 72743G, 2009.
2. R. Navarro, et al, "Extended ATHENA alignment performance and application for the100nm technology node", *Proceedings of SPIE 4344*, pp. 682-694, 2001.
3. J. Huijbrechtse, et al, "Overlay Performance with advanced ATHENA alignment recipes", *Proceedings of SPIE 5038,* pp. 918-928, 2003.

Modeling of CD and placement error in multi-spacer patterning technology

S. Babin, K. Bay

aBeam Technologies, Inc.

5286 Dunnigan Ct., Castro Valley, CA 94546; sb@abeamtech.com

ABSTRACT

The spacer patterning technique is an attractive way to fabricate patterns at resolutions far beyond the limits of traditional optical lithography. In this paper, we have simulated film deposition and dry etch in spacer patterning at 32 nm and 22 nm designs using the commercially available TRAVIT software. Various resist thicknesses and profiles were used, as well as process conditions for film deposition and dry etch. Dynamics of etch profiles, resulting profiles, and critical dimensions (CDs) were extracted, as well as positional errors of features. It was found that the placement error can be significant, especially when using thin resists. Multi-spacer patterning was also simulated. In the multispacer technique, the spacer patterning processes were applied consequently, resulting in the reduction of the lithographic pitch. The fabrication of 11 nm half-pitch lines were simulated using available lithographic techniques at 45 nm.

Keywords: spacer patterning technology, SPT, self aligned double patterning, SADP, dry etch, film deposition, etch profile, CD variation, microloading, placement error

1. INTRODUCTION

The spacer patterning technology (SPT) is based on the fabrication of a spacer by conformal deposition of a material at the edge of a sacrificial layer, followed by anisotropic etch of this material. When the sacrificial layer is removed, the spacer can be used as a hard mask to define structures in underlying layers. The SPT is an attractive way to fabricate patterns at resolutions far beyond the limits of traditional optical lithography. The principle of SPT has been demonstrated in 1983.[1] Progress has recently led to the application of SPT to microelectronic manufacturing.[2-6] The process is sometimes called self-aligned double patterning (SADP).[6]

Simulations of SPT should complement the time-consuming experiments in order to reduce cost and shorten development time. The TRAVIT software tool can accurately simulate multi-step processing of etch and film deposition to accommodate this demand. The software predicts etch profile, linewidth, variation of critical dimensions (CD), and placement errors in SPT.

In this paper, we have simulated film deposition and dry etch in spacer patterning at 32 nm and 22 nm designs. Various resist thicknesses and profiles were used, as well as process conditions for film deposition and dry etch. Dynamics of etch profiles, resulting profiles, and critical dimensions (CDs) were extracted. Multiple examples of the simulations are presented. In addition, multi-spacer patterning (S^nPT) was also simulated. In the multi-spacer technique, the process was applied several times, resulting in the reduction of the lithographic pitch by a factor of 2n, where n is the number of spacer processing steps. Fabricated spacers were used as a sacrificial layer for the following processing.

2. TRAVIT: PROCESS SIMULATION SOFTWARE TOOL

Simulation of microelectronic processing can significantly reduce the process development cost and improve time to market. Two main processes need to be simulated in SPT: film deposition and dry etch.

Simulating a dry etch process is a complex problem that requires a detailed knowledge of plasma physics, the interaction of plasmas with solids, plasma chemistry, kinetics, etc.[7-10] Because of the complexity of the task, it is

Optical Microlithography XXIII, edited by Mircea V. Dusa, Will Conley, Proc. of SPIE Vol. 7640,
764021 · © 2010 SPIE · CCC code: 0277-786X/10/$18 · doi: 10.1117/12.850972

impractical to expect any simulation software to be universally applicable to all dry etch situations. On the other hand, addressing the problem for specific cases using defined boundary conditions with given tolerances is feasible and cost effective; it can deliver usable information.

TRAVIT simulation software uses an analytical model of dry etch. The software tool is focused on the simulation of the dynamics of etched profiles. In addition, it is capable of simulating CDs and CD variations due to dry etch; these are not addressed by currently known models.[11]

The simulation can handle isotropic etch, anisotropic etch, and a combination of the two. The process can be changed on the fly; describing changing recepies or/and conditions of the process. Using etched profiles found by the software, the software then extracts CDs and calculates CD variation. The input parameters for simulation involve the GDSII pattern, the initial profile of a resist or a hardmask, layers and their thicknesses, and etch conditions.

In order to address the accuracy of the model, multiple etch effects have been included in the model. Microloading is automatically simulated; CD variation due to microloading is extracted. Microtrenching and footing effects, as well as angular dispersion, are part of the model. The other target for the model is high simulation speed. The simulation speed should not be affected significantly when additional etch effects are included. Both of these goals were achieved using an analytical model.

In a typical modeling of SPT process, six or more process steps are simulated. In a multiple S^nPT process, the number of simulation steps is multiplied by n.

3. SIMULATION OF SPT

3.1 Example of SPT process at 22 nm

A 22 nm SPT process was simulated. The pattern in the photoresist was four lines at 88 nm pitch; each line was 44 nm wide after the resist development. The stack was silicon with a thick functional layer, a hard mask film, and a resist. The goal was to produce a pattern at 22 nm half-pitch.

The simulations involved:

- setup of a GDSII resist pattern at 88 nm pitch on a hard mask covering functional layer; assigning material thicknesses and resist profile after development
- resist trim;
- conformal deposition of sacrificial layer
- anisotropic etch of sacrificial layer
- etch of the resist;
- etch of a hard mask;
- etch of the functional layer.

The results of the simulations at each step are shown in Figure 1.

In multi-spacer patterning, this full process has been repeated in order to reduce the pitch by a factor of two every other processing.

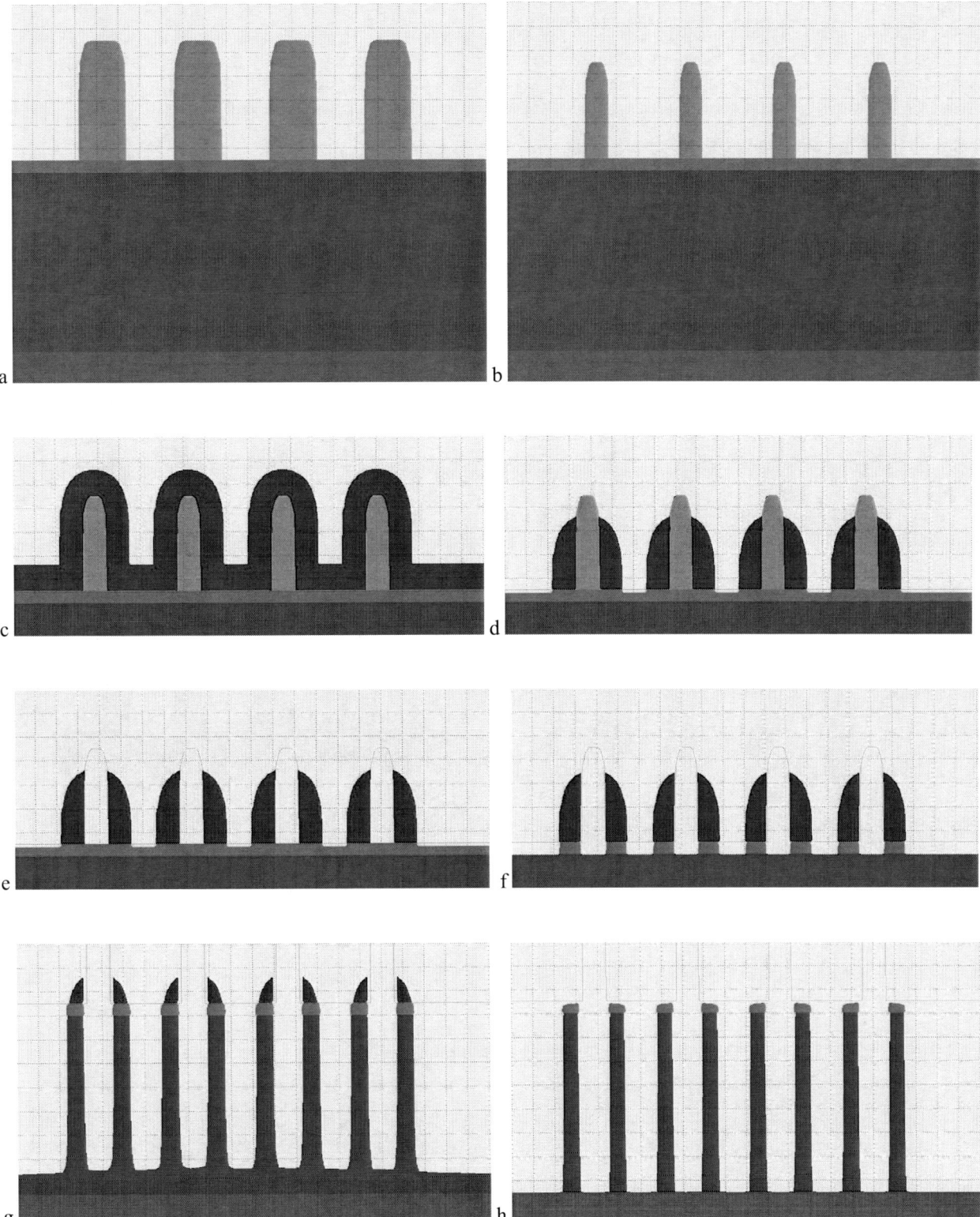

Figure 1. Simulation of the spacer patterning technology process at 22 nm: a) fabrication of a resist pattern at 88 nm pitch, 44 nm lines on a hard mask covering functional layer; b) resist trim; c) deposition of sacrificial layer (only top of the structure is shown); d) etch of sacrificial layer; e) etch of the resist; f) etch of a hard mask; g) and h) etch of the functional layer.

3.2 22 nm half pitch process; effect of resist thickness

A similar process for 22 nm has been simulated in more detail. Here, the effect of the resist thickness was studied. The profile after etch of a hard mask through the sacrificial layer is shown in the Figure. The thickness of the hard mask was 10 nm; the thickness of the photoresist after trimming and before etch was 40 nm, 80 nm, and 160 nm.

The results of etch of the hard mask are displayed in Figure 2.

It was found that the placement error improved for the 80 nm resist compared to the 40 nm thick resist: the placement error dropped from 5.9 nm to 3 nm. However, further increase of the thickness to 160 nm did not bring benefit.

We attributed this effect to the improvement of the "internal" wall angles of the spacer, and partially to additional protection of the hard mask by the sacrificial layer during the etch of the functional material.

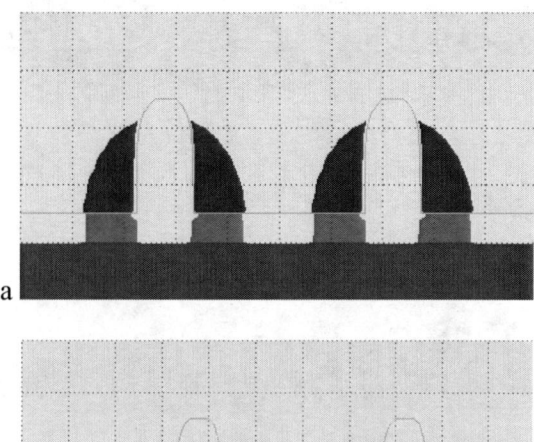

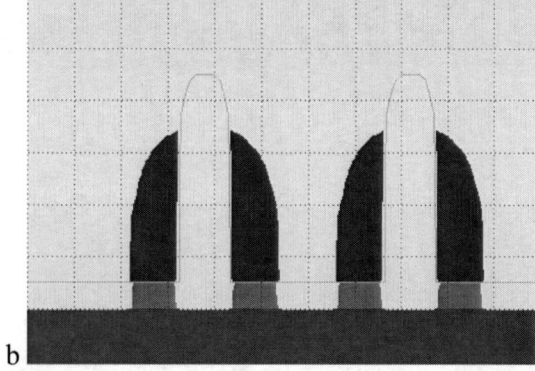

Figure 2. Profiles of sacrificial layer and a hard mask at various resist thickness after trim:
 a) 40 nm; b) 80 nm; c) 160 nm
A significant reduction of placement error was found at 80 nm resist compared to 40 nm, while further increase of thickness did not improve the error at etch conditions used.

3.3 22 nm half pitch process; effect of the resist trim on placement error

Resist trim is an important step in SPT process. The accuracy of the trim directly influences the placement error of features.

Three resist trim processes were used. They resulted in the resist linewidths of 16 nm, 22 nm, and 28 nm; see Figure 3.

The results of simulation of the profile after final etch are displayed in Figure 4. Significant placement errors of 10 nm and 13 nm were found when the trim was not done right. The placement error is well visible in the Figure. The results are summarized in the Table 1.

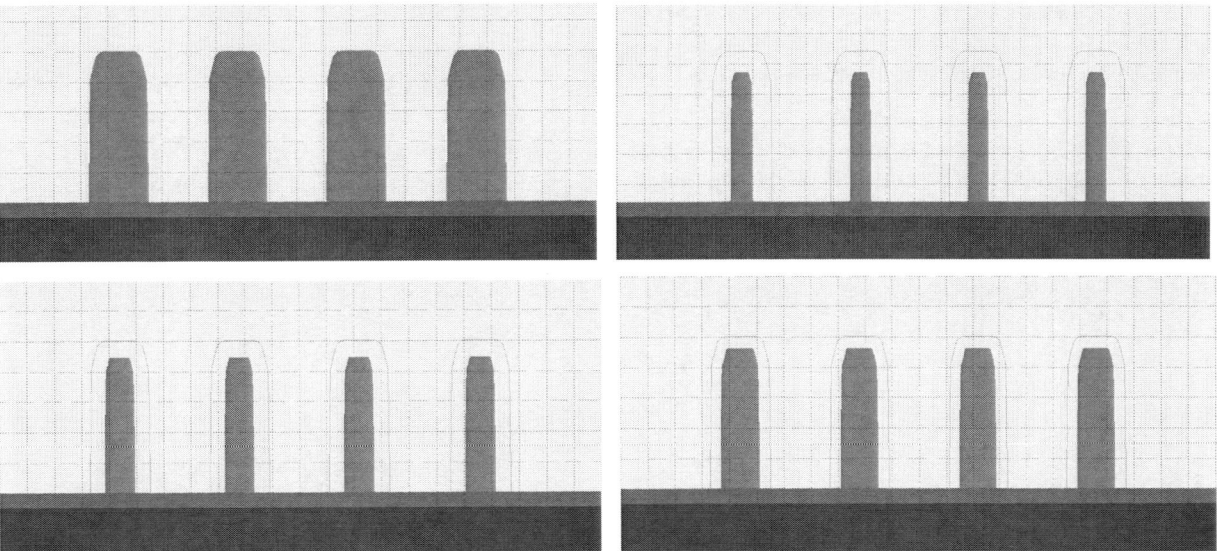

Figure 3. Various trim of photoresist: a) the linewidth was 44 nm after resist development; b)-d) after resist trimming, the resist linewidth was 16 nm, 22 nm, and 28 nm respectively.

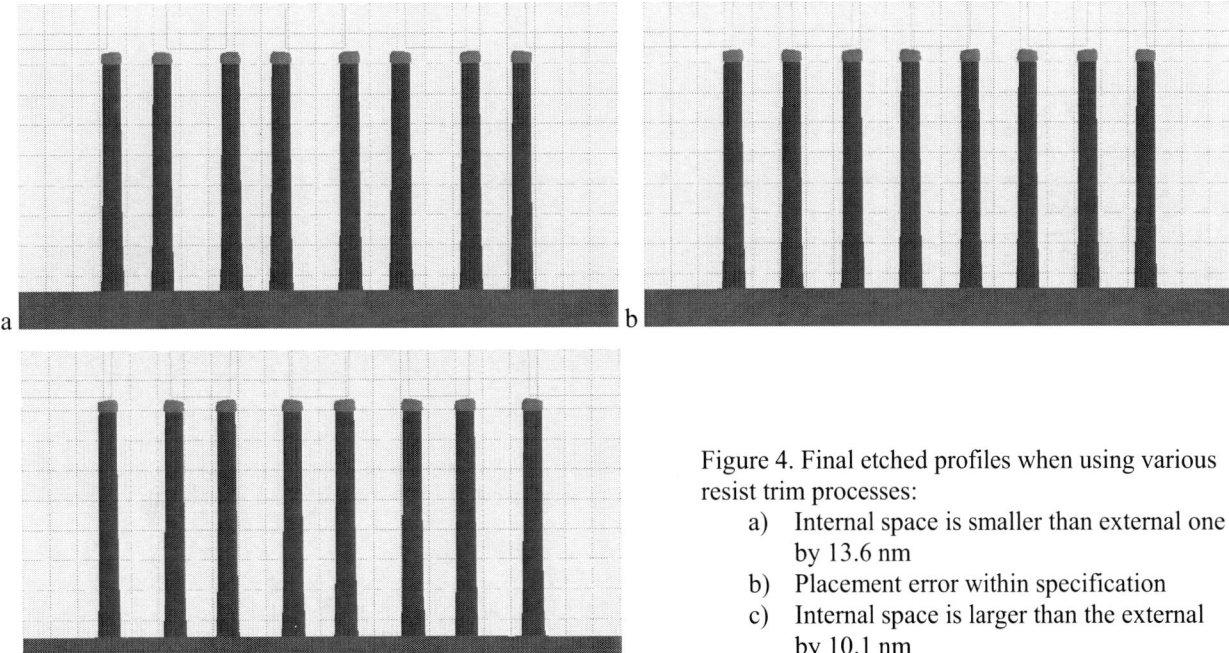

Figure 4. Final etched profiles when using various resist trim processes:

a)   Internal space is smaller than external one by 13.6 nm
b)   Placement error within specification
c)   Internal space is larger than the external by 10.1 nm

Table 1. Placement error at various trim of the resist.

Resist width after trim, nm	Space left, nm	Space right, nm	Placement error, nm
16.3	32.6	19	13.6
22.4	27	24.6	2.4
28.5	21	31.1	-10.1

3.4 32 nm SPT process

SPT process was applied to fabrication of lines at 32 nm half-pitch. A resist pattern at 64 nm linewidth was used for this simulation.

The etch of the functional layer was simulated at various etch durations after the end point detection of the last etch process. The etch profile is displayed in Figure 5. The etch was stopped when the linewidth at the bottom reached 32 nm (a) and when it was overetched to 23 nm (b). As discussed below, overetch results in improved CD variation.

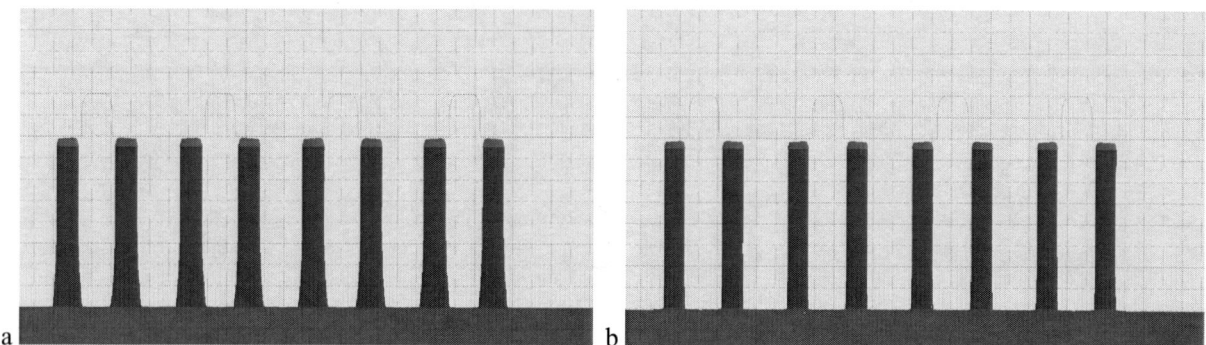

Figure 5. The 32 nm SPT process. The process was stopped at the linewidth 32 nm at the bottom (a) and overetched to get the CDs about 23 nm (b).

In this example, the etch was stopped when the linewidth at the bottom was about 32 nm. Variation of CDs due to the etch process was found to be 1.2 nm; the linewidth varied between 31.6 and 32.8 nm.

It was also found that the placement error is noticeable. The width between lines was 27.9 nm on the left side and 35 nm on the right side of a central line.

The overetch has reduced the CD variation. When the process was stopped at the linewidth of 22.9 nm, it was found that the CD nonuniformity of linewidth due to microloading was decreased to 0.4 nm. However, the distance between lines was varied significantly: it was alternating between 37.9 nm and 44.5 nm. Assuming the placement of the first line was accurate, placement error of 6.6 nm occurred for every second line.

4. MULTI-SPACER PATTERNING FOR 11 nm HALF-PITCH

Multi-spacer patterning is a promised technique to achieve an extremely high resolution without using lithographic tools of exceptional complexity. The process starts with fabrication of resist lines with 45 nm half pitch; this process is available now in modern semiconductor factories.

The simulated process flow is shown in Figure 6. The resist is trimmed (a); a sacrificial layer is deposited (b). The anisotropic etch (c) is followed by the removal of the resist (d). The sacrificial layer is trimmed (e). After that, the spacer patterning is repeated once more using another sacrificial layer which provides good etch selectivity to the hard mask and to the first sacrificial layer. The second layer is deposited (f) and etched (g), and the first sacrificial layer is removed by plasma (h).

The remaining pattern represents a pattern made in the second sacrificial layer. The half-pitch is 11 nm, which is 4 times smaller than initially fabricated in the resist.

The hard mask is etched through this fabricated pattern (i). This hard mask is used to etch the functional layer, see Figure 6 (k). The fabricated pattern after removal of the etch mask is shown in Figure 6(l).

The pattern has been measured. The following parameters of the lines at the bottom have been found:

CDs are ranging from 10.1 nm to 11 nm. The spaces were between 10.2 and 13.1 nm. Compared to the single step SPT, there are two "inside" and "outside" positions for spaces in this process, one for each spacer.

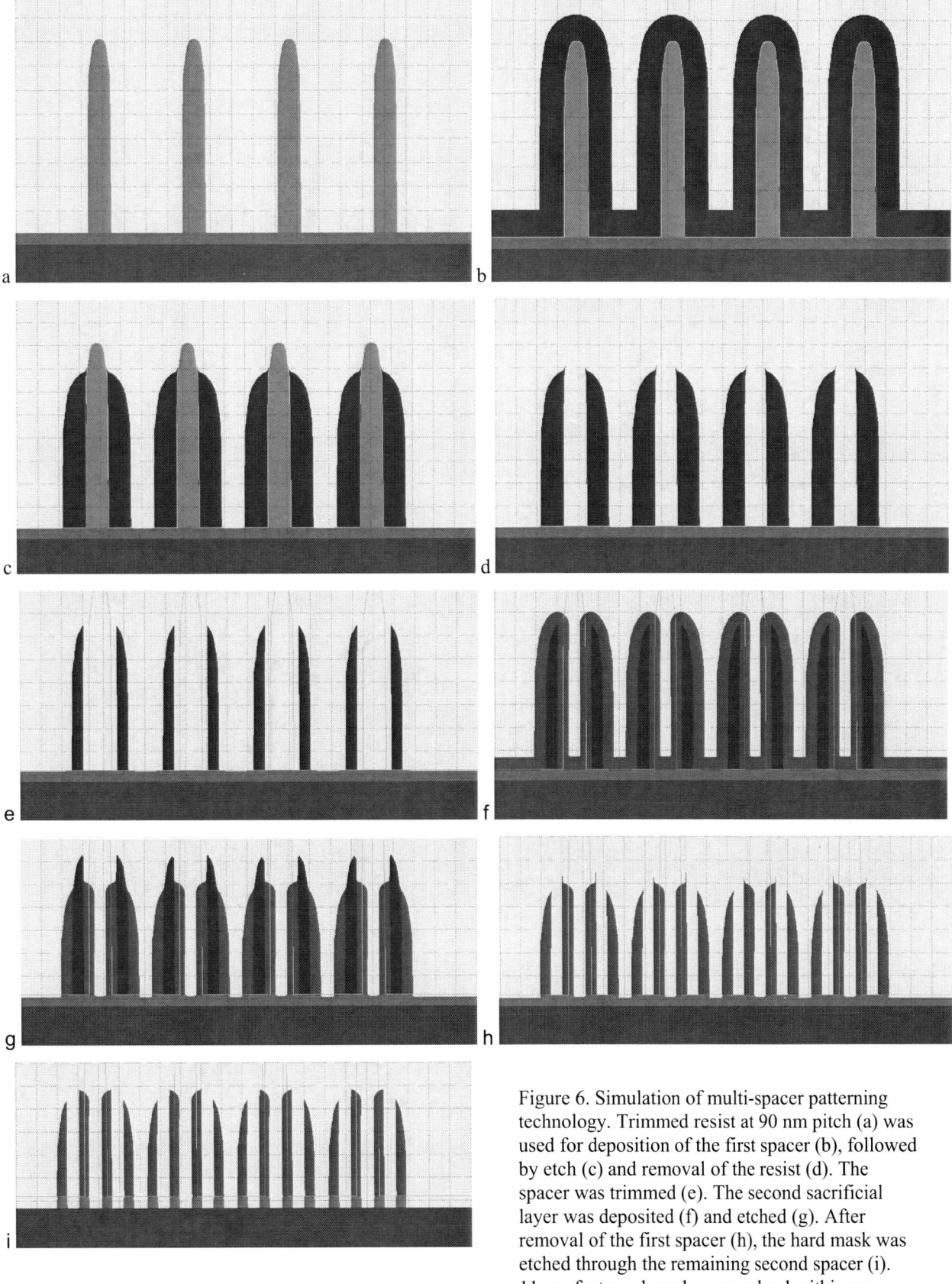

Figure 6. Simulation of multi-spacer patterning technology. Trimmed resist at 90 nm pitch (a) was used for deposition of the first spacer (b), followed by etch (c) and removal of the resist (d). The spacer was trimmed (e). The second sacrificial layer was deposited (f) and etched (g). After removal of the first spacer (h), the hard mask was etched through the remaining second spacer (i). 11 nm features have been resolved within spec.

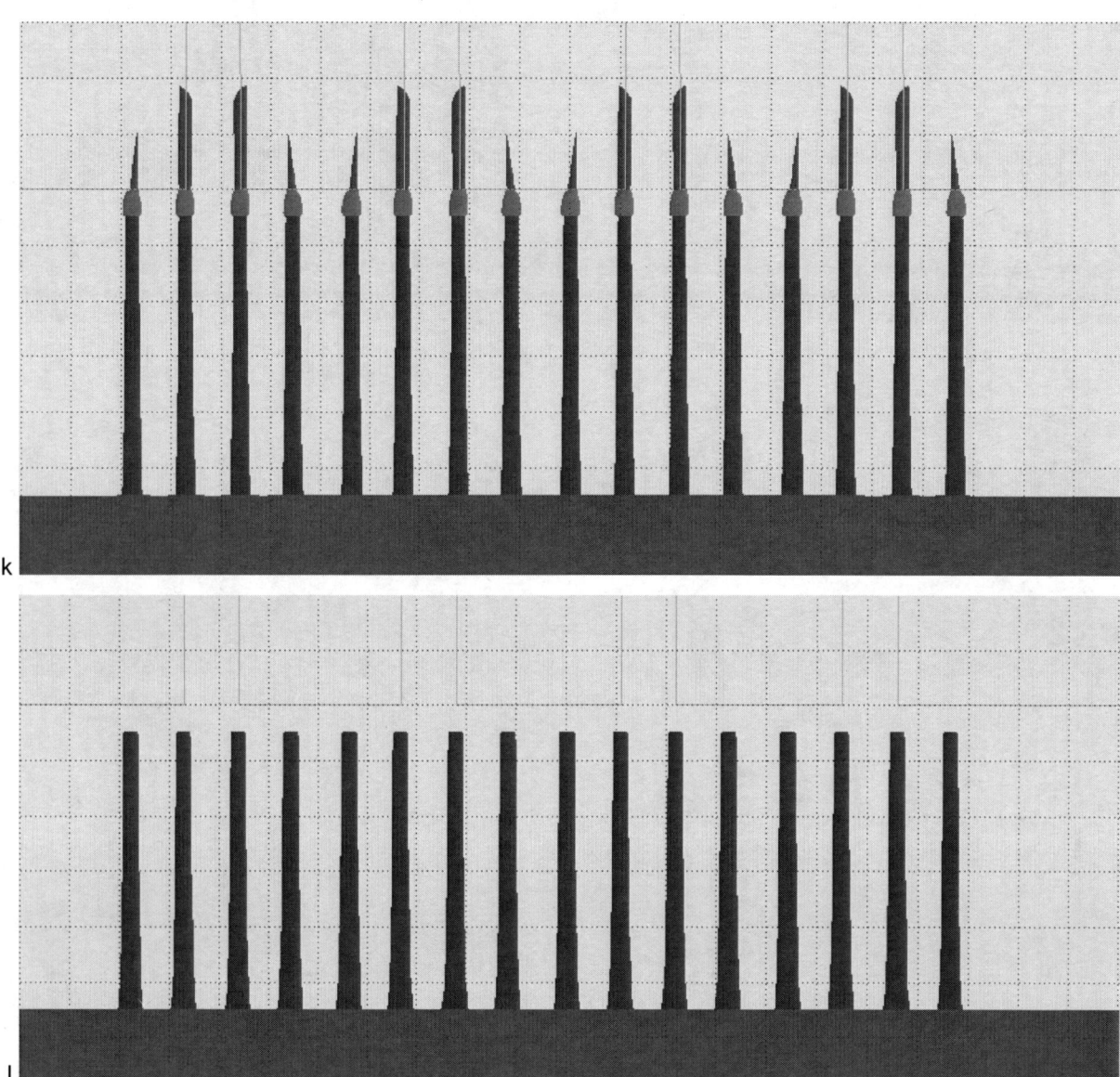

Figure 6. The functional layer has been etched (k) through the hard mask displayed in Fig. 6,i. The hard mask and remaining spacer were etched away (l). The measured features varied in linewidth from 10.1 nm to 11 nm. The spaces between lines were in the range of 10.2 nm and 13.1 nm.

5. CONCLUSION

A software tool, TRAVIT, has been developed to simulate semiconductor processing, including dry etch and film deposition. The software predicts the etch profile, CDs, CD variation, and placement errors.

Examples of simulations have been presented for 22 nm and 32 nm SPT processes. A significant placement error has been found, as well as ways to reduce it as confirmed by simulations. It was found that thicker resist and optimized resist trim processes can improve the placement accuracy.

The simulation of multi-spacer technology has been demonstrated. It was shown that 11 nm lines can be fabricated using 45 nm half-pitch lithography in the resist; the placement error and CD variation were within the required specification.

The software is useful for understanding and optimizing the SPT process. Simulations can reduce the cost of process development and shorten development time for a new process.

REFERENCES

1. Flanders D. C., Efremow N. N., J. Vac. Sci. Technol., B1 (1983) 1105
2. Natelson D., Willett R. L., West K. W., Pfeiffer L. N., Appl. Phys. Lett. 77, (2000) 1991
3. Choi Y-K, King T-J and Hu C, IEEE Trans. Electron Devices 49 (2002) 436
4. Woo-Yung Jung,et al., Proc. SPIE 6156, (2006)
5. Hidefumi Mukai, Eishi Shiobara, Shinya Takahashi and Kohji Hashimoto, Proc. SPIE v. 6924, 692406, (2008)
6. H. Dai, C. Bencher, Y. Chen, S. Sun, X. Xu, C. Ngai, Proc. SPIE v. 7274, 72743G-1, (2009)
7. E. Bogdanov, V. Kolobov, A. Kudryavtsev, L. Tsendin, Proc. IEEE Conf. on Plasma Science, Alberta, Canada (2002) 2P12
8. K. Kwon, S. Kang, S. Park, H. Sung, D. Kim, J. Moon, J. Mat. Sci. Letters, v. 18 (1999) 1197
9. B. Wu, D. Chan, J. Microlith., Microfab., Microsyst., v.2, No. 3 (2003) 200
10. L. Elmonser, A. Rhallabi, M. Gaillard, J. P. Landesman, A. Talneau, F. Pommereau, N. Bouadma, J. Vac. Sci. Technol., A25 1 (2007) 126
11. S. Babin, K. Bay, S. Okulovsky, Proc. SPIE, v.6283, (2006) 62831R

LENS
(Lithography Enhancement Towards Nano Scale) a European Project to support Double Exposure and Double Patterning technology development

Pietro Cantu'[a], Livio Baldi[a], Paolo Piacentini[a], Joost Sytsma[b], Bertrand Le Gratiet[c], Stéphanie Gaugiran[d], Patrick Wong[e], Hiroyuki Miyashita[f], Luisa Rita Atzei[g], Xavier Buch[h], Dick Verkleij[i], Olivier Toublan[l], Francesc Perez-Murano[m], David Mecerreyes[n]

[a] Numonyx, Italy; [b] ASML, Netherlands; [c] STMicroelectronics, France; [d] CEA-LETI, France; [e] IMEC, Belgium; [f] Dai Nipponm Photomask Europe, Italy; [g] Lam Research srl, Italy; [h] JSR, Belgium; [i] FEI, Netherlands; [l] Mentor Graphics, France; [m] Centro Nacional de Microelectrónica; [n] CIDETEC, Spain

ABSTRACT

In 2009 a new European initiative on Double Patterning and Double Exposure lithography process development was started in the framework of the ENIAC Joint Undertaking. The project, named LENS (Lithography Enhancement Towards Nano Scale), involves twelve companies from five different European Countries (Italy, Netherlands, France, Belgium Spain) and includes: IC makers (Numonyx and STMicroelectronics), a group of equipment and materials companies (ASML, Lam Research srl, JSR, FEI), a mask maker (Dai Nippon Photomask Europe), an EDA company (Mentor Graphics) and four research and development institutes (CEA-Leti, IMEC, Centro Nacional de Microelectrónica, CIDETEC).

The LENS project aims to develop and integrate the overall infrastructure required to reach patterning resolutions required by 32nm and 22nm technology nodes through the double patterning and pitch doubling technologies on existing conventional immersion exposure tools, with the purpose to allow the timely development of 32nm and 22nm technology nodes for memories and logic devices, providing a safe alternative to EUV, Higher Refraction Index Fluids Immersion Lithography and maskless lithography, which appear to be still far from maturity.

The project will cover the whole lithography supply chain including design, masks, materials, exposure tools, process integration, metrology and its final objective is the demonstration of 22nm node patterning on available 1.35 NA immersion tools on high complexity mask set.

The paper aims at giving an overview of the LENS project, its activities and results to a broad audience of lithographers.

1. INTRODUCTION

As clearly shown in Figure 1, reporting ITRS 2007 Lithography options for 65nm node and beyond, while water immersion lithography has been widely accepted as patterning technology for the 45nm technology node, solutions for the patterning of 32nm and 22nm technology nodes are not clear yet.

EUV lithography is not available for industrial use, in spite of the impressive progresses registered till now, while multiple beam e beam lithography is still in development. Double patterning seems to be the only viable option to support the development of future process generations in a cost effective way and within the time limits dictated by ITRS roadmap.

Its main advantage consists in enabling the definition of structures beyond resolution capability of existing lithographic tools, without drastic changes in manufacturing infrastructures or huge investments.

Optical Microlithography XXIII, edited by Mircea V. Dusa, Will Conley, Proc. of SPIE Vol. 7640, 764022 · © 2010 SPIE · CCC code: 0277-786X/10/$18 · doi: 10.1117/12.846030

Two alternative approaches are possible, both based on existing immersion scanners:

- Double exposure, which implies two subsequent exposure steps, and the use of different combinations of hard masks or innovative resist materials and development process;

- Pitch doubling based on a single lithography exposure followed by the formation of spacers, by material deposition and etch-back, also in combination with CMP.

Both approaches are being actively investigated, but they are still far from maturity. Among the problems to be solved there are the control of mask to mask alignment, for double exposure, and the control of the thickness of deposited layers, of defects and of profiles, for pitch doubling. Common concerns are cost, size control and the partitioning of the design.

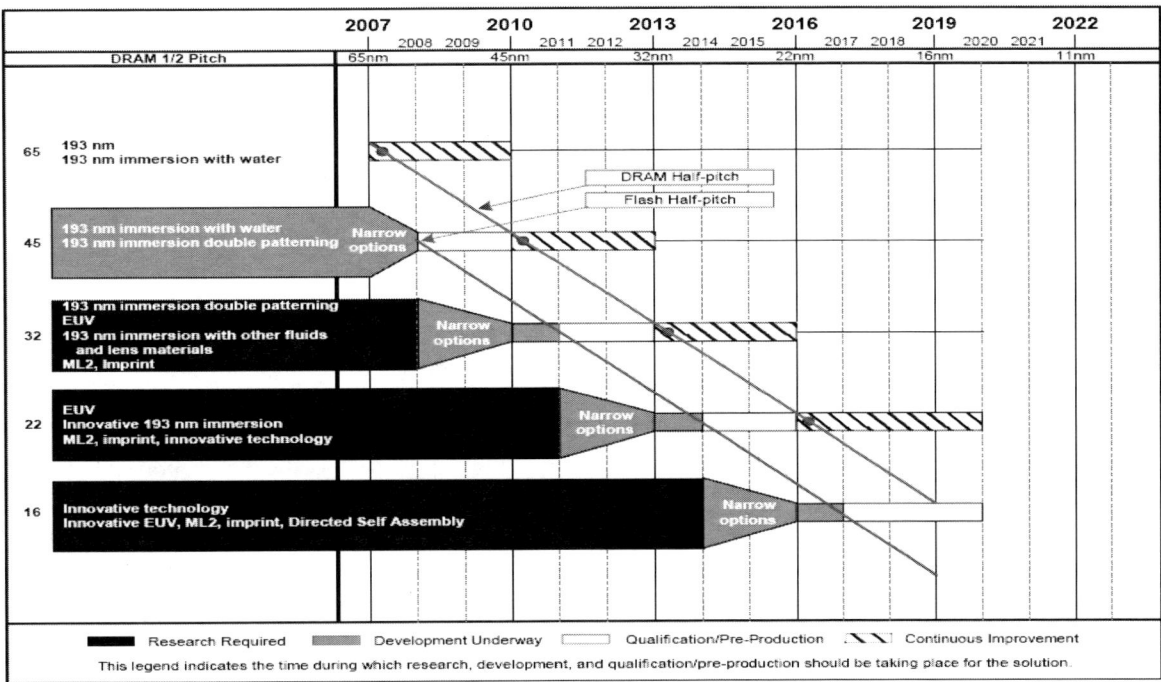

Figure 1: ITRS 2007 Lithography options for 65nm node and beyond

The pitch doubling approach is the more advanced, especially for memories, while double exposure could have a broader application, but will require improvement to equipment.

Although the introduction of double patterning technology will not require a complete change of the manufacturing infrastructure, several issues should be still addressed in order to have this technology mature enough to support industrial production, so research and innovation are required in the whole supply chain.

The LENS consortium includes all required competences to develop all elements of the supply chain required to bring double patterning to industrial maturity, in order to support 32nm and 22nm node mass production.

The 3 years target of LENS project is to build the complete manufacturing infrastructure that will enable the selection of the right technology and its introduction in an industrial environment for the 32nm node, and an assessment of the feasibility of the 22nm node.

2. PROJECT DESCRIPTION

2.1 Project Objectives

LENS project addresses Sub *Programme 8 - Equipment & Materials for Nanoelectronics* of the Annual Work Programme of ENIAC JTI and especially the Target Activity: *Lithography process for beyond 32nm manufacturing*, which requires:

"to support European leadership in lithography equipment and photomask technology for both optical and maskless technologies. An important element in the pre-commercial phase of these technologies is the development of standards, in order to define the interfaces that different suppliers have to use in order create a competitive and future proof platforms. Focus should be on the extension of immersion lithography with improved material and equipment, double exposure and metrology, and new resist and mask concepts."

Aim of the project is to develop and integrate the overall infrastructure necessary to reach patterning resolutions required by 32nm and 22nm technology nodes through the double patterning and pitch doubling technologies supported by conventional immersion exposure tools, with the purpose to allow the timely development of 32 and 22nm technology nodes for memories and logic devices, and to provide a safe alternative to EUV, Higher Refraction Index Fluids Immersion Lithography and maskless lithography, which appear to be still far from maturity.

The project will cover the whole lithography supply chain including design, masks, materials, exposure tools, process integration and metrology.

2.2 Background and motivation

Although water immersion lithography has been widely accepted as patterning technology down to 45nm half-pitch, there is no consensus on solutions for 32 and 22nm nodes yet: immersion lithography would require higher refraction index materials for liquid, but also innovative materials for lenses and resists. The development of lithography systems with numerical apertures higher then 1.35, maximum value achievable with current material and water immersion, is facing troubles and has been continuously delayed by most of the tool manufacturers.

EUV requires radical changes to process and mask technologies. Despite the recent progresses the availability of EUV technology on time cannot be given for granted and, even when ready, would be quite expensive. In this uncertain context double patterning enables going beyond resolution capability of existing lithographic tools without drastic changes in manufacturing infrastructures or huge investments.

Year of Production	2013	2016
DRAM half-pitch (nm) (contacted)	32	22
DRAM half-pitch (nm)	32	23
FLASH half-pitch (nm)	25	18
Contact in resist (nm)	35	25
Contact after etch (nm)	32	23
Overlay (3σ) (nm)	6.4	4.5
CD control (3σ) (nm)	2.6	1.9

Figure 2: ITRS 2007 Lithography requirements for 32 and 22nm nodes

The final objective of the project is the demonstration of 22nm node patterning on available 1.35 NA immersion tools on high complexity mask set. Resolution, alignment and line-width control should be in line with 2007 revision of ITRS requirements for the corresponding technology nodes, which are listed in Figure 2.

As already mentioned special attention will be put on the availability of whole supply chain, including design, masks, exposure tools, materials, process flow and metrology, necessary to support double patterning technologies introduction in a production environment. Logistics, yield, and manufacturability issues will be addressed and, finally, a detailed cost of ownership analysis, with respect to competing technologies, will be carried out.

2.3 Project consortium

As already mentioned LENS consortium includes all required competences to develop all the elements of the supply chain required to bring double patterning to industrial maturity, in order to support 32nm and 22nm node mass production. The project involves twelve companies from five different European Countries: Italy, Netherlands, France, Belgium Spain; and includes: IC makers acting in both memories and logic arenas, Numonyx and STMicroelectronics respectively, a group of equipment and materials companies ASML, Lam Research srl, JSR, FEI, a mask maker Dai Nippon Photomask Europe, an EDA company Mentor Graphics and four research and development institutes CEA-Leti, IMEC, Centro Nacional de Microelectrónica, CIDETEC.

2.4 Project structure

The LENS project work plan is organized into seven major work packages summarized in Figure 3: WP1 project management and coordination, WP2 design, simulation and masks, WP3 metrology tool optimization, WP4 exposure tool optimization and WP5 innovative material research, WP6 double exposure process integration, WP7 double patterning process integration.

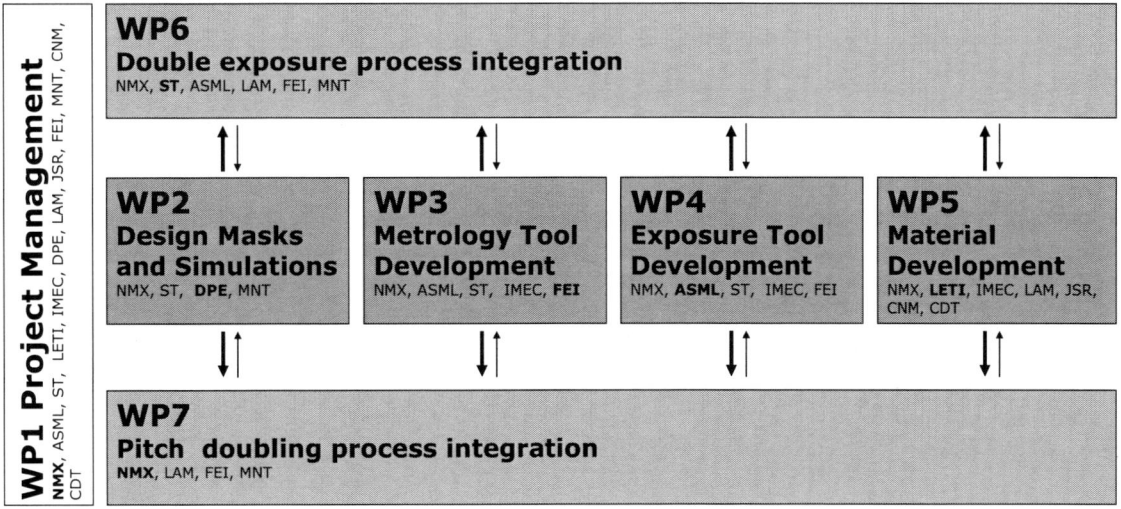

Figure 3: LENS project work packages structure

Four Work Packages (2-5) are clustering the different basic research fields: design, simulation and masks, exposure tool optimization, metrology tool optimization and innovative material research that need to be developed to provide the basic elements of the Double Patterning technology. They will run in parallel but interact with each other, where needed, and are later merged together into two integration work packages (WP 6 and 7) that will integrate their results into a complete process flow. The integration Work Packages will start from the beginning to make a preliminary selection among the different flavors proposed for Double Patterning technologies, and to restrict the spectrum of possibilities, by testing them in Pilot Lines. WP4 will proceed in parallel, assessing new options, which are considered as promising, but still not mature enough for industrial pilot lines. At the end of the second year an assessment will be performed and results transferred to industrial partners, while WP4 will focus on the process and cost control.

Two separate integration Work Packages (WP6 and WP7) are foreseen, because the two different approaches to Double Patterning, that is Double Exposure and Pitch Doubling, require quite different mixtures of basic technologies. The decision to have three transversal basic technology Work Packages has been taken to maintain a strong synergy among the two approaches, and avoid duplication of efforts. At present, there is no clear reason to prefer the one over the other, also because it seems that different applications (e.g. logic versus memory), and different critical layers would require different approaches. However it will be among the tasks of WP1 to monitor the advancement of the project, also taking into account the state-of-the-art worldwide, in order to trim the allocation of resources among the two approaches.

2.5 Work packages breakdown in task and timing

Each Work Package is further broken down into Tasks, for a clearer partitioning of responsibility and to facilitate the monitoring of the progress of activity, also through the link to specific deliverables. Figure 4 summarizes main task for each work package specifying respective timing.

		1st year				2nd year				3rd year			
		Q1	Q2	Q3	Q4	Q1	Q2	Q3	Q4	Q1	Q2	Q3	Q4
WP 1	Project Management												
T1.1	Administrative and financial co-ordination												
T1.2	Project management and technical co-ordination												
T1.3	Project dissemination strategy												
WP 2	Design, Simulation, Masks												
T2.1	Design split methods and data preparation												
T2.2	Mask specifications												
T2.3	Mask writing process												
WP 3	Metrology												
T3.1	Requirements, Specification and interfaces												
T3.2	Industrial grade metrologyTEM/STEM												
T3.3	Test and validation for process development												
WP 4	Exposure Tool												
T4.1	Specifications for overlay and CD budget												
T4.2	Litho tool improvement												
T4.3	Marker development												
T3.5	Double Patterning overlay assessment												
WP 5	Material Development												
T5.1	Materials for double exposure												
T5.2	Materials for pitch doubling												
T5.3	Alternative materials toward 22nm node												
T5.4	Budget and process control												
WP 6	Double Exposure Process												
T6.1	Scalability to 32nm												
T6.2	Validation on 32nm												
T6.3	DP model selection for 22nm												
T6.4	Material selection and integration												
T6.5	Double patterning for 22nm												
WP 7	Pitch Doubling Process												
T7.1	Preliminary screening												
T7.2	Material selection and integration												
T7.2	Morphological validation on 32nm node design												
T7.4	Process scalability down to 22nm node												

Figure 4: Work packages breakdown in tasks and their timing

3. WORK PACKAGES TECHNICAL CONTENT

3.1 Work Package 2: Design, Mask and simulation

Design, Mask and Simulation work package will focus on the design, simulation, fabrication, and qualification of lithographic masks for double patterning (DP). This includes the development and verification of layout splitting algorithms, the development of methods to detect, assess, and compensate for DP specific hot spots and design split conflicts, the definition and assessment of mask specifications for the 32nm and 22nm technology nodes, and the development of improved mask writing processes.

On the design side the activity is focused on the development of scripts (rule based and/or model based) for database splitting, defining rules for the separation of geometries between the masks and providing rule check routines to verify suppression of dense pitches. Figure 5 shows and example of splitting algorithm applied to contact hole layer. The definition of Optical Proximity Correction (OPC) to improve image quality and the following verification, including Sub-Resolution Assist Feature (SRAF), is part of this activity as well.

Mask specifications relevant to double exposure will be developed from the inputs of all partners, defining registration and overlay requirements as well as Critical Dimension Uniformity (CDU) requirements and their potential interactions. The final specification will result from the combination of circuit requirements with the outcome of the modelling of physical mask structure (e.g. optical properties of mask materials, corner rounding, footing effects and line edge roughness of absorber features and mask substrate. Further effects that could impact mask control, like electric field induced migration of chrome, will be studied and specifications for mask usage/storage environment defined.

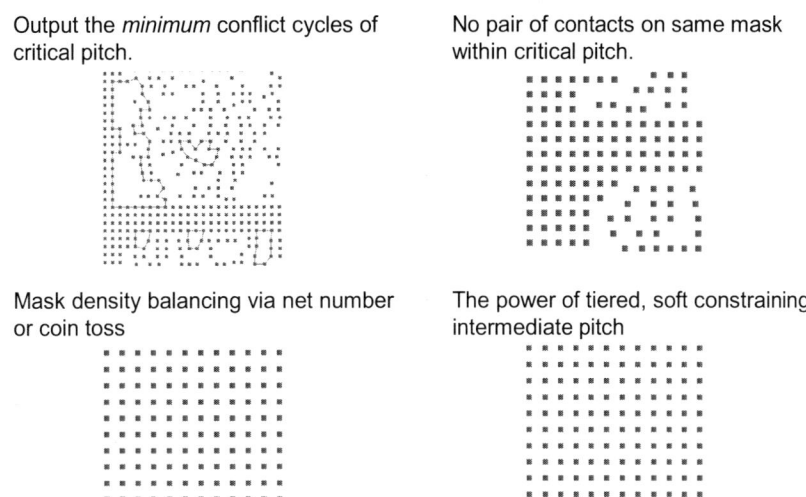

Output the *minimum* conflict cycles of critical pitch.

No pair of contacts on same mask within critical pitch.

Mask density balancing via net number or coin toss

The power of tiered, soft constraining intermediate pitch

Figure 5: Contact holes design splitting example

The overlay improvement methodology on a mask will be developed with a special focus on overlay management during e-Beam writing. Methods will be developed to improve the position accuracy, like writing optimization or specific material development, and the manufacturability of a mask for DP will be assessed for 32nm node and/or 22nm node.

3.2 Work Package 3: Metrology tool development

WP3 aims at the development of metrology tools based on Transmission Electron Microscopy (TEM/STEM) that can be used for specific problems in local areas as well as for calibration and validation of markers, and spacer profiles. TEM/STEM offers the ultimate in resolution in imaging and metrology. In addition, this WP will also develop and apply these metrology-based technologies, together with the semiconductor companies in the consortium, for specific customer process development applications. Especially metrology and characterization of delicate materials with a FAB-like productivity will be a focal point. In the beginning of the project, requirements and specifications for these tools will be defined together with (in-line) metrology partners and end users.

The definition of the requirements for the reference metrology will be done in cooperation with end-users from the semiconductor industry. An important part of the discussion will be the specification of metrology and calibration procedures, and the definition of interfaces needed for sample and data tracking.

Once specifications and requirements will be defined most of the activity of this work package will be focused on the development of new-generation "industrial grade" STEM/TEM with high throughput, reliability and is easy to use.

3.3 Work Package 4: Exposure tool development

In order to meet the demands of Double Patterning lithography in terms of CD control, process control strategies need to be developed on existing equipment for dose, focus and alignment parameters, and close loop control must be established over various processing equipments (scanner, track, etch). With double pattering, the placement accuracy of the two patterns will affect the overall CD in the resulting layer, making CD and alignment parameters interdependent. Control algorithms will be optimized for any combination of lithography parameters.

Existing metrology approaches will need to be improved for fast and reliable data collection. In-line metrology tools need to be calibrated against TEM cross-sections and profile analysis.

In the first part of the project, based on the characterization of the performances of existing equipment and on the analysis of theoretical limits, ASML with the help of IC makers, IMEC and LETI will define specifications for the overlay and CD uniformity budget for double patterning technology in 32nm and 22nm nodes.

ASML will develop new machine actuators to provide higher degrees of freedom in adjusting exposure parameters, so as to increase the potential for process fingerprint compensation. In fact, many exposure tool parameters that are fixed today have a large potential for fingerprint compensation in low-k1 conditions. ASML will investigate and develop innovative scanner control tools to enable the automated and concurrent adjustment of several scanner parameters dynamically. This need to be automated for two reasons: firstly, the parameter space is large enough that humans wouldn't be able to explore it manually; secondly, since the parameters are interdependent, the risk of setting parameter combinations that inadvertently lead to a yield crash is just too high when operating the system manually. There is a large potential improvement to be gained by using multiple parameters dynamically, in the context of fingerprint compensation for both simpler and more complex DPT strategies, like litho-etch-litho-etch (the latter being likely required for logic), as well as more complex patterns than simply lines and spaces. Since the automated scanner control tools will aim at the concurrent adjustment of multiple interdependent control parameters, multi-variant control algorithms will be favored, since those have been developed specifically for such type of applications.

When adopting double patterning approaches measures scatterometry markers, usually placed in the scribe lane between dice, must have characteristics similar to the ones of the actual device structures. Such markers need to be designed, their behavior modeled and their performance optimized, with the support of simulation tools for the modeling. Correlation with device structures needs to be proven in terms of size and profile: for devices where a repetitive unitary cell is not present, by TEM only; where a large repetitive array is available, by TEM and data collected therein by scatterometry.

Finally ASML will optimize the alignment strategy on DPT processes and generate proof material on new DPT dedicated exposure system including APC feedback loops to control CD and overlay performance within requirements. The demonstration of double patterning variance control capability, will be proven through on wafer qualification and characterization for 32 nm node.

In 2009, the "Overlay and CD specifications for 32/28 nm node" were derived and experimentally verified. In particular, for overlay, ASML has developed a test for the scanner, which shows the overlay performance for concrete and realistic wafer exposures. The total on-product-overlay is determined by the baseline tool performance, the customer usage, the processing effects and the control performance. The test addresses the overlay error as relevant for the customer usage: illumination dependence, lens and reticle heating, reticle writing and clamping error, process corrections, etc.

In the realm of new machine actuators for adjusting exposure settings, significant progress was made. Prototyping of several new actuators was successfully completed. The introduction of Flexray (See SPIE 7640-4[1], 7640-7[2], 7640-59[3]), ASML's proprietary illuminator with the on-board flexibility to create free form and very complex illumination pupil shapes, results in larger process windows and maintaining stability of these windows.

The control loops of the scanner are using actuators, the amount by which these actuators are adapted is based on lithography-specific metrology on product wafers. For this, the primary parameters to control are across-wafer and across-field focus and overlay. The required metrology accuracy and precision of focus and overlay were achieved through new developments in the area of marker design and measurement algorithms. This optimization is characteristic to scatterometry. In a typical flow, comparisons are shown between measured versus induced scanner disturbances on monitor wafers. Robustness against process variations and rework was also evaluated. These results are then used in conjunction with a number of innovative developments in the area of control algorithms. The potential stability improvements achieved will be shown in SPIE paper 7640-35[4].

3.4 Work Package 5: Material development

The main goal of the Work Package is to provide advanced lithographic materials allowing cost effective Double Patterning for the 32nm and 22nm nodes. These process modules are promising in terms of performances and cost, but still not mature enough for industrial adoption. A development stage in research centers is therefore needed before the transfer to industrial environment. Two alternative approaches to the conventional Litho-Etch-Litho-Etch (LELE) process will be investigated:

- Double exposure, which eliminates the intermediate etch step and replace it by an "in track" freezing treatment. This technique requires a critical control of mask-to-mask alignment and, at the moment, appears more suitable for logic devices.

- Pitch doubling, which is based on a single lithographic exposure followed by the formation of spacers, by material deposition and etch-back. This technique is mainly adapted to memories.

For what concern the double exposure the main goal is to make a proof of concept of the feasibility of new double patterning approaches that can reduce the cost of double patterning. Alternative materials and processes will be developed to eliminate the intermediate etch step of the conventional Litho-Etch-Litho-Etch double patterning process and replace it by a treatment of the first pattern in the track. Two kinds of freezing processes are currently evaluated: thermal freeze process and spin coating freezing[5]. Concerning the Thermal Cure Process, double patterning performances have been demonstrated both in dry and under high NA immersion lithography with 45nm dense lines patterning in dry lithography and 45nm contact holes patterning in immersion lithography. The Etch transfer feasibility has also been evaluated showing promising results in term of Lines Edge Roughness. This approach will be pushed to higher resolution in the next phase of the project.

 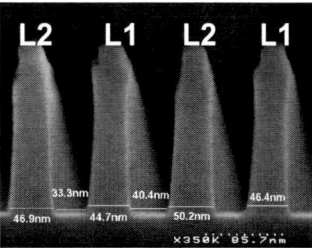

Figure 6: CD sem top view (on the left) and SEM Cross section (on the right) pictures of the 45nm lines after transfer into an underlayer

The overcoat process also reached a good level of maturity with a defectivity level similar to single patterning processes for 32nm dense lines and excellent CDU at 26nm HP. More complex patterns dedicated to logic devices are now under investigation.

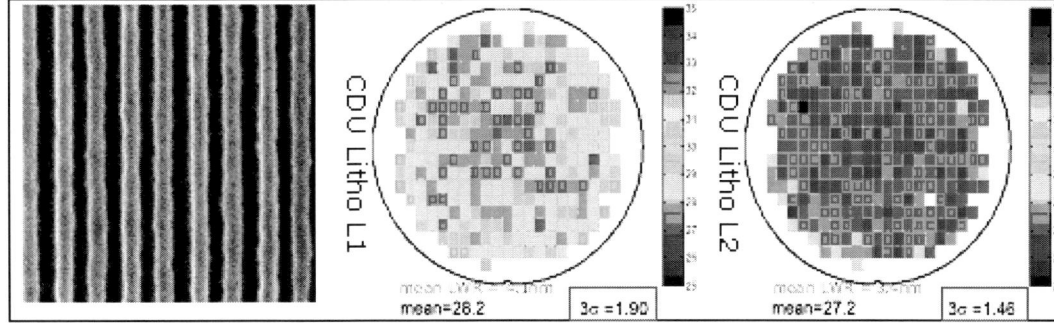

Figure 7: CDU evaluations at 26nm HP with the overcoat process

Block copolymers have recently received much attention for their self-assembly properties. Even if a fully self-assembled process is considered unlikely, block copolymers on structured substrates could be a solution to improve line edge roughness. Basic material properties, like scalability, adhesion to patterned substrates and pattern transfer capability will be assessed[6,7,8,9,10,11,12,13]. Block Co-Polymers synthesis and characterizations have been started and first self assembly has been obtained with Cylindrical PMMA domains perpendicular to the surface.

For what concern the Pitch Doubling, the activity will be focused in developing alternative materials and process flows either to improve the performances or to simplify the conventional spacers define double patterning scheme. Different approaches will be investigated as the use of spin on carbon (SOC) underlayers as an alternative to the conventional double patterning based on spacers; or the spacer deposition on resist patterns which will suppress one etching step and then will reduce the number of process steps and so its cost of ownership. A successful demonstration of 32nm PEALD spacers directly on resist pattern has already been made.

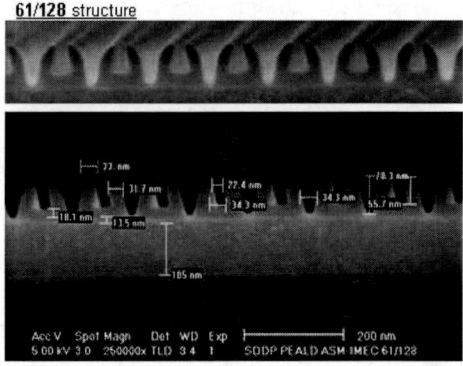

Figure 8: cross-section images of the 32 nm PEALD spacers after oxide etch back and resist + BARC strip

In the same way, the feasibility of 45nm HP PECVD spacers directly on hardened photo-resist patterns has been demonstrated. Process conditions have been optimized to improve the spacer sidewall verticality as well as the deposition conformity.

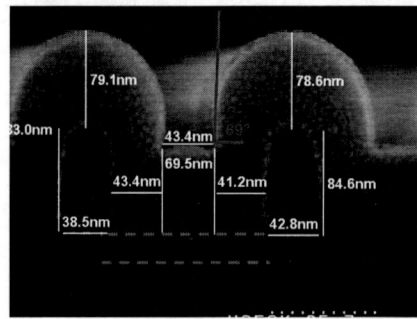

Figure 9: SEM cross section observations of CVD spacer deposition on hardened resist patterns

3.5 Work Package 6: Double exposure process integration

This Work Package will present and investigates the challenges linked to double patterning approach. It will start from requirements coming from already developed double patterning scheme and will outline the challenges that such integration will face when it is extended to 40/32nm node to build up, in a final stage, the strategy that should be used for 22nm node highlighting mandatory items that would make it a success. In 2009 Litho Freeze Litho Etch integration combined with Spin-On carbon/SiARC stack was developed and tested.

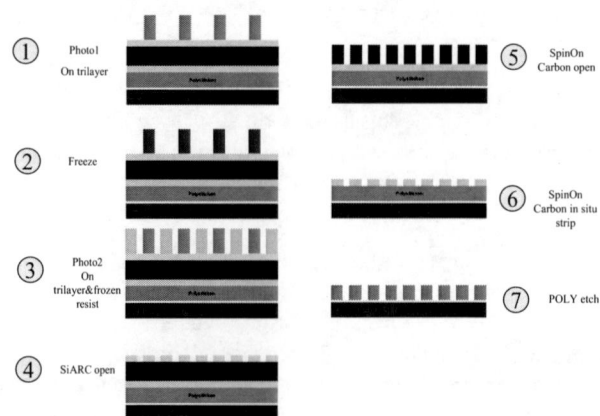

Figure 10: Litho-Freeze-Litho-Etch integration proposal

Figure 10 shows Litho-Freeze-Litho-Etch integration proposal; the main advantage of this scheme is that the resist is frozen by simple thermal curing and can support an overcoat by a second resist. Next figures, Figure 11, present pictures post Litho for Pitch 70nm and 64nm. We see that at pitch 64nm, resist 1 image start deteriorating, new formulation of photo resist 1 and resist 2 are in evaluation to improve this result.

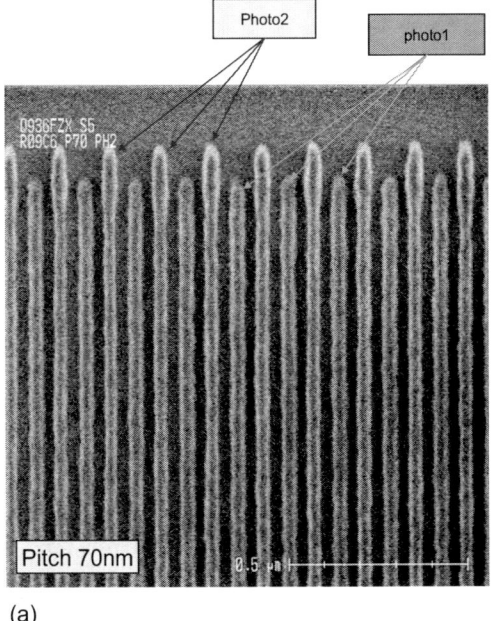

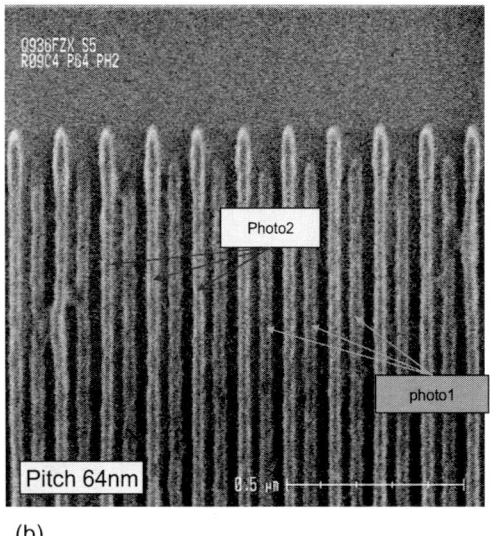

(a)   (b)

Figure 11: Post photolithography 1 and 2 pictures: pitch 70nm (a), pitch 64nm (b)

Some Etch tests have been performed at both 70nm and 64nm pitch; Figure 12 shows patterning results Post Hardmask Etch using Gate etch tool.

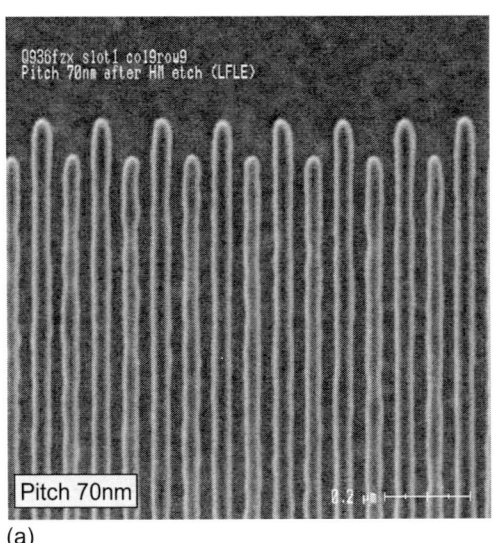

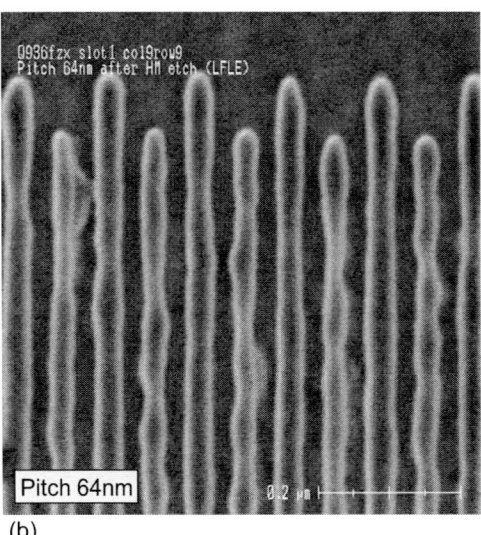

(a)   (b)

Figure 12: Post Hardmask Etch using Gate etch tool: pitch 70nm (a), pitch 64nm (b)

For a complete assessment of 32nm node double patterning and DoseMapper application refer to SPIE Paper 7638-9[14].

3.6 Work Package 7: Double exposure process integration

The work package will assess all the critical issues related to a pitch fragmentation process according to the specification of 32nm and 22nm node devices. Different integration approaches (for example Line-by-Line or Line-by-spaces) will be explored and reviewed. The work will focus on determination of the influence parameters on the morphologically shape and the interaction of the single processes inside the pitch fragmentation process. Different approaches could be used for different layers. Numonyx will develop an integrated pitch fragmentation process, which fulfils the requirements according to CD-Uniformity and geometric shape for the 32nm node. The control of the process-steps of the pitch fragmentation is fundamental. Therefore new Metrology methods will be developed and evaluated. Furthermore the process scalability and limits down to the 22nm node will be evaluated.

In a first phase the activity will be focused to identify the best approach for spacers' definition, basing on the state of the art and new technologies availability. The basic idea is to halve the pitch in a hard mask layer and then transfer the correct pattern to a α-carbon mask and subsequently to the substrate. Different materials will be evaluated for spacer formation (dielectric, photo-resist and polymer) and the process assessment in term of material choice and process flow will be performed. That includes the work to integrate all single processes into the pitch halving approach and exploration of the influence parameters inside the process chain of the pitch halving process and their impact on feature characteristics (CDU, Shape). Two main approaches have been evaluated in 2009, the first one is based on Nitride spacers, its process integration flow is shows in the upper part of Figure 13; the second one is based on low thermal oxide on resist, less mature but much more attractive in term of cost of ownership as requires a much more simplified process integration flow.

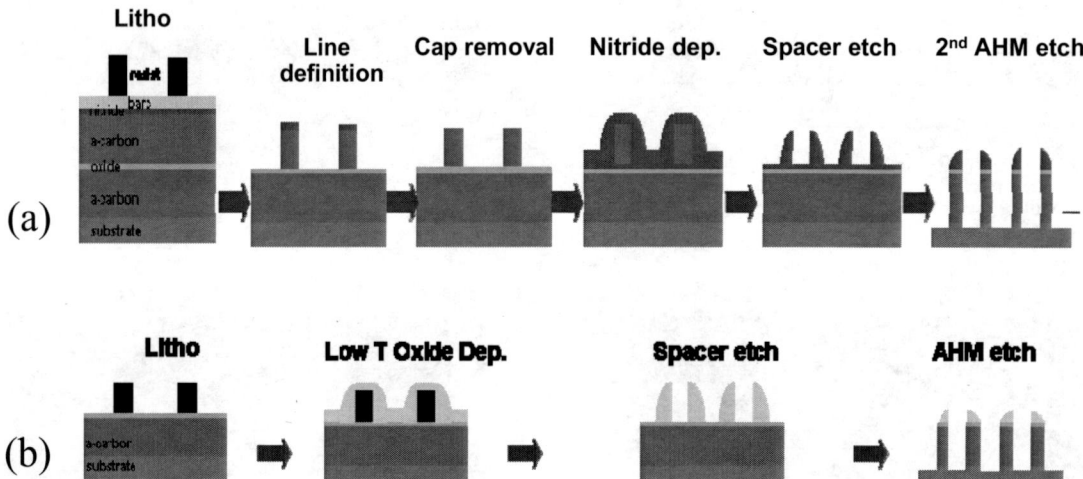

Figure 13: Nitride spacer flow (a) and low thermal oxide on resist flows (b)

In a second phase, innovative materials selected in the ambit of Material development work package will be assessed for maturity and compatibility with industrial pilot lines. Main focus points of the research will be the selection of the materials for sacrificial layers and for spacers, the investigation of the selectivity against RIE and cleaning procedures, the thickness control capability and the defect density.

The best approach selected during the screening phase on 32nm node design will be validated morphologically with respect to the required specification according to CD-Uniformity and feature-shape and then applied to high complexity design for final validation.

The process scalability down to 22nm node will be evaluated morphologically. Feasibility studies will be performed and the process limits investigated. In particular the patterning of the core and array of the cell has to be evaluated. New resist-based approaches developed identified in material development work package will be assessed.

4. CONCLUSION

LENS project was thought to bring double patterning and double exposure techniques developed in the ambit of research and development facilities to industrial maturity able to support high volume manufacturing of complex devices. LENS collected major players in different areas that are working together to improve tools, materials, procedures, methodologies and process integration with the aim to support the patterning required by 32nm and 22nm technology nodes. A single paper can only summarize LENS project content giving an overview of various activities; most of technical results will be presented in dedicated papers.

ACKNOWLEDGMENTS

LENS is a huge project with hundreds of people working behind the scenes on specific technical aspects; we'd like to thank all of them for their precious contribution whose content is going to be presented in several specific technical papers even at this conference. All the activities of LENS project are partially funded under ENIAC JU research contract. Sub *Programme 8 - Equipment & Materials for Nanoelectronics* and especially the Target Activity: ***Lithography process for beyond 32nm manufacturing.***

REFERENCES

[1] Jörg Zimmermann, Paul Gräupner, Dirk Hellweg, Dirk Juergens, Manfred Maul, Bernd Geh, Carl Zeiss SMT AG (Germany); Andre Engelen, Oscar Noordman, Melchior Mulder, ASML Netherlands B.V. (Netherlands), "Generation of arbitrary freeform source shapes using advanced illumination systems in high-NA immersion scanners", Proc. SPIE 7640-4, (2010).

[2] Joost P. M. Bekaert, Bart Laenens, Staf Verhaegen, Lieve Van Look, Darko Trivkovic, Geert Vandenberghe, IMEC (Belgium); Paul J. van Adrichem, Min-Chun Tsai, Orion Mouraille, Koen Schreel, Jo M. Finders, Mircea V. Dusa, ASML Netherlands B.V. (Netherlands); Robert J. Socha, Stanislas Baron, Kai Ning, Stephen D. Hsu, Brion Technologies, Inc. (United States); Jörg Zimmermann, Paul Gräupner, Christoph Hennerkes, Carl Zeiss SMT AG (Germany), "Freeform illumination sources: an experimental study of source-mask optimization for 22-nm SRAM cells", Proc. SPIE 7640-7, (2010).

[3] Drieenhuizen, Cas van Nuenen, Wilfred Endendijk, Jef Verbeeck, Wim Bouman, Robert Kazinczi, ASML Netherlands B.V. (Netherlands); Robert J. Socha, ASML (United States); Dirk Juergens, Jörg Zimmermann, Bastian Trauter, Carl Zeiss SMT AG (Germany); Geert Vandenberghe, Joost P. M. Bekaert, IMEC (Belgium), "Performance of FlexRay: a fully programmable illumination system for generation of freeform sources on high-NA immersion systems", Proc. SPIE 7640-59, (2010).

[4] Peter Vanoppen, Hugo Cramer, Thomas Theeuwes, Martin Ebert, Danu Satriasaputra, ASML Netherlands B.V. (Netherlands), "Lithographic scanner stability improvements through advanced metrology and control", Proc. SPIE 7640-35, (2010).

[5] Patrick Wong, Vincent Wiaux, Diziana Vangoidsenhoven, Mireille Maenhoudt, IMEC (Belgium), "Litho-process-litho for 2D 32-nm hp LOGIC and DRAM double patterning", Proc. SPIE 7640-17, (2010).

[6] Stoykovich, M. P.; Nealey, P. F. "Block Copolymers and Conventional Lithography". Materials Today 9, 20-29 (2006).

[7] Edwards, E. W.; Stoykovich, M. P.; Müller, M.; Solak, H. H.; de Pablo, J. J.; Nealey, P. F. "Mechanism and kinetics of ordering in diblock copolymer thin films on chemically nanopatterned substrates". Journal of Polymer Science Part B: Polymer Physics 43 (23), 3444-3459 (2005).

[8] Nealey, P.F.; Edwards, E. W.; Muller, M.; Stoykovich, M.P.; Solak, H.H.; de Pablo, J.J. "Self-assembling resists for nanolithography". IEEE (2005).

[9] Edwards, E. W.; Montague, M. F.; Solak, H. H.; Hawker C. J.; Nealey, P. F. "Precise Control over Molecular Dimensions of Block-Copolymer Domains Using the Interfacial Energy of Chemically Nanopatterned Substrates". Advanced Materials 16 (15), 1315-1319 (2004).

[10] Stoykovich, M.P.; Muller, M.; Kim, S. O.; Solak, H.H.; Edwards, E. W.; de Pablo, J.J.; Nealey, P.F. "Directed Assembly of Block Copolymer Blends into Nonregular Device-Oriented Structures". Sciencie 308, 1442 (2005).

[11] Kim, S. O.; Solak, H. H.; Stoykovich, M. P.; Ferrier, N. J.; de Pablo, J. J.; Nealey, P. F. "Epitaxial self-assembly of block copolymers on lithographically defined nanopatterned substrates". Nature 424, 411–414 (2003).

[12] Tang, C.; Lennon, E. M.; Fredrickson, G. H.; Kramer, E. J.; Hawker, C. J. "Evolution of Block Copolymer Lithography to Ordered Square Arrays". Science 322, 429 (2008).

[13] Liu, M. Y.; Huang, E.; Russell, T. P.; Hawker, C. "Controlling Polymer-Surface Interactions with Random Copolymer Brushes". Polymer Science 275, 1458 (1997).

[14] Bertrand Le Gratiet, Frank Sundermann, STMicroelectronics (France), "Improved CD control for 45-40 nm CMOS logic patterning: anticipation for 32-28 nm node", Proc. SPIE 7638-9, (2010).

Self-Aligned Double Patterning Process for 32/32nm Contact/Space and beyond using 193 Immersion Lithography

Bencherki Mebarki, Liyan Miao, Yongmei Chen, James Yu, Pokhui Blanco, James Makeeff, Jen Shu, Christoher Bencher, Mehul Naik, Christopher Sui Wing Ngai
Applied Material, 3225 Oakmead Village Drive, M/S 1220, P. O. BOX 58039,
Santa Clara, CA 95054
Phone: (408) 235-4433, Email: bencherki_mebarki@amat.com

Keywords: Contact, Patterning, Lithography, Self-Aligned, Spacer, Core, frequency doubling

ABSTRACT

State of the art production single print lithography for contact is limited to ~43-44nm half-pitch given the parameters in the classic photolithography resolution formula for contacts in 193 immersion tool ($k1 \geq 0.3$, NA = 1.35, and λ = 193nm). Single print lithography limitations can be overcome by (1) Process / Integration based techniques such as double-printing (DP), and spacer based self-aligned double patterning (SADP), (2) Non-standard printing techniques such as electron-beam (eBeam), extreme ultraviolet lithography (EUVL), nano-imprint Lithography (NIL). EUV tools are under development, while nano-imprint is a developmental tool only. Spacer based SADP for equal line/space is well documented as successful patterning technique for 3xnm and beyond. In this paper, we present an adaptation of self-aligned double patterning process to 2-D regular 32/32nm contact/space array. Using SADP process, we successfully achieved an equal contact/space of 32/32nm using 193 immersion lithography that is only capable of 43-44nm resolvable half-pitch contact printing. The key and unique innovation of this work is the use of a 2-D (x and y axis) pillar structure to achieve equal contact/space. Final result is a dense contact array of 32nm half-pitch in 2-D structure (x and y axis). This is achieved on simplified stack of Substrate / APF / Nitride.
Further transfer of this new contact pattern from nitride to the substrate (e.g., Oxide, APF, Poly, Si...) is possible. The technique is potentially extendible to 22/22nm contact/space and beyond.

1. INTRODUCTION

Shrinking integrated circuits (ICs) may result in improved performance, increased capacity and/or reduced cost. Each device shrink requires more sophisticated techniques to form the features. Photolithography is commonly used to pattern features on a substrate. Due to factors such as optics and light or radiation wavelength, however, photolithography techniques have a minimum pitch below which a particular photolithographic technique may not reliably form features. Thus, the minimum pitch of a photolithographic technique can limit feature size reduction. Self-aligned double patterning (SADP) also known as Spacer-based double patterning is one method for extending the capabilities of photolithographic techniques beyond their supposed minimum pitch. This technique is well documented and adopted for density doubling application. Hard mask spacers may be used as an etch mask for patterning the substrate and subsequently removed in a positive tone process flow. Alternatively, in a negative tone process, regions between the spacers is filled with material and used as a mask after the spacers are removed. With either method, the density of features is twice that of the photo-lithographically patterned features, the pitch of hard mask spacers or ribs is half the pitch of patterned features along one dimension. This method was not yet demonstrated to achieve density-doubling for Contact (Via) application in two-dimensions. In this paper, we present an innovative technique to achieve advanced 32nm half pitch Contact (Via) dense array in tow-dimensions (in both X and Y axis).

Optical Microlithography XXIII, edited by Mircea V. Dusa, Will Conley, Proc. of SPIE Vol. 7640,
764023 · © 2010 SPIE · CCC code: 0277-786X/10/$18 · doi: 10.1117/12.862583

2. METHODOLOGY / DATA

Self-Aligned Double patterning process for contact is described in six chronological main steps as follow:

(1) 90nm minimum resolvable pitch dense pillars are first printed using 193 immersion lithography process. As printed diagonal pitch is 128nm.

(2) The resist pattern is trimmed and transferred to nitride hard mask, then to sacrificial APF. After this step, APF pillars CD is ~ 10nm smaller than as printed resist.

(3) Remaining nitride hard mask on top of APF pillars is removed in hot phosphoric wet etch. APF is almost etch-insensitive to hot phosphoric and to other tested wet cleans. There is a slight ~2nm CD bias of APF pillar after nitride wet etch in hot phosphoric.

(4) Nitride spacer is deposited over APF pillars. Nitride spacers overlap in the small 90nm pitch direction, closing the gaps between neighboring APF pillars in 90nm minimum pitch. Meanwhile nitride spacers do not overlap in diagonal 128nm pitch, creating cavities (gaps) between four neighboring APF pillars. Nitride spacer thickness controls the dimension of created cavities.

(5) Once the cavities (contacts) are created, nitrite spacer is etched back to expose APF pillars. This step also clears nitride in created cavities bottoms. Nitride etch back is mostly directional etch process leading to a minimum nitride loss in cavities side walls.

(6) Once APF pillars are exposed, they are stripped away with dry ash process, leaving behind cavities (cores) surrounded by nitride spacers. The final result is a dense array of alternate contacts and spaces of 32 / 32nm.

The method starts by forming a sacrificial structural layer (interchangeably referred to as a core layer) on a substrate. The core layer is followed by a layer of photoresist (Fig. 1), but a hard mask layer may be formed between those two layers, especially when they are etched by similar mechanisms.

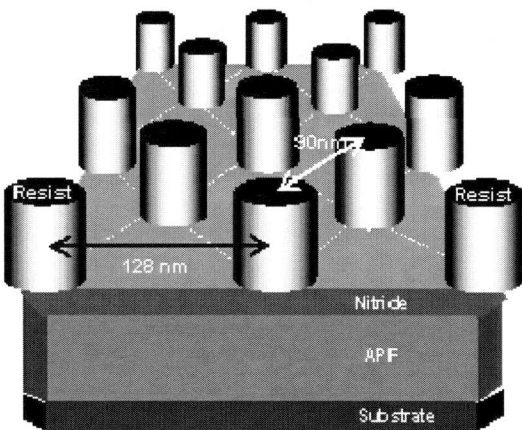

Fig.1: 90nm minimum pitch Pillar print at 193 immersion lithography. Pillar CD~50nm

The photoresist is trimmed then patterned to form a repetitive pattern of pillars with the same pitch in orthogonal directions at or near the resolution limit of a photolithography process (Fig. 2). The pillars may have a diameter of about one quarter of the pitch. The pattern is transferred into the hard mask layer (if used) and the patterned hard mask is then used to pattern the core layer to form a plurality of cores in a separate step where each core corresponds to one of the pillars formed in the photoresist layer and the cores are spaced according to a pitch that is substantially the same as the pitch between the pillars of photoresist.

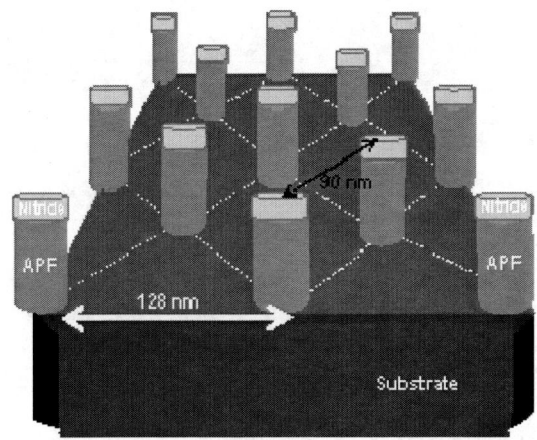

Fig. 2: Resist trim, Pattern transfer to Nitride hard Mask, and to sacrificial APF layer.

Following the remove of the remaining hard mask material, a substrate supports pillars of cores as shown in Fig. 3.

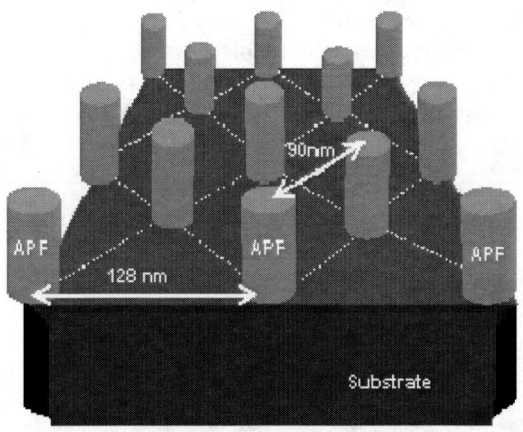

Fig 3: Nitride hard mask cap wet strip

A conformal layer of spacer material is then deposited on the cores and the exposed substrate. Mounds of spacer material accumulate above each core while dimples in the spacer material are created at the center between four adjacent cores (Fig. 4a / 4b). Four adjacent cores form a square with a side equal to the pitch of the core spacing. As the thickness of the conformal layer grows, the width of the dimples is reduced as the spacer material grows away from each adjacent core. As used herein, conformal coverage refers to providing a generally uniform layer of material on a surface in the same shape as the surface, i.e., the surface of the layer and the surface being covered are generally parallel. A person of skill in the art will recognize that the deposited material likely cannot be 100% conformal and thus the term "generally" allows for acceptable tolerances. In this work, conformal layer is a nitride film and is preferably deposited to a thickness such that the diameter of the dimples is near the diameter of the cores.

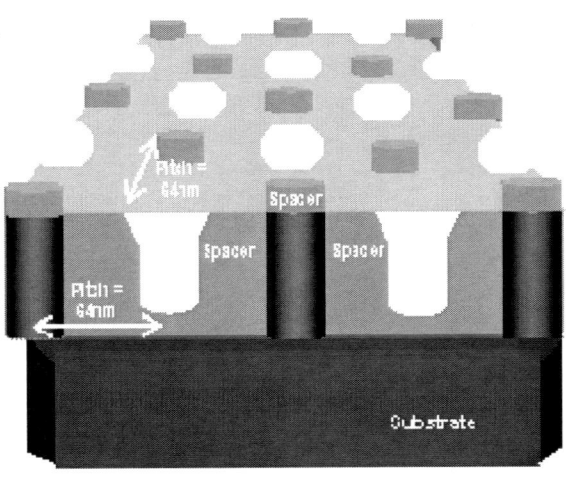

Fig. 4a: Nitride Spacer deposition   Fig. 4b: Top View SEM of Nitride Spacer deposition

The spacer material is selected, in part, so it will withstand the ashing process at least as well as the hydrogenated amorphous carbon in order to retain the patterned spacer layer. The spacer layer is made of silicon nitride. Alternatively, the spacer layer may be made of silicon oxide and the oxygen content in the spacer layer near the interface with the core material can cause ashing. This may compromise the integrity of cores and their ability to support subsequent spacer depositions or withstand an etch step. To avoid ashing the deposition of a silicon oxide spacer layer may begin with a silicon-rich interface and transition to the normal stoichiometry of silicon oxide thereafter. The silicon rich interface has less oxygen content and suppresses ashing of the cores.

The conformal layer is then anisotropically etched (a vertical etch) to expose substrate in the center of each square forming cavities and expose an upper surface of narrow cores (Fig. 5a / 5b).

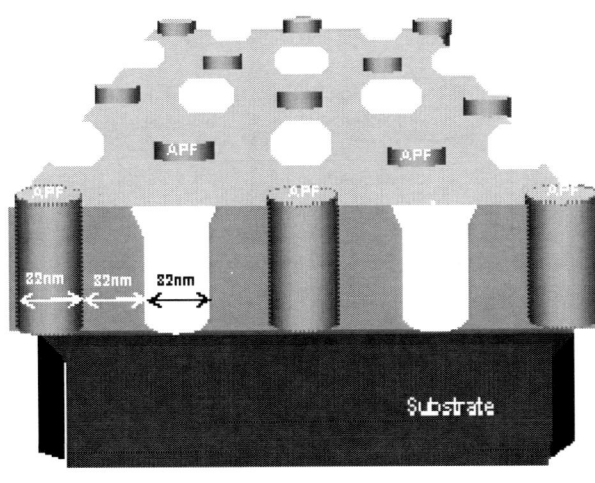
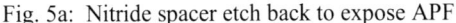

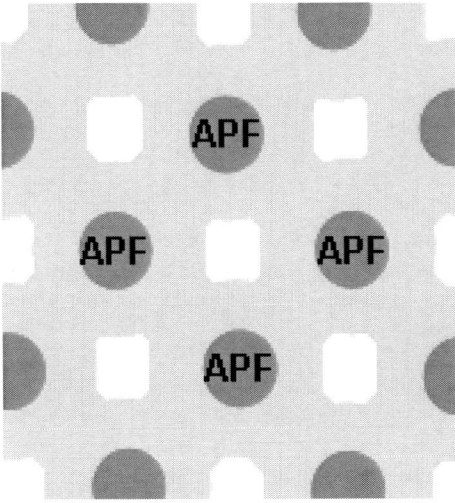

Fig. 5a: Nitride spacer etch back to expose APF   Fig. 5b: Top view: Core APF = Space = Gap = 32nm

3. RESULTS

Finally, exposed core material is removed in an etch step leaving a spacer layer, shown in figure 6, with cavities where the cores were removed and additional cavities near the center of the squares forming a dense contact array of 32nm half-pitch in 2-D structure (x and y axis). The two dimensional density of the cavities is twice the density of the cores. The spacer layer can be used as a hard mask to etch the double density pattern into the underlying substrate or the cavities may be filled with a metal (such as tungsten) to form vias. Another alternative is to continue the process, further increasing the density of the pattern.

The etch step used to remove the core material may exhibit a selectivity in etching the core material relative to the conformal material of the spacers (the spacer material). The core-etch involves ashing the amorphous carbon cores to attain the configurations of FIG. 3B and FIG. 3E. Ashing is often done by introducing O_2 or O_3 into a plasma above the substrate to oxidize the hydrogenated amorphous carbon and pumping the by-products away. The ashing process can also involve halogen-containing gases. During ashing hydrogenated amorphous carbon to form the initial pattern of FIG. 3A, it may be desirable to have a hard mask layer between photoresist and the hydrogenated amorphous carbon since they may have similar ashing rates. Core layers made from another material may not require the hard mask layer.

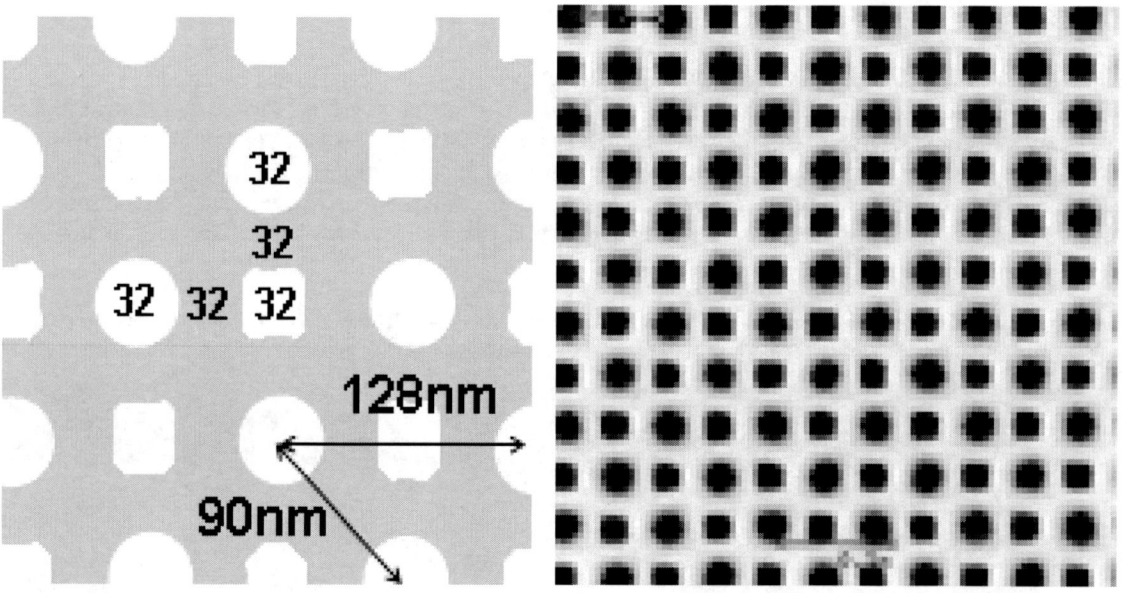

Fig. 6: Top view SEM after core APF strip. Final pitch is 64nm. Equal contact / space of 32 / 32nm

The support substrate may be an insulator or a semiconductor of a variety of doping concentrations and profiles and may, for example, be a semiconductor substrate of the type used in the manufacture of integrated circuits. There is also considerable flexibility in the choice of materials used for the cores and the spacer layer.

The process of forming cavities within the conformal layer of spacer material does not necessarily result in a simple cylindrical shape (Fig. 5a). In this view the cavities are shown to vary from a cylindrical outline. Near the top of each cavity, the diameter may be larger than at the bottom due to the conformal growth pattern of the layer of spacer material and possibly the etch process which creates the cavities. The shape of the opening near the top of the cavities may not be a circle. It is shown in figure 6 as a polygon which may result from the interrupted progression of the spacer material as it grows out from the neighboring cores. These cavities may become more cylindrical during processing steps involving, for example, increased substrate temperature or etching. Reference may be made herein to the diameter of dimples and cavities even when they are irregularly shaped. Such a reference simply means the diameter of a cylindrical approximation to the irregularly shape in a region close to the substrate (where the flaring associated with the conformal growth of the spacer layer has less influence).

4. CONCLUSIONS

Using a Self Aligned Double Patterning process, we successfully achieved an equal contact/space of 32/32nm using 193 immersion lithography that is only capable of 43-44nm resolvable half-pitch contact printing.

This work describes and successfully demonstrates a methodology of forming dense 32nm half pitch Contact array in 2-Dimension (X axis and Y axis) on a substrate having a reduced pitch in two dimensions as compared to what is possible using standard photolithography processing techniques. This work is an adaptation of a well-known frequency doubling self aligned double patterning for line / space. The key and unique innovation of this work is to start with a relaxed 2-D (x and y axis) pillar structure to achieve an aggressive 32nm equal contact/space. Further transfer of this advanced contact pattern into substrate (e.g., Oxide, APF, Poly, Si...) is possible. The technique is potentially extendible to 22/22nm contact/space and beyond.

Novel continuously shaped diffractive optical elements enable high efficiency beam shaping

Yuri V. Miklyaev[1], Waleri Imgrunt[1], Vladimir S. Pavelyev[2], Denis G. Kachalov[3], Tanja Bizjak[1],
Lutz Aschke[1], Vitalij N. Lissotschenko[1]
[1]LIMO Lissotschenko Mikrooptik GmbH
[2]Image Processing Systems Institute of Russian Academy of Sciences
[3]Samara State Aerospace University

ABSTRACT

LIMO's unique production technology is capable to manufacture free form surfaces on monolithic arrays larger than 250 mm with high precision and reproducibility. Different kinds of intensity distributions with best uniformities or customized profiles have been achieved by using LIMO's refractive optical elements. Recently LIMO pushed the limits of this lens production technology and was able to manufacture first diffractive optical elements (DOEs) based on continuous relief's profile.

Beside for the illumination devices in lithography, DOEs find wide use in optical devices for other technological applications, such as optical communications, laser technologies and data processing. Classic lithographic technologies lead to quantized (step-like) profiles of diffractive micro-reliefs, which cause a decrease of DOE's diffractive efficiency. The newest development of LIMO's microlens fabrication technology allows us to make a step from free programmable microlens profiles to diffractive optical elements with high efficiency. Our first results of this approach are demonstrated in this paper. Diffractive beam splitters with continuous profile are fabricated and investigated. The results of profile measurements and intensity distribution of the diffractive beam splitters are given. The comparison between theoretical simulations and experimental results shows very good correlation.

Keywords: diffractive optical elements (DOE), diffractive beam splitters, diffractive continuous microrelief, free form surfaces, beam shaping, homogeneity, microoptics array.

1. INTRODUCTION

Optical lithography with its 193 nm technology is pushed to reach and shift its limits even further. There is strong demand on innovations in illumination part of exposure tools. Current illumination systems consisting of diffractive and refractive optical elements offer numerous benefits such as optimized laser beam shape with extremely high homogeneity and high numerical aperture enabling high efficiency and improved throughput. The illumination should show almost no inhomogeneity and most likely some specially defined far field distribution.

LIMO's technology for manufacture of free form micro-lenses and arrays has been developed and continuously improved for over 18 years [1-5]. These free form micro-optical cylindrical lenses perform variety of beam shaping, enabling special intensity distribution, as well as highly homogeneous illumination with inhomogeneity less then 1% (peak to valley). Micro-optics arrays consisting of cylindrical lens arrays provide separate manipulation of the light distribution in x and y axis. With fill factors close to 100% these optics are extremely efficient.

Optical Microlithography XXIII, edited by Mircea V. Dusa, Will Conley, Proc. of SPIE Vol. 7640,
764024 · © 2010 SPIE · CCC code: 0277-786X/10/$18 · doi: 10.1117/12.846573

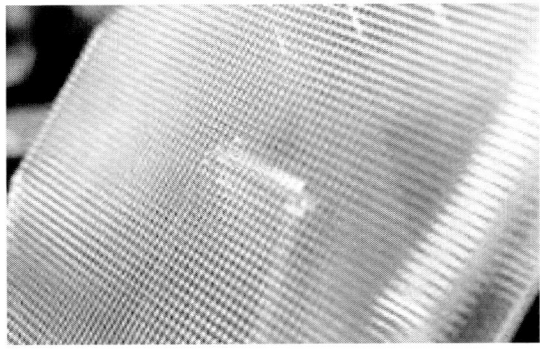 

Figure 1. Typical microlens array (left) produced by LIMO's wafer based technology and corresponding metrology for optical qualification of LIMO's optical solutions (right)

Recently this technology was improved further in order to produce high efficiency free form diffractive micro-optics. These diffractive micro-reliefs are designed to function as diffractive beam splitters. Diffractive beam splitters (multi-order diffractive gratings) are diffraction structure with periodic phase micro-relief and are used to form 1D or 2D set of beams with definite intensity ratio between them. Diffractive beam splitters form one of the most important classes of DOEs. Such a beam splitters have applications in all laser based technologies which need to precisely multiply a number of the incoming laser beams. Examples can be found in laser and illumination part of exposure tools, in optical fiber connectors, optical communications and optical signal processing, optical image multiplications, spectroscopy and technological systems [6].

Owing to the fact that lithographic technology dominates in DOE fabrication techniques, most part of diffractive beam splitters has quantized stepped micro-relief profile. There are many works devoted to design calculation methods of classic quantized micro-relief of multi-order diffractions gratings [7-9]. This profile quantization is a consequence of technological restrictions and it leads to losses in energy efficiency of such a device. Diffractive beam splitters with continuous micro-relief profiles can offer much higher efficiency (>93%). Due to their free form aspheric surface the manufacture of these continuous DOEs is more sophisticated and complicated compared to the quantized ones. Thus the continuously profiled diffractive micro-optics are used less frequently and are still not well known in optical communities [10].

This paper presents first continuous diffractive micro-optics that are manufactured with our production technology. They are designed for beam splitting at 1064 nm and manufactured in BK-7 glass as described in section 2. Experimental results and comparisons to the theoretical values for two different beam splitters are presented in sections 3 and 4. Finally a short conclusion is given.

2. DIFFRACTIVE MICRO-RELIEF PRODUCTION WITH CONTINUOUS PROFILE

In order to get best performance of the beam shaping optics the theoretically optimized surfaces have to be transformed onto a real substrate with minimum deviations between the theoretical and manufactured structures. Therefore, LIMO has developed and continuously improved its unique production technology for the manufacture of high precision asphere single lenses and arrays, where every single lens can be individually shaped. It is based on computer-aided design for the cost-effective and reproducible manufacture of free programmable lens surfaces on wafer basis. According to this technology no etching processes are involved at all. This allows the processing of a broad variety of materials, such as glasses, semiconductors and crystals, e.g. fused silica, BK7, CaF_2 for DUV-NIR applications and Si, Ge and ZnSe for CO_2-lasers in the FIR. Wafer sizes of even more than 200x200 mm² with surface accuracies on the order of 10-100 nm can be produced. The lens apertures cover a range from a few 10 µm to several mm with relative focal length variations well below 1 %. Continuous quality control of the whole wafer is integrated into the production flow by a surface profilometer where the surface of every single lens is measured.

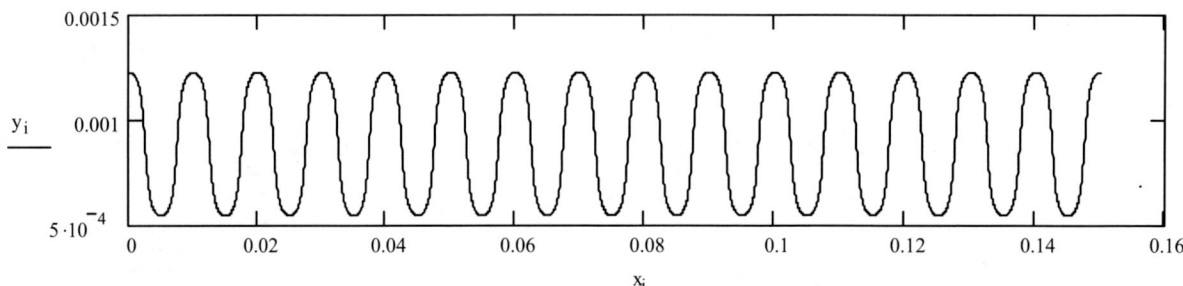

Figure 2. Surface profile of LIMO's continuously shaped diffractive optical element (x and y-axis in mm)

Analysis of microlens profile that are produced by LIMO [3, 4] shows, that this technology can be applied for fabrication of continuous periodic microrelief with pitch of 10 μm or more. This work presents diffractive beam splitters with structure periods of 10 μm and 20 μm. These first continuous diffractive structures were designed as beam splitter at 1064 nm and were manufactured in BK-7 glass. Example for such a continuous structure with pitch of 10 μm is given in figure 2. For continuous profile calculation of such a DOE the calculation procedure based on a beam splitting by phase gratings method is applied [11]. Diffractive beam splitters (multi-order diffractive gratings) are diffraction structure with periodic phase micro-relief and are used to form 1D or 2D set of beams with definite intensity ratio between them. Two different optical devices were made, one with the splitting ratio of 1:3 and one with 1:5 ratio. The experimental results of both beam splitters are demonstrated in the next two sections.

3. EXPERIMENTAL RESULTS OF 1:3 BEAM SPLITTER

Profile of 1:3 ratio beam splitter is described with the following form:

$$h(\mathrm{x}) = \frac{\lambda\varphi(\mathrm{x})}{2\pi(\mathrm{n\text{-}}1)}, \tag{1}$$

where λ is wavelength (in our case λ=1064nm), n is refractive index of BK-7 glass (n=1.507), $\varphi(x)$ is phase function, that was calculated by procedure based on beam splitting by phase gratings method [11]. The calculated and measured profiles are shown in figure 3. Surface profile of the beam splitter was measured with the AFM SolverPro.

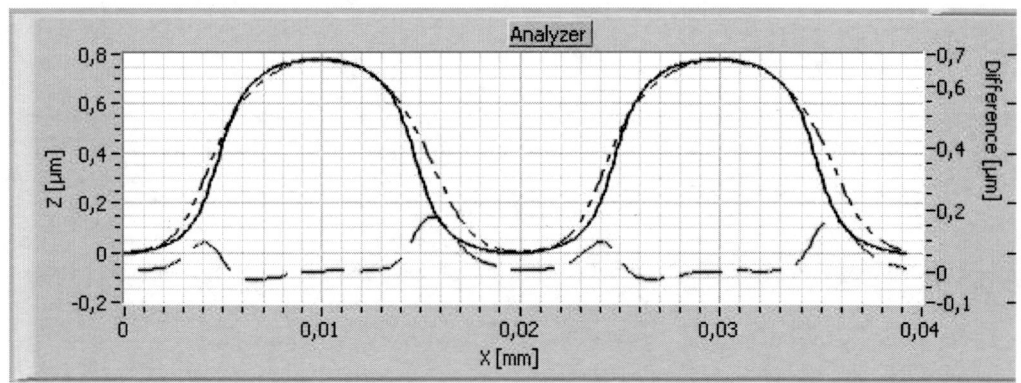

Figure 3. Profile of micro-relief of 1:3 ratio beam splitter with pitch of 20 μm. Solid curve is theoretical profile, dot-dashed curve is measured one and the dashed curve is deviation of fabricated profile from theoretical one.

Figure 4 presents the simulation results of light propagation through theoretical and measured diffractive beam splitters.

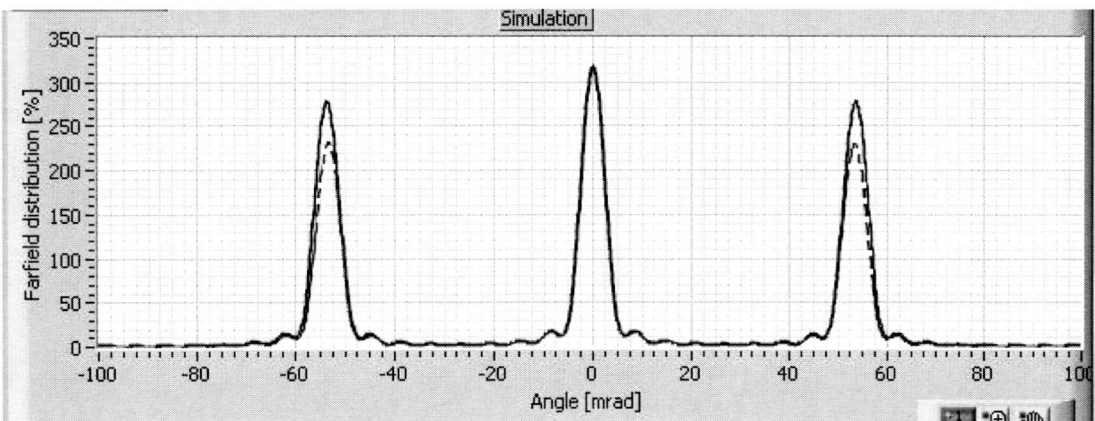

Figure 4. Simulated intensity distribution in far field of the beam splitter 1:3. Solid curve (higher peaks) corresponds to simulation of theoretical profile and the dashed one corresponds to simulation of measured profile.

Optical function of obtained beam splitters was tested with a metrology system which scheme is shown on figure 5.

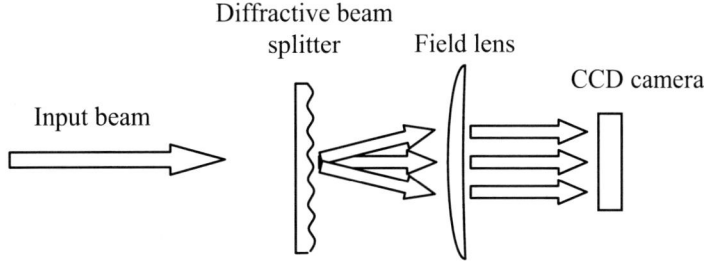

Figure 5. Scheme of the metrology set-up for the qualification of beam splitters.

The results of intensity distribution measurements are shown in figure 6. There is very low level of background and a good correspondence with the simulation of measured profile. This is a good confirmation for our optical simulation based on surface data.

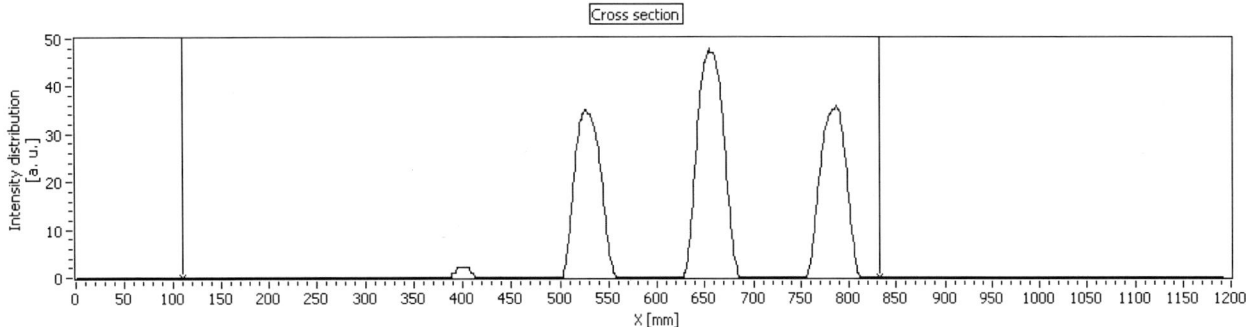

Figure 6. Cross section of intensity distribution in far field of the beam splitter 1:3, illuminated by multimode Nd:YAG laser.

Thanks to analysis of results shown in this chapter (figures 3, 4 and 6) it is possible to develop and improve the structure technology to get high and precise quality of the diffractive micro-optics surface. Further a beam splitter with even more sophisticated form and finer structure is achieved as described in next section.

4. EXPERIMENTAL RESULTS OF 1:5 BEAM SPLITTER

The next step was to manufacture and qualify a diffractive beam splitter with 1:5 splitting ratio. The designed and measured profile of this optical element is shown in figure 7.

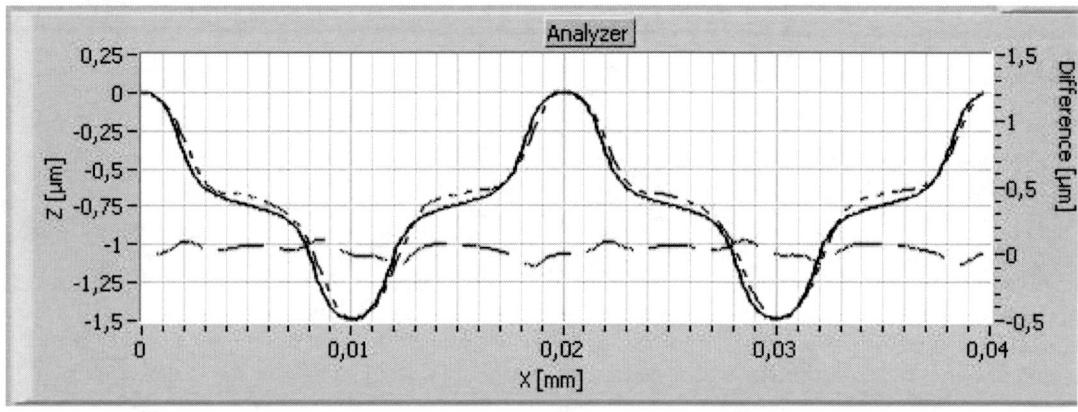

Figure 7. Micro-relief of 1:5 beam splitter with structure period of 20 μm. Solid curve represents theoretical profile, dot-dashed curve a measured one and the dashed curve shows a deviation of fabricated profile from theoretical one.

Functionality of 1:5 ratio beam splitter was tested on the same setup, shown in figure 5. Results of first optical measurements are shown in figure 8.

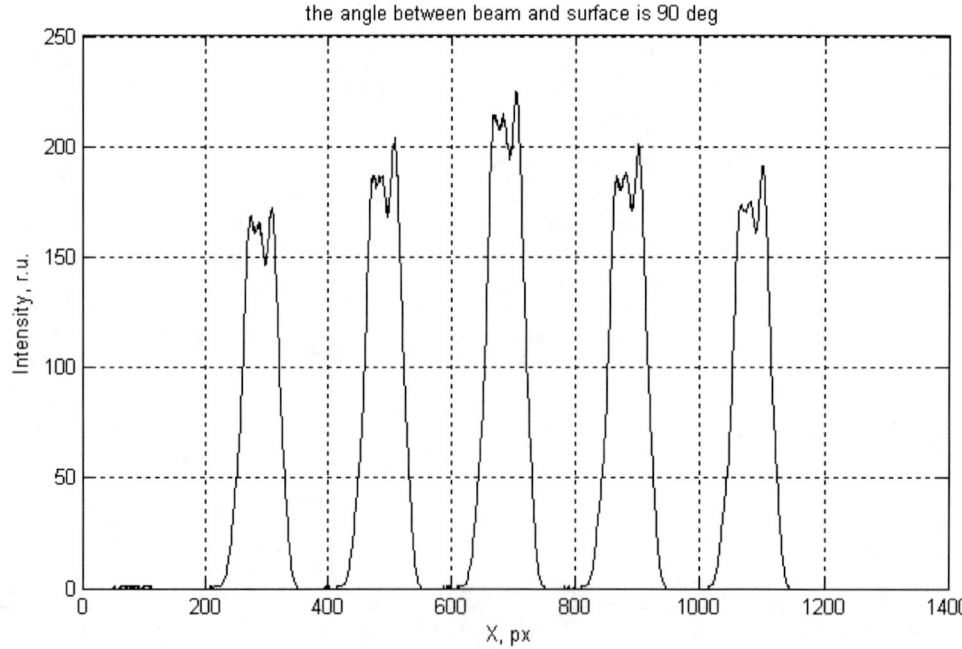

Figure 8. Cross section of intensity distribution in far field of 1:5 beam splitter, illuminated by multimode Nd:YAG laser.

There is some inhomogeneity in intensity distribution between different diffractive orders (figure 8). This inhomogeneity partly results from production technology due to the more complicated shape of the manufactured micro-structure (figure 7), such as lower sag and strong aspheric profile. This inhomogeneity can be compensated by tilting of beam splitter. By tilt of 15° a much better intensity distribution is achieved as shown in figure 9.

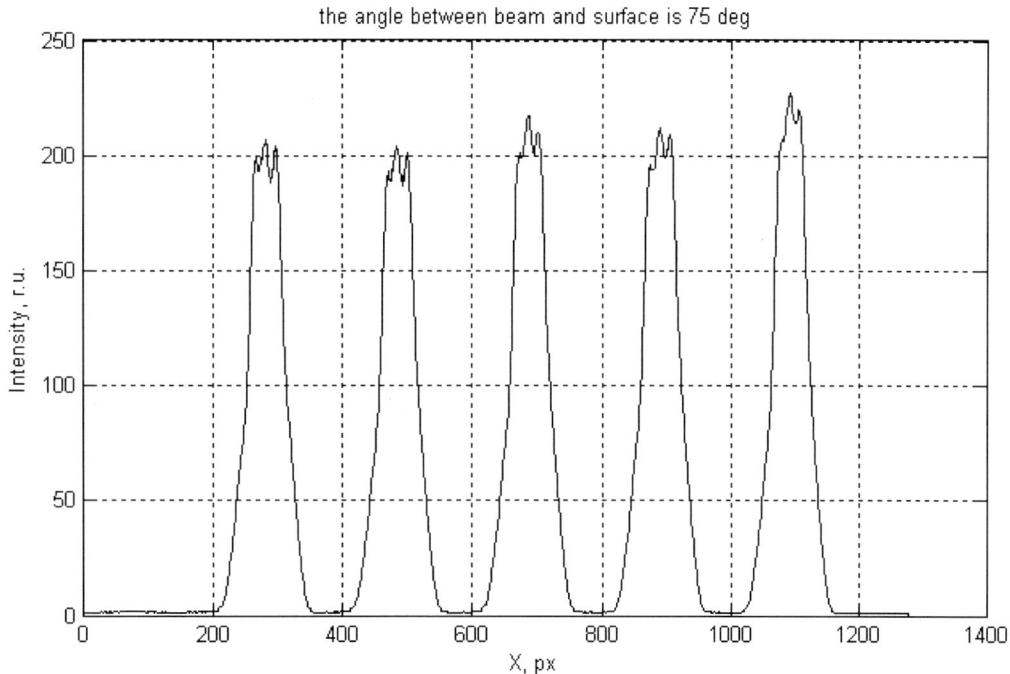

Figure 9. Intensity distribution in far field of 15° tilted beam splitter of 1:5 ratio.

CONCLUSION

LIMO unique production technology is developed and improved to manufacture any kind of free form micro-structure surfaces, not only refractive but also diffractive ones. Thanks to our production novel aperiodic continuously profiled diffractive micro-optical elements can be achieved. Due to their continuous profiles they do not suffer from light scattering on profile steps as the classic quantized DOEs do. Thus high energy efficiency can be obtained. The fabrication and qualification of such a continuous micro-relief in a multi-order diffraction gratings working as a beam splitter is described in this paper. Classic iterative methods for designing DOEs can not be applied for these novel continuous micro-structures. The method [11] of profile design based on beam splitting by phase gratings is used. Further progress in application of this technology for diffractive optical elements manufacturing will require new methods of microrelief design and optimization.

REFERENCES

[1] Bizjak, T., Mitra, T., Aschke, L., "Novel high-throughput micro-optical beam shapers reduce the complexity of macro-optics in hyper-NA illumination systems", *Proc. SPIE 6520* (2007)
[2] Bizjak, T., Mitra, T., Hauschild, D., Aschke, L., "Novel refractive optics enable multipole off-axis illumination", Proc. SPIE 6924, (2008)
[3] Homburg, O., Hauschild, D., Aschke, L., Lissotschenko, V., „Refractive beam shaping: From the solution of the Maxwell equations to products and its applications in laser material processing", Proc. SPIE (2008)

[4] Bizjak, T., Mitra T., Homburg O., Hauschild, D. and Aschke, L., "Innovative refractive optics for exposure tool illumination", Microlithography World, Vol. 17-2 (2008)

[5] Bizjak, T., Mitra, T., Hauschild D., Aschke, L., "Inspection and metrology tools benefit from free-form refractive micro-lens and micro-lens arrays", Proc. SPIE 7272 (2009)

[6] Erwin G. Loewen, Evgeny Popov, Diffraction Gratings and Applications; Marcel Dekker Inc, 601 (1997)

[7] Doskolovich, L.L., Kotlyar, V.V., Soifer, V.A. Chapter 2. "Iterative methods for designing DOEs". In [*Methods for Computer Design of Diffractive Optical Elements*]; Soifer, V.A.; Ed.; John Wiley & Sons, Inc.: New-York, US; 55-158 (2002)

[8] Soifer, V.A., Kotlyar, V.V., Doscolovich, L.L., [*Iterative Methods for Diffractive Optical Elements Computation*]; Taylor & Francis Ltd.: London, (1997).

[9] Wyrowski, F., "Diffractive optical elements: iterative calculation of quantized, blazed phase structures", *J. Opt. Soc. Am. A*, v. 7, no. 6, 961-963 (1990)

[10] Golovashkin, D.L., Kazanskiy, N.L., Soifer, V.A., Pavelyev, V.S., Solovyev, V.S., Usplenyev, G.V., Volkov, A.V., Chapter 4. "Technology of DOE fabrication". In [*Methods for Computer Design of Diffractive Optical Elements]; Soifer, V.A.; Ed.; John Wiley & Sons, Inc.: New-York, US, 267-345 (2002).

[11] Romero, L.A., Dickey, F. M., "Theory of optimal beam splitting by phase gratings. I. One-dimensional gratings", *JOSA A*, Vol. 24, Issue 8, 2280-2295.

Advances in DOE modeling and optical performance for SMO applications

James Carriere, Jared Stack, John Childers, Kevin Welch, Marc D. Himel

Tessera, 9815 David Taylor Drive, Charlotte, NC USA 28262-2369

ABSTRACT

The introduction of source mask optimization (SMO) to the design process addresses an urgent need for the 32nm node and beyond as alternative lithography approaches continue to push out. To take full advantage of SMO routines, an understanding of the characteristic properties of diffractive optical elements (DOEs) is required. Greater flexibility in the DOE output is needed to optimize lithographic process windows. In addition, new and tighter constraints on the DOEs used for off-axis illumination (OAI) are being introduced to precisely predict, control and reduce the effects of pole imbalance and stray light on the CD budget. We present recent advancements in the modeling and optical performance of these DOEs.

Keywords: DOE, Diffractive Optical Element, Diffuser, Off-axis illumination, Source Mask Optimization, Tunable

1. INTRODUCTION

As alternative lithography approaches such as EUV and nanoimprint lithography continue to experience delays in their ability to meet the needs of the industry at the 32nm node and beyond, the requirements placed upon conventional optical lithography are increasing in complexity. The diffractive optical elements (DOEs) used for off-axis illumination (OAI) have additional capabilities that have been relatively untapped for addressing these requirements. SMO has been shown to significantly increase process windows for these critical nodes[1-3]. To take full advantage of SMO, it is important to understand the capabilities DOEs can provide with regard to pupil resolution and grey tone intensity level control for freeform OAI solutions[4] (see Figure 1).

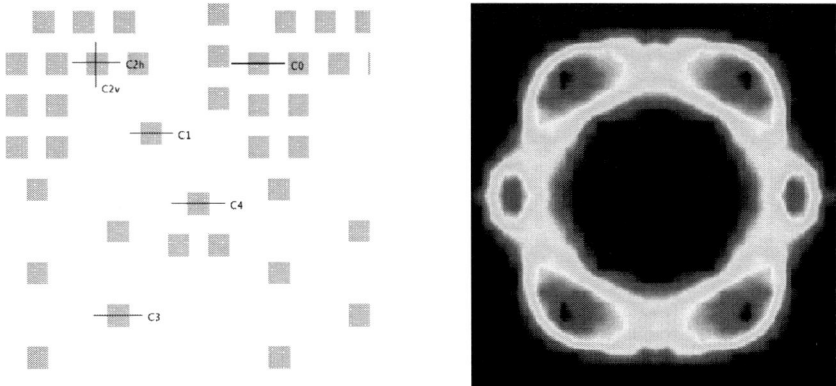

Figure 1. A typical 32nm contact hole pattern and the corresponding optimum freeform illumination source.

In addition, many physical properties of the illumination that were assumed to have negligible impact on performance are becoming significant contributors to the CD budget. Factors like pole imbalance and stray light can be unique to each custom design. Understanding, controlling and minimizing these factors is vital to achieving optimal solutions for the most challenging designs.

To come up with a truly optimized SMO design, an understanding of how the physics and manufacturing tolerances of DOEs can change the illumination pupil from the ideal design is required[5]. It is possible to incorporate these factors and models up front in the iterative SMO design algorithms to find the most robust design. The result is a much needed improvement in the process window of critical layers as well as a reduction of OPC rework required during process

Optical Microlithography XXIII, edited by Mircea V. Dusa, Will Conley, Proc. of SPIE Vol. 7640,
764025 · © 2010 SPIE · CCC code: 0277-786X/10/$18 · doi: 10.1117/12.846619

development thus minimizing development cycle timelines. To aid in these efforts, a comprehensive analysis of DOE constraints and capabilities, including both fundamental design aspects and manufacturing tolerances; as well as recent advancements in the modeling and optical performance of DOEs for OAI is presented.

2. DOE PERFORMANCE PREDICTION MODEL

DOEs are designed to project a specified target pupil to the aerial image plane. For standard deep UV lithographic applications, our target pupil resolutions are ~200x200 pixels, which exceed the resolution limits of conventional scanner illumination systems. Freeform pupils used in SMO applications often require grey toned pixels, which add an extra degree of complexity to the design. Our algorithms can simultaneously optimize 256 grey levels for every pixel in the target pupil, providing a total of 200x200x256 = ~10^7 degrees of freedom in the target pupil.

But to properly define a target pupil, it is also important to suppress illumination where it does not belong. Stray light from diffractive structures is often concentrated around the poles of the target pupil and therefore increases the effective size of each pole. While it isn't possible to completely remove stray light, an accurate prediction model of the stray light performance can be incorporated into the SMO design process to account for the stray light bias of the poles and minimize any detrimental effects it may have. The key to this process is the accuracy of the prediction model algorithm.

Many companies currently provide software to the lithography industry that predict the process performance of a scanner with a given pupil illumination profile. New SMO software packages have also been developed to simultaneously optimize the mask and pupil illumination but assume ideal performance with no effects from DOE stray light. A proper performance prediction model must take into account all elements of the design and manufacturing process. Tessera's 16+ years of experience supplying DOEs to the lithography industry has been leveraged to develop the most accurate DOE performance prediction model available[6]. Figure 2 shows the output of this model for the SMO design in Figure 1.

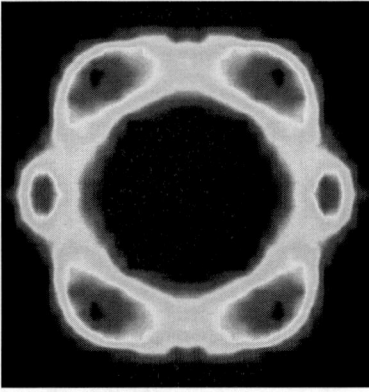 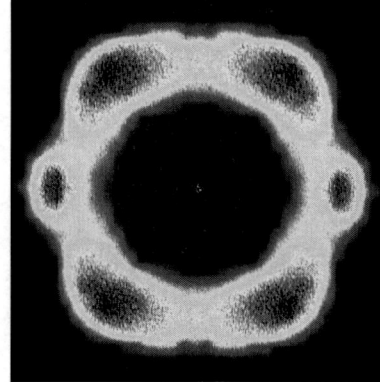

Figure 2. Validation of the performance prediction model for the SMO design of Figure 1. The left image shows the output of the prediction model, the right image shows the measured performance of the manufactured part.

It is easy to see that the tested performance matches the predicted performance within the target region. The speckle apparent in the test data image is due to the high coherence properties of the laser used for this test and would not be visible in standard scanner illumination output where the coherence length is typically too short to produce this level of speckle. As a further demonstration of the prediction model's strength and accuracy, the bright target pixels can be blocked to enhance the stray light in the background as seen in Figure 3.

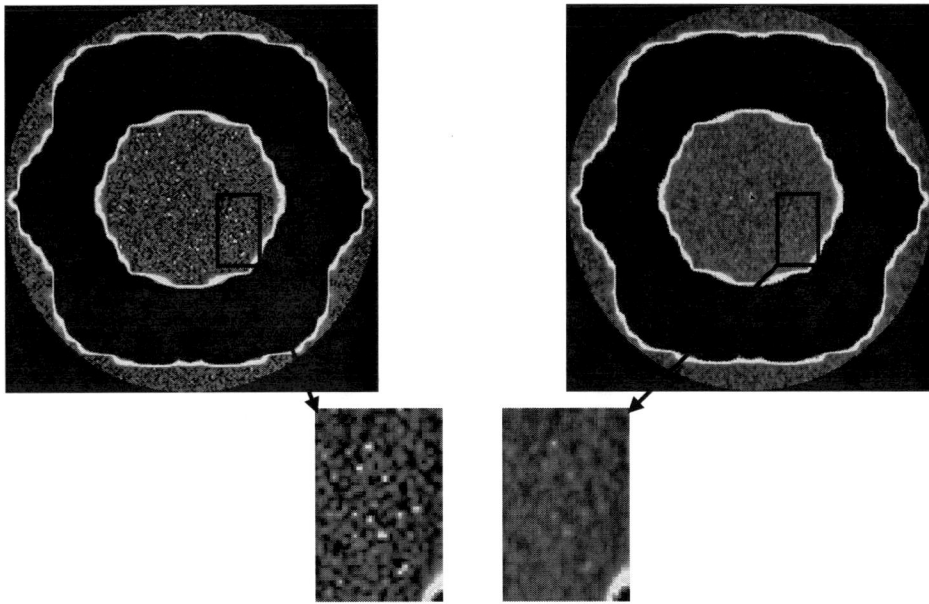

Figure 3. Performance prediction model output (left) and measured performance of the manufactured part (right) with background enhanced. The enlarged portions highlight bright pixels in the non-target region that are accurately predicted by the model.

Even at these relatively low intensity levels, the model accurately predicts both the magnitude and location of the stray light on a pixel-by-pixel basis as demonstrated by the enhanced regions of the figure. To further demonstrate the capability of the model, a horizontal cross-section through the center of the pattern is plotted in Figure 4.

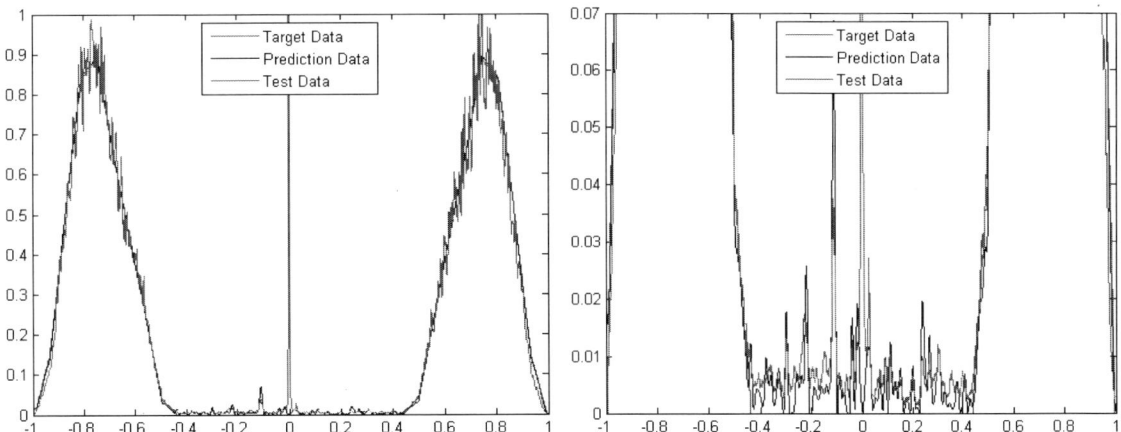

Figure 4. Horizontal cross-section through the center of the images in Figure 2 compared with the ideal target. The right image is magnified to a scale of only 7% of the maximum intensity to show stray light details.

Note that at full scale, the target, prediction and test data all trace the desired pattern quite faithfully; with the deviations in the test data being largely due to speckle from high coherence in the test system as previously mentioned. When expanded to show detail of the stray light region, it is possible to see that individual peaks and valleys predicted by the prediction model are clearly visible at the same location and general magnitude within the test data.

Harnessing the benefits of SMO can be achieved in stages based on the complexity of the required design, but an understanding of the manufactured DOE performance is important at all levels. Beginning with a general concept of what deviations from the ideal target pupil will exist in the manufactured DOE and how much stray light is present in the system; expanding to include effects the location of the stray light would have on the output; to finally examining the stray light contribution on a pixel-by-pixel basis, our performance prediction model can add as much or as little complexity to the SMO optimization algorithm as desired.

3. TUNABLE DOE OUTPUT PERFORMANCE

3.1 Tunable DOE Concept

To increase the capability of DOEs in the lithography industry, Tessera has developed tunable DOE technology where the output of the DOE can be varied from one pattern to another by shifting the position of the beam on the part. For many applications, the optimum illumination pattern can be narrowed down to a family of candidate designs that must be tested to determine which has the best performance. The choice must then be made on whether sufficient time and resources exist to explore each possibility or else choose a likely candidate and perform process optimization as best as possible with limited resources and accept a potentially smaller process window.

With a tunable DOE, multiple patterns can be placed on the same DOE representing each of the candidate designs as seen in Figure 5. By shifting the DOE inside the scanner, the illuminated portion can be adjusted to select one of the candidate designs or a hybrid of multiple designs. Depending on the desired outcome, this can prove to be a powerful tool for dialing in a process before selecting a final design for a production ramp. The following sections describe two examples of how tunable DOEs can be used to optimize the output illumination profile.

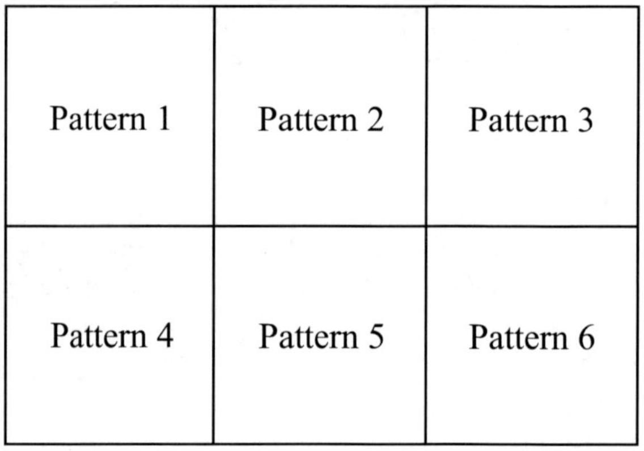

Pattern 1	Pattern 2	Pattern 3
Pattern 4	Pattern 5	Pattern 6

Figure 5. One example of a possible layout for tunable DOEs, in this case there are 6 discrete patterns. Shifting the illumination beam location will vary the output among each of the patterns and produce hybrid patterns in transition areas.

3.2 Tunable Pole Balance Example

Pole balance is an important aspect of scanner illumination patterns. When the energy of the pattern is not evenly distributed among all the poles, an imbalance occurs. If the imbalance is between horizontal and vertical poles, the result is H-V bias. All other imbalances result in telecentricity errors that can be detrimental to process windows. The imbalance is created by asymmetries in the diffractive microstructure profile that preferentially redirect light to one pole over another. We routinely deliver DOEs with < 1% pole imbalance to ensure these effects are minimized.

For most applications, the energy in each pole should be identical and minimizing any pole imbalance is important. Sometimes purposely inducing a pole imbalance is preferential to optimize process performance for different patterned structures on the wafer. In these cases, finding the ideal balance can be tricky. With a tunable DOE, it is possible to achieve both situations in a single part with only 2-3 designs. Choosing the example of a 25° quadrupole as shown in Figure 6, the horizontal and vertical poles can be separately biased in each sub-design. In this case, the bias between X and Y poles was chosen to be a 2:1 ratio.

Figure 6. Sample target pupils used to demonstrate tunable pole balance for a 25° quadrupole. The energy distribution varies from Y-poles having twice the energy as the X-poles (left) to balanced energy in X and Y (center) to X-poles having twice the energy of the Y-poles (right).

When these designs were fabricated on a single tunable DOE part and illuminated with a 6mm diameter circular beam, the output varied as the beam was shifted across the part as seen in Figure 7. When the test beam was placed entirely within one sub-design, the pole balance was relatively constant, closely matching the ideal value for that sub-design, corresponding to the flat regions of the graph. When the beam was shifted into the transition region from one sub-design to another, the output takes on characteristics of both sub-designs with the magnitude of the contribution from each determined by the total area of the beam subtended by each sub-design. The corresponding linear gradient in the transition regions shows that simply by varying the position of the beam across the part, it is possible to dial in any value of pole imbalance between the two sub-design values. If only the two unbalanced sub-designs were chosen for the part, it would be possible to remove the effects of H-V bias completely by varying the beam position until all imbalances were removed.

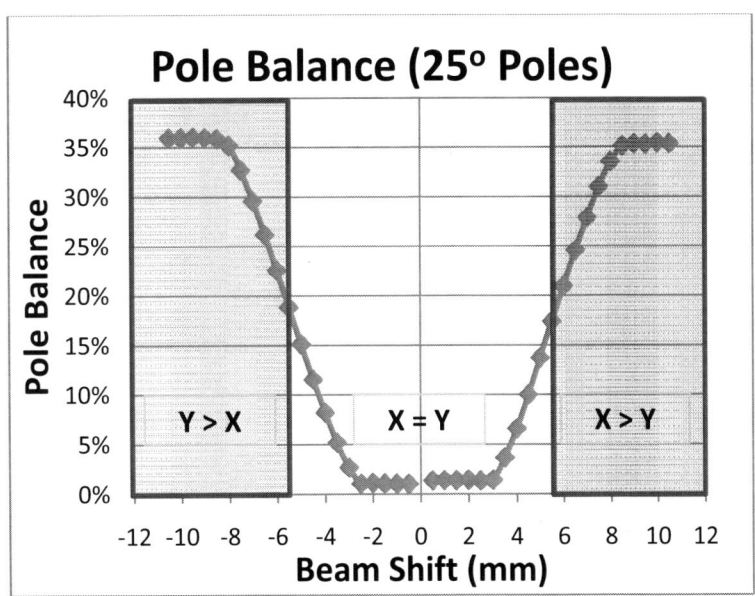

Figure 7. Test data showing pole balance tunability for 25° poles as the beam shifts from Y > X design to balanced to X > Y design. Any desired pole balance from ~0% to ~35% can be selected by shifting the beam location across the part.

Figures 8 and 9 show the corresponding test results to Figures 6 and 7 for the case of a 45° quadrupole design. Once again, the individual sub-design output was achieved by illuminating only that section of the DOE and the output was tuned by shifting the position of the beam across the transition region from one sub-design to another. In principle, this concept can be extended to any arbitrary design where either control or variability of pole balance is desired.

Figure 8. Sample target pupils used to demonstrate tunable pole balance for a 45° quadrupole. The energy distribution varies from Y-poles having twice the energy as the X-poles (left) to balanced energy in X and Y (center) to X-poles having twice the energy of the Y-poles (right).

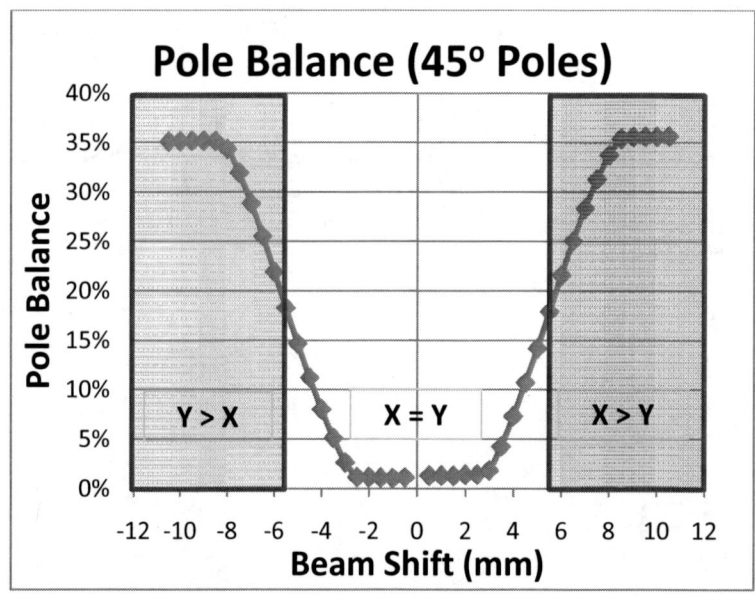

Figure 9. Test data showing pole balance tunability for 45° poles as the beam shifts from Y > X design to balanced to X > Y design. Any desired pole balance from ~0% to ~35% can be selected by shifting the beam location across the part.

3.3 Tunable Pole Widths Example

A second example of how tunable DOEs can be implemented in a scanner illumination system is by varying the output profile between families of designs. Since many common lithography applications print very similar feature structures across the entire chip, often the same type of illumination is used (e.g. dipole, quadrupole, etc). Depending on the specific structures to be printed in a localized region, some modification of the illumination may be optimal. Figure 10 demonstrates the principle of tuning the pole width for a quadrupole design from 25° to 30°. In this case, the transition region is a hybrid with the effective pole width being determined by the amount of energy in the grey transitions around each pole in the center image.

Figure 10. Sample target pupils used to demonstrate tunable pole width for a quadrupole varying from 25° (left) to 30° (right). The center image shows a hybrid of both designs with equal representation of each.

The output test result for this case is shown in Figure 11. To measure the change in effective pole width, the total energy inside a 30.3° pole region was integrated at each beam location and normalized to the maximum pole energy when the test beam is entirely within the 30° sub-section. Note that an additional 0.3° buffer was placed around all dimensions to account for beam divergence effects. As light enters or leaves the pole region corresponding to the change in overlap of the beam with the 30° sub-design, the normalized efficiency will vary accordingly. Since a similar effect would be observed for pole widths between 25° and 30°, the final imaging properties would also be similar.

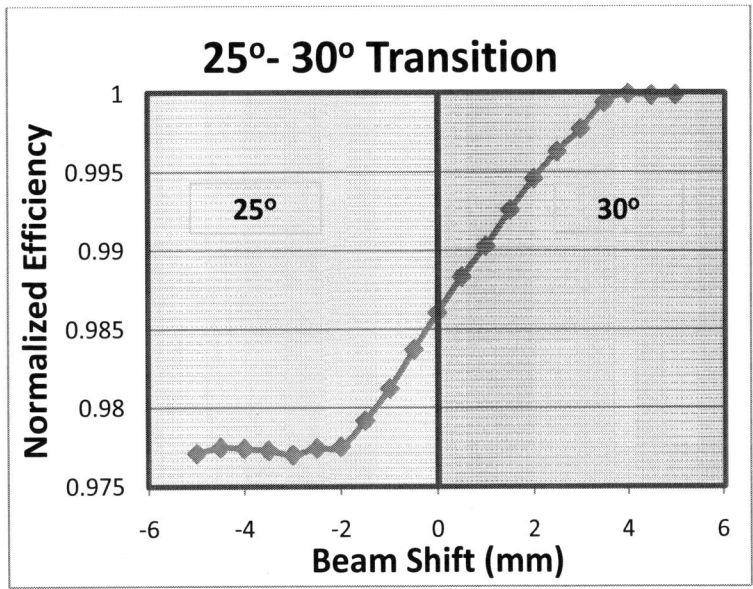

Figure 11. As the beam shifts from one pattern to the other, the normalized efficiency in the poles increases linearly in the hybrid transition region. By selecting an appropriate position, any effective pole width from 25°-30° may be created.

Again, when the 6mm beam is illuminating only one sub-design at a time, the normalized efficiency is constant. As the beam began to cross from one sub-design to the other, a linear gradient was observed. By calibrating this effect, it is possible to dial in any specific effective pole width between the two sub-designs. For the data in Figure 11, there is an apparent shift away from symmetry near the center of the transition region. This is most likely caused by a slight mis-calibration in the absolute positioning of the beam near the transition region.

As with the pole balance example, the concept is equally valid for larger width poles varying from 40° to 45° as shown in Figures 12 and 13. In this case, since the relative change in illuminated area is smaller between 40° and 45° poles than 25° and 30° poles, the magnitude of the change in normalized efficiency is reduced. This effect would make smaller pole designs slightly more sensitive to shifts in the transition region than larger pole designs.

Figure 12. Sample target pupils used to demonstrate tunable pole width for a quadrupole varying from 40° (left) to 45° (right). The center image shows a hybrid of both designs with equal representation of each.

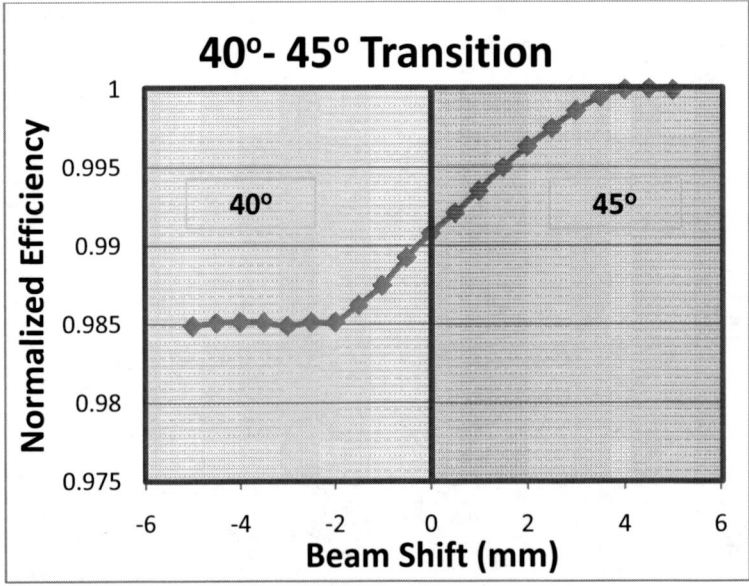

Figure 13. As the beam shifts from one pattern to the other, the normalized efficiency in the poles increases linearly in the hybrid transition region. By selecting an appropriate position, any effective pole width from 40°-45° may be created.

The tunable design output concept can be applied to other examples as well as long as each sub-design is a member of the same "family" of patterns. The two cases presented here were chosen as simple examples to demonstrate a proof of concept. Any practical implementation of these tunable DOEs would only require the additional degree of freedom to adjust the position of the DOE within the beam path of the scanner illumination system.

4. SUMMARY

To obtain truly optimized solutions with the most robust process windows at the 32nm node and beyond, an understanding of the capabilities and limits of the illumination is critical. Small deviations in illumination from the ideal target pupil due to things like pole imbalance, zero order and stray light can have a significant impact if they are not accounted for during the optimization process. Our performance prediction model is robust enough to satisfy the needs of any such optimization process when determining the ideal illumination profile. In addition, recent advances in the area of tunable DOEs can enable in situ variations of the output profile for process development without the need for multiple design iterations. By incorporating some or all of these concepts, it is possible to extend the viability of optical lithography until such time as alternative lithographic approaches are deemed production worthy.

REFERENCES

[1] Lai, K., Rosenbluth, A., et al, "Experimental Result and Simulation Analysis for the use of Pixelated Illumination from Source Mask Optimization for 22nm Logic Lithography Process", Proc. SPIE 7274, 72740A (2009).

[2] Kim, B., Lee, S., et al., "Advanced Resolution Enhancement Technique for 32nm Node Contact Hole Layer Using Source Mask Optimization" Proc. SPIE 6924, 6924-22 (2008).

[3] Bekaert, J., et al., "Freeform illumination sources: Source mask optimization for 22 nm node SRAM", 6th International Symposium on Immersion Lithography Extensions, Prague, O-ILO-03 (2009).

[4] Coskun, T., Sezginer A., et al, "Enabling Process Window Improvement at 45nm and 32nm with Freeform Illumination", Proc. SPIE 7274, 72740B (2009).

[5] Leonard, J., Carriere J., et al, "An Improved Process for Manufacturing Diffractive Optical Elements (DOEs) for Off-Axis Illumination Systems", Proc. SPIE Vol. 6924, 69242O (2008).

[6] Carriere, J., Stack, J., et al., "Advances in Modeling and Optical Performance of DOEs Used for OAI in Immersion Lithography", 6th International Symposium on Immersion Lithography Extensions, Prague, O-ILO-04 (2009).

Abbe-PCA-SMO : Microlithography Source and Mask Optimization Based on Abbe-PCA

Jason Hsih-Chie Chang, and Charlie Chung-Ping Chen,

Graduate Institute of Electronics Engineering, National Taiwan University,

No. 1, Sec. 4, Roosevelt Road, Taipei, 10617 Taiwan (R.O.C).

Lawrence S. Melvin III,

Synopsys Inc., 2025 NW Cornelius Pass Road, Building A, OR 97124.

Abstract

Resolution enhancement technologies (RETs) are so far widely proposed in improving the quality of micro-lithography process. Latest methods such as source mask optimization (SMO) and inverse lithography technology (ILT) are gaining popularity recently. Therefore, high speed simulator is in strong demand for growing computational complexity of RETs. In this paper, we demonstrate that our previously proposed Abbe-PCA is highly efficient for source configuring and pixel-based ILT mask tuning.

1 Introduction

As VLSI process develops, feature size shrinks dramatically with respect to the wavelength of exposure light source. Consequently, diffraction and other non-ideal effects deviate the exposed pattern significantly from the original mask design we expected.

To overcome these non-ideal effects, varieties of resolution enhancement technologies (RETs) have been widely proposed so far in refining the quality and yield of micro-lithography manufacture [6] [7] [8].

Typical RETs are off axial illumination (OAI), phase shift mask (PSM), and optical proximity correction (OPC) Fig.(1). Essentially, the micro-lithography system acts like a low pass filter due to its finite size of exit pupil in optical lenses.

The OAI method tries to configure the shape of illuminator, which consists of exposure light source and condenser lens with illuminator mask. When light source moves off the optical principal axis, the exit pupil shifts correspondingly, which allows more high frequency information passing through and resolution is enhanced.

Besides of source configuring, the mask in object plane is also tuned by PSM and OPC. The PSM method improves printing quality by assigning multiple values to mask but traditional binary block-pass pattern.

In the other hand, the OPC method considers the optical proximity effect of micro-lithography system and preform correction on the mask pattern beforehand. All of these RETs are in favour of improving printing quality [9] [10] [11].

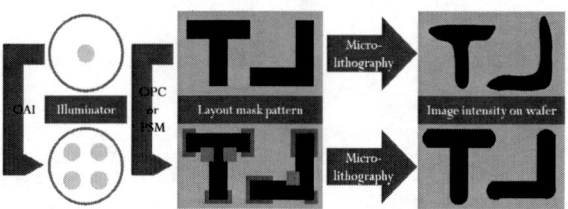

Figure 1: Example of resolution enhancement technologies.

Although these ideas of RET have been proposed for a long time, topics on finer and newer implementation never lose their popularity. Since all these RETs mentioned above are in favour of improving printing quality individually, it is straightforward to try a combination of them.

In these years, the latest theme of RET is source mask optimization (SMO), which tries to combine OAI of source configuring with OPC of mask tuning

Optical Microlithography XXIII, edited by Mircea V. Dusa, Will Conley, Proc. of SPIE Vol. 7640,
764026 · © 2010 SPIE · CCC code: 0277-786X/10/$18 · doi: 10.1117/12.846615

simultaneously.

The SMO method takes not only mask but also source into consideration for an overall optimization, which provides a larger solution space and could lead to a better result than simply OAI or OPC individually [15] [16] [17].

In the other hand, a high speed image intensity simulator is always in strong demand for all kind of RETs since the optimization process is judged by how far is its enhanced image result from the original shape design we expected as a cost function.

In our previous work, we proposed to do principal component analysis (PCA) on Abbe's image kernels, which can derive a set of equivalent compact kernels very efficiently [19] [20] [21].

By modelling the micro-lithography process as a sum of coherent system (SOCS) spanned by the PCA basis of kernel space, we now demonstrate that our Abbe-PCA method works well under the latest SMO framework.

The remain of this paper is organized as following. In section 2, we review basic concepts of micro-lithography image formulation and detailed RET for SMO. In section 3, we introduced the source mask optimization based on our previously proposed Abbe-PCA framework. We provide our simulation results in section 4 and conclude in section 5.

2 Preliminary

In this section, we we review the partial coherent illumination used in micro-lithography system with Abbe's image formulation and detailed RET for further SMO.

2.1 Image Formulation in Micro-lithography System

In modern micro-lithography systems shown in Fig.(2), mask pattern in the object plane is generally illuminated by a partially coherent quasi-monochromatic light source.

The exposure light originates from a coherent point source and collected by a condenser lens with parallel plane wave transmitted. The illuminator may be masked for advanced OAI configuration or source optimization.

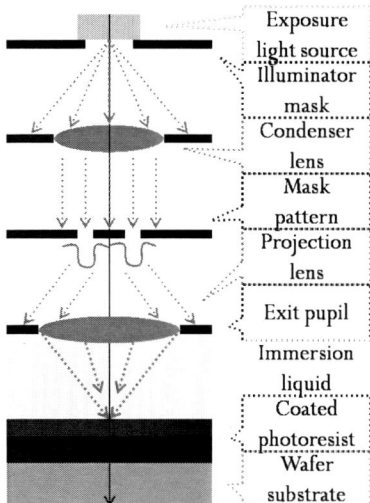

Figure 2: Typical micro-lithography system.

Conventionally, the limit of resolution is determined by the smallest lens in optical system, which was named as the exit pupil in equivalent, acts as a low pass filter and causes image distortion.

Finally, the wafer stack, which acts as a film in micro-lithography system, consisting of coating, photo-resist, and wafer substrate, is exposed by diffracted light thus the mask pattern is printed [1] [2] [3].

In favour of modelling micro-lithography system for mathematical computation, the mask pattern can be viewed as an input while the image intensity on wafer is viewed as an output.

In most cases, the impulse response mapping object plane into image plane can be approximated as a point spread function (PSF) with linear space invariance (LSI) under reasonable assumptions [4] [5].

Thus the image intensity $I(x, y; z)$ with interested depth z in Eqn.(1) can be formulated by corresponding coherent point spread function $H_0(x, y; z)$, mask pattern $O(x, y)$, and mutual intensity $J(x, y)$ with normalizing constant a_{img}.

$$
\begin{aligned}
&I(x, y; z) \\
&= a_{\mathrm{img}} \int\int\int\int J(x' - x'', y' - y'') \\
&\quad H_0(x - x', y - y'; z)O(x', y') \\
&\quad \cdot H_0^*(x - x'', y - y''; z)O^*(x'', y'')dx'dy'dx''dy'' \quad (1) \\
&= a_{\mathrm{img}} \int\int\int\int\int\int \hat{J}(f, g)
\end{aligned}
$$

$$\hat{H}_0(f + f', g + g'; z)\hat{O}(f', g')$$
$$\cdot\, \hat{H}_0^*(f + f'', g + g''; z)\hat{O}^*(f'', g'')$$
$$e^{-i2\pi[(f'-f'')x+(g'-g'')y]}\,df\,dg\,df'\,dg'\,df''\,dg'' \qquad (2)$$

Notations with a hat are in temporal domain with respect to two dimensional Fourier transform, otherwise are in spatial domain.

All the functions shown in Eqn.(1) are properly normalized with respect to wavelength λ of exposure light source, diameter of exit pupil, and numerical aperture (NA).

2.2 Abbe's Image Formulation

Observing the the conjugate symmetry in Eqn.(1), we rewrite it into Eqn.(3) as a squared integration of convolution results with respect to $J(x, y)$.

$$I(x, y; z)$$
$$= a_{\text{img}} \int\!\!\int \hat{J}(f, g) |\int\!\!\int \hat{H}_0(f + f', g + g'; z)$$
$$\hat{O}(f', g')\,df'\,dg'|^2\,df\,dg \qquad (3)$$
$$= a_{\text{img}} \int\!\!\int \hat{J}(f, g) |\int\!\!\int H_0(x - x', y - y'; z)$$
$$e^{-i2\pi[f(x-x')+g(y-y')]}O(x', y')\,dx'\,dy'|^2\,df\,dg \qquad (4)$$
$$= a_{\text{img}} \int\!\!\int \hat{J}(f, g) |(H_0(x, y; z)$$
$$e^{-i2\pi(fx+gy)}) * O(x, y)|^2\,df\,dg \qquad (5)$$

Abbe's image formulation, or discrete point source integration method, was based on approximating the illuminator into sum of discrete point sources $J(f, g) \cong \sum_s^{\text{source}} a_s\delta(f - f_s, g - g_s)$.

These Abbe sources illustrated in Fig.(3) are mutually independent but individually coherent, thus the image intensity can be computed by summing the square of light field caused by each Abbe source written in Eqn.(6).

$$I(x, y; z) \cong a_{\text{img}} \sum_s^{\text{source}} |H_s(x, y; z) * O(x, y)|^2 \qquad (6)$$

Where a_s is the relative intensity of Abbe sources, and $H_s(x, y; z) = H_0(x, y; z)e^{-i2\pi(f_sx+g_sy)}$ is the point spread function caused by an Abbe source located at (f_s, g_s) with shift modulation on H_0.

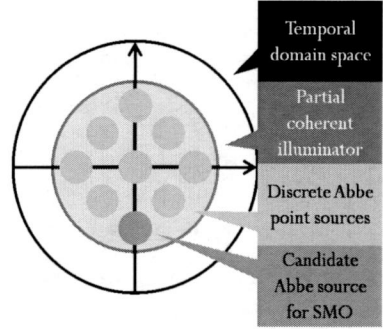

Figure 3: Discrete point sources in Abbe's method.

2.3 Pixel-based Inverse Lithography technology

Conventionally, the most popular RET is model-based OPC shown in Fig.(4), which firstly segments the mask pattern into moving edges and tries to determine each edge should be pulled, pushed, or neither, by treating the image simulation as a black box for feedback control [6] [8].

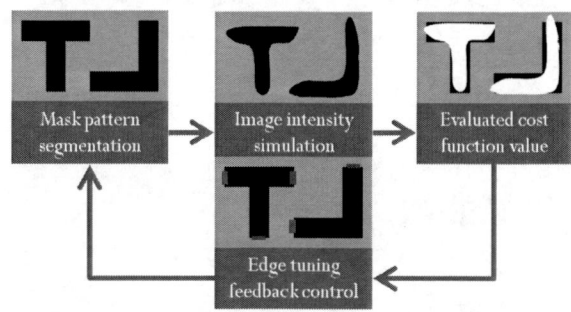

Figure 4: Conventional model-based OPC.

However, issues in MB-OPC such as how to choose a "good" edge segmentation, or the trade-off between smaller segment for better result but together higher complexity, are remained to be solved.

Recently, the latest RET framework named inverse lithography technology (ILT) is proposed, which tries to perform an inverse transform on mask pattern in cancelling the effect of microlithography process [12].

In the other hand, the mask edges pushed or pulled in model-based OPC can be regarded as a sequence of added or deleted mask pixels shown in Fig.(5), are viewed as operation vector in mask optimization.

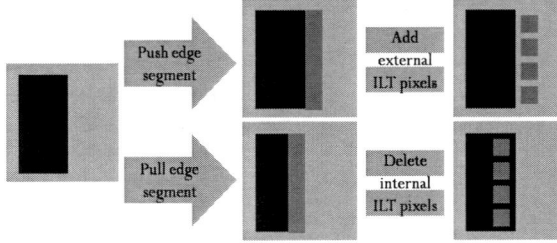

Figure 5: Pixel-based inverse lithography technology.

However, if there is a binary mask with size of $1cm$ by $1cm$ sampled in $1nm$, the solution space is proportional to $2^{10^{14}}$, which is impractical like disaster.

In practical cases, the solution space is reduced by constraint on neighbouring topological invariance or sequentially greedy gradient based search [13] [16].

3 Methodology: Abbe-PCA-SMO

In this section, we firstly introduce the basic concept of Abbe-PCA method involved in our proposed Abbe-PCA-SMO algorithm for microlithography source mask optimization.

3.1 Principle Component Analysis on Abbe's Method

The number PSFs H_s used in Abbe's method is generally ranged from 100 to 1000, depending on the requirement of simulation accuracy. Therefore, it is straightforward to approximate an equivalent compaction on them within acceptable error [19] [20].

As we know, principle component analysis (PCA) is define as an orthonormal basis transforming the original data set into a vector space spanned by its eigenvectors with decreasing order of principal components [18].

Now we collect the Abbe PSFs H_s, or the image convolution kernels, in column vectors. A matrix H of kernel set can be derived with dimension n^2 by k, if there are totally k kernels with each size of n by n.

If there exists $H = USV^*$ defined after the singular value decomposition (SVD) of H, U is the eigenvectors of image kernel space, V is the coefficients of PSF linear combination, S is the singular values.

Thus, the PCA transformation can be written as $K = HV = US$, which mean the PCA space K is a combination of PCA kernels Φ_c in U with weighted λ_c in S.

In the other hand, K can be derived by mapping kernel set H with coefficients in V, which is the eigenvectors of $H^*H = V\Sigma V^*$ written in Eqn.(7).

$$
\begin{aligned}
H^*H &= (USV^*)^*(USV^*) \\
&= V(S^*S)V^* = V\Sigma V^*
\end{aligned} \tag{7}
$$

Each entry in H^*H is the inner product of two Abbe PSFs H_s and H_t, which can be calculated directly by a lookup table of Fourier transformed square of coherent image kernel written in Eqn.(8).

$$
\begin{aligned}
&< H_s, H_t > \\
&= \int \int [H_0^*(x,y)e^{+i2\pi(f_t x + g_t y)}] \\
&\quad [H_0(x,y)e^{-i2\pi(f_s x + g_s y)}]dxdy \\
&= F^2[H_0^*H_0]|_{f=f_s-f_t, g=g_s-g_t}
\end{aligned} \tag{8}
$$

For example, if there are $k = 100$ kernels with size $n = 1000$, the target matrix for compaction will be $n^4 = 10^{12}$ for traditional direct method [8], or a kernel set with size of $kn^2 = 10^8$ for [19], but simply $k^2 = 10^4$ using our Abbe-PCA method [20].

3.2 Applying Abbe-PCA on SMO

Once the PCA space K was built, we can construct a sum of coherent system (SOCS) in Fig.(6) by the compact dominating PCA kernels Φ_c with weighting λ_c in Eqn.(9) for high-speed image simulation.

$$
\begin{aligned}
I &= a_{\text{img}} \sum_s^{\text{source}} a_s |H_s * O|^2 \\
&\cong a_{\text{img}} \sum_c^{\text{compact}} \lambda_c |\Phi_c * O|^2
\end{aligned} \tag{9}
$$

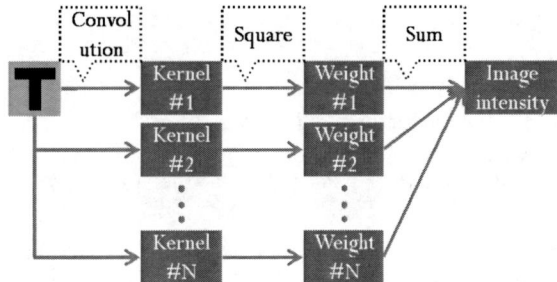

Figure 6: Sum of coherent system built by PCA space.

In SMO, when we try to add (positive a_s) or delete (negative a_s) Abbe sources from the existing illuminator, their impact can be simply derived by updating the weighting in Eqn.(10) due to the superposition property of sources and orthogonality of PCA kernels.

$$\lambda_c = \sum_{s}^{\text{modify}} < a_s H_s, \Phi_c > \tag{10}$$

In the other hand, the PCA compaction process in Eqn.(7) can be performed repetitively after added or deleted enough Abbe sources during the SMO process [21].

3.3 Intensity-based Cost function

As we know, the cost function is critical in optimization process, which affects the final result significantly and should be decided with carefulness.

Traditionally, the cost function is measured by edge placement error (EPE) [8], which calculates the difference between image contour and mask shape we excepted. However, EPE is time-consuming because of full simulation demanded and is hard to handle with mathematical expression.

Recently, an edge intensity error (EIE) based cost function is proposed in saving runtime [11]. It measures the difference of image intensity and image threshold on sampled edge sites shown in Fig.(7).

As long as the EIE is minimized and the number of sampled edge sites is enough, the image contour will approach the mask shape we excepted.

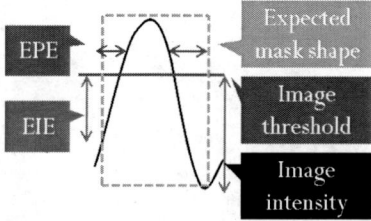

Figure 7: The edge intensity error.

Also, the EIE based cost function can be easily handled via mathematical expression written in Eqn.(11). Where $O_e(x,y) = \sum_e^{\text{edge}} \delta(x - x_e, y - y_e)$ is the function for sampled edge sites, a_{thr} is the image threshold.

$$\text{EIE} = \| I(x,y) O_e(x,y) - a_{\text{thr}} \|$$
$$= \sum_e^{\text{edge}} \| I(x_e, y_e) - a_{\text{thr}} \| \tag{11}$$

However, in spite of minimizing EIE, the edges of image result are also expect to be as sharp as possible. Here we introduce the edge contrast penalty (ECP) to avoid results with bad slopes shown in Fig.(8).

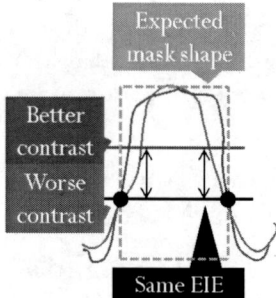

Figure 8: The edge contrast penalty.

Since our cost function in to be minimized, the ECP is defined small for sharp edges, which can be written in Eqn.(12).

$$\text{ECP} \propto^{-1} \| \nabla I(x,y) O_e(x,y) \|$$
$$= \sum_e^{\text{edge}} \| [\frac{\partial I(x_e, y_e)}{\partial x}, \frac{\partial I(x_e, y_e)}{\partial y}] \| \tag{12}$$

3.4 Pixel-based ILT using BFS

As mentioned in previous section, now we start from the pixel-based ILT method, which treats sequence of added or deleted mask pixels in Fig.(5) as operation vector for mask optimization.

Given an initial mask pattern O_0, which is also the image shape we expected, thus the mask pattern during iteration can be presented as $O = O_0 + O_p - O_n$ shown in Fig.(9).

Notation O_p is the function for candidate external pixels to be added, and O_n is the function for candidate internal pixels to be deleted.

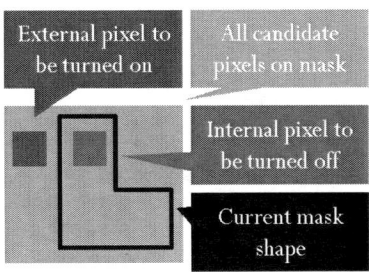

Figure 9: Candidate internal and external pixels.

For greedy optimization, the order of searching sequence would affect the final result significantly and should be decided with carefulness. In this work, we find that added or deleted ILT pixels are mostly chosen near the original mask pattern.

Hence, it is straightforward to perform a breadth first search (BFS) starting from the edges of original mask pattern. Then, candidate external and internal pixels are listed into sequence as propagating wave-fronts both inward and outward iteratively during optimization process.

Furthermore, pixel-based ILT using simply BFS could avoid manufacture unfriendly pixels shown in Fig.(10) without time-consuming topological condition check.

3.5 Abbe-PCA-SMO

To sum up, our Abbe-PCA-SMO algorithm is presented in flowchart of Fig.(11).

4 Simulation Result

Our Abbe-PCA-SMO algorithm is currently implemented under Matlab 7.4 on a PC with Intel Core

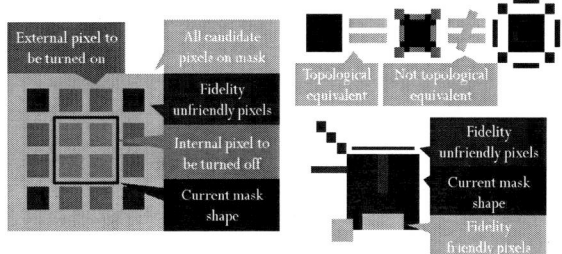

Figure 10: Candidate manufacture friendly pixels.

2 Duo 1.86 GHz processor and 2GB memory.

The environmental parameters for simulation is under typical $45nm$ micro-lithography process [3].

For source optimization, we have an initial source configuration of annular illuminator with coherency $\sigma = 0.8$ while the candidate discrete Abbe sources are shown in Fig.(12).

The target mask pattern is shown in Fig.(13), which has a size of $1\mu m$ by $1\mu m$ and resolution in $10nm$.

The image intensity and comparison during first few iterations of optimization are shown as in Fig.(14), Fig.(15), Fig.(16).

The runtime of full-run kernel compaction compared with single-run Abbe-PCA update during optimization process is shown in Fig.(17), which is consistently 10X speed up using our Abbe-PCA based incremental method.

The cost function is shown gradually decreasing in Fig.(18), while the chosen point sources are shown in Fig.(19).

For mask optimization, the size of pixels considered is $30nm$ by $30nm$ with resolution in $10nm$.

Typical target masks are shown in Fig.(20), Fig.(21), Fig.(22) with their optimized mask shape, modified pixels from original mask, image intensity, and image contour for demonstration.

Chosen added and deleted pixels from candidate external and internal pixels during process are shown in Fig.(23), which demonstrates that our method converges rapidly within few iterations.

The cost function is also shown in Fig.(24), which reveals that our method could provide an estimation measured with efficiency.

Figure 11: Flowchart of Abbe-PCA-SMO.

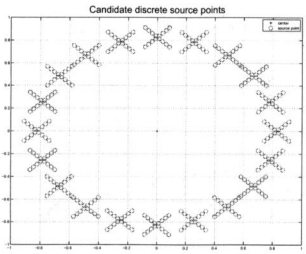

Figure 12: Candidate discrete sources for optimization.

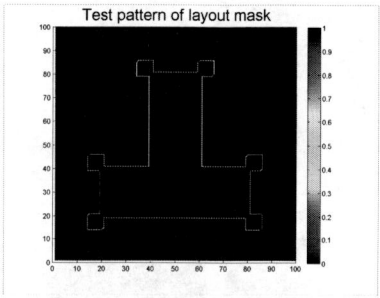

Figure 13: Target mask pattern for source optimization.

5 Conclusion

In this paper, we proposed an efficient Abbe-PCA-SMO algorithm for simultaneous source and mask optimization under Abbe-PCA framework. The image quality is significantly optimized as shown in experimental result.

References

[1] L. F. Thompson, C. G. Willson, and M. J. Bowden, *Introduction to Micro-lithography*, American Chemical Society, Washington, DC, 2 edition, 1994.

[2] Heinrich Kirchauer, *Photo-lithography Simulation*, PhD thesis, TU Vienna, 1998.

[3] A. K. Wong, *Optical Imaging in Projection Micro-lithography*, SPIE Press, 2005.

[4] J. W. Goodman, *Introduction to Fourier Optics*, McGraw Hill, 2 edition, 1996.

[5] Max Born and Emil Wolf, *Principle of Optics*, Press Syndicate of The University of Cambridge, 7 edition, 2005.

[6] A. K. Wong, *Resolution Enhancement Techniques in Optical Lithography*, SPIE Press, 2001.

[7] Y. C. Pati and T. Kailath, "Phase-shifting masks for micro-lithography : automated design and mask requirements," in *Optical Society of America*, 1994.

[8] N. B. Cobb, *Fast optical and process proximity correction algorithms for integrated circuit manufacturing*, the University of Berkeley, 1998.

[9] N. B. Cobb and Y. Granik, "New concepts in OPC," in *Proc. SPIE*, vol. 5377, 2004.

[10] P. Yu, S. X. Shi, and D. Z. Pan, "Process variation aware opc with variation lithography modelling," in *Design Automation Conference*, 2006.

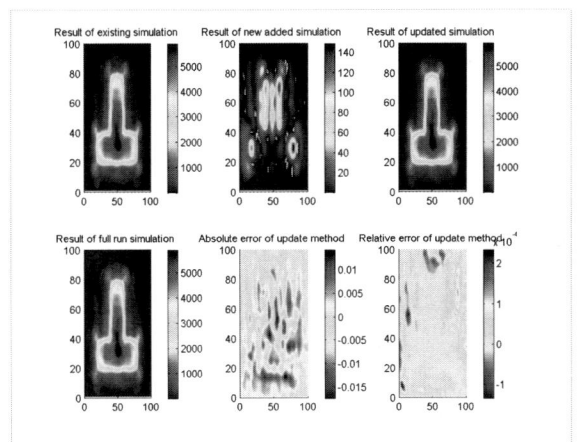

Figure 14: Image intensity during source optimization.

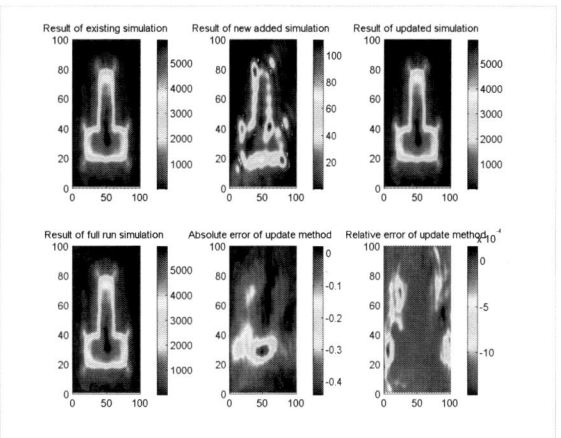

Figure 16: Image intensity during source optimization.

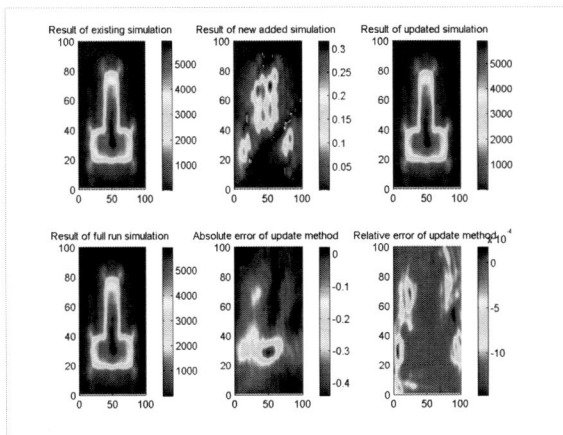

Figure 15: Image intensity during source optimization.

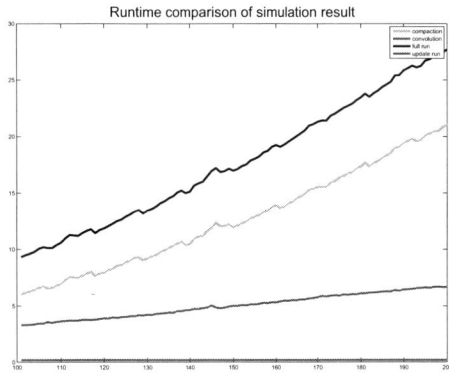

Figure 17: Runtime comparison of source optimization.

[11] P. Yu, D. Z. Pan, "A novel intensity based optical proximity correction algorithm with speedup in lithography simulation," in *Proc. International Conference on Computer Aided Design*, 2007.

[12] A. Poonawala and P. Milanfar, "Mask design for optical micro-lithography - an inverse imaging problem," in *IEEE Trans. on Image Processing*, vol. 16, no. 3, Mar. 2007.

[13] P. Yu, D. Z. Pan, "TIP-OPC: a new topological invariant paradigm for pixel based optical proximity," in *Proc. International Conference on Computer Aided Design*, 2007.

[14] Y. Borodovsky, W. H. Cheng, R. Schenker, V. Singh, "Pixelated phase mask as novel lithography RET," in *Proc. SPIE* , 2008.

[15] R. Socha, X. Shi, D. LeHoty, "Simultaneous source mask optimization (SMO)", in *Proc. SPIE*, 2005.

[16] X. Ma, G.R. Arce, "Pixel-based simultaneous source and mask optimization for resolution enhancement in optical lithography," in *Proc. of Optics Express*, 2009.

[17] J. Zhang, W. Xiong, Y. Wang, Z. Yu, M. C. Tsai, "A highly efficient optimization algorithm for pixel manipulation in inverse lithography

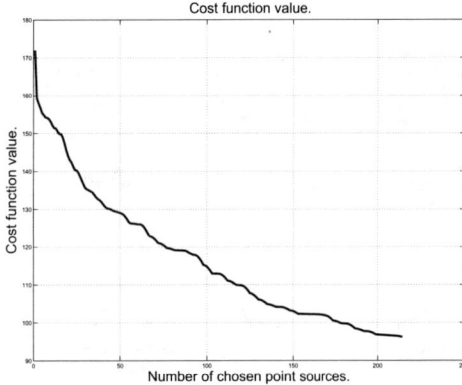

Figure 18: Gradually decreasing cost function in source optimization.

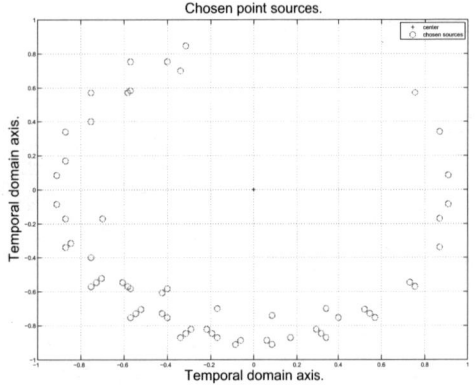

Figure 19: Chosen point sources in source optimization.

technique," in *Proc. International Conference on Computer Aided Design*, 2008.

[18] Jonathon Shlens, *A Tutorial on Principal Component Analysis*, Centre for Neural Science, New York University.

[19] C. C. P. Chen, A. Gurhanli, T. Y. Chiang, and J. J. Hong, "Abbe singular-value decomposition: Compact Abbe's kernel generation for micro-lithography aerial image simulation using singular-value decomposition method," in *Proc. J. Vac. Sci. Technol. B*, Volume 26, November 2008.

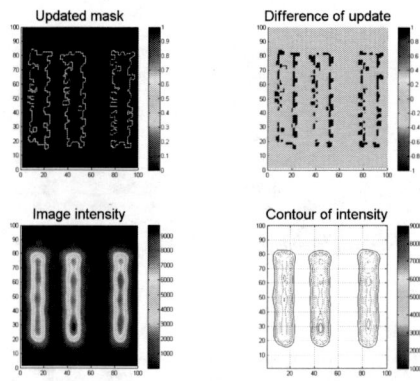

Figure 20: Simulation result of Abbe-PCA-SMO.

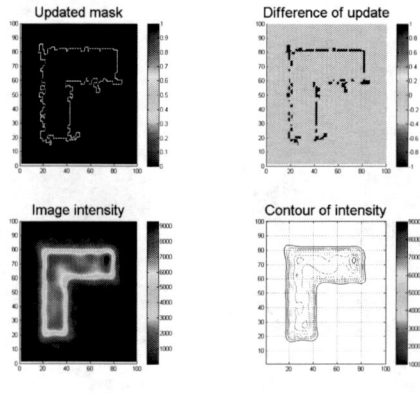

Figure 21: Simulation result of Abbe-PCA-SMO.

[20] M. F. Tsai, S. J. Chang, and C. C. P. Chen, "Abbe-PCA (Abbe-Hopkins): micro-lithography aerial image analytical compact kernel generation based on principle component analysis," in *Proc. SPIE*, Vol. 7274, March 2009.

[21] S. J. Chang, C. C. P. Chen, and L. S. Melvin III, "Abbe-PCA-SMO: micro-lithography simultaneous mask and mask optimization using Abbe-PCA," in *Proc. ASPIE*, November, 2009.

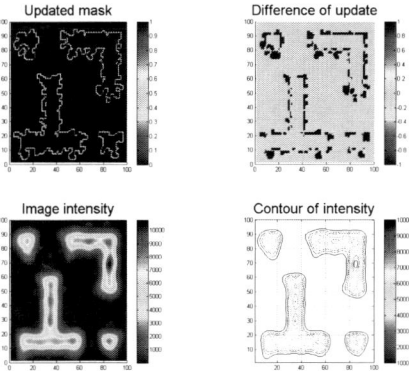

Figure 22: Simulation result of Abbe-PCA-SMO.

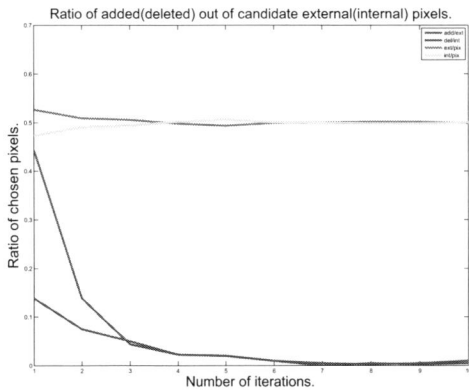

Figure 23: Convergence on chosen pixels w.r.t iteration.

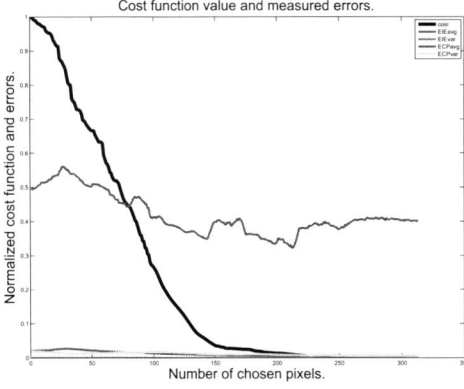

Figure 24: Reducing cost function value w.r.t. chosen pixels.

Optimization on Illumination Source with Design of Experiments

Helen Hu, Yi Zou and Yunfei Deng

GLOBALFOUNDRIES, 1050 E. ARQUES AVE., SUNNYVALE, CA

ABSTRACT

In advanced photolithography process for manufacturing integrated circuits, the critical pattern sizes that need to be printed on wafer are much smaller than the wavelength. Thus, source optimization (SO) techniques play a critical role in enabling a successful technology node. However, finding an appropriate illumination configuration involves intensive computation simulations. EDA vendors have been developing the pixelated source optimization tools that co-optimize both source and mask for a set of patterns. As an alternative approach, we have introduced design of experiments (DOE) methodology for parameterized source optimization to minimize computation efforts while achieving comparable CDU control for given design patterns.

In this paper, we present a Response Surface Methodology (RSM) that simplifies the response function and achieves the optimization goal on multiple responses. Results have shown that the optimal input settings identified by this approach are comparable with the pixelated source optimization results.

Keywords: Source Optimization, Illumination, DOE, Response Surface Methodology

1. INTRODUCTION

Source optimization in general involves finding optimized illuminator shapes for a representative set of design patterns. The objective of finding such optimum illuminator for achieving maximum patterning performance can be defined in many different forms. Construction of PV-band using the process variations information such as dose and focus variations, and mask errors is a technique to quantify the measure for judging layout manufacturability. Layout PV-band width is sufficient as the objective used in our source optimization experiments.

In our illuminator shape optimization flow, a previously optimized mask recipe that we consider to be generic enough for most of illuminators is used, together with a small set of critical design patterns. We implemented Manhattan mask rule constrains (MRC) for optimized mask recipe (vertical/horizontal SRAF constrains, minimal area, writer segment length, etc.)

After a few source candidates are pre-selected, we will include a much large set of design patterns in our simulation. In the simulation verification, we will closely examine decoupled MEEF, DOF, and Process latitude, and any illuminator given the general overall best performance will be chosen as the final source. In many situations, we have to make tradeoffs among many factors, such as the lithographic index (i.e. MEEF, DOF) mentioned, or other practical considerations, such as cost etc.

It should be noted, critical measurement topologies, such as line-ends, edges and etc, are tagged and assigned to different weights according to the importance. Different merit functions are formulated based on the PV-band widths of the topologies and they are co-optimized by using the design of experiments approach.

As discussed by Yi *et al* [4], we continue to use conventional parameterized illuminator for majority of critical layers of advanced technology nodes. Since not all critical layers are at k1 limit that requires very advanced lithographic techniques such as pixelated source optimization or the pixelated source optimization, it becomes important to introduce a methodology that helps reduce intensive computational efforts of commonly used "exhaustive" search for optimizing traditional parameterized source.

Optical Microlithography XXIII, edited by Mircea V. Dusa, Will Conley, Proc. of SPIE Vol. 7640,
764027 · © 2010 SPIE · CCC code: 0277-786X/10/$18 · doi: 10.1117/12.848876

Statistical background on DOEs is reviewed in section 2. Section 3 states the illuminator problem and describes the specific experiment we proposed. Experimental data and statistical analysis follow in section 4. Section 5 compares the result with those from other alternatives. Sensitivity analysis is another benefit of our DOE approach. Section 6 summarizes the conclusion and discusses the future work.

2. STATISTICAL DOES

An experimenter follows the following process in a statistical design of experiments. A good understanding of all the issues is necessary in designing an experiment. The experimenter chooses the factors to be varied and the ranges and specific levels that each factor varies. The response variable is the target parameter of interest. DOE is a method for systematically planning and conducting engineering studies. Controlled changes in puts are made to identify causes and effects in changes of system or process outputs. DOE is much more efficient than the traditional one-variable-a-time approach. The number of runs necessary to perform the experiments is largely reduced by varying multiple variables simultaneously. Different choices of designs are associated different models for analyzing the experimental data. The models provide basis for making valid conclusions over the experimental ranges of the factors, even for settings that are not run. Optimal settings for factors, the effects of individual factors and their interactions are inferred from statistical models. To summarize, DOE is applied in manufacturing to mainly solve four problems: comparison, screening or characterization, modeling and optimization.

Factorial designs are widely used in experiments where the interest is to estimate both main and interaction effects of several factors on a response variable. For k factors, a full-factorial design requires 2^k experimental runs. A 2^k design is often used to fit a first-order response surface model to generate estimates of main and interaction effects. Therefore it is useful for screening experiments where important factors are identified among many. Although each factor takes two levels, the experiment gets large quickly as k increases. When we assume that certain high-order interactions are negligible, then running a fraction of the full factorial design is a more efficient experiment. Fractional factorial designs are among the most popular design in industry for screening experiments. However quite often we observe curvatures in the response surface which requires three levels for factors. A 3^k fractional factorial design can still be expensive when k goes above six, as seen in table 1. Besides, the choice of fractions requires assumptions on the response surface, which brings the risks of neglecting important relationships.

Response surface designs are efficient designs that capture curvature in response surfaces, as shown in conceptual figure 1. They are able to contain quadratic terms in the model equation.

$$
\begin{aligned}
\hat{y} = b_0 &+ b_1 x_1 + b_2 x_2 + b_3 x_3 \\
&+ b_{12} x_1 x_2 + b_{13} x_1 x_3 + b_{23} x_2 x_3 \\
&+ b_{11} x_1^2 + b_{22} x_2^2 + b_{33} x_3^2
\end{aligned}
\tag{1}
$$

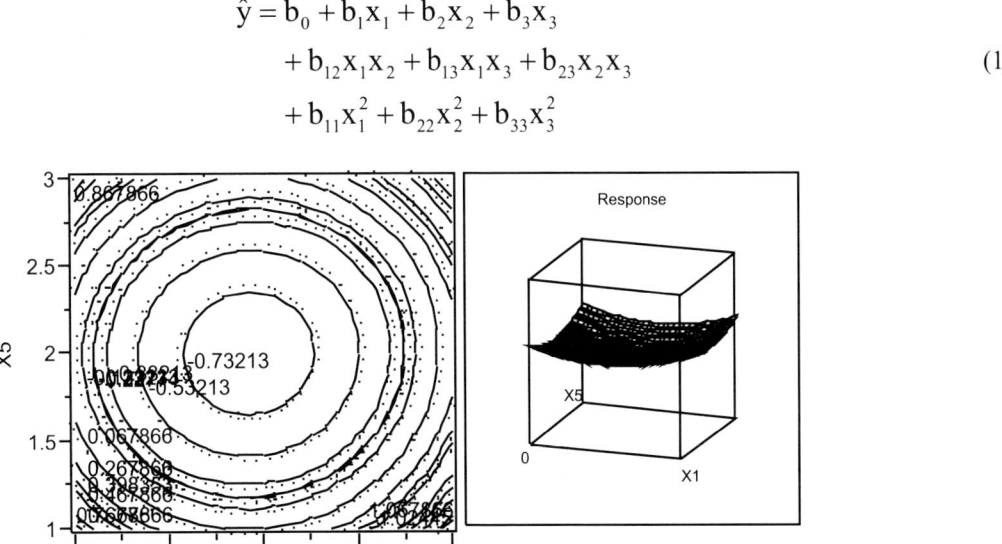

Figure 1. Contour plot and 3-dimensional plot of a response surface that contains curvature.

Two major types of response surface designs are available: Central Composite Design (CCD) and Box-Behnken design. CCDs are based on factorial designs, center runs and axial runs. Central Composite Circumscribed/Inscribed designs have 5 levels for each factor and they are rotatable. Central Composite Face (CCF) designs have axial runs on the faces and hence 3 levels per factor are used. An illustration of a CCF design with factor A, B and C is seen in Figure 2. The efficiency of CCDs relative to factorial designs at 3 levels is shown in Table 1. On the other hand, Box-Behnken designs set factor levels at the center of edges of an experimental space besides the center points. For a complete reference on experimental design and analysis and RSM, see [5-6].

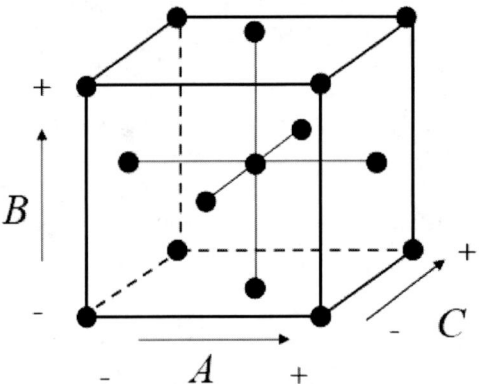

Figure 2. A 3-factor Central Composite Face (CCF) design.

Table 1. A comparison of efficiency between experimental designs.

# factors	Central Composite	3^k
2	13	9
3	20	27
4	31	81
5	32	243
6	53	729
7	92	2187
8	93	6561

3. ILLUMINATOR EXPERIMENTS

The purpose of illuminator optimization is to find optimal input settings, i.e., factors, which generate minimal Process Variation Band (PVBAND). The settings for the eight factors are illustrated in Figure 3, which also specifies the levels experimenters used to set for a full-factorial experiment. For example, we use "X1" to denote "dipole_center_radius", which is set at .77, .82 and .87.

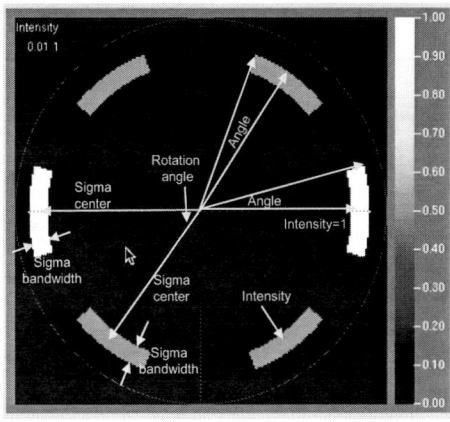

Variable	Illuminator Factor	Levels in Full-Factorial Experiment
X1	dipole_center_radius	{0.77 0.82 0.87}
X2	dipole_sigma_band	{0.1 0.2 0.3 0.4}
X3	dipole_angle_width	{20 30 40}
X4	quasar_rotation_angle	{45 55 65}
X5	quasar_angle_width	{20 30 40}
X6	quasar_center_radius	{0.77 0.82 0.87}
X7	quasar_sigma_band	{0.1 0.2 0.3 0.4}
X8	quasar_dose	{0.4 0.5 0.6 0.7 0.8 0.9}

Figure 3. Input Settings of an illuminator

PVBAND on four layouts were measured. Table 2 shows the six responses from these layouts. Individual PVBANDs should be minimized by choosing optimal settings. Although response surface model can well handle different optimization goals of multiple responses, the problem here is simplified by creating an objective function *ModelAvg* based on the critical dimensions from individual layouts,

$$ModelAvg = AVERAGE(\text{rw}/40, \text{ww}/80, \text{c1\_smooth}/40, \text{c1\_le}/70, \text{c2\_smooth}/50, \text{c2\_le}/50) \qquad (2)$$

Hence, a single response *ModelAvg* is to be minimized as the goal of illuminator experiments.

Table 2. Responses measured in experiment

Layout	Response
Logic Cell 1	edge only: rw
Logic Cell 2	edge only: rw
SRAM Cell 1	edge: c1_smooth; line-end: c1_le
SRAM Cell 2	edge: c2_smooth; line-end: c2_le

The improvement effort for statistical experimental design has the three inherent goals. First, an optimal process setting is to be determined that minimize the overall response. The setting(s) should be comparable with results from industrial standard methods, such as SMO. Second, we achieve the first goal with reduced number of runs, which improves efficiency especially in presence of high dimensions. Third, learning the significance of input factors from a statistical model is important for screening purpose or a sensitivity analysis. The statistical DOE approach involves the design of a response surface (RS) experiment, building a good RS model and the optimization of the RS in search for the optimal parameter settings.

The traditional full-factorial experiment takes the combination of all levels, which requires 3x4x3x3x3x3x4x6 = 23328 runs (refer to Figure 3). A 3^k factorial experiment requires $3^8 = 6561$ runs. We propose a CCF design that contains 2^k factorial runs, $2k$ axial runs and 1 center run, i.e., $2^8 + 2 \times 8 + 1 = 273$. CCF designs are convenient because the axial runs are set on the faces so that each factor only takes on 3 levels (see Figure 2). We have the option to choose an even smaller CCF design such as 145 runs. Since efficiency of the two makes relatively small difference at this magnitude, we opted for the design of 273 runs for the following reasons: First, the larger design explores parameter space at a larger degree. It often results a better model than smaller design. It is not always true, which leads to the need for model selections. However important relationships between input and response can be missed by smaller designs. Second, a moderate number of observations enable us for data filtering, given our specific goal of exploring the region with minimal responses.

4. RESPONSE SURFACE OPTIMIZATION

The RS model takes the function form as in equation (1). Figure 4 illustrates how a response surface is determined for two input factors. The original model, Model A, has a good fit with R^2 at .82, as seen in Figure 5(a). However, a close look at the residual plot in Figure 5(b) suggests two populations exist based on large and small values in response variable. A good statistic model should have random residuals instead of trends and patterns in residual plot, while R^2 is not the primary or only goodness-of-fit statistics. Since our primary goal is to fit the region of small responses, we filter out data highlighted in yellow and only keep the observations in black. The adjusted model, Model B, has a fit of R^2 at .72 and the plot of actual vs. predicted values is shown in Figure 5(c). The residual plot has become much more random than Model A, as seen in Figure 5(d).

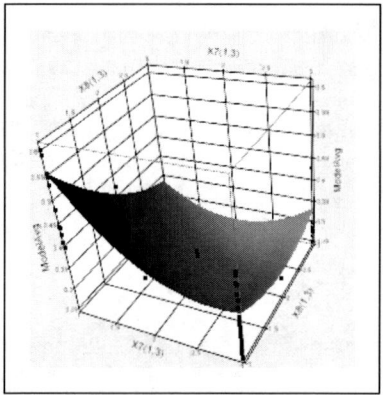

Figure 4. Response surface for two input factors

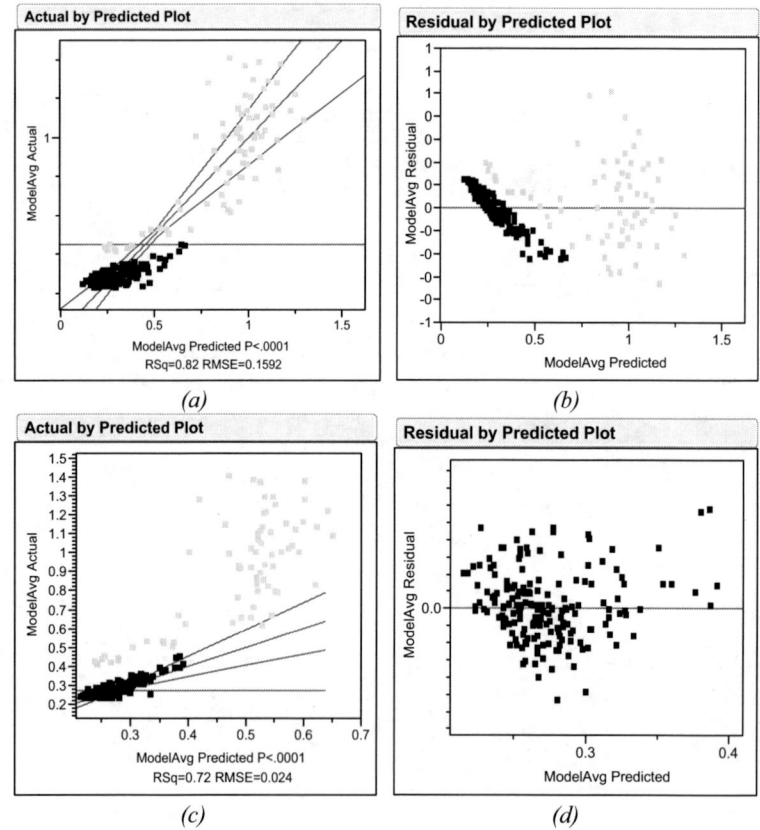

Figure 5. Fit of RS models. *(a)* Actual vs. Predicted plot for Model A. *(b)* Residual plot for Model A.
(c) Actual vs. Predicted plot for Model B. *(d)* Residual plot for Model B.

Based on the adjusted model, Model B, we search for parameter settings that minimize the response *ModelAvg*. Statistical software JMP$^{@}$ provides a platform called desirability profiler [7]. The desirability is defined here so that it is optimized as *ModelAvg* is minimized, shown in Figure 6. The profiler calculates the predicted value based on model for various combinations of input values. We will revisit and interpret the profiler results in a later section. The most flat profiler trace exists in X4, which suggests an insensitivity of the response to input values of X4. Input X7 is between 1 and 2 for optimal result. This leads to several settings listed in Table 3. Table 3 is obtained by taking the inferred DOE input settings and running them through simulation for the actual *ModelAvg* results.

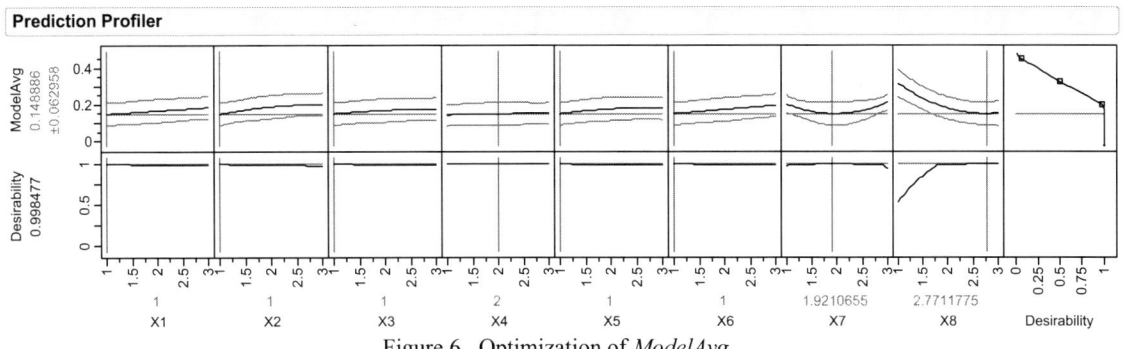

Figure 6. Optimization of *ModelAvg*.

Table 3. Input settings based on DOE optimization

DOE Input Setting	Actual *ModelAvg*
1_1_1_1_1_1_1_3	0.2100
1_1_1_2_1_1_2_3	0.2182
1_1_1_3_1_1_2_3	0.2088

5. ILLUMINATOR RESULTS AND SENSITIVITY

As mentioned in Section 3, the traditional approach is to run an exhaustive search of 23328 runs. It would provide a true solution to the optimization problem. We happened to have the capacity of running such a large simulation and the minimal *ModelAvg* turns out to be .1962. The RS model-based results in Table 3 are yet to reach such a minimum. This is the expense of running an experiment that is 100 times smaller (273 vs. 23328).

Table 4 shows the observed *ModelAvg* and DOE settings from our CCF experiment in an increasing order. We can conclude that the RS model inferred better settings than available runs from the CCF experiment, because all *ModelAvg* values in Table 3 are smaller than that in Table 4, i.e., .2208.

Table 4. Sorted data for CCF experiment

	ModelAvg	DOE
1	0.2208006	1_1_3_1_3_1_1_3
2	0.22342619	1_1_3_1_1_1_3_3
3	0.22399583	1_3_3_1_3_1_3_1
4	0.22543274	3_1_1_1_3_1_3_3
5	0.2266631	1_1_1_3_1_3_3_3
6	0.22674821	1_1_3_1_1_1_3_1

We have compared the five illuminators of (1) the optimal setting from 3k experiment, (2) optimal setting from CCF experiment, i.e., reduced DOE and (3-5) RS model-inferred optimal settings from CCF experiment. Figure 7 shows the actual illuminator configurations with respects to their performance for different designs with different classified features. The illuminator in the first column uses the optimal configuration that gives the minimal ModelAvg among the 3^k full factorial DOE designs. The illuminator in the second column gives the minimal ModelAvg among the reduce 3^k full factorial DOE designs. The last three illuminator shapes are predicted to give near-optimal solutions using the constructed DOE model based on the reduce 3^k full factorial DOE designs. As we can see, although the achieved objective of the three predicted illuminator configurations are not as good as the first one, they're better choices if we compare to the second illuminator.

As previously mentioned, we will need to verify all optimized illuminator candidates by using a much larger set of design patterns. In many situations, it is not necessarily true that the most optimized illuminator gives best performance for all patterns in the verification. In Figure 8, we compare the lithographic performance for an optimized pixelated source, the optimized source from the 3^k full factorial DOE designs and the optimized source from the reduced DOE designs. It can be noted their performance are very close, and no big lithographic benefit for most patterns is observed except for pattern 15.

Illuminator	3^K Factorial Optimized Illuminator	Reduced DOE optimized illuminator	Reduced DOE Predicted illuminator 1	Reduced DOE Predicted illuminator 2	Reduced DOE Predicted illuminator 3
DOE	1 1 2 2 1 2 1 3	1 1 3 1 3 1 1 3	1 1 1 1 1 1 1 3	1 1 1 2 1 1 2 3	1 1 1 3 1 1 2 3
quasar_dose	0.4	0.4	0.4	0.4	0.4
quasar_sigma_band	0.1	0.1	0.1	0.1	0.1
quasar_center_radius	0.82	0.87	0.77	0.77	0.77
quasar_angle_width	30	20	20	30	40
quasar_rotation_angle	45	65	45	45	45
dipole_angle_width	30	20	20	20	20
dipole_sigma_band	0.1	0.1	0.1	0.25	0.25
dipole_center_radius	0.87	0.87	0.87	0.87	0.87
Average	0.196172619	0.220800595	0.210034524	0.218196429	0.208769048

Figure 7. Comparison of optimal settings.

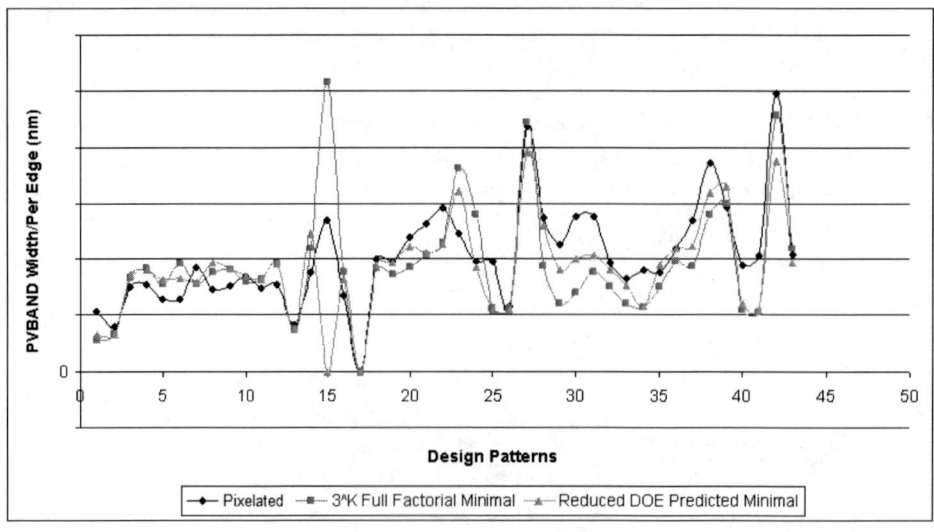

Figure 8. Comparisons between PVBAND performance for the pixelated source and DOE optimized sources.

The RS model approach also provides a benefit of conducting sensitivity analysis. Although not a primary goal for this problem as all illuminator settings are required, the relative sensitivity of input factors does provide some insights on which factors are more important than the others. The profiler in Figure 6 showed different influences each input has on the response. For example, holding other factors constant, X1 is positively correlated with response. The importance of a factor can be assessed to some extent by the steepness of the prediction trace. If the squared terms are significant, the traces may be curved, as with factor X7 and X8.

The estimation of RS model produces F-test statistics for all main, interaction and quadratic terms as shown in Table 5. The effect of a term is said to be statistically significant when the p-value, identified by "Prob > F", is below .05. The relative importance of the terms can be assessed by ranking their p-values from low to high. The conclusions on sensitivities are somewhat consistent with observations from the profiler, though the curvature effects can change the profiler traces depending on what values related factors take. Here X1, X2, X7 and X8 have more influence to response from both profiler and their smaller p-values. Response is relatively insensitive to the setting of factor X4.

Table 5. RS model-based sensitivity analysis

Effect Tests

Source	Nparm	DF	Sum of Squares	F Ratio	Prob > F
X1(1,3)	1	1	0.1191831	4.7039	0.0312*
X2(1,3)	1	1	0.3699291	14.6004	0.0002*
X3(1,3)	1	1	0.0390723	1.5421	0.2157
X4(1,3)	1	1	0.0396774	1.5660	0.2122
X5(1,3)	1	1	0.1842564	7.2723	0.0076*
X6(1,3)	1	1	0.0128288	0.5063	0.4775
X7(1,3)	1	1	9.4344055	372.3581	<.0001*
X8(1,3)	1	1	6.5750477	259.5047	<.0001*
X1*X2	1	1	0.3829920	15.1160	0.0001*
X1*X3	1	1	0.0591551	2.3347	0.1280
X2*X3	1	1	0.0456308	1.8010	0.1811
X1*X4	1	1	0.0124675	0.4921	0.4838

6. CONCLUSION AND FUTHER WORK

In the DOE approach, simulation runs generate multiple responses for design patterns that individually vary on input settings. An objective function of the several responses is constructed based on minimizing averaged relative CD variations, which simplifies the goal to the minimization of a single response. A response surface (RS) design, center composite face (CCF), is proposed to solve the problem with reduced runs. With an input dimension of eight, we choose a CCF of 273 runs, while 6,561 runs are usually run for a 3^k design. This approach identifies several optimal settings by adjusting inputs according to an improved second-order RS model. Verification runs showed that these solutions generate values smaller than the minimum of observed data but larger than the global minimum.

The proposed approach largely improves the efficiency by reducing the number of experiments 30 times. It also enables us to explore a wide range of illuminator options and to understand the impact of illuminator parameters as well as tradeoffs among 1D, 2D and SRAM design patterns.

Future improvements may take several directions. First, sequential experiments can be built based on the CCF screening experiment. However, when the response surface has multiple optimums, the principle of sequential designs following steepest path of decent will not be as accurate. Besides, we often find the optimal settings are near the boundary of factor levels. Another direction is to look into other optimal designs. As shown in Figure 9, Latin-Hypercube designs choose levels that are equally spaced for each factor. As a popular space-filling design, it chooses points that explore the parameter space in a most efficient way. These designs stay small, though the subsequent modeling takes a different path from RSM. Gaussian process model can capture more curvatures in the response surface.

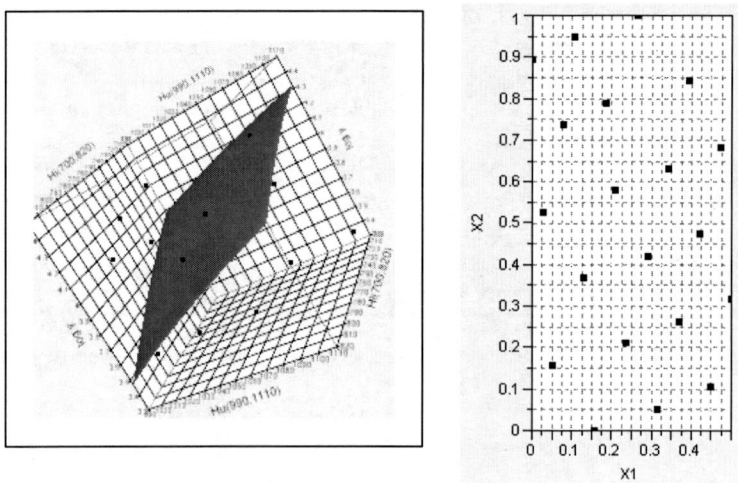

Figure 9. A Latin-Hypercube design

ACKNOWLEDGEMENTS

The authors would like to thank Luminescent Technology, Inc. for its support on simulation experiments.

REFERENCES

[1] Rosenbluth, A.E., Melville, D. O., Tian, K., Bagheri, S., Azpiroz, J.T., Lai, K., Waechter, A., Inoue, T., Ladanyi, L., Barahona, F., Scheinberg, K., Sakamoto, M., Muta H., Gallagher, E., Faure, T., Faure, T., Hibbs, M., Tritchkov, A., and Granik, Y., "Intensive Optimization of Masks and Sources for 22nm Lithography", Proc. of SPIE, Vol. 7274 (2009).
[2] Tolani, V., Hu, P.,Peng, D., Sinn, R., Pang, L. and and Gleason, B., "Source-Mask co-Optimization (SMO) using Level Set Methods", Photomask Technology, Proc. of SPIE, Vol. 7488 (2009).
[3] Deng, Y.F., Zou, Y., Yoshimoto, K., Ma, Y.S., Tabery, C.E., Kye, J., Capodieci, L., Levinson, H.J., "Considerations in source-mask optimization for logic applications", Proc. of SPIE, Vol. 7460-53 (2010).
[4] Zou, Y., Deng, Y.F., Kye, J., Capodieci, L., and Tabery, C.E., "Evaluation of Lithographic Benefits of using ILT Techniques for 22nm-node", Proc. of SPIE, Vol. 7640-20 (2010).
[5] Montgomery, D.C., "Design and Analysis of Experiments", 6th edition, John Wiley and Sons (2005).
[6] Myers, R.H., Montgomery, D.C., "Response Surface Methodology", 2nd edition, John Wiley and Sons (2009).
[7] SAS JMP 8.0.1. User help manual.

Source-mask optimization (SMO): from theory to practice

Thuc Dam*[a], Vikram Tolani[a], Peter Hu[a], Ki-Ho Baik[a], Linyong Pang[a], and Bob Gleason[a]
Steven D. Slonaker[b] and Jacek K. Tyminski[b]
[a]Luminescent Technologies, Inc., 2471 East Bayshore Road, Palo Alto, CA, USA 94303;
[b]Nikon Precision Inc, 1399 Shoreway Road, Belmont, CA, USA 94002-4107

ABSTRACT

Source Mask Optimization techniques are gaining increasing attention as RET computational lithography techniques in sub-32nm design nodes. However, practical use of this technique requires careful considerations in the use of the obtained pixilated or composite source and mask solutions, along with accurate modeling of mask, resist, and optics, including scanner scalar and vector aberrations as part of the optimization process. We present here a theory-to-practice case of applying ILT-based SMO on 22nm design patterns.

Keywords: SMO, ILT, OPC, optical aberrations, scanner aberrations, Zernike aberrations, Jones pupil aberrations

1. INTRODUCTION

Aberrations induce edge placement errors that contribute to variation in sizes and displacement of patterns. Although advanced projection optics have very small aberrations, requirements for total CD control and overlay approaching 2 nm make it important to consider any single component that might contribute as little as 0.5 nm to a CD or overlay budget. Standard practices in common use cover some of the effects of aberrations. OPC models calibrated using SEM measurements of printed patterns easily capture the effects of aberrations on line widths, but aberrations may make it invalid to assume that the measured line widths correspond to equal and opposite deviations of edges from a target. A model based on this invalid assumption will not predict pattern placement properly, and may not extrapolate well to patterns outside its calibration data set. If masks or illuminators are designed to improve depth of focus in addition to putting nominal images on target, performance with respect to aberrations that have certain similarity to defocus will also improve. Any order of spherical aberration or astigmatism is in this class. On the other hand, improving performance in the presence of an asymmetric aberration, such as coma, may require explicit consideration of the aberration during optimization.

The work reported here covers optimization of an illuminator and masks to print a set of test patterns in the presence of either scalar aberrations, represented as a Zernike expansion, or vector aberrations, represented as a Jones pupil. Although it would be instructive to consider the effects of single aberrations, we chose to use sets of many small aberrations because no modern lithographic lens design has significant single aberrations. The aberrations we used do not correspond to an actual scanner. We chose values that are larger than those of real designs because we wanted to explore the limits to which mask and illuminator design can mitigate the effects of aberrations. For the same reason, we did not constrain complexity of mask patterns. Trade-off between lithographic performance and simplification of mask patterns to reduce manufacturing cost is an important topic, but imposing such constraints on the optimization would interfere with our objective of exploring the limits of what is possible in the presence of aberrations. The cases we compare include optimizing both the illuminator and the mask with explicit treatment of aberrations, neglecting aberrations while optimizing the illuminator but including them while optimizing the mask, and neglecting them while optimizing both the illuminator and the mask.

2. EXPERIMENTAL AND RESULTS

2.1 Experimental Setup

The patterns selected for this work are dark-field arrays of square apertures, each of which is well-suited to quadrupole illumination if optimized individually without aberrations. Some perform better if the poles are along the x and y axes, and others if they are rotated by 45°, and the optimum pole locations vary with pattern pitch. The illuminator obtained

Optical Microlithography XXIII, edited by Mircea V. Dusa, Will Conley, Proc. of SPIE Vol. 7640,
764028 · © 2010 SPIE · CCC code: 0277-786X/10/$18 · doi: 10.1117/12.848257

when images of the patterns are optimized collectively is therefore a compromise. Some of the arrays have 90° rotational symmetry. Others have only 180° symmetry, although effects of departure from 4-fold rotational symmetry are mostly confined to shapes near the corners of the arrays. Based on the symmetry of the layouts, we expect that a parametric representation of the illuminator as a pair of quadrupoles will give good results for the unaberrated optimization. Because aberrations can introduce additional asymmetry into the problem, we chose to represent the illuminator as four pairs of dipoles, the lowest illumination symmetry that would not de-center an image out of focus. The SMO algorithm was sequential, alternating between adjusting the illuminator and mask to minimize the same objective function. Details of the method used for mask optimization are published elsewhere. [1,2] Because our objective was to study optical phenomena, we did not use a model for a particular resist in this work, and simply applied a constant threshold to the bulk intensity in the resist film to calculate locations of pattern edges.

Table 2.11 describes the matrix of source mask optimization (SMO) experiments. The parametric SMO experiments consist of two components 1) source and 2) mask optimizations. The source optimization employed is a bounded parametric optimization whereby source parameters inner/outer sigma, fan open angles, and fan intensities are varied. The initial source consists of 4 independent dipoles as depicted in Figure 2.11. The objective function to be optimized includes nominal, -50 nm defocus, and +/- 1 nm mask biased images, each with equal weight. The mask is allowed to be free-form curvilinear with mask rule check (MRC) turned off, consistent with our objectives to investigate the limits of unconstrained optimization. As depicted in table 2.11, SMO was conducted with and without XY polarization, Jones pupil aberrations, and Zernike aberrations. We did not cover the dependence of aberrations on field coordinates in this work because such exhaustive characterization of projection optics is not generally available. Treating the field dependence properly would also involve averaging across the scanning direction, which is unnecessary here because we treat only mean aberrations averaged over the field. Variables in the optimization also included photoresist threshold.

	SMO Conducted with		
	No Aberations	JonesPupil	Zernike
No Polarization	SMO-NN	SMO-JN	SMO-ZN
XY Polarization	SMO-NP	SMO-JP	SMO-ZP

Table 2.11. SMO experiments.

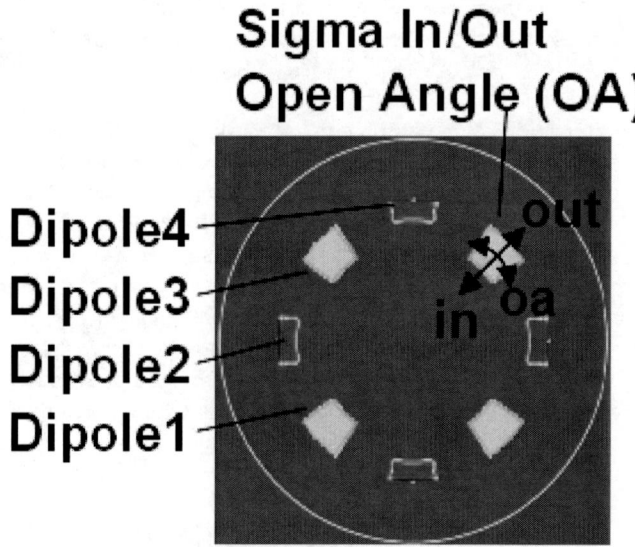

Figure 2.11. Source parameters being varied by parametric optimizer.

Other fixed setup parameters are:

    NA = 1.3
    Resist Filmstack = Typical 193nm resist setup with matched substrate.

Image/Focal Plane = top of film stack
Vector optical model

The following list contains the variables in the optimization, their initial values, and bounds (in brackets).

Dipoles 1 and 3
  Sigma In = 0.5 [0, 0.7]
  Sigma Out = 0.9 [0.7, 1]
  Open Angle = 20° [0 ,40°]
  Starting Angle = 45° & 135° [Fixed]
  Fan Intensity = 0.7 [0, 1]
Dipoles 2 and 4
  Sigma In = 0.6 [0, 0.7]
  Sigma Out = 0.8 [0.7, 1]
  Open Angle = 20° [0 ,40°]
  Starting Angle = 0 & 90 [Fixed]
  Fan Intensity = 1 [0, 1]
Photoresist Threshold = 0.24

Input target patterns used for SMO are described in Figure 2.12. Each pattern has equal weight in the SMO cost function with the pattern region for optimization being focused on line-ends. Pattern corners are not considered in the optimization.

The collected litho performance metrics are line ends edge placement errors (EPE), mask error enhancement factor at +/- 1 nm mask bias (MEEF), and process variation band between images at 0, +50 nm, and -50 nm defocus at 100% dose (PV-band).

type	target CD	pattern pitch	clip#
rect CH	70	160	1
rect CH	70	180	2
rect CH	70	200	3
rect CH	80	160	4
rect CH	80	180	5
rect CH	80	200	6
stagger CH	80	180	7
stagger CH	80	200	8
stagger CH	80	220	9
stagger CH	80	240	10
stagger CH	80	260	11
stagger CH	80	280	12
stagger CH	80	300	13
stagger CH	80	320	14
stagger CH	70	180	15
stagger CH	70	200	16
stagger CH	70	220	17
stagger CH	70	240	18
stagger CH	70	260	19
stagger CH	70	280	20
stagger CH	70	300	21
stagger CH	70	320	22

clip# 1

clip# 15

Figure 2.12. SMO target patterns used for multi-clip optimization.

In the first set of SMO experiments, no aberrations are included during the optimization. The source and masks obtained will be referred to as SMO-NN$_{src}$ and SMO-NN$_{msk}$. The experiment is repeated with XY polarization, and this source mask combination will be referred to as SMO-NP$_{src}$ and SMO-NP$_{msk}$. The forward simulations for these masks and sources are collected, and are considered the ideal results without any aberrations used in either optimizations or forward simulations. In practice scanners, however, do have aberrations, which if unaccounted during SMO optimization could lead to undesirable images during print or simulations. To assess these unaccounted aberrations, the SMO-NN & -NP masks are forward simulated with Jones pupil and Zernike aberrations added to the SMO-NN(P)$_{src}$, and these simulated results will be referred to as SMO-NN(P)$_{msk-fwdJP}$ and SMO-NN(P)$_{msk-fwdZern}$. To assess whether mask optimization alone can correct for these unaccounted aberrations, ILT mask optimization was conducted using SMO-NN(P)$_{src}$ with Jones pupil and Zernike aberrations included in the forward simulation engine. These masks and simulated results will be referred to as SMO-NN(P)$_{msk-ILT\&fwdJP}$ and SMO-NN(P)$_{msk-ILT\&fwdZern}$.

In the second set of SMO experiments, Jones pupil aberrations are included during the optimization. The source and masks obtained will be referred to as SMO-JN$_{src}$ and SMO-JN$_{msk}$. The experiment is repeated with XY polarization, and this source combination will be referred to as SMO-JP$_{src}$ and -JP$_{msk}$.

In the third set of SMO experiments, Zernike aberrations are included during the optimization. The source and masks obtained will be referred to as SMO-ZN$_{src}$ and SMO-ZN$_{msk}$. The experiment is repeated with XY polarization, and this source combination will be referred to as SMO-ZP$_{src}$ and -ZP$_{msk}$.

2.2 Results

Figure 2.13 shows the sources that were obtained at the end of each SMO experiment.

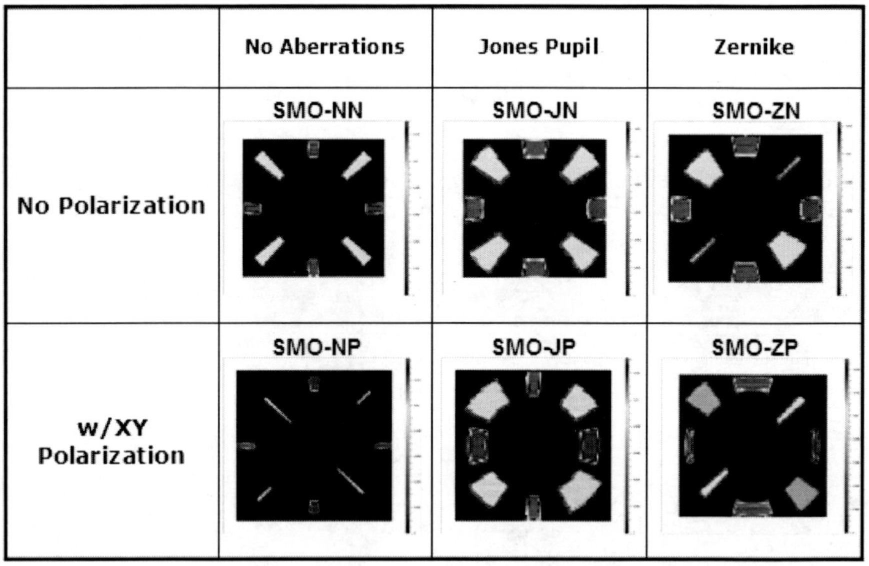

Figure 2.13 Parametric SMO sources obtained from 3 experiments with different aberrations and polarization.

The sources obtained with unaberrated light shows high degree of symmetry and coherent fans. With inclusion of Jones pupil aberrations into SMO, the sources have more asymmetric fans with wider open angles. As Zernike aberrations are applied, the 45° dipole fans shrink significantly and increases the asymmetry in the SMO source.

We selected clip 2 in this study to illustrate how these aberrations affect SMO solutions and their images. The mask and images generated from the first SMO experiments (SMO-NN$_{src}$ and SMO-NN$_{msk}$) are shown in Figure 2.14. Note the absence of perfect four-fold rotational symmetry in these solutions, even though this clip has a four-fold axis. This is due to the illuminator not having exact four-fold symmetry due to its joint optimization with other patterns, such as clip 15 in Figure 2.12. As is discussed below, this degree of asymmetry is small compared to that induced by the aberrations, so it has no effect on our conclusions. In this figure, the green images are generated by simulating the blue masks at nominal and +/50 nm defocus conditions without any aberrations. As one can see, the nominal image converged to the red square target, while the + and – 50 nm defocus images (right side) are smaller than target by ~3 nm. Both defocus images are nearly on top of each other indicating that defocus asymmetry is insignificant.

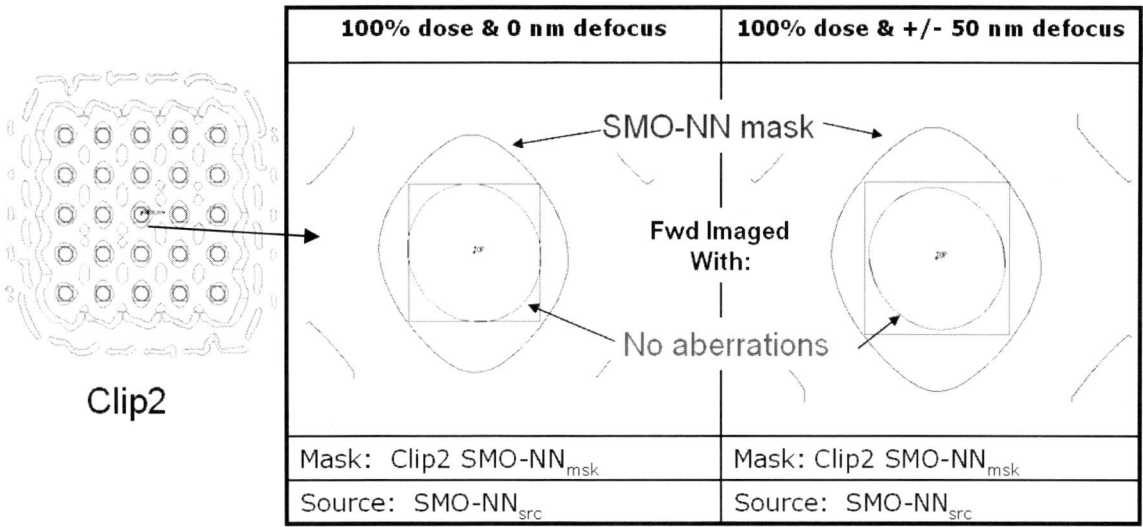

Figure 2.14. Clip2 SMO-NN$_{msk}$ simulated with SMO-NN$_{src}$ without aberrations at nominal and +/- 50 nm defocus.

When Jones pupil aberrations are applied during simulation with SMO-NN$_{src}$ and SMO-NN$_{msk}$ (Figure 2.15), the red nominal and defocus images are smaller than the green, unaberrated images.

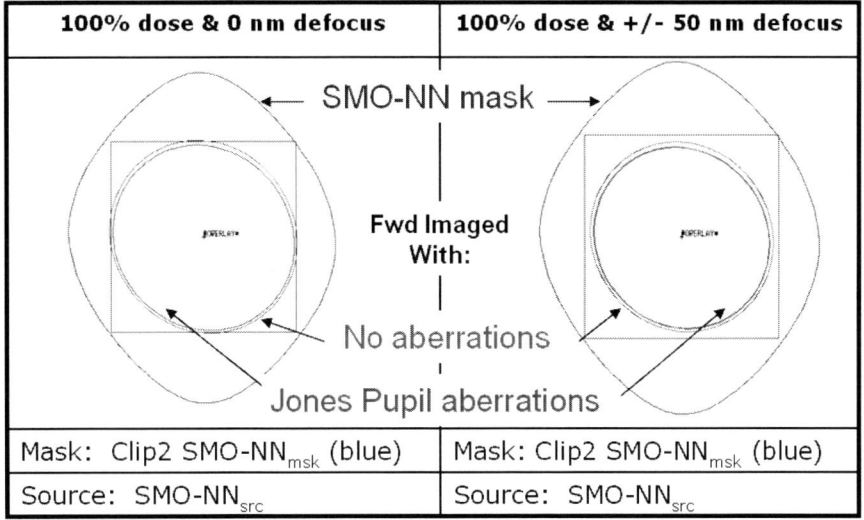

Figure 2.15. Clip2 SMO-NN$_{msk}$ simulated with SMO-NN$_{src}$ with Jones pupil aberration at nominal and +/- 50 nm defocus.

When Zernike aberrations are applied during simulation with SMO-NN$_{src}$ and SMO-NN$_{msk}$ (Figure 2.16), the red nominal and defocus images are shifted 9-10 nm to the right of the green, unaberrated images. The two defocus images (-/+ 50 nm) become asymmetric in defocus and are separated by ~3 nm, where as the green defocus images are nearly on top of one another.

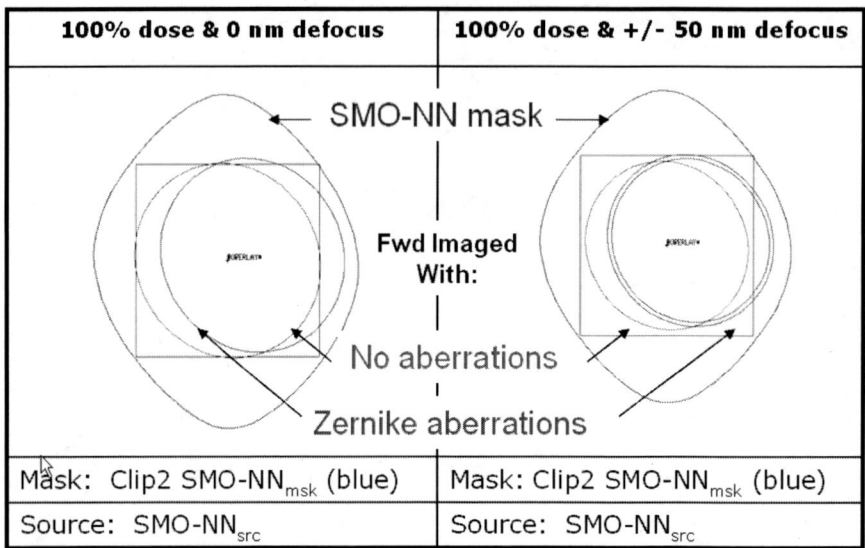

100% dose & 0 nm defocus	100% dose & +/- 50 nm defocus
Mask: Clip2 SMO-NN$_{msk}$ (blue)	Mask: Clip2 SMO-NN$_{msk}$ (blue)
Source: SMO-NN$_{src}$	Source: SMO-NN$_{src}$

Figure 2.16. Clip2 SMO-NN$_{msk}$ simulated with SMO-NN$_{src}$ with Zernike aberration at nominal and +/- 50 nm defocus.

The overall data for the first experiment is summarized for all clips and imaging conditions in Figure 2.17. This result shows the Zernike aberrations affected image quality the most. It shows large EPE deviation (9-10 nm) and larger PV bands for all 22 clips, most of which as demonstrated in the clip2 simulated images are due to lateral image shift. Jones pupil aberrations also show minor EPE & PV shift. Neither aberration affects MEEF very much, and XY polarized light tends to contribute to PV band improvements (Figure 2.17c).

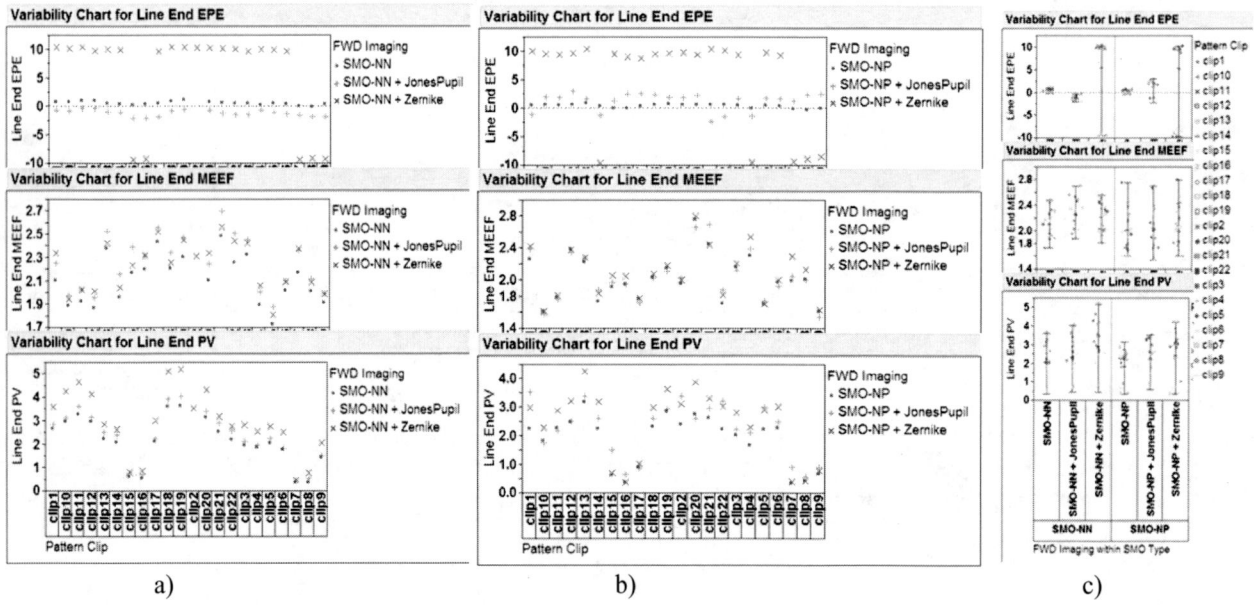

Figure 2.17. a) SMO-NN's line end EPE, MEEF, and PV results. b) SMO-NP's line end EPE, MEEF, and PV results. c) comparison plot between SMO-NN & SMO-NP results.

After documenting the lithographic effects of Jones pupil and Zernike aberrations, we considered two possible ways to mitigate them. One approach is to include aberrations only when optimizing the mask, using the source obtained from the SMO solution without aberrations. The other is to include the aberrations in all SMO steps. In continuation of the first SMO experiment, SMO-NN$_{src}$ is used during mask inversion for clip 2 with aberration. In Figure 2.18, SMO-NN$_{src}$ is used for inversion with Jones pupil aberration. The resulting blue mask (SMO-NN $_{msk-ILT\&fwdJP}$) is slightly larger than

SMO-NN$_{msk}$, and the comparison of the simulated images showed that the blue image contours are converged to the red square target. This indicates that the Jones pupil effects of image shrinkage can be mitigated with mask correction alone.

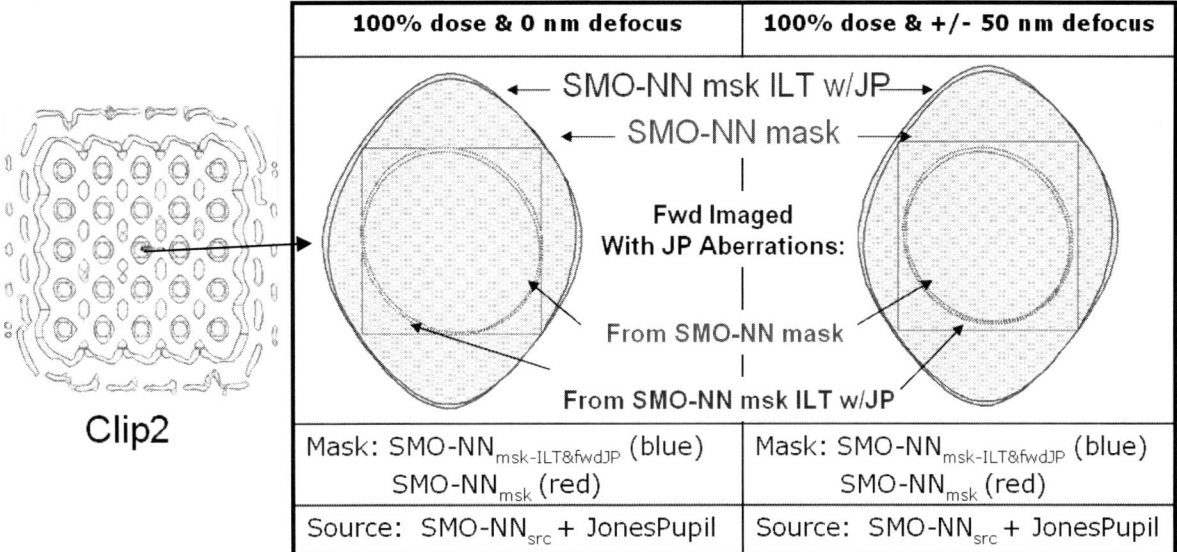

Figure 2.18. With a fixed SMO-NNsrc, mask inversion with Jones pupil can correct image shrinkage caused by Jones pupil aberration.

In Figure 2.19, SMO-NN$_{src}$ is used for inversion with Zernike aberration for clip 2. The resulting blue mask (SMO-NN $_{msk-ILT\&fwdZern}$) is shifted to the left compared to SMO-NN$_{msk}$. The simulated images showed that the blue nominal image contour is better converged to the red square target. This indicates that the Zernike effects of lateral image shift can be mitigated with mask correction. The defocus image contours are also better centered, but the defocus asymmetry between +50 nm and -50 nm defocus remains uncorrectable by mask optimization.

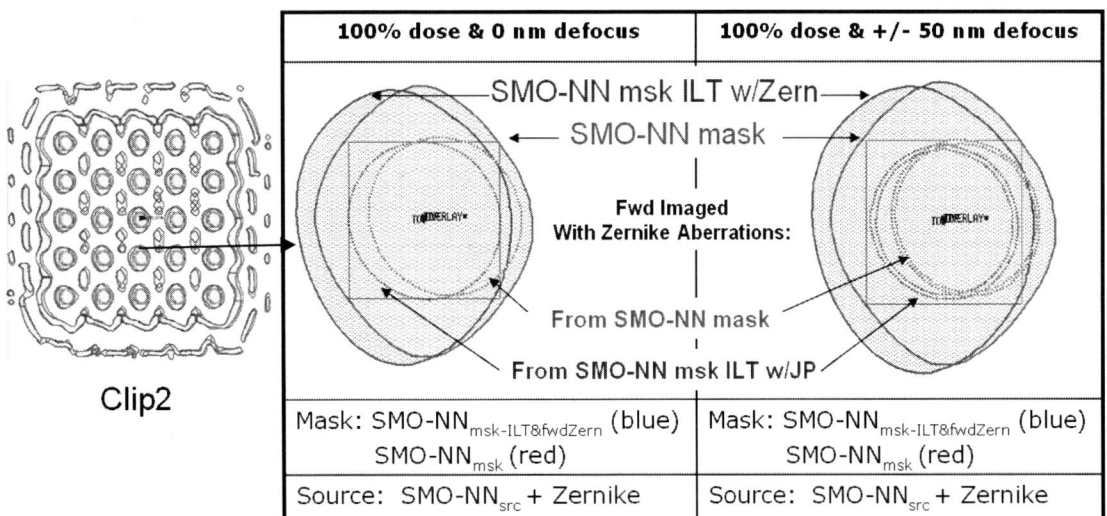

Figure 2.19. With a fixed SMO-NNsrc, mask inversion with Zernike can correct image shift caused by Zernike aberration, but does not seem to be able to correct for the defocus asymmetry.

Figure 2.20 summarizes all the lithographic performance data for mask inversion studies. In general, this data shows that mask inversion with aberrations can correct for the image shift and shrinkage caused by Zernike and Jones pupil aberrations, respectively, without any impact on MEEF. There, however, remains a gap in PV band performance that cannot be recovered by mask inversion alone with the Zernike aberrations. XY polarized light improves PV band performance when Zernike aberrations are present for the conditions and patterns in this study. (Figure 2.20c).

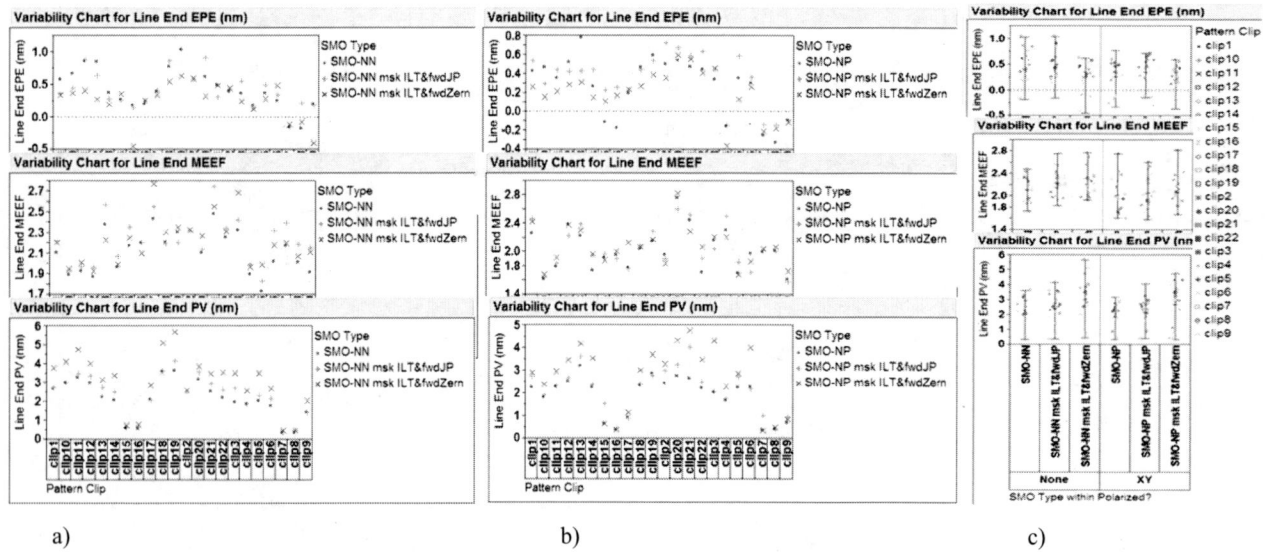

a)   b)   c)

Figure 2.20. a) SMO-NN's line end EPE, MEEF, and PV results for mask inversion that includes aberrations. b) SMO-NP's line end EPE, MEEF, and PV results for mask inversion that includes aberrations. c) comparison plot between polarized and unpolarized results.

In the second set of SMO experiments, Jones pupil aberrations are included in each step of the optimization. The resulting clip2 SMO-JN$_{msk}$ and simulated images are shown in Figure 2.21. Relative to the SMO-NN$_{msk}$, SMO-JN$_{msk}$ is larger, and its resulting simulated images are converged for nominal and are well matched at defocus. It would appear that SMO with Jones pupil accounted for during optimization can result in images that have comparable performance to ideal case.

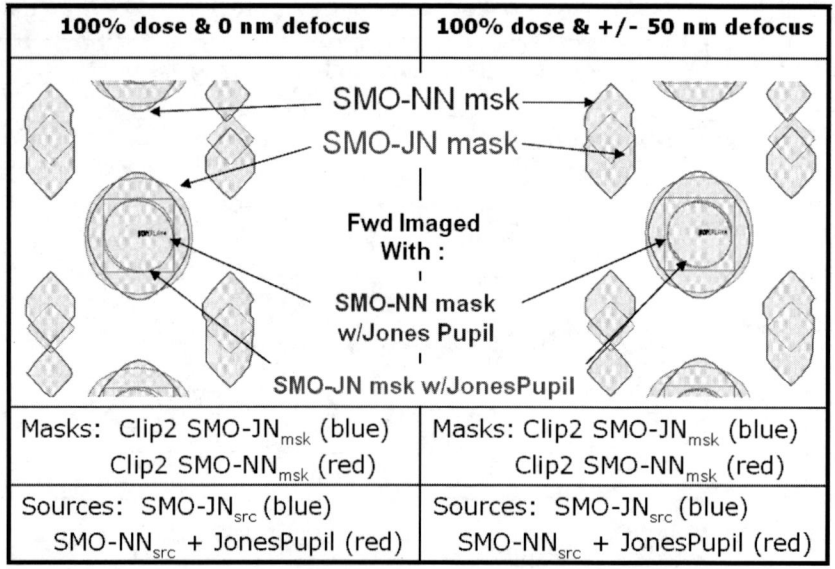

Figure 2.21. Clip 2 SMO-JN$_{msk}$ and simulated images.

In the third set of SMO experiments, Zernike aberrations are included in each step of the optimization. The clip 2 mask SMO-ZN$_{msk}$ and simulated images are displayed in Figure 2.22. The blue SMO-ZN$_{msk}$ is correcting for the Zernike aberration by shifting to the left as was also seen in the mask-only optimization. The other interesting change is the different SMO-ZN$_{msk}$'s SRAF, which occurred mostly likely due to difference in the source. Image centering and targeting are close to those of the unaberrated results in Figure 2.14, although differences between contours corresponding to positive and negative defocus values are not completely eliminated by including the aberrations in the SMO solution.

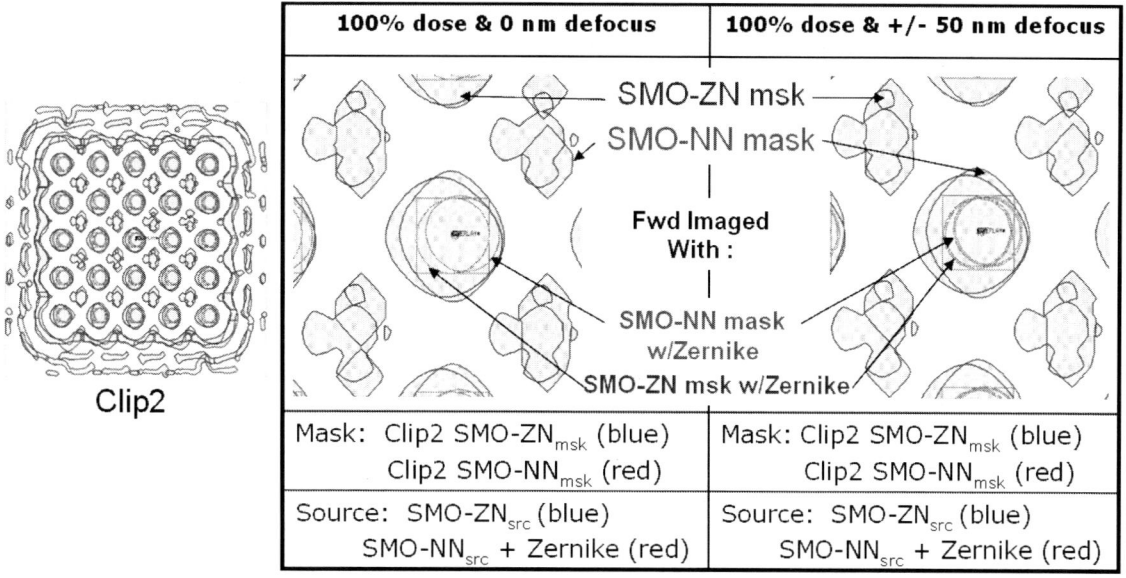

Figure 2.22. Clip 2 SMO-ZN$_{msk}$ and simulated images.

The compilation of litho performance metric for SMO conducted with Jones pupil and Zernike aberrations are summarized in Figure 2.23. This data indicates that SMO with aberrations can lead to converged images (EPE close to 1 nm for all cases), and very little impact on MEEF. PV is best for the unaberrated case, while the Zernike aberrations appear difficult to correct, most likely due to the asymmetry in defocus images. XY polarization had little effect on SMO conducted with aberrations for the patterns in this study.

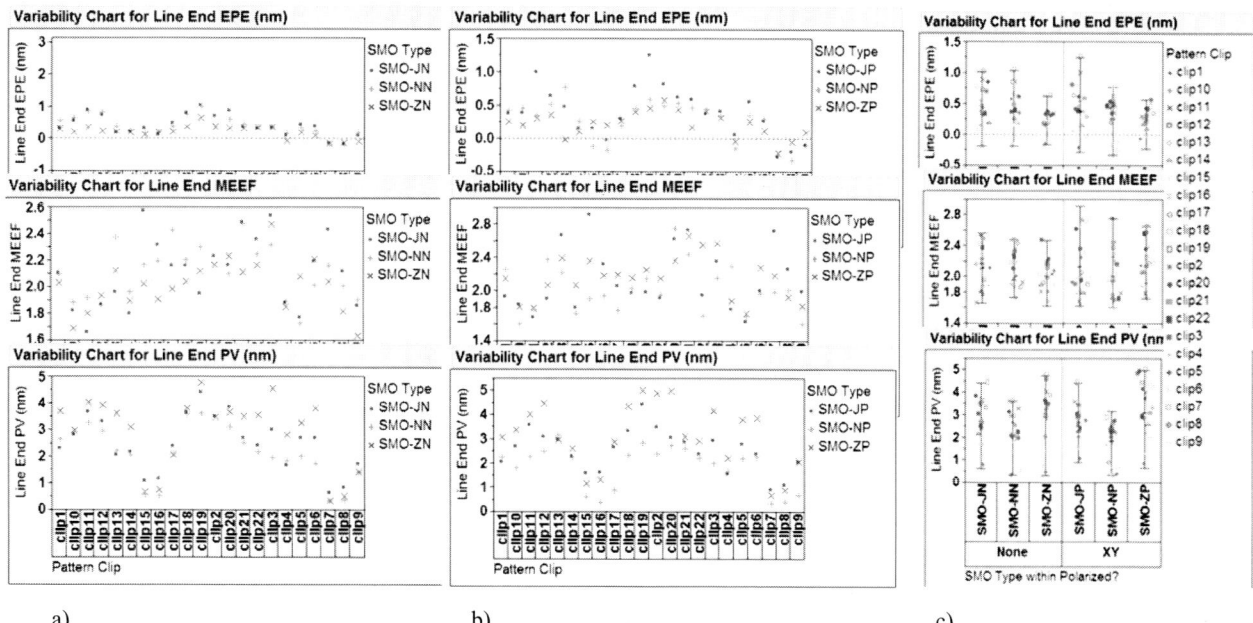

a)   b)   c)

Figure 2.23. a) Line end EPE, MEEF, and PV results from SMO with aberrations included and with unpolarized light. b) Line end EPE, MEEF, and PV results from SMO's with aberrations included and with XY polarized light. . c) comparison plot between XY polarized and unpolarized results

3. DISCUSSIONS & CONCLUSION

Accounting for aberrations during each step of the SMO process can produce solutions whose lithographic performance is close to that of the ideal unaberrated case, even with aberrations that are larger than those of any modern projection optics. For the types of patterns in this study, dark-field arrays of square apertures, the dominant effects of aberrations showed in nominal image fidelity and depth of focus. MEEF performance was a much less sensitive indicator. We observed improvement in image quality for both scalar and vector aberrations, and for unpolarized and polarized illumination. Although the scalar aberrations we chose to study showed poorer lithographic performance than the vector aberrations, this was simply due to the presence of a large image shift that happened to be in the Zernike coefficients, and says nothing about the relative importance of using the Jones pupil formalism to cover dependence of aberrations on polarization. One of the most important conclusions from this work is that even with such a large aberration, which shifts and distorts the nominal image contour, we obtain good solutions both in and out of focus by properly accounting for it during SMO. This aberration is a good example of one that affects both overlay and CD variation. Even if it were practical to calibrate an OPC model using measurements of individual edge positions rather than CDs in an attempt to account for pattern displacements, it would be impossible to determine whether the displacement of this particular pattern was due to simple distortion (pattern-independent) or an aberration such as coma (pattern-dependent), solely from measurements performed on its image. This is why explicit treatment of aberrations in the SMO process provides better extrapolation outside of the patterns in the calibration data set, and better prediction of off-nominal image properties. These considerations will become more important as we push total budgets for overlay and CD control down to the 2 nm level.

REFERENCES

[1] Daniel Abrams and Lingyong Pang, Fast Inverse Lithography Technology, Proceedings of the S.P.I.E. 6154-55, 2006

[2] Daniel Abrams, Danping Peng, and Stanley Osher, Method or Time-Evolving Rectilinear Contours Representing Photo Masks, U.S. Patent 7,124,394, 2006

Partial spatial coherence in an excimer-laser lithographic imaging system

Arlene Smith, Anna Burvall and Christopher Dainty

School of Physics, National University of Ireland, Galway, Ireland

ABSTRACT

We have recently explored the Elementary Function method, previously presented by Wald et al (Proc. SPIE 59621G, 2005), and we have demonstrated under what circumstances this method can be used to reduce the propagation calculations of partially coherent light to two dimensions. In this paper, we examine the methods used to measure the spatial coherence of a light source in the literature. We present a method based on work previously shown by Mejía et al (Opt Comm 273 (428-434), 2007) which uses an array of pinholes with one degree of redundancy. We discuss the design of the pinhole array and present the results of some simulations.

Keywords: partial coherence, excimer, elementary function, imaging

1. INTRODUCTION

Excimer lasers operating in the Deep UV (193nm and 248nm) have been used in lithographic systems for the last couple of decades. The output from a pulsed excimer laser is spatially partially coherent. This adds complexity to simulations of light propagation through excimer imaging systems as partially coherent calculations require four-dimensional calculations. The reason for this is the correlation between fields at different points in space must be taken into account, so integration must be performed not just over all points of a two-dimensional field distribution, but over all pairs of points.

We have recently explored the Elementary Function method[1,2] previously presented by Wald et al[3] and, in the space-time and space-frequency domains, by Vahimaa and Turunen[4] and Friberg et al.[5] We have demonstrated under what circumstances this method can be used to reduce the propagation calculations of partially coherent light to two dimensions. This increases the speed of the calculations and reduces the need for high memory capacities.

In this paper we examine the methods used to measure the spatial coherence of a light source previously published in the literature. We present a method based on work previously shown by Mejía and Gonzáles[6] which uses an array of pinholes with one degree of redundancy. We discuss the design considerations for the pinhole array. We present the results of simulations for the fully coherent and partially coherent cases.

2. THEORY

Spatial coherence is the study of the randomness of a field over spatial coordinates, usually in directions perpendicular to the propagation of the optical field. The measurement of spatial coherence is well explored in the literature. In lithographic systems, it is usually necessary to narrow the bandwidth of the laser. The excimer laser is operated in the pulse mode with a broad linewidth, so both the temporal and spatial coherences are low. However, the process of line narrowing can increase the spatial coherence of the laser quite dramatically. Also, many lithographic systems employ homogenizing optics to smooth the intensity profile of the beam. It leads on from this that precise information about the degree of spatial coherence in the source is important for the development of accurate models of lithographic imaging systems.

Author contact: arlene.smith@nuigalway.ie

Optical Microlithography XXIII, edited by Mircea V. Dusa, Will Conley, Proc. of SPIE Vol. 7640,
764029 · © 2010 SPIE · CCC code: 0277-786X/10/$18 · doi: 10.1117/12.846349

2.1. Measuring Spatial Coherence: Existing Methods

Many methods to measure spatial coherence of a light source are based on the Young interferometer. In 1938, Zernike determined the degree of coherence directly from the visibility of the interference fringes formed in a Young interferometer. The spatial coherence of KrF excimer lasers was explored by Kawata et al[7] using a shearing interferometer with double gratings. The visibility of the interference fringes at the shear of the two first-order diffracted beams gives the coherence length of the laser.

Methods based on the measurement of fringe visibility have involved masks containing two apertures, but this has the disadvantage that the mask must be moved laterally to sample all parts of the beam. Also, if the phase and modulus of the complex degree of spatial coherence are shift variant, a single interferogram is not sufficient to fully characterise the light field. Nugent and Trebes[8] and Castañeda[9,10] proposed an alternative: using a mask with multiple apertures spaced evenly (uniformly redundant array) and analysing the Fourier spectrum. Mejía and Gonzáles[6] showed experimental results of a method involving a nonredundant array. In his case, the array has five apertures and each spacing is unique and an integer multiple of the smallest spacing. The mask is illuminated by a coherent laser source (633nm) decohered using a piece of rotating ground glass. When the array of pinholes is illuminated by the source, a pattern is produced which is the result of the intererence of each aperture pair adding together. Fourier analysis of the far-field interferogram follows. The degree of spatial coherence can de deduced from the visibility of the interference fringes multiplied by a factor dependent on the intensity at each aperture as shown by Wolf[11],[6]

2.2. Pinhole array with one degree of redundancy

The source to be measured is a 248nm KrF excimer laser. In our design, we considered the advantages of the pinhole array over two-pinhole set-up. One measurement using a mask with multiple pinholes would provide sufficient data to deduce the spatial coherence of the source. The sizes of the pinholes were chosen to allow the mask to be used with various sources of different wavelengths. Figure 1 shows the design of the pinhole plate. The layout of the pinholes is suitable for measurements of spatial coherence in both the horizontal and vertical direction. Figure 2 shows the spacing of the pinholes in the array. The spacings chosen form an array with one degree of redundancy i.e. one aperture pair spacing occurs more than once. In this case, a spacing of five units occurs twice. This will affect the distribution of intensity among the interference fringes. The number of pinholes (five) will give nine classes of aperture pair. Under spatially partially coherent illumination, the effective contributing classes will be chosen by the modulus of the complex degree of spatial coherence, in such a way that its magnitude specifies the weight of the contribution and its support the number of contributing classes. From this, the modulus and complex degree of coherence can be determined.

Pair	Class	Separation (mm)
2,3	1	0.08
1,2	2	0.16
1,3	3	0.24
3,4	4	0.32
2,4	5	0.40
4,5	5	0.40
1,4	7	0.56
3,5	9	0.72
2,5	10	0.80
1,5	12	0.96

Table 1. Classes of aperture pairs yielded by the mask of Figure 2.

3. SIMULATION & RESULTS

Initial simulations have involved testing for the case of fully coherent light. Figure 3(a) is the intensity cross section of the interference pattern produced for the five pinhole array with spacing as in Figure 1. The pinhole

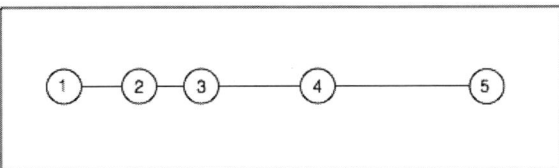

Figure 1. Layout of five-pinhole array. Spacings: $1{-}2 = 0.16mm$, $2{-}3 = 0.08mm$, $3{-}4 = 0.32mm$, and $4{-}5 = 0.40mm$.

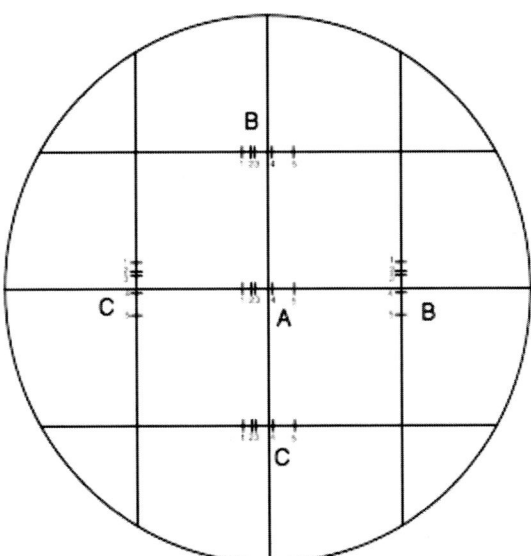

Figure 2. The finished design of mask with arrays of five pinholes. The pinhole diameters are $A = 5\mu m$, $B = 10\mu m$, and $C = 15\mu m$.

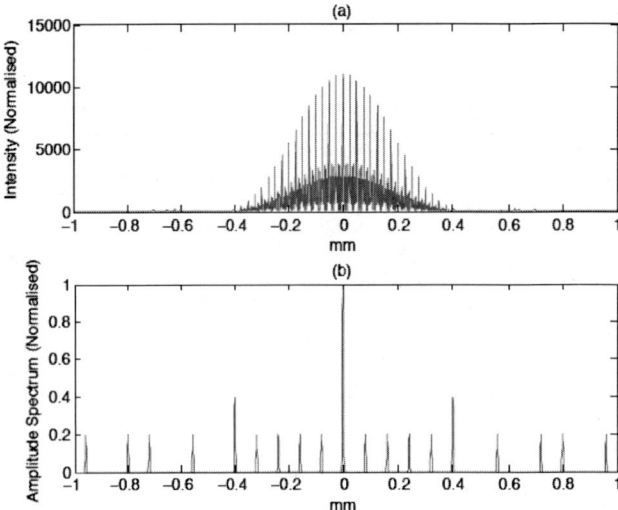

Figure 3. (a) Intensity cross section of interference fringes. (b) Plot of the Fourier Transform of the power spectrum for pinhole mask illuminated with coherent light (normalised).

diameter is $5\mu m$. The interference pattern is produced by way of a Fourier transform of the pinhole array to give the image amplitude; the intensity is the square modulus of the amplitude. Taking that intensity cross-section and performing a Fourier transform will give an output as in Figure 3(b). We can see a contribution from nine aperture pairs: classes 1 - 5, class 7, classes 9, 10 and 12. Classes 6, 8 and 11 are not possible with this arrangement, and do not contribute. As discussed in 2.2, an aperture pair of class 5 occurs twice. Thus we see a larger (double) intensity contribution at class 5.

The spatial coherence of the source can be inferred from these results using a general form of Wolf's equation.[6] The height and phase of the peaks in the Fourier spectrum determines the modulus and phase of the complex degree of spatial coherence respectively.

4. CONCLUSION

We have presented a method to measure the spatial coherence of a light source. This method can be used to determine the spatial coherence of a pulsed laser source operating in the UV domain, but is also suitable for other partially coherent sources. The use of an array of pinholes gives a usable result in one measurement, instead of several measurements from individual pinhole pairs moved laterally over the beam. The results of the spatial coherence measurements can be used in models of partially coherent imaging systems to give precise information on that system. This is particularly useful in lithographic systems in which the coherence of the source may be affected by various elements in the system, such as beam homogenization optics and line narrowing optics. The effect of the homogenization of the beam on the spatial coherence will be explored in a future publication.

ACKNOWLEDGMENTS

This research was supported by the Irish Research Council for Science, Engineering and Technology (IRCSET) and by Science Foundation Ireland under Grant No. SFI/07/IN.1/I906

REFERENCES

1. A. Burvall, A. Smith, and C. Dainty, "Elementary functions: Propagation of partially coherent light," *Journal of Optical Society of America A* **26(7)**, pp. 1721–1729, 2009.

2. A. Smith, A. Burvall, and C. Dainty, "Partially coherent image computation using elementary functions," *Proc. SPIE* **7274**, p. 727434, 2009.

3. M. Wald, M. Burkhardt, A. Pesch, H. Gross, and J. Greif, "Design of a microscopy illumination using a partial coherent light source," *Proc. SPIE* **5962**, pp. 420–429, 2005.

4. P. Vahimaa and J. Turunen, "Finite-elementary source model for partially coherent radiation," *Opt. Express* **14(4)**, pp. 1376–1381, 2006.

5. A. T. Friberg, H. Lajunen, and V. Torres-Company, "Spectral elementary-coherence-function representation for partially coherent light pulses," *Opt. Express* **15(8)**, pp. 5160–5162, 2007.

6. Y. Mejía and A. I. González, "Measuring spatial coherence by using a mask with multiple apertures," *Optics Communications* **273**, pp. 428–434, 2007.

7. S. Kawata, I. Hikima, Y. Ichihara, and S. Watanabe, "Spatial coherence of KrF excimer lasers," *Appl. Opt.* **31(3)**, pp. 387–396, 1992.

8. K. A. Nugent and J. E. Trebes, "Coherence measurement technique for short-wavelength light sources," *Rev. Sci. Instrum.* **63(4)**, pp. 2146–2151, 1992.

9. R. Castañeda and Z. Jaroszewicz, "Determination of the spatial coherence of schell-model beams with diffraction gratings," *Opt. Comms.* **173**, pp. 115–121, 2000.

10. J. Garcia-Sucerquia and R. Castañeda, "Full retrieving of the complex degree of spatial coherence: theoretical analysis," *Opt. Comms.* **228**, pp. 9–19, 2003.

11. E. Wolf, "A macroscopic theory of interference and diffraction of light from finite sources," *Proc. Roy. Soc. Lond. A* **225**, pp. 96–111, 1954.

Flexible and reliable high power injection locked laser for double exposure and double patterning ArF immersion lithography

Masaya Yoshino*, Hiroshi Umeda, Hiroaki Tsushima, Hidenori Watanabe,
Satoshi Tanaka, Shinich Matsumoto, Takashi Onose, Hiroyuki Nogawa,
Yasufumi Kawasuji, Takashi Matsunaga, Junichi Fujimoto and Hakaru Mizoguchi
Gigaphoton Inc., 400 Yokokura-Shinden, Oyama-shi, Tochigi, JAPAN 323-8558

ABSTRACT

ArF immersion technology is spotlighted as the enabling technology for the 45nm node and beyond. Recently, double exposure technology is also considered as a possible candidate for the 32nm node and beyond. We have already released an injection lock ArF excimer laser, the GT61A (60W/6kHz/10mJ/0.30pm) with ultra line-narrowed spectrum and stabilized spectrum performance for immersion lithography tools with N.A.>1.3, and we have been monitoring the field reliability data of our lasers used in the ArF immersion segment since Q4 2006.

In this report we show field reliability data of our GigaTwin series – twin chamber ArF laser products. GigaTwin series have high reliability. The availability that exceeds 99.5% proves the reliability of the GigaTwin series.

We have developed tunable and high power injection-lock ArF excimer laser for double patterning, GT62A (Max90W/6000Hz/Tunable power with 10-15mJ/0.30pm (E95)) based on the GigaTwin platform. A number of innovative and unique technologies are implemented on GT62A.
 - Support the latest illumination optical system
 - Support E95 stability and adjustability
 - Reduce total cost (Cost of Consumables, Cost of Downtime and Cost of Energy & Environment)

Keywords: 32nm node, ArF excimer laser, Injection Lock, line narrow, 193nm lithography, Immersion, spectrum bandwidth, high power

1. INTRODUCTION

193nm ArF light sources are widely used in semiconductor mass production from the 90 nm node and beyond. And the ArF immersion technology is even spotlighted as the enabling technology for the 45nm node and beyond. In addition, double patterning is considered to be most promising technology to meet the requirement of the next generation 32nm node. To achieve this, market demands for ArF light source are getting more severe, for example, higher power and narrower spectral bandwidth are required for higher throughput and higher NA lithography respectively.

We have already released an injection lock ArF excimer laser with high output power and high repetition rate for higher throughput and higher NA first immersion tool: GT60A (60W/6000Hz/0.5pm (E95)) to the ArF immersion market in Q1 2006[1]. In the technology for 45nm and beyond, a light source is required to offer a narrower spectrum and high average laser power. We succeeded in releasing the next generation model, GT61A (6kHz/60W/0.30pm (E95)) with narrower spectral bandwidth used for high-NA lithography at the 45nm node in 2007[2]. Both a newly developed high-precision E95 measuring module and a stabilization control system are provided as standard features, allowing a highly stable spectrum performance throughout the entire product lifetime. The higher throughput model, GT62A (6kHz/90W/0.30pm (E95)) with the higher power was developed for double patterning lithography at the 32nm node[3]. For the GT62A, a variety of technologies to reduce the running cost of laser is introduced, which is applicable backward for the previous GigaTwin series lasers[4]. In addition, the latest generation model GT62A-1SxE is the laser matching the enhancement technology of advanced exposure systems. For example, in order to provide illumination power optimum for resist sensitivity, it has extendable power from 60W to 90W. All laser systems are built on the GigaTwin platform, a common and reliability-proven platform. (Table 1)

*masaya_yoshino@gigaphoton.com; phone +81-285-28-8416; fax +81-285-28-8439; http://gigaphoton.com.

Optical Microlithography XXIII, edited by Mircea V. Dusa, Will Conley, Proc. of SPIE Vol. 7640,
76402A · © 2010 SPIE · CCC code: 0277-786X/10/$18 · doi: 10.1117/12.846337

In this paper, we report on the innovative technology of GT62A-1SxE and reliability data of the GigaTwin series in the field.

Technology Node (typical)	Main driver	Requirement for ArF Laser light source	Power	GT model
32 nm	double patterning higher throughput (advenced system)	6kHz//0.3pm(E95)	60 - 90W	GT62A-1SxE
32 nm	double patterning higher throughput	6kHz/0.3pm(E95)	60W	GT62A-1S
45 nm	higher NA	6kHz//0.3pm(E95)	60W	GT61A
50 nm	higher throughput higher NA	6kHz//<0.5pm(E95)	60W	GT60A
65 nm	higher throughput	4kHz//<0.5pm(E95)	45W	GT40A

Table 1. Technology nodes and required performance for ArF light sources

2. FEATURES AND MAJOR SPECIFICATIONS OF THE GT SERIES

2.1 Gigaphoton injection lock system

Gigaphoton's injection lock (MOPO) system consists of a Master Oscillator (MO) and a Power Oscillator (PO). Low energy and highly spectrally narrowed bandwidth seed light is produced by the MO and is amplified by the PO. We adopt injection lock system for the following reasons[3].

Merits	Benefits
1) Higher efficiency	Easy to get higher power
2) Narrow spectral bandwidth	Easy to get narrower spectrum
3) Wide tolerance of synchronization timing	Better stability and 2-charger system
4) Very small seed light energy	Low Cost of Ownership (CoO) from low optical load
5) Long pulse duration	Low CoO from low optical load

By making use of these injection lock characteristic, output power has been changed tunably from 60W to 90W without having negative impacts on major laser performances, including spectrum and wavelength stability

2.2 GigaTwin series major specifications
Gigaphoton's technological advance allows semiconductor industry to challenge not only for higher throughput but for the shrinking of IC design geometry. Major specifications of the GigaTwin series are shown in table 2.
The latest generation model GT62A-1SxE has extendable power from 60W to 90W tunably without upgrading. In MOPO, extension power is achieved easily. The MO condition does not necessarily change, because PO has higher gain. Therefore spectrum and wavelength stabilities stay unaffected. In addition, the GT62A-1SxE has inherited proven high reliability and low running cost on GigaTwin platform.

ArF model		GT40A	GT60A	GT61A	GT62A-1S	GT62A-1N	GT62A-1SxE
Wavelength	nm	193	193	193	193	193	193
Power	W	45	60	60	60	90	60 - 90
Pulse energy	mJ	11.25	10	10	10	15	10 - 15
Max. rep rate	Hz	4000	6000	6000	6000	6000	6000
FWHM	pm	0.2	0.2	N.A	N.A	N.A	N.A
E95	pm	<0.5	<0.5	0.3	0.3	0.3	0.3
Durability (Expected)							
MO Chamber	Bpls	40*	40*	40*	40*	40*	>40***
PO Chamber	Bpls	40*	40*	40*	40*	40*	>40***
LNM / MO LNM	Bpls	60**	60**	60**	60**	60**	60**
MM	Bpls	30	30	30	30	30	30
FM / PO FM	Bpls	30	30	30	30	30	30
PO RM	Bpls	30	30	30	30	30	30

*   GRYCOS technology
**   MPL (Multi Positioning LNM)
*** Durability can be extendable @ <90W

Table 2.   Major specifications of the GigaTwin series.

3.   MAJOR PERFORMANCE OF THE GT62A-1SxE

We tested the major performances and they were confirmed to meet design targets.
Its conditions are as follows:
    Output power: tuned power in the step of 10W from 60W to 90W to 60W
    Repetition rate: 6kHz
    Measured performances: output pulse energy, energy dose stability, wavelength stability, spectral bandwidth,
                pulse duration, beam profile, beam divergence, beam position and beam pointing
                at the same time at each power
    Time for changing target power: three minutes each time
The results are described below.

3.1 Output pulse energy and output power
Tunable output power provides illumination power optimum for resist sensitivity. Fig.1 shows the pulse train of output
energy at 60W, 70W, 80W and 90W. We have confirmed that the tunable range of output power is from 60W to 90W.

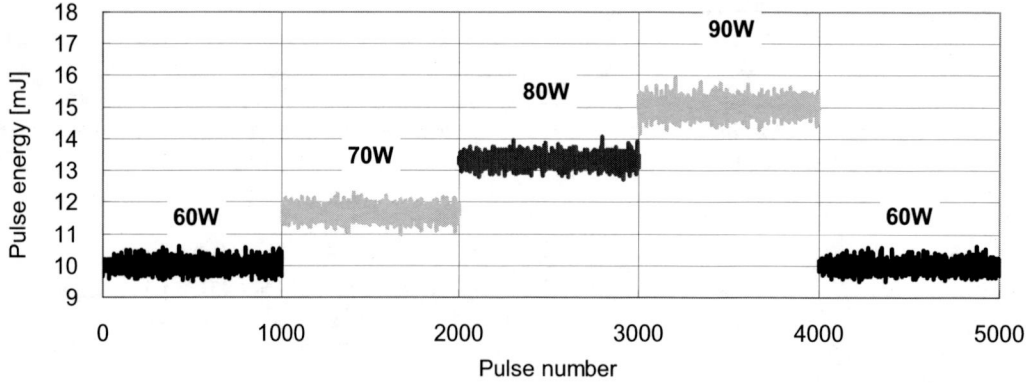

Figure1. Pulse energy and output power

3.2 Dose stability

Dose stability is an important property of laser output because it affects CD control. Fig.2 shows the trend of energy dose stability at 60W, 70W, 80W and 90W. These data was calculated by integrating the energy over the specified moving window. We have confirmed that there is no difference from 60W to 90W operation.

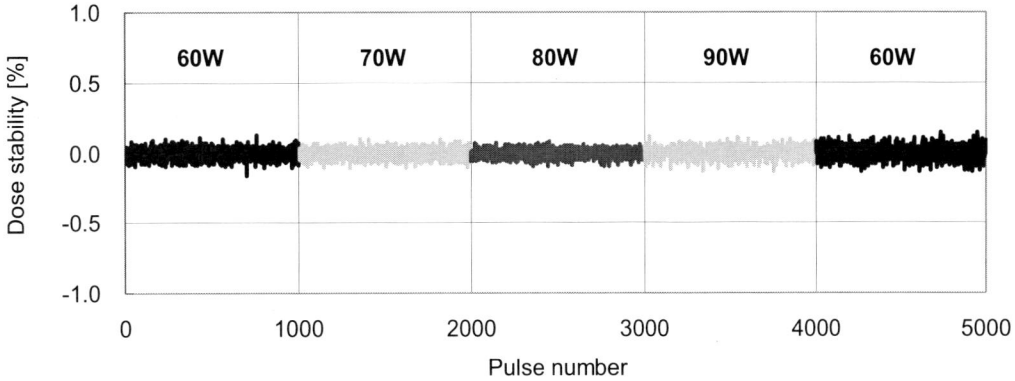

Figure 2. Dose stability from 60 to 90W

3.3 Wavelength stability

Changes of wavelength cause defocus, so the stability of the wavelength is important. Fig.3 and Fig.4 show the dependency of wavelength error and wavelength stability sigma with wavelength control on output power levels at 60W, 70W, 80W and 90W. These data were calculated by statistically treating the wavelength error averaged over the specified moving window. We have confirmed that wavelength control accuracy is independent of output power.

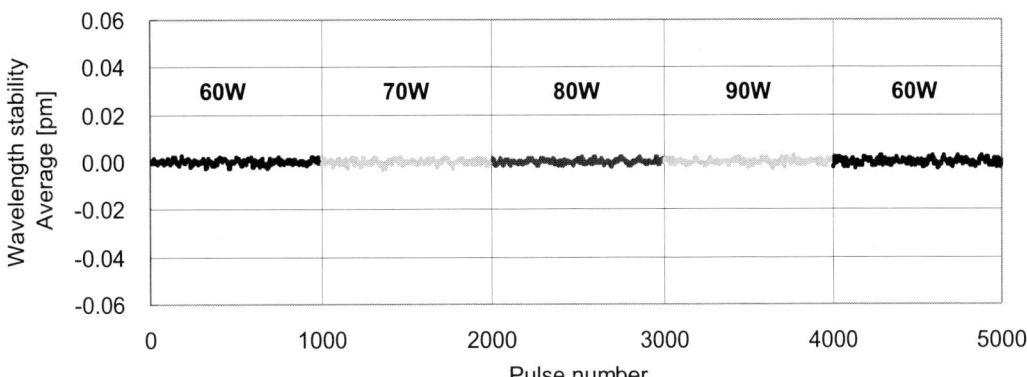

Figure 3. Wavelength stability error from 60 to 90W

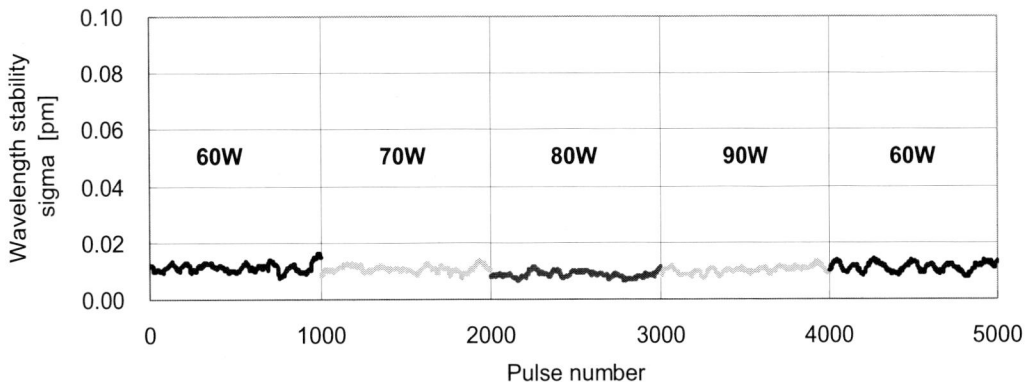

Figure 4. Wavelength stability sigma from 60 to 90W

3.4 Spectral bandwidth

The spectral bandwidth of laser is an important factor for imaging ability and CD control. Fig. 5 shows the data of spectral bandwidth of 95% energy concentration (E95) with spectral bandwidth control with E95 set point 0.3pm at 60W, 70W, 80W and 90W. Fig. 6 shows the spectral profile shape at 60W, 70W, 80W and 90W. We have confirmed that spectral bandwidth control accuracy and spectral profile shape are independent of output power.

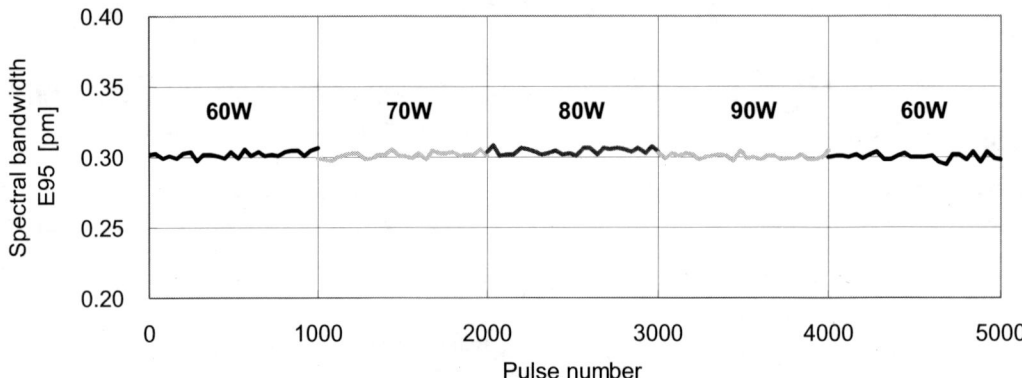

Figure 5. Spectral bandwidth from 60 to 90W

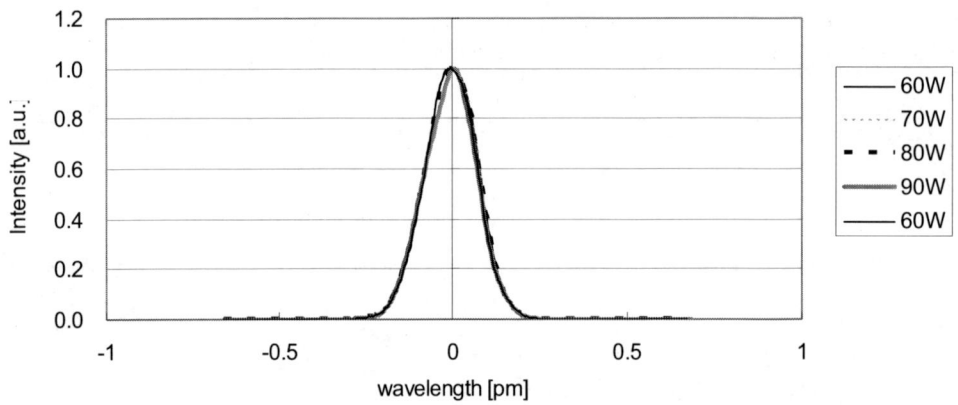

Figure 6. Spectral profile shape from 60 to 90W

3.5 Pulse duration

Long pulse duration is important because it lowers CoO. This is because the peak power intensity of laser pulses affects the lifetime of optical components inside scanners. In additional, long pulse duration is able to reduce the line edge roughness[5]. Fig.7 shows the laser pulse shape and the pulse duration T_{IS} (Time Integrated Square) at 60W, 70W, 80W and 90W. We have confirmed that pulse duration keeps more than 150 nsec under output power from 60W to 90W

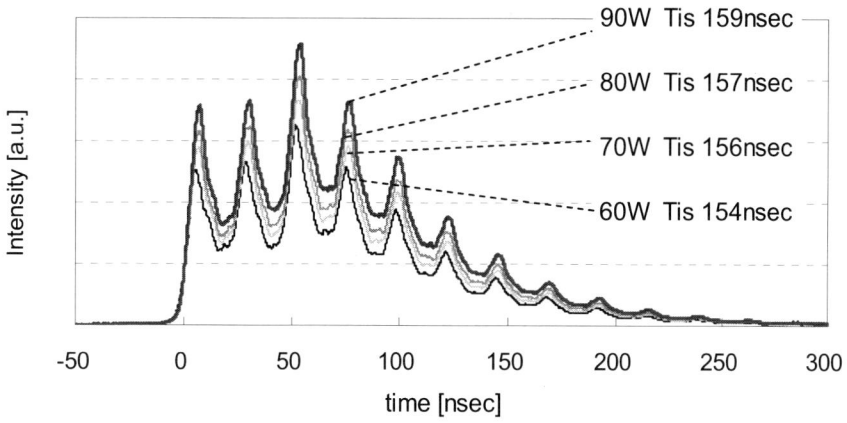

Figure 7. Laser pulse shape and the pulse duration

3. 6 Beam profile and divergence

New illumination system like a double patterning lithography requires ArF laser with more stable optical performances. Fig. 8 and 9 show the dependency of the fluctuation of beam profile and divergence on output power levels of 60W, 70W, 80W and 90W, respectively. These data were normalized at 60W data. We have confirmed that beam profile and divergence are stable from 60W to 90W.

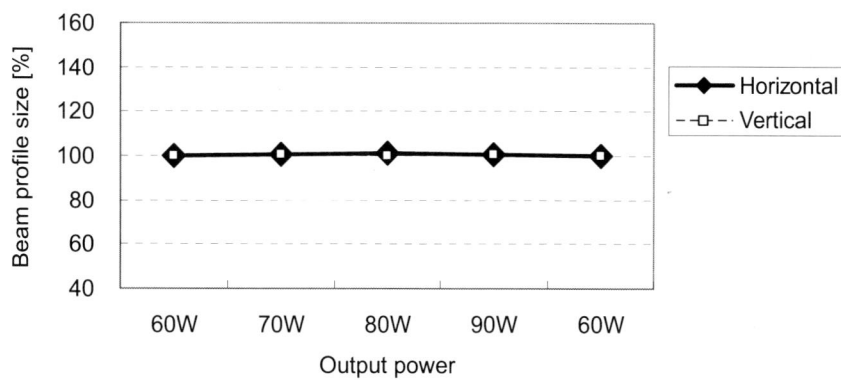

Figure 8. Output power dependency of beam profile size

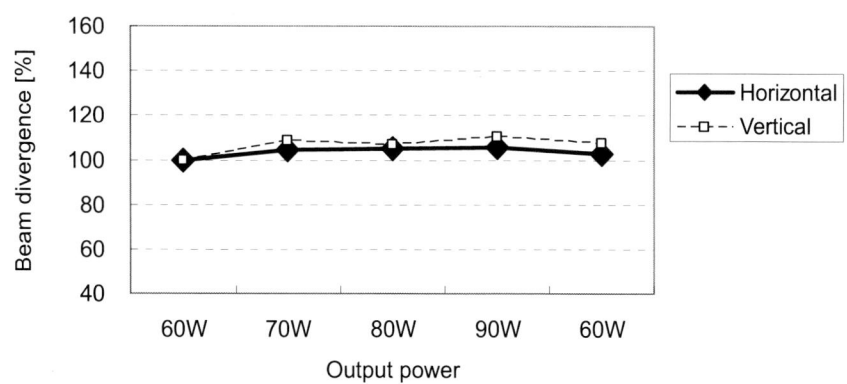

Figure 9. Output power dependency of beam divergence

3.7 Beam position and pointing

As in preceding section, new illumination system like a double patterning lithography requires ArF laser with more stable optical performance. Fig. 10 and 11 show the output power dependency of beam position and pointing at 60W, 70W, 80W and 90W, respectively. These data were calculated to initial 60W data. We have confirmed that beam position and pointing are stable from 60W to 90W.

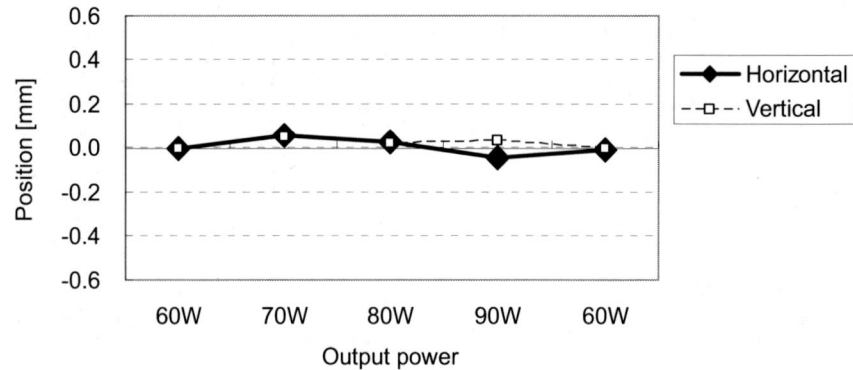

Figure 10. Output power dependency of beam position

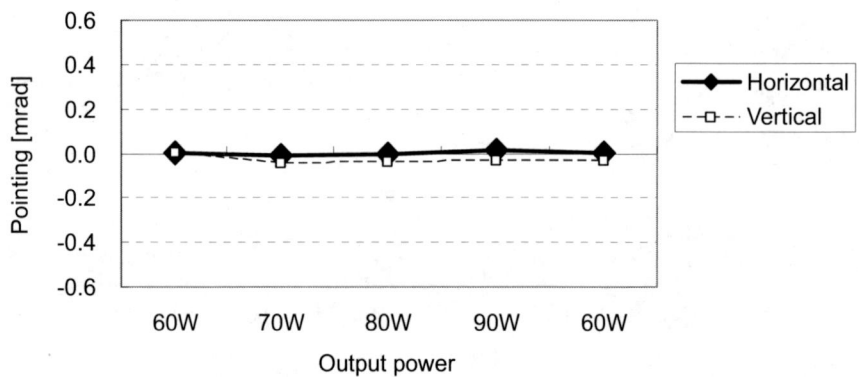

Figure 11. Output power dependency of beam pointing

4. OTHER FEATURES OF THE GT SERIES

4.1 Reliability

Now ArF lithography moves into high volume production, and reliability of the laser is industry's common request. We evaluated reliability by "Availability" as reliability indicators. "Availability" shows system available time by percentage of total time. The definition of Availability in this report is shown as follows.

Availability = [Total Hour – (Scheduled Downtime + Unscheduled Downtime)] / [Total Hour]

Availability of GigaTwin series up to Q4 2009 is shown at Fig.12. GigaTwin series have high reliability performance. Various technologies used for GigaTwin series are contributing high reliability. The availability that exceeds 99.5% proves the reliability of the GigaTwin series.
GT62A-1SxE has proven reliability by inheriting the GigaTwin platform.

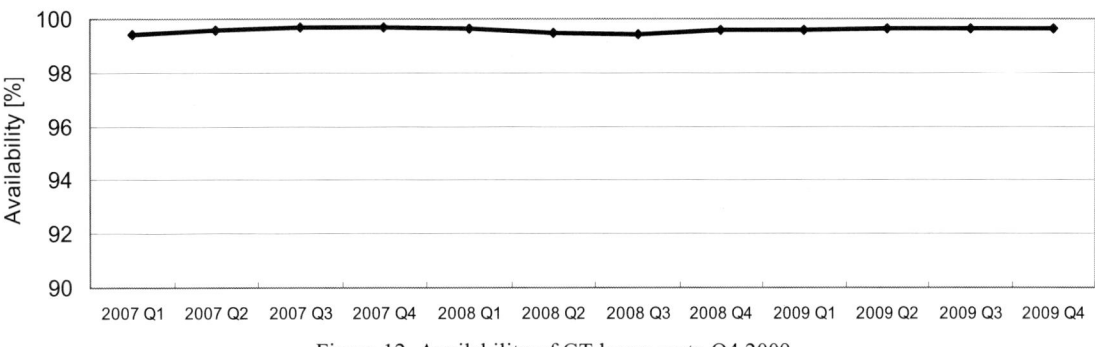
Figure 12. Availability of GT lasers up to Q4 2009

4.2 Reduction of running cost

In the semiconductor industry, price competition has become more intense than ever. The reduction of the equipment running cost, therefore, is one of the major concerns. We have developed a variety of technologies to reduce the laser running cost, or the cost of operation (CoO). A Number of innovative and unique technologies are implemented on the GigaTwin series in order to reduce the running cost of laser. The improvements are:

1) Chamber lifetime extension: The Gigaphoton Recycled Chamber Operation System (GRYCOS)
    20 billion pulses → 40 billion pulses
2) LNM lifetime extension: Multi Positioning LNM (MPL)
    30 billion pulses → 60 billion pulses
3) Gas lifetime extension: Total Gas Manager (TGM)
    3days → 15days: 24times/year

These three technologies can be applied to all the laser types built on GigaTwin platform. The details of the technologies were reported in the previous paper[4]
Inheriting the GigaTwin platform, GT62A-1SxE features the reduced running costs.

5. CONCLUSION

Gigaphoton has developed the tunable output power (60W - 90W) laser GT62A-1SxE.
It is designed to support the requirement of process parameter flexibility of exposure tool and end customer.
    -Optimized illumination power for various resist sensitivities.
    -Meeting the advanced illumination system.
    -Contribution to reduce optics deterioration and the line edge roughness.
    -Well maintained CD variation at all power range.
Inheriting the GigaTwin platform, it features proven reliability and the reduced running costs by GRYCOS, MPL and TGM.

6. REFERENCES

1. H. Mizoguchi, T.Inoue, J. Fujimoto, T. Suzuki, T. Matsunaga, S. Sakanishi, M. Kaminishi, Y. Watanabe, T. Nakaike, M. Shinbori, M. Yoshino, T. Kawasuji, H.Nogawa, H. Umeda, H. Taniguchi, Y. Sasaki, J. Kinoshita, T. Abe, H. Tanaka, H. Hayashi, K. Miyao, M. Niwano, A. Kurosu, M. Yashiro, H. Nagano, T igarashi, T. Mimura, K. Kakizaki: "High Power Injection Lock 6kHz 60W Laser for ArF Dry/Wet Lithography", Proc. SPIE Vol. 6154, 615425(2006)
2. T. Suzuki, K. Kakizaki, T. Matsunaga, S. Tanaka, Y. Kawasuji, M. Shimbori, M. Yoshino, T. Kumazaki, H. Umeda, H. Nagano, S. Nagai, Y. Sasaki, H. Mizoguchi: "Ultra line narrowed injection lock laser light source for higher NA ArF immersion lithography tool", Proc. SPIE Vol. 6520, 652024(2007)

3. M. Yoshino, H. Nakarai, T. Ohta, H. Nagano, H. Umeda, Y. Kawasuji, T. Abe, R. Nohdomi, T. Suzuki, S. Tanaka, Y. Watanabe, T. Yamazaki, S. Nagai, O. Wakabayashi, T. Matsunaga, K. Kakizaki, J. Fujimoto, H. Mizoguchi: "High-power and high-energy stability injection lock laser light source for double exposure or double pattering ArF immersion lithography", Proc. SPIE Vol. 6924, 6924-199 (2008)

4. Hiroaki Tsushima, Masaya Yoshino, Takeshi Ohta, Takahito Kumazaki, Hidenori Watanabe, Shinichi Matsumoto, Hiroaki Nakarai, Hiroshi Umeda, Yasufumi Kawasuji, Toru Suzuki, Satoshi Tanaka, Akihiko Kurosu, Takashi Matsunaga, Junichi Fujimoto and Hakaru Mizoguchi: "Reliability report of high power injection lock laser light source for double exposure and double patterning ArF immersion lithography", Proc. SPIE Vol. 7274, 7274-156 (2009)

5. Katsuhiko Wakana, Hiroaki Tsushima, Shinichi Matsumoto, Masaya Yoshino, Takahito Kumazaki, HidenoriWatanabe, Takeshi Ohta, Satoshi Tanaka, Toru Suzuki, Hiroaki Nakarai, Yasufumi Kawasuji, Akihiko Kurosu, Takashi Matsunaga, Junichi Fujimoto and Hakaru Mizoguchi: "Optical performance of laser light source for ArF immersion double patterning lithography tool", Proc. SPIE Vol. 7274, 7274-153 (2009)

Laser bandwidth effect on overlay budget and imaging for the 45 nm and 32nm technology nodes with immersion lithography

Umberto Iessi [a], Michiel Kupers[b], Elio De Chiara[a] Pierluigi Rigolli[a], Ivan Lalovic[c],G. Capetti[a]

[a] Numonyx, Via C. Olivetti 2, Agrate Brianza (MI) 20041, Italy
[b] Cymer B.V., De Run 4312B, 5503 LN Veldhoven, Netherlands
[c] Cymer, Inc., 17075 Thornmint Court, San Diego, CA 92127

ABSTRACT

The laser bandwidth and the wavelength stability are among the important factors contributing to the CD Uniformity budget for a 45 nm and 32nm technology node NV Memory. Longitudinal chromatic aberrations are also minimized by lens designers to reduce the contrast loss among different patterns. In this work, the residual effect of laser bandwidth and wavelength stability are investigated and quantified for a critical DOF layer. Besides the typical CD implications we evaluate the "image placement error" (IPE) affecting specific asymmetric patterns in the device layout. We show that the IPE of asymmetric device patterns can be sensitive to laser bandwidth, potentially resulting in nanometer-level errors in overlay. These effects are compared to the relative impact of other parameters that define the contrast of the lithography image for the 45nm node. We extend the discussion of the contributions to IPE and their relative importance in the 32 nm double-patterning overlay budget.

Keywords: CD Uniformity, Overlay Budget. Lens Aberrations, Laser bandwidth, Image Placement error

1. INTRODUCTION

As the industry heads towards ever smaller feature sizes and complex pattern layouts, sources of imaging and overlay errors that were neglected in the past are now becoming increasingly significant. In this paper we will investigate the impact of laser bandwidth on CD and the image placement error of asymmetric device patterns. Although the CD imaging effects have been explored before, for example in a paper by Bisschop et. al. (see reference [3]), here we will apply methods to specifically quantify the image placement effects for specific asymmetric device patterns for NV Memory (see reference [1]). This pattern was printed with a 6%-attenuated Phase Shift Mask at NA=1.20 using dipole illumination at 193nm wavelength. The dipole was horizontally oriented and had 60-deg opening angle, radially delimited by SigmaInner=0.65, SigmaOuter=0.85. The exposure-light was polarized in y-direction.

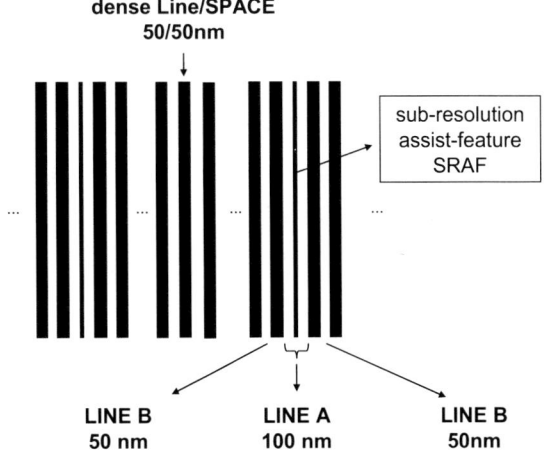

Figure 1: Pattern to be printed: Dense lines and spaces with pitch-interruption. Investigated are Critical Dimensions CD of 'LINE A', 'LINE B' and 'SPACE'. Dimensions indicated in the figure are target-values.

Optical Microlithography XXIII, edited by Mircea V. Dusa, Will Conley, Proc. of SPIE Vol. 7640,
76402B · © 2010 SPIE · CCC code: 0277-786X/10/$18 · doi: 10.1117/12.846552

Furthermore, we will use simulation software to study the impact of laser bandwidth on CD as well as on the Image Placement Error (IPE) and the relative contributions of the higher order Zernike coefficients. A recent update in the Panoramic™ lithography simulation software now makes it possible to directly calculate the impact of the higher-order chromatic aberrations. The resulting aerial image is calculated by sampling the spectrum at discrete wavelength points. A set of aerial images is obtained with each image corresponding to a single wavelength sample. Finally, the images are weighted by the intensity in the laser spectrum at the corresponding wavelength and summed together. This computation method has been described before, for example references [3,4], and is described schematically in the figure below. In our case, the individual aerial images are computed with different aberration levels as defined by a set of Z4-Z37 Zernike coefficients corresponding to each discrete wavelength sample.

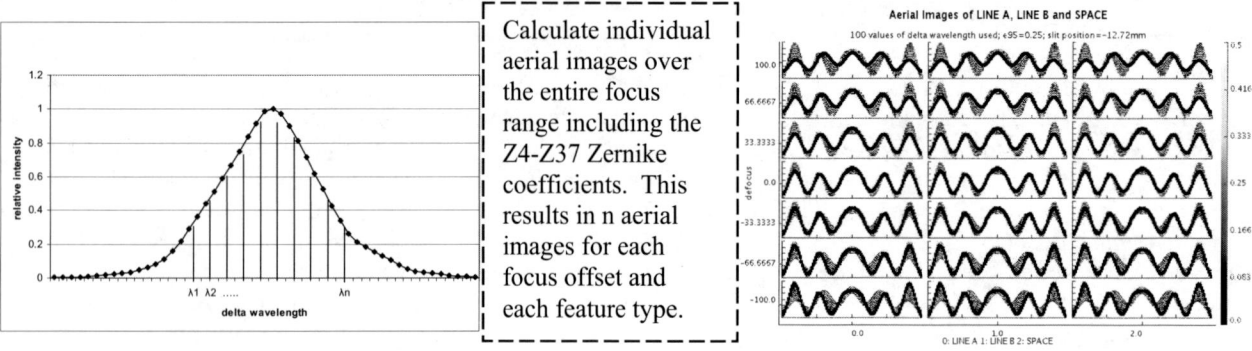

In this study, the aggregate aerial image is calculated using a typical spectrum from a Cymer XLA360 laser, which is shown in Figure 2.

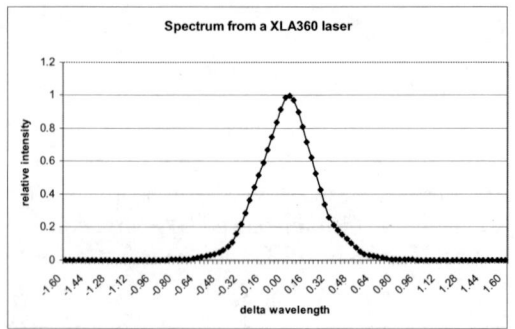

Figure 2 XLA360 spectrum used as a weighting function for the simulations.

A previous study has quantified the effect of IPE as a function different overlay marks (see reference [2]). Although these effects are quite small, we have extended this work to study the effect of the higher order chromatic aberrations for the AIM overlay marks used for this particular layer

1.1 Impact of higher order aberrations on IPE

Higher order aberrations are commonly defined as the Z5 to Z37 Zernike polynomial terms that describe deformations from spherical wavefronts (see for example reference [6]). In prior studies, it has been shown that the IPE is mostly affected by the lower order terms and it was assumed that higher order terms are small and can be neglected. Our simulation will quantify the magnitude of the contributions of the higher order terms for the particular features described above.

2. SIMULATIONS

2.1 IPE simulations for 45 nm node

For our simulations we will use Hyperlith™ from Panoramic Inc. The following inputs were used:

- GDS data from the reticle using 0.5 nm of simulation grid
- NA: 1.2 ; sigma inner=0.85 sigma outer=0.65, using X 60° dipole (Y polarized)
- Used -350 nm/pm as chromatic sensitivity

- Used measured Zernike data at 3 wavelengths (0, -0.5pm, +0.5pm) from an ASML XT1700i scanner in the Numonyx Fab. Due to confidentiality reasons we will not disclose the exact numbers.
- +/- 10% and +/- 100 nm of aerial image threshold and focus variation applied respectively.
- The image threshold is anchored to the SPACE feature

The results are shown in the plots of Figure 3 and Figure 4. Note that the plots are a function of focus and slit position. Except for LINE A, the variation of the IPE in the slit due to the higher order Zernikes is quite small. LINE B is more sensitive to focus variations than LINE A. In Figure 4 we show the variation of the IPE through the slit for the different types of pattern. These values are confirmed by an overlay experiment, and we discuss these results in section 3.4.

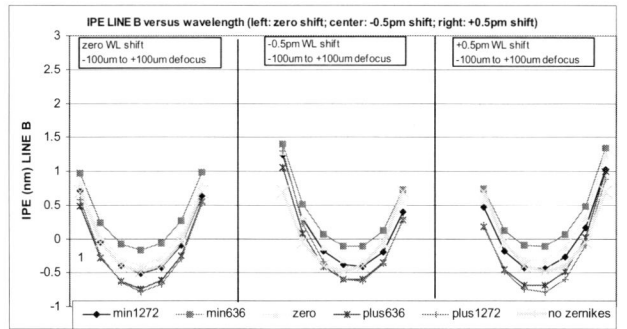 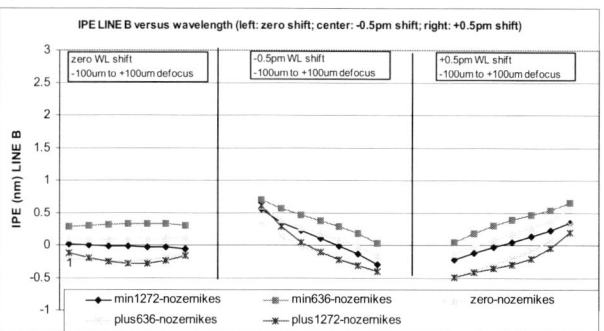

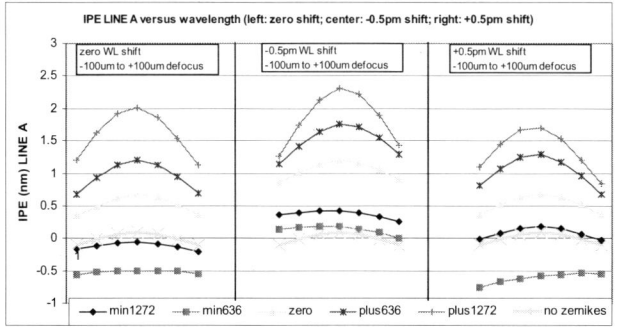 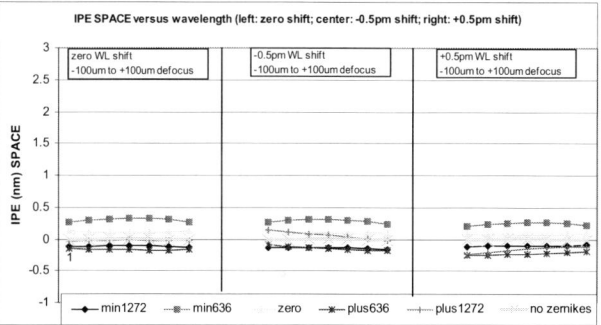

Figure 3: IPE (nm) through focus, slit position, wavelength and feature type. The wavelength of the laser was changed by -0.5pm and +0.5pm. From left to right and top to bottom: IPE LINE B through focus and wavelength; the change in IPE through focus when the no Zernike case is subtracted; IPE LINE A through focus and wavelength; IPE SPACE through focus and wavelength

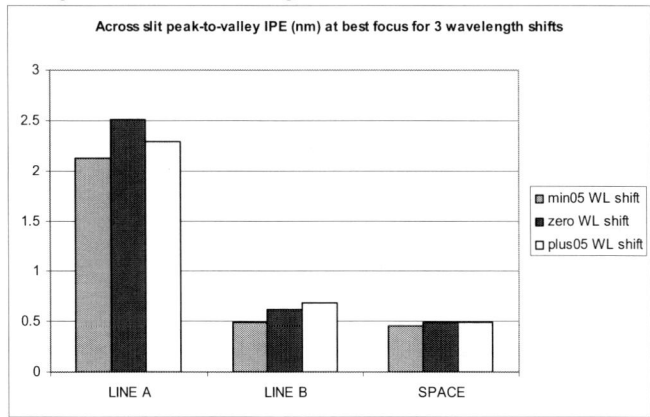

Figure 4 : total IPE variation over the slit for 3 types of patterns

To investigate the effect of chromatic aberrations and finite laser bandwidth on IPE, we will now generalize this approach and consider the aggregate aerial images computed over a range of wavelengths. To do this we must obtain the Zernike coefficients as a function of wavelength. The Zernike sensitivities to wavelength for this optical lithography scanner where determined experimentally, by measuring the aberrations at several wavelength set-points. These Zernike coefficients are typically linear with the wavelength, over a range of several picometers of wavelength offsets from nominal, which means we can perform interpolation to determine a range of intermediate values. Previous studies (reference [3] and [7]) have discussed the sampling requirements in order to accurately simulate the effect of bandwidth when using the defocus or Z4 aberration term only. Since it is rather easy to program the number of interpolations in the simulation software we have chosen to interpolate and calculate the aggregate aerial image over 100 samples of wavelength shift. . We also consider 6 values of the E95 laser bandwidth: 0.00pm, 0.25pm, 0.32pm, 0.38pm, 0.50pm and 1.00pm. We have included the e95=1.0pm value in order to assess the effect of bandwidth significantly beyond the usual operating range of lasers such as the XLA360 (see section 3.4 for a discussion on this).

To get an accurate description of the behavior of IPE versus bandwidth with and without the higher order Zernikes, we will compute the IPE at 5 different positions in the slit as well as through focus (-100um to +100um). The results for LINE B are plotted in Figure 5. Similar plots can be made for the other features. Figure 6 and Figure 7 show the IPE at best focus for LINE B, LINE A and SPACE for all the E95 values as well as the case where the Zernike coefficients are absent from the simulation. We see that LINE B is most sensitive to bandwidth whereas LINE A is mostly sensitive to the higher order Zernikes. In Figure 8 we show the variation of IPE over the slit as a function of bandwidth when we also consider the impact of the lower order Zernikes (Z2, Z3, Z4). From Figure 5 we see that the IPE for the most sensitive pattern, LINE B, changes by 0.9nm to 1.5 nm (depending on focus) over the 1pm change in BW. The other patterns are not sensitive to bandwidth. Note that the IPE effect is significantly lower over the typical bandwidth operating ranges of the XLA 360 laser, namely about 0.2nm to 0.5nm at extremes of defocus. The effect is even lower for laser systems that feature active bandwidth stabilization, such as the XLR 560i.

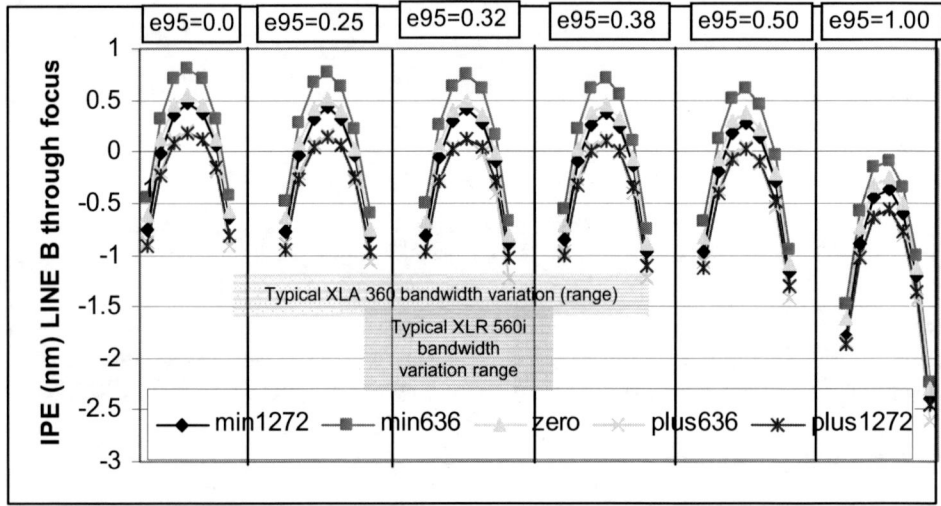

Figure 5: IPE as a function of laser bandwidth for LINE B through focus (-100nm,-66nm,-33nm, 0, +33nm, +66nm, +100nm)

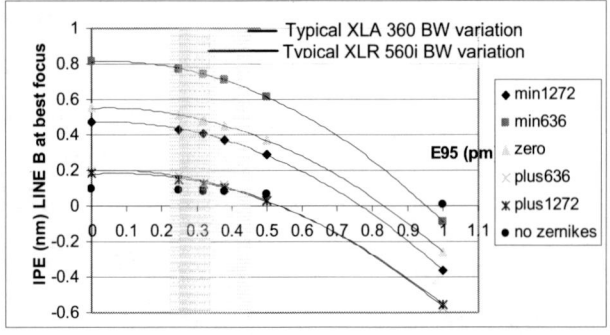

Figure 6: IPE as a function of bandwidth for LINE B. The bandwidth variation range for the XLA 360 laser type used in these experiments is shown in the shaded area; the typical bandwidth variation for XLR 560i systems featuring advanced bandwidth stabilization is also included for comparison

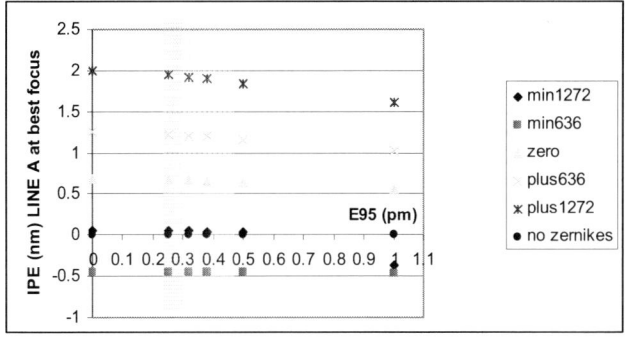 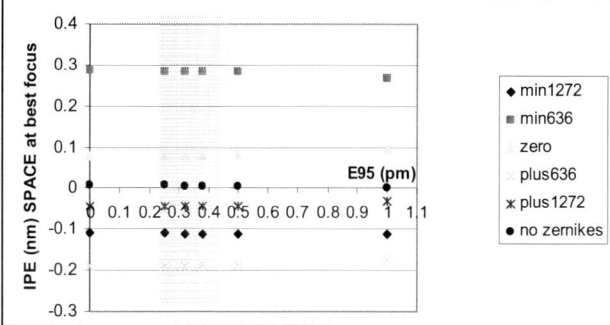

Figure 7: IPE as a function of laser bandwidth for LINE A and SPACE. The bandwidth variation range for the XLA 360 laser type used in these experiments is shown in the shaded area

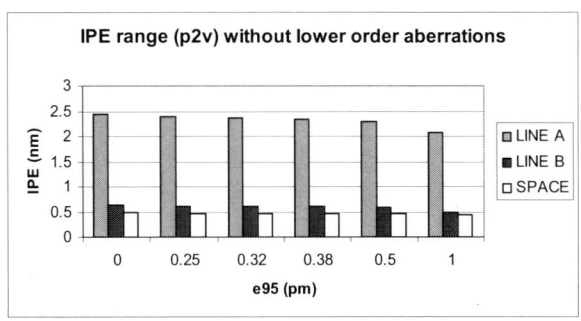 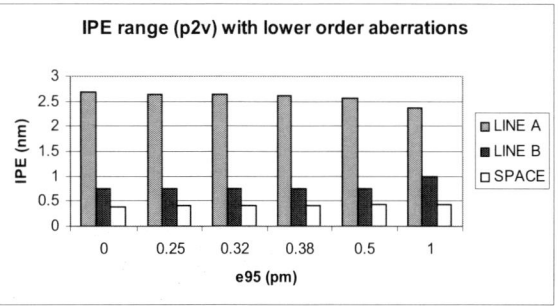

Figure 8: Across-slit IPE range at best focus with and without the lower order aberrations as a function of laser bandwidth for LINE A, LINE B and SPACE

2.2 IPE simulations for 32 nm node

The 32 nm lithography step for NV Memories require the introduction of Self Align Double Patterning (SADP) methods. With this technique, it could be necessary to print asymmetric patterns in order to obtain the final desired structure after spacer definition. In our case study we evaluate the IPE of a pattern named "L" (see Figure 9) with different lens NA :1.2 immersion and 0.93 dry.

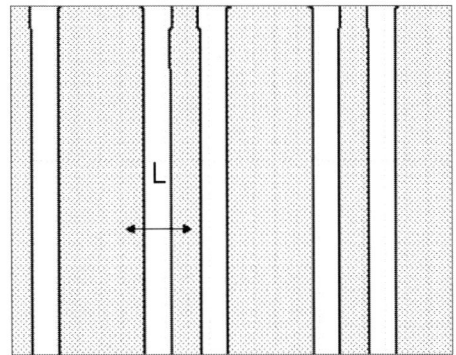

Figure 9: asymmetric layout for a 32nm technology node SADP memory device

The lithography pitch for a 64 nm SADP device is the half of the final patterned one so that dry lithography can be sufficient. Figure 10 shows that IPE is less than 0.2 nm and it is not sensitive to defocus. The relaxed k1 factor for both dry and immersion process (0.31 and 0.4 respectively) is protecting this technology node from IPE issues.

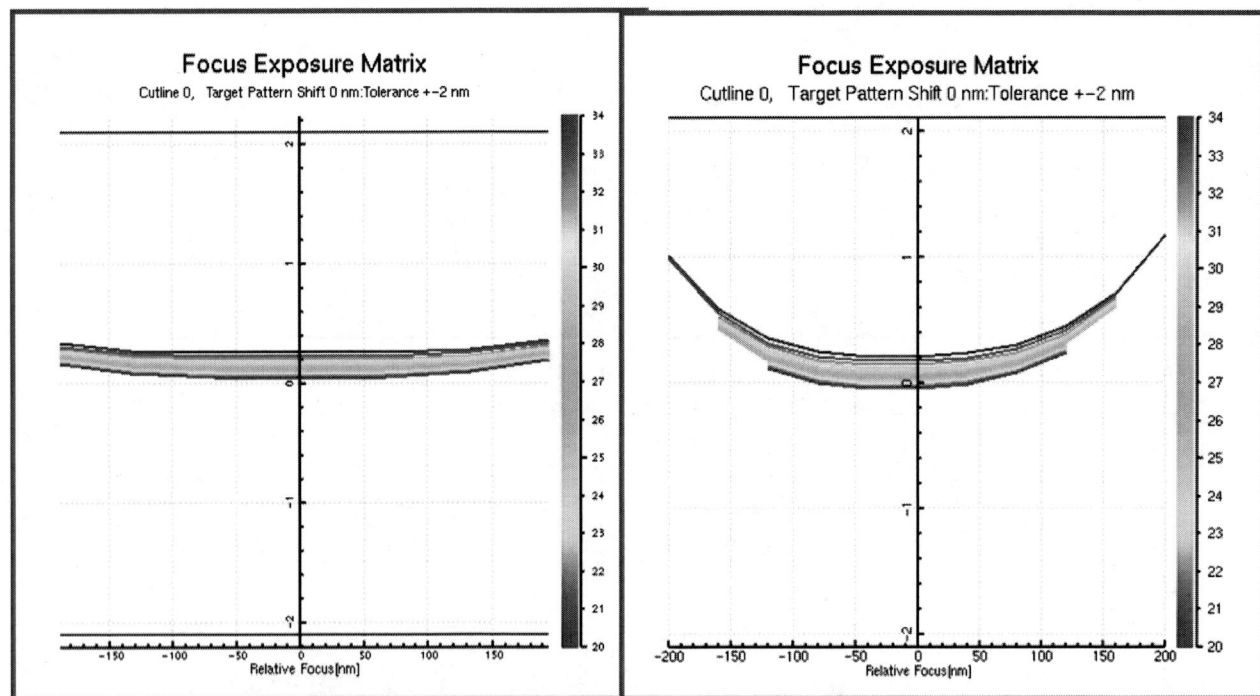

Figure 10: IPE(nm) through focus for the "L" pattern at different exposure doses

2.3 CD Uniformity simulations for 45 nm node

In order to evaluate the impact of laser bandwidth on our critical patterns we use ASML LithoCruiser Software to run simulations including in the model the appropriate scanner lens and illuminator fingerprint and a calibrated resist model.

This software first calculates the sensitivities of critical patterns to the main scanner parameters and than estimates a CDU budget for each of this main contributors in an early immersion lithography litho cell.

	Line A 3σ (nm)	Line B 3σ (nm)	Space 3σ (nm)
All	**3.05**	**2.78**	**2.30**
Scanner	0.52	0.83	0.26
Laser	**0.91**	**0.07**	**0.44**
Reticle	2.27	2.00	1.41
Process	1.51	1.51	1.51
Other	0.88	0.87	0.87

Table 1: CD uniformity for LINE A, LINE B and SPACE for each main contributor

The laser bandwidth contribution sensitivity is calculated by varying the bandwidth FWHM around the nominal value of 0.12 pm (equivalent to e95=0.25pm) and considering the defocus term only.

The laser bandwidth FWHM is varied over a range of 0.04 pm (0.08pm to 0.16 pm) and the model of the spectrum is a Modified Lorentzian with "n" factor equal to 3.6.

From Figure 11 we can conclude that, depending on the pattern, the laser contribution to the total CDU budget can be about 10%. LINE B is less affected by the laser contribution than the other patterns.

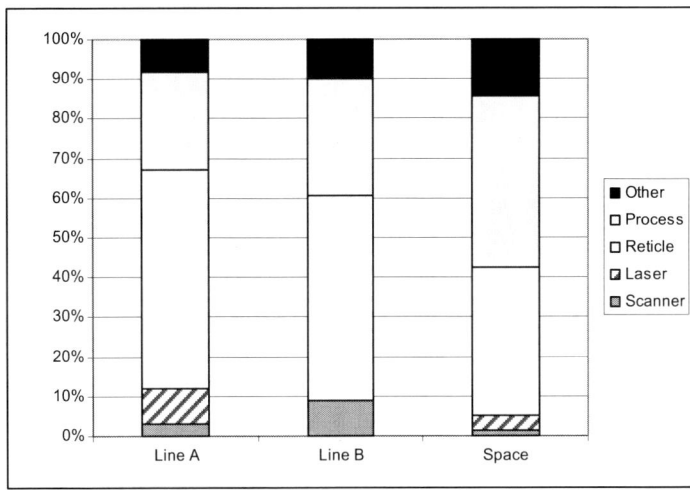

Figure 11: CDU budget for the different patterns and contributors

2.4 DOF simulation vs Laser Bandwidth

We determined the process window for the 3 different patterns for different typical conditions of the laser bandwidth. The depth of focus and exposure latitude of the overlapping process window is shown in **Table 2**.

Setting	FWHM (pm)	E95 (pm)	DOF (nm)	EL at 10% CD (nm)
A	0.12	0.25	86	19
B	0.22	0.37	64	18
C	0.33	0.5	34	6

Table 2: Elliptical DOF and EL for different laser bandwidth setting of FWHM and E95

Conditions "A" and "B" can be considered as the range of bandwidth used for normal operation. Within this range a consistent DOF reduction of about 14% can be explained by the high sensitivity of the LINE B pattern to defocus as can be seen in Figure 12. We will perform an experiment to confirm these results.

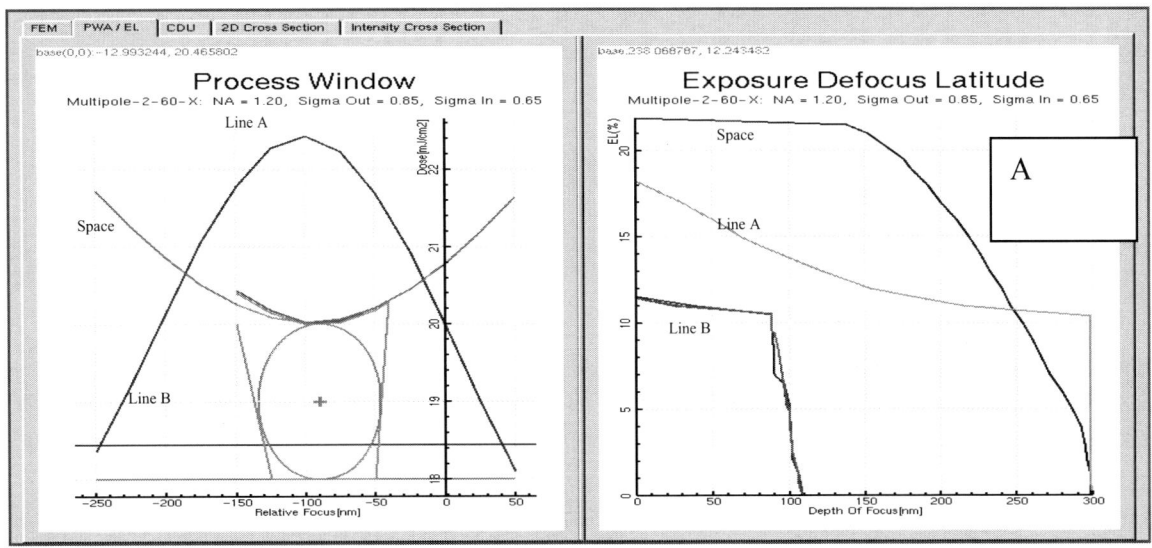

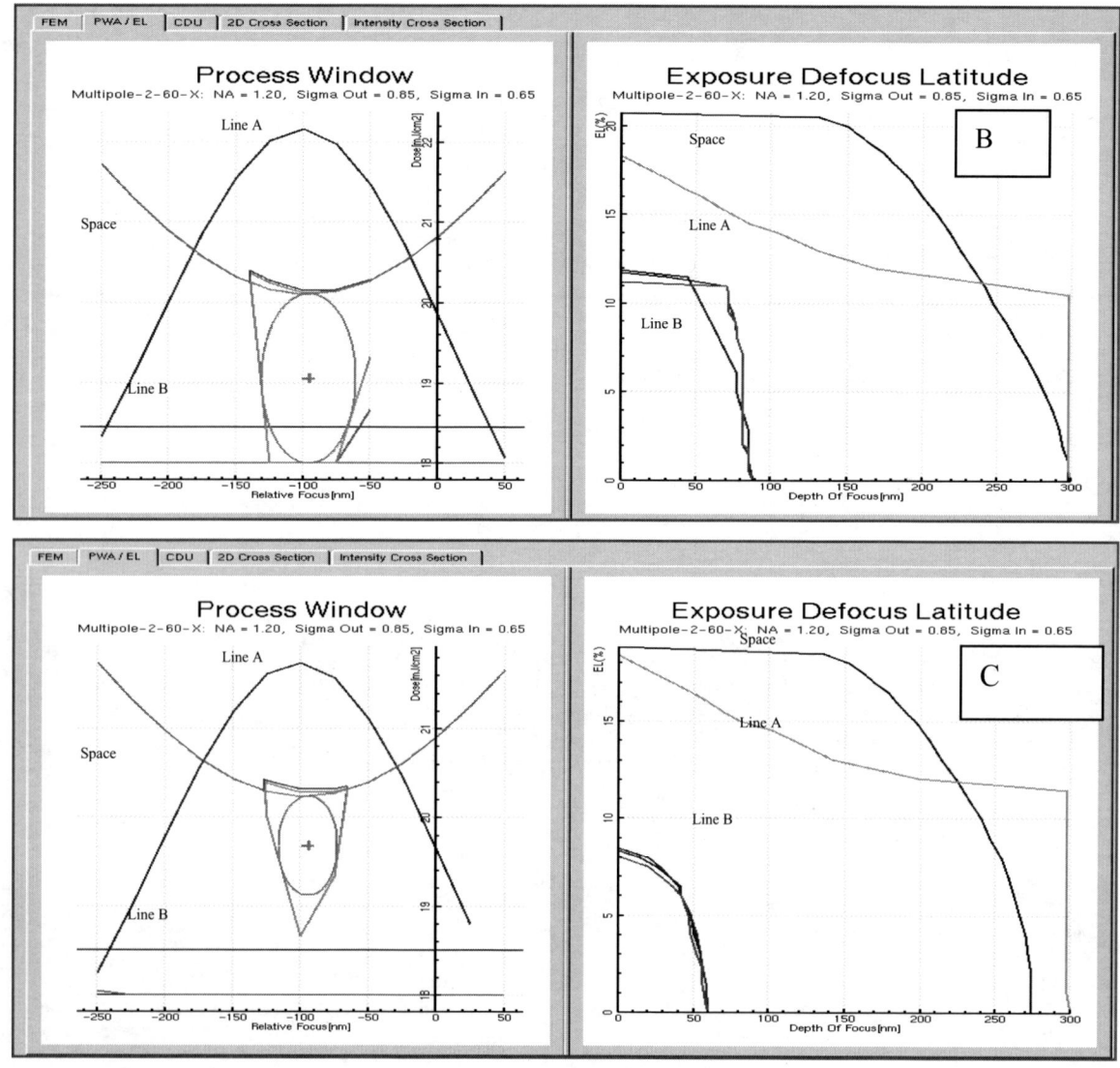

Figure 12: process window plots at different bandwidth settings

3. EXPERIMENTAL RESULTS

3.1 Overlay experiment

In order to confirm the simulations, we ran a 2-pass overlay test, where the first (reference) layer is exposed with the nominal wavelength and the second one with a wavelength set-point offset at two WL offset settings (+0.5pm, -0.5pm) in addition to the baseline exposure. According to the simulations, we expect a maximum effect of about 2.5nm through the slit (see Figure 8) for the most sensitive feature, LINE B. The bandwidth of the laser for this experiment is set to e95=0.25.

Ideally the overlay targets should be representative of standard process monitoring conditions for this technology.

In particular, different types of overlay mark segmentation may exhibit a different sensitivity to chromatic aberrations compared to non-segmented marks. At the same time, the defocus due to longitudinal chromatic aberration may begin to result in contrast loss of segmented marks particularly for higher WL offsets.

The purpose of this test is to extrapolate the shift behavior of different targets with the wavelength in order to convolute the results simulating a bandwidth impact. We will compare the experimental results with the simulations.

3.2 Overlay target and sampling description

The experiment described in the previous section was carried out using Advanced Imaging Metrology (AIM) targets with different features size as reported in Figure 13 a) and measured on Archer™ 100IS metrology tool. Dedicated overlay targets sampling plan was introduced in order to catch distortion map across the field and exposure slit as depicted in Figure 13 b) and c) respectively. Full wafer coverage has been considered.

AIM Targets	Size [µm]	Pitch [µm]	Feature Size [µm]	
			Inner Grating	Outer Grating
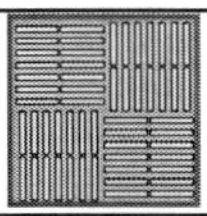	35 x 35	2.4	1.1	1.1
	13 x13	1.2	0.3	0.3
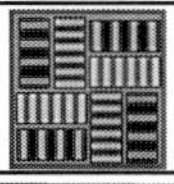	27 x 27	Optical Grating	0.2	0.2
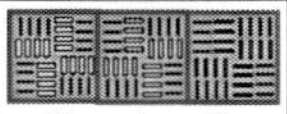 G A E	15x15	1.8	G=1 A=0.7 E=0.4	0.2

a)

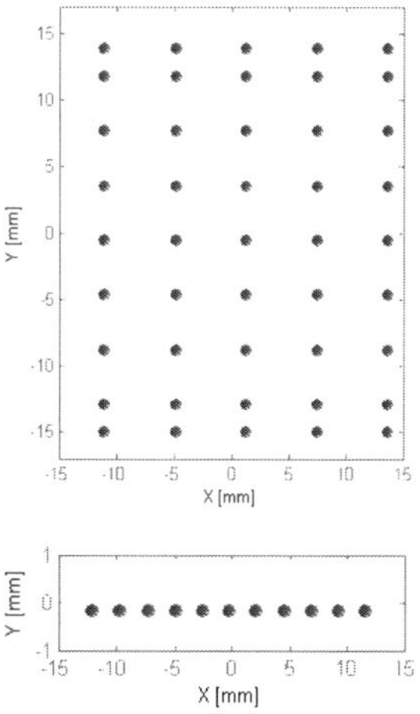

Figure 13: starting from the top: the first 3 rows are AIM targets used for distortion calculation, 2-pass overlay test. The last row shows targets {G, A, E} which are suitable targets to evaluate the imaging fingerprint across the slit. b) Field sampling for 2-pass overlay test. c) Slit sampling for IPE calculation

Except for the {G, A, E} targets, the metrology targets shown in Figure 13 are 2 layers targets where inner and outer gratings of AIM are printed at two different exposure runs. On the contrary, {G, A, E} are single layer AIM targets and printed at the same time in a single exposure run.

In order to have a reliable characterization of metrology measurements we have estimated the maximum error contribution in terms of Total Measurement Uncertainty (TMU) and random error propagation. The result is $3\sigma \leq$ 0.4nm. This is the maximum error bar for both X and Y measurements.

3.3 Discussion of the results

The results of metrology measurements, conducted using three different AIM targets as described in the previous section, are presented. As shown later, all measurements using different AIM targets are well matched one to another. Therefore, in Figure 14 only the Non Correctible Errors (NCE) of 13x13 µm AIM targets have been depicted. This match means that the 2-run overlay test is independent from target used for overlay characterization, in particular from feature size of the bar assembling the grating. From Figure 14 , it's clearly visible that both NCE in X and Y are a function of the wavelength shift $\Delta\lambda$. The effects might be ascribed to non-linear overlay contributions. In particular, the NCE Y appears as a 2nd order distortion across the slit whereas the NCE X appears as a 3rd order distortion across the slit, see Figure 14 b), d) respectively. As is obvious, the reference state $\Delta\lambda=0$ does not have large high order components (however, a small residual 3rd order NCE results from the specific illumination mode used). For what concerns the range of variation in terms of peak to valley (p2v) as a function of wavelength shift with respect to the reference state ($\Delta\lambda=0$), we found p2v of NCE X and NCE Y of around 2nm. From the simulation results we also found errors of around 2nm.

Figure 14: a) NCE in the Y direction across the field and through wavelength shift. b) Orthogonal projection of NCE Y onto the exposure slit. c) NCE X direction across the field and through wavelength d) Orthogonal projection of NCE X onto the exposure slit. The NCE is calculated by subtracting the linear model terms from measured data

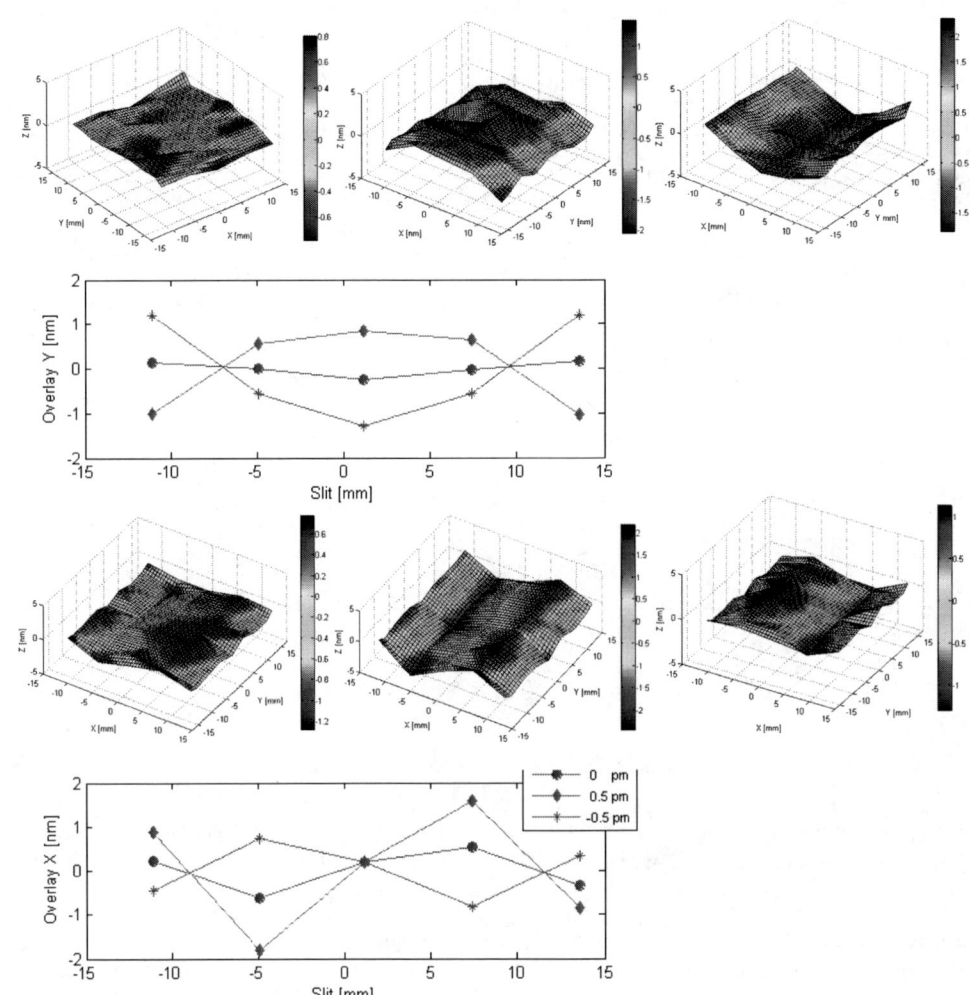

3.4 IPE discussion

Considering two different set of variables, wavelength $\equiv \{\Delta\lambda=0,\ \Delta\lambda=0.5,\ \Delta\lambda=-0.5\}$ and IPE targets $\equiv \{G, A, E\}$ we are able to evaluate the relative maximum contribution to Image Placement Error. Taking a look at Figure 15 and selecting the most sensitive target the relative maximum contribution to IPE as a function of wavelength is ≤ 0.1nm. These experimental results are in good agreement with the simulations shown in Figure 4 for the different product patterns (LINE A, LINE B and SPACE). Keeping the wavelength constant, the relative maximum contribution to IPE as a function metrology targets is ≤ 0.7nm.

As a comparison, a 0.5pm wavelength variation used in these experiments and simulations is over an order of magnitude greater than the laser wavelength stability specification for the XLA 360 generation lasers, and up to two orders of magnitude greater than actual performance for the latest generation XLR lasers.

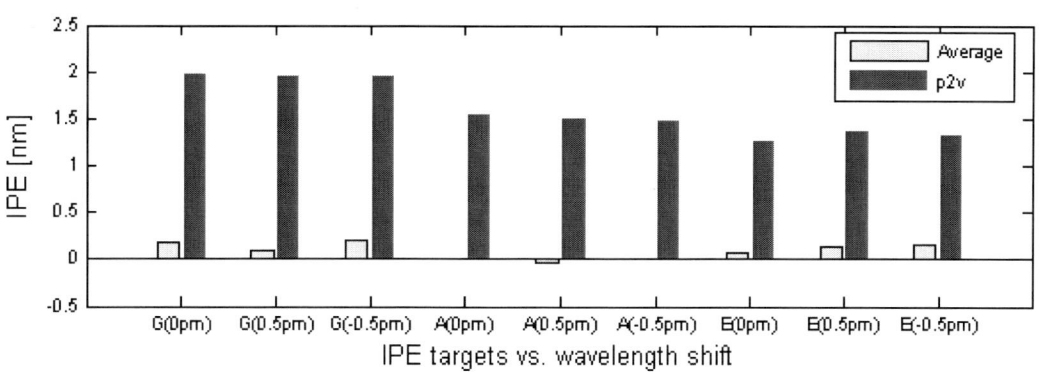

Figure 15: Image Placement Error as a function of metrology targets {G, A, E}   and wavelength shift {Δλ=0, Δλ=0.5, Δλ=-0.5}

3.5 IPE simulation with measured Zernikes

To understand the wavelength setpoint effect on overlay we simulate the AIM target shift with the full sets of lens Zernikes collected after each wavelength offset.

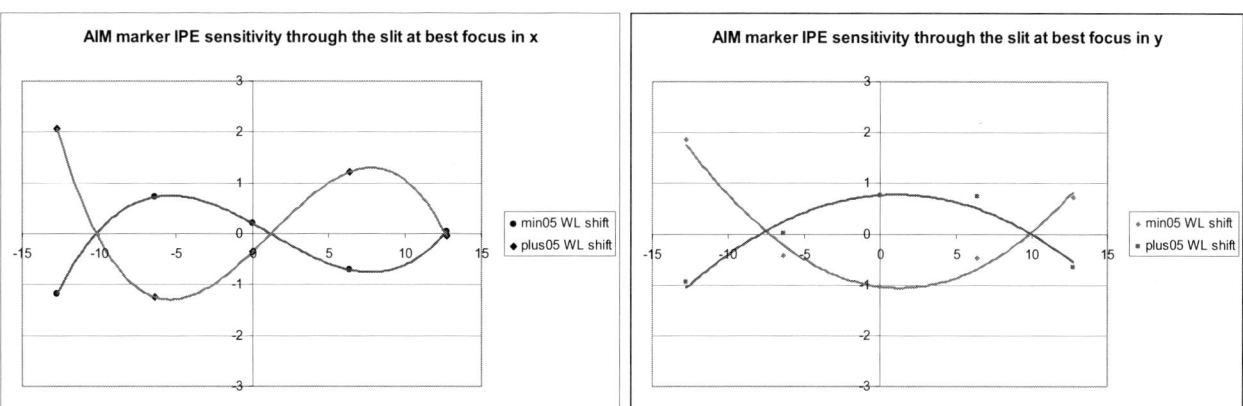

Figure 16: AIM marker IPE sensitivity through the slit

Figure 16 confirms the second order shift in Y and the third order shift in X coming mainly by Z2 and Z3 wavelength sensitivity in the slit.

This effect is not automatically compensated during the exposure inducing a third order in X and second order in Y lens distortion.

These misalignments are anyway perfectly symmetrical with respect to the reference wavelength hence we can conclude that a bandwidth drift considered as a convolution of three wavelengths here analyzed may not significantly impact the lens distortion.

4. SUMMARY AND CONCLUSIONS

The work in this paper has shown the IPE resulting from the higher order Zernike terms are small but depend on the feature type. We simulated the effect and followed up with an experiment to confirm the result.

The maximum across slit IPE that we simulated and subsequently confirmed by an experiment is about 2.5nm, 0.7nm and 0.4nm for the LINE A, LINE B and SPACE features respectively. A simulation found that the effect of a bandwidth variation of 0.5pm is about 0.2 nm to 0.5nm for the most sensitive pattern, LINE B, at the defocus extremes. Note that 0.5pm bandwidth variation modelled in this work is significantly higher than controlled in current-generation lasers.

We found that the DOF of the most sensitive pattern, LINE B, is reduced by about 14% within the standard operating range of the laser. This will be confirmed by an experiment. Current-generation lasers are able to control the bandwidth such that the risk of this type of effect is reduced. The CD Uniformity contribution from the laser to the total CD uniformity budget is estimated by dedicated simulations to be less than 10%.

CDU uniformity as a function of bandwidth depends on the type of pattern and the laser contribution is highest for the LINE A feature. These two results show that asymmetric patterns can be sensitive to bandwidth. We note that the XLR 560i and XLR 660ix both offer lower nominal bandwidth operation and significantly lower bandwidth variability due to bandwidth stabilization technology and a new wavelength controller.

Additionally we showed that the IPE is insensitive to the type of overlay measurement mark. We investigated 3 types of marks and found that the maximum difference is less than 0.1nm. This was also confirmed by a simulation.

An investigation of the 32nm pattern found that it is less sensitive to IPE than the pattern shown in Figure 1.

5. REFERENCES

[1] P. Rigolli et al., "AIM technology for Non-Volatile Memories microelectronics devices", SPIE2006, 6152-175

[2] E. Hendrickx, A. Colina, A. van der Hoff, J. Finders, G. Vandenberghe, Image placement error: closing the gap between overlay and imaging, J. Microlith., Microfab., Microsyst. 4(3), 033006 (Jul–Sep 2005)

[3] P. de Bisschop, I. Lalovic, F. Trintchouk, Impact of finite laser bandwidth on the critical dimension of L/S structures, J. Micro/Nanolith. MEMS MOEMS 7(3), 033001 (Jul–Sep 2008)

[4] M. Smith, J. Bendik, I. Lalovic, N. Farrar, W. Howard, C. Sallee, "Modeling and Performance Metrics for Longitudinal Chromatic Aberrations, Focus-drilling, and Z-noise; Exploring excimer laser pulse-spectra," *Proc. SPIE Optical Microlithography XX* **6520** -127 (2007).

[5] M.Terry, I. Lalovic, G. Wells, A. Smith, Behavior Of Lens Aberrations As A Function Of Wavelength On KrF and ArF Lithography Scanners, Proceedings of SPIE Vol. 4346 (2001)

[6] V.N. Mahajan, Optical Imaging and Aberrations-II: Wave Diffraction Optics, SPIE Press, Bellingham (2001)

[7] I. Lalovic, O. Kritsun, S. McGowan, J. Bendik, M. Smith, N. Farrar, "Defining a physically-accurate laser bandwdith input for optical proximity correction (OPC) and modeling", *Proc. BACUS XXII Photomask Technology Symposium* **7122** -62, (2008).

Laser Spectrum Requirements for Tight CD Control at Advanced Logic Technology Nodes

R. C. Peng*[a], H. J. Lee[a], John Lin[a], Arthur Lin[b], Allen Chang[c], and Benjamin Szu-Min Lin[c]

[a]MTC, Taiwan Semiconductor Manufacturing Corp., Hsinchu Science Park, Hsinchu, Taiwan 300-77, R.O.C.; [b]KLA-Tencor Corp., Chupei City, Hsinchu, Taiwan, R.O.C.; [b]Cymer Inc., Kuang Fu Rd. HsinChu, Taiwan, R.O.C.

ABSTRACT

Tight circuit CD control in a photolithographic process has become increasingly critical particularly for advanced process nodes below 32nm, not only because of its impact on device performance but also because the CD control requirements are approaching the limits of measurement capability. Process stability relies on tight control of every factor which may impact the photolithographic performance. The variation of circuit CD depends on many factors, for example, CD uniformity on reticles, focus and dose errors, lens aberrations, partial coherence variation, photoresist performance and changes in laser spectrum. Laser bandwidth and illumination partial coherence are two significant contributors to the proximity CD portion of the scanner CD budget. It has been reported that bandwidth can contribute to as much as 9% of the available CD budget, which is equivalent to ~0.5nm at the 32nm node. In this paper, we are going to focus on the contributions of key laser parameters e.g. spectral shape and bandwidth, on circuit CD variation for an advanced node logic device. These key laser parameters will be input into the photolithography simulator, Prolith, to calculate their impacts on circuit CD variation. Stable though-pitch proximity behavior is one of the critical topics for foundry products, and will also be described in the paper.

Keywords: CD control, laser parameter, laser spectrum, photolithography simulation

1. INTRODUCTION

As critical dimension (CD) is shrunk following Moore's Law, tight circuit CD control becomes more and more difficult, because of not only the reduced CD tolerance, but also hitting the equipment control limitations and CD measurement limitations. There are many factors that impact CD variations, for example, CD uniformity on reticles, focus errors, lens aberrations, partial coherence variation, photoresist performance and laser spectrum. Laser bandwidth and illumination partial coherence are two of the largest contributors to the proximity CD portion of the scanner CD budget, in the other words, Iso-Dense Bias (IDB) or through-pitch performance. IDB performance can be attributed to numerous factors that generate changes in image contrast or induce focus blur, for examples, illumination condition adjustment and laser light source spectral bandwidth (E95%). These factors are necessary to be controlled to fully compensate for IDB variation to the level required at advanced process nodes.[1]-[5]

In previous studies, the sensitivity of IDB and through-pitch performance with regards to laser spectral bandwidth were reported for both 45nm Node logic device[6] and 32nm Node logic device.[7] These results showed that for IDB change of 1nm, E95% variation needs to be controlled to 69fm and 45fm, for 45nm and 32nm logic devices respectively. Increasingly higher spectral bandwidth stability is required for each successive technology node. This indicates that the 22nm Node will demand even tighter bandwidth stability and the added ability to set spectral bandwidth with both high accuracy and flexibility.

Factors from light sources impacting product yields can be divided into two groups, CD control and overlay. Figure 1 shows the laser parameters that contribute to CD control and overlay. Contrast and intensity are two key factors that impact CD control. With further breaking down, the key laser parameters contributing CD control are bandwidth, spectral shape, ASE, wavelength stability, beam stability, energy stability, laser coherence and polarization. In this manuscript, the effects of laser parameters, specifically bandwidth and spectral shape, on CD control will be discussed.

Optical Microlithography XXIII, edited by Mircea V. Dusa, Will Conley, Proc. of SPIE Vol. 7640,
76402C · © 2010 SPIE · CCC code: 0277-786X/10/$18 · doi: 10.1117/12.846516

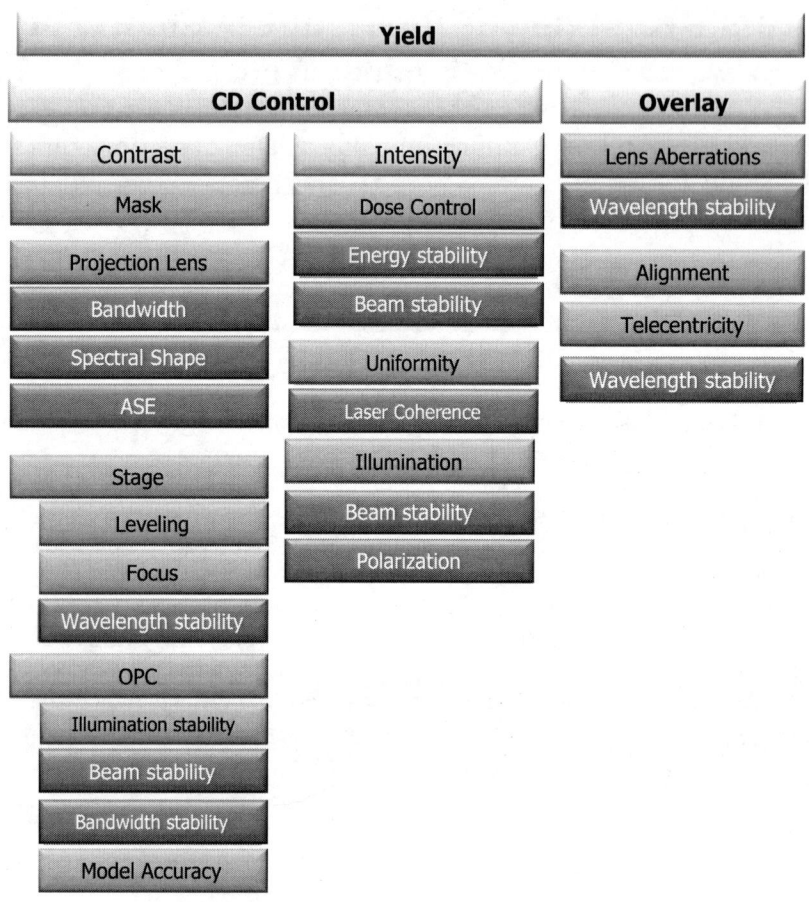

Figure 1. The key laser parameters contributing to product yield (CD control + overlay).

2. SIMULATION CONDITIONS

Figure 2 illustrates the flow chart used for simulation. An L/S layer from an advanced node logic device was chosen as the target layout for this simulation work. Through-pitch 1-D L/S patterns from the minimum pitch to full isolated features were simulated. No sub-resolution assistant features (SRAFs) were used, but only mask biases were applied to compensate the optical proximity effects at a certain chosen laser condition. Original through-pitch 1-D L/S patterns and illumination conditions (193nm Immersion + Dipole illumination + Y-Polarization + Att. PSM) were the 1st set of input parameters to calculate the Eop to meet the nominal target CD size for the feature with the minimum pitch and then calculate mask biases for all the other features. Prolith version Ver. X3 was used for all the simulation work in this manuscript. Only aerial image CDs were taken into account for simplifying the simulation work. Varied laser parameters, including E95, were used as the 2nd set of input parameters to check through-pitch DOF (@ 5% exposure latitude), proximity CDs, and CDU. Sigma fine-tuning was the 3rd set of input parameters to improve DOF at the forbidden pitches; optimum center sigma and radius sigma values were selected to achieve enough process margins for those forbidden pitches. Laser requirements were then derived from simulated CD results, and E95 CD sensitivity to meet the target CD/CDU criteria.

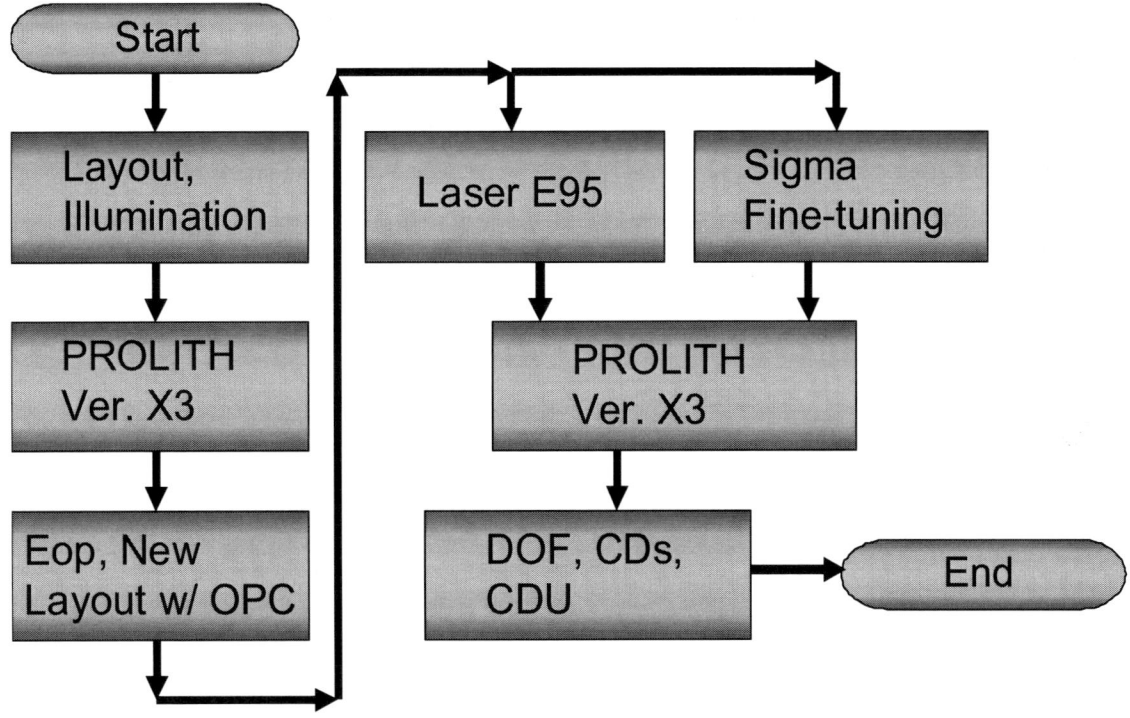

Figure 2. Simulation Flow Chart.

3. RESULTS AND DISCUSSIONS

3.1 DOF and CDU

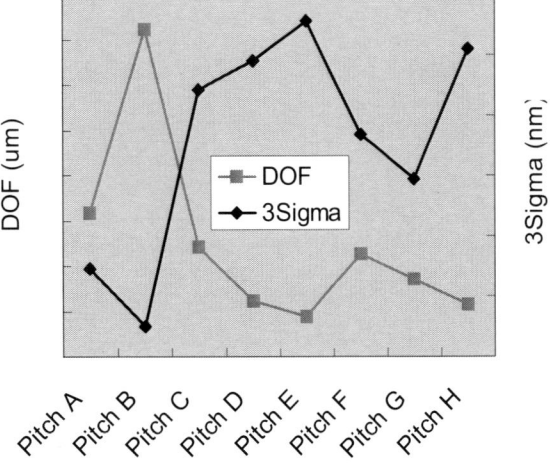

Figure 3. The through-pitch simulation results of DOF and CDU with assigned illumination conditions and built-in Cymer XLA light source parameters. The square line represents the through-pitch DOF trend while the diamond line the through-pitch CDU trend.

Figure 3 illustrates the through-pitch simulation results of DOF and CDU with assigned illumination conditions and built-in Cymer XLA300 light source parameters. The square line represents the through-pitch DOF trend while the diamond line the through-pitch CDU trend. No SRAFs were added but optimal mask biases were applied to through-pitch 1-D L/S patterns to compensate the OPC effects. Dipole illumination enhances the DOF and CDU values at the tighter pitches, however, the forbidden pitches (Pitch D and E) and isolated (Pitch H) L/S features show less DOF and expectedly worse CDU. It was believed that SRAFs could definitely enhance DOF and CDU for those isolated features, but would not improve DOF and CDU for the forbidden-pitch features because of insufficient space to insert any SRAF, under the reticle manufacturability constraints. Under the current illumination conditions and process settings, CD variations at forbidden pitches should be considered as the worst cases instead of CD behaviors at isolated features as usual due to their narrower process margins.

3.2 Bandwidth CD Dependency

Figure 4 shows the proximity curves in terms of CD differences exposed at different Cymer XLA laser bandwidths. E95%, instead of FWHM (full-width-half-maximum, default tunable parameter in Prolith Ver. X3), was used to represent the laser bandwidth changes here. The simulated E95% range in this study is from 355fm to 555fm. Proximity CD difference curve at an E95% of 435fm is flat because the through-pitch optimal mask biases were calculated at this condition. Based on the same mask bias settings, proximity CD difference curves for different E95% values were then calculated. Larger proximity CD difference variations occur as E95% is changed for the forbidden-pitches (Pitch D and E) and isolated (Pitch H) patterns. Even without SRAF, the proximity CD difference variation as a function of E95% at Pitch E is higher than that at isolated patterns. This indicates that forbidden-dense bias (FDB) may be a more relevant measure of bandwidth sensitivity than isolated-dense bias (IDB).

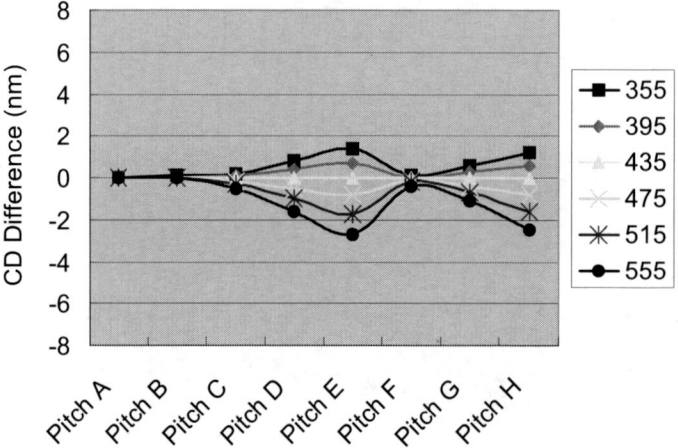

Figure 4. The proximity curves in terms of CD differences exposed at different Cymer XLA laser bandwidths. E95%, ranged from 355fm to 555fm, instead of FWHM (full-width-half-maximum, default tuning parameter in Prolith Ver. X3) and was used to indicate the laser bandwidth here. Proximity CD difference curve under E95% of 435fm is flat because the through-pitch optimal mask biases was calculated at this condition.

The E95% CD sensitivities of different pitches in the E95% range from 355fm to 555fm were calculated in Figure 5. The range of E95% - CD sensitivities is from 0.23nm/100fm to 2.04nm/100fm. The CD variation is derived as 0.09nm ~ 0.81nm based on typical Cymer XLA performance resulting in E95% 3 sigma of 40fm without active bandwidth control. In this case, CD variation can be improved down to 0.05nm ~ 0.47nm by Cymer new XLR platform, which features advanced bandwidth stabilization (ABS) technology, with typical long term E95% stability of 3 sigma of 23fm as shown in Figure 6. Pitch C and Pitch F patterns have the lowest E95% CD sensitivities of 0.35nm/100fm and 0.23nm/100fm, respectively. The highest E95% CD sensitivity occurs at forbidden pitch (Pitch E), while the 2nd highest E95% CD sensitivity occurs for isolated (Pitch H) patterns. If SRAF can be optimally implemented for isolated patterns and use of patterns with forbidden pitch (Pitch E) is restricted, the maximum E95% CD sensitivity can be improved by 42% (from

2.04nm/100fm to 1.19nm/100fm), which means 0.27nm CD variation under typical performance of Cymer's XLR platform long term stability of 23 fm E95%.

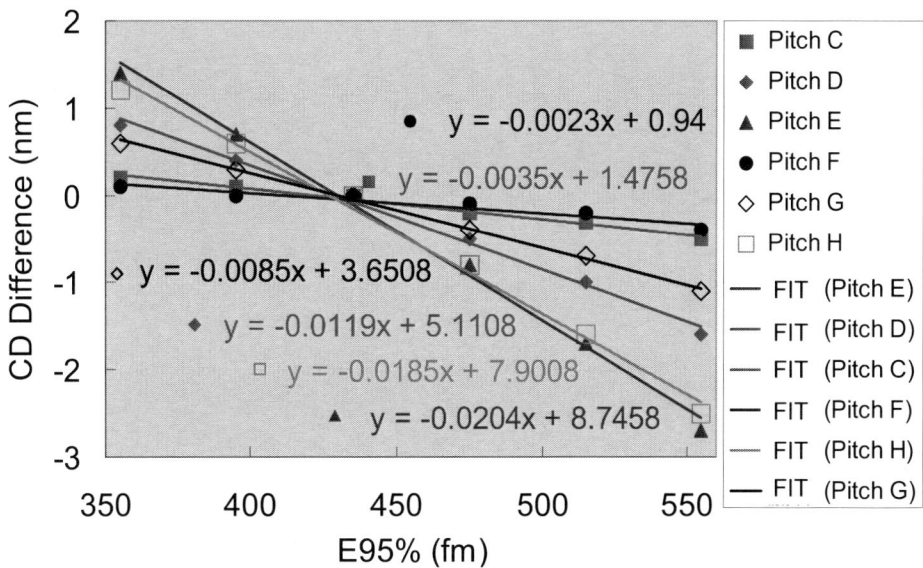

Figure 5. The E95% CD sensitivities of different pitches in the E95% range from 355fm to 555fm were calculated. The range of E95% CD sensitivities are from 0.23nm/100fm to 2.04nm/100fm.

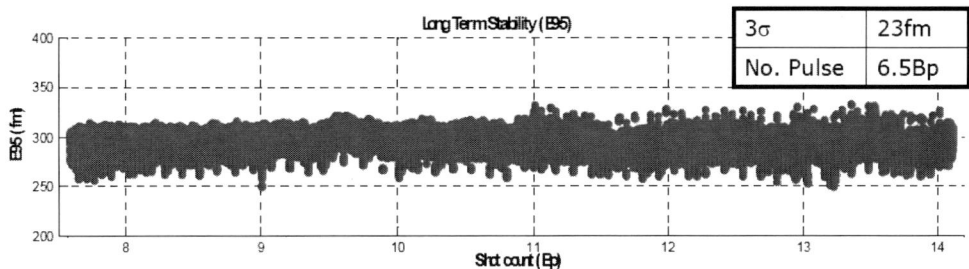

Figure 6. Typical long term E95% stability (3 sigma of 23fm) of Cymer new XLR platform with pulse numbers over 6 Bp. CD variation can be further improved as 0.05nm ~ 0.47nm by this Cymer new XLR platform.

Spectral Shape Effects

Besides Cymer XLA300 spectrum as a built-in light source spectrum in Prolith for accurate simulation, a Lorentzian spectrum is also available for parametric and fast simulation. However, Lorentzian spectral shape looks very different from the Cymer XLA300 spectral shape. Power law coefficient (PLC) in Prolith Ver. X3 is therefore created to vary the Lorentzian spectral shapes into what is referred to as modified Lorentzian. Figure 7 demonstrates the spectral shapes of (a) Lorentzian spectral shape with PLC=1, (b) Lorentzian spectral shape with PLC=2, and (c) Cymer XLA300 spectral shape. It is obvious that the peak of the modified Lorentzian shape becomes rounded and the tails are suppressed as PLC increases from 1 to 2, which result in the modified Lorentzian spectral shape more closely fitting the Cymer XLA300 spectral shape. The effects of using the modified Lorentzian or Gaussian analytic approximations to actual laser spectra were also discussed previously.[8]

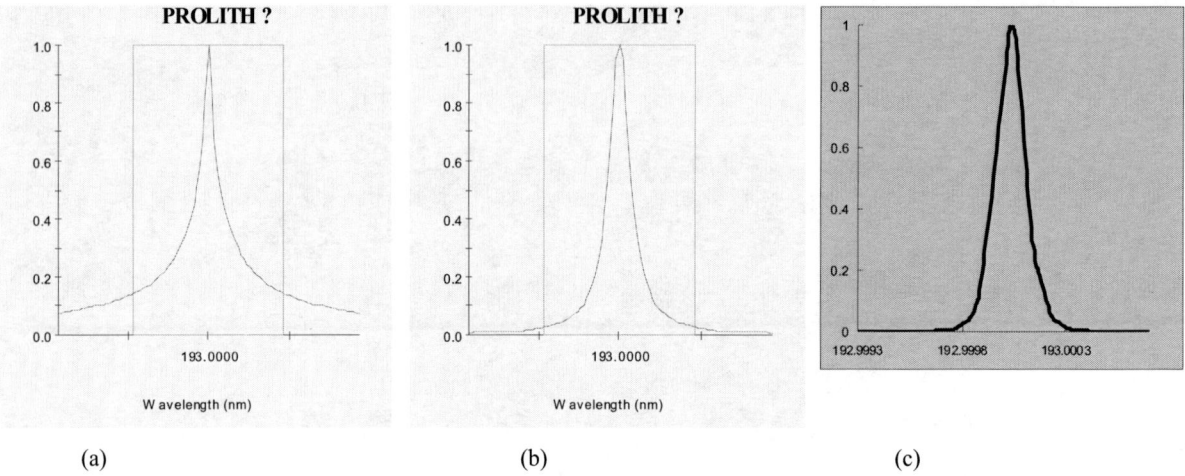

(a)   (b)   (c)

Figure 7. The spectral shapes of (a) Lorentzian spectral shape with PLC=1, (b) Lorentzian spectral shape with PLC=2, and (c) Cymer XLA300 spectral shape.

The proximity curves, in terms of their CD differences, can be drawn in Figure 8(a) and Figure 8(b) with Lorentzian spectral shape with PLC=1 and Lorentzian spectral shape with PLC=2, respectively. CD differences of the case of PLC=1 is apparently higher than that of the case of Cymer XLA300, while CD differences of the case of PLC=2 much closer to the case of Cymer XLA300. Even though, bandwidth CD sensitivity of the case of PLC=2 is calculated as 2.8nm/100fm, which is about 40% larger than bandwidth CD sensitivity of Cymer XLA300 as 2.04nm/100fm.

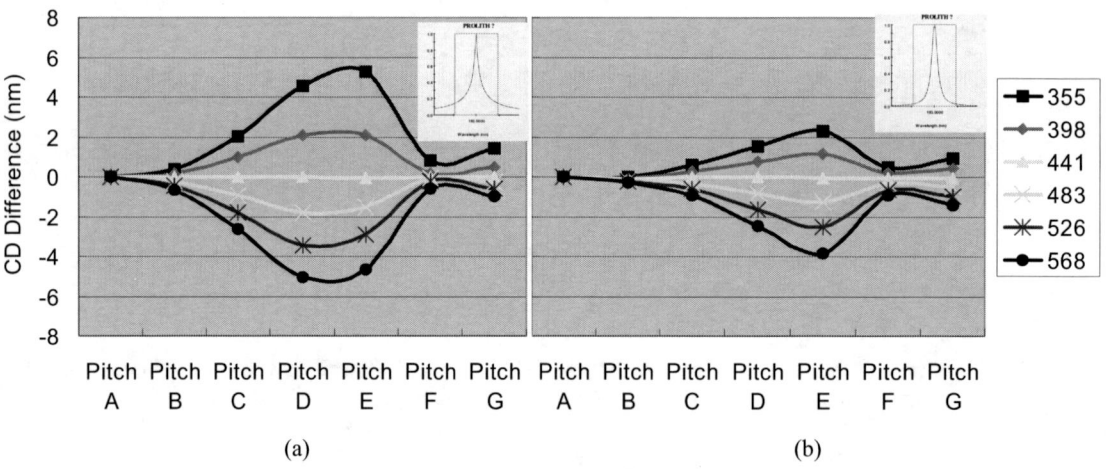

(a)   (b)

Figure 8. The proximity curves in terms of CD differences with (a) Lorentzian spectral shape with PLC=1 and (b) Lorentzian spectral shape with PLC=2.

In order to matching modified Lorentzian spectral shapes and actual Cymer XLA300 spectral shape more closely, modified Lorentzian spectra with higher PLCs were studied. From the log scale overlapping spectral chart (Figure 9), it can be told that modified Lorentzian spectral shape with PLC=4 matches actual Cymer XLA spectral shape, except those low level noises at the actual spectral footings.

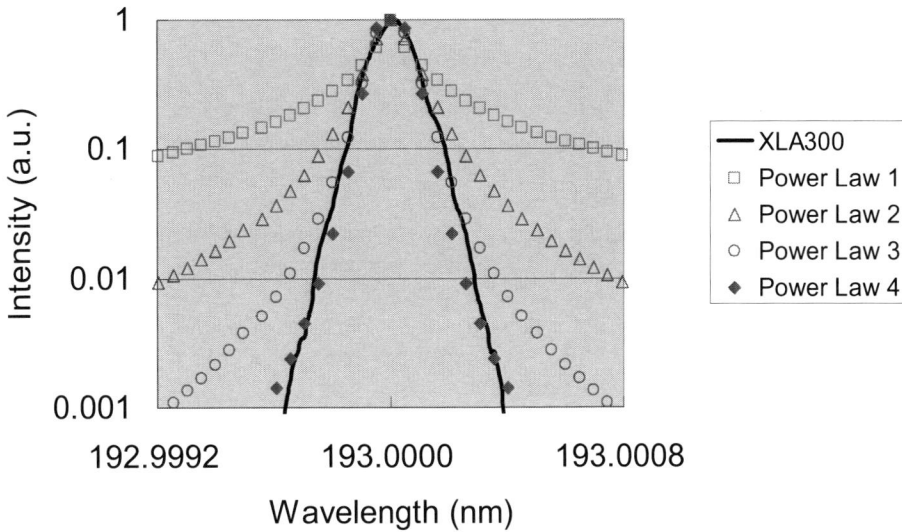

Figure 9. Lorentzian spectral shapes with PLC from 1 to 4 overlapped with actual Cymer XLA300 spectral shape.

Illumination Optimization

Illumination optimization is a common and well-known way to improve DOF and CD variation of the forbidden-pitch patterns. However, it will suffer the process margin loss for those patterns with tighter pitches. Table 1 shows the relative DOF values (100% DOF @ Pitch E pattern under dipole center of 0.9 and dipole radius of 0.05) at different dipole center sigma and radius sigma combinations for Pitch A and Pitch E patterns. Dipole illumination generally favors tighter pitches and any off-optimal condition will result in a reduction of dense pattern process margins. Table 1(a) shows relative DOF values for Pitch A patterns, and the maximum DOF value occurs at the condition of dipole center of 0.8 and dipole radius of 0.05. Although DOF for Pitch E patterns, shown in Table 1(b), also achieves the maximum value under the Pitch A preferred condition, it may soon drop to zero if dipole center or dipole radius shifts to a lower value. In order to improve the process control margins, suggested dipole conditions could be (1) dipole center of 0.85 and dipole radius of 0.1 and (2) dipole center of 0.85 and dipole radius of 0.05, depending on the controllability of dipole center and dipole radius.

DOF	Dipole Radius Sigma			
Dipole Center Sigma	0.05	0.1	0.15	0.2
0.6	133%	170%	146%	212%
0.65	201%	211%	272%	399%
0.7	323%	373%	497%	505%
0.75	836%	972%	715%	507%
0.8	1596%	943%	684%	485%
0.85	632%	845%	671%	520%
0.9	266%	296%	439%	515%

(a)

DOF	Dipole Radius Sigma			
Dipole Center Sigma	0.05	0.1	0.15	0.2
0.6	0%	0%	0%	127%
0.65	0%	0%	0%	0%
0.7	0%	0%	0%	0%
0.75	0%	0%	0%	0%
0.8	204%	0%	0%	0%
0.85	145%	153%	154%	0%
0.9	100%	119%	126%	135%

(b)

Table 1. The relative DOF values (100% DOF @ Pitch E pattern under dipole center of 0.9 and dipole radius of 0.05) at different dipole center sigma and radius sigma combination for (a) Pitch A and (b) Pitch E patterns.

4. CONCLUSION

In this manuscript, key laser light source parameters, e.g. bandwidth and spectral shape, have been studied to determine their contributions to CD variation and proximity variation. E95% forbidden-dense bias (FDB) should be characterized in addition to the usual IDB, because FDB has a higher E95% CD sensitivity due to the narrower process margins. Bandwidth dependent CD variation of 0.47nm for the advanced node logic device can be achieved with typical Cymer XLR long term E95% bandwidth performance of 3 sigma of 23fm, which is enabled by advanced bandwidth stabilization and control technologies. Bandwidth dependent CD variation can be further improved down to 0.27nm by adding SRAFs to isolated patterns and avoiding the use of the forbidden pitch (Pitch E in this case) in the pattern design. Illumination optimization can also increase DOF for those forbidden-pitch patterns, which can also improve E95% FBD sensitivity. In this study, we find that the Modified Lorentzian spectrum with PLC=4 can match the imaging results obtained using a Cymer XLA300 spectral shape at best focus.

REFERENCE

[1] M. Terry et al., "*Behavior of lens aberrations as a function of wavelength on KrF and ArF lithography scanners*", Optical Microlithography XIV, SPIE 4346-41 p.15-24 (2001).

[2] T. Brunner et al., "*Laser bandwidth and other sources of focus blur in lithography*", Optical Microlithography XIX, SPIE 6154-31 (2006).

[3] Kevin Huggins et al., "*Effects of laser bandwidth on OPE in a modern lithography tool*", Optical Microlithography XIX, SPIE 6154-36 (2006).

[4] Feder Trintchouk et al., "*XLA 300: the Forth-Generation ArF MOPA Light Source for Immersion Lithography*", Optical Microlithography XIX, SPIE 6154-76 (2006).

[5] T. Oga et al., "*Challenging to meet 1nm Iso- Dense Bias (IDB) by controlling Laser Spectrum*", Advanced Lithography XXI, SPIE 6520-144 (2007).

[6] K. Yoshimochi et al., "*45nm Node Logic Device OPE Matching between Exposure Tools Through Laser Bandwidth Tuning*", Optical Microlithography XXI, SPIE 6924-92 (2008).

[7] K. Yoshimochi et al., "*32nm Node Device Laser-bandwidth OPE Sensitivity and Process Matching*", Optical Microlithography XXI, SPIE 7274-115 (2009).

[8] Lalovic et al., "*Defining a physically-accurate laser bandwidth input for optical proximity correction (OPC) and modeling*", Proc. BACUS XXII Photomask Technology Symposium 7122-62, (2008).

Lithography Light Source Fault Detection

Matthew Graham, Erica Pantel, Patrick Nelissen, Jeffrey Moen, Eduard Tincu,
Wayne Dunstan, Daniel Brown
Cymer Inc.

ABSTRACT

High productivity is a key requirement for today's advanced lithography exposure tools. Achieving targets for wafers per day output requires consistently high throughput and availability. One of the keys to high availability is minimizing unscheduled downtime of the litho cell, including the scanner, track and light source. From the earliest eximer laser light sources, Cymer has collected extensive performance data during operation of the source, and this data has been used to identify the root causes of downtime and failures on the system. Recently, new techniques have been developed for more extensive analysis of this data to characterize the onset of typical end-of-life behavior of components within the light source and allow greater predictive capability for identifying both the type of upcoming service that will be required and when it will be required.

The new techniques described in this paper are based on two core elements of Cymer's light source data management architecture. The first is enhanced performance logging features added to newer-generation light source software that captures detailed performance data; and the second is Cymer OnLine (COL) which facilitates collection and transmission of light source data. Extensive analysis of the performance data collected using this architecture has demonstrated that many light source issues exhibit recognizable patterns in their symptoms. These patterns are amenable to automated identification using a Cymer-developed model-based fault detection system, thereby alleviating the need for detailed manual review of all light source performance information. Automated recognition of these patterns also augments our ability to predict the performance trending of light sources.

Such automated analysis provides several efficiency improvements for light source troubleshooting by providing more content-rich standardized summaries of light source performance, along with reduced time-to-identification for previously classified faults. Automation provides the ability to generate metrics based on a single light source, or multiple light sources. However, perhaps the most significant advantage is that these recognized patterns are often correlated to known root cause, where known corrective actions can be implemented, and this can therefore minimize the time that the light source needs to be offline for maintenance. In this paper, we will show examples of how this new tool and methodology, through an increased level of automation in analysis, is able to reduce fault identification time, reduce time for root cause determination for previously experienced issues, and enhance our light source performance predictability.

Keywords: light source, fault detection and classification, availability

Optical Microlithography XXIII, edited by Mircea V. Dusa, Will Conley, Proc. of SPIE Vol. 7640,
76402D · © 2010 SPIE · CCC code: 0277-786X/10/$18 · doi: 10.1117/12.852758

1. Introduction

The increasing cost sensitivity of semiconductor manufacturing is driving significant attention towards equipment uptime and availability. Particularly in leading edge lithography processes, the demand for uptime and availability cascades down to the litho tools which are the designed constraint in the manufacturing workflow. Improving uptime and availability means reducing equipment downtime, where downtime includes preventative maintenance and replacement of consumables, see SEMI E10 Standard [1] illustrated in Figure 1. Furthermore, one of the key elements in maximizing the availability efficiency is to minimize unscheduled downtime. While one solution would be to proactively replace consumables well before they cause unscheduled down time, the cost associated with premature replacement will negatively impact the overall operating costs.

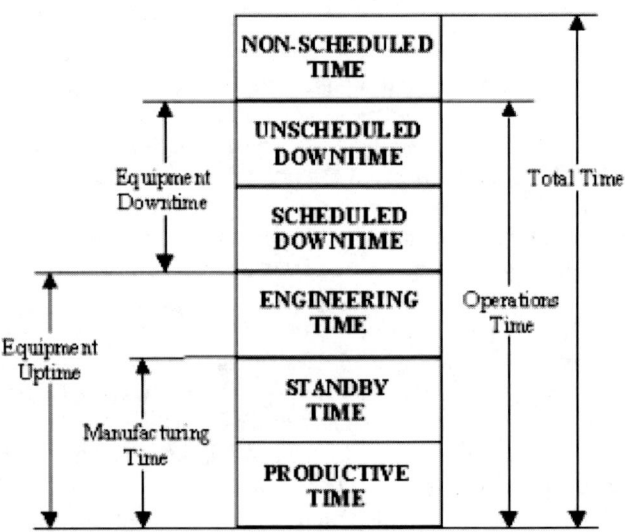

Figure 1: Breakdown of SEMI E10 standard (*Specification for Definition and Measurement of Equipment Reliability, Availability, and Maintainability.*)

Minimizing unscheduled downtime requires data to provide detailed analysis about tool performance to extend the consumable use and avoid premature replacements. In order to provide up-to-date information for troubleshooting and scheduling maintenance, while minimizing impact to the manufacturing process, tool performance data is collected on-line under production conditions. From the earliest excimer light sources, Cymer has collected extensive performance data from light sources that have been used to identify root cause for failures and create action plans for preventative maintenance. Recently, new techniques have been developed for more advanced automated data analysis to characterize the onset of end-of-life behavior of consumables and allow greater predictive capability for identifying when and what type of service will be required.

Comprehensive analysis of the performance data collected using this architecture has demonstrated that many light source issues exhibit recognizable patterns in their symptoms. These patterns are amenable to automated identification using a Cymer-developed model-based fault detection and classification (FDC) system, thereby alleviating the need for detailed manual review of all light source performance data. Furthermore, automated recognition of these patterns also augments the ability to predict performance trend of light sources.

Such automated analysis provides several efficiency improvements for light source troubleshooting and preventative maintenance scheduling by providing more content-rich standardized summarization of light source performance along with reduced time-to-identification for previously classified faults. Automation provides the ability to generate light source performance metrics based on the performance of a single light source, or of multiple light sources. However, perhaps the most significant advantage is that these recognized patterns are often correlated to a known root cause, where known corrective actions can be implemented. This can minimize the time that the light source needs to be offline for maintenance.

This paper presents the Cymer model-based FDC tool used to enhance light source service strategy. An example of the tool use is provided to demonstrate the FDC methodology. Through an increase level of automation in data analysis, Cymer is able to realize reduced troubleshooting time for previously experienced issues and hence improve light source performance predictability.

2. Data Management Architecture for Fault Detection and Classification

Fault detection identifies that an operating condition is different from normal, while classification identifies what caused the difference. The fault detection and classification techniques described in this paper are based on two core elements of the Cymer light source data management architecture. The first is enhanced performance logging features added to current-generation light source software that captures detailed performance data synchronous with detection of fault conditions. The on-board logging capability is responsible for the on-line monitoring of performance. During normal operation, on-board logging mostly consists of time-based average performance data. However if a fault condition is detected, on-board logging captures detailed performance data around the fault condition.

The second core element of the Cymer light source data management architecture is Cymer OnLine (COL), which facilitates the collection, transmission, and storage of light source performance data for classification analysis. Most performance data is processed outside the light source software in order to provide the highest quality data analysis, since expert service engineers consider all current information regarding known operating conditions and performance issues. Similarly, the automated classification analysis is performed on a data analysis server to take advantage of central processing and rapid access to updated information. By relying on a centralized performance analysis structure, ongoing improvements and refinements of the FDC tools can be quickly implemented without disturbing the fielded light sources. A diagram of the current service work flow, including automated data analysis, is presented in Figure 2.

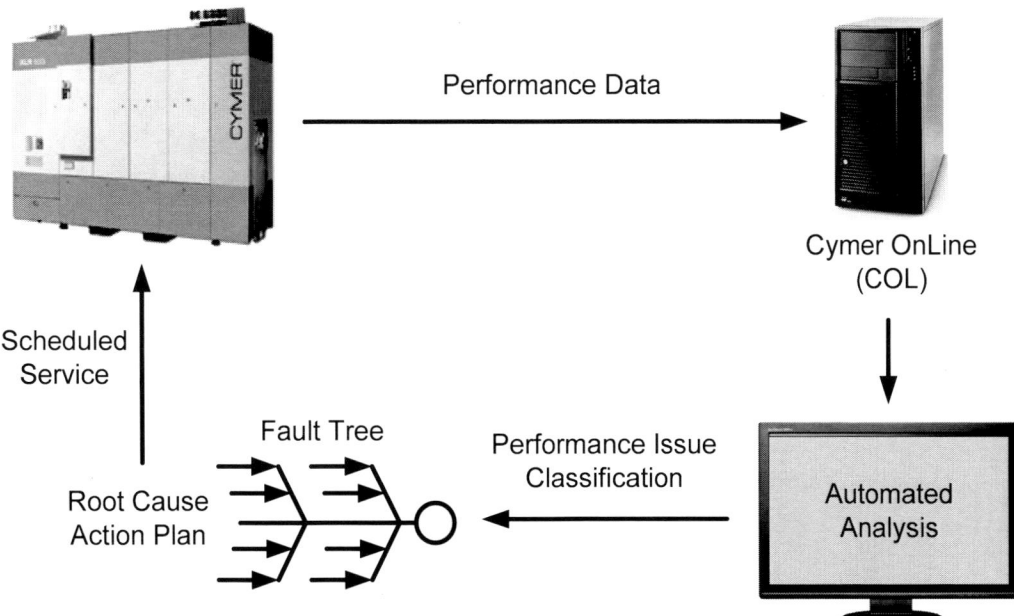

Figure 2: Flow diagram for Cymer FDC architecture. Performance data is created on-board the light source then transmitted and collected by COL. Automated analysis classifies the performance issue, which is then used to assign root cause and determine the appropriate action plan for service.

Prior to the existence of the automated data analysis discussed in this paper, performance data was directly accessed by service engineers who interpreted information contained in the data. Automated data analysis enhances this service strategy by interpreting raw performance data into information about the health of the light source. This additional information is readily available to service engineers for developing actions and scheduling maintenance activities. The Cymer FDC was developed such that whenever applicable a one-to-one relationship is established between the performance issue and final root cause (see fault tree presented in Figure 3). Several levels of classification analysis exist, the primary level designates the performance issue to a category while subsequent levels narrow down the performance issue to the eventual root cause. In case a root cause can not be determined, perhaps due to lack of fault excitation and observability in the data set, then an issue category is assigned and manual troubleshooting can be directly applied to resolve that section of the fault tree.

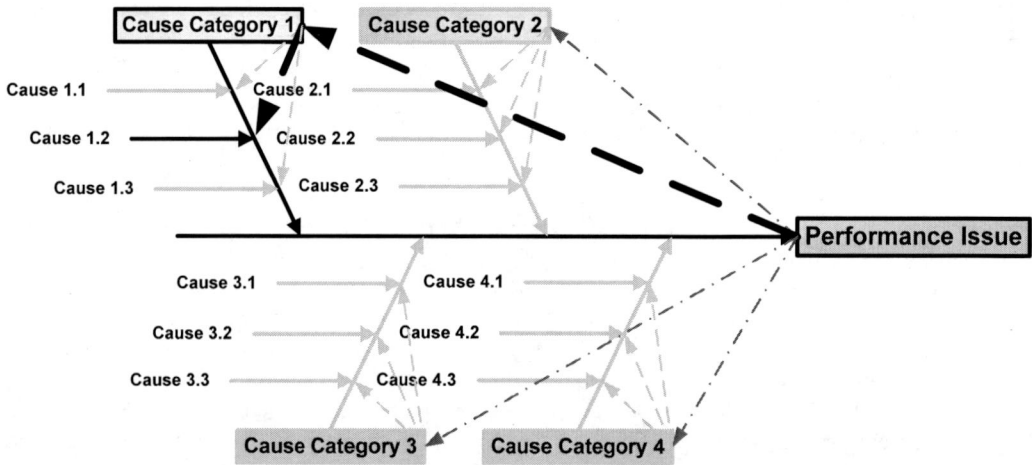

Figure 3: Diagram of a fault tree documenting possible root causes for performance issues. Several category levels within the fault detection and classification provide a relationship that leads between the performance issue and root cause.

Finally, automated analysis is capable of processing much larger quantities of data, enabling a greater understanding of light source performance with respect to population statistics as well as rapid development of issue resolution strategies. Furthermore, this augments the ability to predict the trend of light source performance and therefore minimize downtime and resources required for maintenance.

3. Fault Detection and Classification Methodology

Different approaches for fault detection and classification using mathematical models have been developed over many years [2]. These methods augment the more general statistical pattern recognition techniques, see [3] and references therein, particularly where information about the process is known and useful for constructing a process model. Model-based approaches use prescribed dependencies between different measured signals that are expressed in mathematical models in order to select relevant variables in the performance data. Feature extraction is the process of mapping original measurements into more useful features. Relevant features might correspond to the operation of sensors, actuators or subsystems of a process.

Important features in light source performance data are typically non-linear mapping functions of the original measurements. The feature extraction mappings are based on mathematical models of light source operation, developed to amplify characteristics that are expected under certain fault conditions. An example nonlinearity included into feature extraction models is signal processing logic that explicitly considers performance transients between operating modes. In most cases, the signal processing involves constructing multivariate signals to form effective feature extraction mappings that are metrics of light source performance.

An effective feature extraction mapping is one that correctly indicates the likelihood of the presence of recognizable symptoms of a particular fault condition. The likelihood of a fault above a specified threshold determines the detection and classification of the fault condition. The Cymer fault analysis software is built upon a flexible infrastructure that enables rapid development and deployment of feature extraction mappings. Training data sets are used to evaluate feature extraction mappings over numerous light source operating conditions for effective and reliable identification of a particular fault condition. Once validated, newly developed features are easily deployed to the mainstream automated analysis software. Since the architecture relies on a centralized analysis function, software that resides on the fielded light sources does not need to be modified, and the implementation can be immediate.

For each light source performance data set received, multiple features can be extracted. The automated data analysis classifies the data set into a fault condition by observing all the extracted features and quantifying any discrepancies from expected behavior. Some fault conditions are amenable to single features that directly relate to root cause, while others require correlation of several features in order to recognize a pattern that indicates root cause. In either case, the automated analysis provides consistent information that reduces the reliance on expert service engineer judgment in identifying the correct resolution for a performance issue.

3.1 Case Study

The following case study provides an example for the methodology of the Cymer model-based FDC. The maximum pulse-to-pulse energy exceeds a specified threshold and triggers the fault condition (see Figure 4). Once the fault condition is detected, light source performance data is collected and transmitted for analysis as described in Section 2. Performing single-variate analysis, that is considering one signal at a time, the cause for the performance issue often can not be identified. Indeed several key signals, shown in Figure 5 – a and 5 – b, are well within their normal variation leading up to the fault condition.

Here it may be interesting to note that the fault condition presented in this case study is considered a rare occurrence. Particularly in known but rarely occurring fault events, detailed analysis beyond the readily available performance data (single-variate signals) would have required time and resources before issue resolution. Automating the data analysis expedites the troubleshooting processing by quickly delivering the relevant information about the fault symptoms regardless of how often the fault condition actually occurs.

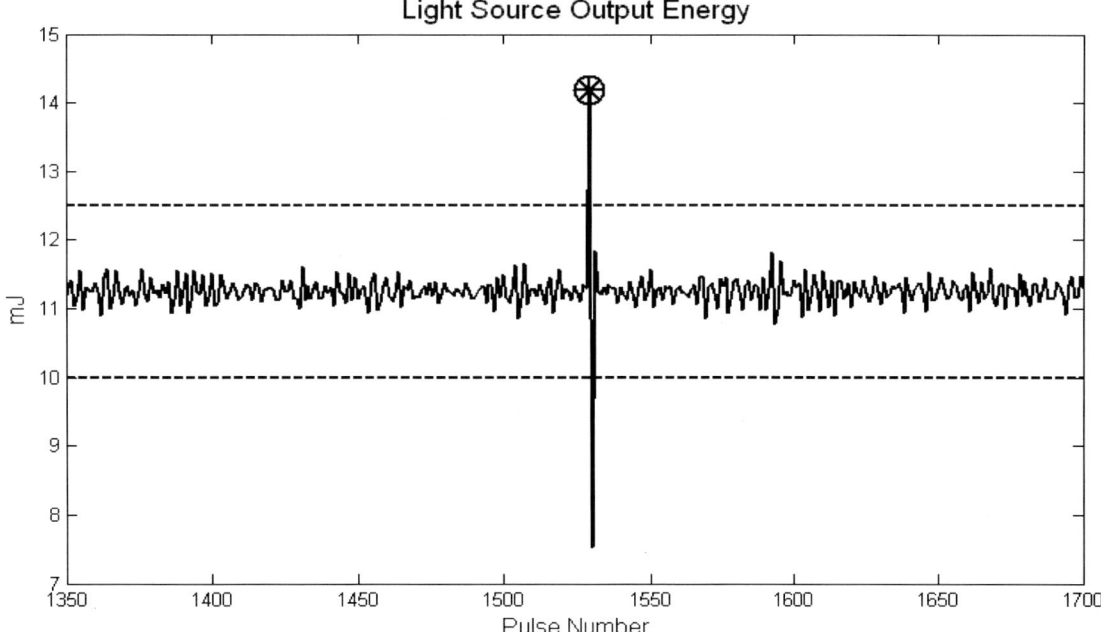

Figure 4: Light source measured output energy (solid line) is plotted against the pulse-to-pulse energy fault condition (dotted line). A pause in light source operation is indicated by the circle (O) which coincides with the first pulse to exceed the threshold designated by the marker (*).

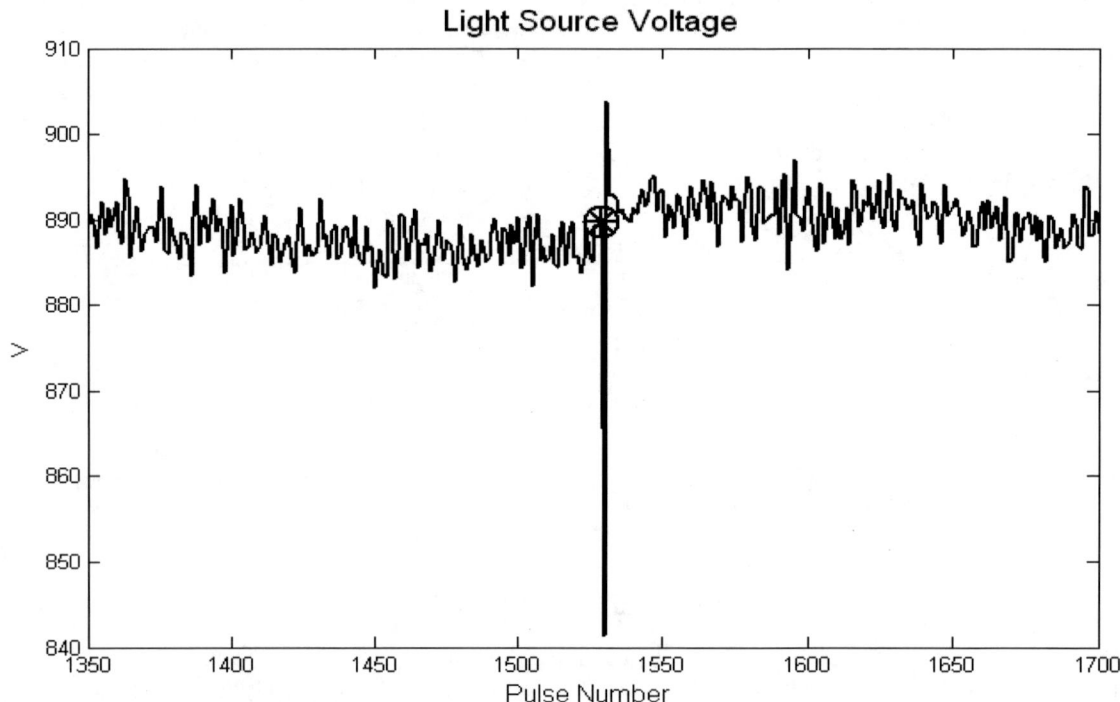

Figure 5 – a: Light source measured input voltage (solid line). A pause in light source operation is indicated by the circle (O) which coincides with the first pulse to exceed the threshold designated by the marker (*).

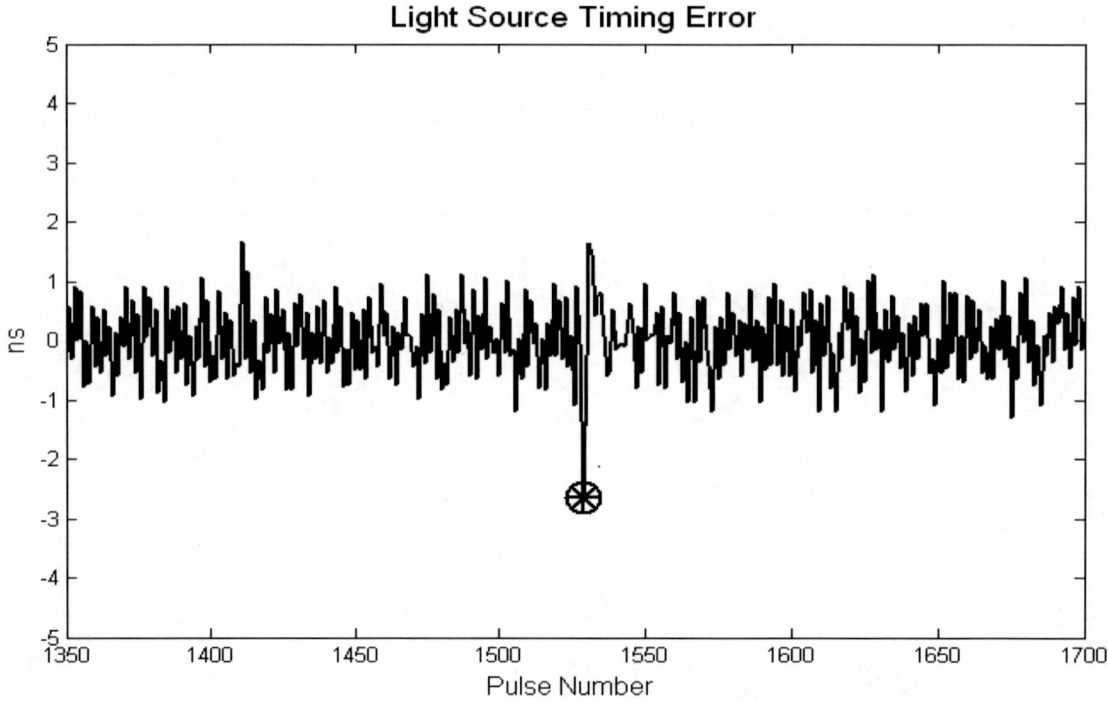

Figure 5 – b: Light source measured timing error (solid line). A pause in light source operation is indicated by the circle (O) which coincides with the first pulse to exceed the threshold designated by the marker (*).

The collected fault condition performance data is evaluated over multiple feature extraction models of the light source that are built in to the FDC system. For this case study example, one extracted feature exhibits sufficient information to identify root cause. This feature relates the signals shown in Figures 5–a and 5–b via mathematical model of the light source physics.

The constructed multivariate signal, plotted in Figure 6 as a function of light source voltage, clusters the expected light source performance along a single line. Discrepancies from expected performance are clearly visible from this representation of this visualization of the data, however it does not yet indicate a root cause for the fault condition. The relevant feature for this particular fault condition is a projection of anomalous data points onto the light source model. The projected value is the expected voltage driving the light source inferred from other measurements of light source performance, i.e. that there is a discrepancy between the reported voltage and the one actually applied to the light source. The projected discrepancy indicates a rare occurrence on legacy software versions, whereby an internal software deadline is not met, and the applied voltage is incorrect. The issue is quickly resolved by recognizing the root cause for the fault condition and upgrading software.

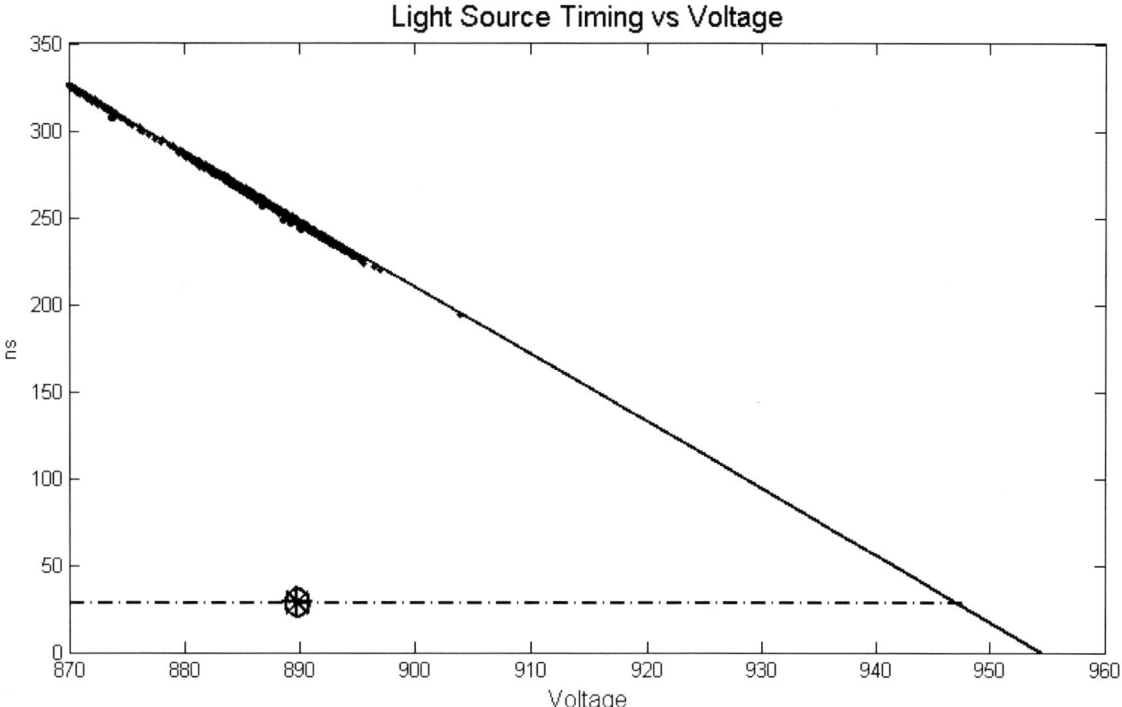

Figure 6: Light source measured output energy (solid line) is plotted against the pulse-to-pulse energy fault condition (dotted line). A pause in light source operation is indicated by the circle (O) which coincides with the first pulse to exceed the threshold is designated by the marker (*).

4. Fault Detection and Classification Performance

The Cymer FDC system was first tested and released in January 2009. Currently the automated data analysis is processing performance data from more than 300 fielded light sources around the world and approximately 1 [Gbyte] per day. On average, 60% of the performance issues detected by each light source are classified according to the symptoms displayed in the performance data during the fault condition, see Figure 7. Although not each classification implicates the fault condition root cause, the analysis adds information by identifying or ruling out symptoms of the fault condition.

Fault Detection and Classification Performance

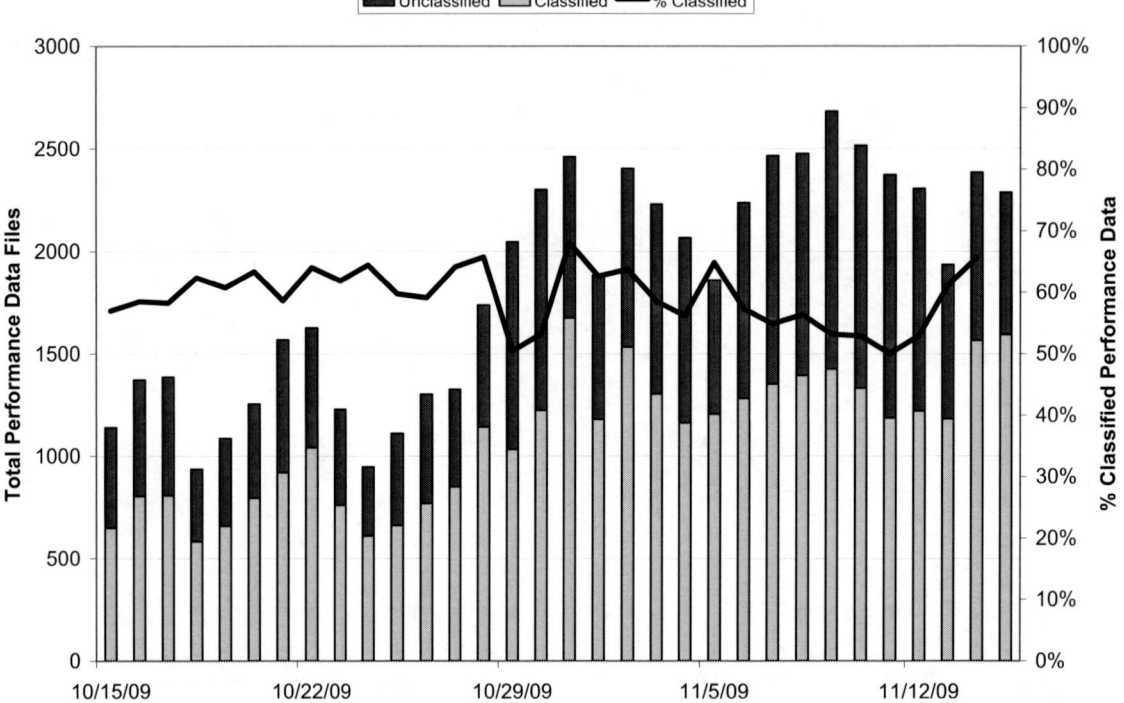

Figure 7: FDC performance in monitoring all fielded Cymer light sources. On average, 60% of performance issues are classified.

As mentioned in Section 2, the Cymer FDC system enhances current service strategy and augments the ability to predict the performance trending of light sources. Many fault conditions progress with operation of the light source, thus by monitoring the performance data issues can be recognized and addressed more accurately and with more time to prepare scheduled maintenance action plans. Consider for example the performance issue presented in Figure 8. The fault condition was automatically identified prior to customer impact and the appropriate action plans were developed and ready for the next scheduled maintenance opportunity. The performance issue caused by the fault condition was resolved during scheduled maintenance, reducing the impact to production.

As a part of an overall service strategy, automated FDC augments the ability to predict the performance trending of light sources. The impact of augmenting performance predictability can be observed in comparing scheduled versus unscheduled downtime. Cymer Technical Support deployed a Proactive Service Program (PSP) in conjunction with FDC in December 2008. The FDC system adds information performance summaries for each light source. Figure 9 shows the monthly scheduled versus unscheduled events for one customer participating in this program. Shortly after deployment, the ratio between scheduled versus unscheduled service improved by 10%. While there are many factors that contribute to this improvement, FDC enhances the capabilities and value of the Cymer PSP program.

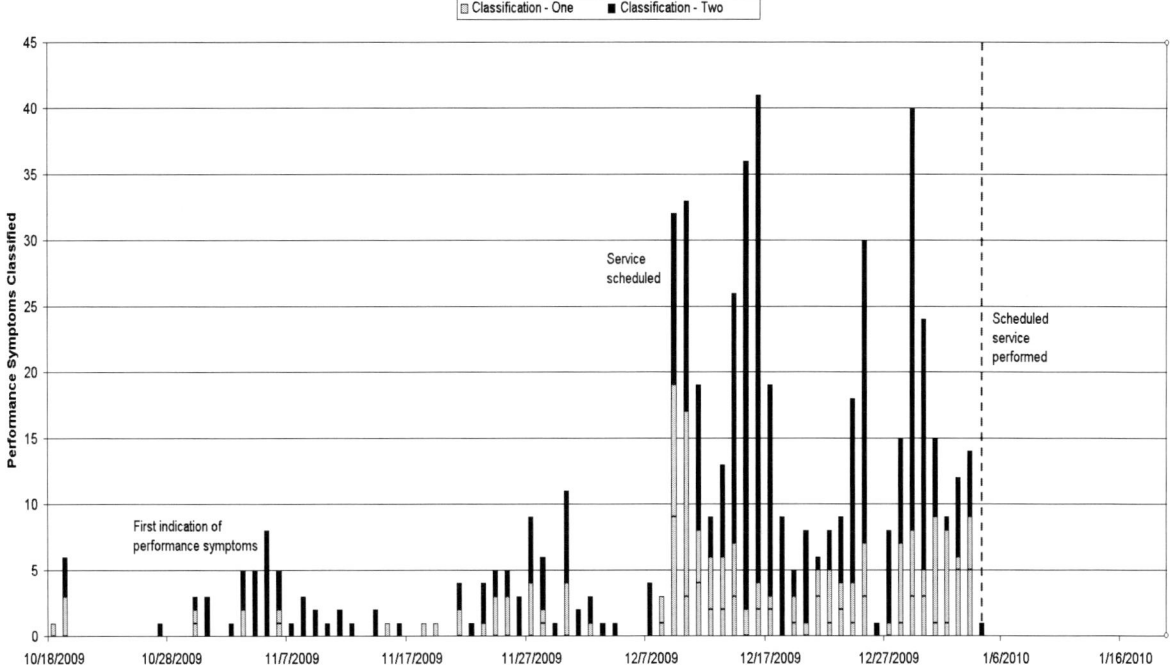

Figure 8: FDC summary for a single light source during the progression of performance symptoms. The performance symptoms begin with small low rate of events that increase in frequency over time. Two classification types, which are related to a single light source subsystem, have been identified. In this case, automated classification enabled awareness of the performance issue root cause with sufficient time to schedule maintenance with two week horizon.

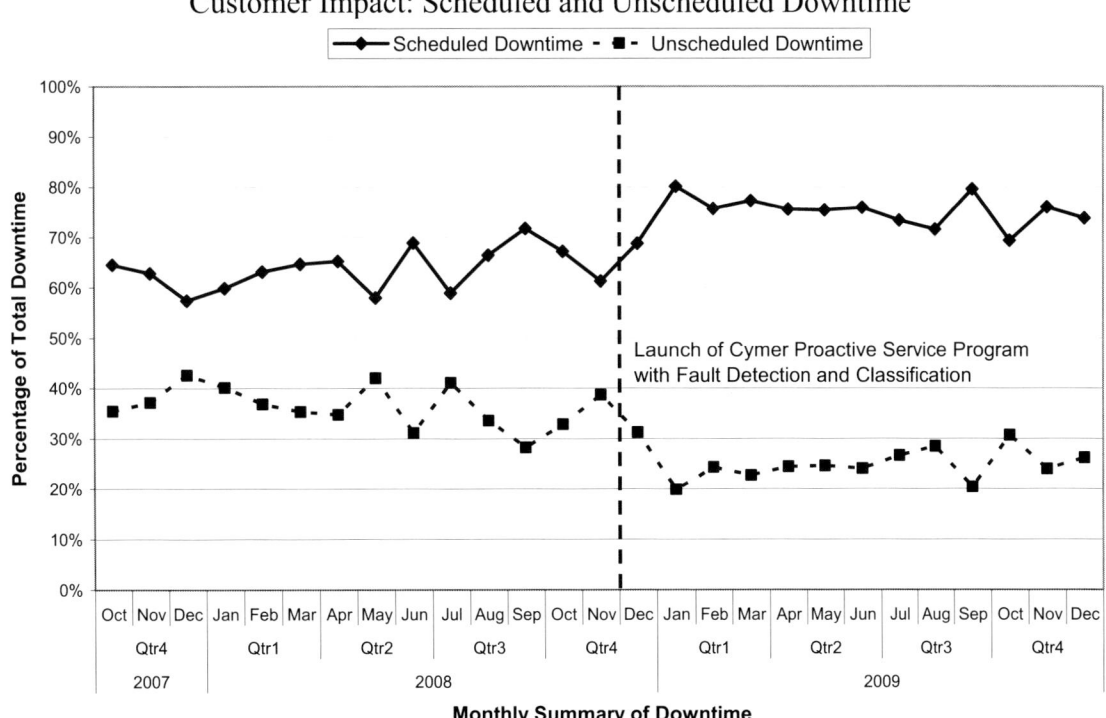

Figure 9: Single customer comparison of scheduled versus unscheduled downtime that shows the improvement in scheduled service with the deployment of Proactive Service Program (PSP) in conjunction with FDC.

5. Conclusions and Future Work

Cymer FDC provides several efficiency improvements for light source troubleshooting by providing more content-rich standardized summaries of light source performance along with reduced time-to-identification for previously classified faults. Perhaps the most significant advantage is that performance issues that exhibit recognized patterns are often correlated to known root cause, where known corrective actions can be implemented. This is therefore used to schedule service activities and minimize the impact of the duration of time that the light source is offline for maintenance. Through an increased level of automation in analysis, Cymer is able to reduced fault identification time, reduce time for root cause determination for previously experienced issues, and enhance light source performance predictability.

References

[1] SEMI E10 - for Definition and Measurement of Equipment Reliability, Availability, and Maintainability.

[2] Isermann, R. "Model-based fault detection and diagnosis – status and applications" Annual Reviews in Control, 29(1), 71-85, (2005).

[3] Jain, A.K., Duin, R.P.W., Mao, J. "Statistical pattern recognition: A review" IEEE Transactions on Pattern Analysis and Machine Intelligence, 22 (1), 4-37, (2000).

Pattern deformation caused by deformed pellicle
with ArF exposure

*Jee-Hye You[1], Ilsin An[2] and Hye-Keun Oh[1]

[1]Lithography Laboratory, Department of Applied Physics, Hanyang University,
[2]Nano-View Co. Sa 1- dong, Sangrok-gu, Ansan, Kyunggi-do, 426-791, S. Korea
Tel: +82-31-400-4137, Fax: +82-31-406-1777, E-mail: jeehye.u@hanmail.net

ABSTRACT

It will directly affects pellicle degradation, at the irradiated part of the pellicle, and make a sloped pellicle surface and will act like a prism before change of phase or transmittance occurs, because the energies of C, F and O single bondings composing the ArF pellicle film is quite smaller than the energy of 193 nm ArF. Thus, outgoing light has information of smaller space than mask size. In order to offer some tip to find the appearance of pellicle thinning caused defect, several types of pattern deformation caused by pellicle degradation is studied.

1. INTRODUCTION

The wavelength of exposure light in photolithography has tendency to become shorter because of the demand of better resolution in photolithography, and ArF laser was introduced to perform the critical photolithography process.[1] Recently, extensive studies that are willing to seek for the solutions about the haze have been performed by various researches with especially photo mask cleaning process improvement. However, because a phase shift mask (PSM) is still weak to haze formation, a binary mask is preferred in ArF lithography. But, a new problem was happened after binary mask introduction that is critical dimension (CD) variation. As the accumulated dosage is increased, CD variation in a lithography shot is appeared in a certain time and is gradually grown in progress of exposure process. And finally, CD variation grown considerably causes defect in wafer level. It can be inferred that CD variation is closely related to the transmittance of the reticle because the transmittance of the reticle where CD variation is detected has some differences from the initial level.[2]

Due to its high photon energy, ArF laser accelerates photochemical reactions generating a non-organic crystal defect called as haze, so that it makes a cleaning cycle of photomask quite short. Moreover, high energy of ArF laser affects pellicle degradation in direct because the energy of C, F and O single bonding composing ArF pellicle film is quite smaller than ArF laser photon energy as

Optical Microlithography XXIII, edited by Mircea V. Dusa, Will Conley, Proc. of SPIE Vol. 7640,
76402E · © 2010 SPIE · CCC code: 0277-786X/10/$18 · doi: 10.1117/12.846483

shown in Figure 1.[3] As a result of broken C, F and O single bonding, the pellicle of the exposed open pattern is locally getting thinner with exposure process. In some papers, even less than 4nm pellicle membrane thickness reduction causes 2% light transmission rate.[4, 5]

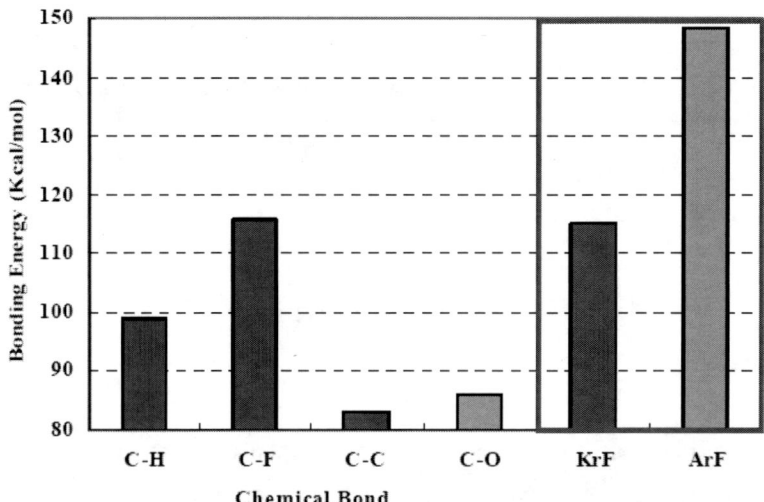

Figure 1: Comparison between chemical bonding energy and laser photon energy

Through the experiment, we found that thickness of pellicle get thinner during the exposure process as shown in Figure 2. At the exposed area on the pellicle membrane, it will be making a sloped pellicle deformation like a concave lens. These thinner parts will be act like prism; outgoing light has information of smaller space than mask size.

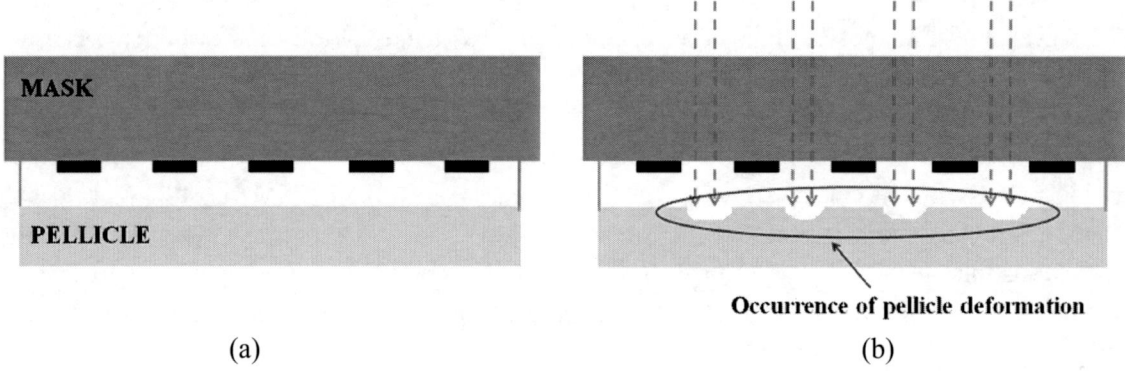

Figure 2: Schematic diagram of (a) mask and attached pellicle before exposure and (b) occurrence of local sloped pellicle deformation due to exposure

As shown in Figure 3, and slopes that made by high photon energy are different with space CD on mask and pellicle deformed depth, so it is very important to very smaller sized pattern. And it will more affect if we use phase shift mask (PSM), modified off-axis illumination (OAI), high numerical aperture (NA), etc. In order to see the appearance of pellicle thinning caused defect, several types of pattern deformation caused by pellicle degradation is studied by using commercial simulation EM-SUITE by Panoramic.

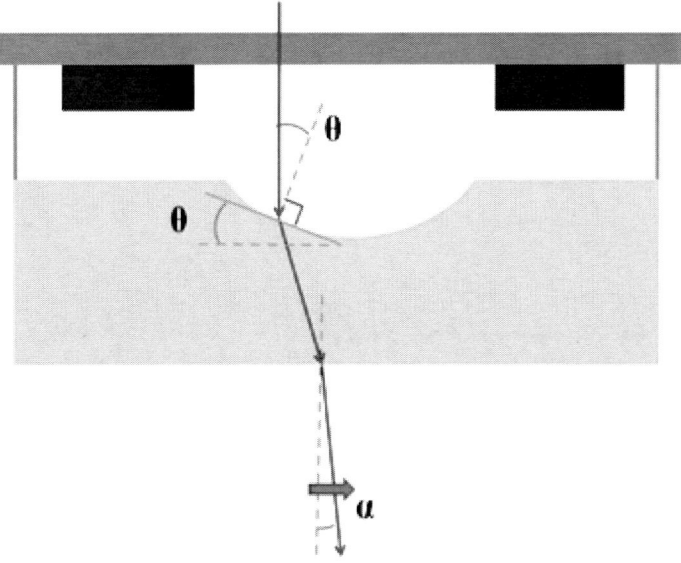

Figure 3: Ray off to inside because of the difference of index of refraction ($n_{air} = 1$, $n_{pellicle}=1.389$)

2. RESULTS

(1) Width and Depth of Pellicle Deformed Area

We simulated the effect of width and depth of pellicle deformed area. Target CD of 90 nm is chosen by using binary Cr mask and pole radius of illumination is 0.3 pole angel is 0°with dipole. Pellicle thickness is 833 nm.[6] Figure 4 shows the aerial images with different width of pellicle deformed area. Pellicle deformed width range is varied from 20 to 200 nm in mask size and deformed depth is fixed to 50 nm to all cases. It should that there is some difference of intensity and CD variation is not much different for this.

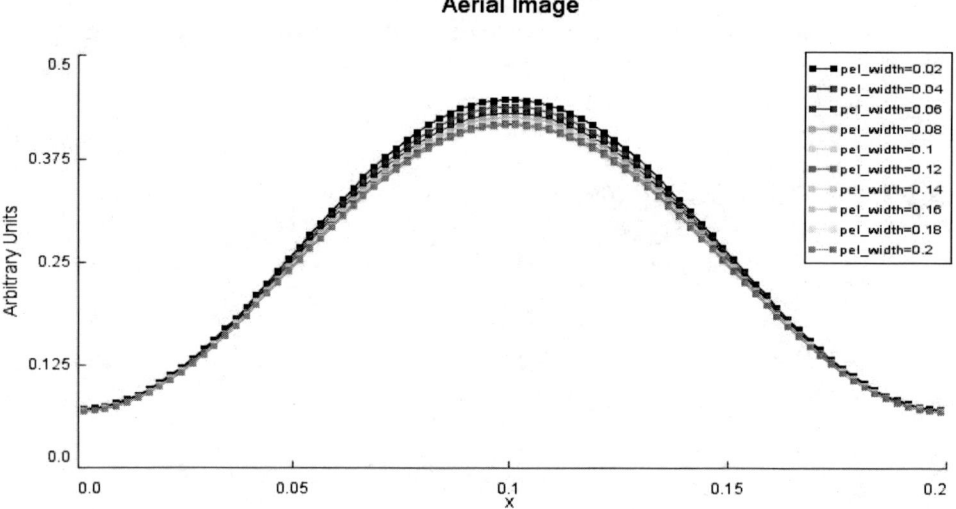

Figure 4: Aerial Images with the different widths of pellicle deformed area for 90 nm

Figure 5 shows the aerial image with different depth of pellicle deformed area. Deformed depth range is from 35 nm to 60 nm in mask size and deformed width is fixed to 360 nm for all the cases. The effect about depth of pellicle deformed area is larger effect to aerial image and resist pattern CD than about width of pellicle deformed area. Also the same tendency was found for the 45 nm and 65 nm target CD. Consequently, pattern deformation on the wafer can be occurring during exposure process although exposure process keeps within pellicle lifetime.

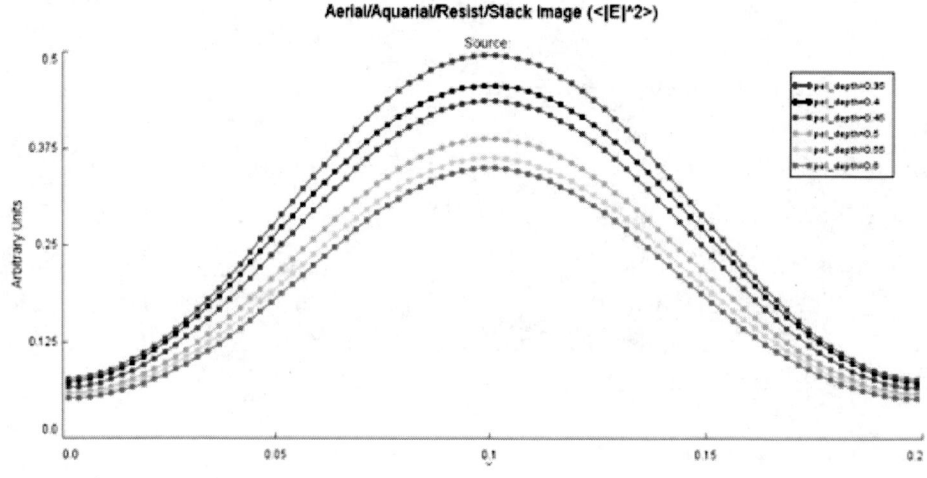

Figure 5: Aerial Images with the difference of pellicle deformed depth for 90 nm CD

(2) Different pitches

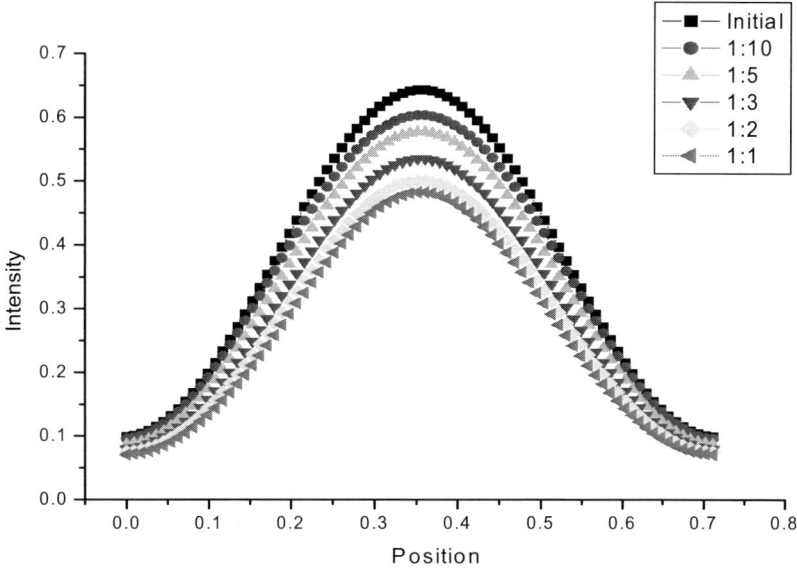

Figure 6: Aerial images for 90 nm pattern with various pitches

We simulated the same effect with different CDs and pitches. Figure 6 shows the result for 90 nm pattern with 1:1, 2, 3, 5 and 10 pitches. For 90 nm target CD, we found the some difference at aerial image, or intensity, and CD variation is ignorable numerical value, about 1.5 nm.

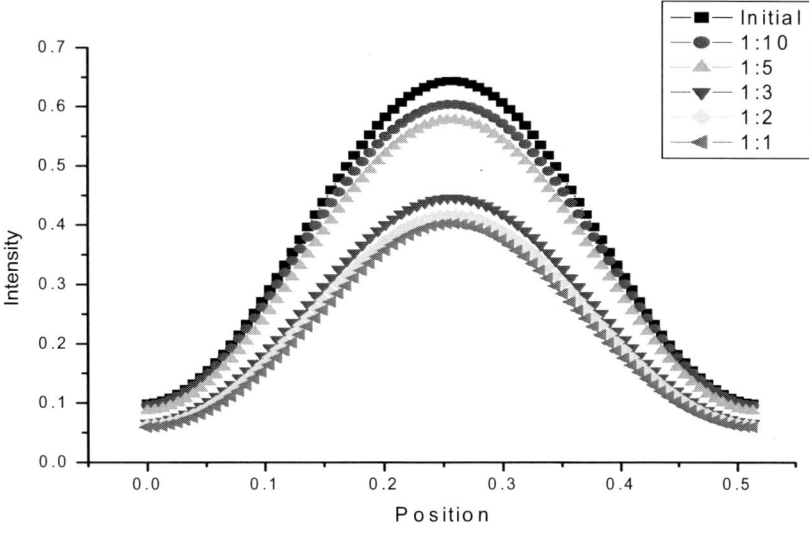

Figure 7: Aerial images for 60 nm pattern with various pitches

Figure 7 shows the result for 60 nm pattern with 1:1, 2, 3, 5, 10 pitches. For 60 nm target CD, there is a large gap with difference pitches; CD difference on the wafer is about 2.6 nm. There is large difference to ~ 1:3 pitches, but 1:5 and 1:10 pitch aren't specific difference on the wafer.

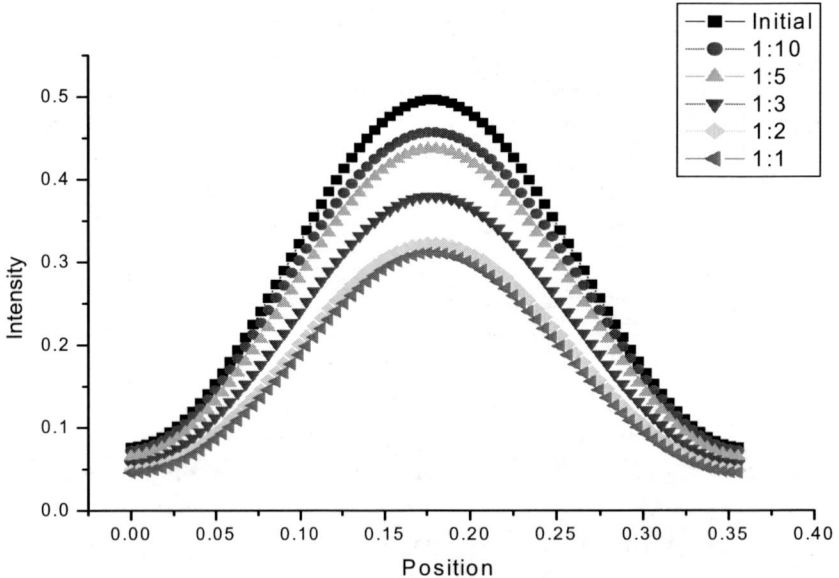

Figure 8: Aerial images for 45 nm pattern with various pitches

Figure 8 shows the result for 45 nm pattern with 1:1, 2, 3, 5 and 10 pitches. For 45 nm target CD, there are differences with difference pitches; pattern deformation on the wafer is about 5 nm. This value is very large and it will more affect if we use PSM, OAI and NA etc. For the smaller sized pattern, as shown in Figure 3, degree of pellicle slope will be steep very largely, and pattern deformation on the wafer could be very serious problem and the pellicle exchange cycle has to be shorten than the present period.

3. SUMMARY

Pellicle is very important to protect from dust and must have item for 193 nm lithography. But ArF source occur pellicle deformation because of high photon energy than C, F and O bonding, and patterns on the wafer be seriously. We simulated several cases with simulation tool, and found that that pattern deformation is affected by depth of pellicle deformed area and smaller sized patterns.

REFERENCE

1) I. Sakurai, T. Shirasaki, M. Kashida and Y. Kubota, Proc. of SPIE Vol. 3748 **177** (1999)

2) H. Choi, Y. Ahn, J. Ryu, Y. Lee, B. An, S. Lee, Proc. of SPIE Vol. 6518 **1** (2007)

3) H. Choi, Y. Ahn, J. Yoon, Y. Lee, Y. Cho, J. Kim, Proc. of SPIE Vol. 6730 **2** (2007)

4) S. Shigematsu, M. Kondo, H. Nakagawa, Proc. of SPIE Vol. 3748 **168** (1999)

5) J. Miyazaki, M. Uematsu, T. Ogawa, Proc. of SPIE Vol. 3679 **534** (1999)

6) R.de Kruif, T. van Rhee, E. van der Heijden, Proc. of SPIE Vol. 7470 **1** (2009)

Study for lithography techniques of hybrid mask shape of contact hole with 1.35NA polarized illumination for 28nm-node and below logic LSI

Yuji Setta, Katsuyoshi Kobayashi, Tatsuo Chijimatsu, and Satoru Asai,
Fujitsu Microelectronics Limited (Japan)

1500, Mizono, Tado-cho, Kuwana-shi, Mie-ken, 511-0192 Japan
Email: y-setta@jp.fujitsu.com

ABSTRACT

In this presentation, the advantage in the use of combination of polarized illumination and technique of optimum shape mask for contact-hole lithography will be discussed. Both simulation and experimental work were carried out to characterize performance of this technique. We confirmed that some polarized illuminations show improvement in image contrast, MEEF, and DOF for nested contact-hole than non-polarized condition. In addition, certain shape mask shows more improvement. Totally 63% DOF improvement from traditional square shape with non-polarized condition was confirmed. In final single exposure era for contact-hole, this result with techniques of hybrid mask shape and polarized illumination is very attractive.

Keywords, low k1 lithography, contact-hole, polarization

INTRODUCTON

Demand for scaling of integrated circuit device has continued, however, the backup of improvement of RET (resolution enhancement technique) and supply of high NA (Numerical Aperture) exposure tool has made it possible. In recent years, in order to improve the resolution, different approaches compared with traditional way such as double patterning and EUV (Extreme Ultra Violet) lithography are focused. Also, computational lithography such as SMO (source mask optimization) and ILT (inverse lithography technology) has been explored.

Dry ArF exposure tool has become general at the mass production of 90nm to 65nm node. Those day demands of minimum pitch are about 1.1 to 1.5 times larger than the light source wavelength (Figure 1). Patterning of contact hole is one of difficult issue for optical lithography and this minimum pitch have an influence to shrink design rule. Even those situations were very difficult but litho engineer could achieve the process for mass production with research and enthusiasm. On the other hand, demand of minimum pitch for 28nm node has reached a half of exposure wavelength. Our strategy as lithographers is to push it costly and robust than before.

One of characterizations in lithography is to use reticle for making pattern on wafer. As DR (design rule) scaling, demand of mask cost of ownership has been sever. In this situation, data preparation time, writing time, and inspection time become a big component in the mask cost. Lithographer should concern some approach to reduce it.

In the 28nm and below, polarized illumination are discussed and explored. This technique has some merit to get higher image contrast and reduce the MEEF for nested region[1], [2] ,[3]. The situation that cannot increase NA of immersion exposure tool has made this technique the last stronghold. Among contact-hole lithography, we would like to use this advantage as possible as we can.

In this presentation, the advantage in the use of combination of polarized illumination and technique of optimum mask shape for contact-hole lithography will be discussed. It is said that quadrupole illumination is suitable for nested pattern of contact hole/via. But that's only said in textbooks because actual configuration of LSI pattern for contact hole/via is extremely numerous. Setting of quadrupole illumination with polarization is often difficult because each aperture of quadrupole is located on slit position. So several polarized condition with annular were explored. We confirmed that some polarizations show improvement in image contrast, MEEF, and DOF than non-polarized condition. In addition, octagonal mask shape shows more improvement. Totally 63% DOF improvement from traditional rectangle shape with non-polarized condition was confirmed.

Optical Microlithography XXIII, edited by Mircea V. Dusa, Will Conley, Proc. of SPIE Vol. 7640,
76402I · © 2010 SPIE · CCC code: 0277-786X/10/$18 · doi: 10.1117/12.848316

Effect of polarized illumination for contact-hole

As a means of getting high RET, high NA exposure tool cannot be expected. In addition, EUV lithography cannot meet our schedule. Recently the effect of polarized illumination is aggressively explored; line and space pattern is an example to this effect. In order to take this advantage of polarization for contact-hole lithography, we compared several conditions with polarization.

On the beginning of this evaluation, in order to understand the merit of polarized illumination effect, the comparison of effect by simulation was done. Figure 2 shows image intensity of same pitch pattern. Figure 2a is for line and space pattern, and Figure 2b is for contact hole. This result shows that the usage of polarized illumination for line and space pattern leads 25% improvement in NILS (Normalized Intensity Log Slope), also 15% improvement in MEEF (Mask Enhanced Error Factor). These improvements were very attractive and promising in low k1 lithography era. On the other hands, the effect of polarized illumination for contact hole was unfortunately smaller than that of line and space. The improvement rate of NILS and MEEF is 3.6% and 9.2%, respectively. This effect was small as we expected, but we have to manage to drive with this situation.

According to nested regular array contact hole patterns, it is generally said that quadrupole illumination is desirable. The quadrupole illumination consists of 4 openings located on 45-degree, 135-degree, 225-dgeree, and 315-degree, respectively. In addition, the thoughts of SMO is recently discussed and explored. But there is concern about the optimization for all patterns in LSI chip, it might create weak configuration layout. That is, the illumination for contact hole lithography until 28nm generation might be conservative annular illumination if possible. To complicate matters, however, even annular "4 segments" polarized illumination has certain slits in those octagonal positions. The slits are required for switching of polarized illumination lights direction. These slits are about 5% area loss.

Figure 3a-c show illumination maps that we used on this evaluation. This experiment was done with several annular illuminations. Figure 3a is simple annular illumination without polarization. We named it "non polarization". Figure 3b is XY polarization with annular illumination, and this aperture consists of 4 main segments. So we called it "4 segments" polarized illumination. Figure 3c is a kind of tangential polarized illumination; 8 segments polarizing elements divide this illumination into 8 equal parts. So we named it "8 segments" polarized illumination. Also the direction of polarized lights was descried in each figure.

Figure 3d shows the DOF comparison with above-mentioned three illumination conditions. Compared with "non-polarization" condition, improvement of DOF for both polarized illuminations was confirmed as we expected. DOF of "4 segment" polarized illumination was improved by 43%, and that of "8 segments" polarization was 39%. Each effect of polarization is very helpful, however, in case of comparison of "4 segments" with "8 segments" polarized illumination using regular array contact hole, effect of "8 segments" polarized condition did not show further improvement than "4 segments". The results were worse by 6% compared with "4 segments" condition. This result falls short of our expectations. However another contact hole array might be better for this characteristic "8 segment" polarized illumination

Then, CD variations with each illumination were evaluated. Compared with CD variation of "non-polarization" condition, both CD variation of each polarized condition were improved by 7%. Regarding CD variation, the effect of polarization was almost the same value.

Octagonal shape mask

After evaluation of each polarization states, in order to achieve further improvement of DOF, MEEF, and CD-variation, the shape of mask pattern was focused. That is why single exposure is must key item for low cost mass production. Therefore next step of our work went to stage of optimization of mask.

In 28-nm node generation target CD become small as design scaling, but MEEF for 28-nm node estimated in ITRS roadmap is as same as previous generation. That is, further CD control as much as we can is required.

Figure 4a shows the study of image contrast of nested contact hole patterns by simulation. That is comparison of traditional square shape mask with our proposal octagonal shape mask. The image contrast of traditional square shape mask was higher than that of octagonal shape mask in the region of 0.12um pitch and wider. But below 0.12 um pitch shows that the image contrast of octagonal shape mask was higher than that of square mask. That is, usage of octagonal shape mask below 0.12um pitch might have possibilities to reduce MEEF, CD-variation. And then these pitches correspond to contacted pitch in 28nm generation. So we began to confirm the merit of octagonal shape mask with a combination of polarized illumination.

From the point of view of optics, circle shape mask shows better optical performance. On our previous fundamental work, free form SMO pattern was investigated. However, unfortunately the numbers of reticle writing data become huge in spite of clipped layout. (Figure 4b) Therefore, instead of circle shape mask, we evaluated effect of octagonal shape mask. In case of octagonal shape mask, the volume of reticle writing data might be reduced.

Figure 5 shows the result of DOF comparison in these experiments. The point is that each results of octagon shape mask were better than that of square shape mask. Comparing with above-mentioned results, the rank order of improvements of DOF for octagonal shape mask was as same as that of square shape mask.

Regarding CD variation and MEEF, the result of this experiments shows that those of octagonal shape mask were better than those results of square shape mask (Figure 6). Usage of octagonal shape mask provides 25% improvement of CD variation. In addition, resist collapse test using FEM (focus exposure matrix) shows wide process window (Figure 7).

CONCLUSION

We investigated the promising solution for contact hole lithography toward 28 nm node and below. The pattern pitch we evaluated in this paper was almost the half of wavelength. The result of this experiment was that usage of "4 segments" XY polarized illumination shows the DOF improvement by 43%, and CD-variation reduction by 7%. But in case of "8 segments" tangential polarization further improvement was not unfortunately confirmed.

On the other hands octagonal shape mask shows better image contrast in narrow pitch. In order to achieve further improvement of narrow region we evaluated octagonal shape mask with "4 segments" XY polarization. Consequently, comparing with traditional square shape mask shape, a combination of "4 segments" XY polarized illumination with octagonal shape mask shows the DOF improvement by 6-13%, and CD-variation reduction by 25%.

These improvement results might be helpful for us to mitigate the control value of overlay for Diffusion-Contact and Poly-Contact. In low-k1 lithography era we would like to manage to do these consciousnesses. So we propose to use both traditional square shape and octagonal shape mask as the situation demands. We named it Hybrid shape mask.

REFERENCES

[1] Seiji Matsuura, et al, "Intrinsic Problem Affecting Contact Hole Resolution in Hyper NA Era", Japanese Journal of Applied Physics, vol.44, No.7B, 2005, pp.5489-5495.
[2] Ki-Yeop (Chris) Park, et al, "Lithography performance and simulation accuracy at different polarization states for sub 40nm node", proceeding of SPIE, vol.7140-132, 2008.
[3] Sohan Singh Mehta, et al, "C-Quad polarized illumination for back end thin wire: Moving beyond annular illumination regime", proceeding of SPIE, vol.7274-128, 2009

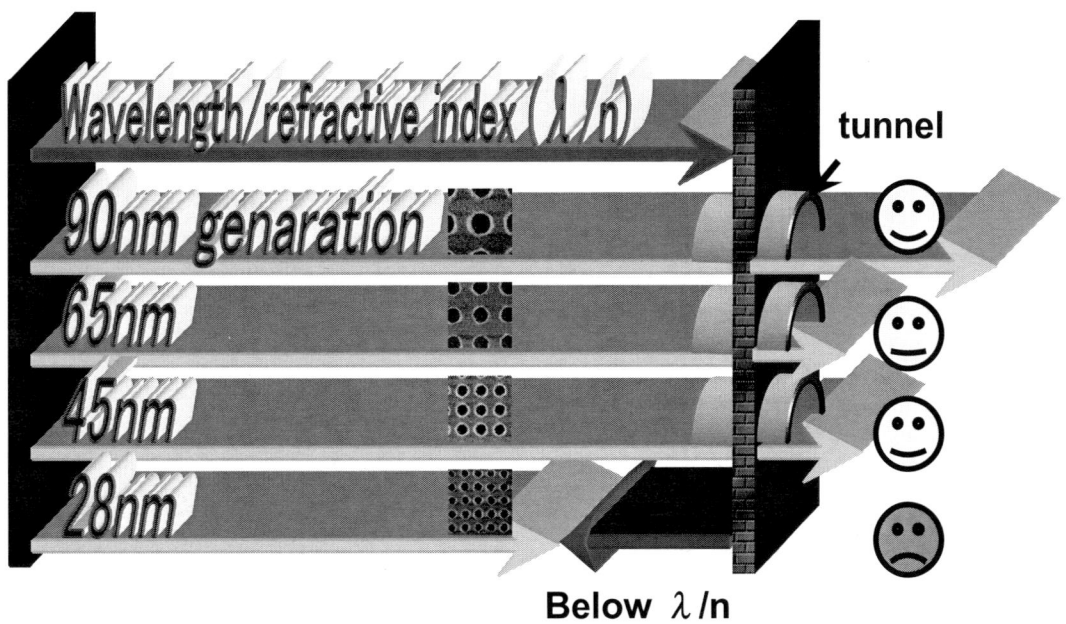

Figure 1 : relation between wavelength and critical pitch. Critical pattern pitch are becoming smaller than wavelength.

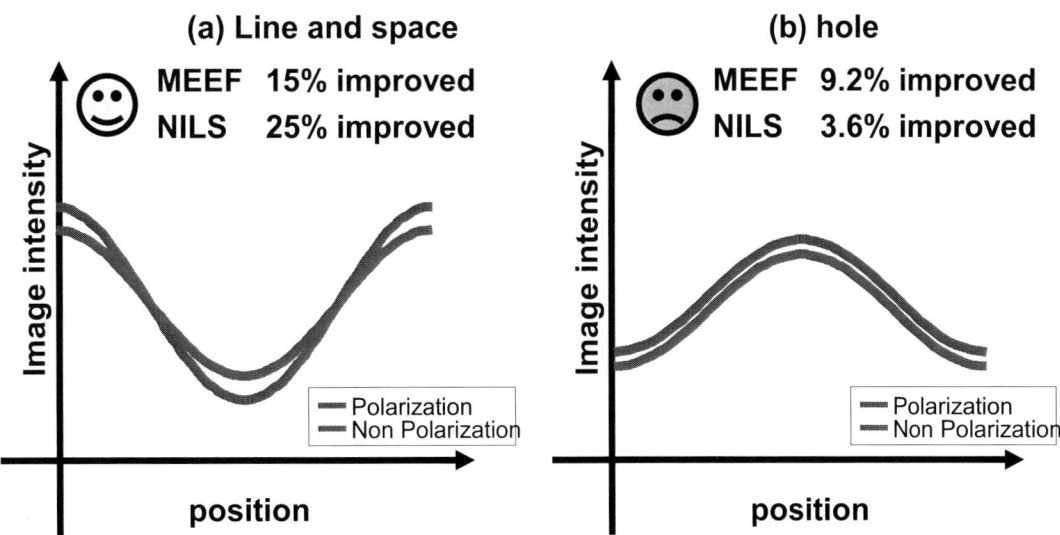

Figure 2 : comparison of polarization effect in image intensity

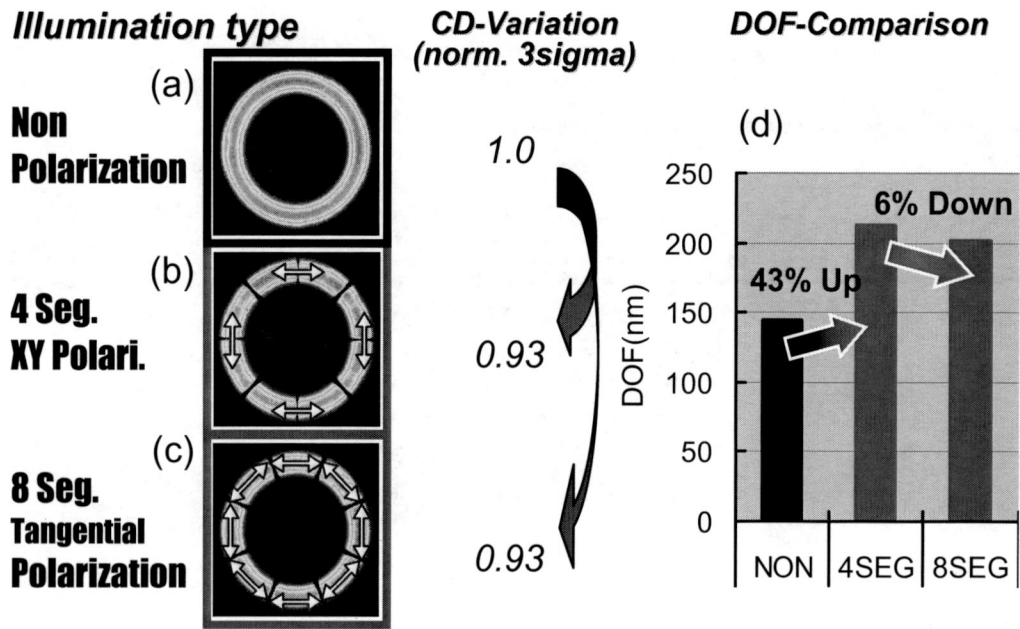

Figure 3 : Comparison of experimental DOF and CD-variation

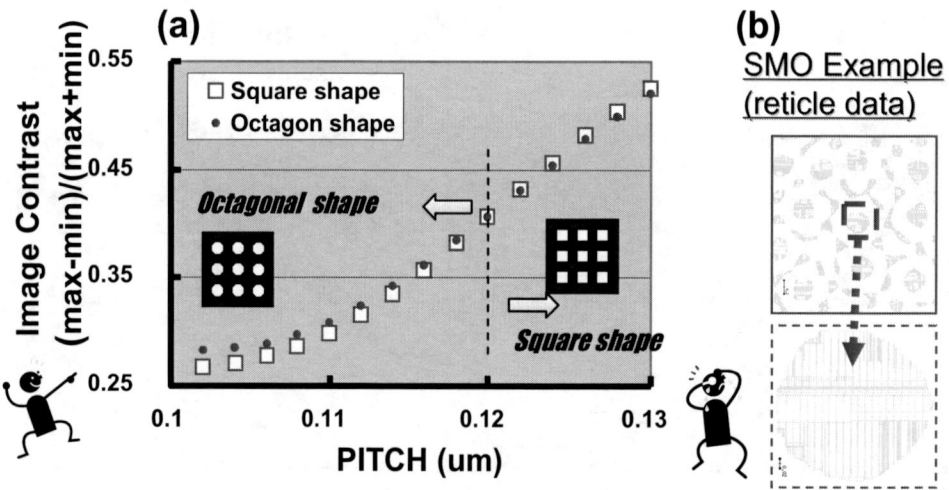

Figure 4 : (a)pitch dependency of image contrast. (b)example of reticle writing data for freeform MRC

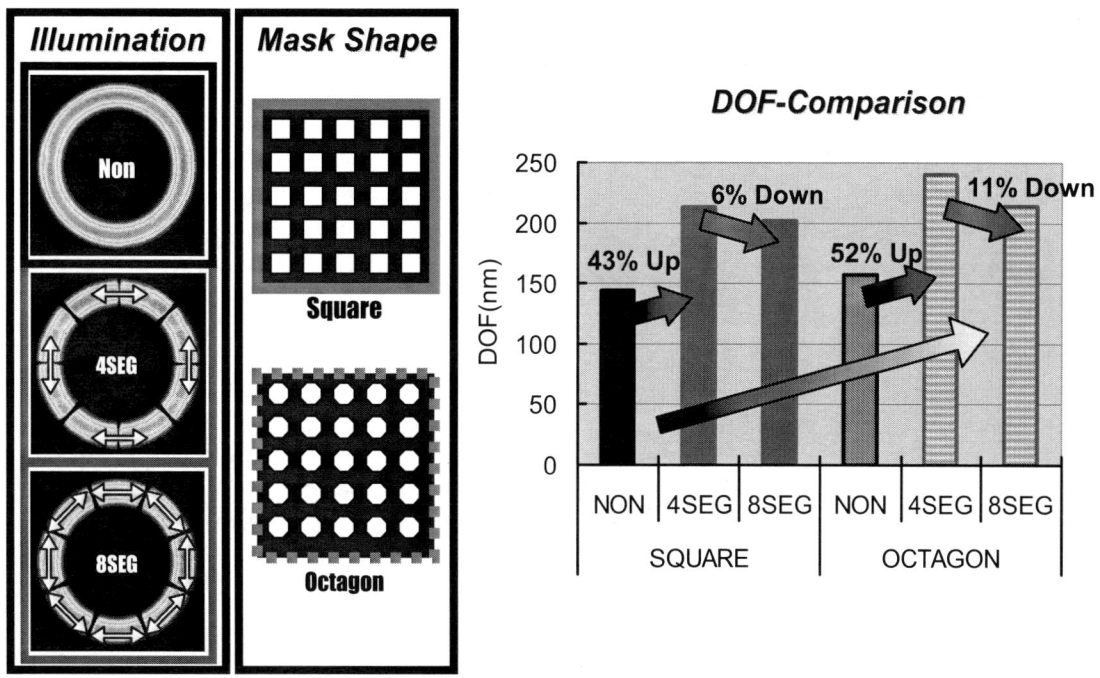

Figure 5 : evaluation for combination of polarization and Octagon shape mask

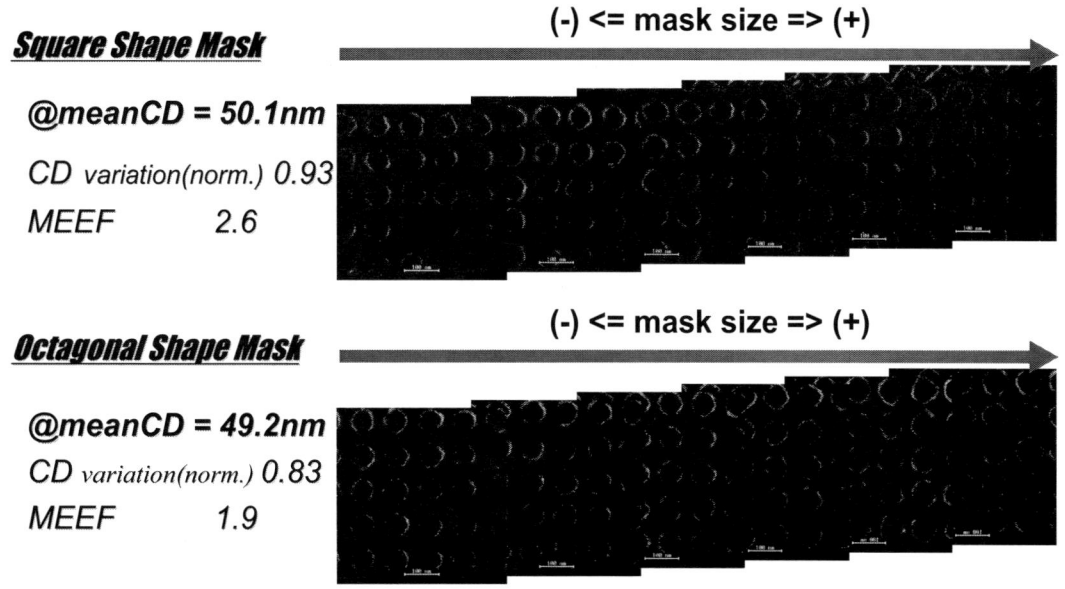

Figure 6 : SEM images for MEEF and CD-variation

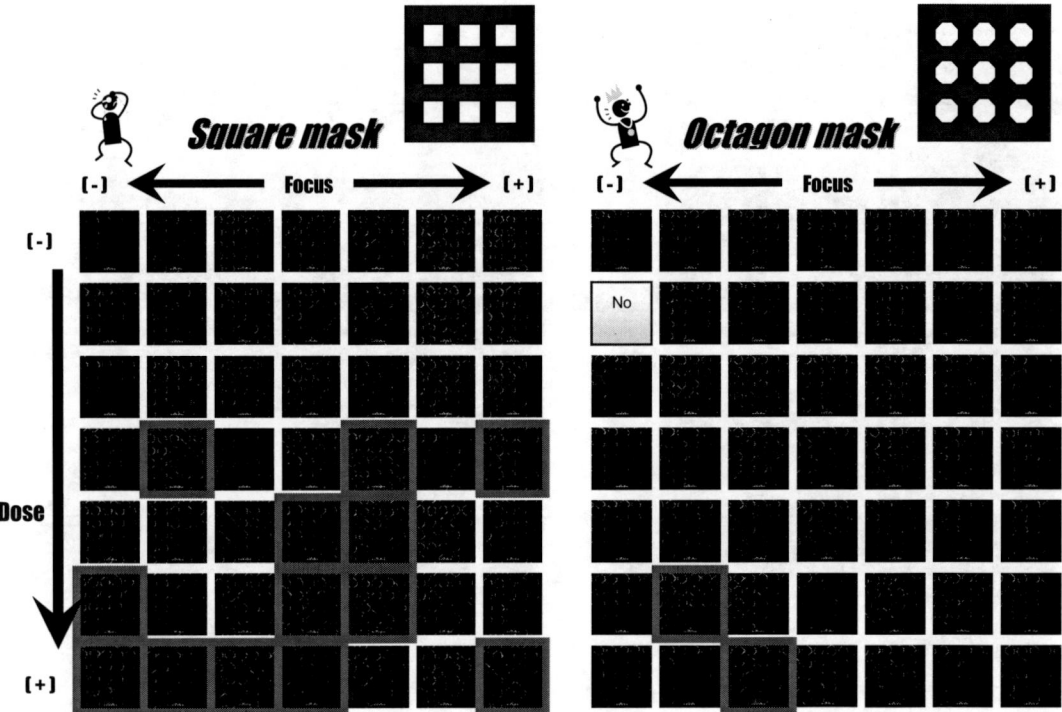

Figure 7 : image gallery for FEM. Marker " □ " shows collapse of resist pattern like short. Octagonal shape mask has wider process window.

Applications of MoSi-based Binary Intensity Mask
for Sub-40nm DRAM

Tae-Seung Eom*, Eun-Kyoung Shin, Eun-Ha Lee, Yoon-Jung Ryu, Jun-Taek Park,
Sunyoung Koo, Hye-Jin Shin, Seung-Hyun Hwang, Hee-Youl Lim, Sarohan Park,
Kyu-Tae Sun, Noh-Jung Kwak and Sungki Park

R&D Division, Hynix Semiconductor Inc.
San 136 -1, Ami-ri, Bubal-eub, Ichon-si, Kyoungki-do, 467-701, Korea

ABSTRACT

In this paper, we will present applications of MoSi-based binary intensity mask for sub-40nm DRAM with hyper-NA immersion scanner which has been the main stream of DRAM lithography. Some technical issues will be reported for polarized illumination and mask materials in hyper-NA imaging. One att.PSM (Phase Shift Mask) and three types of binary intensity mask are used for this experiment; those are ArF att.PSM (MoSi:760Å , transmittance 6%), conventional Cr (1030Å) BIM (Binary Intensity Mask), MoSi-based BIM (MoSi:590Å , transmittance 0.1%) and multi layer (Cr:740Å / MoSi:930Å) BIM. Simulation and experiment with 1.35NA immersion scanner are performed to study influence of mask structure, process margin and effect of polarization. Two types of DRAM cell patterns are studied; one is a line and space pattern and the other is a contact hole pattern through mask structure. Various line and space pattern is also through 38nm to 50nm half pitch studied for this experiment. Lithography simulation is done by in-house tool based on diffused aerial image model. EM-SUITE is also used in order to study the influence of mask structure and polarization effect through rigorous EMF simulation. Transmission and polarization effects of zero and the first diffraction orders are simulated for both att.PSM and BIM. First and zero diffraction order polarization are shown to be influenced by the structure of masking film. As pattern size on mask decreases to the level of exposure wavelength, incident light will interact with mask pattern, thereby transmittance changes for mask structure. Optimum mask bias is one of the important factors for lithographic performance. In the case of att.PSM, negative bias shows higher image contrast than positive one, but in the case of binary intensity mask, positive bias shows better performance than negative one. This is caused by balance of amplitude between first diffraction order and zero diffraction order light.[1]

Process windows and mask error enhancement factors are measured with respect to several types of mask structure. In the case of one dimensional line and space pattern, MoSi-based BIM and conventional Cr BIM show the best performance through various pitches. But in the case of hole DRAM cell pattern, it is difficult to find out the advantage of BIM except of exposure energy difference. Finally, it was observed that MoSi-based binary intensity mask for sub-40nm DRAM has advantage for one dimensional line and space pattern.

Key words : Polarization, Hyper-NA Immersion Lithography, MoSi-based BIM, Att.PSM, DRAM

1. INTRODUCTION

Optical Microlithography XXIII, edited by Mircea V. Dusa, Will Conley, Proc. of SPIE Vol. 7640,
76402J · © 2010 SPIE · CCC code: 0277-786X/10/$18 · doi: 10.1117/12.846728

Since advanced lithography has been a driving force for high density memory device implementations, lithography community has concerned on the resolution improvement by developing viable imaging solutions. In order to improve the limitation of optical lithography, resolution enhancement technology (RET) such as hyper-NA immersion lithography, polarized illumination, double patterning technology (DPT) and spacer patterning technology (SPT) have been examined intensively. Currently all technical efforts might be focused on the sub-40nm patterning by applying hyper-NA (NA>1) immersion lithography system. As long as the features on the mask were closed to the exposure light wavelength, degree of polarization would be changed dramatically by 3D mask structures on the mask. Since bigger NA means larger incident angles on mask, three dimensional consideration of illumination light becomes more important especially in the area of NA larger than 1. It is well known that binary intensity mask acts as TE (Transverse Electric) polarizer and attenuated phase shift mask acts as TM (Transverse Magnetic) polarizer. Recently, many papers are reporting better performance from MoSi-based BIM compared to conventional Cr BIM and attenuated PSM for sub-40nm half pitch with respect to process window, CD uniformity and mask-making consideration by using hyper-NA immersion scanner.

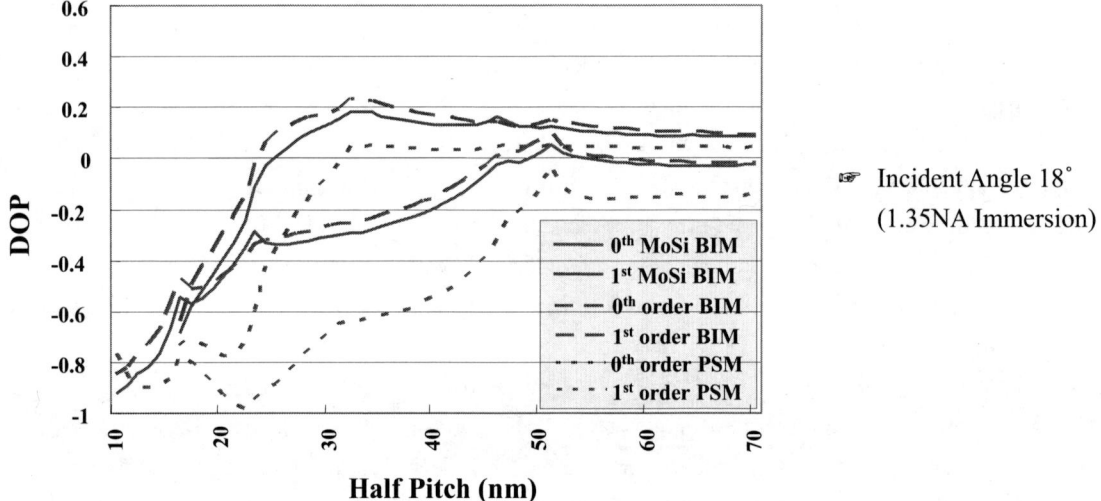

Fig. 1 Degree of polarization according to mask structure. First and zero diffraction order polarization are shown to be influenced by the structure of masking film. BIM has advantage of transmittance of polarization for sub-40nm.

As mentioned above, as the features on the mask were closed to the exposure light wavelength, degree of polarization would be changed dramatically by 3D mask structures on the mask. In scalar imaging theory, complete interference is assumed between each diffracted beams, but it can not be happened in reality and the difference grows if NA increased further above 1. There is an inappreciable difference of DOP between MoSi-based BIM and conventional Cr BIM, however diffracted light drives toward TM (Transverse Magnetic) at att.PSM for sub-40nm especially. Therefore, the image contrast of att.PSM is lower than BIM as imbalance of amplitude between zero diffraction order and first diffraction order beam. As many lithographers have been interested in this phenomenon, many researchers are reporting better performance from binary intensity mask compared to att.PSM. Some papers are reporting binary intensity mask has better contrast for sub 40nm half pitch due to difference of transmittance of binary intensity mask compared to att.PSM by using hyper-NA scanner. Figure 1 shows the degree of polarization according to the mask structure. First and zero diffraction order are shown to be influenced by the structure of masking film. To verify this phenomenon, simulation and experiment are needed to thoroughly understand the influence of mask material and polarization state on process margin for sub-40nm node DRAM cells. As you can see in figure 1, MoSi-based BIM shows similar transmittance comparing conventional Cr BIM for sub-40nm half pitch.

2. EXPERIMENTS AND SIMULATION

For this experiment, simulation was conducted by using in-house tool (Diffused Aerial Image Model) and EM-SUITE. Simulations with various mask structure and pattern pitches were performed by EM-SUITE with rigorous EMF simulation. Experiment was performed by using 1.35 NA ArF scanner with polarized illumination. Resist of 0.09 μm thickness was coated on bottom anti-reflective (BARC) materials in order to reduce the thin film interference effect of photo resist thickness on DOF. Wafer CD was measured by using Hitachi CD SEM at the same field position to avoid measurement error. To avoid photo resist shrinkage during CD measuring, we measured CD by using off-site focusing method. Four types of mask were made for this experiment. One is a ArF attenuated phase shift mask which has 6% transmittance, others are three types of binary intensity mask; those includes MoSi-based BIM (transmittance 0.1%), conventional Cr BIM and multi layer BIM. All types of mask are applied for the line and space patterns through 38nm to 50nm design rule. And two types of DRAM cell patterns were studied such as line and space and contact hole.

Items	Conditions
Exposure	ArF 1.35NA Immersion Scanner
Illumination	Crosspole and Dipole w/ Polarization
Resist	ArF 0.09 μm Thickness on Organic BARC
Mask	6% Att.PSM and 3 Types of BIM
Simulation	HOST (Diffused Aerial Image Model)
	EM-SUITE (rigorous EMF Simulation)
Pattern Size	38nm ~ 50nm Line & Space
	Sub-40 m DRAM Patterns

Table 1 Experimental conditions

Table 1 and Figure 2 shows mask structure that we made for this experiment. They are ArF att.PSM (MoSi:760Å , transmittance 6%), conventional Cr (1030Å) BIM (Binary Intensity Mask), MoSi-based (MoSi:600Å , transmittance 0.1%) BIM and multi layer (Cr:740Å / MoSi:930Å) BIM.

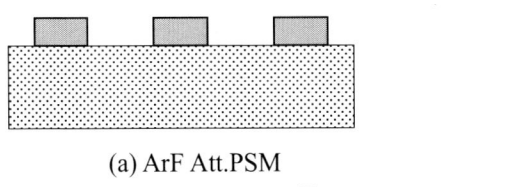

(a) ArF Att.PSM

Shifter (MoSi) : 760□

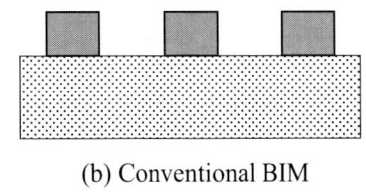

(b) Conventional BIM

Cr : 1030□

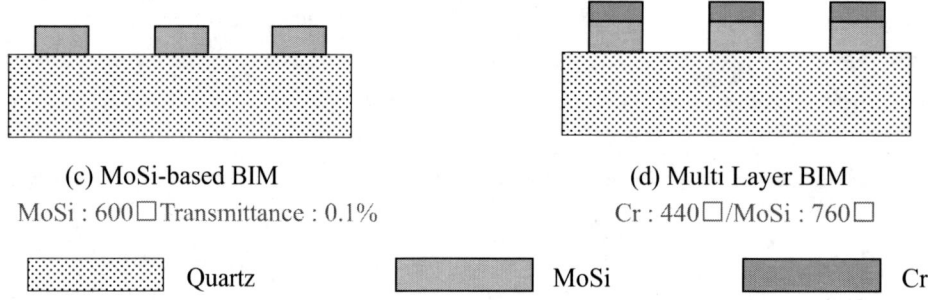

(c) MoSi-based BIM   (d) Multi Layer BIM

MoSi : 600□ Transmittance : 0.1%   Cr : 440□/MoSi : 760□

▦ Quartz   ▬ MoSi   ▬ Cr

Fig. 2 Mask structure for experiment.

3. RESULTS AND DISCUSSION

Since only a few primary diffraction orders can contribute to form an intensity image in sub-wavelength lithography, it is sufficient to describe the mask diffraction field with these primary diffraction orders. In the case of att. PSM with various transmittance and duty ratio of lines and spaces (L/S) pattern, zero and first diffraction orders can be calculated as:

$$\left|Mag\right|_{0th-order} = \left[1 + \sqrt{T}\right](s/p) - \sqrt{T} \; : \; A_0$$
$$\left|Mag\right|_{1st-order} = \left[1 + \sqrt{T}\right]\left|\frac{\sin(\pi s/p)}{\pi}\right| \; : \; A_1 \qquad\qquad [1]$$

where T is transmittance of the phase shifters, s and p is the space and the pitch respectively. These zero and first order beams are reconstructed at the wafer plane and the interference between those beams has resulted in the intensity modulation. In addition, the illumination source of lithographic system has poor spatial coherence, interference can be taken place only between the diffraction beams leaving from same source point. As the features become smaller and smaller, some fraction of 1st diffraction order goes outside the aperture and the zero order beam corresponding to this fraction of lost 1st order will remains as background intensity. Now, the fraction of 1st order beam which are captured by the lens system is defined as 1st order efficiency. Then the image contrast by two beams having magnitude A0 and A1 and 1st order efficiency "r" can be derived as :

$$Contrast = \frac{2rA_0A_1}{A_0^2 + rA_1^2} \qquad\qquad \text{Unpolarized Case} \qquad\qquad [2]$$

$$Contrast = \frac{2rA_{0\_TE}A_{1\_TE}}{A_{0\_TM}^2 + A_{0\_TE}^2 + rA_{1\_TE}^2} \qquad\qquad \text{Polarized Case (Crosspole Illumination)} \qquad [3]$$

$$Contrast = \frac{2rA_{0\_TE}A_{1\_TE}}{A_{0\_TE}^2 + rA_{1\_TE}^2} \qquad\qquad \text{Polarized Case (Dipole Illumination)} \qquad [4]$$

According to eq.[1], [2], [3] and [4], image contrast depends on illumination condition for polarized case. Therefore, we have to consider vector model. Since bigger NA means larger incident angles on mask, polarization consideration of illumination light becomes more important especially in the area of NA larger than 1.[5]

Figure 3 shows aerial image contrast with respect to mask bias. In figure 3, binary intensity masks show higher contrast than att.PSM through pitch if they have optimum mask bias. It is obvious that image contrast difference between att.PSM and BIMs at the optimum mask bias has increased with half pitch size shrinks as shown on figure 3. It means that BIM is more suitable for patterning as the half pitch decreases towards wavelength of illumination light. Image contrast of MoSi-based BIM shows similar trend to conventional Cr BIM through mask bias.

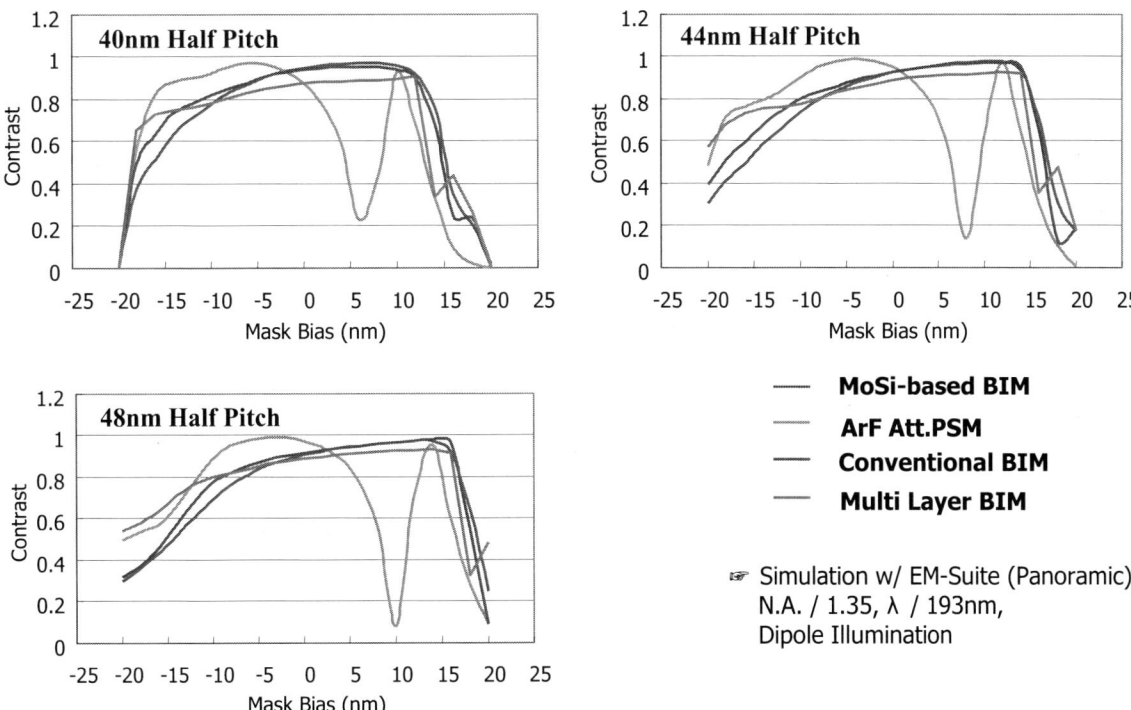

Fig. 3 Image contrast on mask bias through half pitch (40nm ~ 48nm)   In the case of att.PSM, negative bias shows higher image contrast than positive one, but in the case of binary intensity mask, positive bias shows better performance than negative one.

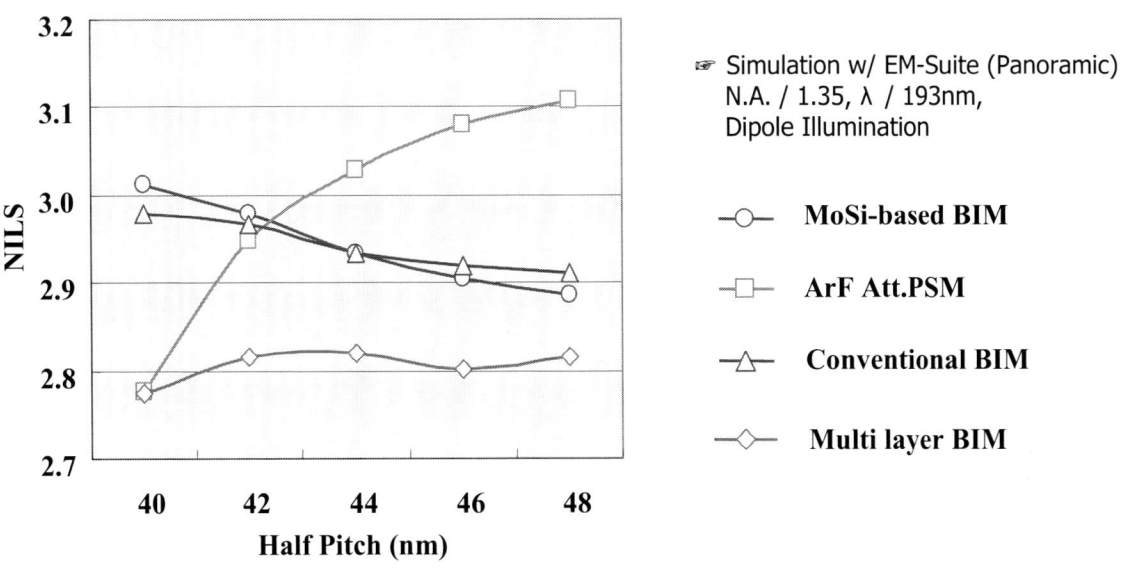

Fig. 4 Simulated NILS value according to mask structure through half pitch (40nm ~ 48nm)
NILS value of MoSi-based BIM shows similar trend to conventional Cr BIM through mask bias.

Figure 4 shows the NILS values with respect to half pitches for various mask structures. MoSi-based BIM and conventional Cr BIM show higher NILS value than att.PSM especially at sub 42nm half pitches. The NILS value of MoSi-based BIM shows similar trend to conventional Cr BIM through mask bias. We can expect to get same performance between MoSi-based BIM and conventional Cr BIM for sub-40nm DRAM.

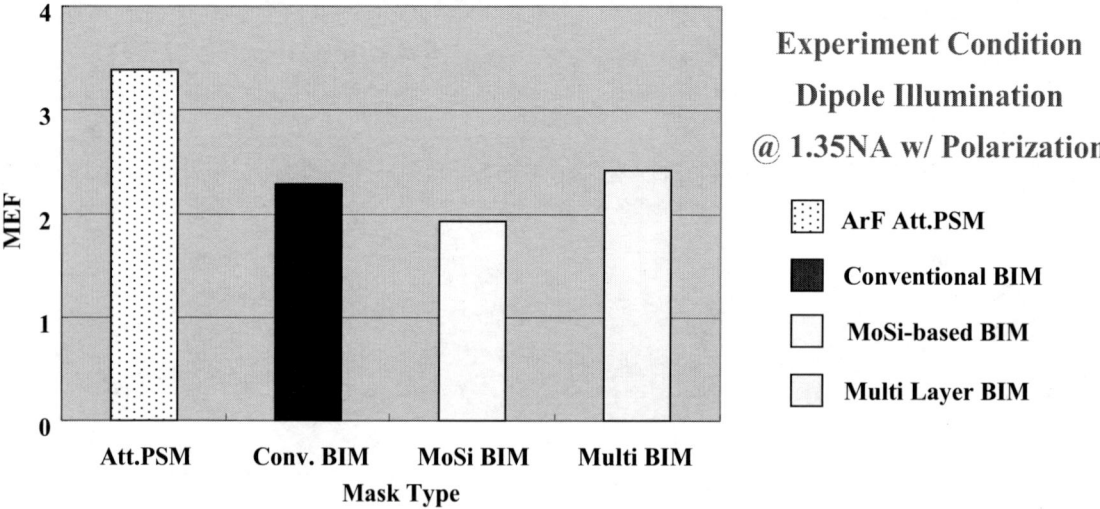

Fig. 5 Experimental verification of effect of mask structure through pitch
ArF 1.35NA with dipole illumination is applied.

Figure 5 is an experimental data of effect of mask structure. As you can see, MoSi-based BIM has an advantage in terms of mask error factor. This is caused by balance of amplitude between first diffraction order and zero diffraction order light. MEF result is well matched with NILS result. If MEF value is amplified above 2, it is difficult to have sufficient process margin for wafer patterning. Therefore, applying of MoSi-based BIM is necessary for sub-40nm DRAM fabrication.

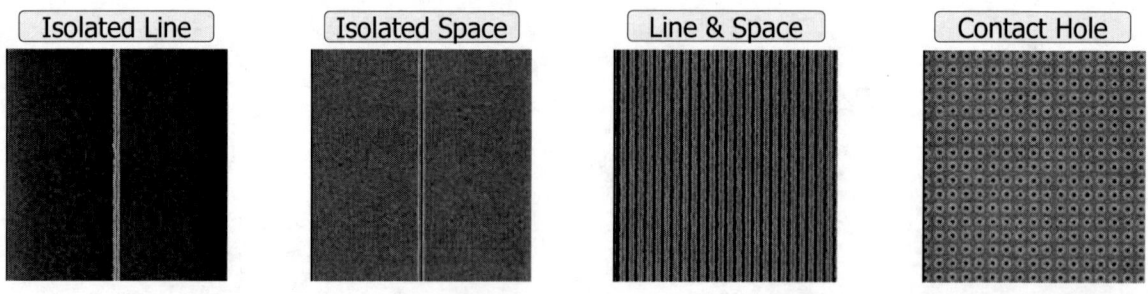

Fig. 6 MoSi-based BIM SEM images of various patterns for sub-40nm DRAM

Blank mask with MoSi absorber and AR layer was prepared for this study after optimizing the mask-making process. As shown in figure 6, relevant pattern fidelity on mask and vertical absorber profile was achieved through various DRAM patterns. The mask used for this study is made by company's own captive maskshop. Figure 7 shows mask CD linearity of MoSi-based binary intensity mask. CD linearity was controlled within 10% CD variation across all the pattern sizes.

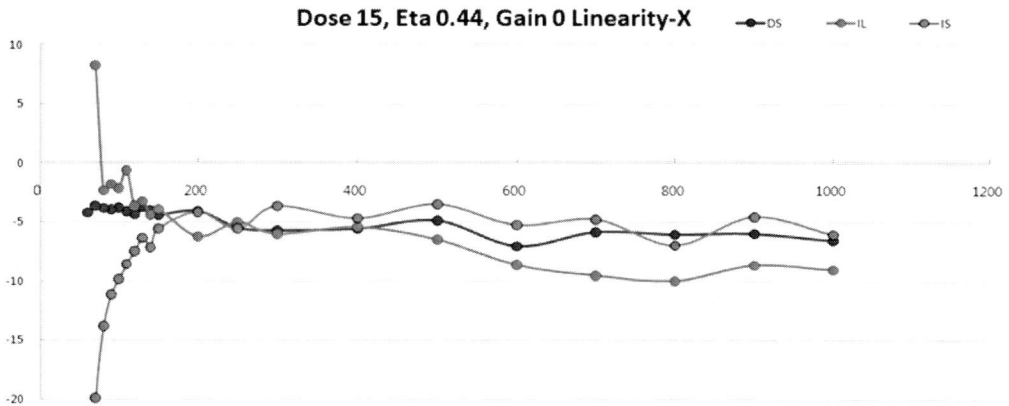

Fig. 7 Mask CD linearity of MoSi-based BIM

Experiment was performed on actual wafer to compare the difference between the masks for device application perspective. One att.PSM and three types of binary intensity mask were used for; those including ArF att.PSM (MoSi:760Å), conventional Cr (1030Å) BIM, MoSi-based (600Å) BIM. Lithographic characteristics such as DOF and EL were investigated for D2X DRAM pattern. Figure 8 shows experimental result of exposure latitude for line & space pattern for D2X DRAM pattern. It is obvious that applying to BIM for sub-40nm is a good choice to improve process window.

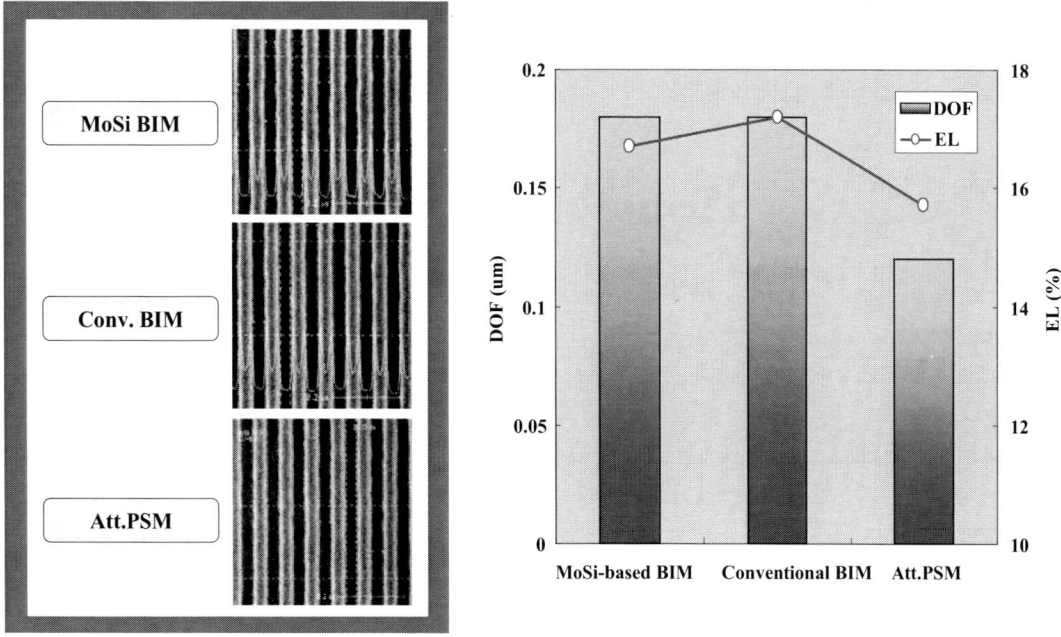

Fig. 8 SEM images and process window of D2X DRAM cell with line and space pattern
Experiment is performed by using 1.35NA ArF scanner with polarized dipole illumination

For the perspective of device application, we applied DRAM cells with peri circuit and contact hole pattern as shown in figure 9. On the contrary to line and space results both cells showed nearly no difference in exposure latitude. It is not clear why one dimensional and two dimensional pattern shows different exposure latitude dependency. In my opinion 2-dimensional pattern has a different transmittance behavior to that of 1-dimensional line and space pattern.[1-4]

We also found that transmittance of mask is difference between attenuated PSM and MoSi-based binary intensity mask. Difference dose to size between MoSi-based BIM and Att.PSM means polarization dependent transmittance behavior for mask structure.

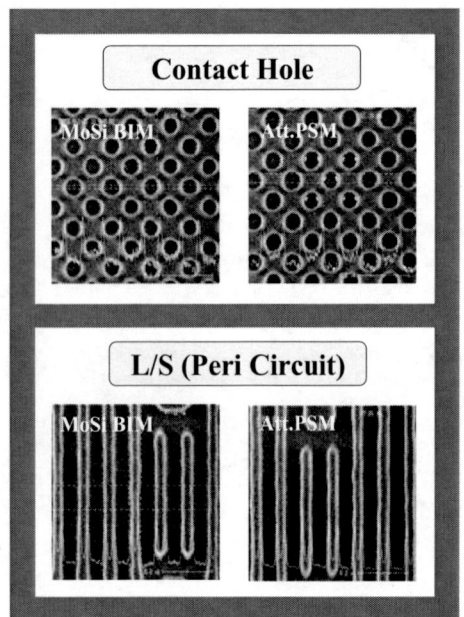 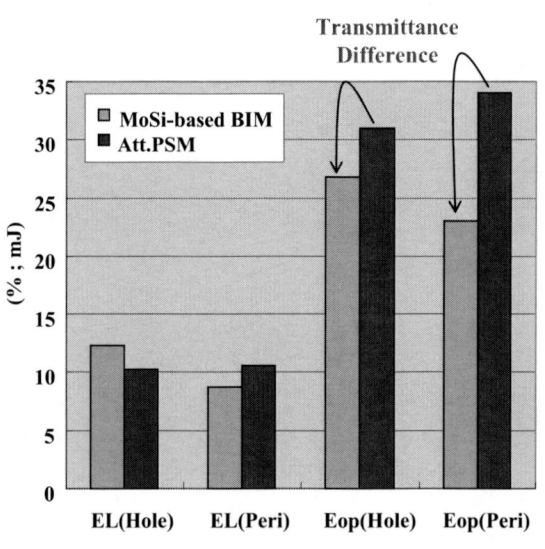

Fig. 9 SEM images and process window of D2X DRAM cell with contact hole and peri circuit
Experiment is performed by using 1.35NA ArF scanner with polarized dipole illumination

4. SUMMARY & DISCUSSION

We have investigated polarization effects of binary intensity masks and attenuated phase shift mask in hyper NA immersion lithography. Conventional Cr (1030Å) BIM, MoSi-based (600Å) BIM, multi layer (Cr:740Å / MoSi:930 Å) BIM and att.PSM (MoSi:760Å) were used for the experiment.

In the case of one dimensional line and space pattern, experimental data were matched well with simulation in terms of MEF and exposure latitudes. We can conclude that applying MoSi-based BIM for one dimensional pattern can be a good choice for sub-40nm region to improve process window.

We tested also 2-dimensional peri circuit and contact hole patterns of real DRAM structures. There is no difference for the exposure latitude between attenuated PSM and MoSi-based binary intensity mask. Although dose to size showed little difference in 2-dimensional pattern, mask types did not affect exposure latitude of 2-dimensional pattern in general.

We believe that transmittance behavior of 2-dimensional pattern was quite different from that of 1-dimensional pattern and further investigation will be needed.

5. ACKNOWLEDGEMENTS

The authors would like to thank mask development team for supporting and contributing to this work.

6. REFERENCES

[1] Tae-Seung Eom et. al., "Comparative Study of Binary Intensity Mask and Attenuated Phase Shift Mask using Hyper-NA Immersion Lithography for Sub-45nm Era", Proc. SPIE 6924, (2008).

[2] Chang-Moon Lim et. al., "Hyper-NA Polarized Imaging of 45nm DRAM", Proc. SPIE 6520, (2007)

[3] Thomas Faure et. al., "Characterization of Binary and Attenuated Phase Shift Mask Blanks for 32nm Mask Fabrication", Proc. SPIE 7122, (2008)

[4] Kazuya Sato et. al., "Mask 3D Effect on 45nm Imaging using Attenuated PSM", Proc. SPIE 6520, (2007)

[5] Tae-Seung Eom et. al., "Comparative Study of Chromeless and Attenuated Phase Shift Mask for 0.3 k1 ArF Lithography of DRAM" Proc. SPIE 5040, (2003)

OMOG Mask Topography Effect on Lithography Modeling of 32nm Contact Hole Patterning

Lei Yuan[1*], Wenzhan Zhou[1], Larry L. Zhuang[2], Kwang Sub Yoon[3], Qun Ying Lin[1], Scott Mansfield[2]

[1.] GLOBALFOUNDRIES Singapore, 60 Woodlands Industrial Park D, Singapore 738406

[2.] IBM Microelectronics, 2070 Route52, Hopewell Junction, NY 12533, USA

[3.] Samsung Electronics Co., Ltd., 2070 Route52, Hopewell Junction, NY 12533, USA

[*]Tel: 845-894-4413 Email: yuan@us.ibm.com

ABSTRACT

The topography effect of Opaque MoSi on Glass (OMOG) mask on 32nm contact hole patterning is analyzed by examining the difference of image intensity profile between thin mask approximation and rigorous electro-magnetic field (EMF) simulation. The study shows that OMOG topography results in more than a 20% decrease of image intensity. The impact of OMOG mask topography on lithography modeling of a 32nm contact hole process is explored by fitting lithography simulation with experimental results for both thin mask model and EMF model. This study shows that thin mask modeling is a good approximation of EMF modeling for a contact pitch larger than 120nm, but yields about 10nm prediction error for a 110nm contact pitch. Thin mask modeling is shown to be inaccurate in predicting critical dimension of contact arrays with sub-resolution assistant feature (SRAF). In addition, thin mask modeling is too pessimistic in predicting SRAF printability. In contrast, EMF model shows good prediction of contact arrays with and without sub-resolution feature. A modified thin mask modeling technique utilizing an effective SRAF size is proposed and verified with experimental results.

Keywords: opaque MoSi on glass, MoSi binary mask, OMOG, lithography modeling, mask topography, contact hole

1. INTRODUCTION

Opaque MoSi on Glass (OMOG) mask is a new type of binary mask developed by Toppan and ShinEtsu [1]. OMOG mask is fabricated by first printing patterns on a very thin chrome film through E-beam lithography followed by pattern transfer into MoSi stack through dry etch using the thin chrome film as hard mask. The thin chrome film is about10nm thick and will be removed afterwards. The application of the thin chrome film enables thinner Ebeam resist and thus offers better mask resolution [1-2]. Furthermore, the thin chrome hard mask reduces both loading effect and process bias for dry etching process, thus improving CD uniformity. Other OMOG mask advantages that have been demonstrated include better cleaning and radiation durability, better linearity, better mask flatness and faster cycle time [3-4]. Provided all these advantages, OMOG mask has been adopted by 32nm technology and will be extended to 22nm node.

With continuously shrinking feature dimension, mask topography effects have become more and more significant, generating serious challenges to full chip lithography simulation as required by optical proximity correction (OPC) [5-6]. The same concern applies to the application of OMOG mask. Although study has shown that the topography effect of OMOG mask is better than chrome on glasss (COG) binary mask and attenuated PSM mask, a more quantitative analysis of OMOG mask topography effects remains necessary. In this study, a 32nm contact lithography process will be used to investigate OMOG topography effects and their impact on lithography modeling.

This paper consists of five sections. Section 2 introduces experimental and modeling conditions that are used in this study. In section 3, OMOG mask topography effect will be demonstrated by comparing image intensity profile provided by both rigorous electro-magnetic field (EMF) simulation and thin mask simulation. In section 4, more quantitative analysis will be done for contact arrays with and without sub-resolution assistant feature (SRAF) by comparing lithography simulations with silicon results. A modified thin mask modeling by utilizing effective SRAF size will be presented and verified with silicon results. Finally, a conclusion will be given in section 5.

Optical Microlithography XXIII, edited by Mircea V. Dusa, Will Conley, Proc. of SPIE Vol. 7640,
76402K · © 2010 SPIE · CCC code: 0277-786X/10/$18 · doi: 10.1117/12.846547

2. EXPERIMENTAL AND MODELING CONDITION

All experiments in this paper are done by applying a 32nm contact hole lithography process, wherein, wafers are exposed on ASML TWINSCAN 1900i scanner and a trilayer resist film stack is coated and processed on Lithius-i track. Annular illumination is applied and an OMOG mask blank is used. The OMOG blank consists of about 40nm opaque MoSi film and about 20nm anti-reflective MoSi layer on glass substrate as shown in Fig. 1.

Fig. 1 Opaque MoSi on Glass (OMOG) mask stack

Lithography simulations in this paper are done by use of Synopsys lithography software — Sentauraus-Litho version D-2009.12, where a vector imaging model is applied to calculate an optical image profile in resist film stack and a reaction-diffusion system is assumed to model the post-exposed bake (PEB) process. Both a thin mask model (Kirchhoff mask model) and rigorous mask topography model (EMF model) are available in Sentauraus-Litho. In the thin mask model, 0.1% transmission rate is applied. In the EMF simulation, a finite difference time domain (FDTD) algorithm is used to solve Maxwell equations in the three-dimensional mask structure.

The accuracy of EMF simulation will depend on the grid size of mask discretization. To determine the grid size, various grid sizes have been applied to simulate image intensity of 120nm contact pitch. As shown in Fig. 2, a grid size of 2nm will be adequate. In this paper, a grid size of 2nm will be applied in all EMF simulations unless explicitly specified. It is necessary to point out that the grid size here is on wafer scale that corresponds to 8nm on mask scale.

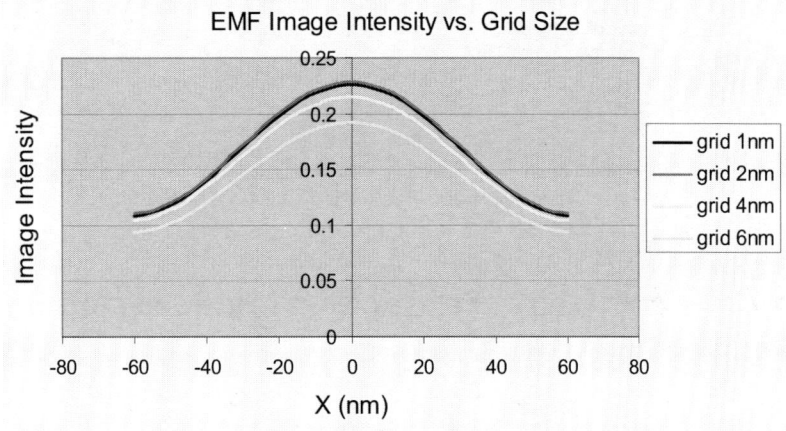

Fig. 2 EMF image simulation by grid size

3. OMOG TOPOGRAPHYIC EFFECT ON IMAGE INTENSITY

The thin mask model is an approximation of the rigorous EMF model and its accuracy is dependent on the significance of mask topography effects. The more significant mask topography effects are, the more difference can be observed between thin mask simulation and EMF simulation. Therefore, the magnitude of mask topography effects can be reflected by the differences between thin mask simulation and EMF simulation.

In the following, image intensity of multiple contact pitches at best focus is simulated using both EMF model and thin mask model, where best focus is determined by examining Bossung plot simulation for each mask model. As shown in Fig.3, thin mask simulation predicts maximum image intensity 20%~30% higher than that of EMF simulation.

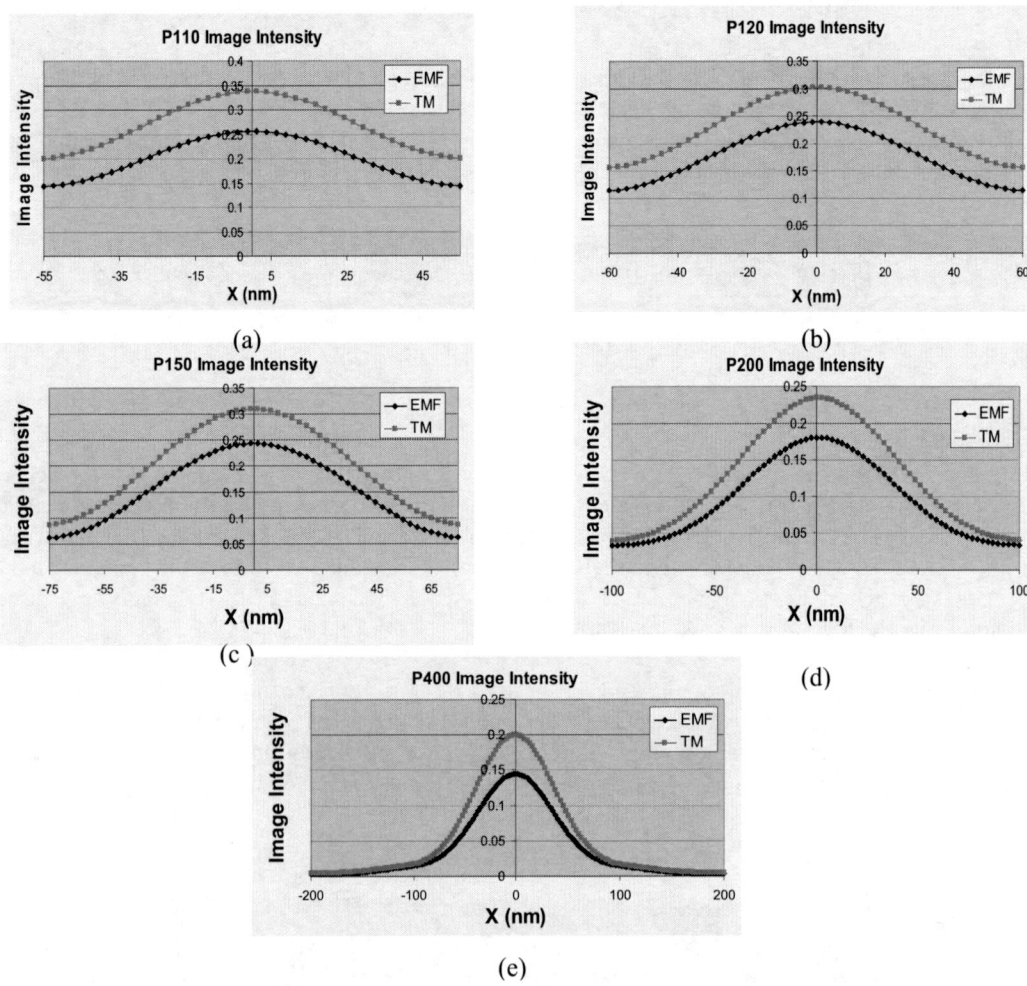

Fig. 3 Image intensity profile by EMF model and thin mask model

4. OMOG TOPOGRAPHY EFFECT ON LITHOGRAPHY MODELING

Thin mask simulation is currently the only practical solution of full chip lithography simulation since EMF simulation is very expensive. Given the impact of OMOG topography effect on image intensity results, the effectiveness of thin mask modeling of a 32nm lithography process requires careful investigation.

In the following, the image intensity profile as simulated by Sentauraus-Litho is normalized with respect to the maximum image intensity of a 120nm contact pitch for each mask model. The normalization will not affect the accuracy of lithography modeling since it is equivalent to applying a different image threshold value.

Sec. 4.1 Contact array without SRAF

The normalized image intensity profiles of multiple contact pitches are shown in Fig. 4 for both mask models. It shows that, for a contact pitch of 120nm ~ 200nm, the image profile of the thin mask simulation is very close to that of EMF simulation, while there is slight discrepancy for a contact pitch of 400nm. However, for a very dense pitch of 110nm, the thin mask model will result into about 10nm prediction error of critical dimension (CD). This result implies that thin mask modeling is a good approximation for a pitch of 120nm and above, but inaccurate for 110nm pitch.

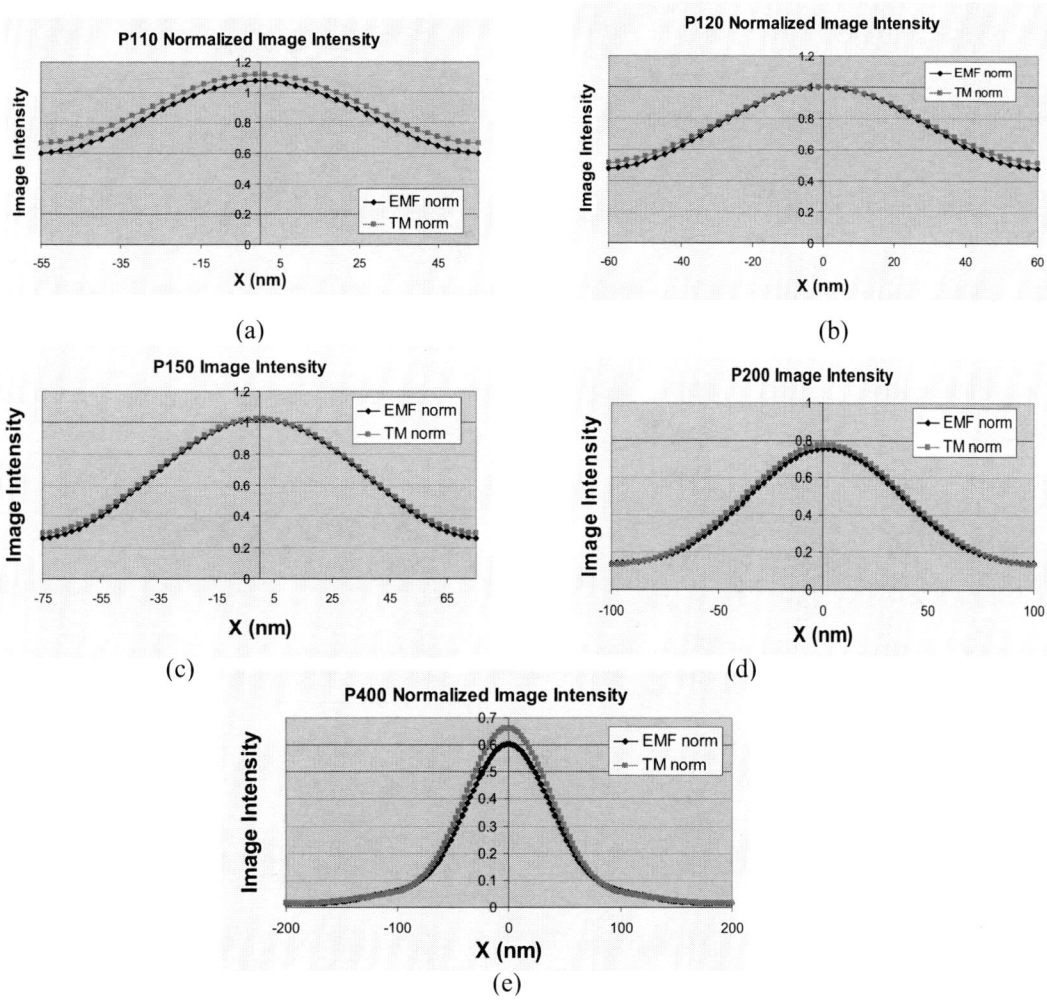

Fig. 4 Normalized image intensity by EMF model and thin mask model

Sec. 4.2 Contact array with sub-resolution assistant feature

To explore the impact of EMF on sub-resolution assist feature performance, contacts of 200nm pitch with SRAF of 25nm x 130nm and 270nm pitch with SRAF of 30nm x 130nm are simulated. The normalized image intensity is shown in Fig. 5a and 5b, while Fig. 5c and 5d represent the cross section image. It can be seen that the thin mask model over-estimates the image intensity of sub-resolution features, although it gives reasonable image intensity for main features. Sub-resolution features show more significant mask topography effects because MoSi height to space ratio of sub-resolution feature is much higher than that of main features.

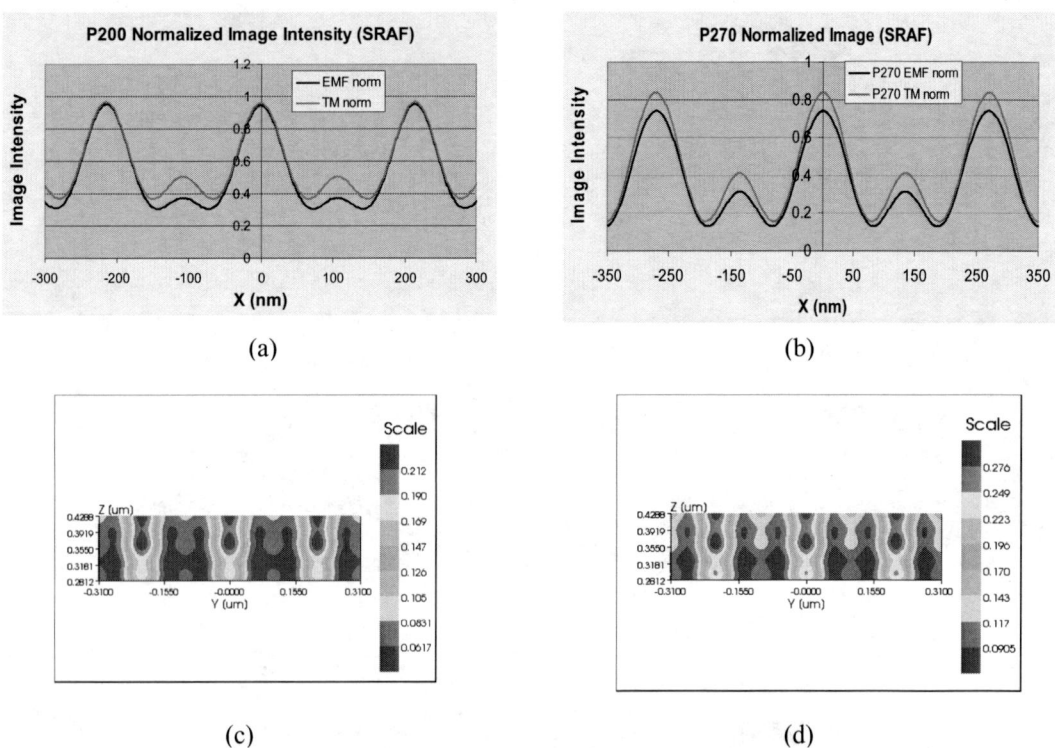

(a)

(b)

(c)

(d)

Fig. 5 (a) Normalized image intensity of 200nm contact pitch with SRAF of 25nm x 130nm (b) normalized image intensity of 270nm contact pitch with SRAF of 30nm x 130nm (c) cross section image of 200nm contact array with SRAF of 25nm x 130nm (d) cross section image of 270nm contact array with SRAF of 30nm x 130nm

Sec. 4.3 Experimental verification of lithography modeling

In the following, rigorous resist modeling will be integrated into lithography simulation to predict experiment results. The resist parameters are characterized by fitting FEM experimental data from a 200nm contact pitch for each mask model, which means that each mask model will have its own set of resist parameters. The FEM conditions of best focus and best dose are -0.005um and 29.3mJ/cm^2, while the focus and dose step are 0.025um and 1.2mJ/cm^2 respectively. The fitting results, shown in Fig. 6a and 6b, demonstrate that both mask models are able to fit the FEM experimental data well.

Next, the same set of resist parameters is used to predict FEM results for the 240nm contact pitch to validate the prediction capability of the lithography model. The prediction results, shown in Fig. 6c and 6d, demonstrate the good prediction capability for both mask models.

Lithography simulation is then compared with the FEM experimental data of contact arrays with SRAF. Two contact arrays that are exposed include a 200nm pitch and a 240nm pitch with SRAF of 25x130nm. The same FEM condition as used for resist characterization is also applied here. As shown in Fig. 7, EMF simulation is able to provide accurate prediction of FEM experimental results. Conversely, thin mask simulation over-predicts critical dimension by about 10nm due to the significant mask topography effects of sub-resolution features.

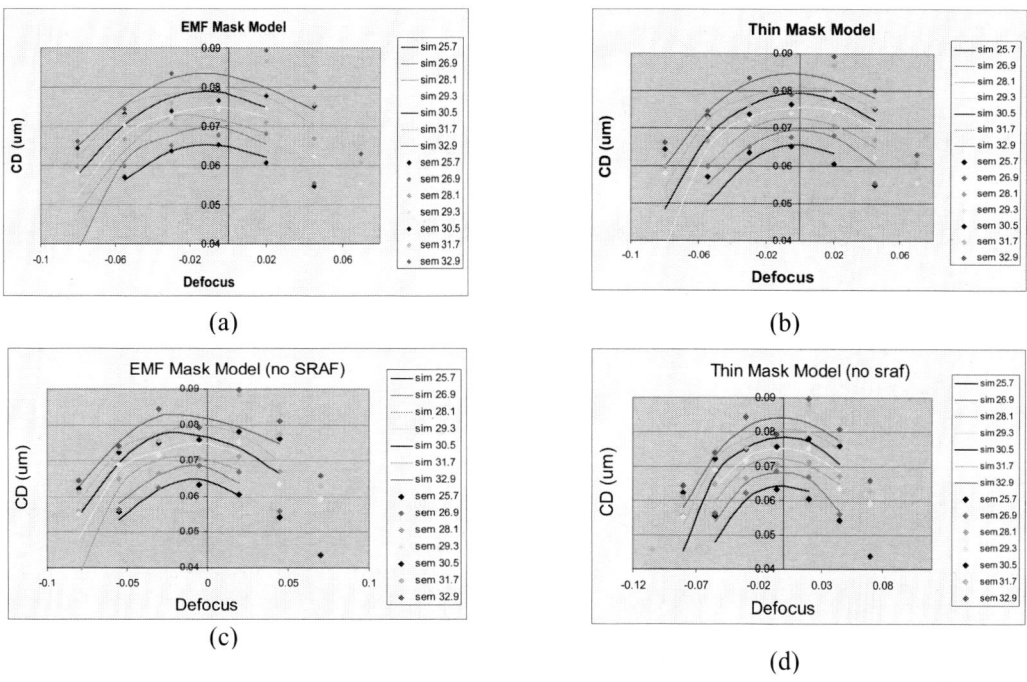

Fig. 6 FEM experiment and simulation (a) EMF model, 200nm contact pitch (b) thin mask model, 200nm contact pitch (c) EMF model, 240nm contact pitch (d) thin mask model, 240nm contact pitch

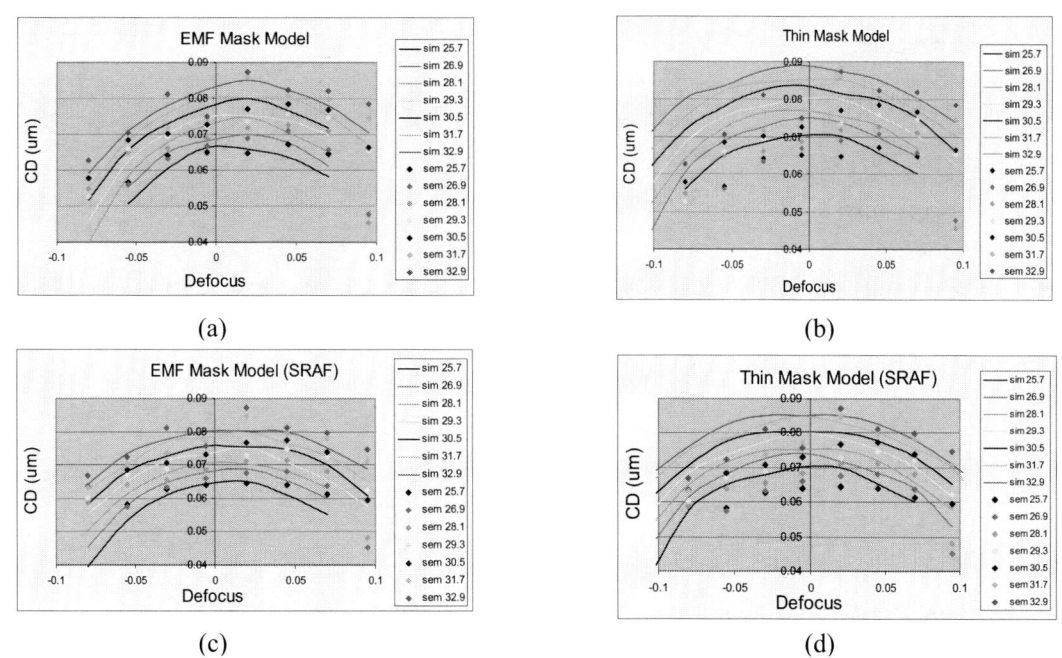

Fig. 7 Prediction of FEM experiment (a) 200nm contact array, SRAF 25x130nm, EMF model (b) 200nm contact array, SRAF 25x130nm, thin mask model (c) 240nm contact array, SRAF 25x130nm, EMF model (d) 240nm contact array, SRAF 25x130nm, thin mask model

Both the EMF model and thin mask model are also used to predict the experiment of SRAF printability, where a 200nm contact pitch having SRAF of 25x130nm is exposed with the dose range of 26.9mJ/cm^2 ~ 32.9mJ/cm^2 at the best focus, -0.005um. The top down SEM images in Fig. 8a show no SRAF printing up to a dose of 32.9mJ/cm^2. EMF simulation of the same exposure condition predicts silicon observation very well, as shown in Fig. 8b. On the other hand, thin mask modeling predicts SRAF printing starting from dose 29.3mJ/cm^2, as shown in Fig. 8c.

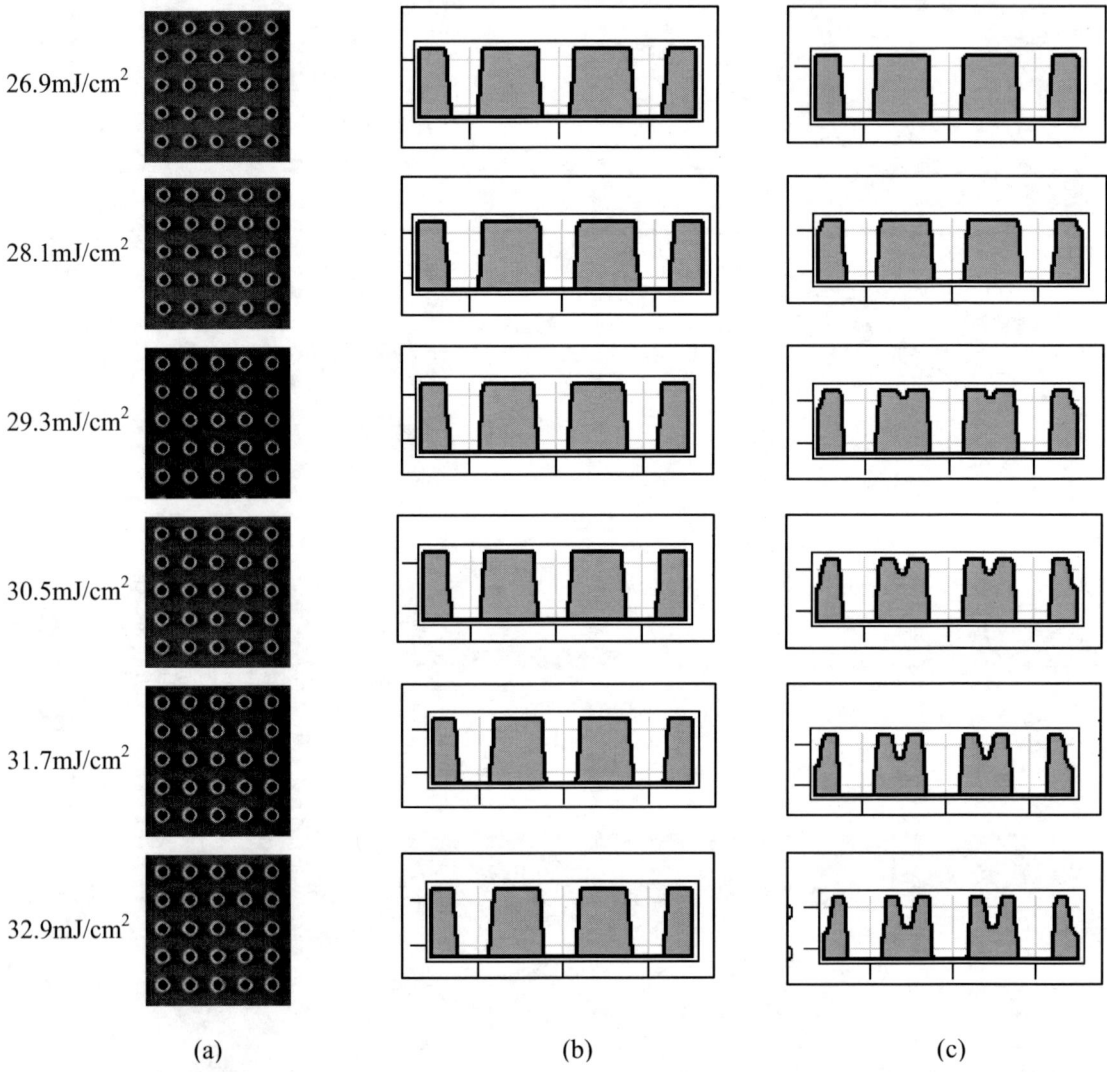

(a)   (b)   (c)

Fig. 8 200nm contact pitch with SRAF 25nm x 130nm, dose range 26.9~32.9 mJ/cm^2 (a) top down SEM image (b) cross section profile by EMF simulation (c) cross section profile by thin mask modeling

Sec. 4.4 Improving thin mask modeling utilizing an effective SRAF size

It is apparent from the EMF simulation that the effective size of clear SRAF features are smaller than their design size, due to the light absorption on the sidewall of OMOG mask stack. A method to improve thin mask SRAF modeling is thus to utilize an effective SRAF size instead of the actual design size. The effective SRAF size can be determined by either matching thin mask simulation of SRAF image intensity to rigorous EMF simulation or best fitting experimental FEM results.

To explore the effective SRAF size, a 200nm contact pitch with SRAF of 25nm x 130nm is simulated by thin mask modeling. Various SRAF sizes are used and the resulting image profiles are compared with that of rigorous EMF simulation, as shown in Fig. 9a. It can be seen that the effective SRAF size of 20nm, 5nm smaller than design size, is able to generate accurate SRAF image intensity.

Next, FEM experiments of two contact pitches of 200nm and 240nm, both of which have SRAF of 25nm x 130nm, are simulated using the thin mask model assuming an effective SRAF size of 18nm — 7nm smaller than the design size. The FEM simulations are then compared with the experimental measurements, as shown in Fig. 9b and 9c, where a more accurate prediction is achieved compared with using the actual design size as presented in Fig. 7c and 7d.

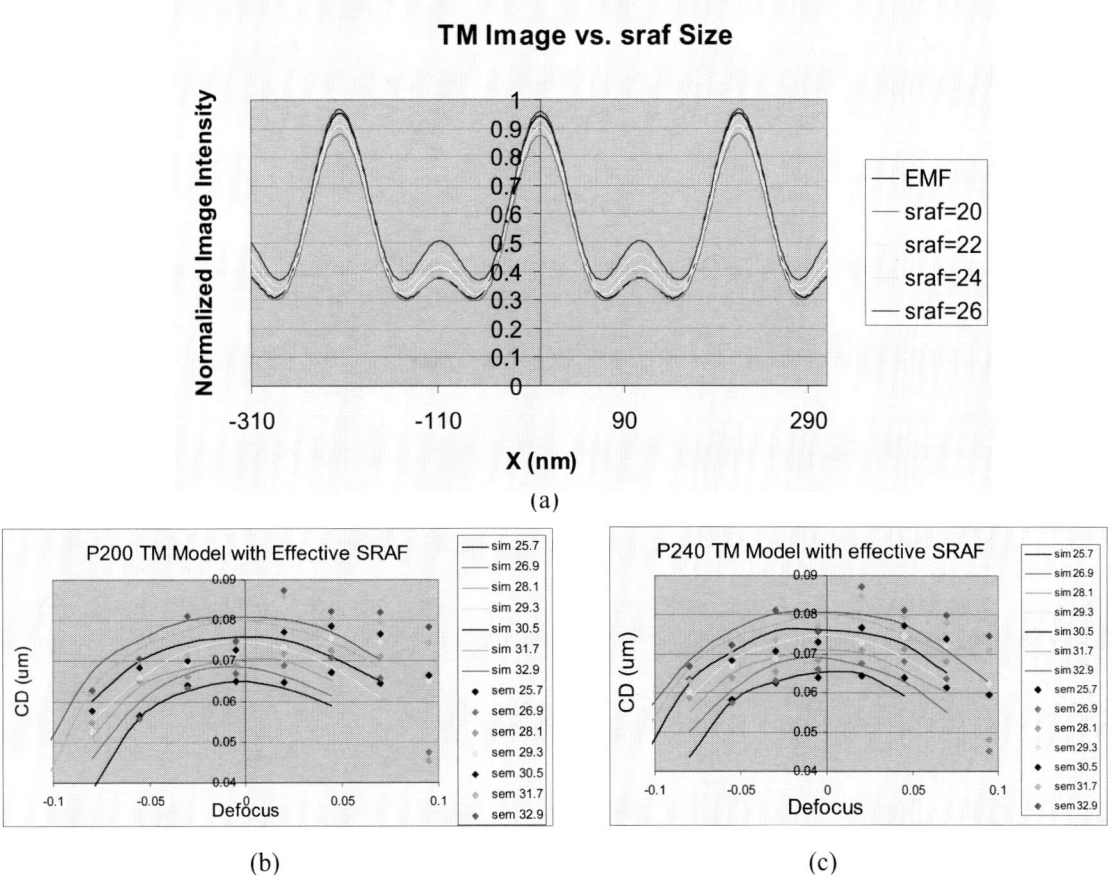

Fig. 9 (a) Thin mask simulation vs. SRAF size, FEM experiment and simulation by thin mask model with an effective SRAF size of 18nm for (b) 200nm contact array with SRAF of 25x130nm (c) 240nm contact array with SRAF of 25x130nm

5. CONCLUSION

The topography effects of an OMOG mask have been studied by comparing EMF modeling and thin mask modeling of a 32nm contact lithography process. It was found that the OMOG mask topography effects result in a 20~30% decrease of image intensity.

The impact of OMOG mask topography effects on lithography modeling of 32nm contact hole patterning has been carefully investigated. For contact arrays without SRAF, thin mask modeling is a good approximation of rigorous mask topography simulation for contact pitches of 120nm and above. However, for a dense pitch of 110nm, thin mask simulation may cause 10nm CD prediction error. The thin mask model also proved to be inaccurate in predicting contact printing with sub-resolution assist features, as a 10nm CD prediction error was observed. In addition, thin mask modeling is too pessimistic in predicting SRAF printability and this study showed that utilizing an effective SRAF size of 5~7nm smaller than the design size can effectively improve the thin mask simulation.

ACKNOWLEDGEMENTS

The authors would like to thank Derren Dunn of IBM, East Fishkill NY for assistance on Sentauraus-Litho application, John Lewellen of Synopsis for efficient technical support, and Binbin Yan of GLOBALFOUNDRIES Singapore for providing part of CD SEM results. The authors are grateful to Dr. David Medeiros for his valuable advice. This work has been supported by the independent Bulk CMOS and SOI technology development projects at the IBM Microelectronics Division, Semiconductor Research & Development Center, Hopewell Junction, NY 12533.

REFREENCES

1. Yamana, M., Lamantia, M., Philipsen, V., Wada, S., Nagatomo, T., and Tonooka, Y., "Comparison of Lithographic Performance between MoSi Binary Mask and MoSi Attenuated PSM", Proc. of SPIE Vol. 7379, 2009
2. McIntyre, G., et al, "Lithographic qualification of new opaque MoSi binary mask blank for the 32-nm node and beyong", Journal of Micro/Nanolith, MEMS, MOEMS, 2010 (in press)
3. Faure, T., Gallagher, E., Hibbs, M., Kindt, L., Racette, K., Wistrom, R., Zweber, A., Wagner, A., Kikuchi, Y., Komizo, T., and Nemoto, S., "Characterization of Binary and Attenuated Phase Shift Mask Blanks for 32nm Mask Fabrication", Proc. of SPIE Vol. 7122, 2008
4. Badger, K., Kodera, Y., Gallagher, E., Lawliss, M., "Impact of the OMOG Substrate on 32nm Mask OPC Inspectability, Defect Sensitivity and Mask Design Rule Restrictions", Proc. of SPIE Vol.7122, 2008
5. Shim, S., Kim, Y., Lee, S., Choi, S. and Han, W., "Study of the mask topography effect on the OPC modeling of hole patterns", Proc. of SPIE, Vol. 6924, 2008
6. Mimotogi, A., Itoh, M., Mimotogi, S., Sato, K., Sato, T. and Tanaka, S., "Mask topography effects of hole patterns on hyper-NA lithography", Proc. Of SPIE, Vol.6607, 2007

Fast-Converging Iterative Gradient Decent Methods for High Pattern Fidelity Inverse Mask Design

Jue-Chin Yu and Peichen Yu

Department of Photonics and Institute of Electro-Optical Engineering,
National Chiao Tung University, 1001 Da-hsueh Rd., Hsinchu 30050, Taiwan
E-mail address: yup@faculty.nctu.edu.tw

Abstract

Convergence speed and local minimum issue have been the major issues for inverse lithography. In this paper, we propose an inverse algorithm that employs an iterative gradient-descent method to improve convergence and reduce the Edge Placement Error (EPE). The algorithm employs a constrained gradient-based optimization to attain the fast converging speed, while a cross-weighting technique is introduced to overcome the local minimum trapping.

Keywords: optical proximity correction, inverse lithography, image gradient

1. Introduction

Semiconductor fabrication is the cornerstone of the current IC (Integrated Circuit) industry. With recent advances in microlithography now pushing towards nano-scale features, the problem of printing circuit layouts on wafers has become more intricate and convoluted. Optical Proximity Correction (OPC) is a resolution enhancement technique that modifies mask layout designs in order to minimize their distortion when transferred to silicon. A good OPC implementation may prove sufficient for a given process technology, precluding the need for a more expensive alternative like Double Patterning, Alternating Phase-Shift Masks (AltPSM), Immersion Lithography and so on. Clearly, OPC has obvious advantages in terms of efficiency and manufacturing cost.

Segment-based OPC has been the general industry approach and has proven successful through many CMOS generations. Because it only modifies existing edges in the layout, segment-based OPC has the advantage of being easy to implement, particularly in iterative algorithms. However, as the Critical Dimension (CD) becomes ever smaller, this type of edge-only compensation is not flexible enough to exploit the full range of possible mask corrections. Therefore, inverse mask design, also called Inverse Lithography Technology (ILT), has been suggested as an alternative due to its more relaxed constraints and full-mask approach. However, inverse calculation is faced with several problems, including bad convergence and the existence of local minima. To overcome these issues, many approaches have been proposed, such as pixel-flipping, gradient strategies and so on. Still, these approaches need to be further developed and refined to become the next-generation OPC.

In this paper, we propose an inverse algorithm that employs an iterative gradient-based method to improve convergence and reduce the Edge Placement Error (EPE). The algorithm achieves fast convergence by defining a digitized gradient vector and a cross-weighting matrix to determine the corresponding weighting factors. The digitized gradient vector of the cost function depicts an optimized step direction for the iteration, while the cross-weighting matrix is a tensor expression that takes the correlations of EPEs of different edges into account for the weighting factors.

2. Methodology

The Köhler's illumination model [1, 2, 3] is widely used in optical lithography. Figure 1 shows the configuration of an exposure system. The condenser lens $\mathbf{L_c}$ collimate the radiative light from the illumination source which is a quasi-monochromatic light source with a central wavelength $\bar{\lambda}$ and imaged on the pupil plane by a lens $\mathbf{L_1}$. Moreover, 193 nm ArF excimer laser is current exposure illumination source. With the nature of the quasi-monochromatism, the partially coherent image formation is applied to evaluate the image intensity on the wafer. The aerial image intensity of position (x, y) on image plane can be expressed as [1, 2]

$$I(x,y) = \int\int\int\limits_{-\infty}^{\infty}\int J(x_o^{'} - x_o^{"}, y_o^{'} - y_o^{"})m(x_o^{'}, y_o^{'})m^*(x_o^{"}, y_o^{"})$$

$$H(x - x_o^{'}, y - y_o^{'})H^*(x - x_o^{"}, y - y_o^{"})dx_o^{'}dy_o^{'}dx_o^{"}dy_o^{"} \tag{1}$$

Optical Microlithography XXIII, edited by Mircea V. Dusa, Will Conley, Proc. of SPIE Vol. 7640,
76402L · © 2010 SPIE · CCC code: 0277-786X/10/$18 · doi: 10.1117/12.846568

where $J(x_o^{'} - x_o^{''}, y_o^{'} - y_o^{''})$ is the mutual intensity which indicates the interference by two object points, $(x_o^{'}, y_o^{'})$ and $(x_o^{''}, y_o^{''})$, $m(x_o, y_o)$ is the mask function, $H(x_o, y_o)$ is the impulse response of the optical image system. The asterisk * denotes the complex conjugate.

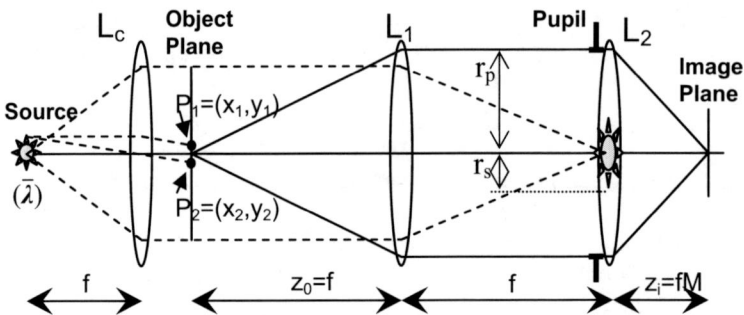

Figure 1. Configuration sketch of an optical lithography imaging system.

However, such four-fold integration like eq. (1) is a time-consuming calculation. To enhance the calculation efficiency, a numerical method named singular value decomposition (SVD) [4] are early proposed to decompose the eq.(1) to the summation of eigenvalues and eigenvector multiplication [5, 6, 7] that is

$$I(x, y) = \sum_{q=1}^{Q} \lambda_q \left| h_q(x, y) \otimes m(x, y) \right|^2, \tag{2}$$

where λ_q is the eigenvalue of the eigenvector h_q. Only Q dominant eigenvalues and their eigenvectors are used to synthesize the aerial image. The Eq.(2) shows the aerial image formation as a sum of coherent system [6]. Furthermore, such light intensity expression in Eq.(2) can be represented by the field form where

$$\begin{cases} E_q(x, y) = h_q(x, y) \otimes m(x, y), & (3) \\ I(x, y) = \sum_{q=1}^{Q} \lambda_q \left| E(x, y) \right|^2 & (4) \end{cases}$$

that is consistent with the time-average electromagnetic wave intensity. Finally a cost function is defined as following to evaluate the image performance in our optimization that is

$$F(m(x, y)) = \left\| I_t(x, y) - I(x, y) \right\|^2 \tag{5}$$

where I_t is the target image which is configured by the drawn patterns. $\|\cdot\|$ denotes the *Euclidean* length. Moreover, the mask is constrained in the range of $0 \leq m(x,y) \leq 1$.

Our approach is first based on the Frank and Wolfe method [8, 9] which transforms a nonlinear problem into an approximate linear optimization. In our inverse problem, we would like to find the minimum of a cost function F,

$$F(m(x, y)) \rightarrow \min., \tag{6}$$

where $m(x, y)$ is the mask function that spans in the x and y directions. Our approach can be expressed as an iterative calculation,

$$\nabla F^k (m^k(x, y)) \cdot m'(x, y) \rightarrow \min., \tag{7}$$

where k is the iteration number and

$$\left| m'(x, y) \right| \leq 1. \tag{8}$$

To calculate the iteration step, we establish the solution of $m'(x, y)$ as

$$m'(x, y) = \hat{m}^k(x, y). \tag{9}$$

Moreover, to satisfy Eq.(2) and Eq.(3) $\hat{m}^k(x, y)$ is given the following expression,

$$\hat{m}^k(x, y) = sign\left(\nabla F^{k-1}\left(m^{k-1}(x, y)\right)\right), \tag{10}$$

where

$$sign(x) = \begin{cases} 1, & when\ x > 0 \\ 0, & when\ x = 0\ , \\ -1, & when\ x < 0 \end{cases} \tag{11}$$

and the step direction is given by

$$\Delta m^k(x, y) = \hat{m}^k(x, y) - m^{k-1}(x, y). \tag{12}$$

To properly scale the step length through the iterations, we define a parameter $\kappa \in [0,1]$ that specifies the ratio between the configurations $m^{k-1}(x, y)$ and $m^k(x, y)$. This parameter can be used as

$$m^k(x, y) = m^{k-1}(x, y) + \kappa \Delta m^k(x, y). \tag{13}$$

This shows a linear exploration with origin in $m^{k-1}(x, y)$ that moves in the direction of $\Delta m^k(x, y)$ by the amount of $\kappa \left| \Delta m^k(x, y) \right|$. Therefore, the minimization problem becomes

$$F^k\left(m^{k-1}(x, y) + \kappa \Delta m^k(x, y)\right) \to min, \tag{14}$$

therefore a line search approach is employed to find the optimal κ every iteration.

Moreover, the optimization is accelerated by appropriate weighting factors for the gradient cost function, which are calculated using a cross-weighting matrix that takes into account the correlations among different edges, such matrix can be expressed as follows

$$\begin{bmatrix} w_{Inner} \\ w_{Outer} \\ w_{Side} \end{bmatrix} = \begin{bmatrix} C_{11} & C_{12} & C_{13} \\ C_{21} & C_{22} & C_{23} \\ C_{31} & C_{32} & C_{33} \end{bmatrix} \begin{bmatrix} EPE_{Inner} \\ EPE_{Outer} \\ EPE_{Side} \end{bmatrix}, \tag{15}$$

where w_{Inner}, w_{Outer} and w_{Side} are the weighting factors for the gradient vectors of inner, outer, and side edges, respectively, and C_{ij} are coefficients that take into account the effects of EPEs of different edges on the weighting factors. Figure 2 shows the exemplary definitions of the edges for two closely-placed vias and one via near a vertical bar. Such approach classifies the edges that have their own detecting points into three kinds of groups. Therefore the correction of the eight edge segments in figure 2 (a) originally described by an 8×8 matrix can be reduced to a 3×3 matrix. As the same reason a 14×14 matrix in figure 2 (b) can also be reduced to a 3×3 matrix. So the computing time is reduced especially in large size template. Furthermore other demands can be added by extending the dimension of the cross-weighting matrix.

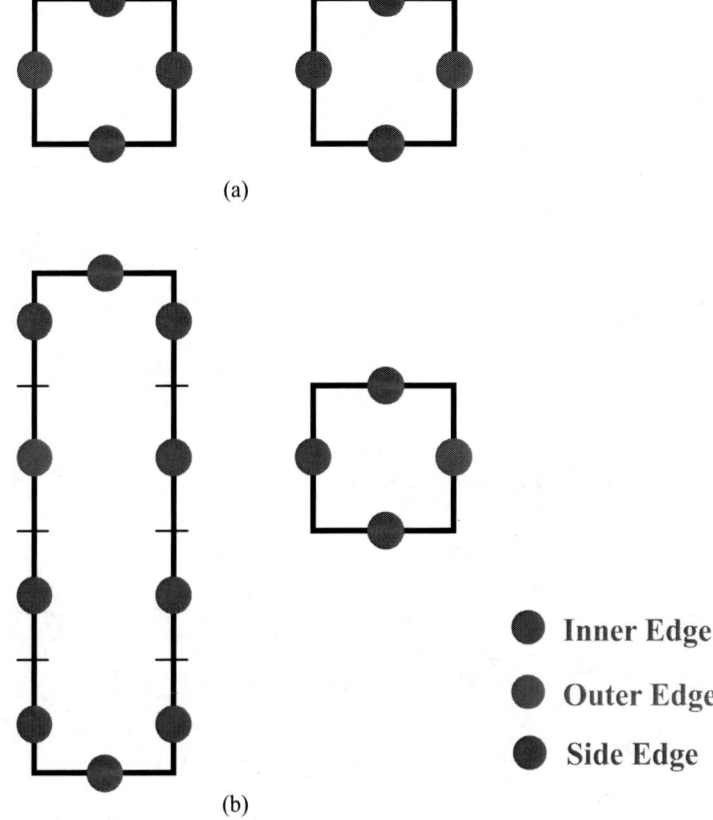

(a)

(b)

● Inner Edge

● Outer Edge

● Side Edge

Figure 2. Two exemplary definitions of the edges for (a) two closely-placed vias and (b) one via near a vertical bar.

4. Results and Discussion

In this section we demonstrate our algorithm under λ=193nm, $NA = 0.7$ and partial coherent illumination modeled with eight kernels. For comparison, the drawn mask of two 90 nm×90 nm vias with 100 nm separation placed on a 1.28 μm×1.28 μm template composed of 10 nm×10 nm pixels is first calculated under different weighting strategies. Three weighting approaches are first applied. One is non weighting optimization where the eq. (15) becomes

$$\begin{bmatrix} w_{Inner}^k \\ w_{Outer}^k \\ w_{Side}^k \end{bmatrix} = \begin{bmatrix} 0 & 0 & 0 \\ 0 & 0 & 0 \\ 0 & 0 & 0 \end{bmatrix} \begin{bmatrix} \overline{EPE}_{Inner}^k \\ \overline{EPE}_{Outer}^k \\ \overline{EPE}_{Side}^k \end{bmatrix}, \tag{16}$$

another is constant weighting where the eq. (15) can be expressed as

$$\begin{bmatrix} w_{Inner}^k \\ w_{Outer}^k \\ w_{Side}^k \end{bmatrix} = \begin{bmatrix} 1 & 0 & 0 \\ 0 & 1 & 0 \\ 0 & 0 & 1 \end{bmatrix} \begin{bmatrix} 25 \\ 25 \\ 25 \end{bmatrix}, \tag{17}$$

and the other is dynamic allocating the weighting value by the EPE of every iteration where the eq. (15) can be adapted

to

$$\begin{bmatrix} w^k_{Inner} \\ w^k_{Outer} \\ w^k_{Side} \end{bmatrix} = \begin{bmatrix} 15 & 0 & 0 \\ 0 & 15 & 0 \\ 0 & 0 & 20 \end{bmatrix} \begin{bmatrix} \overline{EPE}^k_{Inner} \\ \overline{EPE}^k_{Outer} \\ \overline{EPE}^k_{Side} \end{bmatrix},$$ (18)

where k is the iteration number and $\overline{(\cdots)}$ means the average operation in eq. (16), (17) and (18). Finally because the correlations between different kinds of the edges are included in our cross-weighting approach, the non-zero elements do not only locate along the diagonal entries of the matrix. So in our simulation the eq. (15) has the formation as

$$\begin{bmatrix} w^k_{Inner} \\ w^k_{Outer} \\ w^k_{Side} \end{bmatrix} = \begin{bmatrix} 20 & -3 & -2 \\ -3 & 20 & -2 \\ -1.5 & -1.5 & 23 \end{bmatrix} \begin{bmatrix} \overline{EPE}^k_{Inner} \\ \overline{EPE}^k_{Outer} \\ \overline{EPE}^k_{Side} \end{bmatrix}$$ (19)

where k is the iteration number. $\overline{(\cdots)}$ means the average operation.

Continuously before running the simulation, the detecting points for evaluating the EPE of every edge should be first decided. Subsequently the edges' types which are belonging to inner, outer or side are also defined. Figure 3 (a) shows the settings of the detecting points and (b) displays the classification of inner, outer and side edges that are labeled by red, green and blue spots respectively.

In figure 4 (a) and (b) respectively show the optimized gray level mask and its aerial image under non-weighting IL correction. Obviously, there is no threshold contours. (c) and (d) show the constant weighting results. (e) and (f) show the results under dynamic weightings allocated by every iteration's EPEs. (g) and (h) show our proposed cross-weighting results where the weightings are governed by eq. (19). Moreover, the contours at a threshold value of 0.5 in figure 4 (b), (d), (f) and (h) are displayed with a deep green line.

Figure 5 shows the EPEs as recorded in every iteration. The results under constant weighting are shown in figure 5 (a). And figure 5 (b) shows dynamic weightings which are allocated by the EPE when every iteration. Then the results calculated with our proposed cross-weighting approaches are shown in figure 5 (c). On all graphs, the fluctuations in the beginning are caused by searching the direction of the local minimum. Starting around iteration ten and for about twenty iterations, the EPEs converge stably and keep their trend toward zero. Then, just after the thirtieth iteration there is a clear change in the weight settings that further pushes the EPEs closer to zero. The final EPEs are approximately 8.5 nm in figure (a), 8.2 nm in figure (b) and 1.5 nm in figure (c).

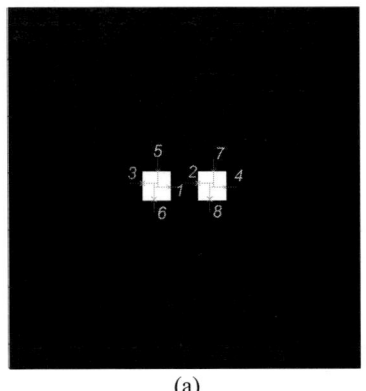

(a)

(b)

Figure 3. Two closely placed vias which are 90 nm×90 nm in area are aligned in a 1.28 μm×1.28 μm template, where (a) shows the detecting point setting of the two vias. (b) displays the classification of the edges where the red spots denote the inner edges, the greens denote the outer edges and the blues denote the side edges.

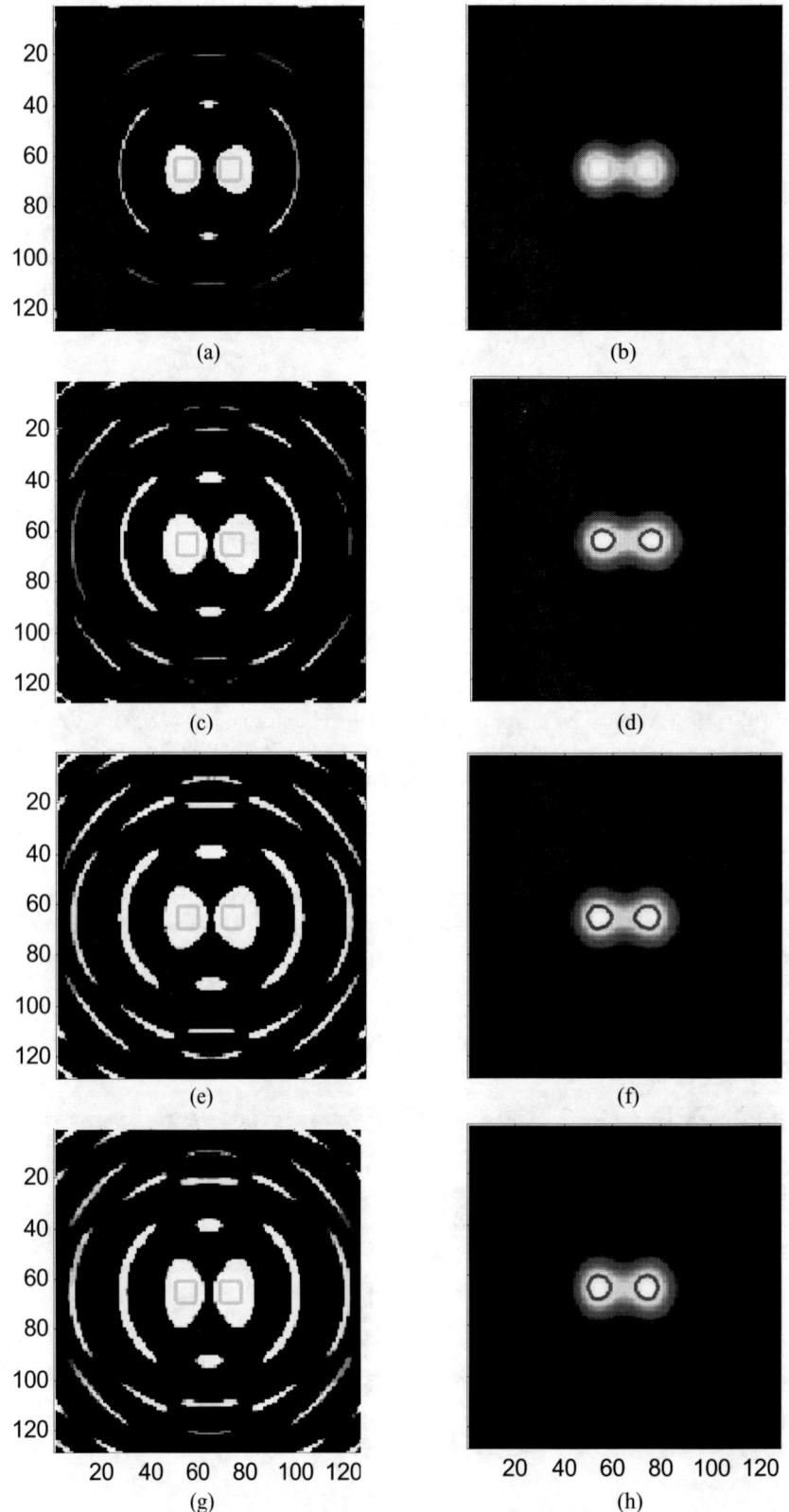

Figure 4. The corrected gray-level mask patterns of the two closely-placed vias under different weighting approaches. Part (a), (c), (e) and (f) show the optimized gray level mask corresponding to the varying transmittance from 0 to 1. Part (b), (d), (f) and (h) show the aerial image with contours at a threshold equal to 0.5 in deep green. The cyan lines in both plots show the drawn patterns.

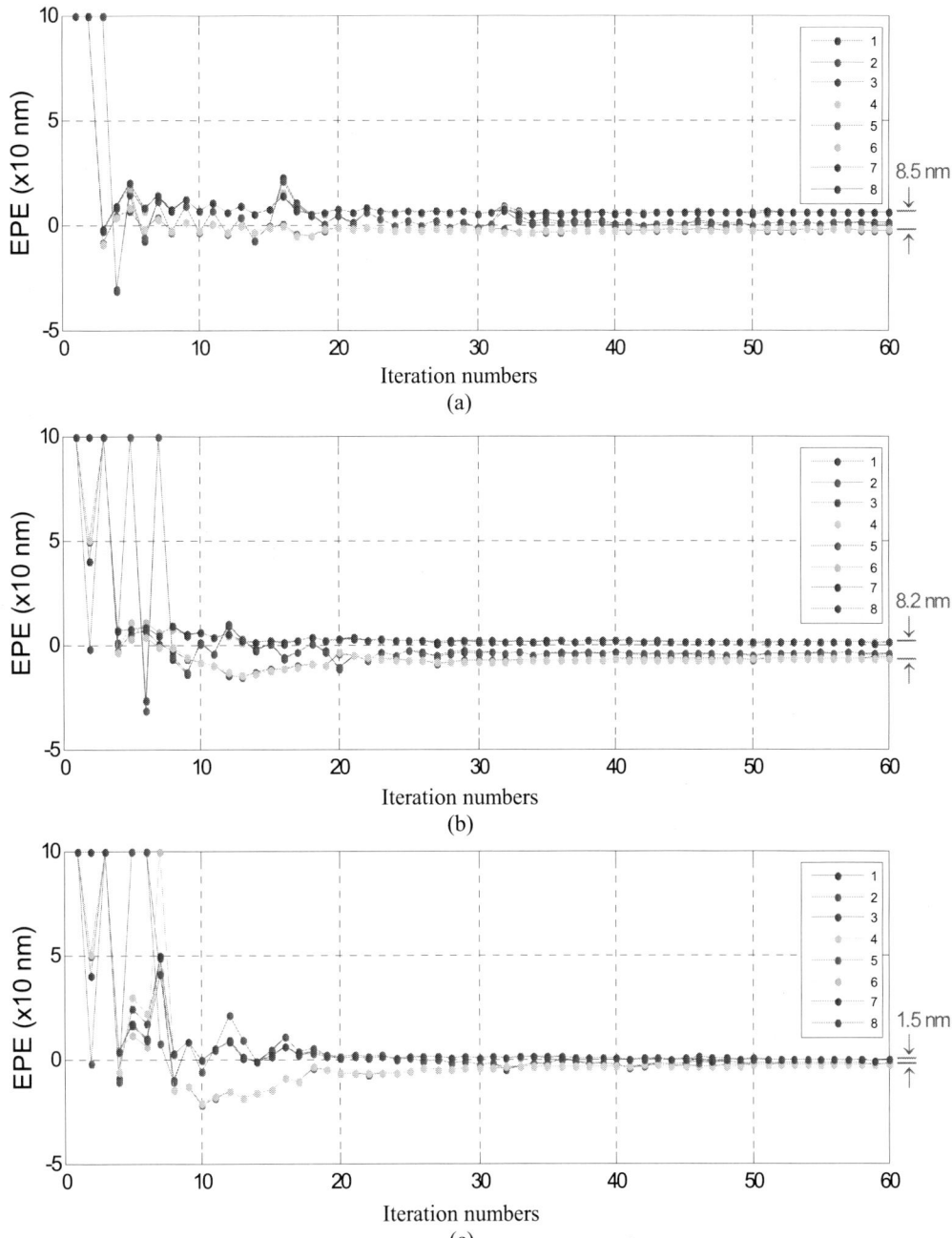

Figure 5. Edge Placement Error (EPE) vs. Iteration Number. (a) Constant weighting. (b) The weighting is re-assigned in every iteration according to the measured EPE of individual anchor points. (c) Cross-weighting technique. There are eight anchor points on the drawn edges. The horizontal axis shows the iteration numbers and the vertical axis is the EPE in number of pixels with a unit of 10 nm.

In summary because the feature size and the configuration are both far beyond the diffraction limit which is governed by *Rayleigh* criterion where R = 0.61λ/NA ~ 170 nm, the IL correction has no threshold contour without weighting treatment. Moreover duo to the sub-wavelength feature size and configurations, the sever diffraction effects cause complex correlations between different location. The weighting methods only incorporing with single edge are not sufficient to obtain the promising EPEs. Therefore, the optimization incorporating the interference between different locations should be considered. Comparing to figure 5 (a) and (b), our proposed cross-weighting technique that had taken such phenomena into account present the better performance as shown in figure 5 (c).

Similarly, figure 6 (a) and (b) show the optimized gray level mask and aerial image of an arbitrary SRAM contact array. The two middle vias array are composed of 190 nm×190 nm vias and the two side vias array are composed of 120 nm×120 nm vias. The template composed of 10 nm×10 nm pixels has size of 2.56 μm×2.56 μm. Moreover the final EPEs are still aggressive as shown in figure 7 where the EPEs at the every iteration are plotted and finally converged to 1.6 nm by our IL calculation incorporating with cross-weighting technique.

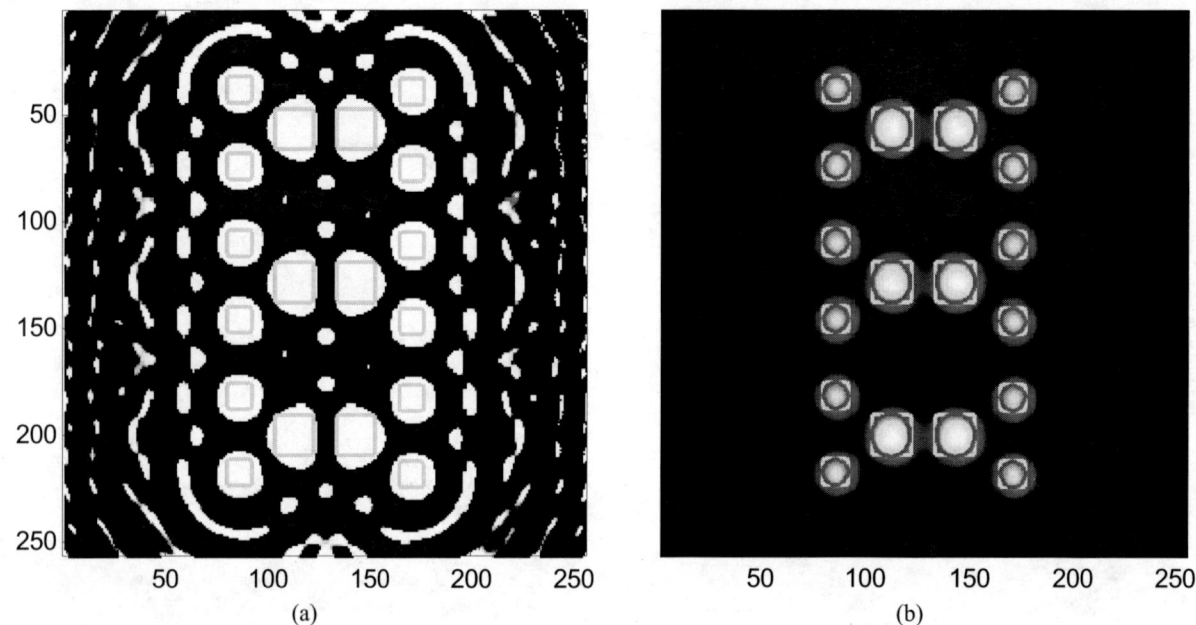

(a)   (b)

Figure 6. The corrected gray-level mask patterns of the two closely-placed vias under different weighting approaches.

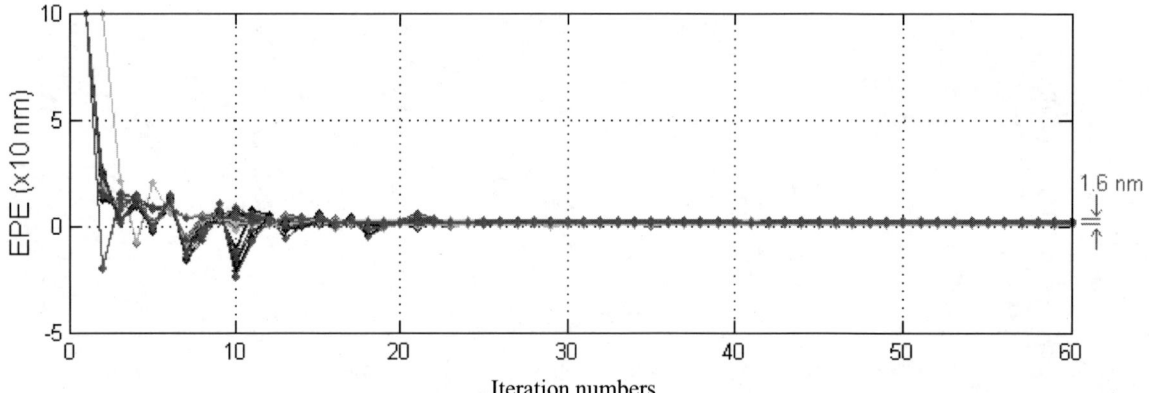

Iteration numbers

Figure 7. Edge Placement Error (EPE) vs. Iteration Number.

5. Conclusion

We successfully demonstrate the inverse lithography approach by using the constrained gradient approach iteratively. By incorporating with the proposed cross-weighting technique, the optimization calculations hardly suffer from local minimum trapping. The drawn patterns with sub-wavelength feature size and configuration receive the sizing corrections that are similar with segment-based OPC. Moreover, the surrounding assist features are simultaneously generated by inverse lithography calculation and with irregular geometries which is not achievable in conventional OPC. Furthermore, the fast convergence results are obtained where there are less than thirty iterations for converging in all above simulation.

Reference

[1] M. Born and E. Wolf, *Principles of Optics*, Pergamon, 7th ed., Cambridge University, (1999).

[2] J. W. Goodman, *Statistical Optics*, John Wiley and Sons, (1985).

[3] Alfred Kwok-kit Wong, *Optical Imaging in Projection Microlithography*, SPIE, Washington, (2005).

[4] Steven J. Leon, *Linear Algebra with applications*, 6th ed., Prentice-Hall, (2002).

[5] B. E. A. Saleh and M. Rabbani, "Simulation of partially coherent imagery in the space and frequency domains and by modal expansion," Applied Optics **21**, 15066-15079 (1982).

[6] N. B. Cobb, *Fast optical and process proximity correction algorithms for integrated circuit manufacturing*, University of California at Berkeley, Berkely, California, (1998).

[7] Edmund Y. Lama and Alfred K. K. Wong, "Computation lithography: virtual reality and virtual virtuality," Optics Express **17**, 12259-12268 (2009).

[8] Yuri Granik, "Fast pixel-based mask optimization for inverse lithography," J. Microlith., Microfab., Microsyst. **5**, 043002 (2006).

[9] M. Minoux, *Mathematical programming in Theory and Algorithms*, Wiley, New York, (1986).

Appendix

To calculate the gradient of the cost function F, we first apply the operator ∇, which is defined as

$$\nabla = \begin{bmatrix} \dfrac{\partial}{\partial m_{(x_1,y_1)}} & \cdots & \dfrac{\partial}{\partial m_{(x_1,y_n)}} & \cdots & \dfrac{\partial}{\partial m_{(x_1,y_N)}} \\ \vdots & \ddots & \vdots & \ddots & \vdots \\ \dfrac{\partial}{\partial m_{(x_m,y_1)}} & \cdots & \dfrac{\partial}{\partial m_{(x_m,y_n)}} & \cdots & \dfrac{\partial}{\partial m_{(x_m,y_N)}} \\ \vdots & \ddots & \vdots & \ddots & \vdots \\ \dfrac{\partial}{\partial m_{(x_M,y_1)}} & \cdots & \dfrac{\partial}{\partial m_{(x_M,y_n)}} & \cdots & \dfrac{\partial}{\partial m_{(x_M,y_N)}} \end{bmatrix}, \tag{A1}$$

where $m_{(x_m,y_n)}$ is the (m,n) pixel value and (x_m, y_n) the coordinate of this pixel. $m_{(x_m,y_n)}$ covers the region $\left[x_m - \dfrac{\delta x}{2}, x_m + \dfrac{\delta x}{2} \right] \cap \left[y_n - \dfrac{\delta y}{2}, y_n + \dfrac{\delta y}{2} \right]$ where the δx and δy are the differentials of the pixel in the x and y directions respectively. We may assume the value of $m_{(x_m,y_n)}$ is independent to the other pixels. This means

$$\frac{\partial m(x,y)}{\partial m_{(x_m,y_n)}} = \delta_d (x - x_m, y - y_n) = \begin{cases} \dfrac{1}{\delta x \delta y}, & x = x_m \ and \ y = y_n \\ 0, & x \neq x_m \ or \ y \neq y_n \end{cases}, \tag{A2}$$

where $\delta_d (x - x_m, y - y_n)$ is a discrete delta function. So, if $\delta x \to 0$ and $\delta y \to 0$, $m(x,y)$ is a continuous function and $\nabla m(x,y)$ becomes

$$\begin{bmatrix} \delta(x - \dfrac{M}{2}\delta x, y - \dfrac{N}{2}\delta y) & \cdots & \delta(x - \dfrac{M}{2}\delta x, y) & \cdots & \delta(x - \dfrac{M}{2}\delta x, y + \dfrac{N}{2}\delta y) \\ \vdots & \ddots & & \ddots & \vdots \\ \delta(x, y - \dfrac{N}{2}\delta y) & \cdots & \delta(x,y) & \cdots & \delta(x, y - \dfrac{N}{2}\delta y) \\ \vdots & \ddots & \vdots & \ddots & \vdots \\ \delta(x + \dfrac{M}{2}\delta x, y - \dfrac{N}{2}\delta y) & \cdots & \delta(x - \dfrac{M}{2}\delta x, y) & \cdots & \delta(x + \dfrac{M}{2}\delta x, y + \dfrac{N}{2}\delta y) \end{bmatrix}, \tag{A3}$$

where M, N are the total number of pixels in the x and y directions respectively and are both infinite. The delta function $\delta(x, y)$ has the property

$$\delta(x,y) = \begin{cases} \infty, & x = y = 0 \\ 0, & x \neq 0 \ or \ y \neq 0 \end{cases}, \tag{A4}$$

$$\int_{0^-}^{0^+} \int \delta(x,y)dxdy = 1. \tag{A5}$$

According to the above description, if $m(x,y)$ is a mask with infinitesimal pixels, we may express $\nabla m(x',y')$ as

$$\nabla m(x',y') = \delta(x-x', y-y')\Big|_{x' \in \left[-\frac{M}{2},\frac{M}{2}\right], \ y' \in \left[-\frac{N}{2},\frac{N}{2}\right]; \ M \to \infty, N \to \infty}. \tag{A6}$$

The cost function F may be expressed by the Manhattan norm of the difference between the real image $I(m,x,y)$ and ideal image $I_0(x,y)$ as

$$F(m) = \left\|I(m,x,y)-I_0(x,y)\right\|^2 = \iint (I(m,x,y)-I_0(x,y))^2 dxdy, \tag{A7}$$

then we derive F by using the ∇ operator, where

$$\nabla F(m) = \nabla \left\|I(m,x,y)-I_0(x,y)\right\|^2$$

$$= \nabla \iint (I(m,x,y)-I_0(x,y))^2 dxdy$$

$$= \iint \nabla (I(m,x,y)-I_0(x,y))^2 dxdy$$

$$= \iint 2(I(m,x,y)-I_0(x,y))\nabla I(m,x,y)dxdy. \tag{A8}$$

We may substitute $I(m,x,y)-I_0(x,y)$ by ΔI and define that the phasor formation of the electric field $E(x,y)$ is equal to $h(x,y) \otimes m(x,y)$ where $h(x,y)$ is the optical system response function and $\otimes$ denotes the convolution operation. Therefore the intensity $I(m,x,y)$ can be expressed as $I(m,x,y) = E(x,y)E(x,y)^* = (h(x,y) \otimes m(x,y)) \cdot (h(x,y) \otimes m(x,y))^*$. Then, the equation can be adapted to

$$\nabla F(m) = \iint 2\Delta I(x,y)\nabla \left(\sum_{k=1}^{K} \lambda_k (h(x,y) \otimes m(x,y))(h(x,y) \otimes m(x,y))^* \right) dxdy$$

$$= \sum_{k=1}^{K} \lambda_k \iint 2\Delta I(x,y)\left(E(x,y)(h(x,y) \otimes \nabla m(x,y))^* + (h(x,y) \otimes \nabla m(x,y))E^*(x,y) \right) dxdy,$$

$$= 4\sum_{k=1}^{K} \lambda_k \iint \Delta I(x,y) \operatorname{Re}\left[E^*(x,y)\big(h(x,y) \otimes \nabla m(x,y)\big)\right] dxdy$$

$$= 4\sum_{k=1}^{K} \lambda_k \iint \operatorname{Re}\left[\Delta I(x,y) E^*(x,y)\left(\iint h(x-x',y-y')\nabla m(x',y')dx'dy'\right)\right] dxdy$$

$$= 4\sum_{k=1}^{K} \lambda_k \iint\iint \operatorname{Re}\left[\Delta I(x,y) E^*(x,y) h(x-x',y-y')\nabla m(x',y')\right] dxdydx'dy'$$

$$= 4\sum_{k=1}^{K} \lambda_k \iint \operatorname{Re}\left[\iint \big(\Delta I(x,y) E^*(x,y)\big) h(x-x',y-y')dxdy\right] \nabla m(x',y') \ dx'dy'$$

$$= 4\sum_{k=1}^{K} \lambda_k \iint \operatorname{Re}\left[\big(\Delta I(x',y') \cdot E^*(x',y')\big) \otimes h(-x',-y')\right] \nabla m(x',y') \ dx'dy'$$

$$= 4\sum_{k=1}^{K} \lambda_k \iint \operatorname{Re}\left[\big(\Delta I \cdot E^*\big) \otimes h(-x',-y')\right] \delta(x''-x',y''-y') \ dx'dy'$$

$$= 4\sum_{k=1}^{K} \lambda_k \operatorname{Re}\left[\big(\Delta I \cdot E^*\big) \otimes h(-x'',-y'')\right] \tag{A9}$$

Radial segmentation approach for contact hole patterning in 193 nm immersion lithography

Moh Lung Ling, Gek Soon Chua[a] , Sia Kim Tan[a], Cho Jui Tay, Chenggen Quan, Qunying Lin[b]
Department of Mechanical Engineering
National University of Singapore
9 Engineering Drive 1, Singapore 117576

[a] Advanced Litho Enablement, Mask Technology
[b]Advanced Module Technology Development, Lithography
GLOBALFOUNDRIES Singapore Pte. Ltd.
60 Woodlands Industrial Park D, Singapore 738406

ABSTRACT

In this paper, a novel optical proximity correction (OPC) method for contact hole patterning is demonstrated. Conventional OPC for contact hole patterning involves dimensional biasing, addition of serifs, and sub resolution assist features (SRAF). A square shape is targeted in the process of applying conventional OPC. As dimension of contact hole reduces, features on mask appear to be circular due to strong diffraction effect. The process window enhancement of conventional OPC approach is limited. Moreover, increased encounters of side lobes printing and missing contact holes are affecting the process robustness. A new approach of changing the target pattern from square to circular is proposed in this study. The approach involves a change in shape of mask openings and a radial segmentation method for proximity correction. The contact holes patterns studied include regular contact holes array and staggered contact holes. Process windows, critical dimension (CD) and aerial image contrast is compared to investigate the effectiveness of the proposed contact holes patterning approach relative to conventional practice.

Keywords: Contact holes patterning, immersion lithography, mask aperture, circular aperture, side lobe printing, optical proximity correction, resolution enhancement techniques, segmentation

1. INTRODUCTION

The patterning of contact holes in immersion lithography has become one of the most critical issues in enabling the continual scaling of integrated circuit density. In the past five years, immersion lithography has enabled the extension of 193 nm lithography for sub 90 nm technology nodes by lowering resolution limits through larger numerical aperture (NA). However, moving beyond 45 nm technology node, the process windows gain by applying immersion lithography for contact holes becomes limited. Difficulties in the patterning of contact holes can be attributed to the two dimensional layout contact layer. The low-pass frequency filtering effect of optical lithography system results in the lost of high frequency diffraction components [1]. Consequently, the corners appear to be rounded and patterned features become circular as dimensions reduce. There are several resolution enhancement techniques (RET) that have been proposed to enable lower k_1 patterning of contact holes. These includes the use of phase shifting masks for enhancing the aerial image log slope [2, 3], double patterning (DP) [4, 5] for extremely dense contact holes array, interference mapping lithography (IML) for assist features placement [6, 7], contact holes shrink process using chemical shrink materials [8, 9], and more recently simultaneous source and mask optimization (SMO) [10]. These techniques are capable of improving the process window and printability of contact holes. However, IML and SMO are computationally intensive and the solution is layout dependent.

Optical Microlithography XXIII, edited by Mircea V. Dusa, Will Conley, Proc. of SPIE Vol. 7640,
76402M · © 2010 SPIE · CCC code: 0277-786X/10/$18 · doi: 10.1117/12.848431

In this paper, a new approach of contact hole patterning is proposed. This involved targeting a printed contact hole as circular shape instead of a square shape used in conventional approach. The mask apertures adopted in this method is circular. For regular contact holes array, dimensional biasing by changing the radius of apertures is applied to enable patterning of different pitches. For random contact hole pattern, such as staggered contact holes pattern, a radial segmentation approach is proposed to enhance the aerial image contrast. The imaging performances such as critical dimension (CD) control, depth of focus (DOF), and normalized image log slope (NILS) resulted from the proposed method is compared with that obtained from conventional approach.

2. THEORETICAL ANALYSIS

The Fraunhofer diffraction field of an isolated square hole with sides a is given by [11]

$$E(f, g) = C\ sinc\ (kfa)\ sinc\ (kga) \qquad (1)$$

In Eqn. (1), C is the constant depends on the source and point of observation, k is the wave propagation number equivalent to $\frac{2\pi}{\lambda}$, f and g are the spatial frequencies in x and y directions respectively. The $sinc$ function above is given by

$$sinc\ (a) = \frac{\sin a}{a}$$

For an isolated circular hole with radius r, the Fraunhofer diffraction field is given by

$$E(w) = D\frac{2J_1(krw)}{krw} \qquad (2)$$

In Eqn. (2), w is the spatial frequencies in polar coordinates such that $w = \sqrt{f^2 + g^2}$ and J_1 is the Bessel function of the first kind. D is the constant depends on the source and point of observation.

In Fig. 1 , it is seen that the diffraction pattern generated by a circular aperture features a broader central peak and has lower secondary peaks compared with the square opening. Thus, a circular aperture has more diffracted energy at lower spatial frequencies. As CD of the contact holes reduces, light is diffracted at larger angle after it passes through the mask. Since optical lithography is a low-pass frequency filtering process, conventional square aperture is likely to suffer more image contrast degradation as it has more light diffracted towards higher spatial frequencies.

The mask function that represents circular contact holes array is convolution of a *circ* function and *comb* function in x and y directions, given by [12]

$$m(x, y) = circ(\frac{q}{r}) \otimes comb(x - np)comb(y - mp) \qquad (3)$$

In Eqn. (3), $circ(\frac{q}{r})$ is a function that characterizes a circular aperture with radius r and q is the position in radial direction, $q = \sqrt{x^2 + y^2}$. $comb(x - np)comb(y - mp)$ is a collection of delta directions function in x and y directions, with distance of p which is the pitch of the array. n and m are positive integers.

For square contact holes array, the mask function is similar to that Eqn (3), given by

$$m(x, y) = rect(\frac{x}{a}) \otimes comb(x - np)comb(y - mp) \qquad (4)$$

In Eqn. (4), $rect(\frac{x}{a})$ is a function that characterizes a square aperture, with sides a.

For $circ(\frac{q}{r})$, the Fourier spectrum is bounded by Bessel function of first kind, whereas for $rect(\frac{x}{a})$, the Fourier spectrum is bounded by *sinc* function. Due to broader central peak of Bessel function, the magnitude of first order light collected at the pupil plane using circular apertures is higher than that of square apertures. Assuming only first and zero order light collected, higher first order light will result in better image contrast. Thus circular apertures mask is capable of improving the image contrast for contact holes array.

3. METHODOLOGY

3.1 Circular aperture and radial segmentation approach

A new approach for contact hole patterning is proposed. Instead of using a square as the target shape for printing, a circular printed target is used. The process flow of the proposed circular aperture and radial segmentation approach is presented in Fig. 2. As shown in Fig. 2, the procedures for pattern generation in this new approach can be summarized as:
1.   Circle OPC target is generated based on design information.
2.   The circular OPC target is segmented.
3.   Certain segments of the segmented circle will be biased by extending the arc in the radial direction. This is done with reference to nearby feature distribution.
4.   Mask GDS out data is generated based on shape produced in step 3.

In this new approach, the parameters to be considered when determining the size of an aperture on mask are radius (r), radial extension (Δr) of segment, and opening angle (θ) of segment, as shown in Fig. 3. The radius of the opening is dependent on the target CD that is desired. For regular contact holes array, dimensional biasing by increasing the radius of the aperture is adopted. For random contact holes layout, radial segmentation approach is applied. The radial extension and opening angle of the segment can be varied to achieve optimal imaging results.

3.2 Simulation conditions

Simulation based on LithoCruiser[TM] is carried out to examine the effectiveness of the proposed method for enhancement in contact hole patterning. The off-axis illumination (OAI) source shape applied in this study is quadrupole illumination with opening angle of 45 degree, as shown in Fig. 4. Attenuated phase-shifting mask (AttPSM) with 6% transmission is used for the simulation. For regular contact holes array, illumination conditions are optimized to the smallest pitch of 130 nm. Dimensional biasing on the mask is used to achieve common process window and CD control for the pitch range that is studied. The targeted CD is 80 nm and the pitch size study ranges from 130 nm to 500 nm. These dimensions are chosen in accordance to ITRS roadmap specification for 45 nm technology node. Results are compared with that of optimized square openings. Table 1 is the summary of illumination conditions and pitch range that are examined in this study.

Table 1 Simulation conditions for regular contact holes array

	Conventional Square Apertures	Proposed Circular Apertures
OAI source shape	Quadrupole	
Numerical aperture (NA)	0.97	0.93
Partial Coherence factor (σ_{in}, σ_{out})	(0.81 , 0.97)	(0.80 , 0.97)
Minimum pitch (nm)	130	
Maximum pitch (nm)	500	
OPC method	Dimensional biasing	

Staggered contact holes (as shown in Fig. 5) are simulated to study the effects of radial segmentation approach on random contact holes layout. The OAI source shape and mask simulated are identical to that used for regular contact holes array. The illumination settings are optimized for the smallest staggered contact holes array with staggered spacing d of 100 nm. Radial segmentation is applied at the diagonal direction since image intensity is lower in this direction as a result of the smaller spacing. In this study, the opening angle θ of the extended segment is fixed at 45 degree. The radial extension Δr is varied according to the minimum spacing between neighboring features. The imaging result is compared with that produced by conventional square apertures. Simulation conditions for staggered contact holes in this study is summarized in Table 2.

Table 2 Simulation conditions for staggered contact holes

	Conventional Square Apertures	Proposed Circular Apertures
OAI source shape	Quadrupole	
Numerical aperture (NA)	1.26	
Partial Coherence factor (σ_{in}, σ_{out})	(0.6 , 0.9)	
Minimum d (nm)	100	
Maximum d (nm)	200	
OPC method	Dimensional biasing	Radial segmentation

4. RESULTS AND DISCUSSION

4.1 Regular contact holes array

Comparison between simulated CD through pitch for masks with square and circular apertures is shown in Fig. 6. It is noted that only dimensional biasing is applied on the mask features to achieve desired CD control. In general, the CD control for both square and circular openings are within 10 % of the nominal CD. The maximum CD error resulted by mask with circular openings is around 4 % at 250 nm pitch. The maximum CD fluctuation from one pitch to another is 6 % which occurs from 250 nm to 300 nm pitch. Dimensional biasing is sufficient to improve the image contrast and enable the patterning of contact holes array from 130 nm to 500 nm pitch. For mask with square opening, the maximum CD error resulted is 10 % at 140 nm pitch. The maximum CD fluctuation is 10 % from 130 nm to 140 nm pitch. It is noted that there is no CD shown in Fig. 6 for 450 nm and 500 nm pitch. This is because the image contrast cannot be further improved by applying dimensional biasing. As a result, there is no sufficient process window at these pitches. Thus, mask with circular openings is capable of achieving a better CD control and reduced CD fluctuation. The improvement in CD through pitch performance can be attributed to enhanced image contrast resulted of the mask with circular openings. From Eqn. (3), the magnitude of first order light collected using circular openings is higher than that of square openings since Bessel function has a broader central peak. Thus at image reconstruction, higher first order light will improve the image contrast.

The DOF resulted from implementing circular aperture is compared with that of conventional square aperture in Fig. 7. Depth of focus is the highest at the smallest pitch of 130 nm because illumination settings are optimized to produce the largest process window at this pitch. However at larger pitch, lower DOF is observed since some of the diffracted light collected does not interfere constructively during image reconstruction and results in poorer image contrast. Mask with circular openings has the highest DOF of 800 nm at 130 nm pitch. The DOF reduces as pitch increases. At 500 nm pitch, DOF drops to 170 nm. However, the DOF through pitch performance is sufficient for application, considering the resist thickness coated is in the range of 100 nm to 150 nm. For mask with square openings, the DOF is 540 nm at 130 nm pitch. The DOF reduces to only 57 nm at 300 nm pitch. At 450 nm and 500 nm pitch, there is no DOF since dimensional biasing is insufficient to improve the image contrast. Also, it is seen that the DOF at larger pitch drops below 150 nm for pitch larger than 250 nm. The average process window improvement provided by application of mask with circular openings is around 15 % for pitch ranging from 150 nm to 220 nm. Since image contrast is directly proportional to DOF, it can be concluded that mask with circular openings has better image contrast than that of square openings. This can be attributed to the improvement in the magnitude in diffraction order collected, as suggested by Eqn. (3)

The normalized image log slope (NILS) comparison between masks circular and square apertures for pitch ranging from 150 nm to 220 nm is reported in Fig. 8. For mask with circular openings, it is seen that NILS is the best at 170 nm pitch. As pitch increases to 220 nm, NILS drops to the lowest. Similarly, mask with square openings has the best NILS at 160 nm pitch but it drops to the lowest at 220 nm pitch. At 170 nm pitch, the best NILS for circular opening mask has improved by 35 % compared to square opening mask. For square openings mask, the best NILS at 160 nm pitch and it has improved by 9 % using circular openings mask. The average improvement in NILS for 150 nm to 220 nm pitch is 15 %. From the results, it is observed that by using circular apertures on mask for contact hole patterning, the image contrast is enhanced. The improvement in CD control and DOF observed in Fig. 6 and Fig. 7 is a result of increasing image contrast. The gain in NILS has indicated that the image contrast is improved and this is due to the improvement in the magnitude of higher order diffracted light, which agrees well with Eqn. (3).

The aerial image intensity plot for contact holes array of 150 nm, 170 nm, and 190 nm pitch is shown in Fig. 9. The image intensity is normalized to the peak intensity of each curve respectively. From the image intensity plot, it is seen that circular mask openings results in better aerial image contrast than that of square openings. For 170 nm pitch, the image contrast for square openings mask is around 0.71. It improves by 10 % to 0.78 by applying mask with circular openings. It is seen that the average improvement is more than 10 % for the three pitches mentioned above. Since aerial image contrast is directly related to the NILS, better image contrast will result in higher NILS, which agrees well with the observation from the simulated results.

4.2 Staggered contact holes

The DOF for staggered contact holes of different staggered spacing d resulted from circular and square apertures is compared in Fig. 10. It is seen that DOF is the largest at the smallest d of 100 nm since illumination settings are optimized to produce maximum process window at this spacing. For subsequent spacing, DOF reduces gradually. In general, the behavior DOF variation is similar for square and circular aperture. The largest DOF for staggered contact holes patterned using circular aperture with radial segmentation is around 140 nm at d of 100 nm. The smallest DOF at d of 200 nm is 26 % less than d of 100 nm. For conventional square aperture, the largest DOF at d of 100 nm is 124 nm. Similarly, DOF reduces as d increases. The smallest DOF at d of 200 nm is around 25 % less than that at optimum spacing. From the simulated DOF performance, an average improvement of around 10 % is achieved by the implementation of circular aperture with radial segmentation. This indicates that circular apertures with radial segmentation does not only benefit regular contact array but also applicable for random contact holes. As dimension of contact holes reduces, the spacing between neighboring features becomes smaller. By implementing the proposed method for contact hole patterning, the process window can be enhanced to enable random contact holes patterning for smaller dimensions.

Aerial image is examined to understand the influence of circular aperture. The NILS comparison for different d studied is plotted in Fig. 11. It can be observed that the behavior of NILS variation is similar for square and circular aperture. At the smallest spacing, NILS achieved for circular aperture is 1.64. For subsequent spacing, the maximum NILS at d of 160 nm is 10 % higher. For square aperture, NILS at the smallest spacing is around 1.50. The maximum NILS at 160 nm is 5 % higher. From Fig. 11, it is seen that staggered contact holes patterned using circular aperture have higher NILS. On average, the improvement is around 10 % throughout different spacing. The improvement in NILS has contributed to the process window enhancement.

The aerial image intensity plots for staggered contact holes pattern with d of 160 nm and 200 nm are shown in Fig. 12. The difference in the maximum and minimum image intensity is higher for contact holes printed using circular aperture. Besides, it is seen that at these spacing values, the aerial image experiences the formation of secondary peaks in between neighboring features. These peaks are the result of constructive interference between light diffracted from the aperture into the space between neighboring holes. If the intensity of these peaks is high enough, undesired side lobes will be printed and affecting the process yield. At d of 160 nm, the secondary peak resulted by implementing circular aperture is 35 % of the maximum intensity. The secondary peak for square aperture is 10% higher that that of circular aperture. At d of 200 nm, the secondary peak formed by circular aperture is around 25 % of the maximum image intensity. However, the image formed by square aperture resulted in a secondary peak that is 20 % higher. Thus circular aperture is capable of improving the aerial image contrast and minimizing the risk of side lobe printing.

5. CONCLUSIONS

A new approach for contact holes patterning is demonstrated in this study. This approach involves targeting a circular printed feature. Circular apertures on mask and radial segmentation method are implemented. Comparison between the proposed approach and conventional approach of square apertures is made by simulating patterning of regular contact holes array and staggered contact holes pattern. From the results, it is observed that the proposed method has achieved 10 % average improvement in DOF compared with that resulted from conventional square aperture approach. Similar gain in aerial image NILS of around 10 % on average compared with conventional approach is achieved. In addition, the risk of side lobe printing is minimized as the image contrast of main feature is enhanced. In conclusion, this study has successfully verified the feasibility circular aperture and radial segmentation approach for contact holes resolution enhancement.

REFERENCES

[1] A. K. Raub, A. M. Biswas, Y. Borodovsky, G. Allen, S. R. J. Brueck, "Simulation of dense contact hole (k1=0.35) arrays with 193-nm immersion lithography", Proc. SPIE Vol. 6154, pp.61542U1-61542U10 (2006).

[2] M. D. Levenson, "Improving resolution in photolithography with a phase-shifting mask", IEEE Transaction on Electron Devices, Vol. ED-29(12), pp. 1828 – 1836 (1982).

[3] M. D. Levenson, D. S. Goodman, S. Lindsey, P. W. Bayer, H. Santini, "The Phase-Shifting mask II: Imaging simulations and submicrometer resist exposures", IEEE Transaction on Electron Devices, Vol. ED-31(6), pp. 753 -763, (1984).

[4] A. Vanleenhove, D. Van Steenwinckel, "A litho-only approach to double patterning ", Proc. SPIE 6520, 65202F (2007).

[5] H. Nakamura, M. Omura, S. Yamashita, Y. Taniguchi, "Ultra-low k1 oxide contact hole formation and metal filling using resist contact hole pattern by Double L&S formation method", Proc. SPIE Vol. 6520, 65201E1 – 65201E10 (2007).

[6] R. Socha et. al., "Contact hole reticle optimization by using interference mapping lithography (IML)", Proc. SPIE 5446, 516 (2004).

[7] M. Hsu, J. F. Chen, D. V. D. Broeke, S. E. Tszng, J. Shieh, "RET masks for patterning 45 nm node contact hole using ArF immersion lithography". Proc. SPIE 6283, 628317-1 – 628317-12 (2006).

[8] J. W. Park, S. Shu, I. Kim, Y. Kang, "Robust double exposure flow for memory", Proc. SPIE 6154, 61542E (2006).

[9] M. Terai et. al., "Below 70-nm contact hole pattern with RELACS process on ArF resist", Proc. SPIE 5039, 789 (2003).

[10] A. E. Rosenbluth et. al., "Intensive optimization of masks and sources for 22nm lithography" ,Proc. SPIE 7274, 727409 (2009).

[11] E. Hecht, [Optics], Addison Wesley (2002).

[12] J.W. Goodman, [Introduction to Fourier optics], Mcgraw-Hill (1996).

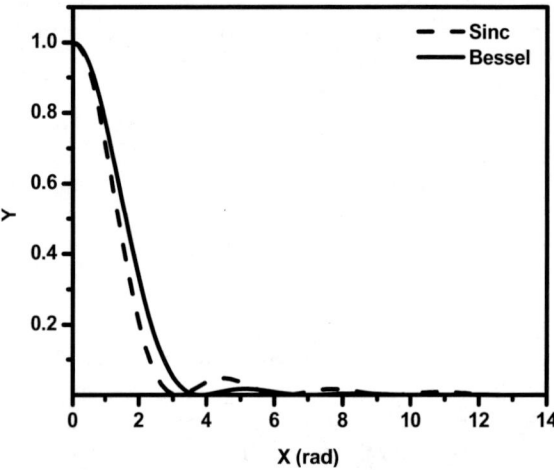

Fig. 1 Sinc and Bessel function

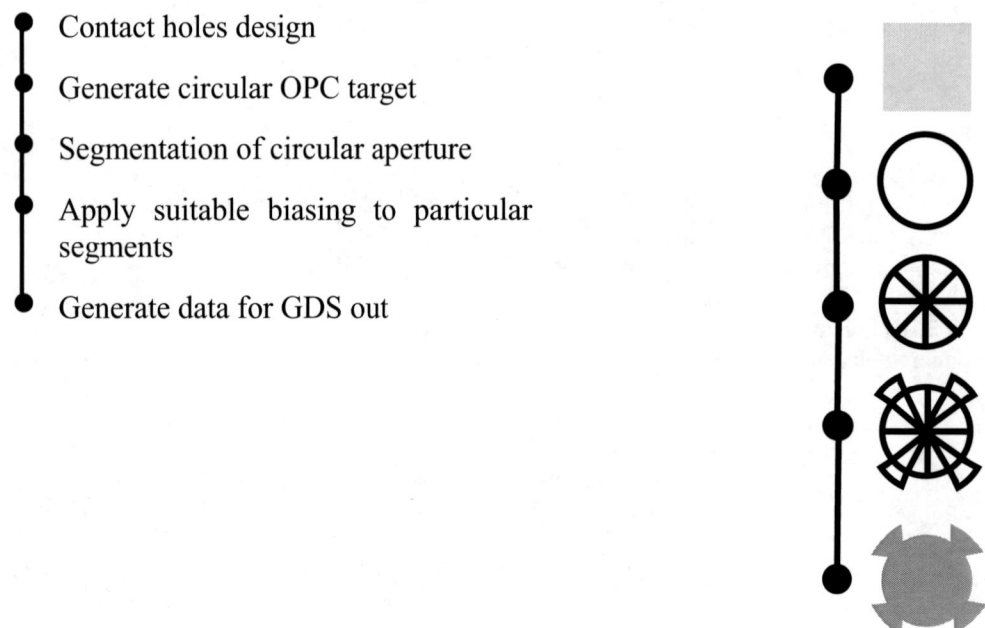

- Contact holes design
- Generate circular OPC target
- Segmentation of circular aperture
- Apply suitable biasing to particular segments
- Generate data for GDS out

Fig. 2 Flow chart for non rectangular mask pattern generation

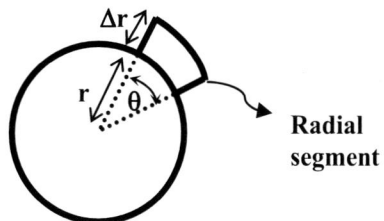

Fig. 3 Parameters for radial segmentation

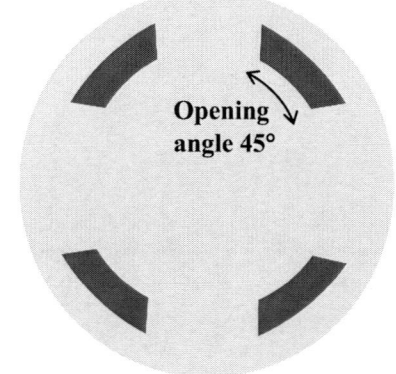

Fig. 4 Quadrupole off-axis illumination

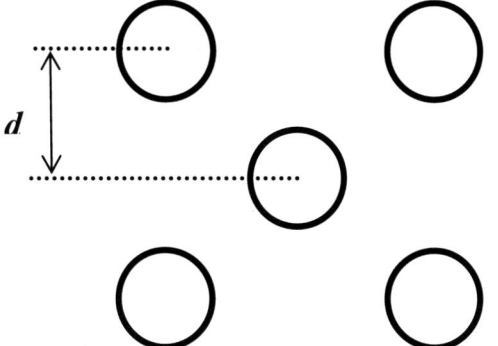

Fig. 5 Staggered contact holes pattern

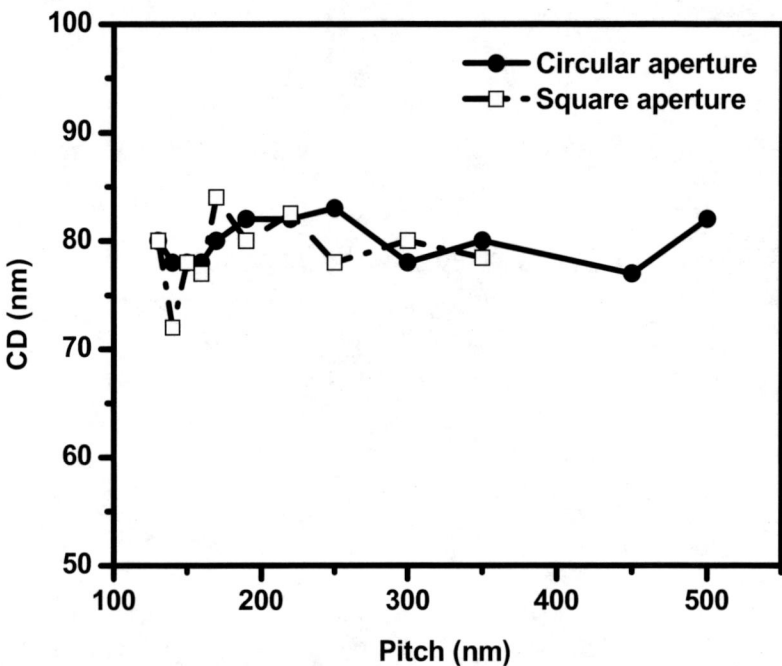

Fig. 6 CD through pitch comparison between circular and square aperture

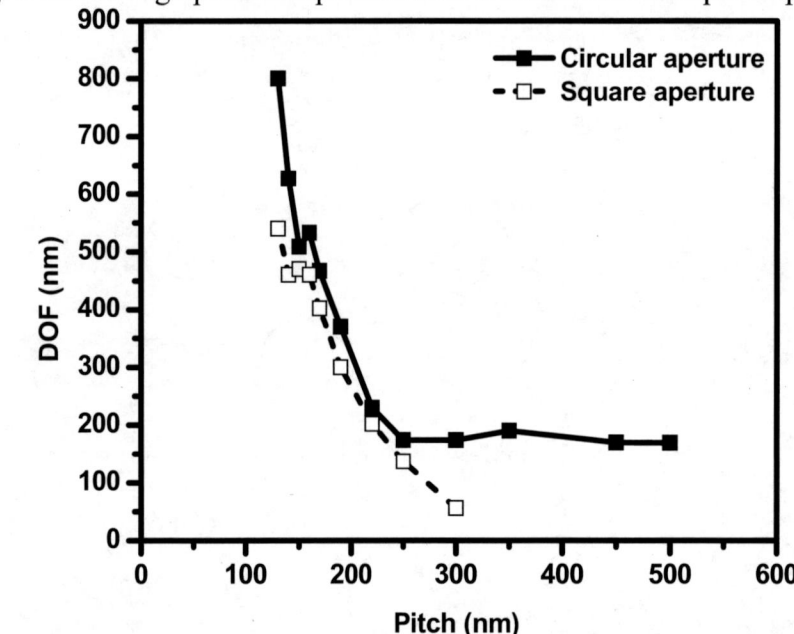

Fig. 7 DOF through pitch comparison between circular and square aperture

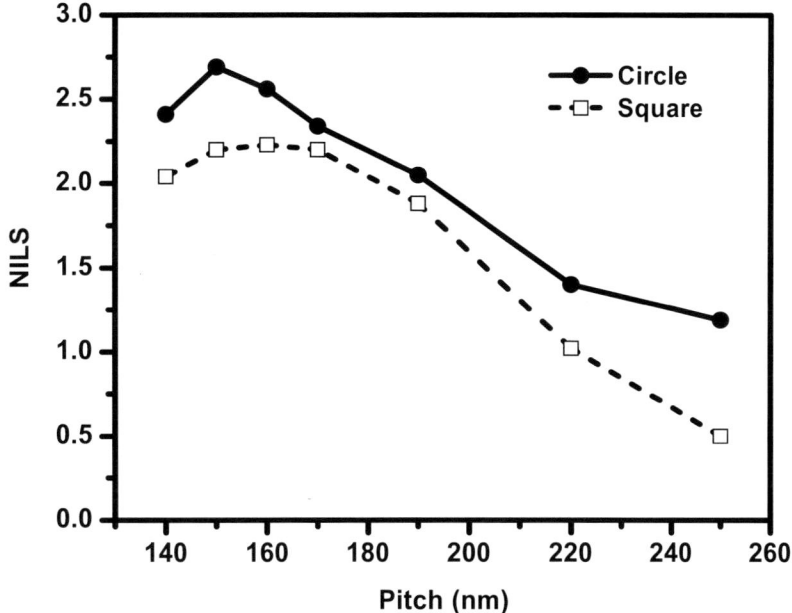

Fig. 8 NILS comparison between circular and square aperture for contact hole array of 150 nm to 220 nm pitch

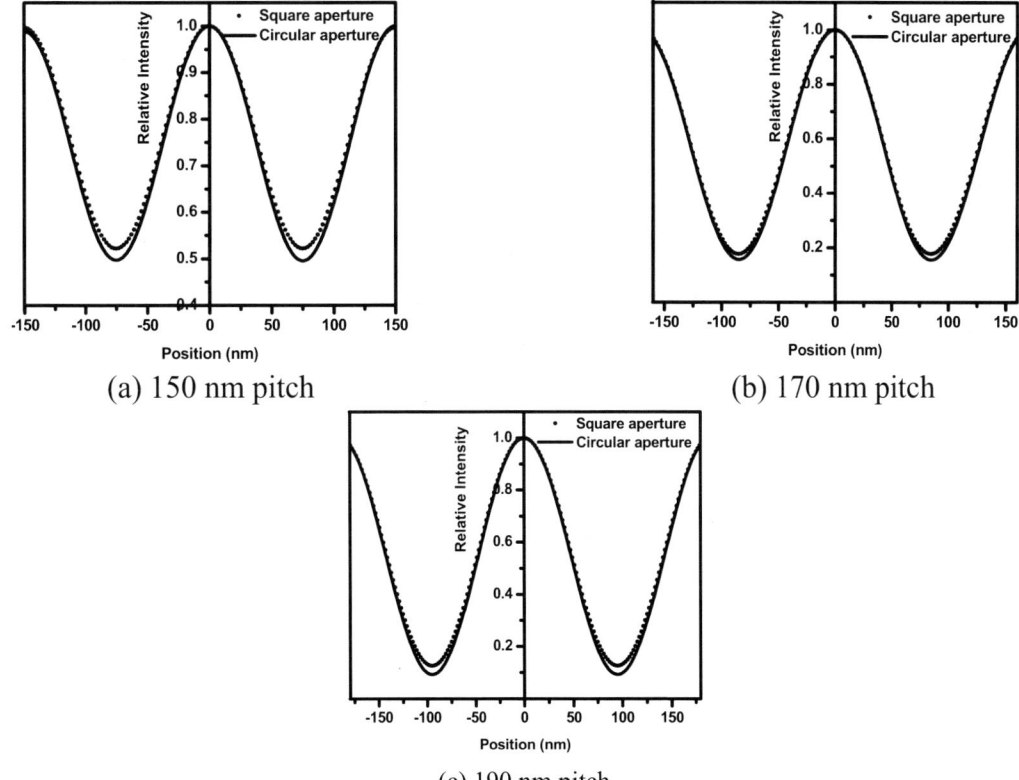

(a) 150 nm pitch   (b) 170 nm pitch

(c) 190 nm pitch

Fig. 9 Aerial image intensity plot comparison between circular and square aperture

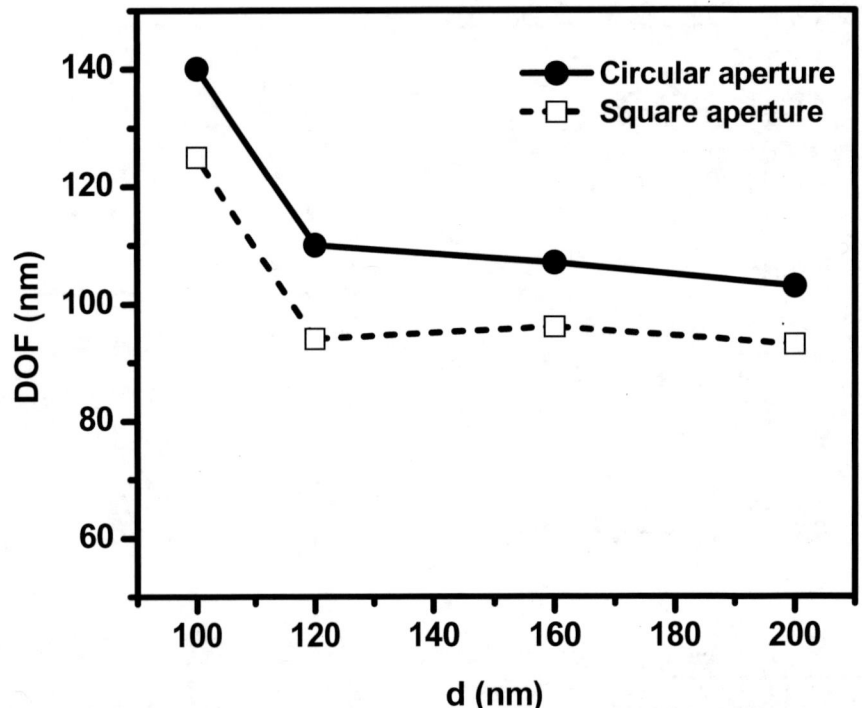

Fig. 10 DOF for staggered contact holes pattern of different spacing (d)

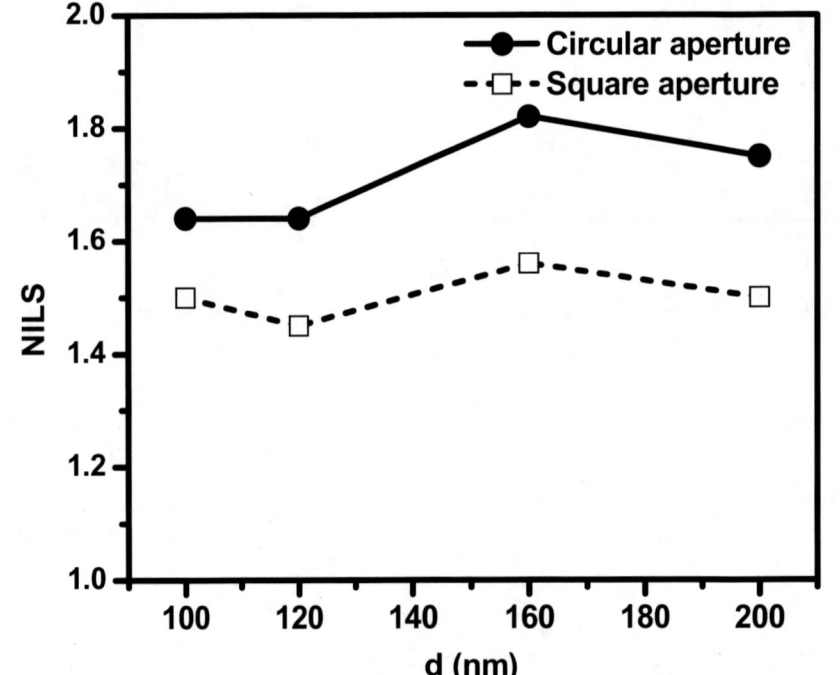

Fig. 11 NILS for staggered contact holes pattern of different spacing (d)

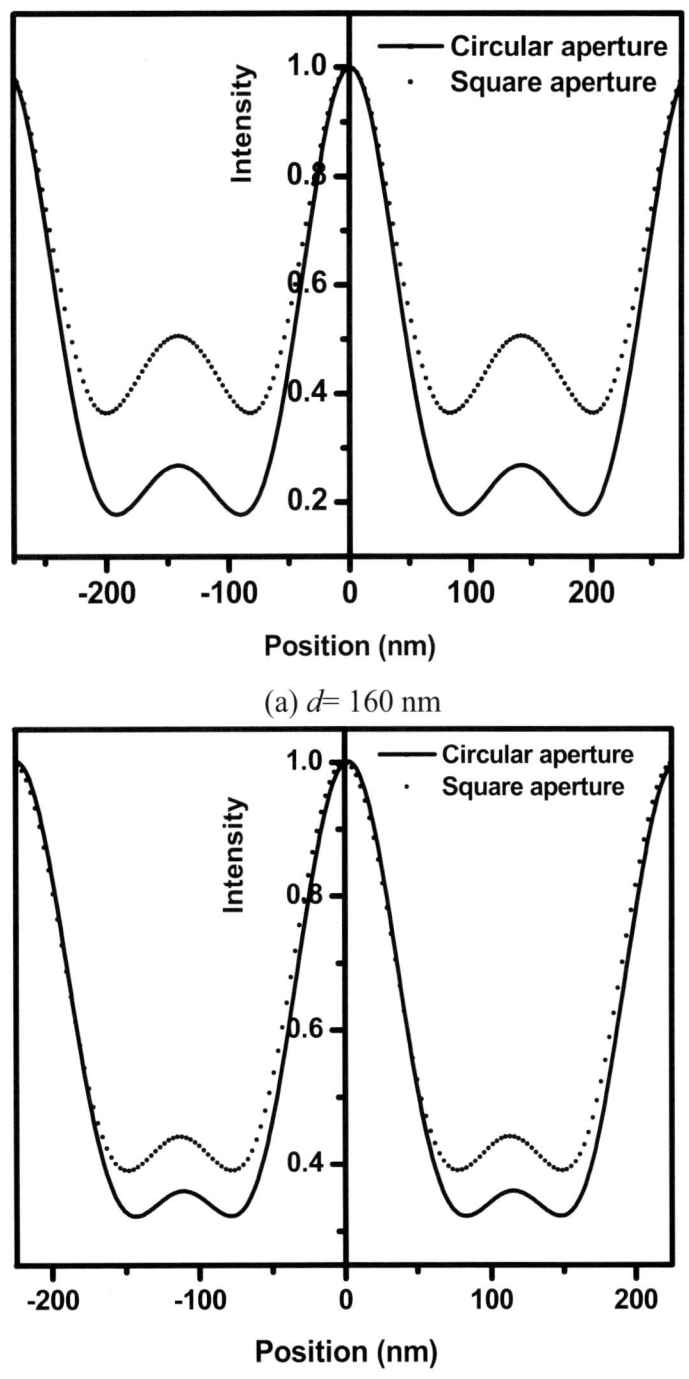

(a) $d = 160$ nm

(b) $d = 200$ nm

Fig. 12 Aerial image intensity for staggered contact holes pattern

Binary Mask Optimization for Forward Lithography based on Boundary Layer Model in Coherent Systems *

Xu Ma and Gonzalo R. Arce

Department of Electrical and Computer Engineering, University of Delaware, Newark, DE,

U.S.A.

ABSTRACT

Recently, a set of generalized gradient-based optical proximity correction (OPC) optimization methods have been developed to solve for the forward and inverse lithography problem under the thin-mask assumption, where the mask is considered a thin 2-D object. However, as the critical dimension printed on the wafer shrinks into the subwavelength regime, thick-mask effects become prevalent and thus these effects must be taken into account in OPC optimization methods. OPC methods derived under the thin-mask assumption have inherent limitations and perform poorly in the subwavelength scenario. This paper focuses on developing model-based forward binary mask optimization methods which account for the thick-mask effects of coherent imaging systems. The boundary layer (BL) model is exploited to simplify and characterize the thick-mask effects, leading to a computationally efficient OPC method. The BL model is simpler than other thick-mask models, treating the near field of the mask as the superposition of the interior transmission areas and the boundary layers. The advantages and limitations of the proposed algorithm are discussed and several illustrative simulations are presented.

Keywords: Binary mask, thick mask, forward lithography, coherent imaging

1. INTRODUCTION

Due to the resolution limits of optical lithographic systems, the electronics industry has relied on resolution enhancement techniques (RET) to compensate and minimize mask distortions as they are projected onto semiconductor wafers.[1] Resolution in optical lithography obeys the Rayleigh resolution limit $R = k\frac{\lambda}{NA}$, where λ is

---

* This work was supported by the Intel corporation.

Optical Microlithography XXIII, edited by Mircea V. Dusa, Will Conley, Proc. of SPIE Vol. 7640,
76402N · © 2010 SPIE · CCC code: 0277-786X/10/$18 · doi: 10.1117/12.836858

the wavelength, NA is the numerical aperture, and k is the process constant which can be minimized through RET methods.[2-5] In optical proximity correction (OPC), mask amplitude patterns are modified by the addition of sub-resolution features that can pre-compensate for imaging distortions.

Several approaches to forward and inverse lithography have been proposed in the literature. Sherif, et al. derived an iterative approach to generate binary masks in incoherent diffraction-limited imaging systems.[6] Liu and Zakhor developed a binary and phase shifting mask (PSM) design strategy based on the branch and bound algorithm and simulated annealing.[7] Pati-Kailath developed sub-optimal projections onto convex sets for PSM designs.[8] In addition, Erdmann proposed automatic optimization of the mask and illumination parameters with a genetic algorithm.[9] Linyong Pang, et al. gave an overview of ILT and provided some simulations to demonstrate the benefit of ILT.[10] Granik described and compared solutions of inverse mask problems.[11]

Poonawala and Milanfar developed a pixel-based optimization framework for inverse lithography, well suited for gradient-based search.[12] Ma and Arce generalized this algorithm so as to admit multi-phase components having arbitrary PSM patterns.[13, 14] However, both of the algorithms focus on the coherent illumination system. Recently, Ma and Arce used the sum of coherent systems (SOCS) model and average coherent approximation model to develop effective and computationally efficient binary mask design algorithms for inverse lithography under partially coherent illuminations.[15, 16] Subsequently, Ma and Arce develop a PSM design algorithm based on an SVD model under partially coherent systems.[17, 18] In addition, they extended their work to allow for the joint optimization of the source and the mask.[19]

All of the algorithms above, however, have been developed under the thin-mask assumption, where Kirchhoff's boundary condition is directly applied to the mask topology and consequently the mask is treated as a 2-D object.[20, 21] As the critical dimension (CD) printed on the wafer shrinks into the subwavelength regime, the thick-mask effects become very pronounced such that these effects should be taken into account in the mask optimization. Thick-mask effects include polarization dependence due to the different boundary conditions for the electric and magnetic fields, transmission error in small openings, diffraction edge effects or electromagnetic coupling and so on.[20] The thick-mask effects can be rigorously represented by the near-field pattern of the mask, which is different from the Kirchhoff approximation of the mask topography. Two decades ago, Wong and

Neureuther discovered the intensity imbalance of alternating PSM, and applied the finite-difference time-domain method (FDTD) to study the mask topography effects in the projection printing of PSM.[22, 23] This phenomenon was proved by experimental results later.[24] Yuan exploited the waveguide method (WG) to model the light diffraction of 2D phase shifting masks,[25] which was subsequently generalized by Lucas to the 3D topography.[26] Erdmann, et al. evaluated and compared the FDTD method and the WG method for the simulation of typical hyper NA ($NA > 1$) imaging problem.[27] Adam and Neureuther introduced domain decomposition methods for the simulation of photomask scattering.[28] Nevertheless, these approaches are too complex to be applied in model-based binary mask design.

Recently, Azpiroz et al. introduced a novel boundary layer (BL) model for fast evaluation of the near-field of a thick mask.[20, 21] Different from other computationally complex and resource consuming rigorous mask models, the BL model treats the near field of the mask as the superposition of the interior transmission areas and the boundary layers, which have fixed dimensions and determined locations. The BL model effectively compensates for the inaccuracy of Kirchhoff's approximation, which is attributed to thick-mask effects, different polarizations, and phase errors. The simplicity and accuracy of the BL model enables the formulation of a computationally efficient optimization algorithm for binary masks. This paper thus focuses on the formulation of a computationally effective model-based forward binary mask optimization algorithm based on the BL model to take into account the thick-mask effects under coherent illumination. This is accomplished as follows: First, the optical lithography process under coherent illumination is formulated as the combination of the BL model and the Hopkins diffraction model. The cost function of the binary mask optimization problem is formulated as the square of the l^2-norm of the difference between the real aerial image and the desired pattern on the wafer. Then the gradient of the cost function, referred to as the cost sensitivity function is developed and used to drive cost function in the descent direction during the optimization process. Topological constraints of the binary mask is introduced and used to limit the minimum opening size of the optimized mask pattern.

The remainder of the paper is organized as follows: The Hopkins diffraction model of optical lithography systems is summarized in Section 2. The BL model of coherent imaging system is summarized in Section 3. The lithography preliminaries and the cost sensitivity function are developed in Section 4. The binary mask optimization algorithm based on BL model under coherent illumination is described in Section 5. Simulation results are illustrated in Section 6. Conclusions are provided in Section 7.

2. HOPKINS DIFFRACTION MODEL OF THE OPTICAL LITHOGRAPHY SYSTEM

According to the Hopkins diffraction model, the light intensity distribution exposed on the wafer, referred to as the aerial image under partially coherent illumination (PCI) is bilinear and described by[29]

$$I(\mathbf{r}) = \int \int M^*(\mathbf{r_1}) M(\mathbf{r_2}) \gamma(\mathbf{r_1} - \mathbf{r_2}) h^*(\mathbf{r} - \mathbf{r_1}) h(\mathbf{r} - \mathbf{r_2}) d\mathbf{r_1} d\mathbf{r_2}, \tag{1}$$

where $\mathbf{r} = (x, y)$, $\mathbf{r_1} = (x_1, y_1)$ and $\mathbf{r_2} = (x_2, y_2)$. $M(\mathbf{r})$ is the mask pattern, $\gamma(\mathbf{r_1} - \mathbf{r_2})$ is the complex degree of coherence, and $h(\mathbf{r})$ represents the amplitude impulse response of the optical system. The amplitude impulse response $h(\mathbf{r})$ is defined as the Fourier transform of the circular lens aperture with cutoff frequency NA/λ;[30,31] therefore,

$$h(\mathbf{r}) = \frac{J_1(2\pi r NA/\lambda)}{2\pi r NA/\lambda}. \tag{2}$$

The complex degree of coherence $\gamma(\mathbf{r_1} - \mathbf{r_2})$ is generally a complex number, whose magnitude represents the extent of optical interaction between two spatial locations $\mathbf{r_1} = (x_1, y_1)$ and $\mathbf{r_2} = (x_2, y_2)$ of the light source.[1] The complex degree of coherence in the spatial domain is the inverse 2-D Fourier transform of the illumination shape. In general, this equation is tedious to compute, both analytically and numerically.[32] The system reduces to simple forms in the two limits of complete coherence or complete incoherence. For the completely coherent case, the illumination source is at a single point, thus, $\gamma(\mathbf{r}) = 1$. In this case, the aerial image in Eq. (1) is separable on $\mathbf{r_1}$ and $\mathbf{r_2}$, and thus

$$I(\mathbf{r}) = |M(\mathbf{r}) \otimes h(\mathbf{r})|^2, \tag{3}$$

where $\otimes$ is the convolution operation. For the completely incoherent case, the illumination source is of infinite extent and thus, $\gamma(\mathbf{r}) = \delta(\mathbf{r})$. In this case, the aerial image reduces to

$$I(\mathbf{r}) = |M(\mathbf{r})|^2 \otimes |h(\mathbf{r})|^2. \tag{4}$$

This paper focuses on the binary mask optimization based on the BL model under coherent illumination. The schematic of an optical lithography system with coherent illumination is illustrated in Fig. 1.

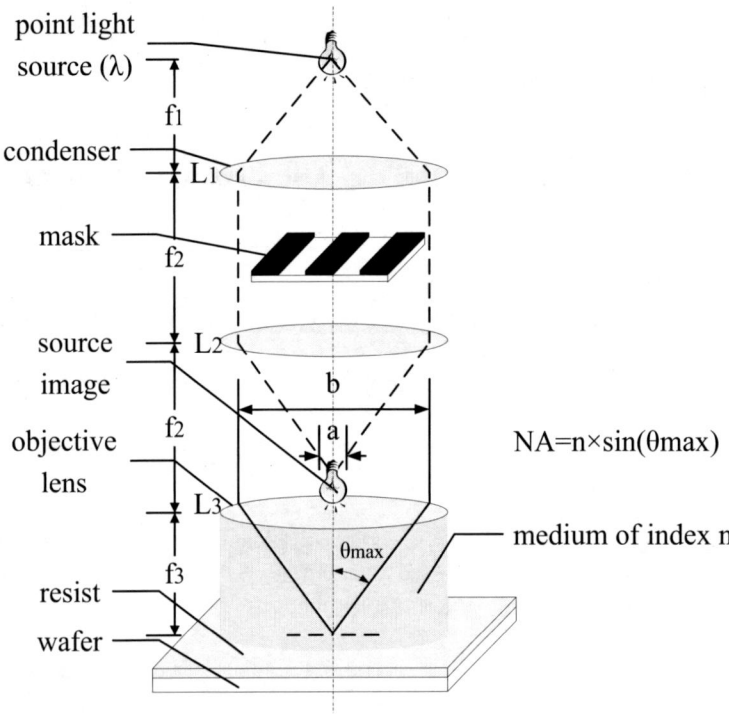

Figure 1. Optical lithography system with coherent illumination

3. BOUNDARY LAYER MODEL OF COHERENT IMAGING

The Kirchhoff approximation has been extensively used in the development of OPC methods, where the mask thickness is assumed to be infinitesimal and the mask is considered a 2D object. As the CD printed on the wafer shrinks into the subwavelength regime, however, the thick-mask effects have become significant and thus these need to be taken into account in the design of OPC methods. Although numerous rigorous mask models simulating the 3D electromagnetic field of the mask were developed, these models are resource consuming and too complex to be applied in the model-based binary mask design for forward lithography.

Recently, Azpiroz et al. introduced a novel boundary layer (BL) model for the fast evaluation of the near-field of the thick mask, where the near field is modelled as the superposition of the interior transmission areas and the boundary layers with fixed dimensions and determined locations.[20, 21] The concepts of the BL model under coherent illumination are illustrated in Fig. 2, where the polarization of the impending electric field **E** is assigned to be in the horizontal direction. Figure. 2 shows a typical rectangular opening of the binary mask with width equal to a, and height equal to b. The harmonic mean of the area's width a and height b is $d = \frac{2ab}{a+b}$. . The

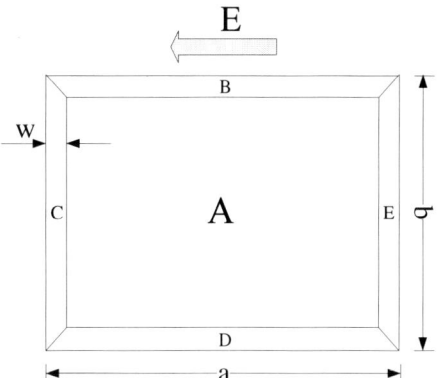

Figure 2. BL model under coherent illumination, where the polarization of the electric field is assigned to be in the horizontal direction. w is the width of the boundary areas. a and b are the width and height of the entire opening area respectively.

near field of the opening is divided into five areas: A, B, C, D and E. A is the interior transmission area with transmission coefficient $\eta_I = 1$ for the binary mask. The transmission coefficients of the boundary layers depend on the polarization of the electric field of the impending light. Since the polarization of the electric field is assigned to be in the horizontal direction, B and D are the tangential boundary areas with width w and transmission coefficient η_T. C and E are the normal boundary areas with width w and transmission coefficient η_N. In the BL model, the relative error of amplitude of the electric field on the wafer produced by the thin-mask approximation is measured by the deviation of its real component from the rigorously FDTD calculated EM field value. Experimental results show that the relative error of amplitude is in proportion to the width of the boundary layer w and inversely proportional to the harmonic mean d, represented as

$$Re\{\frac{\Delta\mathbf{E}}{\mathbf{E}}\} = \frac{4w}{d} = \frac{(2a+2b)w}{ab} = \frac{Boundary\ Layer\ Area\ (real\ part)}{Total\ Area}, \tag{5}$$

where $Re\{\cdot\}$ denotes the real part of the argument. $\mathbf{E}$ is the total electric field from the rigorously FDTD calculated EM field value. $\Delta\mathbf{E}$ is the electric field error from the thin-mask assumption, and $\frac{\Delta\mathbf{E}}{\mathbf{E}}$ is the the relative error of amplitude. Given a and b, w can be calculated from Eq. (5). The deviation of the real component is compensated by the opaque boundary layers surrounding all openings on the mask, whose transmission coefficients are zero. Given the value of w, it is shown experimentally that the relative error of phase is in proportion to w and transmission coefficient magnitude $|\eta_T|$, and inversely proportional to the height of the opening b, represented as

$$Im\{\frac{\Delta \mathbf{E}}{\mathbf{E}}\} = |\eta_T|\frac{2w}{b} = |\eta_T|\frac{2aw}{ab} = \frac{Boundary\ Layer\ Area\ (imaginary\ part)}{Total\ Area}, \tag{6}$$

where $Im\{\cdot\}$ denotes the imaginary part of the argument. Subsequently, the relative error of phase is compensated by the boundary layers with complex transmission coefficient η_T and width of w on the opening edges tangent to the electric field of impending light. The inaccuracy of the thin-mask approximation is effectively offset by the superposition of complex-valued boundary layers. The real values of the boundary layers are zero (opaque) around the area A. The complex values are η_T on the tangential direction and zero on the normal direction. These values for η_T and η_N have been shown in[20, 21] to effectively compensate for the thin-mask distortion. The transmission coefficient of the tangential boundary areas, η_T is calculated from the slope of the linear relation described in Eq. (6). However, the relationships described in Eq. (5) and Eq. (6) are not accurate when the opening size reduces below the wavelength. For the validity of the BL model, the minimum size of the opening is constrained to be larger than the wavelength. The simplicity and accuracy of the BL model are suitable for the model-based binary mask optimization algorithms.

Azpiroz et al. studied two types of optical lithography systems. The first one is a 4X projection system with $NA = 0.68$ and $\lambda = 248nm$, while the second one is with $NA = 0.85$ and $\lambda = 193nm$. For the first type of optical lithography system, $w = 24.8nm$, $\eta_T = 0.0i$, $\eta_N = 0$, and the minimum allowed opening size on the mask (the width of the area A) is $248nm$. For the second type, $w = 14.5nm$, $\eta_T = 0.8i$, $\eta_N = 0$, and the minimum allowed opening size is $200nm$. In our work, we will use the two types of lithography systems described by Azpiroz et al. to develop binary mask optimization algorithms.

4. LITHOGRAPHY PRELIMINARIES AND COST SENSITIVITY FUNCTION

Let $M(x,y)$ be the input binary mask to an optical lithography system $T\{\cdot\}$ with coherent illumination. The system $T\{\cdot\}$ includes two steps. The first step is the evaluation of the near field of the thick mask, which is based on the BL model. The second step is the optical imaging system leading to the aerial image on the wafer, which is approximated by the Hopkins diffraction model. The output aerial image on the wafer is denoted as $Z(x,y) = T\{M(x,y)\}$. Given a $N \times N$ desired output pattern $\tilde{Z}(x,y)$, the goal of OPC mask design is to find the optimized $M(x,y)$ called $\hat{M}(x,y)$ such that the distance

$$D = d(Z(x,y), \tilde{Z}(x,y)) = d(T\{M(x,y)\}, \tilde{Z}(x,y)) \tag{7}$$

is minimized, where $d(\cdot,\cdot)$ is the square of the l^2-norm criterion. The OPC optimization problem can thus be formulated as the search of $\hat{M}(x,y)$ over the $N \times N$ real space $\Re^{N \times N}$ such that

$$\hat{M}(x,y) = arg \min_{M(x,y) \in \Re^{N \times N}} d(T\{M(x,y)\}, \tilde{Z}(x,y)). \tag{8}$$

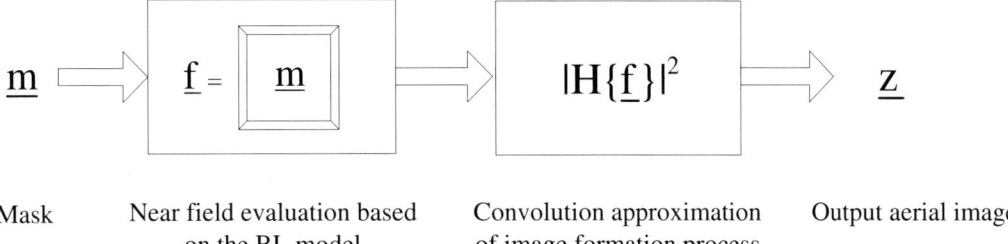

| Mask | Near field evaluation based on the BL model | Convolution approximation of image formation process | Output aerial image |

Figure 3. Approximated forward imaging process based on the BL model under coherent illumination, where the polarization of the electric field is assigned to be in the horizontal direction.

The forward imaging process is illustrated in Fig. 3. The electric field propagating through the thick-mask pattern is affected by the 3D topography of the mask, forming the near field which is then influenced by the diffraction and mutual interference in the optical imaging system. Light that is transmitted through the optical system reaches the photoresist and forms the aerial image. In Fig. 3, $|\cdot|$ is the element-by-element absolute operation, and the output of the convolution and the absolute operation model is the intensity distribution of the aerial image.

Following the definitions above, the following notations are used:

1) The $M_{N \times N}$ matrix represents the mask pattern, with entry values equal to 0 or 1 for the binary mask. The $N^2 \times 1$ equivalent raster scanned vector representation is denoted as $\underline{m}$.

2) The $\Gamma_{N \times N}(M)$ matrix, with all entry values equal to 0 or 1, represents the interior transmission area pattern of the near field corresponding to the mask M. Its vector representation is denoted as $\underline{\gamma}$.

3) The $\Gamma \uparrow$ and $\Gamma \downarrow$ represent the shifted version of Γ by shifting Γ along the vertical direction (up and down) by one pixel, respectively. Their vector representations are denoted as $\underline{\gamma} \uparrow$ and $\underline{\gamma} \downarrow$.

4) The $F_{N \times N}(M)$ matrix represents the near-field corresponding to the mask M, with complex entry values. Its vector representation is denoted as $\underline{f}$. Let the polarization of the impending electric field $\mathbf{E}$ be in the horizontal direction. For the binary mask in the first type of optical lithography system, all the boundary layers are opaque, with transmission coefficient of 0. In order to represent all the features on the mask by integral number of pixels, the pixel size is assigned equal to the greatest common divisor between the boundary layer width and the minimum opening size, thus, the pixel size is set to be 24.8 nm. The minimum opening size is $248nm = 10 \times pixel\ size$. Thus, the near-field is the same as the interior transmission area. Therefore,

$$\underline{f}_p = \underline{\gamma}_p, \quad p = 1, 2, \ldots, N^2. \tag{9}$$

For the binary mask in the second type of optical lithography system, the normal boundary layers are opaque and the tangential boundary layers have complex transmission coefficient of $0.8i$. The pixel size is set to be 14.5 nm. In order to represent the minimum opening size by integral number of pixels, the minimum opening size is increased to be $203nm = 14 \times pixel\ size$. Therefore

$$\underline{f}_p = \begin{cases} 0.8j & : \quad (\underline{\gamma}_{p-N} = 1 \text{ and } \underline{\gamma}_p = 0) \text{ or } (\underline{\gamma}_{p+N} = 1 \text{ and } \underline{\gamma}_p = 0) \\ \underline{\gamma}_p & : \quad \text{otherwise} \end{cases} \tag{10}$$

Eq. (10) can be rewritten as

$$\underline{f}_p = 0.8j(1 - \underline{\gamma}_p)\underline{\gamma}_{p-N} + 0.8j(1 - \underline{\gamma}_p)\underline{\gamma}_{p+N} + \underline{\gamma}_p, \quad p = 1, 2, \ldots, N^2, \tag{11}$$

where $\underline{\gamma}_p = 0$, if $p < 1$ or $p > N^2$.

5) A convolution matrix H is a $N^2 \times N^2$ matrix with an equivalent two-dimensional filter h.

6) The desired $N \times N$ binary output pattern is denoted as $\tilde{Z}$. It is the desired aerial image sought on the wafer. Its vector representation is denoted as $\underline{\tilde{z}}$.

7) The initial interior transmission area of the optimization process is $\tilde{Z}$. The corresponding initial mask pattern is $\tilde{M}$.

8) The output aerial image is the $N \times N$ matrix denoted as:

$$Z = |H\{F\}|^2. \tag{12}$$

The equivalent vector is denoted as $\underline{z}$.

9) The optimized mask denoted as $\hat{M}$ minimizes the distance between Z and $\tilde{Z}$, ie,

$$\hat{M} = arg \min_{M} d(|H\{F\}|^2, \tilde{Z}). \tag{13}$$

Given the output aerial image $\underline{z} = |H\{\underline{f}\}|^2$, the pth entry in this vector can be represented as

$$\underline{z}_p = |\sum_{q=1}^{N^2} h_{pq}\underline{f}_q|^2, \quad p = 1, \dots N^2, \tag{14}$$

where h_{pq} is the $(p, q)th$ entry of the filter. The cost function is the L_2 norm of the difference between $\underline{z}$ and $\tilde{\underline{z}}$. Therefore,

$$D = \|\tilde{\underline{z}} - \underline{z}\|_2^2 = \sum_{p=1}^{N^2} (\tilde{\underline{z}}_p - \underline{z}_p)^2, \tag{15}$$

where $\underline{z}_p$ in Eq. (15) is represented in Eq. (14).

In the following, the sensitivity of the cost function D with respect to the change of the interior transmission area will be used to guide the optimization process. The sensitivity of the cost function D is $\frac{\partial D}{\partial \underline{\gamma}}$. For the first type of optical lithography system,

$$\frac{\partial D}{\partial \underline{\gamma}} = -H^T[(\tilde{\underline{z}} - \underline{z}) \odot H(\underline{\gamma})], \tag{16}$$

where T is the conjugate transposition, and $\odot$ is the element-by-element multiplication operation. For the second type of optical lithography system,

$$\begin{aligned}
\frac{\partial D}{\partial \underline{\gamma}} &= -4\text{Re}\{H^T[(\tilde{\underline{z}} - \underline{z}) \odot H(\underline{f})] \odot (0.8i\underline{\gamma} \uparrow +0.8i\underline{\gamma} \downarrow +1) \\
&+ H^T[(\tilde{\underline{z}} - \underline{z}) \odot H(\underline{f})] \downarrow \odot 0.8i(1 - \underline{\gamma} \downarrow) \\
&+ H^T[(\tilde{\underline{z}} - \underline{z}) \odot H(\underline{f})] \uparrow \odot 0.8i(1 - \underline{\gamma} \uparrow)\}
\end{aligned} \tag{17}$$

where $\text{Re}\{\cdot\}$ denotes the real part of the argument. $\uparrow$ and $\downarrow$ are shifting operations by shifting the $N \times N$ equivalent matrix of the vector in the argument along vertical direction (up and down) by one pixel, respectively.

5. THE BINARY MASK OPTIMIZATION ALGORITHM BASED ON BL MODEL UNDER COHERENT ILLUMINATION

5.1. Topological Constraint

According to the BL model summarized in Section 3, the interior transmission area has a one-to-one correspondence to the mask. Therefore, the proposed OPC mask design algorithm directly optimizes the interior transmission area, from which the mask can be easily reconstructed. The BL model constrains the minimum size of the openings on the binary mask.[20, 21] In order to meet the requirements, some topological constraints are imposed in the optimization process of the interior transmission area.[33, 34] In the following, some definitions and operations for shape topologies are listed.

Definition 1 (White block and black block). Any pixel in the interior transmission area can have either a value 0 or 1. A white block is a square area with all pixels values equal to 1, while a black block has all of its pixels equal to 0.

Definition 2 (Flipping-on and flipping-off operations). Turning a pixel value from 0 to 1, and from 1 to 0 are referred to as flipping-on and flipping-off a pixel. In general, flipping-on and flipping-off operations of a block means to turn the block to a white block and to a black block.

Definition 3 (Type I singular pixel). A type I singular pixel is one that does not belong to any $L \times L$ white block on the interior transmission area pattern Γ, where L depends on the minimum opening size of the BL model.

Definition 4 (Type II singular pixel). A type II singular pixel is one that does not belong to any 3×3 black block on the interior transmission area pattern Γ. Since the openings on the optimized binary mask contain the additional surrounding boundary layers compared to the corresponding interior transmission areas, the type II singular pixel introduces the mergence of adjacent openings on the mask.

Definition 5 (Cost sensitivity matrix of a block). The cost sensitivity function corresponding to a block G on the interior transmission area pattern, calculated by Eq. (16) or Eq. (17) is $\nabla D(G)$ defined as the cost sensitivity matrix of the block G.

Definition 6 (Changeable block). A $K \times K$ changeable block is a block whose cost sensitivity matrix contains K positive or negative values. If the cost sensitivity matrix contains K positive values, the block is defined as a positive changeable block. Vice versa, it is defined as a negative changeable block.

In our binary mask optimization approach, only the positive or negative changeable blocks are considered to be flipped-off or flipped-on. These topological constraints guarantee that the features of the optimized binary mask are larger than the minimum opening size.

5.2. The binary mask optimization algorithm based on BL model under coherent illumination

Following the topological constraints, the proposed binary mask optimization algorithm is shown in Table 1, where the parameters K in **Step 3** and L used in *Definition 3* depend on the minimum opening size of the BL model.

6. SIMULATIONS

To prove the efficiency of the proposed algorithm, the optimization method described in Table 1 is used to design a mask targetting the desired aerial image shown in Fig. 4. In Fig. 4, p is the pitch width. For the first type of optical lithography system, $p = 223.2nm$, and the system parameters are: $NA = 0.68$ and $\lambda = 248nm$. Since the system is a 4X projection system, the pitch width of the initial interior transmission area pattern $\tilde{\Gamma}$ of the optimization process is $892.8nm$. In the simulation, the initial mask pattern has the dimension of $2.23\mu m \times 2.23\mu m$. The pixel size is $24.8nm \times 24.8nm$, which is the same as the boundary width. The convolution kernel shown in Eq. (2) is assumed to vanish outside the area A_{h1} defined by $x, y \in [-1.5\mu m, 1.5\mu m]$. The parameters of the optimization algorithm are $K = L = 8$. The simulations results using the algorithm depicted in Table 1 for the first type of optical lithography system are shown in Fig. 5. Top row (from left to right) shows: the initial mask pattern and the corresponding output aerial image, with output pattern error of 1200.1. Middle row (from left to right) shows: the optimized binary mask using the algorithm depicted in Table 1 based on thin-mask approximation and the corresponding output aerial image, with output pattern error of 1039.4. Bottom row (from left to right) shows: the optimized binary mask based on BL model and the corresponding output aerial image, with output

Table 1. The Binary Mask Optimization Algorithm

Step 1	Initialization of the interior transmission area pattern: $\tilde{\Gamma} = \tilde{Z}$. The corresponding initial mask pattern is $\tilde{M}$.
Step 2	Calculate the cost sensitivity function using Eq. (16) or Eq. (17).
Step 3	Scan the cost sensitivity matrix from top to bottom and from left to right. Find the first encountered $K \times K$ changeable block G.
Step 4	Flip-on or flip-off G if it is a negative or positive changeable block.
Step 5	If (flipping operation has introduced type I or type II singular pixel) flag=1.
Step 6	If (flag==1) or (cost function D is increased) or (any pixel value $\neq 0$ or 1) restore G to its original values.
Step 7	Clear the cost sensitivity matrix of G: $\nabla D(G) = \mathbf{0}$.

Step 8	If $\nabla D \neq \mathbf{0}$
	    Go to step 3.
	Otherwise
	    If no block is flipped in the current iteration
	        End.
	    Otherwise
	        Go to step 2.

pattern error of 972.3. Black and white represent 0 and 1 respectively. It is shown that optimization of the binary mask based on thin-mask approximation reduces the output pattern error by 13.4%. On the other hand, algorithm based on BL model reduces the output pattern error by 19.0%. Figure 6 illustrates the intersections of the aerial images on the $45th$ row. The solid line, dashed line and dotted line represent the intersections corresponding to the initial mask, the OPC based on the thin-mask assumption, and OPC based on the thick mask assumption, respectively.

For the second type of optical lithography system, $p = 137.8nm$, the system parameters are: $NA = 0.85$ and $\lambda = 193nm$. The pitch width of the initial interior transmission area pattern $\tilde{\Gamma}$ of the optimization process is $551.0nm$. In the simulation, the initial mask pattern has the dimension of $1.38\mu m \times 1.38\mu m$. The pixel size is

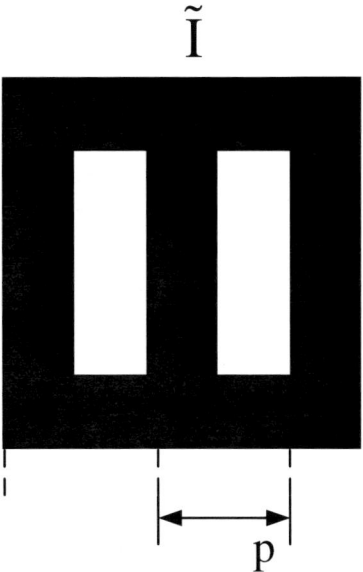

Figure 4. Desired pattern of the aerial image searched on the wafer.

$14.5nm \times 14.5nm$, which is the same as the boundary width. The convolution kernel is assumed to vanish outside the area A_{h2} defined by $x, y \in [-1.0\mu m, 1.0\mu m]$. The parameters of the optimization algorithm are $K = L = 12$. The simulations results for the second type of optical lithography system are shown in Fig. 7. Top row (from left to right) shows: the initial mask pattern and the corresponding output aerial image, with output pattern error of 1352.4. Middle row (from left to right) shows: the optimized binary mask using the algorithm depicted in Table 1 based on thin-mask approximation and the corresponding output aerial image, with output pattern error of 1135.9. Bottom row (from left to right) shows: the optimized binary mask based on BL model and the corresponding output aerial image, with output pattern error of 1089.6. Black and white represent 0 and 1 respectively. It is shown that optimization of the binary mask based on thin-mask approximation reduces the output pattern error by 16.0%. On the other hand, algorithm based on BL model reduces the output pattern error by 19.4%. Figure 8 illustrates the intersections of the aerial images on the $48th$ row. The solid line, dashed line and dotted line represent the intersections corresponding to the initial mask, the OPC based on the thin-mask assumption, and OPC based on the thick mask assumption, respectively. As shown in Fig. 5 and Fig. 7, the proposed binary mask optimization algorithm effectively reduces the output pattern errors and obtains more desirable aerial images. The performance differences between optimizing mask based on thin-mask

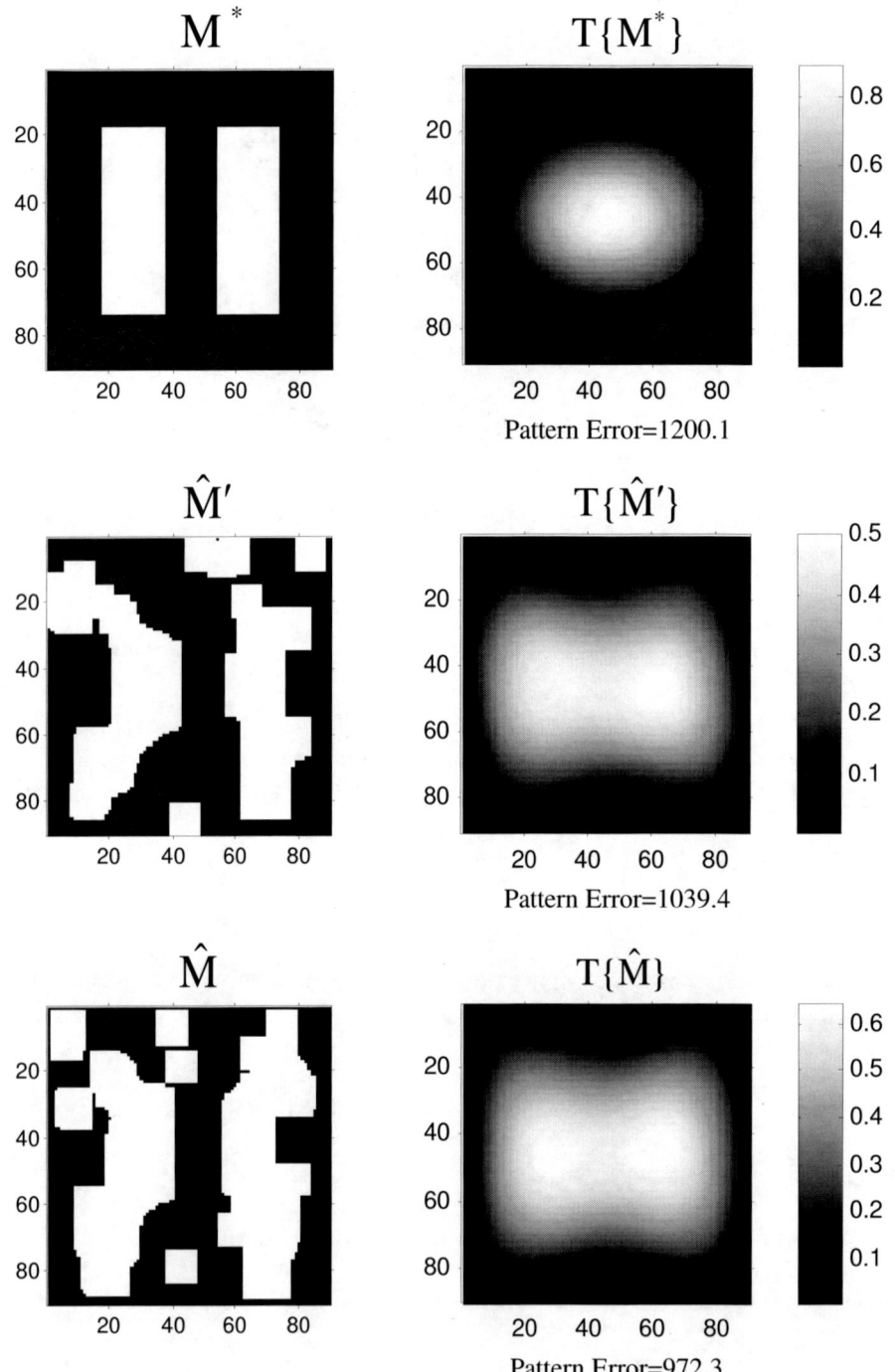

Figure 5. Binary mask optimization based on BL model for the first type of coherent optical lithography system. $\lambda = 248nm$. Top row (from left to right) shows: the initial mask pattern and the corresponding output aerial image. Middle row (from left to right) shows: the optimized binary mask based on thin-mask approximation and the corresponding output aerial image. Bottom row (from left to right) shows: the optimized binary mask based on BL model and the corresponding output aerial image. Black and white represent 0 and 1 respectively.

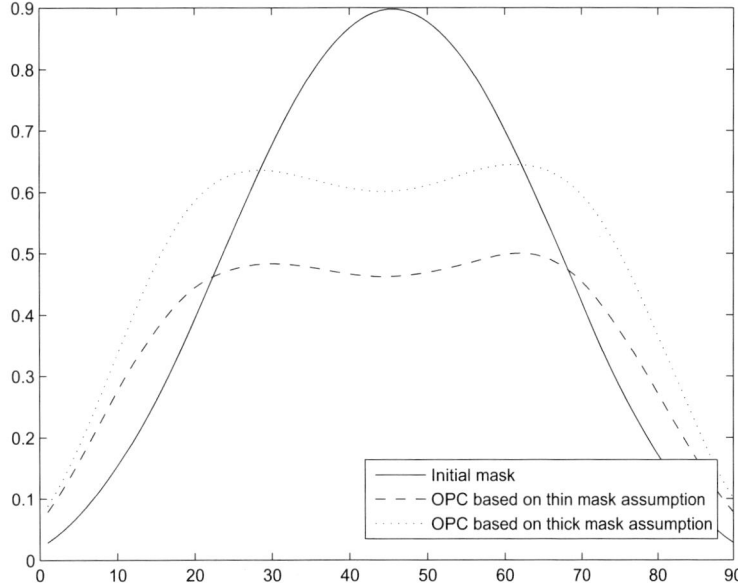

Figure 6. Intersections of the aerial images shown in Fig. 5 on the 45*th* row.

approximation and BL model show the necessity of the proposed algorithms taking into account the thick-mask effect. These results are consistent with those obtained in other simulations with different desired patterns we have ran.

7. CONCLUSION

This paper studies binary mask optimization for model-based forward lithography taking into account the thick-mask effects under coherent illumination. The BL model is applied to evaluate the near field of the thick mask. Based on this model, the binary mask optimization algorithm is proposed for two typical kinds of optical lithography systems. Topological constraints are applied in the optimization framework to limit the minimum feature size on the mask. Simulations illustrate that our approach is effective and practical.

8. ACKNOWLEDGEMENTS

We wish to thank Christof Krautschik, Yan Borodovsky and the TCAD group at the Intel corporation for their comments and support.

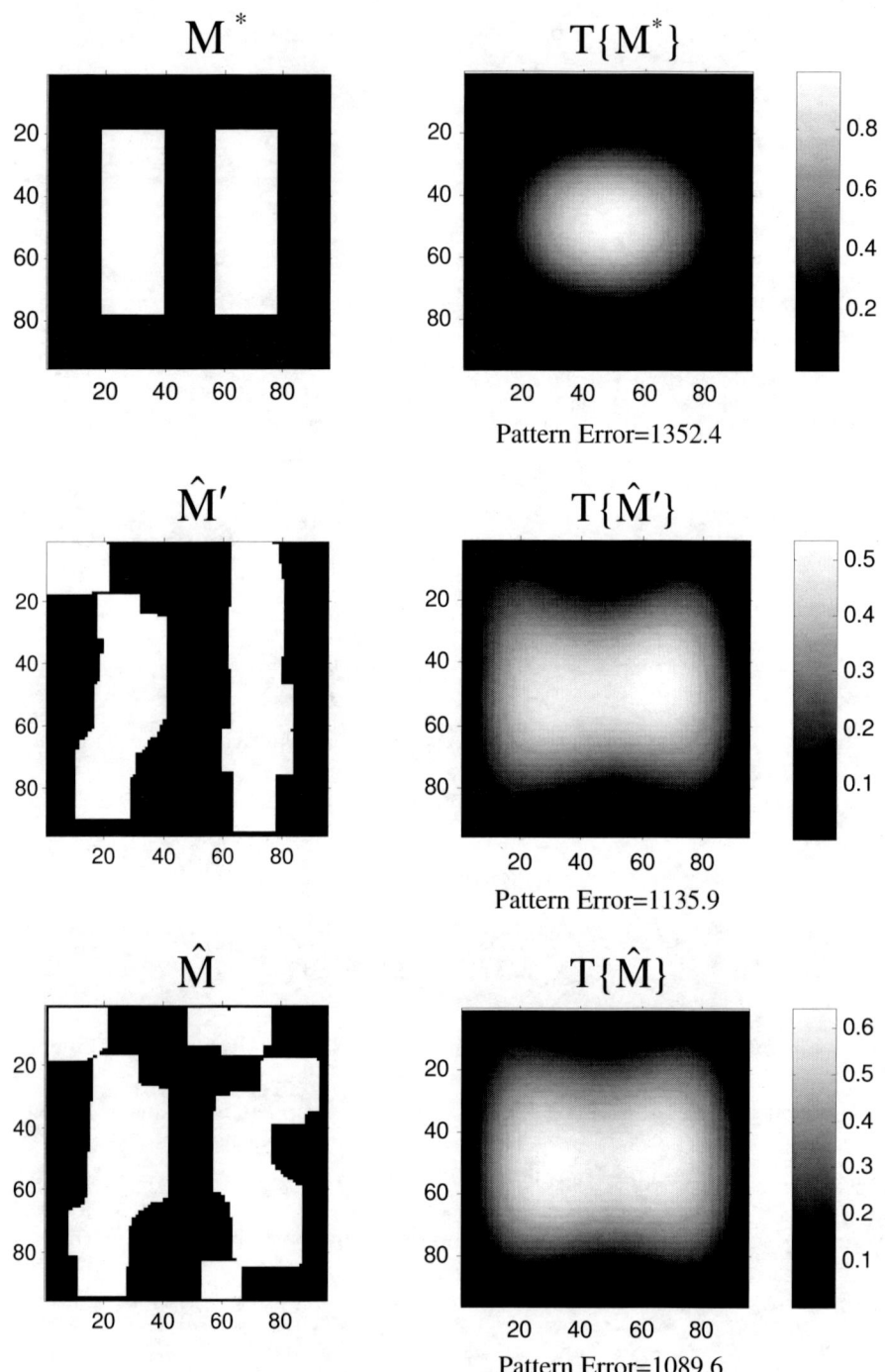

Figure 7. Binary mask optimization based on BL model for the second type of coherent optical lithography system. $\lambda = 193nm$. Top row (from left to right) shows: the initial mask pattern and the corresponding output aerial image. Middle row (from left to right) shows: the optimized binary mask based on thin-mask approximation and the corresponding output aerial image. Bottom row (from left to right) shows: the optimized binary mask based on BL model and the corresponding output aerial image. Black and white represent 0 and 1 respectively.

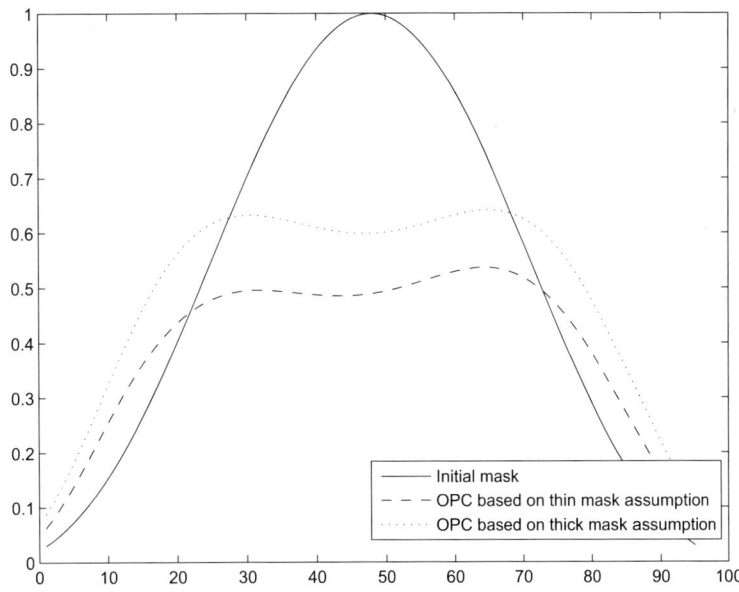

Figure 8. Intersections of the aerial images shown in Fig. 7 on the 48*th* row.

REFERENCES

1. A. K. Wong, *Resolution enhancement techniques*, vol. 1, SPIE Press, 2001.

2. S. A. Campbell, *The science and engineering of microelectronic fabrication*, Publishing House of Electronics Industry, Beijing, China, 2nd ed., 2003.

3. F. Schellenberg, "Resolution enhancement technology: The past, the present, and extensions for the future, optical microlithography," in *Proc. SPIE* **5377**, pp. 1–20, May 2004.

4. F. Schellenberg, *Resolution enhancement techniques in optical lithography*, SPIE Press, 2004.

5. L. Liebmann, S. Mansfield, A. Wong, M. Lavin, W. Leipold, and T. Dunham, "Tcad development for lithography resolution enhancement," *IBM Journal of Research and Development* **45**, pp. 651–665, Sep. 2001.

6. S. Sherif, B. Saleh, and R. Leone, "Binary image synthesis using mixed integer programming," *IEEE Trans. on Image Proc.* **4**, pp. 1252–1257, Sep. 1995.

7. Y. Liu and A. Zakhor, "Binary and phase shifting mask design for optical lithography," *IEEE Trans. on Semiconductor Manufacturing* **5**, pp. 138–152, May 1992.

8. Y. C. Pati and T. Kailath, "Phase-shifting masks for microlithography: Automated design and mask requirements," *Optical Society of America* **11**, pp. 2438–2452, Sep. 1994.

9. A. Erdmann, R. Farkas, T. Fuhner, B. Tollkuhn, and G. Kokai, "Towards automatic mask and source optimization for optical lithography," *Optical Microlithography, Proc. SPIE* **5377**, pp. 646–657, 2004.

10. L. Pang, Y. Liu, and D. Abrams, "Inverse lithography technology (ilt): What is the impact to the photomask industry?," in *Proc. SPIE* **6283**, May 2006.

11. Y. Granik, "Illuminator optimization methods in microlithography," in *Optical Microlithography, Proc. SPIE* **5524**, pp. 217–229, Oct. 2004.

12. A. Poonawala and P. Milanfar, "Mask design for optical microlithography - an inverse imaging problem," *IEEE Trans. on Image Proc.* **16**, pp. 774–788, Mar. 2007.

13. X. Ma and G. R. Arce, "Generalized inverse lithography methods for phase-shifting mask design," in *Optical Microlithography XX, Proc. SPIE* **6520**, pp. 65200U, (San Jose, CA), Mar. 2007.

14. X. Ma and G. R. Arce, "Generalized inverse lithography methods for phase-shifting mask design," *Optics Express* **15**, pp. 15066–15079, 2007.

15. X. Ma and G. R. Arce, "Binary mask optimization for inverse lithography with partially coherent illumination," in *Proc. SPIE* **7140**, pp. 71401A, (Taiwan), Nov. 2008.

16. X. Ma and G. R. Arce, "Binary mask optimization for inverse lithography with partially coherent illumination," *J. Opt. Soc. Am.* **25**(12), 2008.

17. X. Ma and G. R. Arce, "Psm design for inverse lithography using illumination with samll partial coherence factor," in *Proc. SPIE* **7274**, (San Jose, CA), Feb. 2009.

18. X. Ma and G. R. Arce, "Psm design for inverse lithography with partially coherent illumination," *Optics Express* **16**(24), pp. 20126–20141, 2008.

19. X. Ma and G. R. Arce, "Pixel-based simultaneous source and mask optimization," *Optics Express* **17**(7), pp. 5783–5793, 2009.

20. J. Tirapu-Azpiroz, P. Burchard, and E. Yablonovitch, "Boundary layer model to account for thick mask effects in photolithography," in *Optical Microlithography, Proc. SPIE* **5040**, pp. 1611–1619, 2003.

21. J. Tirapu-Azpiroz and E. Yablonovitch, "Fast evaluation of photomask near-fields in sub-wavelength 193nm lithography," in *Optical Microlithography, Proc. SPIE* **5377**, pp. 1528–1535, (Bellingham, WA), 2004.

22. A. Wong, *Rigorous three-dimensional time-domain finite difference electromagnetic simulation.* PhD thesis, University of California, Berkeley, 1994.

23. A. Wong and A. R. Neureuther, "Mask topography effects in projection printing of phase shift masks," *IEEE Trans. Electron Devices* **41**, pp. 895–902, 1994.

24. C. Pierrat, A. Wong, and S. Vaidya, "Phase-shifting mask topography effects on lithographyic image quality," in *Tech. Dig. - Int. Electron Devices Meet.*, pp. 53–56, 1992.

25. C. M. Yuan, "Calculation of one-dimension lithographyic aerial images using the vector theory," *IEEE Trans. Electron Devices* **40**(1604), 1993.

26. K. Lucas, H. Tanabe, and A. J. Strojwas, "Efficient and rigorous three-dimensional model for optical lithography simulation," *J. Opt. Soc. Am.* **A13**(2187), 1996.

27. A. Erdmann, P. Evanschitzky, G. Citarella, T. Fühner and P. D. Bisschop, "Rigorous mask modeling using waveguide and FDTD methods: An assessment for typical hyper NA imaging problems," in *Photomask and Next-Generation Lithography Mask Technology XIII, Proc. of SPIE* **6283**(628319), (Yokohama, Japan), Apr. 2006.

28. K. Adam and A. R. Neureuther, "Domain decomposition methods for the rapid electromagnetic simulation of photomask scattering," *Journal of Microlithography, Microfabrication, and Microsystems* **1**(253), 2002.

29. B. E. A. Saleh and M. Rabbani, "Simulation of partially coherent imagery in the space and frequency domains and by modal expansion," *Applied Optics* **21**, Aug. 1982.

30. M. Born and E. Wolfe, *Principles of optics*, Cambridge University Press, United Kingdom, 1999.

31. R. Wilson, *Fourier Series and Optical Transform Techniques in Contemporary Optics*, John Wiley and Sons, New York, 1995.

32. B. Salik, J. Rosen, and A. Yariv, "Average coherent approximation for partially cohernet optical systems," *J. Opt. Soc. Am. A* **13**, Oct. 1996.

33. L. Lam, S. W. Lee, and C. Y. Suen, "Thinning methodologies-a comprehensive survey," *IEEE Trans. Pattern Anal. Mach. Intell.* **14**(9), pp. 869–885, 1992.

34. P. Yu and D. Z. Pan, "Tip-opc: a new topological invariant paradigm for pixel based optical proximity correction," in *Proc. ACM/IEEE International Conference on Computer-Aided Design (ICCAD)*, (San Jose, CA), Nov. 2007.

Improvement in Process Window Aware OPC

Xiaohai Li[a], Yasushi Kojima[b], Hironobu Taoka[b], Akemi Moniwa[b], Matt St. John[a], Yang Ping[a],
Randall Brown[a], Robert Lugg[a], Sooryong Lee[a]

[a]Synopsys Inc., 2025 NW Cornelius Pass Road, Hillsboro, OR 97124, USA;
[b]Renesas Technology Co., 4-1, Mizuhara, Itami-shi, Hyogo, 664-0005, Japan

ABSTRACT

In this paper, we present some important improvements on our process window aware OPC (PWA-OPC). First, a CD-based process window checking is developed to find all pinching and bridging errors; Secondly, a rank ordering method is constructed to do process window correction; Finally, PWA-OPC can be applied to selected areas with different specifications for different feature types. In addition, the improved PWA-OPC recipe is constructed as sequence of independent modules, so it is easy for users to modify its algorithm and build original IPs.

Keywords: optical proximity correction (OPC), process window, process window aware OPC (PWA-OPC)

1. INTRODUCTION

The use of optical lithography for 32nm node and beyond requires hyper-NA immersion optics in combination with advanced illumination and mask technology, which, however, leads to a significant shrinkage of tolerable lithography process window.[1] This is exacerbated by the fact that 1.35-NA, the limit of water-based immersion lithography, has a focus window of less than 150nm. The exposure dose window is also decreasing because contrast decreases as semiconductor manufactures attempt to maximize the benefit of modern scanners. As a result, a small process variation (focus and exposure dose variation) may cause some catastrophic failures at these more advanced technology nodes. Traditional Optical Proximity Correction (OPC) only uses nominal model (nominal focus and nominal exposure dose), which may result in an unacceptably small process window. Therefore process window aware OPC (PWA-OPC) is becoming more and more important to improve process robustness and enhance yield.

In our previous study[2], we have demonstrated the effectiveness of PWA-OPC at full chip scale. In this method, traditional model-based OPC with a nominal model is applied to the pattern for the first several iterations; then the process window checking and correction are added to find and fix any unacceptable edge placement error (EPE) under all the models being considered. During the process window checking, if a violation of EPE limits with any model is found, a constraint is added. During process window correction, a cost function comprised of the weighted sum of constraints is minimized to find the best possible edge position. Figure 1 illustrates the flow diagram of this PWA-OPC method.

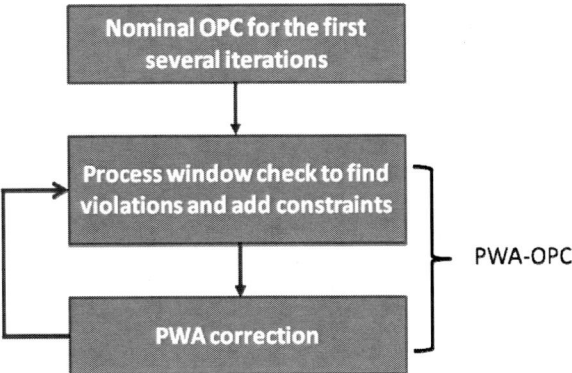

Figure 1. Flow diagram of PWA-OPC.

Optical Microlithography XXIII, edited by Mircea V. Dusa, Will Conley, Proc. of SPIE Vol. 7640,
76402O · © 2010 SPIE · CCC code: 0277-786X/10/$18 · doi: 10.1117/12.849904

In this paper, we present several important improvements on this PWA-OPC approach in terms of quality and flexibility. The remainder of this paper is organized as follows: Section 2 describes the new improvements, which include a CD-based process window checking to find all pinching or bridging errors; a novel rank ordering method for doing process window correction; the flexibility of applying PWA-OPC at selected areas with different requirements for feature types. Section 3 demonstrates the effectiveness of the improved PWA-OPC by applying it to some test patterns. And finally we conclude in Section 4.

2. IMPROVEMENTS ON PWA-OPC

2.1 CD-based process window checking

PWA-OPC usually involves setting up EPE limits for all applicable models. By trading off the EPE at the nominal condition, the printing fidelity is improved across process variations. As showed in fig. 2, simulated resist contours from all models for edge A are required to fall in the region between the outer EPE limit A and the inner EPE limit A to prevent pinching or bridging errors. The EPE tolerance can be directly specified by users or determined according to user input specifications for minimum space and width. In the latter case, the distance between the outer EPE limit A and the outer EPE limit B, or the inner EPE limit A and the inner EPE limit C is equal to the minimum required space or width respectively. In our PWA-OPC, the worst EPE on edge A is then calculated using all process models. If it falls outside the EPE limits, a constraint is added at the corresponding location. Later the process window correction takes this constraint into account and attempts to bring the contours into the specification limits.

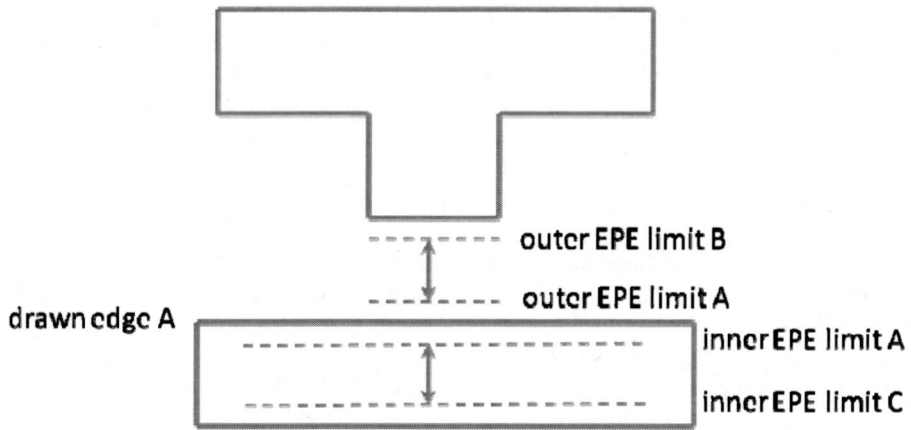

Figure 2. In PWA-OPC, the EPE limits are set up to prevent pinching and bridging errors.

However, in some cases such EPE-based process window checking can be too conservative. Figure 3 shows that even the contour for edge A is outside the outer EPE limit A, it is still possible to meet the minimum space requirement. Therefore given user input specifications for minimum space and width, an alternative way for finding pinching and bridging errors is to directly base on the actual contour space and width. On a segment, the minimum contour space or width is measured for all models. In our implementation, any-angle contour space or width check is supported as showed in Fig. 3. If the measured contour space or width from any model violates the user specifications, a constraint is then added for that segment. Such CD-based process window checking usually requires more model simulations than the EPE-based method, but has the advantage of achieving better correction results, e.g., better nominal EPE, or better chance to resolve the conflicts between minimum space and width requirements for the same segment.

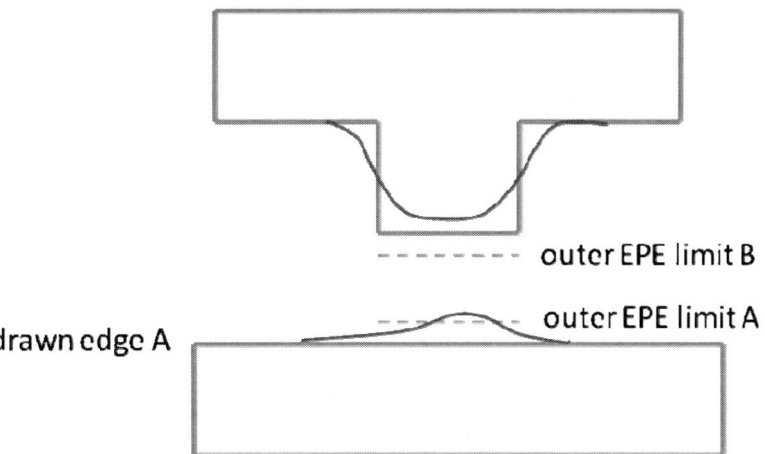

Figure 3. The contour for edge A can exceed the outer EPE limit A, but the contour space is still within the specification for the minimum space.

2.2 Rank ordering method for process window correction

After constraints are added, in our previous paper we demonstrate a correction method by minimizing a cost function. By placing a weight on the error for each constraint, the cost function becomes the weighted sum of all errors. In this paper, we present a different correction method based on priority. Considering the situation where users are in favor of preventing pinching errors since it causes more detrimental effects, assigning weights to different requirements may not be a trivial task. Instead anther way is to prioritize those requirements. If requirement A has higher priority than requirement B, we always try to achieve requirement A before requirement B. We do not even think about requirement B before requirement A is achieved.

2.3 Reduction of PWA-OPC areas

Compared to the traditional OPC with nominal model, PWA-OPC is usually computational expensive. To reduce the computational cost, it is often desirable to apply PWA-OPC when necessary. There are numerous metrics that can be used to identify a layout region that should be corrected using PWA-OPC, such as feature widths less than a specified length, image contrast less than a specified intensity, edge shift of more than a specified value for a test PW condition; nominal OPC unable to correct all values to a "satisfactory" result. Some of these metrics are rule based; others are model based. Furthermore, users may have different requirements for different feature types. With the highly-programmable interface provided by Proteus OPC from Synopsys, all these can be easily integrated into a OPC recipe by users.

3. RESULTS

We compare our PWA-OPC with nominal OPC on a test pattern that is generated to mimic a metal layer design. The PWA-OPC uses CD-based processing window checking and rank ordering method for correction. The user input specifications for minimum space and width are 65 nm and 50 nm respectively, and they are set to have the same priority. Nominal model and one defocus model are used for OPC. Notice that in real practice users are allowed to pick the desired number of models they want. The defocus model used here is probably outside the corners of the actual process window in a real process. Nevertheless, this experiment is aimed to demonstrate the benefits of our PWA-OPC.

Figure 4 depicts all the pinching and bridging errors from the lithographic check for nominal OPC. It can be seen that there are quite a few pinching and bridging errors. From Fig. 4(a), we can even see several catastrophic defects where the printed lines are completely broken. In comparison, Figure 5(a) shows that by using PWA-OPC, the number of pinching errors is dramatically reduced, and the worst pinching error is also mitigated. Figure 6 is an example where using nominal OPC there is a complete pinching, while using PWA-OPC both the minimum width and space specifications can be met. We also notice that using PWA-OPC the number of bridging errors is slightly increased as showed in Fig. 5(b), however the worst bridging error and the average space that are outside the minimum space specification are improved. The increased number of bridging errors arises from the balance between the minimum space requirements and the minimum width requirements.

To demonstrate how the priority setting affects the correction results when using rank ordering method in our PWA-OPC, two additional runs are performed. In Fig. 7, the minimum width requirement takes the higher priority than the minimum space requirement, the lithography checking shows that the number of pinching errors is further reduced compared to fig. 5(a) at the expense of more bridging errors than in Fig. 5(b). The reverse trend occurs in Fig.8 where the minimum space requirement has the higher priority. Figure 9 illustrates an example how the simulated resist contours after PWA-OPC are affected by the priority setting.

Figure 10 shows an example where the CD-based process window checking yields better results than the EPE-based checking. In this example, the middle line has large space on its right, by using the CD-based process window checking, the PWA-OPC automatically shift the contour from the defocus model slightly to the right so that both the minimum space and width requirements can be met.

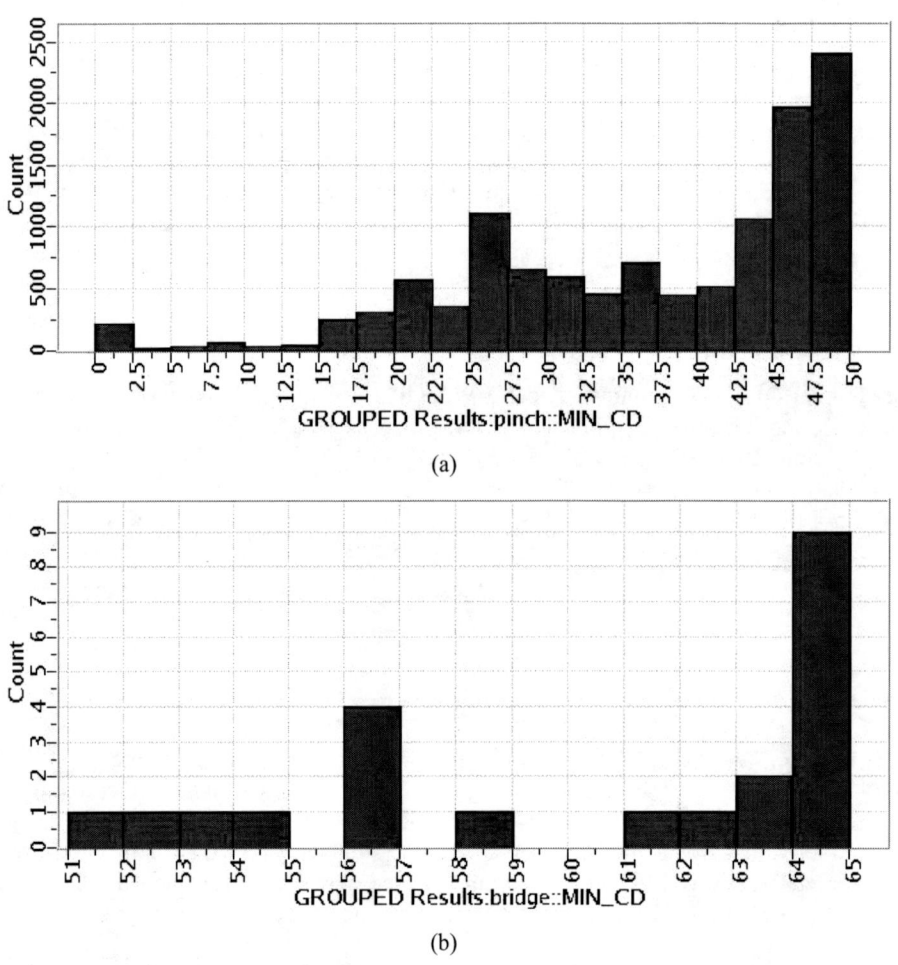

(a)

(b)

Figure 4. Histogram of pinching and bridging errors after nominal OPC.

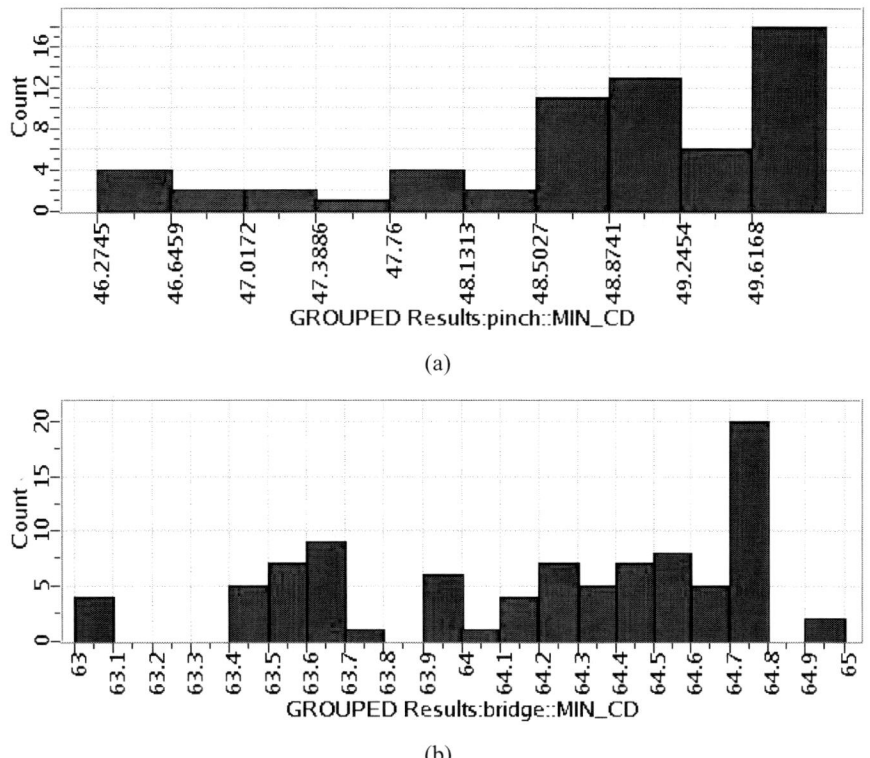

(a)

(b)

Figure 5. Histogram of pinching and bridging errors after PWA-OPC.

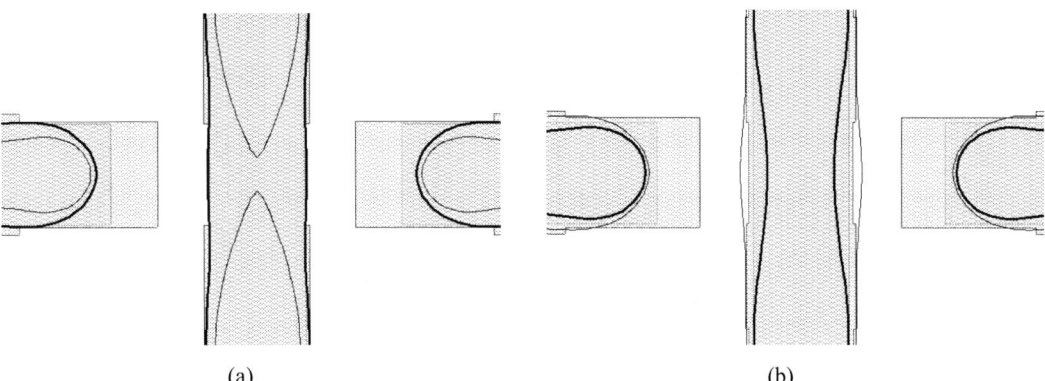

(a)   (b)

Figure 6. (a) There is a complete pinching using nominal OPC. (b) At the same location, both contour space and width are within specification using PWA-OPC.

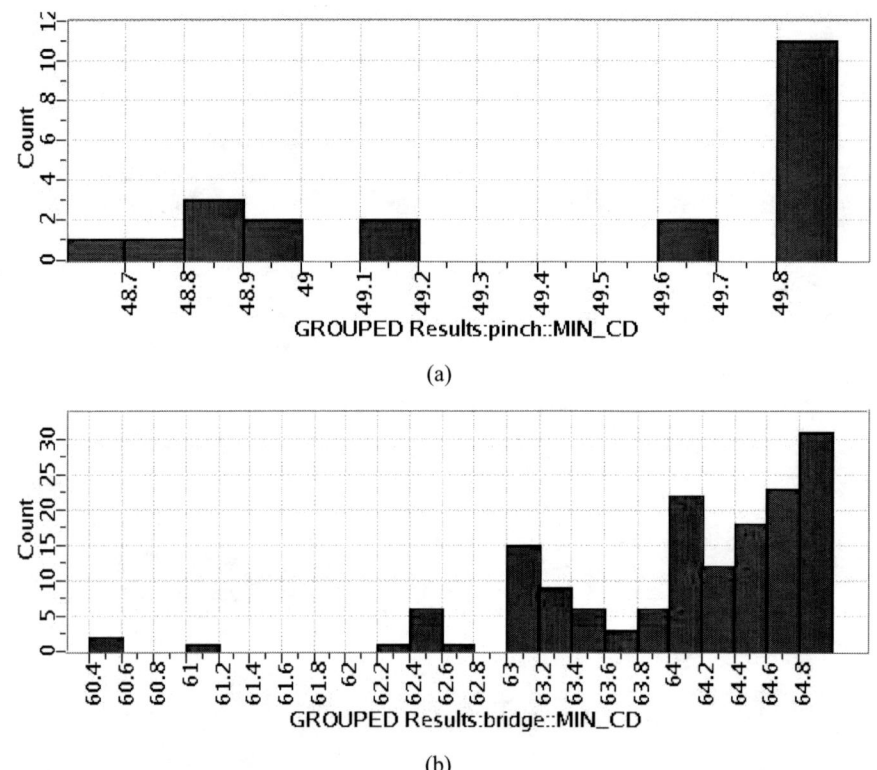

(a)

(b)

Figure 7. Histogram of pinching and bridging errors after PWA-OPC. The minimum width requirement has higher priority than the minimum space requirement.

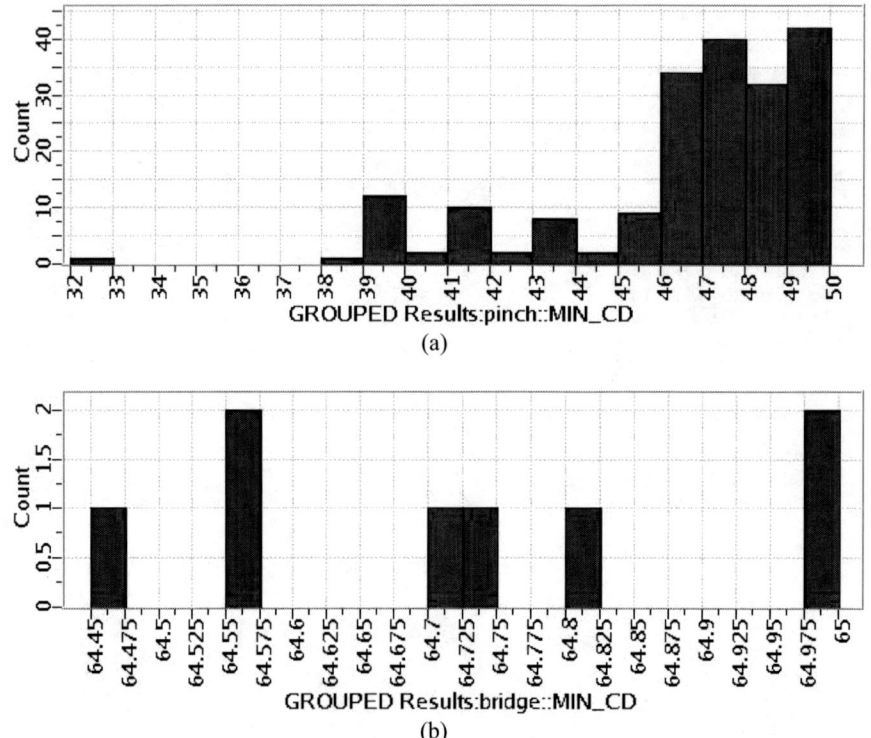

(a)

(b)

Figure 8. Histogram of pinching and bridging errors after PWA-OPC. The minimum space requirement has higher priority than the minimum width requirement.

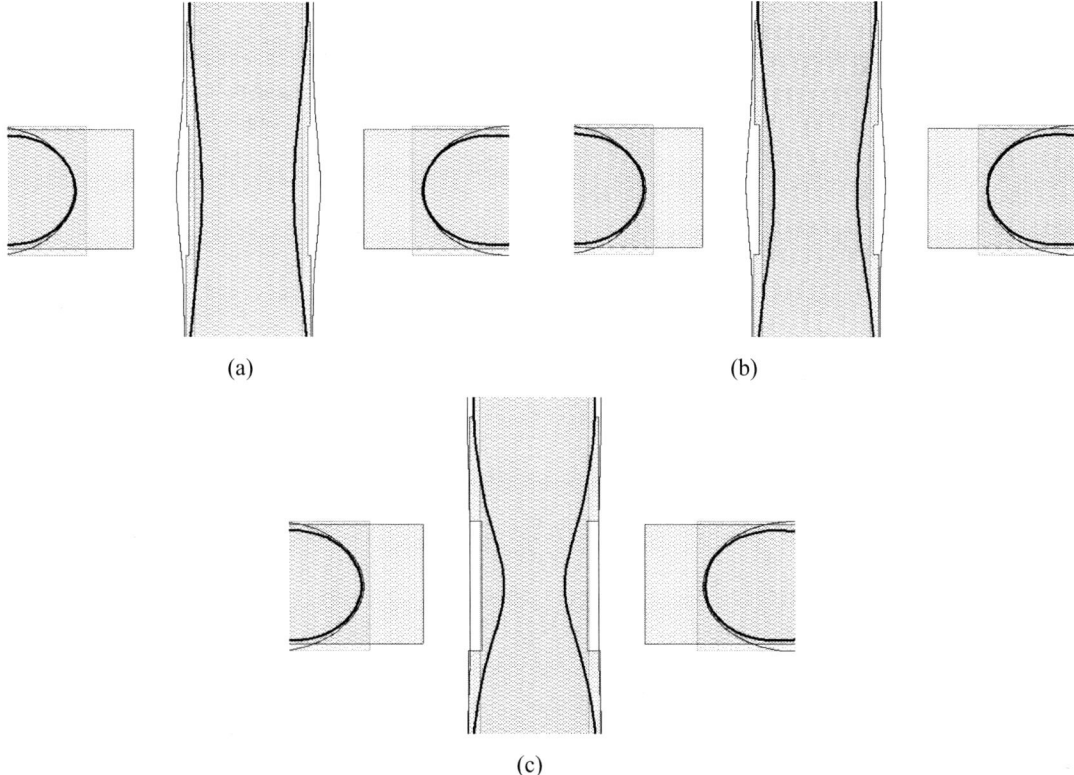

Figure 9. Contours are affected by the priority setting. For the middle feature, (a) if the minimum space requirement and minimum width requirement have the same priority, both pinching and bridging errors exist; (b) if the minimum width requirement has the higher priority, pinching error is eliminated and bridging error remains; (c) if the minimum space has the higher priority, bridging error is eliminated and pinching error remains.

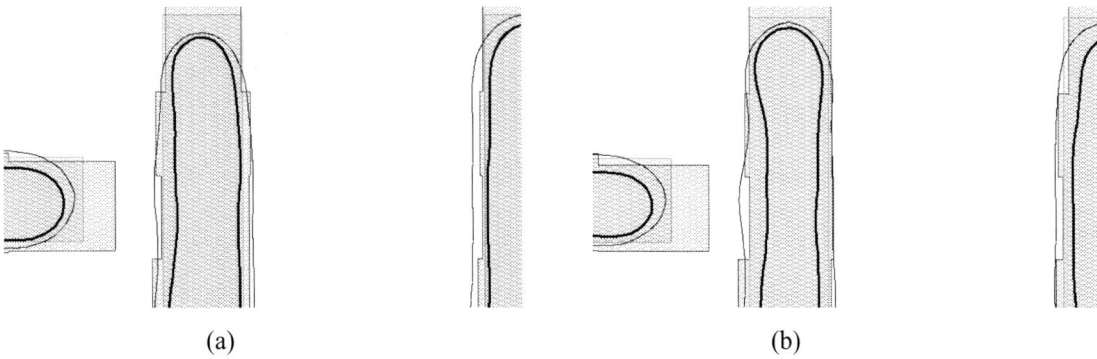

Figure 10. (a) CD-based process window checking is used. (b) EPE-based process window checking is used. For middle feature, using CD-based process window checking, both contour space and width are within the specifications.

4. CONCLUSIONS

In conclusion, in this paper we have presented several import improvements on our previous process window aware OPC method in terms of quality and flexibility. By applying this improved PWA-OPC method to a test pattern, we have demonstrated its benefits over nominal OPC. CD-based process window checking can better address the conflicts between different requirements; rank ordering method for correction makes the correction results more controllable and predictable. In addition, the improved PWA-OPC recipe is constructed as sequence of independent modules, so users can easily modify its algorithm and build original IPs.

REFERENCES

[1] Zhang, Q., Yan, Q., Zhang, Y. and Lucas, K., "Continuous process window modeling for process variation aware OPC and lithography verification," Proc. SPIE 6925, 69251A (2008).

[2] Lugg, R., St. John, M., Zhang, Y., Yang, A. and Adrichem, P. V., "Full-chip process window aware OPC capability assessment," Proc. SPIE 6730, 67302U (2007).

A Non-Delta-Chrome OPC Methodology for Process Models with Three-Dimensional Mask Effects

Philip C. W. Ng,[a] Kuen-Yu Tsai,[a*] Chih-Hsien Tang,[a,b] and Lawrence Melvin[c]

[a]National Taiwan University, Department of Electrical Engineering, Taipei 10617, Taiwan
[b]Synopsys Taiwan Ltd., Hsinchu 30288, Taiwan
[c]Synopsys, Inc. Hillsboro OR 97124, United State

ABSTRACT

Delta-chrome optical proximity correction (OPC) has been widely adopted in lithographic patterning for semiconductor manufacturing. During the delta-chrome OPC iteration, a predetermined amount of chrome is added or subtracted from the mask pattern. With this chrome change, the change of exposure intensity error (IE) or the change of edge placement error (EPE) between the printed contour and the target pattern is then calculated based on standard Kirchhoff approximation. Linear approximation is used to predict the amount of the proper chrome change to remove the correction error. This approximation can be very fast and effective, but must be performed iteratively to capture interactions between chrome changes. As integrated circuit (IC) design shrinks to the deep sub-wavelength regime, previously ignored nonlinear process effects, such as three-dimensional (3D) mask effects and resist development effects, become significant for accurate prediction and correction of proximity effects. These nonlinearities challenge the delta-chrome OPC methodology. The model response to mask pattern perturbation by linear approximation can be readily computed but inaccurate. In fact, computation of the mask perturbation response becomes complex and expensive. A non-delta-chrome OPC methodology with IE-based feedback compensation is proposed. It determines the amount of the proper chrome change based on IE without intensive computation of mask perturbation response. Its effectiveness in improving patterning fidelity and runtime is examined on a 50-nm practical circuit layout. Despite the presence and the absence of nonlinear 3D mask effects, our results show the proposed non-delta-chrome OPC outperforms the delta-chrome one in terms of patterning fidelity and runtime. The results also demonstrate that process models with 3D mask effects limit the use of delta-chrome OPC methodology.

Keywords: Optical proximity correction; three-dimensional mask effects; nonlinearity; feedback compensation.

1. INTRODUCTION

Traditionally, process models used for optical proximity correction (OPC) are approximately linear or quadratic to the mask pattern. A property of approximately linear or quadratic process models is that the model response to some tiny change in the mask pattern can be readily computed [1]–[3], without much computational effort. This makes the so-called delta-chrome OPC methodology [4]–[9] possible which has been widely adopted in lithographic patterning. During the delta-chrome OPC iteration, a predetermined amount of chrome is added or subtracted from the mask pattern. The exposure intensity error (IE) or the edge placement error (EPE) between the printed contour and the target pattern is then calculated with this chrome change based on standard Kirchhoff approximation, and that IE or EPE is used to calculate the change of IE or the change of EPE with respect to the change in chrome. This approximation can be very fast and efficient.

As integrated circuit (IC) design shrinks to the deep sub-wavelength regime, high-NA immersion lithography and more aggressive resolution enhancement techniques are widely adopted. In the mean while, OPC process models also have to take previously ignored non-linear effects into account, such as three-dimensional (3D) mask effects [10] and resist development effects [11]. The nonlinearity of such model components imposes a big challenge to the delta-chrome OPC methodology. The model response to mask pattern perturbation by linear approximation can be readily computed but inaccurate. In fact, computation of the mask perturbation response becomes complex and expensive.

---

*E-mail: kytsai@cc.ee.edu.tw, Tel: +886-2-33663689, Fax: +886-2-23671909

Optical Microlithography XXIII, edited by Mircea V. Dusa, Will Conley, Proc. of SPIE Vol. 7640,
76402P · © 2010 SPIE · CCC code: 0277-786X/10/$18 · doi: 10.1117/12.846687

A non-delta-chrome OPC methodology with IE-based feedback compensation is proposed. It determines the amount of the proper chrome change based on IE without intensive computation of mask perturbation response. Its effectiveness in improving patterning fidelity and runtime is examined on a 50-nm practical circuit layout comprising of seven critical layers with a minimum pitch size of 125 nm. Despite the presence and the absence of nonlinear 3D mask effects, our results show the proposed non-delta-chrome OPC outperforms the delta-chrome one in terms of patterning fidelity and runtime. The results also demonstrate that process models with 3D mask effects limit the use of delta-chrome OPC methodology.

2. OVERVIEW OF OPC PROCESS MODELING

Any OPC process models can be described as a function $T[\cdot]$,

$$W(x,y) = T\big[M(x,y)\big], \tag{1}$$

which transfers the mask transmission function $M(x,y)$ to the printed resist contour function $W(x,y)$. The (x,y) in $M(x,y)$ and $W(x,y)$ denotes the locations on the mask and the wafer respectively.

2.1 Imaging Formation

The way used to computed the aerial image intensity $I(x,y)$ by the kernel decomposition [3] can be classified into the *sparse* simulation method [1]–[3] and the *dense* simulation method [12]–[14]. First method is used to fast compute the image intensity in the spatial domain based on point by point basic using lookup tables as

$$I(x,y) = \sum_{n=1}^{N} \sigma_n \big| K_n(x,y) \otimes M(x,y) \big|^2 . \tag{2}$$

Second method optimized for circuit layout with high vertex density is used to efficiently compute the image intensity of full field in the frequency domain using fast Fourier transform (FFT) as

$$I(x,y) = \sum_{n=1}^{N} \sigma_n \big| \mathcal{F}\{\mathcal{F}^{-1}\{K_n(x,y)\} \mathcal{F}^{-1}\{M(x,y)\}\} \big|^2 , \tag{3}$$

where $K_n(x,y)$'s are the convolution kernel functions, which optimally approximate lithography system information with the effects of resist and etch. The runtime of this approximate is closely proportional to the number of kernels used. $\mathcal{F}\{\cdot\}$ and $\mathcal{F}^{-1}\{\cdot\}$ are the Fourier and inverse Fourier transform operators.

2.2 Resist Development

In general, resist model used for OPC is simply constant or variable threshold to rapidly delineate the printed resist contour on a wafer. The printed resist contour function from Eq. (1) can be rewritten as

$$W(x,y) = R\big[I(x,y)\big], \tag{4}$$

where the function $R[\cdot]$ has the following form for constant threshold resist (CTR) model,

$$R\big[I(x,y)\big] = \begin{cases} 1 & I(x,y) \geq I_T \\ 0 & I(x,y) < I_T \end{cases} \tag{5}$$

3. ANALYSIS OF DIFFICULTIES IN USE OF DELTA-CHROME OPC METHODOLOGY

Mask transmission function is required for the computation of the image intensity. This transmission function is conventionally based on Kirchhoff approximation. Since it assumes that the mask is infinitively flat, this approximation is sometimes called thin mask approximation. When the mask opening is significantly larger than the thickness of the absorber, the Kirchhoff approximation is reasonable. Any 3D mask effects can be neglected. In this case, the mask transmission function is approximate to the circuit layout $L(x, y)$ with assigned transmission and phase values. Hence, the mask perturbation response can be readily computed based on *sparse* simulation method in Eq. (2) as

$$I(x, y) + \Delta I(x, y) = \sum_{n=1}^{N} \sigma_n \left| K_n(x, y) \otimes \left[M(x, y) + \Delta M(x, y) \right] \right|^2 , \qquad (6)$$

where $\Delta M(x, y)$ is the perturbation transmission function with respect to the chrome change of the mask. It is proportional to the change of circuit layout, i.e. $\Delta M(x, y) \propto \Delta L(x, y)$.

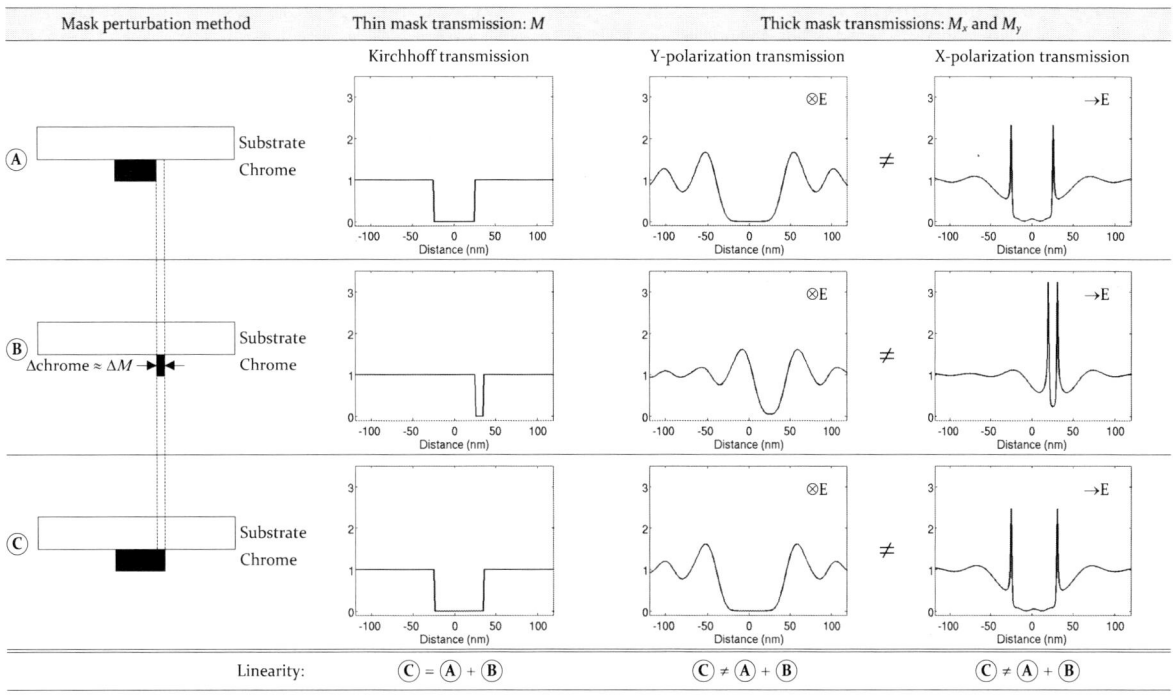

Fig. 1 Linearity of mask perturbation method.

However, as the feature size approaches the absorber thickness, electromagnetic field effects become significant contributions to the mask pattern. Kirchhoff approximation is no longer reasonable. Fig. 1 shows the linearity of mask perturbation method for the cases of thin mask transmission and thick mask transmission (also known as 3D mask transmission). By the Kirchhoff approximation, the mask transmission function $C = A + B$. However, the mask transmission function $C \neq A + B$ extracted from Sentaurus Lithography [15] by the finite-different time-domain (FDTD) method for solving Maxwell's equation. In addition, significant differences to the Kirchhoff mask transmission function as well as polarization effects are obvious. These differences have significant contribution on the correction result. Hence, the perturbation transmission function with respect to the chrome change of the mask is not linear to the change of circuit layout, i.e. $\Delta M_x(x, y) = f_x\left[\Delta L(x, y)\right]$ or $\Delta M_y(x, y) = f_y\left[\Delta L(x, y)\right]$. This implies that the model response to mask pattern perturbation by linear approximation such as MEEF or slope can be readily computed but inaccurate. MEEF is the mask error enhancement factor that is defined as

$$\text{MEEF} = \frac{\Delta \text{EPE}}{\Delta M(x, y)} , \qquad (7)$$

and the slope is defined as

$$slope = \frac{\Delta IE}{\Delta M(x,y)},$$ (8)

where ΔEPE is the change of edge placement error and ΔIE is the change of intensity error. In fact, computation of the mask perturbation response becomes complex and expensive because of nonlinearity. Hence, a new OPC methodology is necessary that does not rely on the computation of the mask perturbation response. But it still exhibits good correction convergence properties.

4. A PROPOSED NON-DELTA-CHROME OPC METHODOLOGY

Fig. 2 shows a standard delta-chrome OPC correction flow. The full-field intensity within the ambit is computed based on *dense* simulation method in Eq. (3). The EPE for each segment is then calculated on the target point from IE, where IE is the difference between the intensity and the desired value. In constant resist threshold model, the desired value is a predetermined constant threshold. If the EPE for any segment is not satisfied, the computation of the mask perturbation response to predict the amount of the proper chrome change is performed based on segment by segment basic. A predetermined amount of chrome is added or subtracted from the segment. It is performed iteratively to adjust the chrome change until the EPE is satisfied. After the chrome change for each segment is determined, the amount of the proper chrome change is then calculated. It is multiplied with a constant damping value or a variable damping value in order to prevent the divergence of correction. The variable damping value is defined according to the iteration. The mask pattern within the ambit is updated with those proper chrome changes. In addition, classical control theory can be used to optimize the convergence [6][7][16][17]. The correction process will stop if the convergence requirement or the predefined number of iterations is achieved. The total OPC runtime is approximately proportional to the number of iterations required.

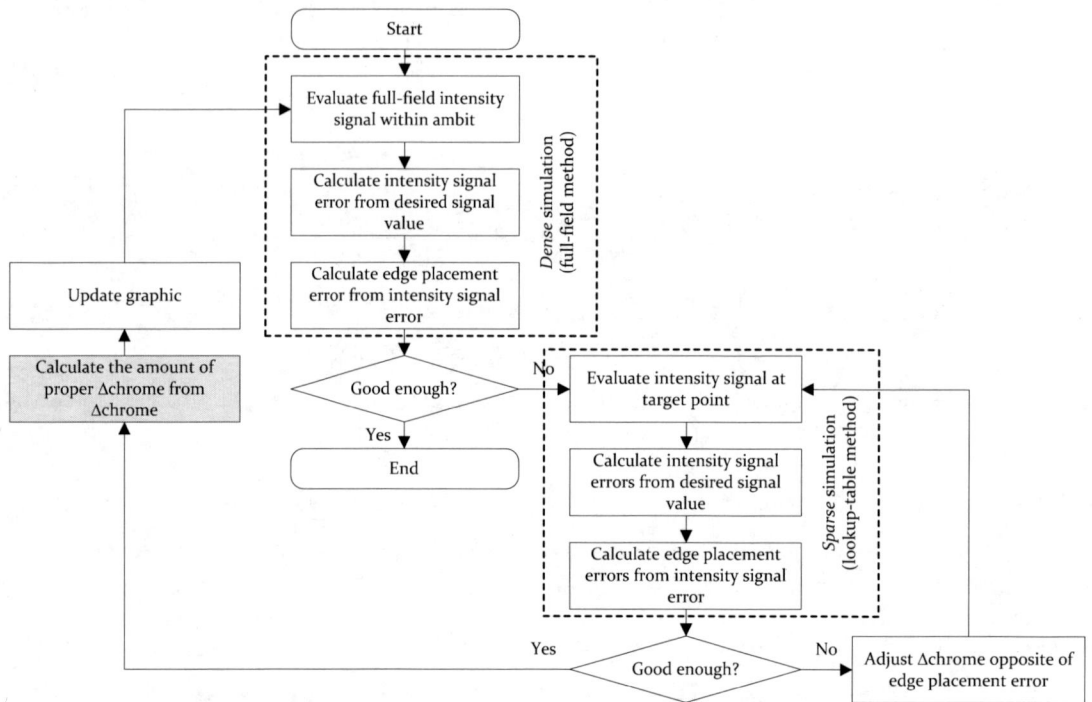

Fig. 2 A standard delta-chrome OPC correction flow.

Instead, a non-delta-chrome OPC methodology is proposed to determine the amount of the proper chrome change based on IE without intensive computation of the mask perturbation response as shown in Fig. 3. It is expected that this proposed methodology can possibly have a fast OPC runtime. The algorithm used to compute the amount of the proper chrome change is based on feedback control. The feedback controller can be a classical proportional (P), proportional-integral (PI), proportional-derivative (PD), or proportional-integral-derivative (PID) controller. After the IE for each segment is calculated, the amount of the proper chrome change can be computed as

$$u_k = u_{k-1} + K_p e_k + K_i \sum_{n=1}^{k} e_n + K_d \left(e_k - e_{k-1} \right) \tag{9}$$

where u_k is the amount of the proper chrome change for k iteration, e_k is the amount of IE for k iteration, K_p is the proportional parameter, K_i is the integral parameter, and K_d is the derivative parameter. A heuristic approach for tuning the K_p, K_i, and K_d parameters to achieve the convergence requirement has been proposed [17].

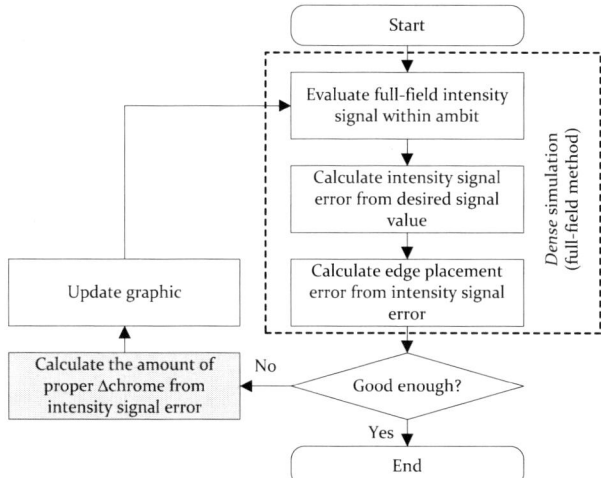

Fig. 3 A proposed non-delta-chrome OPC correction flow.

5. VALIDATION OF NON-DELTA-CHROME OPC METHODOLOGY

The proposed non-delta-chrome OPC methodology has been implemented in Proteus OPC engine [18]. The following simulations were performed on dual Intel Xeon 2.27-GHz central processing units (CPU) with the main memory of 24 gigabytes (GB). Two OPC process models generated from Progen [19] were examined. First was a 50-nm half-pitch-equivalent process model without nonlinear 3D mask effects. The optical system has a 1.1 NA, 193 nm wavelength, an outer sigma of 0.84 outer and an inner sigma of 0.64 inner. The constant resist threshold is 0.23. Second was a 40-nm half-pitch-equivalent process model with nonlinear 3D mask effects, which was calibrated based on IMEC data [10]. The optical system has a 1.2 NA c-Quad illumination, 193 nm wavelength, an outer sigma of 0.96 and an inner sigma of 0.71. The c-Quad light source is X-Y polarized. The constant resist threshold is 0.17. A simulation grid size of 0.25 nm, a segment length of 30 nm, 8 iterations, and EPE tolerance of ±2 nm were used for the correction. In the delta-chrome OPC, variable damping values of 0.6, 0.8, and 0.9 were used for first, second, and other iterations respectively. A PD controller was used for the proposed non-delta-chrome OPC. The P and D parameters were 140 and −20 respectively, which were heuristically tuned for correction convergence.

A 50-nm practical circuit layout comprising of seven critical layers was examined. The comparison of the OPC quality and the runtimes at different process models are shown in Table 1. "# Seg" columns denote the total number of segments assigned to each layer. "$\mu_{|EPE|}$" columns denote the mean of the EPE absolute value. "$\sigma_{|EPE|}$" columns denote the standard deviation of the EPE absolute value. "# OTS" columns denote the number of out-of-tolerance segments (where the EPE absolute value > 2 nm). "RT" columns denote the OPC runtime. "# OTS Reduc" columns denote the reduction of out-of-tolerance segment numbers. "RT Reduc" columns denote the OPC runtime reduction. The averages over the seven layers for the appropriate quantities are shown at different process models.

From the results, it is clear that the proposed non-delta-chrome OPC gives significant reduction in terms of runtime. It is about 23% runtime reduction. This is because the computation of mask perturbation response is not required. In the absence of 3D mask effects, non-delta-chrome OPC results are comparable with the delta-chrome OPC results. In fact, the non-delta-chrome OPC still guarantees almost the same standard deviation of |EPE|, although it increases the mean of |EPE| a little. For the fourth layer, the non-delta-chrome OPC gives a negative percentage of the out-of-tolerance segment numbers reduction but decreases the mean and standard deviation of |EPE|. However, there is no significant average reduction. In the presence of 3D mask effects, our results show the proposed non-delta-chrome OPC gives a good OPC quality. Better OPC quality can be seen as more advanced process model is used for the 50-nm circuit layout. However, the delta-chrome OPC failed to make a good correction results. This is due to the use of mask perturbation estimation by linear approximation in a nonlinear process model is no longer valid. The results demonstrate that the proposed non-delta-chrome OPC is indeed a promising method.

Table 1. OPC quality and runtimes comparison

		50-nm half-pitch-equivalent process model without 3D mask effects									
		Delta-chrome OPC				Non-delta-chrome OPC				# OTS Reduc (%)	RT Reduc (%)
Layer #	# Seg	$\mu_{|EPE|}$ (nm)	$\sigma_{|EPE|}$ (nm)	# OTS	RT (s)	$\mu_{|EPE|}$ (nm)	$\sigma_{|EPE|}$ (nm)	# OTS	RT (s)		
1:0	30702	0.32	0.35	8	36	0.32	0.34	7	31	0.00	13.89
2:0	59255	0.43	0.67	1559	58	0.42	0.70	997	52	0.95	10.34
3:0	14436	0.13	0.11	0	88	0.28	0.19	0	22	0.00	75.00
4:0	69894	0.57	0.80	2592	69	0.53	0.65	3235	55	−0.92	20.29
5:0	3448	0.12	0.11	0	33	0.24	0.17	0	16	0.00	51.52
6:0	26317	0.36	0.39	1	33	0.36	0.38	1	28	0.00	15.15
8:0	22641	0.43	0.41	1	31	0.45	0.38	4	28	−0.01	9.68
Average		0.34	0.41			0.37	0.40			0.00	27.98
		40-nm half-pitch-equivalent process model with 3D mask effects									
		Delta-chrome OPC				Non-delta-chrome OPC				# OTS Reduc (%)	RT Reduc (%)
Layer #	# Seg	$\mu_{|EPE|}$ (nm)	$\sigma_{|EPE|}$ (nm)	# OTS	RT (s)	$\mu_{|EPE|}$ (nm)	$\sigma_{|EPE|}$ (nm)	# OTS	RT (s)		
1:0	31096	26.0	81.8	31087	100	0.20	0.25	8	88	99.95	12.00
2:0	58866	1434.7	863.1	58524	132	0.20	0.24	5	103	99.41	21.97
3:0	14436	1964	0.0	14436	55	0.07	0.06	0	43	100.00	21.82
4:0	68938	16.4	72.7	64826	137	0.26	0.32	359	83	93.51	39.42
5:0	3448	1964	0.0	3448	36	0.07	0.05	0	33	100.00	8.33
6:0	26292	28.5	106.3	25781	61	0.18	0.17	0	54	98.06	11.48
8:0	22856	26.1	50.0	22841	86	0.15	0.19	15	75	99.87	12.79
Average		779.9	167.7			0.16	0.18			98.69	18.26

6. CONCLUSIONS

A proposed non-delta-chrome OPC methodology has been successfully implemented in Proteus OPC engine [15]. Its effectiveness in improving runtime and patterning fidelity with the presence and the absence of nonlinear 3D mask effects has been demonstrated. Despite the presence and the absence of nonlinear 3D mask effects, our results show the proposed non-delta-chrome OPC outperforms the delta-chrome one. The results also reveal that process models with 3D mask effects limit the use of delta-chrome OPC methodology. It can be concluded that the proposed methodology is indeed a promising method. Further study on resist development effects is currently ongoing.

ACKNOWLEDGEMENTS

The authors would like to acknowledge Synopsys for the funding support, the donation of software licenses, and providing the test patterns and model with 3D mask effects. Also thank Eason Su and Chadwick Lin from Synopsys for their helpful support and discussions.

REFERENCES

[1] N. Cobb and A. Zakhor, "Fast, low-complexity mask design," *Proc. SPIE* **2440**, 313–327 (1995).
[2] N. Cobb and A. Zakhor, "Fast sparse aerial image calculation for OPC," *Proc. SPIE* **2621**, 534–545 (1995).
[3] N. Cobb, "Fast optical and process proximity correction algorithms for integrated circuit manufacturing," Ph.D. dissertation, University of California at Berkeley (1998).
[4] N. Cobb and Y. Granik, "Model-based OPC using the MEEF matrix," *Proc. SPIE* **4889**, 1281–1292 (2002).

[5] Y. Granik and N. Cobb, "MEEF as a matrix," *Proc. SPIE* **4562**, 980–991 (2002).

[6] B. Painter, L. Melvin, and M. Rieger, "Classical control theory applied to OPC correction segment convergence," *Proc. SPIE* **5377**, 1198–1206 (2004).

[7] S. Choi, A. Je, J. Hong, M. Yoo and, J. Kong, "MEEF-based correction to achieve OPC convergence of low-k1 lithography with strong OAI," *Proc. SPIE* **6154**, 61540P (2006).

[8] Y. Chen, K. Wu, Z. Shi, and X. Yan, "A feasible model-based OPC algorithm using Jacobian matrix of intensity distribution functions," *Proc. SPIE* **6520**, 65204C (2007).

[9] P. Yu and D. Pan, "A novel intensity based optical proximity correction algorithm with speedup in Lithography simulation," *Proc. ICCAD*, 854–859 (2007).

[10] L. Melvin, T. Schmoeller, C. Kalus, and J. Li, "Three-dimensional mask effects in OPC process model development from first principle simulation," *Proc. SPIE* **6792**, 6792D7 (2008).

[11] Y. Fan, L. Zavyalova, Y. Zhang, C. Zhang, K. Lucas, B. Falch, E. Croffie, J. Li, L. Melvin, and B. Ward, "Resist development modeling for OPC accuracy improvement," *Proc. SPIE* **7274**, 727418 (2009).

[12] N. Cobb, "Flexible sparse and dense OPC algorithms," *Proc. SPIE* **5853**, 693–702 (2005).

[13] N. Cobb and Y. Granik, "Dense OPC for 65nm and below," *Proc. SPIE* **5992**, 599259 (2005).

[14] N. Cobb and D. Dudau, "Dense OPC and verification for 45nm," *Proc. SPIE* **6154**, 61540I (2006).

[15] Sentaurus Lithography is a commercial product of Synopsys Inc., http://www.synopsys.com.

[16] G. Franklin, J. Powell, and M. Workman, *Digital Control of Dynamic Systems* 3rd Ed., Addison Wesley, Menlo Park (1998).

[17] Y. Su, P. Ng, K. Tsai, and Y. Chen, "Design of automatic controllers for model-based OPC with optimal resist threshold determination for improving correction convergence," *Proc. SPIE* **6924**, 69243Z (2008).

[18] Proteus is a commercial product of Synopsys Inc., http://www.synopsys.com.

[19] Progen is a commercial product of Synopsys Inc., http://www.synopsys.com.

[20] International technology roadmap for semiconductors, "Lithography," Chap. in *ITRS 2008 Update*, ITRS (2008).

A New Etch-Aware After Development Inspection (ADI) Technique for OPC Modeling

Jing Xue [*a], Jason Huang [a], Aram Kazarian [a], Brad Falch [b]

[a]Synopsys Inc., 700 East Middlefield Road, Mountain View, CA 94043, USA
[b]Synopsys Inc., 1301 S MO Pac Expy, Austin, TX 78746

ABSTRACT

This paper presents a new etch-aware after development inspection (ADI) model with an inverse etch bias filter. We model the etch bias as a function of pattern geometry parameters, and we introduce it to the ADI model by means of an inverse bias matrix that works in conjunction with an ADI specification related matrix. The inverse bias filter tunes the ADI model to be highly correlated to the etch effects and provides simplified and designable inputs to the after etch inspection (AEI) model and hence improves its performance over the staged modeling flow. In addition, the inverse bias filter creates a model based rule table for design retargeting. Some of the etch effects are corrected by the inverse bias filter as the lithography model is calibrated, thus speeding up and simplifying the etch AEI model, while maintaining lithography ADI model with a good accuracy.

Keywords: Etch Bias, After Development Inspection (ADI), After Etch Inspection (AEI), Staged model, Retarget model, OPC modeling

1. INTRODUCTION

Two main processes, lithography and etch, have traditionally been subjected to Optical Proximity Correction (OPC) modeling. There is now a recognized need to extend this to the more general Process Proximity Correction (PPC), in order to address the critical patterning steps in the Integrated Circuit (IC) manufacturing. OPC as well as PPC require compact, tunable and accurate modeling of the processes in question. Lithography and etch therefore have to work closely to achieve the required Critical Dimension (CD) control targets. However, non-uniform etch bias, defined as the CD difference in resist patterns after development and in the related permanent patterns after etching, leads to difficulties in the traditional etch models and degrades the overall lithography-etch modeling capabilities. This problem becomes one of the main challenges in the OPC/PPC modeling, especially at advanced technology nodes where various complex patterning schemes are used to improve printing resolution [1] [2].

Nowadays, lumped, retargeted, and staged models are primarily used for etch modeling. Ignoring intermediate process steps, the lumped model generally uses optical and resist model components and Gaussian load kernels to directly predict the etched CD. Depending on the etch bias and the combined model form, the lumped model may distort the optical image surface and mutual coherence properties, and fail to preserve lithography process window information [2]. The retarget model has the advantage of one-stage correction flow while maintaining the lithographic model properties, however a retarget rule table based on the prior measurement data is required, and therefore model accuracy and fidelity are limited by the previously measured pattern structure and metrology quality [3]. The staged model considers lithography and etch separately, and allows tracking and verification of process behaviors at each stage [4] [5]. But the traditional staged model calibrates the ADI model without considering the AEI model properties. Since the resist measurement is generally noisy and the ADI model is the input to the etching model on the staged flow, the quality and accuracy of the etch model are limited by the traditional ADI model.

In this paper, we introduce a new etch aware lithography ADI model with an inverse bias filter. The etch bias is introduced into the ADI model by an inverse bias matrix and an ADI specification related matrix. The concept of the

---

* E-mail: Jing.Xue@synopsys.com, Telephone: 650 584 2893, Address: Synopsys Inc., 700E Middlefield Road, Mountain View, CA 94043

Optical Microlithography XXIII, edited by Mircea V. Dusa, Will Conley, Proc. of SPIE Vol. 7640,
76402Q · © 2010 SPIE · CCC code: 0277-786X/10/$18 · doi: 10.1117/12.846683

etch-aware lithography model is described in section 2. The implementation of the inverse bias filter is discussed in section 3. The conclusions are presented in section 4.

2. ETCH-AWARE LITHOGRAPHY MODEL CONCEPT

Semiconductor manufacturing includes a number of processes which involve complicated physical and chemical interactions. In many cases, rigorous first principle models that truly predict process behavior are still open research topics. On the other hand, first principle models, even when available, generally cannot be applied at full-chip scale because of their computational complexity. This is especially true in OPC/PPC applications where a computationally efficient model, or a semi-empirical model, is typically needed. OPC/PPC models must account for etch effects that occur due to structure etch and other additional etch steps following photolithography. These etch effects are determined by a complex series of reactions in the etch chamber and governed by etch recipes, which control plasma composition and density, electric field configuration, discharge energy, pressure, temperature, overall step timing and sequence, and other factors. The geometry of the actual device layout also significantly impacts the etch effects. Different feature sizes, line to space ratios and aspect ratios can cause local loading, shadowing, sidewall coating, and other effects, which result in varying etch bias across pattern geometry, layout context, and wafer topography. For the purpose of fast OPC/PPC simulations, we assume that the process recipe is repeatable, and that the aspect ratio of the trench reduces to an effective trench width at the wafer plane. The etch bias can then be represented by a series of pattern geometry parameters, e.g. line width, trench width, pattern density, polygon density, area density [6], etc. Eq. (1) shows an example where etch bias is based on a pattern geometry parameter x within m discrete layout contexts:

$$bias_i = a_{i0} + a_{i1}x + a_{i2}x^2 + \cdots + a_{ik}x^k \cdots + a_{in}x^n \qquad \text{Eq. (1)}$$

where a_{ik} with $i = 0, 1, \ldots, m$ and $k = 0, 1, \ldots, n$ are used as the fitting coefficients for each of the m bias functions, and n is the polynomial order. Each of the m bias functions corresponds to a different layout context (also viewed as the second additional pattern geometry parameter). Consequently, Eq. (1) computes the etch bias as a function of the first pattern geometry parameter x when the value of the second pattern geometry parameter remains a constant. Eq. (1) can be written in a matrix form as:

$$\hat{bias} = A_{bias} \cdot \hat{x}_{bias} \qquad \text{Eq. (2)}$$

where,
$$A_{bias} = \begin{bmatrix} a_{00} & a_{01} & \cdots\cdots & a_{0n} \\ a_{10} & a_{11} & \cdots\cdots & a_{1n} \\ a_{20} & a_{21} & \cdots\cdots & a_{2n} \\ & & \cdot & \\ & & \cdot & \\ & & \cdot & \\ a_{m0} & a_{m1} & \cdots\cdots & a_{mn} \end{bmatrix}, \text{ and } \hat{x}_{bias} = \begin{bmatrix} 1 \\ x \\ x^2 \\ \cdot \\ \cdot \\ \cdot \\ x^n \end{bmatrix}.$$

After constructing the set of bias functions, the coefficients a_{ik} can be determined by fitting the set of the bias functions to the measured etch bias data. Note that the number of terms n in each bias function can be decided during the fitting process to control fitting accuracy while avoiding over fitting. Generally n of 3 to 5 is sufficient for a typically complex case. In a similar way, the bias can also be extended to a function of multiple geometry parameters. For example, if three parameters (one represented by x, and the other two represented by the indices i and j) are involved in the etch model, Eq. (1) becomes:

$$bias_{ij} = a_{ij0} + a_{ij1}x + a_{ij2}x^2 + \cdots + a_{ijn}x^n \qquad \text{Eq. (3)}$$

where a_{ijk} are the coefficients of the bias function $bias_{ij}$ with $i = 0, 1, \ldots, m$, $j = 0, 1, \ldots, l$, and $k = 0, 1, \ldots, n$. The matrix formulation in Eq. (2) is still valid in the case of multiple parameters, while the matrix components should be adjusted accordingly. It is noted that not all of the pattern geometry parameters are independent. One parameter may be derived from one or more of other pattern parameters. Hence, the etch bias function does not have to include many pattern parameters. Two or three parameters may be sufficient for general applications.

The lithography ADI model used in OPC/PPC applications is typically a semi-empirical model. In order to achieve the desired performance, the ADI model is fitted to the resist measurement data in order to capture the practical lithographic behavior. Therefore one of the main goals of the lithography ADI model is to minimize the residuals between model fitting CDs and resist measurement CDs:

$$\{p_1, p_2, \cdots p_n\} = \arg\min \sum_i ((M_i(p_1, p_2, \cdots p_n) - W_i)^2 \qquad \text{Eq. (4)}$$

where W_i is ith wafer measurement result, $\{p_1, p_2, \cdots p_n\}$ represents the tunable model parameters, and the ADI model is a function of the model parameters: $M(p_1, p_2, \cdots p_n)$. To simplify the following annotations, we denote the lithographic measurement results as W_{ADI}, the ADI modeling results as M_{ADI}, and the residuals between wafer measurement and ADI modeling results as R_{ADI}, so:

$$R_{ADI} = M_{ADI} - W_{ADI} \qquad \text{Eq. (5)}$$

In the etch-aware lithography ADI model, we include the inverse of the etch bias model into the lithography ADI model fitting target. Also, in order to track lithographic behavior with a good accuracy, a second term of the programmed constant $\hat{C}$ is added to the inverse etch bias so that the sum of the two terms yields a "zero-centered" fitting target:

$$f_{ADI,i} \equiv -(A_{bias} \cdot \hat{x}_{bias})_i + C_i \qquad \text{Eq. (6)}$$

where $\hat{C}$ is a programmed constant bias vector. For example, if the first line-width corresponds to an inverse bias $-bias_0$ between (-20nm, -30nm), we can choose the programmed constant C_0 to be around 25nm, so that the shifted fitting target for the first line-width is between (-5nm, 5nm). C_0 can also be adjusted in order to obtain a substantially zero mean. One of the goals of normalizing the reversed etch bias with the programmed constants is to ensure that the ADI model calibration results are within a reasonable range. Then the lithography model calibration target is adjusted to be:

$$\{p_1, p_2, \cdots p_n\} = \arg\min \sum_i ((M_i(p_1, p_2, \cdots p_n) - W_i) - f_{ADI,i})^2 \qquad \text{Eq. (7)}$$

By including the inverse bias filter, as described in Eq. (2), (6) and (7), into the ADI model calibration, the ADI model residuals become highly correlated to the etch bias effects, and the solution to the etch AEI model becomes significantly simpler. If the mean of the first term in Eq. (6) has no large offset from zero, the ADI model simply predicts the etch process behavior. As an additional advantage, one can perform measurement noise analysis by comparing the measurement data and the simulation data. If the first term has large shifts from a reasonable range, the second term is used to adjust the calibration target, which provides a pre-bias value for the retarget model. It is noted that the retarget value is obtained from the current process data, and it could be visible before model calibration. The flow chart of the new retarget model is shown in Figure 1 [7]. In Figure 1(a), the post etch measured data are corrected by the inverse bias filter rule table, and then entered into the lithography model in order to generate the retarget model. Both the retarget model and the inverse bias filter rule table are used as inputs to the next OPC correction step. In Figure 1(b), the post lithography measured data are loaded into the lithography model, and the inverse bias filter is used in the model calibrations.

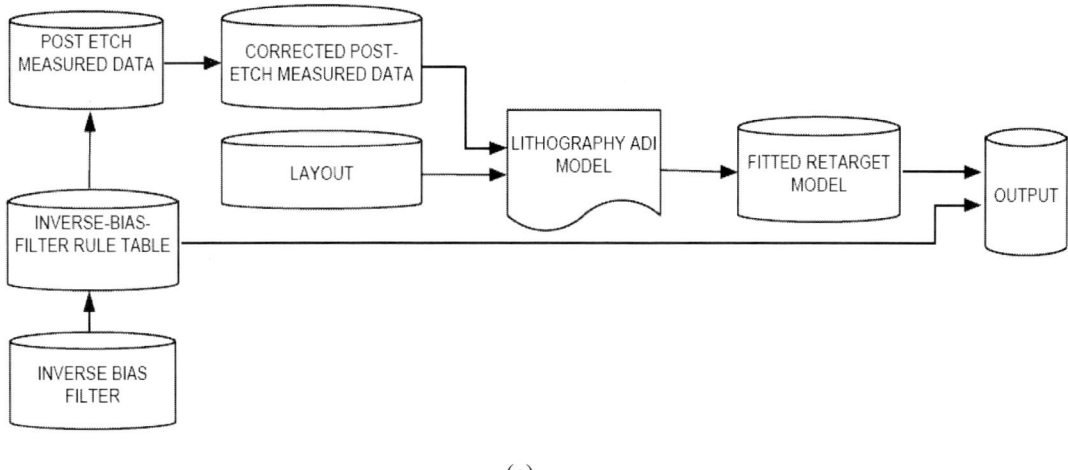

(a)

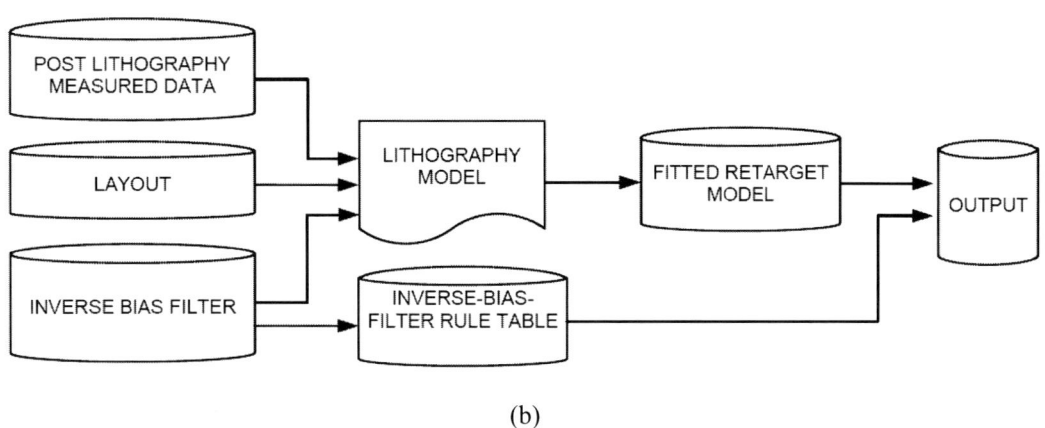

(b)

Figure 1 Illustration of the retarget model flow with the inverse bias filter. (a) The corrected post etch measured data by the inverse bias filter rule table are entered into the lithography model to generate the retarget model. (b) Post lithography measured data and the inverse bias filter are used as inputs into the lithography model in order to create the retarget model.

Next, for the staged flow, the ADI model output CD is the input to the AEI model. We note that by including the inverse bias filter into the ADI model, the AEI model initial residual is:

$$R_{AEI} = M_{ADI} - W_{AEI} = (M_{ADI} - W_{ADI}) + (W_{ADI} - W_{AEI}) = Invbias + bias = C + \varepsilon \qquad \text{Eq. (8)}$$

where *Invbias* represents the inverse of the bias model used in the ADI model calibrations, and ε is a small refinement to the ADI model, representing the residual errors due to the ADI model fitting residuals in addition to the difference between etch bias model and etch measurement data. Eq. (8) indicates that by using the inverse bias filter, the impact of the first layout parameter x has been accounted for, so the AEI model becomes a simple discrete step function of the remaining layout parameters. Some common etch bias effects, such as iso-dense bias have been corrected at the lithography ADI model stage. The degrees of freedom and the complexity of the AEI model are significantly reduced. The flow chart of the inverse bias filter related staged modeling is shown in Figure 2 [7].

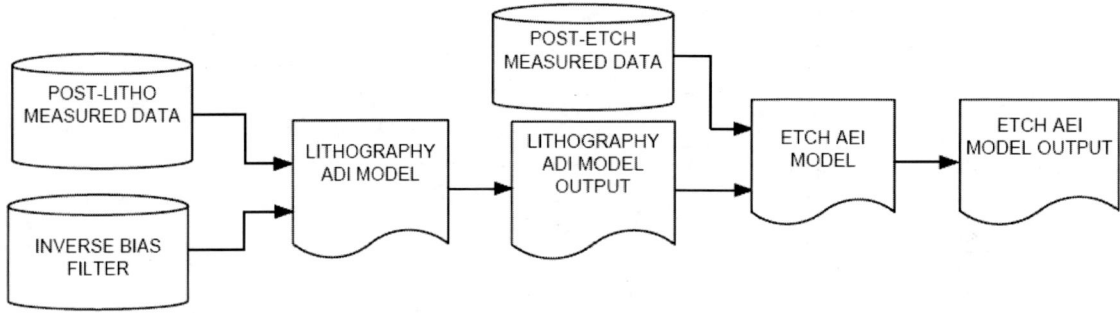

Figure 2 Illustration of the staged modeling flow with the inverse bias filter.

3. IMPLEMENTATIONS OF THE INVERSE BIAS FILTER

A dataset of lithography and etch in a DRAM process is used to illustrate the implementations of the inverse bias filter. In this example metal gate stack material is being etched. The experimental data is fed into the etch bias model parameter fitting procedure, as shown in Figure 3. Two pattern geometry parameters, space width and line width (given in seven discrete levels), are the variables in the five order polynomial model. The fitting results are indicated by the solid curve in Figure 3(a) with the minimum R^2 of 0.93. By using Eq. (2), the simulated etch bias is reversed, as shown in Figure 3 (b), where the inverse bias is in the range of -5 to -45 units. The next step is to zero-center the inverse etch bias mean by adding the constant terms depicted in Eq. (6), so that the means of the inverse bias are around zero, as depicted in Figure 3(c).

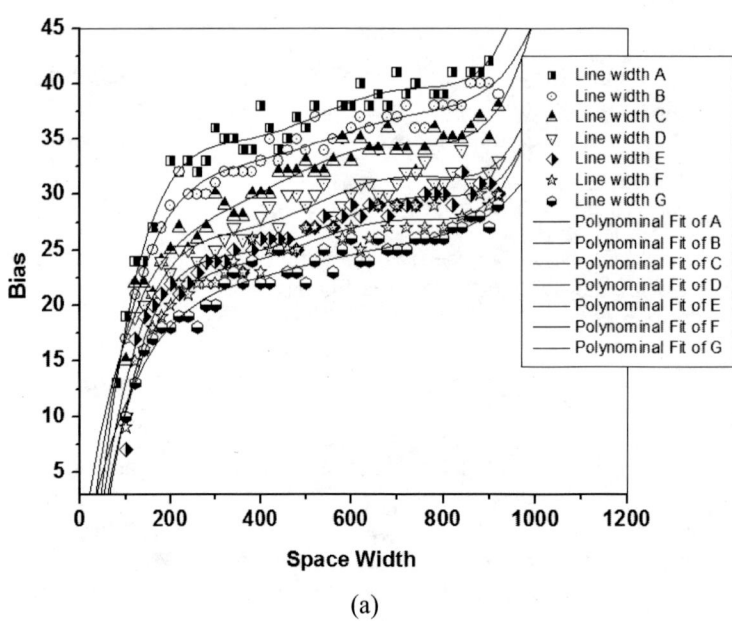

(a)

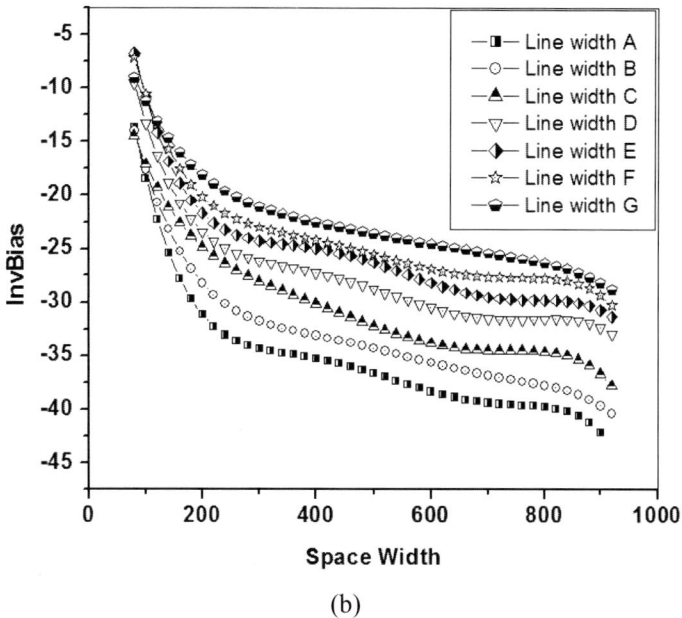

(b)

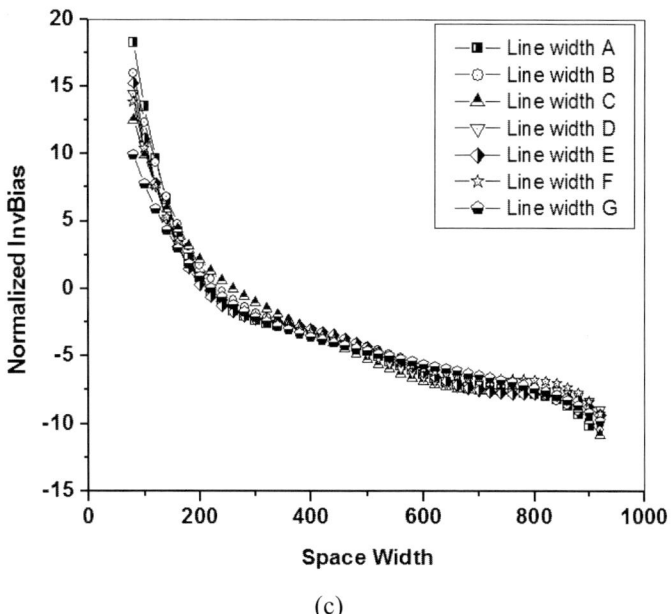

(c)

Figure 3 Illustration of the implementation of the inverse bias filter. (a) Etch bias measured data and etch bias polynomial model as a function of space width and line width; (b) Inverse of the etch bias model; (c) Zero-mean-centered inverse etch bias.

Next, the AEI model initial residuals are calculated in Figure 4 (a) with Eq. (8). The spread in each group is randomly distributed around their mean with the maximum standard deviation of 1.25. Therefore the two dimensional etch model is transformed to the one dimensional discrete step function of line-width. Some etch effect such as iso-dense bias as a function of space width is eliminated. For one of the applications, the mean of each curve is counted as a retarget pre-bias value and generates the etch retarget rule table, as shown in Figure 4 (b). When appropriate, the discrete function can be replaced with a continuous function through interpolation and/or extrapolation as shown in Figure 4 (b). For the staged modeling flow, the simple discrete functions significantly simplify the AEI model computations.

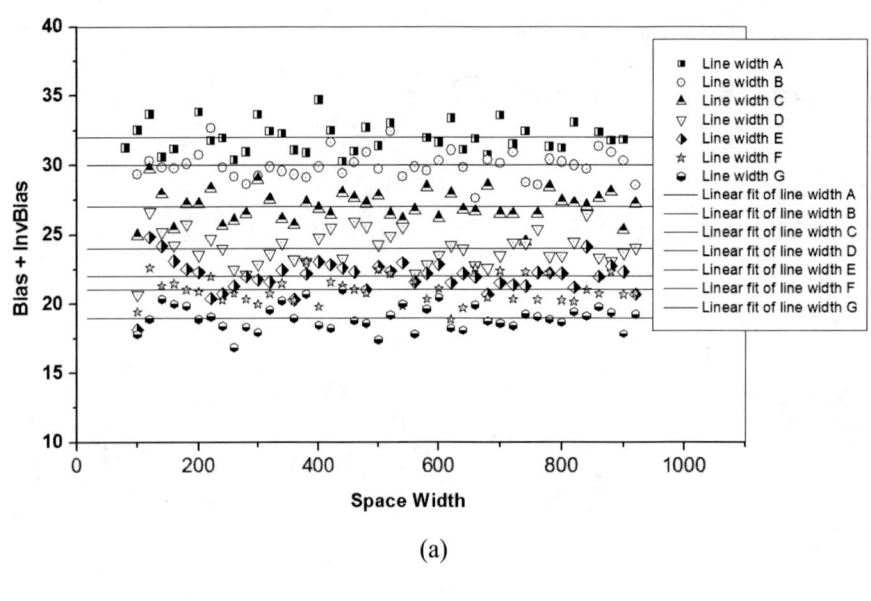

(a)

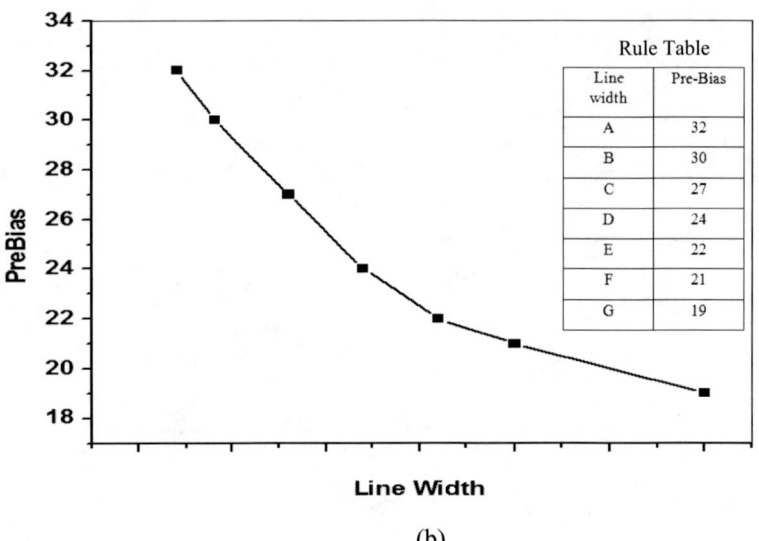

(b)

Figure 4 Illustrations of the inverse bias based rule table (a) Sum of etch bias and normalized inverse bias; (b) The inverse bias predicted pre-bias value, presented as a continuous function created through simple interpolation. The discrete rule table is summarized at the upper right of the graph.

4. CONCLUSIONS

This paper introduces an etch-aware ADI model with an inverse etch bias filter. By applying the inverse bias filter to the ADI model, the ADI model calibration target becomes highly correlated to the etch effect, and some of the etch effects such as iso-dense bias can be addressed in the lithography model, while the lithography model accuracy level is well maintained. The inverse bias filter can generate a model based rule table for design retargeting, which speeds up and simplifies etch model and OPC corrections. In addition, it creates a discrete function as an input to the AEI model in the staged flow, where the solution to the etch AEI model is significantly simplified. The new retarget etch model and staged litho-etch model flow are presented accordingly.

Least-squares polynomial fitting using actual data is used to implement the etch bias model and create the inverse bias filter. However, both of parametric and non-parametric model fitting methods are compatible with the inverse bias filter and the new retarget and staged modeling methods. Similarly, not only one dimensional pattern geometry parameters such as line width, space width, etc., but also two dimensional variables like area density or polygon density, etc., can be incorporated in the inverse bias filter related modeling applications.

The etch-aware lithography model bridges the gap between etch and lithography processes in the OPC modeling. It can be applied in the various patterning techniques, including double and multiple patterning sequences, likely to be required in the upcoming advanced technology nodes.

REFERENCES

1. International Technology Roadmap for Semiconductors (ITRS), 2008
2. D. F. Beale, J. P. Shiely, L. L. Melvin, and M. L. Rieger, "Advanced Model Formulations for Optical and Process Proximity Corrections", *Proc. of* SPIE, Vol. 5377, pp.721, 2004
3. M. Drapeau, and D. Beale, "Combined Resist And Etch Modeling And Correction For The 45nm Node", *Proc. of* SPIE, Vol. 6349, 634921, 2006
4. D. F. Beale, J. P. Shiely, M. L. Rieger, "Multiple Stage Optical Proximity Correction", *Proc. of* SPIE, Vol. 5040, pp1202, 2003
5. S. Shang, Y. Granik, and M. Niehoff, "Etch Proximity Correction by Integrated Model-Based Retargeting and OPC Flow", *Proc. of* SPIE, Vol. 6730, 67302G, 2007
6. D. F. Beale, "Method and apparatus for determining a proximity correction using a visible area model", *US* patent 7340713, 2008
7. J. Xue and J. Huang, "Etch Aware OPC Model Calibration by Using an Etch Bias Filter", *Synopsys Inc.* patent 1205, 2010

Wafer LMC Accuracy Improvement by Adding Mask Model

Wei Cyuan Lo[1], Yung Feng Cheng[1], Ming Jui Chen[1], Peter Huang[1], Stephen Chang[2], Eiji Tsujimoto[3]

1. Advanced OPC2 Department, ATD Pattern.
United Microelectronics Corporation, No. 18, Nan-Ke Rd. 2, Science-Based Industrial Park,
Sinshih Township, Tainan County 741, Taiwan, ROC

2. Brion Technologies, Santa Clara, CA, USA

3. Dai Nippon Printing Co., Saitama, Japan

ABSTRACT

Mask effect will be more sensitive for wafer printing in high-end technology. For advance only using current wafer model can not predict real wafer behavior accurately because it do not concern real mask performance (CD error, corner rounding..).

Generally, we use wafer model to check whether our OPC results can satisfy our requirements (CD target). Through simulation on post-OPC patterns by using wafer model, we can check whether these post-OPC patterns can meet our target. Hence, accuracy model can help us to predict real wafer printing results and avoid OPC verification error.

To Improve simulation verification accuracy at wafer level and decrease false alarm. We must consider mask effect like corner rounding and line-end shortening...etc in high-end mask. UMC (United Microelectronics Corporation) has cooperated with Brion and DNP to evaluate whether the wafer LMC (Lithography Manufacturability Check) (Brion hot spots prediction by simulation contour) accuracy can be improved by adding mask model into LMC verification procedure. We combine mask model (DNP provide 45nm node Poly mask model) and wafer model (UMC provide 45nm node Poly wafer model) then build up a new model that called M-FEM (Mask Focus Energy Matrix model) (Brion fitting M-FEM model). We compare the hotspots prediction between M-FEM model and baseline wafer model by LMC verification. Some different hotspots between two models were found. We evaluate whether the hotspots of M-FEM is more close to wafer printing results.

Keywords: mask model, poly, 45nm, verification, simulation

Optical Microlithography XXIII, edited by Mircea V. Dusa, Will Conley, Proc. of SPIE Vol. 7640,
76402S · © 2010 SPIE · CCC code: 0277-786X/10/$18 · doi: 10.1117/12.846903

1. INTRODUCTION

As usual photolithography process, UMC use mask to print pattern on wafer. Generally speaking, wafer printing performance will depend on two factors. One is photolithography performance and another is mask performance. Today we focus on mask and study the mask effect.

For the advance photolithography technology, we use hyper NA and strong σ process. Therefore OPC (Optical Proximity Correction) become more complicated and mask manufacturing has more challenge. Actually, a good mask still exist some mask error like mask corner rounding, CD error, and uniformity. Owing to the implementation of hyper NA and strong σ process, the mask effect will become obvious.

UMC always do some verification before mask tape-out. But the verification is based on wafer model. It means that the verification only consider wafer side and assume the mask is perfect. As pervious explanation, a good mask still exist some mask error. If we want to improve simulation verification accuracy we must consider mask error effect. UMC has cooperated with Brion and DNP to improve the verification accuracy with mask error effect. The verification will be introduced in detail in this paper.

2. MASK MODEL

Mask model is the most important for this research. DNP cooperated with Brion to build up mask model. Mask model is very similar with wafer model. It need to collect a lot of mask CD data and put these CD data into fitting. So mask model can predict some mask error after mask manufacture. Although mask model can predict mask error, it still exist some issues. Mask performance is based on writing tool. We think different writer should have different mask model. This will be a big impact for mask house because they need build a lot of mask models for different type writing tool in the beginning. Mask model also can be used on mask house to compensate mask error during mask making. Here we just focus on the benefits of mask model on wafer predict accuracy.

3. M-FEM MODEL ACCURACY VERIFICATION

We use Brion LMC to be our post-OPC verification tool. In order to compensate mask effect, we use mask model which was built by DNP. However, the verification tool can't simulate with wafer model and mask model at the same time. To solve this problem Brion help us to combine two model to one model. We call this combinative model as M-FEM model.

To guarantee M-FEM model accuracy, we need to verify model accuracy. Generally speaking, we need exposure wafer and collect wafer data to verify the model accuracy. DNP provided 45nm node Poly mask model. Therefore, we choose one 45nm product poly layer mask which was made by DNP and exposed this mask on wafer. The verification patterns include proximity (1-D pattern) and device patterns (2-D pattern). We simulate contour and measure contour CD by baseline model (not include mask model) and M-FEM model. Figure 1 shows the difference of wafer CD and simulation contour CD on proximity pattern. All proximity CD error RMS (root mean square) of baseline model is 1.54nm and M-FEM model is 1.35nm. From proximity result, M-FEM model accuracy is a little better than baseline model.

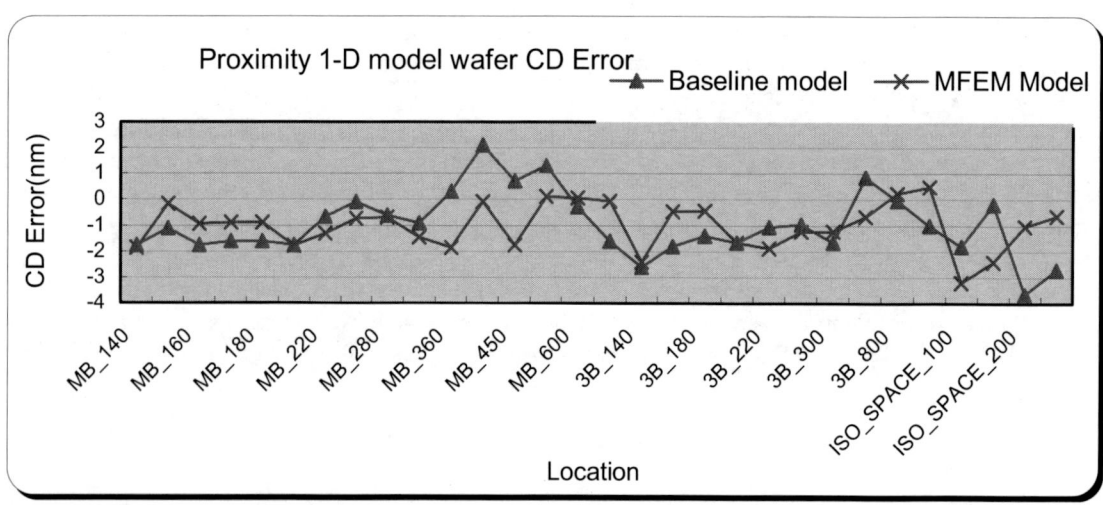

Figure 1. Proximity 1D pattern wafer verification result of baseline model and M-FEM model.

For device pattern we also simulate contour and measure contour CD by baseline model and M-FEM model. Figure 2 shows the difference of wafer CD and simulation contour CD on device pattern. Device pattern CD error RMS of baseline model is 1.92nm and M-FEM model is 1.75nm. From proximity result, M-FEM model accuracy is a little better than baseline model. Both proximity and device patterns result show that M-FEM model accuracy is a little better than baseline and CD error RMS is less then 2nm.

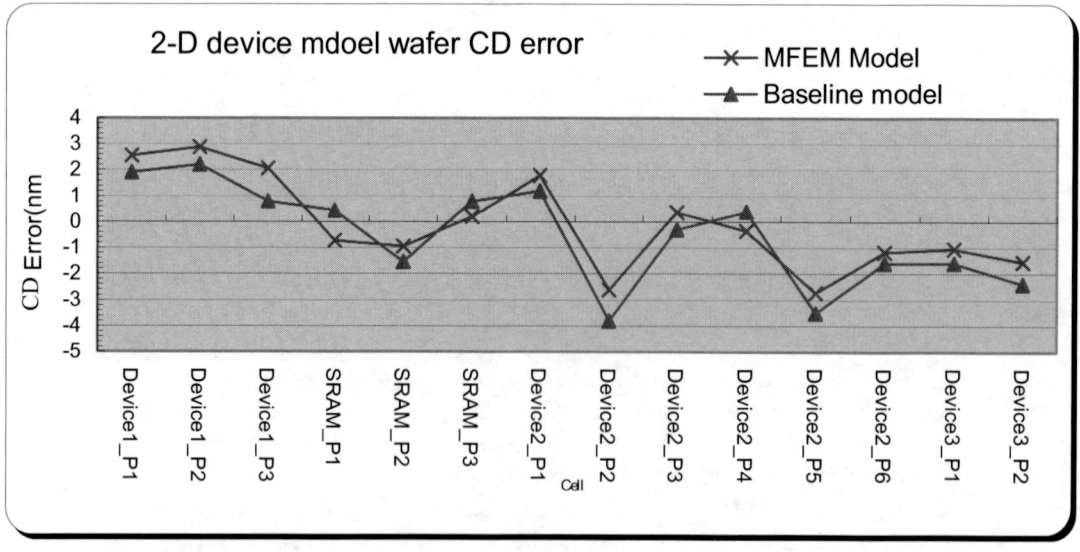

Figure 2. Device 2D pattern wafer verification result of baseline model and M-FEM model.

4. LMC SIMULATION ON M-FEM MODEL AND BASELINE MODEL

LMC is a software which can simulate contour and predict weak point. We set some criteria for LMC simulation. These criteria are based on our process capability. If the simulation result is out of criteria LMC will highlight some hotspots. Model accuracy is very important because LMC simulation is based on model. We had checked model accuracy at pervious section. After the M-FEM model accuracy verification complete, we can start using M-FEM model to do LMC simulation. We choose 3 devices and simulate by baseline model and M-FEM model. After comparing two LMC hotspots of baseline model and M-FEM model, we found that most of differences are bridge and necking type. Figure 3 is an example that LMC catches necking hotspot on baseline model and contour CD is 58.6nm but LMC don't catch necking hotspot on M-FEM model and contour CD is 62.55nm.

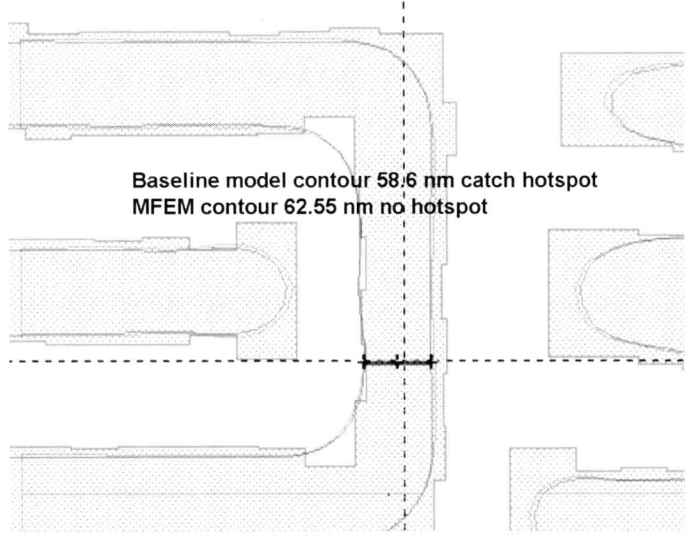

Figure 3. LMC necking type hotspot different

Figure 4 is an example that LMC catches bridge hotspot on baseline model and contour CD is 53.0nm, LMC also catch bridge hotspot on M-FEM model and contour CD is 40.8nm. Although baseline model and M-FEM model LMC result catch bridge in figure 4 but simulation contour have 12nm difference.

We summarize different hotspots in Table 1. The most difference between M-FEM model and baseline model is 34.7nm. The average of contour CD difference is around 7nm. The difference is obvious. We try to analyze the root cause of huge difference. We check post-OPC layout of these locations and we find that most of difference come form corner rounding and line-end. From Table 1 we can obviously catch that the difference on corner rounding is the most serious. The difference which come from corner round is around 12nm, line-end is around 6nm and CD error is around 1~2nm. It also proved that M-FEM model has compensated corner rounding, line-end error and CD error.

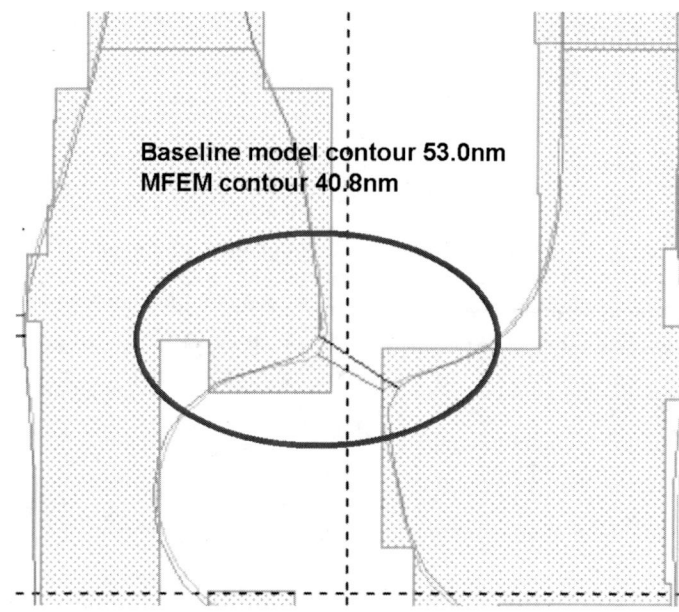

Figure 4. LMC bridge type hotspot different

Table 1. LMC simulate contour different between M-FEM and baseline model.

	Type	MFEM CD(nm)	Baseline CD(nm)	M-FEM-Baseline(nm)
Device1_P1	Line-end	52	57.5	-5.5
Device1_P2	Line-end	48.1	54	-5.9
Device1_P3	Line-end	50	59	-9
Device1_P4	Line-end	48	53	-5
Device1_P5	Line-end	48	54	-6
Device1_P6	CD error	59	56.7	2.3
Device1_P7	Corner rounding	62.55	58.6	3.95
Device2_P1	Corner rounding	35.2	46.7	-11.5
Device2_P2	CD error	55	53	2
Device2_P3	CD error	55	54	1
Device3_P1	Corner rounding	40.8	53	-12.2
Device3_P2	Corner rounding	51.4	58.5	-7.1
Device3_P3	Corner rounding	0	34.7	-34.7
Device3_P4	Corner rounding	50.9	64.2	-13.3
Device3_P5	Corner rounding	51.5	56.5	-5

5. WAFER VERIFY ON LMC SIMULATION

Base on LMC model verification result (figure 1 & figure 2) we can summarize that M-FEM model can compensated some mask error. But, we need to concern whether the accuracy of simulation result are good enough. Wafer printing verification can help us to check the simulation accuracy. We start wafer exposure and measure wafer CD by CD-SEM. Figure 5 shows wafer data and LMC simulation result. We can discover that M-FEM simulation trend is close to wafer CD trend. But, baseline model simulation trend have some different with wafer CD trend. M-FEM model CD error RMS is 3.85nm and baseline model CD error RMS is 12.26nm. It means that the contour prediction of M-FEM model is more accuracy than baseline model. Figure 6 shows LMC simulation contours and wafer SEM images. The contour (line-end ,corner and CD) prediction of M-FEM mode is more close to wafer image.

Wafer data proves that the LMC simulation of M-FEM model has better accuracy. However, using M-FEM model on LMC simulation will cost more runtime. For our experience, M-FEM model LMC simulation runtime will be 2~4 times of baseline. The additional runtime come form additional mask contour calculation. Using M-FEM model to simulate wafer contour is different with normal flow. M-FEM model simulation will generate mask contour firstly and use mask contour as source mask to simulate wafer contour. It means M-FEM model simulation include two times simulations. It is predictable that M-FEM model runtime will be twice than baseline model at least.

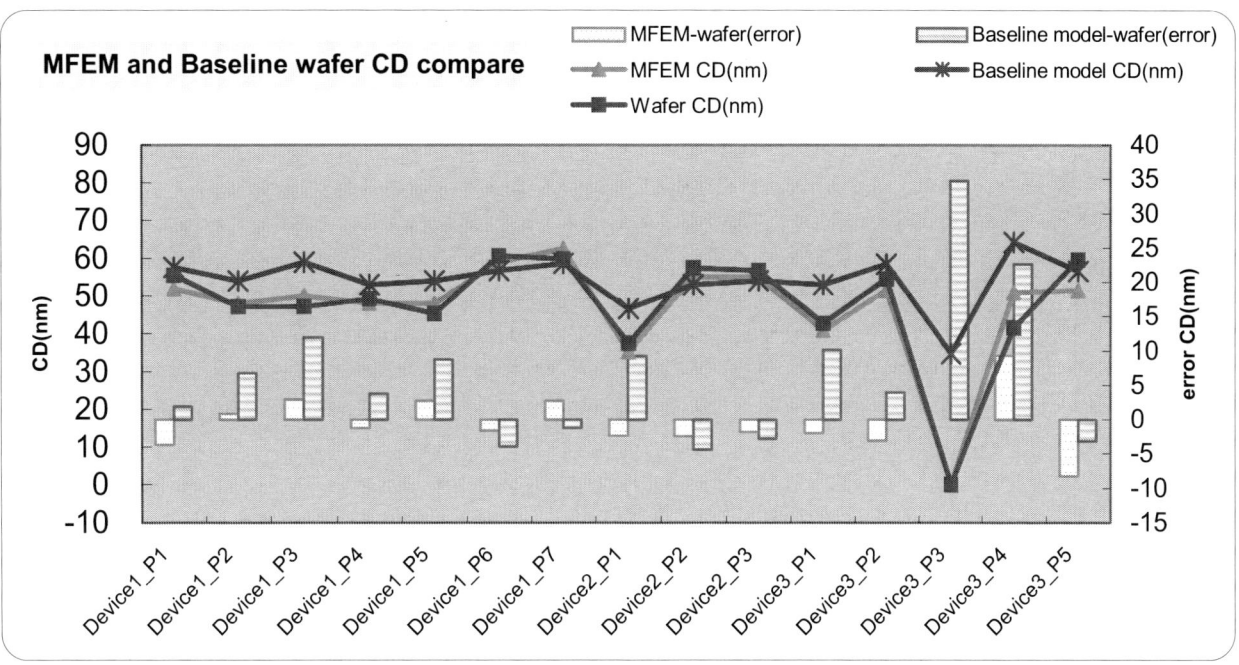

Figure 5. LMC simulation result compare with wafer data

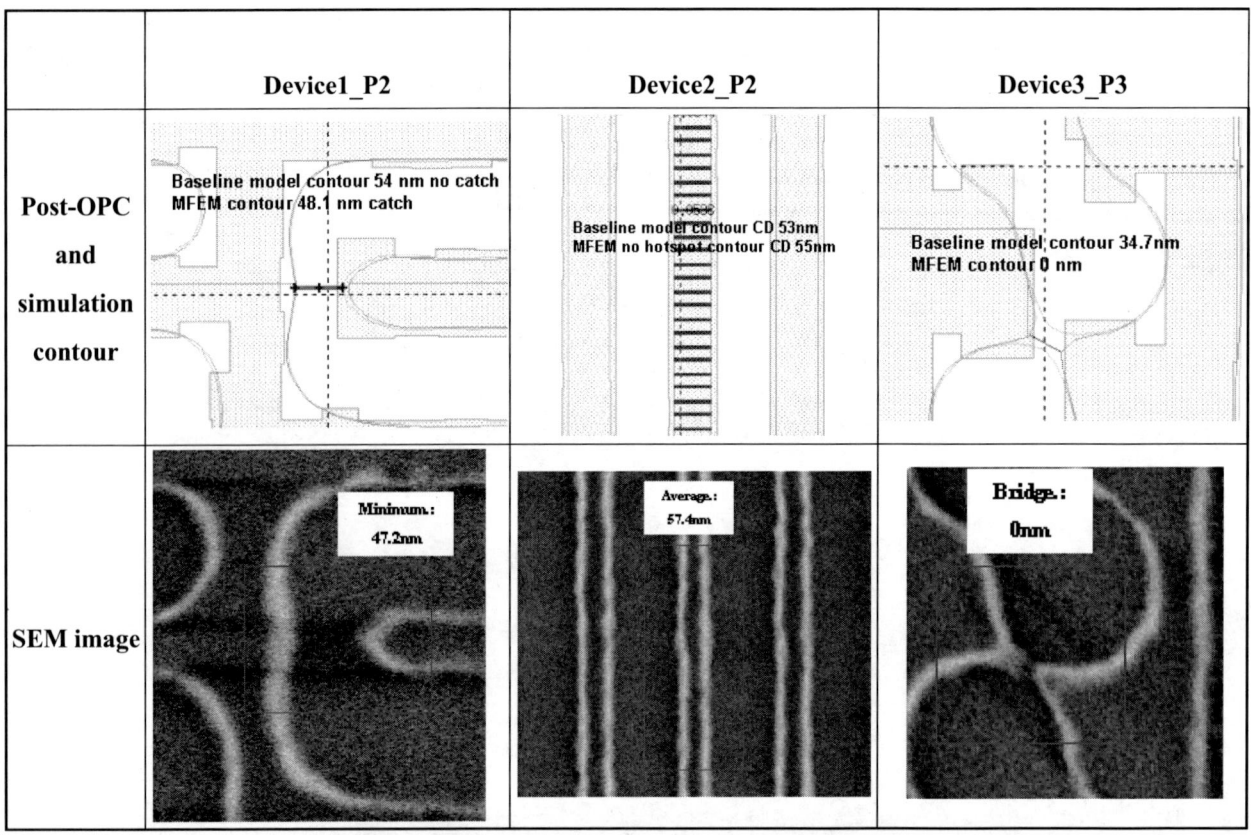

Figure 6. LMC simulation contours and wafer SEM images

6. SUMMARY

Usually we use wafer model to verify our post-OPC data. However in smaller generations, hyper NA and strong σ will cause photolithography more sensitivity. Mask effect on photolithography become more and more serious. We must consider mask effect on our post-OPC verification to avoid some false defects and missing defects. In order to reduce the impact of mask effect, mask model is a good way to help us to improve verification accuracy. UMC cooperated with DNP and Brion to improve wafer LMC accuracy by adding mask model.

If we want to use mask model to do wafer LMC verification, we need combine wafer model with mask model into one model. The combinative model is M-FEM model. At first, we must verify new model before we use it. The verification pattern include proximity (1-D pattern) and device patterns (2-D pattern). The verification result shows M-FEM model CD error RMS is 1.35nm on proximity and 1.75nm on device patterns. M-FEM model accuracy is good enough and better than baseline model. After M-FEM model verification, we can use LMC simulation on our verification step. We compare simulation result of baseline mode and M-FEM model. We discover that there are obvious difference on mask corner and line-end. So far, we need wafer data to check which one is correct. From wafer CD-SEM image result, we know M-FEM contour is more similar with wafer image than baseline model contour. M-FEM contour CD error RMS is 3.85 better than baseline model 12.26nm on device chip. The mask model can improve wafer LMC accuracy. The mask model help us to get better wafer prediction by simulation. However, the runtime is an issue and need to improve it.

7. REFERENCES

1. P. Liu, et al., "Validation of a fast and accurate 3D mask model for SRAF printability analysis at 32nm node" Proc. SPIE, Vol. 6730, 2007.

2. Vicky Philipsen, et al., "Impact of alternative mask stacks on the imaging performance at NA1.20 and above," Paper 6730-57, SPIE Photomask Technology 27th Annual Symposium, 2007.

Study of Model based etch bias retarget for OPC

Qingwei Liu[1*], Renqiang_Cheng[1], Liguo Zhang[2]

[1]Semiconductor Manufacturing International Corp. (China),
[2]Mentor Graphics Corp. (China)
*18 Zhangjiang Road, PuDong New Area, Shanghai, P.R.C Zip:201203
Phone: +86-21-5080-2000 ext:16199; fax:+86-21-5080-4010; E-mail: Qingwei_Liu@smics.com

ABSTRACT

Model based Optical proximity correction is usually used to compensate for the pattern distortion during the micro-lithography process. Currently, almost all the lithography effects, such as the proximity effects from the limited NA, the 3D mask effects due to the shrinking critical dimension, the photo resist effects, and some other well known physical process, can all be well considered into modeling with the OPC algorithm. However, the micro-lithography is not the final step of the pattern transformation procedure from the mask to the wafer. The etch process is also a very important stage. It is well known that till now, the etch process still can't be well explained by physics theory. As we all know, the final critical dimension is decided by both the lithography and the etch process. If the etch bias, which is the difference between the post development CD and the post etch CD, is a constant value, it will be simple to control the final CD. But unfortunately this is always not the case. For advanced technology nodes with shrinking critical dimension, the etch loading effect is the dominate factor that impacts the final CD control. And some people tried to use the etch-based model to do optical proximity correction, but one drawback is the efficiency of the OPC running will be hurt. In this paper, we will demonstrate our study on the model based etch bias retarget for OPC.

KEYWORD: Model-Based, Etch bias, RET, OPC, Calibre

INTRODUCTION

An OPC (Optical Proximity Correction) model is always used to represent the real process effects including the micro-lithography, the resist behavior and even the etching effects. Basically, an OPC model usually can be decomposed into two parts, the Optical part and the resist part. Nowadays, the procedure followed for generating an OPC (Optical Proximity Correction) model generally starts off with a test reticle that contains test cells, which attempt to span both the design rule and the process space. OPC model generation involves taking SEM test cell measurements of resist or etched images of features exposed at the desired illumination conditions represented on the reticle layout. An exhaustive matrix of measurements obtained from this exercise represents the entire optical "region of interest" consisting of structures of various dimensions, pitches, separations, orientations, and so on. Through all these test patterns, the transmission factor of different spatial frequencies for the certain partial coherent optical system can then be sampled. All these measurements are then fed into the appropriate software package, where a mathematical engine generates a set of coefficients, which form the basis for the final model. Usually the Optics should dominant the whole model, but to achieve satisfied prediction of the wafer level behavior, the resist part is also very important. Most OPC models describe the resist behavior through an empirical approach, which means just calibrating the resist model with some mathematical tricks. One drawback of this approach is that the mathematical treatment is based on the sampling data sets, which is always with finite quantity. So while set up an OPC model, the sampling parameter space coverage should be promised. Anyway, after a long period of study, current OPC technology can cover the optical and resist effects with high accuracy. But as we all know, the pattern transfer procedure from the design layout the wafer includes not only the optics and the photo resist effects, the wafer etch is also a very critical stage. To pursue the final critical dimension control ability, sometimes the etch process effect mush also be considered into the correction stage before the mask is prepared.

Optical Microlithography XXIII, edited by Mircea V. Dusa, Will Conley, Proc. of SPIE Vol. 7640,
76402T · © 2010 SPIE · CCC code: 0277-786X/10/$18 · doi: 10.1117/12.845962

Generally there are several kinds of etch effect which will have impact on the CD variation during pattern transfer procedure. They are known as the CD-dependent etch-bias variation, the space-dependent etch-bias variation, the line end shortening induced by etch and some other effects. All these effects combined with the pattern on the resist will generate the pattern after Etch. As we all know, due to the etch-bias variation induced by the loading effect, or in other words, the pattern density variation, it is usually the case that the etch bias will not be a constant value for all kind of features. That means, providing the identical CD on the resist, different CD after etch may be achieved. It is not acceptable to the chip manufacturers. And with the dramatic shrinkage of the critical dimension in the chip design, the foundries are facing big challenge in the wafer CD variation control. So some solutions are need to compensate for the CD variation in the etch stage.

Usually there are two approaches that can be adopted to solve this problem. One is to setup etch model based on resist model to mathematically present the etch-bias variation induced by the pattern density. Another one is to use rule-based compensation combined with accurate resist model to handle with the etch-bias variation. Compared to the etch-bias model, rule based compensation is a proved to be a more effective and direct approach. That's because compared to the lithography process, till now the etch process still cannot be described with high accurate physical model. And another drawback of the etch model is that, though very easy to be implemented, the additional model kernels used to describe the etch process and the corresponding image convolution will bring large punish in the total OPC running time.

In this work, we will demonstrate our study in the VEB (variable etch bias) model and the MER (model based etch bias retarget) flow. The platform we adopted is the Mentor Graphics Calibre tool family. Result with high accuracy and satisfied running time is illustrated.

ALGORITHM

As we mentioned above, two factors that dominate the etch-bias variation are the CD-dependence and the space-dependence. Most etch model tools adopt simple gaussian function to represent both the two etch effects. Calibre VEB model, however, use different types of kernels to describe the two different effects. One is called the shifted-density kernel, as is shown in Figure 1.

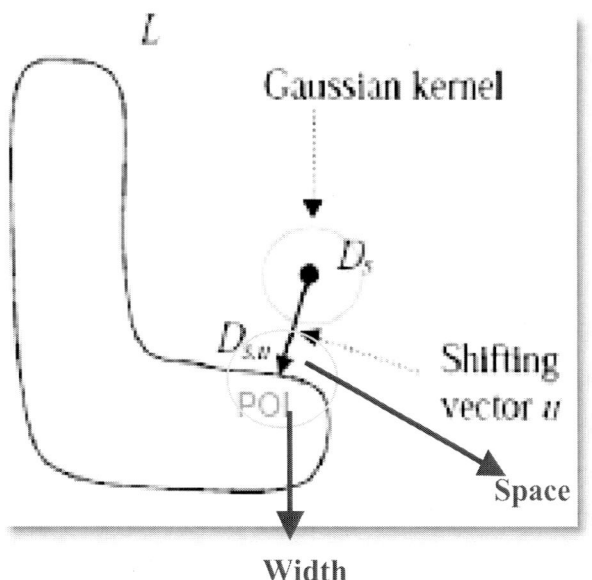

Figure 1. The shift gaussian kernel

With the Shifted-Gaussian Density kernel, the resist layer R(x,y) is first convolved with a Gaussian kernel of diffusion s nm G(s;x,y). Then the result is shifted by u nm in the direction of the gradient of resist contour. When the shift is positive, the kernel is shifted to the open are outside the polygon, thus the result is skewed towards characterizing the separation between polygons, and vice versus. Another kind of kernel is the Visible kernel, as is shown in Figure 2.

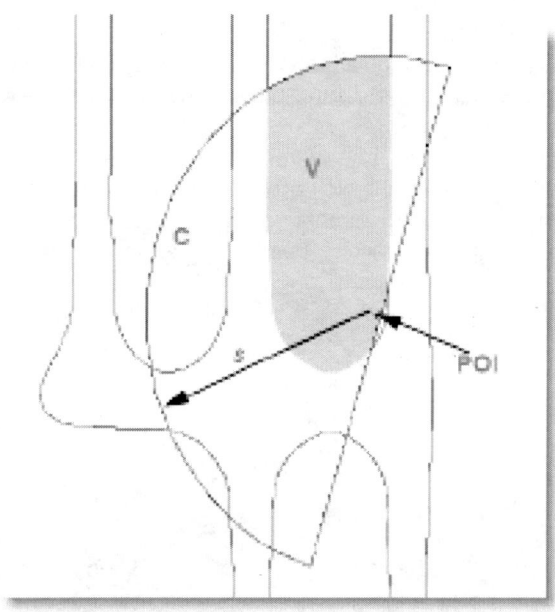

Figure 2. The visible kernel

With the Visibility Density kernel, we can determine by the area that is directly visible from the point of interest (POI), if we "look" in the direction perpendicular to the polygon edge (where POI is), as far as diffusion length s. The shift u is used as an offset from the resist contour s that POI can be positioned in the opening between polygons.

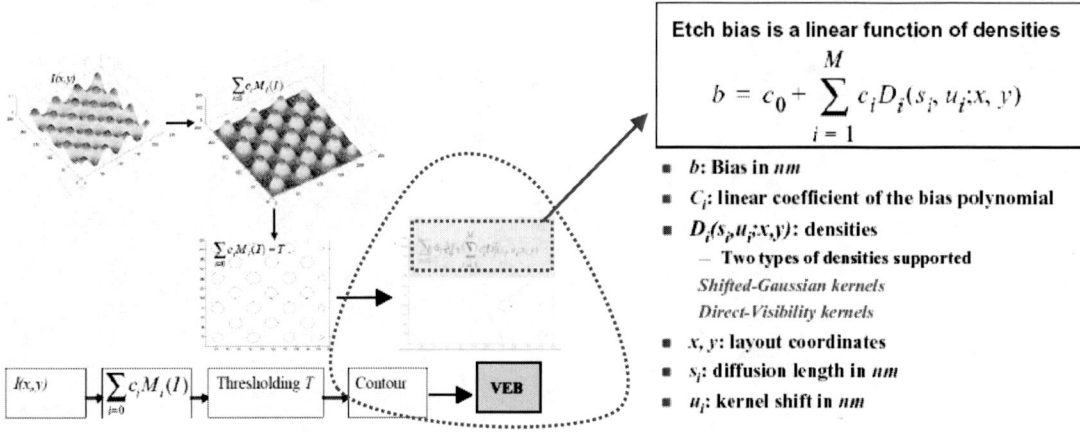

Figure 3. The calculation flow of the VEB model.

Figure 3 shows the calculation flow of the VEB model. The intensity is first achieved through the Optical model. And combined with the resist model and the threshold, the resist contour is then achieved. The VEB model is combinations of different kernels for etch effects. VEB uses CM1 resist model simulation contour as input. VEB moves vertices of input resist contour in the direction perpendicular to the resist contour. And the movement amount, or the bias varies along the contour, is determined by etch bias polynomial. Compared to the etch model with only simple gaussian kernels, the VEB model has relatively more physical description and fitting freedom, which will provide relatively better model accuracy.

MODEL BASED ETCH BIAS RETARGET

As we mentioned in the introduction session, one drawback of the etch model for OPC is, compared to the rule based etch bias compensation, more kernels need to be considered and convolved with the resist contour. As we all know that OPC is an iteration-based procedure. So the convolution between the etch kernels and the resist contour will happen during every iteration of OPC, this will be a time-consuming job. The correction with VEB model, however, is different with the traditional etch model. Variable Etch Bias (VEB) is a process that allows us to separately account for etch effects. This methodology differs from previous OPC practices, where all three modeling effects (optical, resist, and etch) are calculated at the end, using only the post-etch target data. This VEB model-based retargeting process enables us to separate out the modeling of etch bias effects from optical and resist effects, also separate out the correction of etch effects and correction of optical and resist effects. An etch-corrected design based on the VEB model can then be derived. It uses fragmentation and edge movement calculations, similar to making OPC corrections to a layout.

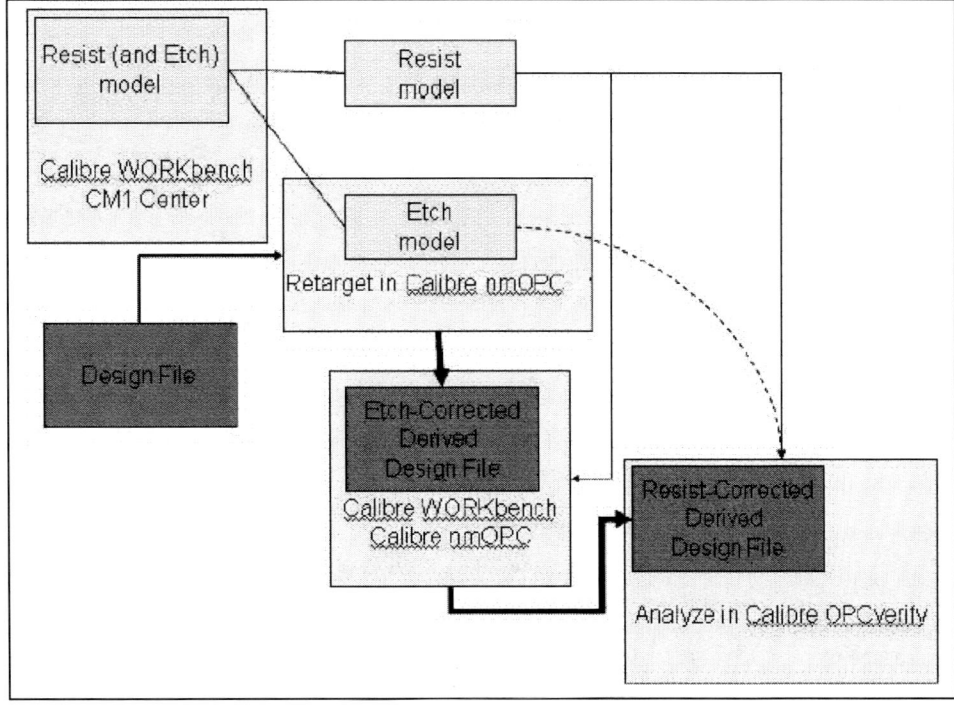

Figure 4. The VEB model correction flow.

Figure 4 illustrates the retargeting flow. This methodology differs from previous OPC practices, where all three modeling effects (optical, resist, and etch) are calculated at the end, using only the post-etch target data. Separating out the post-etch data and using it to retarget the design can reduce the run-time of etch bias correction, since the corrections are applied to the fragment edges on the original target rather than applied to the resist contours. This flow also provides us better control over OPC corrections through separating the resist and etches effect requirements. We can verify post-OPC corrections at the resist level against the resist target, enabling detection of problems in the resist phase.

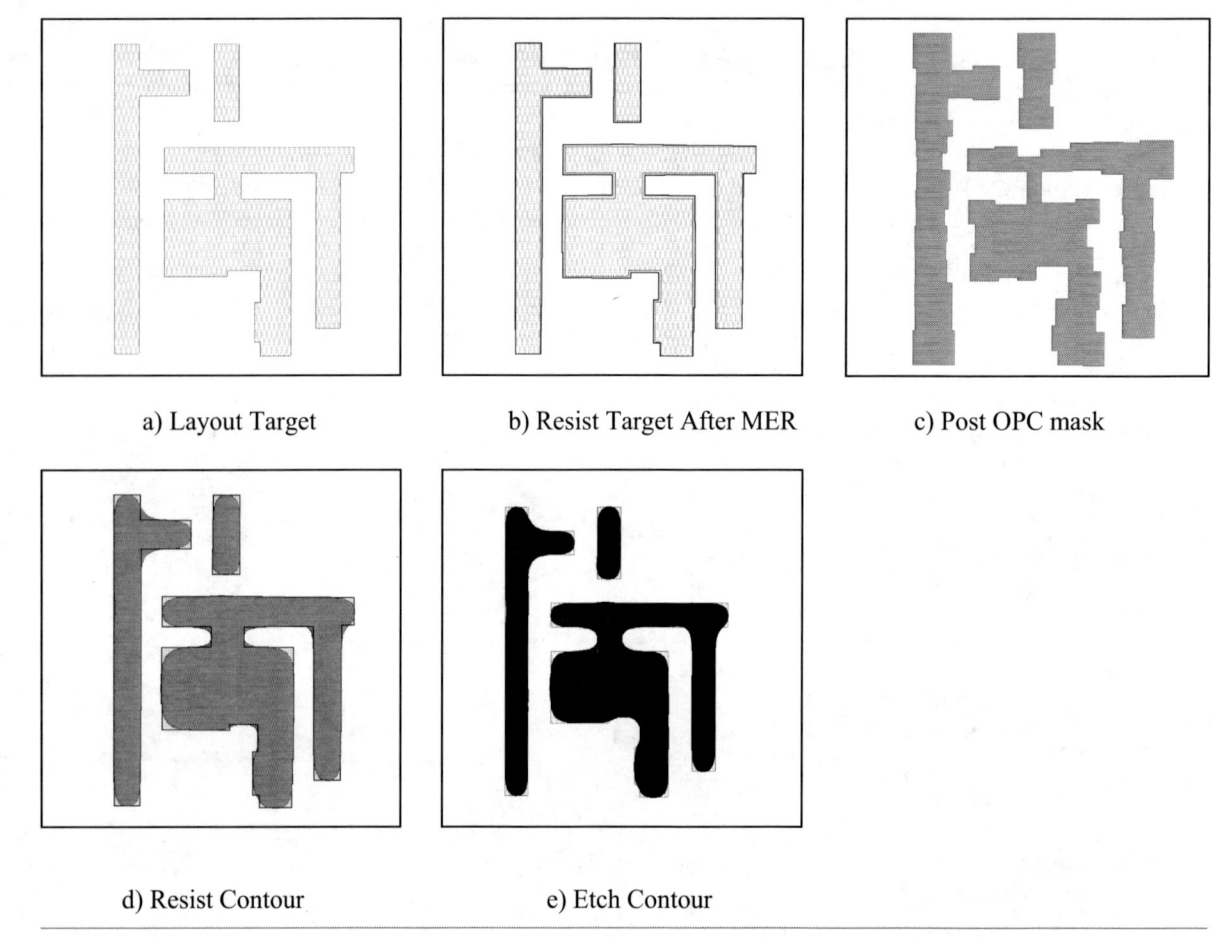

a) Layout Target   b) Resist Target After MER   c) Post OPC mask

d) Resist Contour   e) Etch Contour

Figure 5. An example of using VEB on a test structure._

In figure 5, we demonstrate an example of using VEB model for OPC on a test structure. From the design target, which should also be the target of the post-etch, we can derive the ADI target with the implement of the VEB model. Since this stage is based on the design target, not the resist contour, it can be finished within 2 to 3 iterations. And based on the model-based etch bias retarget layout, we further do OPC with optical and resist model. After finishing the OPC, we will check the resist contour and post-etch contour with the resist and etch model. This is to make sure that the model-based etch bias retarget is done correctly. Due to that the etch calculation and the lithography calculation are separated, the punishment of run time for considering etch effect can be managed.

CONCLUSION

In this work, we demonstrate our study in the VEB (variable etch bias) model and the MER (model based etch bias retarget) flow. The platform we adopted is the Mentor Graphics Calibre tool family. The correction with VEB model, is different with the traditional etch model. Variable Etch Bias (VEB) is a process that allows us to separately account for etch effects. This methodology differs from previous OPC practices, where all three modeling effects (optical, resist, and etch) are calculated at the end, using only the post-etch target data. This VEB model-based retargeting process enables us to separate out the modeling of etch bias effects from optical and resist effects, also separate out the correction of etch effects and correction of optical and resist effects. According to our test result, the punishment of run time for considering etch effect can be managed with the flow of Model based etch bias retarget.

REFERENCE

1. "Calibre WorkBench User's Manual", Chapter 13, 15 Calibre 2009.3
2. Alfred Kwok-Kit Wong, Resolution Enhancement Techniques in Optical Lithography, Vol. TT47 SPIE PRESS.

Intra Field CD Uniformity Correction by Scanner Dose Mapper™ using Galileo® Mask Transmission Mapping as the CDU Data Source

Gek Soon Chua[a#], Chason Eran[b], Sia Kim Tan[a], Byoung IL Choi[a],
Teng Hwee Ng[a], Poh Ling Lua[a], Ofir Sharoni[b], Guy Ben-Zvi[b]
[a]GLOBALFOUNDRIES Singapore Pte. Ltd. 60 Woodlands Ind. Park D St 2. Singapore 738406
[b]Carl Zeiss SMS, 44 Maale Camon, 21613 Karmiel, Israel
[#]corresponding author Tel (65) 64137646 Email chuags@globalfoundries.com

ABSTRACT

Intra-field CD variation can be corrected through wafer CD feedback to the scanner in what is called the Dose Mapper (DOMA) process. This will correct errors contributed from both reticle and scanner processes. Scanner process errors include uncorrected illumination non uniformities and projection lens aberration. However, this is a tedious process involving actual wafer printing and representative CD measurement from multiple sites. A novel method demonstrates that measuring the full-field reticle transmission with Galileo® can be utilized to generate an intensity correction file for the scanner DOMA feature. This correction file will include the reticle transmission map and the scanner CD signature that has been derived in a preliminary step and stored in a database. The scanner database is periodically updated after preventive maintenance with CD from a monitoring reticle for a specific process. This method is easy to implement as no extra monitoring feature is needed on the production reticle for data collection and the new reticle received can be immediately implemented to a production run without the need for wafer CD data collection. Correlation of the reticle transmission and wafer CD measurement can be up to 90% depending on the quality of CD data measurements and repeatability of the scanner signature. CD mapping on the Galileo® tool takes about 20 minutes for 1500 data points (there is no limit to the number of measurement point on the Galileo®), which is more than enough for the DOMA process. Turn Around Time (TAT) for the whole DOMA process can thus be shortened from 3 Days to about an hour with significant savings in time and resources for the fab.

Keywords: Intra-field, Reticle Transmission, Dose Mapper, DOMA, CD CDU, Galileo, Scanner signature, Mask Transmission Mapping

1. INTRODUCTION

In today's manufacturing processes, and especially for 45nm node and below, CDU becomes a major contributor to yield and consequently a big concern for the lithography engineers. Main contributors to the wafer CDU are:

1. The mask CDU, being affected by the mask manufacturing process (film, writing and etching signatures) and CDU degradation due to DUV exposure in the scanner and mask cleaning,

2. The scanner CDU signature and

3. The resist process magnifying the CDU of the aerial image in the wafer plan.

Contributors to the CDU, that show a distinct signature across the mask, even on the exposed resist, can be compensated using a localized dose adjustment in the scanner while exposing the wafer. This feature is incorporated in ASML scanners as the Dose Mapper.

In this paper, the results of a collaborative study between GLOBALFOUNDRIES in Singapore and Carl Zeiss SMS (Pixer Technology) in Israel are presented. Using the Galileo® to accurately scan a reticle's transmission, converting the transmission map into a Critical Dimension (CD) map and to use it to feed forward to the ASML Dose Mapper, 50% improvement in final CDU can be achieved.

Additionally, in order to take the scanner signature into account, the Carl Zeiss Galileo® tool can be used (with the help of a reference tool for reticle CD measurement or wafer CD measurement) to extract the scanner CDU signature for CDU prediction and improvement.

Optical Microlithography XXIII, edited by Mircea V. Dusa, Will Conley, Proc. of SPIE Vol. 7640,
76402U · © 2010 SPIE · CCC code: 0277-786X/10/$18 · doi: 10.1117/12.852819

2. BACKGROUND

2.1 Process flow of Galileo® Transmission measurement tool feeding the ASML Scanner DoseMapper

As shown in other publications, reticle intensity measurements can be used as a source for reticle CDU feedback to the scanner [3-7]. The method works only if reticle CDU is systematic and is a major source of the wafer CDU contributions. Current studies shows that with the advanced development in mask technology in CDU control, scanner contribution to the wafer CDU have also become one of the major sources.

In this paper we present a method of correcting wafer CDU using Reticle Transmission (RT) uniformity measurements and a scanner signature, derived from Reticle Transmission (RT) uniformity measurements. Reticle Transmission (RT) measurement is a macro view of the transmitted light intensity. By using this RT measurement, long range proximity effects can be corrected. In regions where small pitch line and space features are clustered, the variation will be more significant compared to isolated features. This observation is closely related to the MEEF. A correlation setup is done with a monitoring reticle to determine the constant (k) from the linear relation of RT with respect to CD.

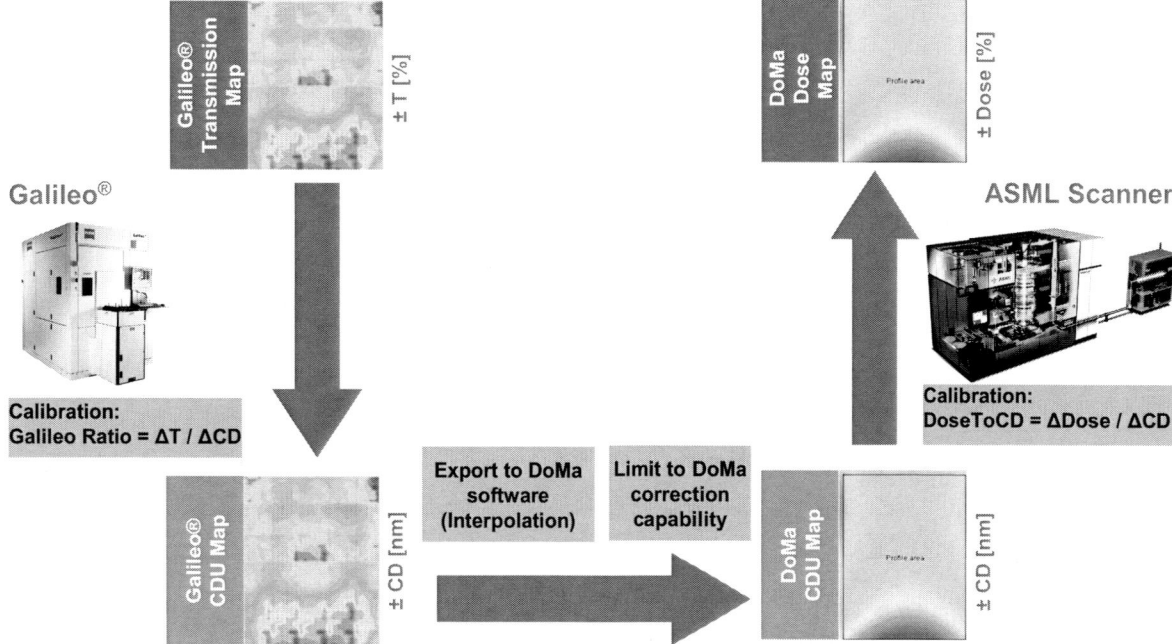

Figure 1 – Process flow of Carl Zeiss Galileo® Transmission measurement tool feeding the ASML Scanner DoseMapper

Figure 1 shows the work flow of compensating CDU for resist CD using Galileo® Transmission measurement tool to feed forward to the ASML scanner DoseMapper. This will save on wafer printing on scanner with high Work-In-Progress (WIP) and long hours of CD measurement (typically about 5-8 hrs) that are needed to setup the wafer CD recipe and CD data collection for DOMA. Furthermore in the current practice, measurement features with suitable requirement design pitch and CD are needed to be identified on MEBES for each unique tapeout. There are occasions where only limited suitable features can be found on the devices which create situations of unrealistic representation of the mask systematic error. Henceforth, CD mapping on the Galileo® tool presents a faster Turn-Around-Time (TAT) for the whole DOMA process with repeatable accuracy. It is easy to implement on the production reticle as no production wafer is needed for CD measurement. Correction can be applied on first production wafer whereby no unique/extra monitoring feature is needed on the production reticle for data collection. The new reticle received can be immediately implemented onto production run without the need for wafer CD data collection.

2.2 Work flow of Galileo® Transmission measurement tool to generate a correction file for the scanner

In this paper we report measuring a full-field reticle transmission map using Galileo® Transmission measurement tool and utilizing it to generate a correction file for the scanner. This correction file includes the reticle transmission information map obtained. The scanner process signature for a specific layer was derived after subtracting its reticle transmission map from its wafer CDU map. This scanner signature can further be stored in a database. The scanner database will be periodically updated after scanner Preventive Maintenance (PM) using the CD of a monitoring reticle for that specific process.

Figure 2 shows the novel flow for obtaining the scanner signature using a monitoring reticle. Before commencing production runs for a new device tapeout, various scanner signatures will be set up using a monitoring reticle. Full field wafer CD measurement data will be collected for various scanners. Correlation between reticle transmission map and wafer CD map will be performed and a specific scanner signature will be derived. Thereafter the specific scanner signature for a specific layer and wafer process will be stored into the database for production runs. Periodic update of the scanner signature will be conducted as and when necessary and after every PM. For new production runs, the new production reticle will undergo RT measurement on Galileo® during the in-coming check. Combining the RT map with its respective scanners signature, a specific correction file for that reticle will be derived and stored in the scanner database for wafer printing.

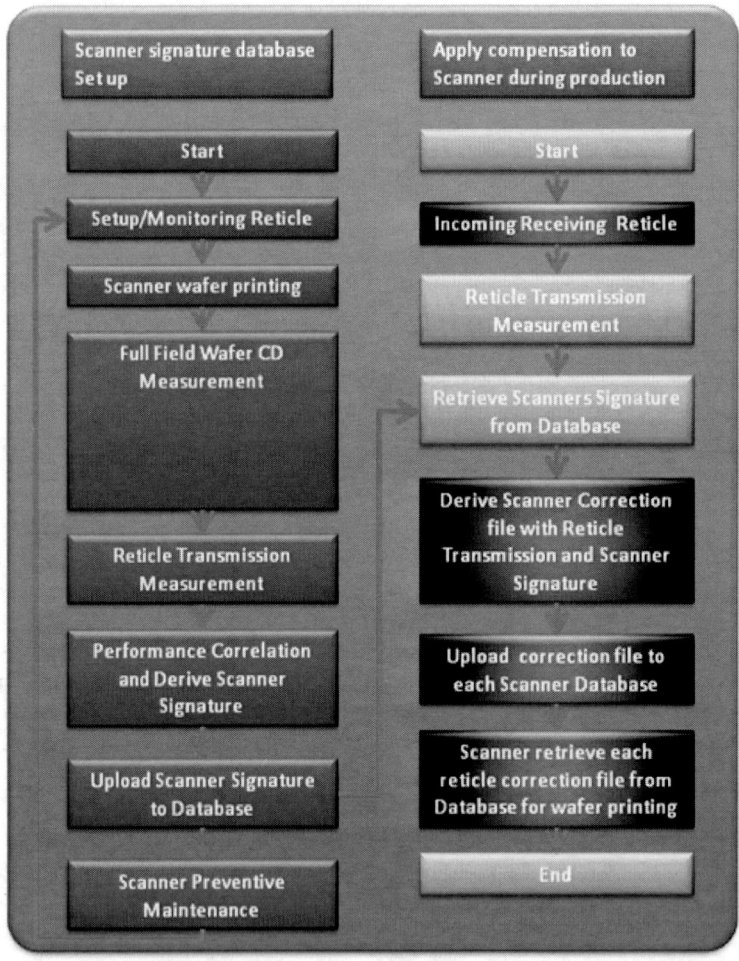

Figure 2 – Scanner signature setup and production reticle correction file generation flow for resist CDU improvement

2.3 Introducing the Galileo® Reticle Transmission measurement system

The Galileo® is a non-imaging transmission measurement system with high transmission repeatability and extremely high signal to noise (SNR) ratio. The Galileo® uses a deep UV lamp as the light source, which helps to get a more stable, low noise and "low cost performance" illumination as it is required for very high sensitivity transmission measurements. The Galileo® scans various measurement locations across the mask using a DUV spot of 180μm-600μm diameter (FWHM), resulting in a transmission value on each location which is averaged over the area of the spot. A reasonably resolved mask transmission map is generated within 30 minutes using the Galileo®. However, higher lateral resolution can be generated if needed.

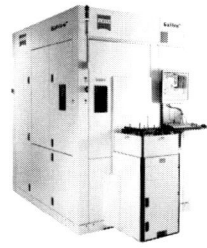

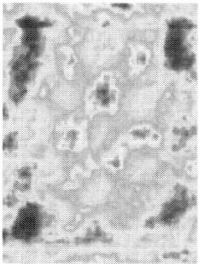

Figure 3 – Galileo® Reticle Transmission Measurement   Figure 4 – Galileo® Transmission Map measured on reticle

2.4 CDU measurement using Galileo®

The transmission uniformity map (Figure 4) correlates to the wafer CDU with a calibration factor, which needs to be calibrated either using mask or wafer CD data generated by an alternative CD measurement system (e.g. CD SEM or optical CD). In order to generate a CDU map as seen in Figure 4, the critical features covering the averaging area of the spot needs to be spread over the mask on several locations.

2.5 Transmission maps and the correlation to wafer CD

The CDU distribution on the wafer can be described by equation (1):

$$\Delta CD_{Wafer} = \left(\left(\frac{\Delta CD_{Mask}}{4} \right) + \Delta CD_{Scanner} \right) \times MEEF$$

Equation (1)

To transfer the Galileo® transmissions maps to the CDU maps (where the units of each measurement point is in nm) a one-time calibration process (per design rule/layer) is needed correlating the Galileo® data to a reference tool measurement, usually wafer CD, OCD or Mask CD.

Under the assumption, that the CDU range is mainly contributed by the mask, the slope of the Galileo® Transmission vs. Wafer CD correlation graph is used to generate the calibration factor. This process already takes the resist MEEF (Mask Error Enhancement Factor) into the calibration.

Converting the Galileo® map into CD map

In this step, the wafer CD data is used to convert the Galileo® RT map into a CD map. Both sets of data are plotted and correlated to each other. The Galileo® data is of high resolution and consist of many measurement points. The wafer CD/OCD or Mask CD are usually limited by the number of measurements to around 150 points per field only. Due to that problem the Galileo® application is used to interpolate the original data to the same coordinates and number of points as of the reference tool's wafer CDU data. This interpolation can be seen in Figure 4 vs. Figure 6. Both Figure 4 and Figure 6 show the same CDU measurement grid where Figure 6 is a Galileo® map plotted in low resolution for the correlation purpose.

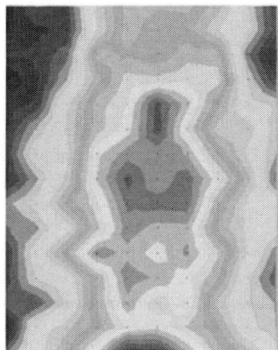

Figure 5 – Wafer Measurement

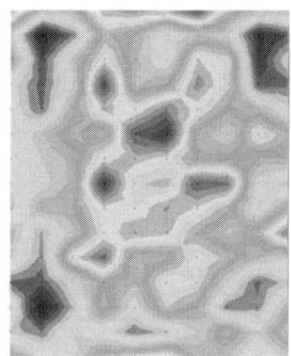

Figure 6 – Galileo® scan (interpolated)

2.6 Scanner signature extraction

To account for the additional influence of the scanner signature on the wafer CDU, the scanner signature can be derived in first approximation using a mask with a small CDU signature and wafer prints. The scanner signature can then be used to predict the wafer CDU more accurately. Figure 7 shows the process of extracting the scanner signature and then using it to create a DoseMapper correction map.

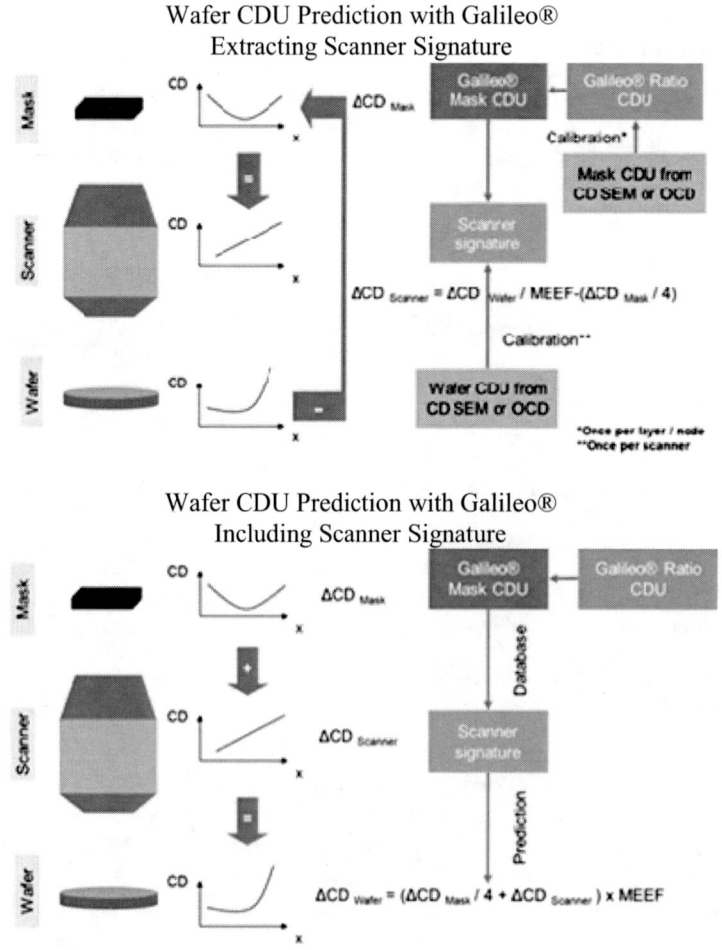

Figure 7 - Wafer CDU prediction with Galileo®. Extracting and using the scanner signature

3. RESULTS

3.1 Wafer CDU map correlation to Galileo® TRU (Transmission Uniformity) map

The Galileo® Transmission map measurements show the mask signature whilst wafer CD results show the mask signature + Residual (mainly scanner non uniformity). Henceforth, in order to have the most significant wafer CDU correction by DOMA the scanner signature should be added to the mask signature as measured by the Galileo®.

The mask non-uniformity correlation to the scanner signature is presented in the following formula in equation 2:

$$Mask\ Signature\ \times \left(\frac{CD}{Dose}\right) = (Wafer\ CD) - (Scanner\ non\ uniformity)$$

Equation (2)

The $\frac{CD}{Dose}$ is estimated in the Correlation graph in Figure 8 and is reported in nm/%dose. All CD data for mask and scanner in this paper is provided in nm at wafer level.

Figure 8 shows the slope of the formula in Eq (2) that represents the factor for conversion of the Galileo® map. In this case, the correlation between the CD data and the Galileo® data is 71% (R = √0.505 = 0.71) and the derived factor is 1.434nm/%.

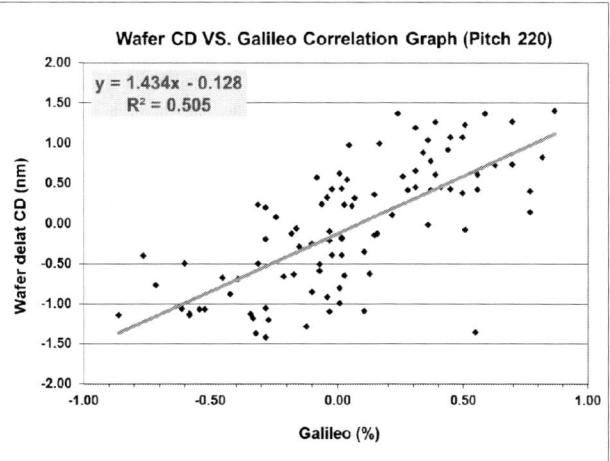

Figure 8 – Correlation graph of wafer CD vs. Galileo measurement

The Galileo® transmission map is basically a dose map. With this factor, it is now possible to convert the Galileo® transmission map from Figure 9 to a CDU map. By applying the CD/dose factor of ~1.45 (1.434 from Figure 7), the mask CD contribution was calculated to be approximately 2.5nm.

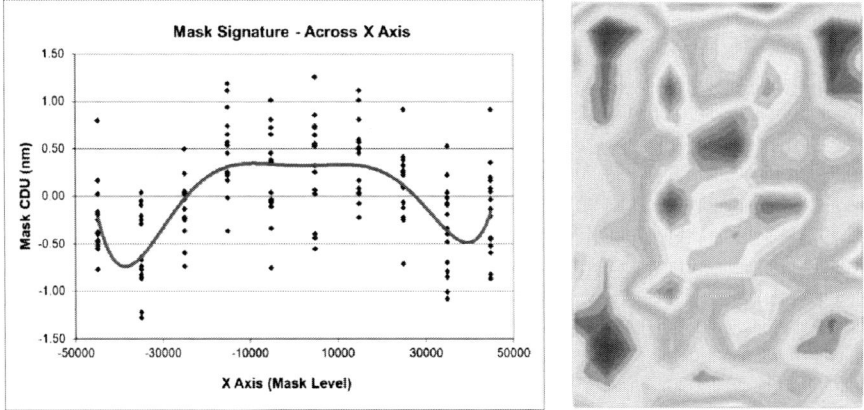

Figure 9 – Galileo® mask signature and its CDU map

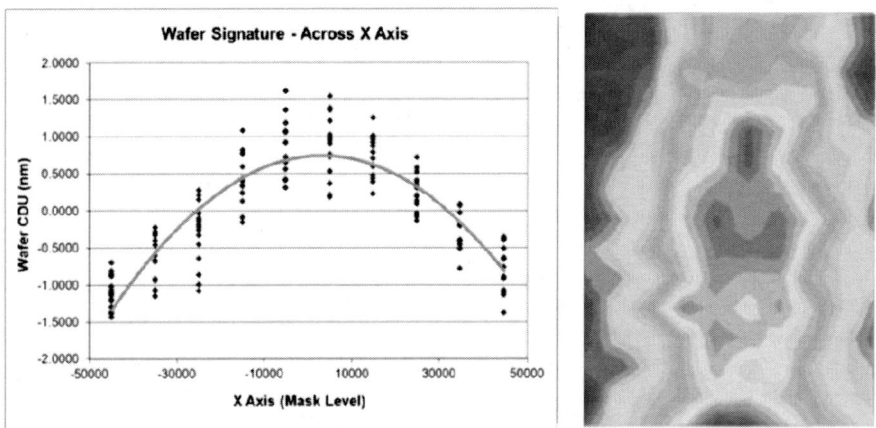

Figure 10 – Wafer CD signature and its CDU map

Figure 10 shows the wafer intra-field CDU as measured by CD-SEM system.

3.2 Correlation with & without scanner signature

From Eq. (1) that represents the wafer CDU, the scanner signature can be calculated. Since the Galileo® was calibrated against wafer data where the MEEF is already included and accounted for; Eq (3) signifies the extraction for scanner signature:

$$\Delta CD_{Scanner} = \Delta CD_{Wafer} - \Delta CD_{Mask}$$

Equation (3)

Figure 11 shows the graphical representation of the scanner signature. With the Galileo® application it is also possible to interpolate the scanner signature into a high resolution map that can be used with any new reticle.

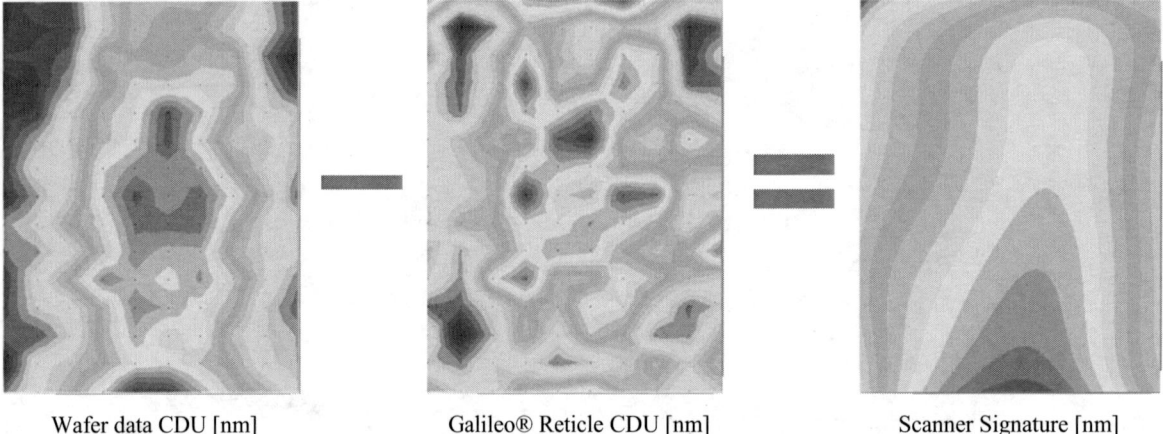

Wafer data CDU [nm]   Galileo® Reticle CDU [nm]   Scanner Signature [nm]

Figure 11 – Graphical representation of scanner signature extraction

Figure 12 shows a good correlation between the scanner signature derived based on the method shown in Figure 11 and the actual scanner signature obtained from standard across-slit uniformity measurement at wafer level. In this case, the correlation is close to 80%.

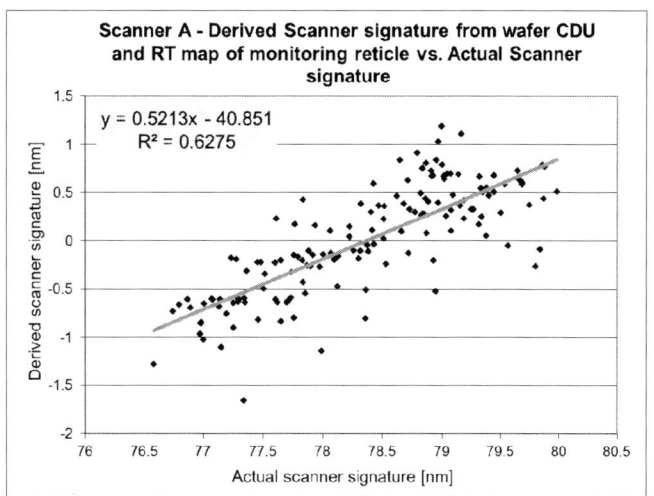

Figure 12 – Derived scanner signature from wafer CDU and RT map of monitoring
reticle in correlation to actual scanner signature

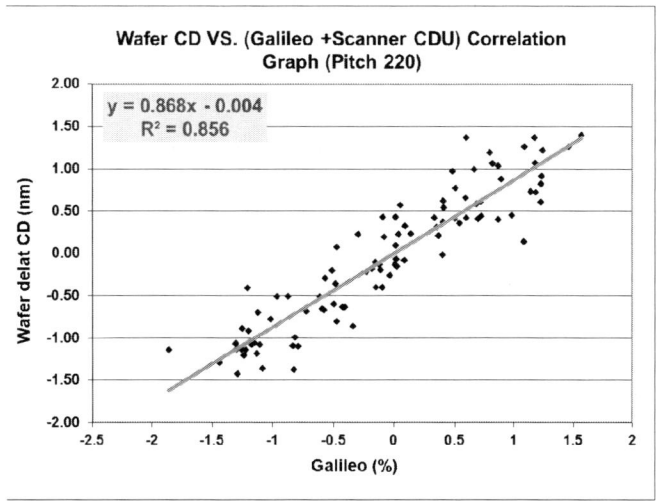

Figure 13 – Correlation graph of wafer CD vs. Galileo® measurement with scanner signature (220nm pitch)

Figure 13 shows an improved correlation between wafer CD data and Galileo® data, which includes the scanner signature for the same mask as in Figure 8. After including the scanner signature the correlation factor is improved to 92.5% (R = $\sqrt{0.856}$ = 0.925) as compared to 71% in Figure 8.

Having derived the scanner signature, it can be easily used to estimate the final wafer CD uniformity by interpolating and overlaying it on any reticle scanned on the Galileo® for the same scanner signature on similar layer and wafer process. Combining the Galileo® RT map with its respective scanners signature, a specific DOMA correction file for that reticle can be derived and used for the DOMA CDU application (instead of using wafer CD).

Cases where mask CDU signature is insignificant compared to the scanner signature

A different case where mask CDU contribution to wafer CD is minimal and scanner signature is very significant is shown in Figures 14 and 15. Figure 14 shows the wafer correlation to Galileo® CDU for this mask without scanner signature whilst Figure 15 exhibits the wafer correlation to Galileo® CDU with scanner signature. There is almost no correlation to wafer data without the scanner signature in Figure 14 which would render only noise for DOMA application. However, a useful 60% correlation is observed with scanner signature applied on the Galileo® scan as shown in Figure 15.

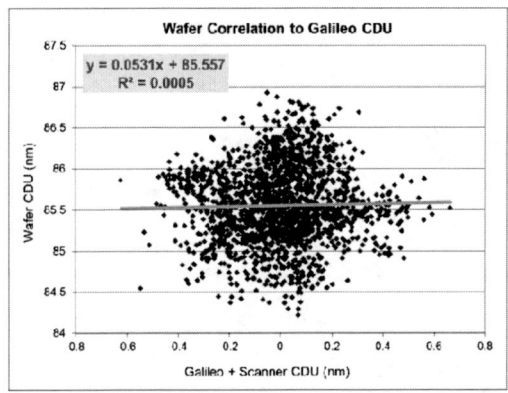

Figure 14 – Wafer correlation to Galileo® CDU
without scanner signature

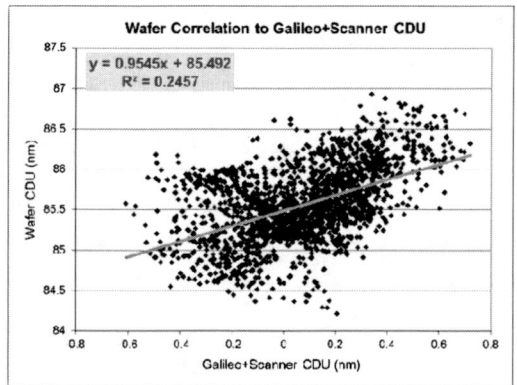

Figure 15 – Wafer correlation to Galileo® CDU
with scanner signature

It is clear from the above example that applying our technique will provide valuable DOMA correction files in both cases, either if the mask contribution to the wafer CDU is large relative to the scanner signature as well as in cases where the mask CDU is very good and the main contributor to wafer CDU is the scanner signature

3.3   CDU improvement with Galileo® & scanner signature

As the scanner signature was extracted based on one feature type (i.e Line feature of a specific design pitch, multiple array dies) of a monitoring reticle, there is a need to convert its "range" when using it on other type of reticles (e.g. for relaxed design rule or single-die reticle). Hence, the averaged transmission of the calibration and the new reticle will be used in order to calculate the factor to compensate the scanner CDU.

The dose ratio between the two masks (dose ratio = transmission ratio) is to be taken into account in order to use it to convert the scanner signature to the new mask. In the progress of this test, the scanner signature was derived from a test reticle with the average transmission of about 20%. The average transmission of the reticle to be corrected was 45% and in order to derive the correct range the map was multiplied by the ratio of the transmission ranges of $\frac{20\%}{45\%} = 0.44$

<u>DOMA Simulation and production results</u>

With the scanner signature adapted to the new scanned reticle, the scanner map can now be interpolated/ extrapolated to fit the new reticle. From Eq. (1) that represents the wafer CDU, the combined map will describe the final wafer CDU ($Wafer_{CDU} = Mask_{CDU} + Scanner_{CDU}$). Using the scanner signature generated from a specific design pitch, CDU improvement for that pitch will be improved while other pitches will have "lower" improvement only, but will still be improved all together. Henceforth, we can generate a signature based on more features, giving a different weight for each critical pitch as needed. Furthermore, scanners can be easily tracked for changes in signature (due to PM, usage etc) and DOMA application will be updated with a new signature promptly with new file generated per reticle for specific scanner as needed. Table 1 provides simulated results of running the Galileo DOMA process on several production masks in the fab.

Device description	Die size	3sigma from wafer before correction [nm]	Corrected 3sigma with Galileo® & scanner signature [nm]	CDU Improvement (DOMA simulation)
2x3 layout (6 dies)	X Large	1.134	0.585	48.4%
3x4 layout (12dies)	Large	1.886	0.853	54.7%
3x5 layout (15 dies)	Large	0.827	0.653	21.0% (*)
4x5 layout (20 dies)	Medium	0.662	0.469	29.1%
4x6 layout (24 dies)	Medium	1.385	0.995	28.1%
6x7 layout (42 dies)	Small	1.371	0.951	30.6%

(*) This reticle correction map was tested on a production wafer and yielded 17% real CDU improvement

Table 1: CDU improvement (simulation and production) summary table in terms of layout and die size

3. DISCUSSION

Galileo® metrology data is streamed in a feed-forward approach through DoseMapper into the scanner, to create a dose compensation recipe which improves the overall CDU performance. From Table 1 it can be seen, that by applying the described process of using the Galileo® transmission measurement for reticle CDU prediction and scanner signature extraction to feed the ASML Dose Mapper, significant CDU improvements have been achieved on several masks

Other than reticle transmission of new reticles and design pitches anchored, die sizes and layout will also contribute to the amount of CDU improvement. Table 1 shows the CDU improvement summary in terms of layout and die sizes. The results show that larger die sizes are more sensitive to MaskCDU and ScannerCDU. Hence CDU improvement is greater for larger die size array. The results discuss that intra-field wafer CDU gains for single-die reticle by using conventional method of correcting through wafer CD feedback to the scanner using DoseMapper is negligible. This is because there are larger non-uniformities in reticle manufacturing process than in a multiple-die reticle. The resist coating processes and etching processes non-uniformities are more dominant for single-die reticle as compared to multiple-die reticle. Henceforth, a single-die reticle will normally show a prevailing radial effect with tilt than multiple-die reticle. As a result, Galileo® metrology data will be able to correct for these sources of CDU errors and improve the wafer CDU for larger die size devices tremendously.

4. CONCLUSIONS

In this paper, it has been demonstrated that a full-field reticle transmission map measured using Galileo® can be utilized to generate a correction file for the scanner and efficiently reduce intra-field CDU on printed wafers. The scanner process signature for a specific layer can be derived by subtracting its reticle transmission map from its wafer CDU map. After extracting the scanner signature, it can be easily used to estimate the final wafer CD uniformity by interpolating and overlaying it on any reticle scanned on the Galileo® for the same scanner signature on similar layer and wafer processes. Our conclusions from this study are:

1. This feed-forward method can effectively replace the feedback concept using send-ahead wafers and extensive CDSEM measurements.

2. It is easy to implement in production as no production wafer is needed for CD measurement.

3. Correction can be applied already on the first production wafer.

4. No unique monitoring feature is needed on the production reticle and no wafer CD or after-etch CD measurement is needed on the production reticle.

5. The result is a significant cost and resource saving and will improve fab productivity.

5. ACKNOWLEDGEMENTS

The authors would like to thank and acknowledge Ching Win Kong and Lim Hui Kow (GLOBALFOUNDRIES Singapore Pte. Ltd.) for their support in bringing the tool Galileo® onsite for evaluation.

REFERENCES

1. DoseMapper^{TM} and DOMA^{TM} are trademarks of ASML.

2. Galileo® is a registered trademark of Pixer Technology Carl Zeiss SMS

3. Ben Yishai,et.al. "An IntenCD map of a reticle as a feed-forward input to DoseMapper" Proc. SPIE Vol. 7028, pp. 70283H-70283H-11 (2008)

4. Mangan Shmoolik,et.al. "Novel lithography approach using feed-forward mask-based wafer CDU correction increase fab productivity and yield" Proc. SPIE Vol. 7272, pp. 72722B-72722B-12 (2009)

5. J. van Schoot,et.al. "CD uniformity improvement by active scanner corrections" Proc. SPIE Vol. 4691, pp.304-314 (2002)

6. Rainer Pforr,et.al. "Performance comparison to techniques for intra-field CD control improvement" Proc. SPIE Vol. 6730,pp. 673032 (2007)

7. H. van der Laan,et.al. "Etch, Reticle, and Track CD Fingerprint Corrections with Local Dose Compensation" Proc. SPIE Vol. 5755, pp. 107-118 (2005)

8. G. Ben Zvi,et.al. "Mask CD Control (CDC) with Ultrafast Laser for Improving Mask CDU Using AIMS™ as the CD Metrology Data Source" Proc. SPIE Vol. 6730, pp. 67304X (2007)

9. Guy Ben-Zvi,et.al. "CD control (CDC) using AIMS as the CD metrology data source" Proc. SPIE Vol. 7028, pp. 7028-47 (2008)

Metamaterials for enhancement of DUV lithography

Andrew Estroff, Neal V. Lafferty, Peng Xie, Bruce W. Smith

Microsystems Engineering, Rochester Institute of Technology, 82 Lomb Memorial Drive, Rochester, NY, 14623

ABSTRACT

The unique properties of metamaterials, namely their negative refractive index, permittivity, and permeability, have gained much recent attention. Research into these materials has led to the realization of a host of applications that may be useful to enhance optical nanolithography, such as a high pass pupil filter based on an induced transmission filter design, or an optical superlens. A large selection of materials has been examined both experimentally and theoretically through wavelength to verify their support of surface plasmons, or lack thereof, in the DUV spectrum via the attenuated total reflection (ATR) method using the Kretschmann configuration. At DUV wavelengths, materials that were previously useful at mid-UV and longer wavelengths no longer act as metamaterials. Composites bound between metallic aluminum and aluminum oxide (Al_2O_3) exhibit metamaterial behavior, as do other materials such as tin and indium. This provides for real opportunities to explore the potential of the use of such materials for image enhancement with easily obtainable materials at desirable lithographic wavelengths.

Keywords: Surface Plasmon Resonance, Attenuated Total Reflection, DUV, Metamaterials, Superlens, Pupil Filter, Optical Lithography

1. INTRODUCTION

The refractive index of a material is conventionally considered to have a positive value. There is, however, no physical requirement that its corresponding electric permittivity (ε) and/or magnetic permeability (μ) must also be positive.

For the case when both the permittivity and permeability are negative, the negative root must be chosen ($n = \pm\sqrt{\varepsilon\mu}$), resulting in a negative refractive index. This phenomenon was first described by Veselago[1] in 1967.

Veselago demonstrated that Snell's Law is still obeyed for negative index materials (NIM) by showing that an incident wave propagating through a positive index material (PIM) with $n = 1$ into a negative index material with $n = -1$ results in a flow of power which still bends towards the normal. Contrary to convention, however, it refracts to the same side of normal as the incident wave and the wave vector points backwards towards the interface of the two materials. This leads to the notion that a simple slab of a NIM can behave as a lens. If a point source is at a distance d in front of a NIM slab, it will be brought back to a focus and imaged at the same distance d beyond the NIM slab. Assuming the PIM is surrounded by vacuum, and that the NIM has an equal but opposite refractive index of the PIM, a "perfect" lens is created; i.e. there is no reflection at the interfaces and all of the radiation is transmitted. Veselago described this, but also knew that there were no naturally occurring materials possessing a negative index.

Some materials can have a negative permittivity but exhibit a positive permeability. These are most generally metals at optical frequencies. Pendry showed that a lens can be created out of a slab of negative permittivity material (NPM), assuming the dimensions of the system are smaller than the wavelength of light (e.g. a photomask) and that the incident radiation is TM-polarized[2]. This NPM slab, or superlens, is capable of amplifying and refocusing the evanescent waves which are usually lost in a diffraction limited system and restoring their contribution to the final image. This leads to the realization of a near field imaging approach for nanolithography[3-5].

Furthermore, Berning and Turner[6], and many others[7-9], noticed that classically opaque metal films exhibit enhanced transmission when a thin layer of this metal possessing negative permittivity is sandwiched between two dielectrics. In this structure, termed an induced transmission filter, normally incident radiation typically experiences a decrease in transmission, with greater transmission observed at more oblique angles, while additionally being sensitive to both polarization and wavelength. A frequency band pass pupil filter could also be conceived from this film stack which could be applied to a far-field projection imaging system[10].

Optical Microlithography XXIII, edited by Mircea V. Dusa, Will Conley, Proc. of SPIE Vol. 7640, 76402W · © 2010 SPIE · CCC code: 0277-786X/10/$18 · doi: 10.1117/12.848427

It is understood that surface plasmons (SPs) are important to the enhancement and refocusing of evanescent waves in superlensing applications as well as providing enhanced transmission characteristics at oblique angles in induced transmission filters. However, very little SP research has been performed in the deep ultraviolet (DUV) portion of the electromagnetic spectrum[11-13], a region of interest for current nanolithography practices. The goal of this paper is to explore materials and determine those that have SP resonances in this region using ATR experiments and simulations.

2. SURFACE PLASMONS

A surface plasmon is an electromagnetic wave that travels along the boundary between a conductor and a dielectric[14,15,16]. The SP dispersion relation can be derived from Maxwell's Equations and is shown in Equation 1, where β is the propagation constant, k_0 is the wave vector of the incident radiation, and ε_m and ε_d are the dielectric functions of the conductor and dielectric:

$$\beta = k_0 \sqrt{\frac{\varepsilon_m \varepsilon_d}{\varepsilon_m + \varepsilon_d}} \, . \tag{1}$$

It should be noted that this dispersion relation is only valid for TM polarization (also known as p polarization) states; surface plasmons are not excited by TE polarization.

Equation 1 shows that the propagation constant becomes asymptotic when the dielectric functions are equal and opposite in sign. This condition is known as the surface plasmon resonance (SPR). Figure 1 shows the plasmon dispersion relation for several cases. The air light line represents the dispersion line for radiation in free space. Of interest is what occurs at the Al/air interface. At frequencies far below the SPR frequency, the SP propagation constant is relatively small, meaning that the SP wavelength is only slightly smaller than the incident radiation. As the frequency of the incident radiation increases to the surface plasmon resonance frequency, the propagation constant increases and approaches infinity at the SPR frequency, meaning that the SP wavelength becomes infinitesimally small. This is the mechanism that allows for radiation to be squeezed into very small holes, as well as providing for deep sub-wavelength resolution.

These dispersion curves do not become entirely asymptotic, and at frequencies above the SPR frequency the dispersion curve sharply moves back to a smaller propagation constant. This is because the complex dielectric function was used, and the imaginary part of the dielectric function acts as a damping term.

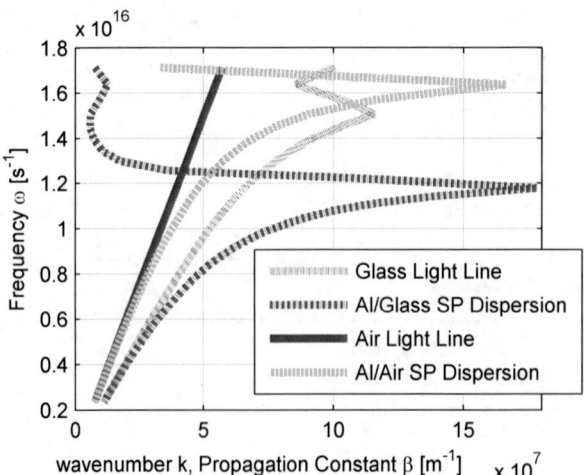

Figure 1: Surface plasmon dispersion curves for Kretschmann ATR configuration.

It is apparent from the dispersion curves that a surface plasmon on a simple NPM-dielectric interface cannot be excited by light due to the SP propagation constant always being larger than the wave vector of light in the dielectric. However, incident light can be phase-matched with the SP by means of prism coupling, a technique referred to as Attenuated Total Reflection (ATR), using either the Otto configuration[17], in which a dielectric prism and metal are separated by an air gap, or the Kretschmann configuration[18] (used in this paper), which has the prism and metal film in direct contact. In both configurations, phase matching is accomplished by varying the angle of incidence of the TM polarized radiation, resulting in a sharp decrease in the reflected intensity at some angle beyond the critical angle. The only propagation constants available are those that lie between the air and prism light lines. For ATR to be successful, the NPM film used needs to be sufficiently thin so that the incident radiation can couple from the prism to the SP propagating on the opposite side. ATR profiles are highly dependent upon film thickness and optical properties.

3. MATERIAL CONSIDERATIONS

The surface plasmon dispersion equation shows that a negative permittivity material is needed in conjunction with a dielectric to support surface plasmons. Figure 2 shows materials to be explored for use in plasmonic applications at 193nm and 248nm[19]. These materials are represented in both the dielectric function space and optical parameter space. The relationship between these two spaces is shown in Equations 2-4.

$$\varepsilon_1 = n^2 - k^2 \tag{2}$$

$$\varepsilon_2 = 2nk \tag{3}$$

$$\varepsilon = \varepsilon_1 + i\varepsilon_2 \tag{4}$$

The real part of the dielectric function, ε_1, needs to be negative to support SPs. Materials exhibiting larger negative permittivity values (i.e. a smaller refractive index and a larger extinction coefficient) are more conductive and should exhibit a stronger plasmon resonance with lower damping and absorption because of the small imaginary part of the dielectric function, ε_2. Knowing this, it is observed from Figure 2 that material candidates for use in plasmonic applications at 193nm could be Al, Sn, In, W, Si, Mo, Rh, and Cr, while at 248nm the same materials could be useful with the addition of Ni and Pd. Missing from this list are materials such as Au and Ag, which support SPs at mid-UV and longer wavelengths, but exhibit less desirable optical characteristics in the DUV spectrum.

For the perfect lens described by Veselago[1], the plasmonic superlens described by Pendry[2], and demonstrated by Fang[4], it was mentioned that the dielectric medium surrounding the NPM lens should have a permittivity of equal value but opposite in sign to reduce reflections at the interfaces. The plots in Figure 2 have a line in the positive permittivity space where a host of transparent dielectric materials exist, as well as a line in the opposite negative permittivity space where desirable NPM might exist. The material SL denotes the optical parameters for Ag at 365nm, as this material was works well as a superlens at that wavelength. At that wavelength it has low absorption, lower than most all materials displayed in Figure 2. In superlensing applications, absorption dictates the minimum resolution possible as well as greatly reducing an already small depth of focus[20]. In and Sn are close to this point at 193nm, but exhibit a higher imaginary permittivity component, resulting in more absorption. At 248nm, the only materials close to the desired permittivity space are Pd and Ni, but they exhibit even higher absorption.

It is easily observed that the optical properties of Al and Al2O3 lie on the opposite sides of the desired permittivity space. This fact suggests that a combination of the two materials, when treated as an optically homogeneous new material, may satisfy the requirements for the superlensing material. In the simple case of a planar film stack with all material boundaries parallel to the electric field, the properties of the combined Al/Al2O3 stack would reduce to the spatial average of the constituents, i.e. lie on the line connecting Al and Al2O3[21]. Such a material could be produced using sputtering techniques. The optical properties of a composite material comprised of ~85% Al and ~15% Al2O3 at 193 nm exhibit similar behavior to those of Ag at 365 nm.

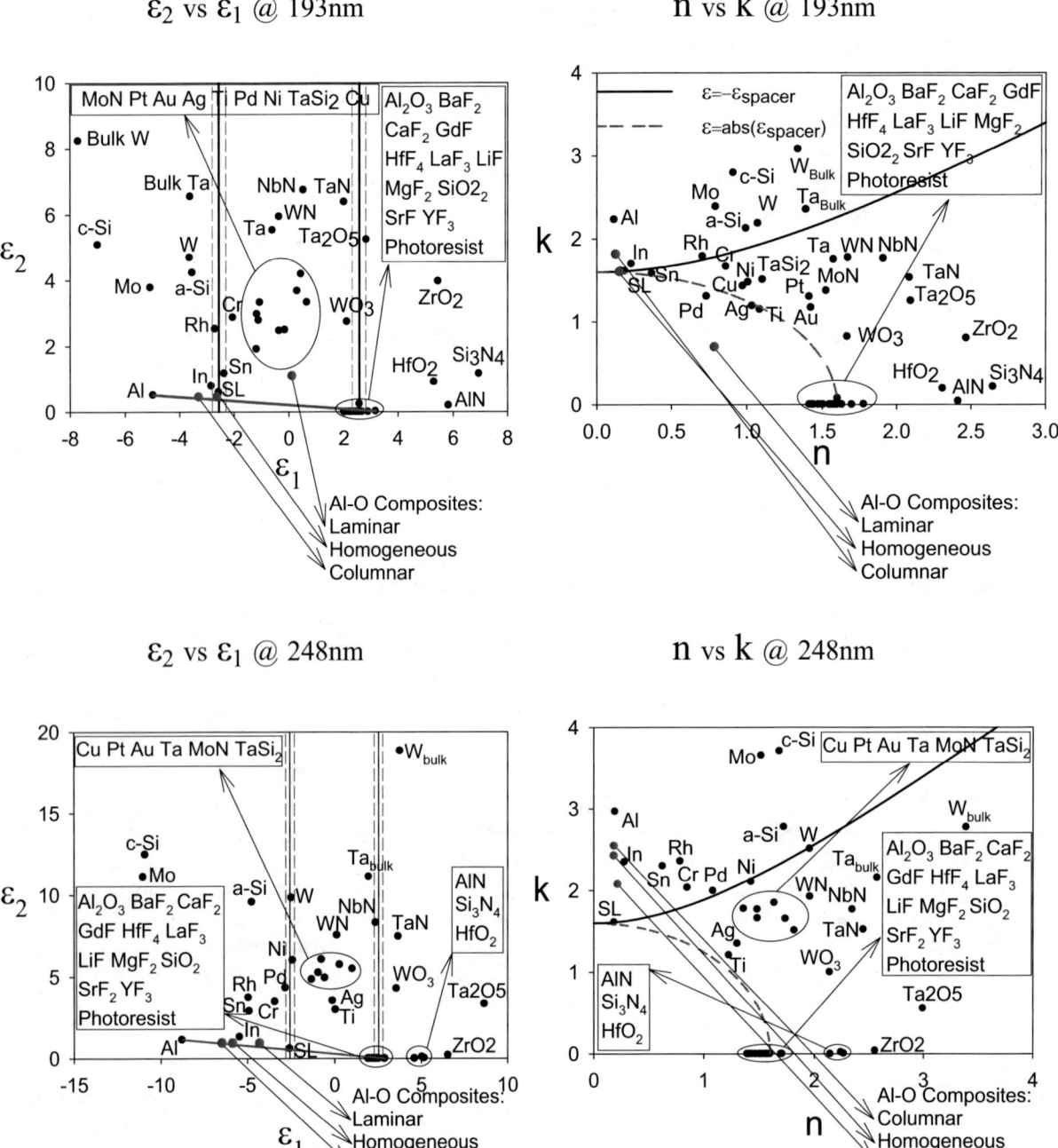

Figure 2: Material space to be explored for support of surface plasmons in the DUV spectrum. The legend for the 193nm n vs k plot applies to the 248nm case as well.

It has also been shown by Ramakrishna and Pendry[22] that the dielectric materials cladding the NPM slab lens do not need to exhibit a permittivity equal but opposite in sign to the NPM, and that this asymmetric lens can help counteract absorption effects from the NPM. This therefore removes some constraints on picking a SP supporting NPM for superlensing, and makes Al a more attractive material to as it is easily integrated with standard semiconductor processing techniques whereas materials such as Sn or In would be problematic in their unbound states.

4. SPECTRAL ANALYSIS OF SP SUPPORTING MATERIALS VIA ATR CALCULATIONS

As described previously, surface plasmons are governed by Maxwell's equations. Standard thin film transmission and reflection matrix calculations are also governed by Maxwell's equations. This matrix formalization can be used to show the interaction between the incident radiation and SPs. Subsequent modelling was carried out and a simulator was created based on these calculations to examine the SP resonance induced decrease in reflectivity at a certain range of angles and conductor thicknesses. It was also designed to examine the peak angle at which the resonance occurs as well as the resonance's width at any wavelength.

The matrix methods outlined by Macleod[23] offer an elegant way of calculating reflectance from an ATR setup. Equations 5 though 11 were used to calculate reflection for TE and TM polarization states for a given wavelength, varying the incident angle and NPM thickness.

$$
\begin{bmatrix} B \\ C \end{bmatrix} = \left\{ \prod_{j=1}^{q} \begin{bmatrix} \cos \delta_j & \dfrac{i \sin \delta_j}{\eta_j} \\ i\eta_j \sin \delta_j & \cos \delta_j \end{bmatrix} \right\} \begin{bmatrix} 1 \\ \eta_{sub} \end{bmatrix}
\tag{5}
$$

$$
y_j = n - ik
\tag{6}
$$

$$
\eta_{j_s} = \sqrt{n_j^2 - k_j^2 - n_0^2 \sin^2 \theta_0 - 2in_j k_j}
\tag{7}
$$

$$
\eta_{j_p} = \frac{y_j^2}{\eta_{j_s}}
\tag{8}
$$

$$
\delta_j = \frac{2\pi d}{\lambda} \sqrt{n_j^2 - k_j^2 - n_0^2 \sin^2 \theta_0 - 2in_j k_j}
\tag{9}
$$

$$
\rho = \frac{y_0 B - C}{y_0 B + C}
\tag{10}
$$

$$
R = \rho \rho^*
\tag{11}
$$

For the ATR technique there is only one layer, the NPM, so q and j are both 1. The substrate is air, and the initial medium is the quartz prism. The characteristic admittance of the NPM is represented as y_j and consists of the optical parameters of the NPM. The oblique optical admittance is represented by η and must be calculated for each polarization state. The film phase factor is δ. The amplitude reflection coefficient ρ can be calculated from the characteristic matrix of the film stack (Equation 5), and from this result, the reflectance can be obtained (shown in Equation 11). With oblique incidence and absorbing materials, care must be taken to ensure that the solutions for η for TE polarization states and δ are in the fourth quadrant to obtain the correct solution (positive real, negative imaginary).

Figure 3 shows the reflectance calculated for an Al film ($n = 0.179$, $k = 1.95$) in the Kretschmann configuration at a wavelength of 248nm. As expected, the incident radiation for TM polarization states is largely absorbed into the SP resonance at a range of angles between ~42-60 degrees and for a range of film thicknesses to ~15-50nm; no resonance is observed for TE polarization states.

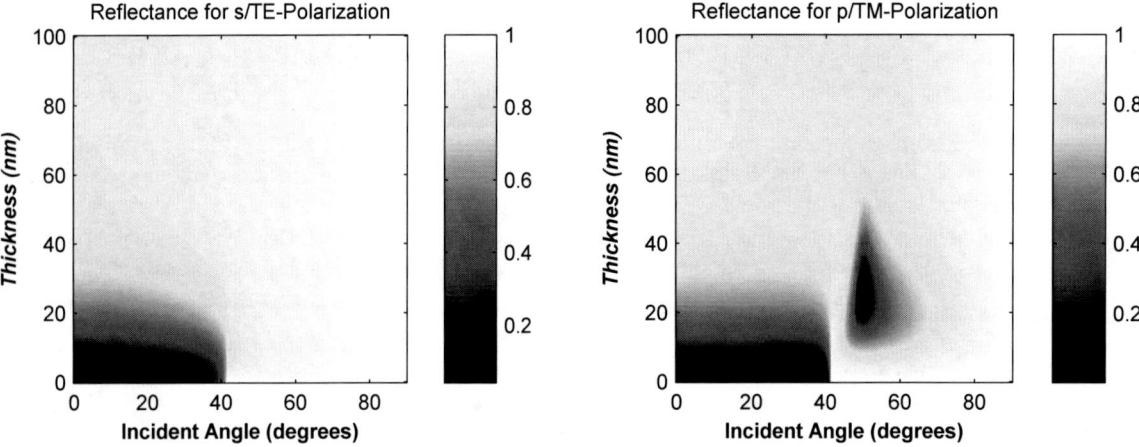

Figure 3: Calculated reflectance from an aluminum film in the Kretschmann configuration at 248nm.

To determine the spectral range over which materials from Figure 2 support surface plasmon resonances, an array was setup containing each material and its optical properties at all wavelengths at which data had been collected. For each material at each wavelength, the location of the center of the ATR peak was determined at the optimal thickness. The optimal thickness for each material case varies with wavelength. For more absorbing materials, the reflectance minimum is expected to be shallower and broader. To determine which materials are least absorptive, the filtering criteria were set to only calculate the ATR peak for materials exhibiting reflectance reaching 10% or lower, with a peak breadth at a reflectance value of 10% no greater than 10 degrees.

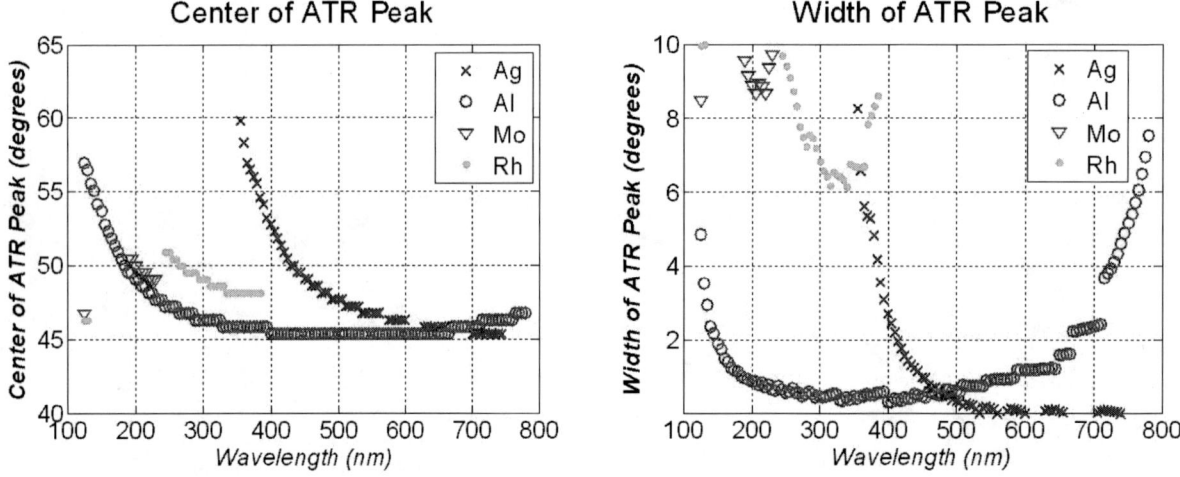

Figure 4: Calculated ATR peak location and width for selected materials.

Figure 4 shows the calculated location of the center of the ATR peak and its width for several materials of interest through wavelength. Ag has been used extensively for plasmonic applications in the mid-UV wavelengths and beyond, but these plots clearly show that it is not a useful plasmonic material below ~350nm. Alternatively, Al exhibits plasmonic behavior from the DUV through the end of the visible spectrum. Only below 150nm and above 700nm does its ATR peak begin to broaden. Mo and Rh appear to support SPs at 193nm and 248nm respectively, however, they are highly absorptive and have a very limited useful wavelength range. Sn and In which are of interest for plasmonic applications in the DUV are not shown here for several reasons. Firstly, there was little through wavelength refractive index data in the source used. Additionally, for data that was available[24], it behaved similarly to Al.

5. EXPERIMENTAL ATR RESULTS IN THE DUV

Attenuated total reflectance experiments were performed on sputtered Al and Mo films deposited on fused silica substrates (with refractive index $n_{193nm} = 1.56$ and $n_{248nm} = 1.51$). The Al film was deposited using a PE4410 RF sputterer with the following process parameters: base pressure=5.5E-7Torr, process pressure=4mTorr, argon flow rate=40sccm, sputter power=200W, deposition time=10min. The same sputter tool was used to deposit the Mo film with the following parameters: base pressure=6.0E-6Torr, process pressure=4mTorr, argon flow rate=40sccm, sputter power=700W, deposition time=2.5min. Profilometry results showed that the Al thickness was near 25nm and the Mo thickness near 15nm. A Woollam variable angle spectroscopic ellipsometer was used to obtain the thickness and optical constants of Al at 248nm, where $n = 0.329$, $k = 2.09$, and thickness=25.4nm. The experimental ATR setup is shown in Figure 5.

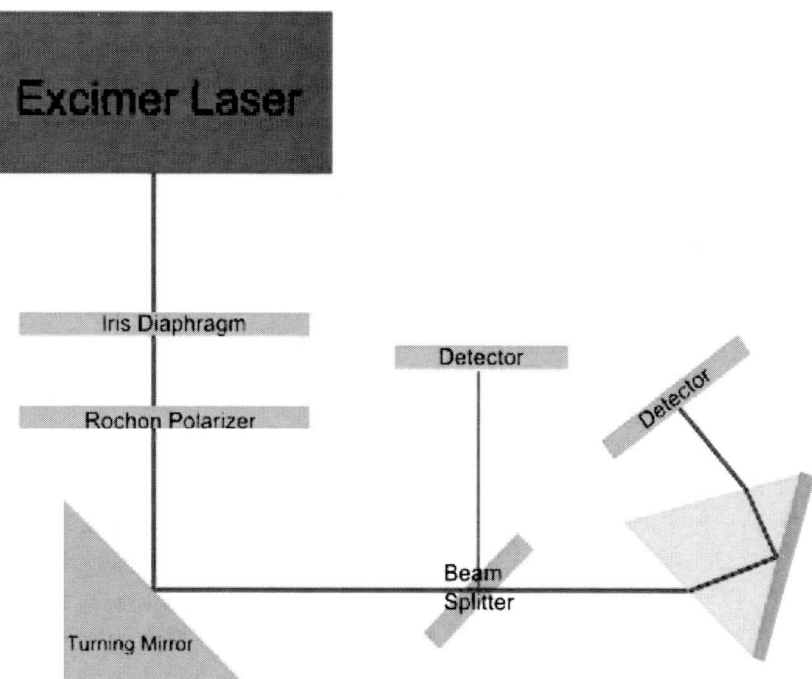

Figure 5: Experimental ATR setup.

A GAM KrF 248nm excimer laser and a GAM ArF 193nm excimer laser were used as sources, and a Coherent J5-09-0A0-5M detector was used to monitor the laser power and to measure the reflected intensity. The fused silica substrates were mated to a fused silica prism with water as an index matching fluid, with the Mo/Al films interfacing with the air. The prism was placed on a goniometric stage, which was aligned using retro-reflectance. It was estimated that the alignment was accurate to one degree.

ATR measurements were taken in one degree increments around the region where the SP resonance peak was expected to occur. A significant amount of beam spreading occurred at more oblique angles of incidence limiting the range of angles that could be measured. The ATR measurement results are shown in Figure 6.

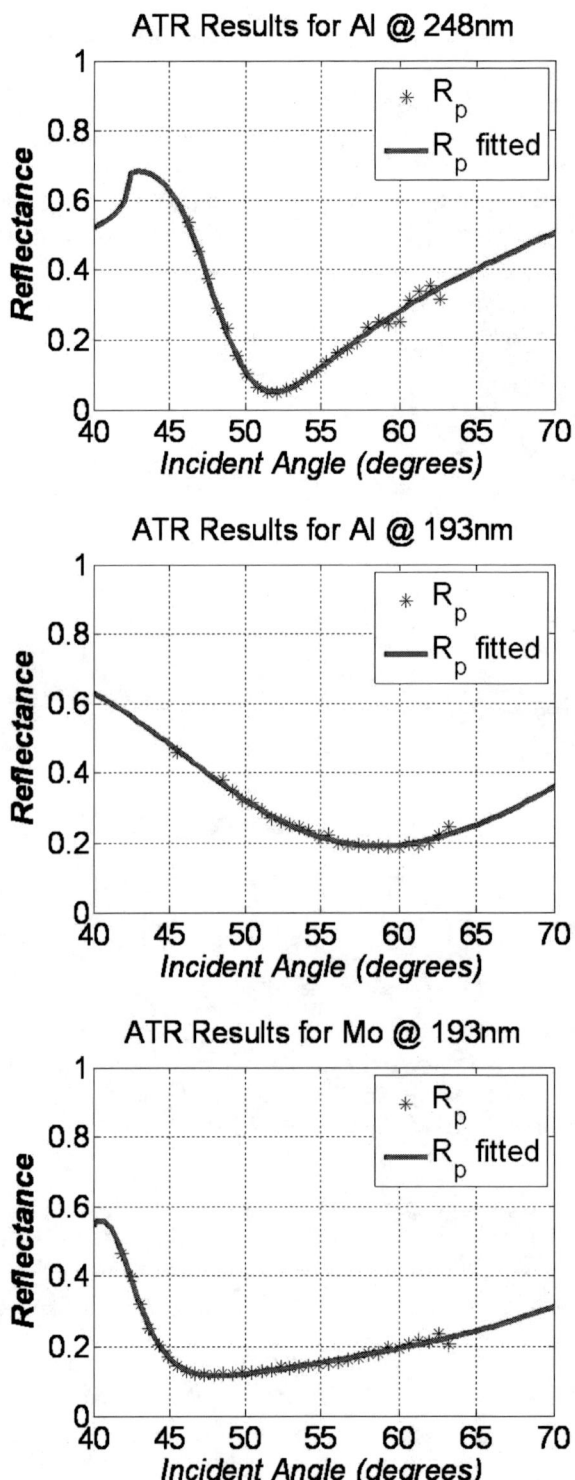

Figure 6: ATR experimental results with calculated ATR data fit by least squares optimization.

After performing the ATR experiments, a nonlinear least-squares optimization (Equation 12) was used to fit the calculated ATR data from the thin film matrix formalization (Equation 11) to the experimental results. The parameters allowed to vary to fit the data were refractive index, absorption coefficient, film thickness, experimental angular error (a global lateral shift in the data), detector offset (a global vertical shift in the data), and a scaling factor to match the two datasets. The results from this modeling routine are shown in Table 1. This optimization routine is quite useful, as it is capable of providing the optical constants for a single thin conducting film. If there is a secondary film involved, such as a surface oxide, an effective index and thickness for the entire stack can be arrived at, but not the properties of the individual films.

$$\min_x \sum_i (experimental_i - R_i(parameters))^2 \tag{12}$$

Table 1: Parameters determined by least-squares optimization of calculated ATR data to experimental.

Parameter	Al @ 248nm	Al @ 193nm	Mo @ 193nm
n	0.413	0.246	1.005
k	1.871	1.182	2.266
Thickness (nm)	24.3	24.4	16.3
Scaling factor	18.705	129.2	27.7
Angular offset (deg)	0.988	-2.99	-0.190
Detector offset	0.0276	0.0005	0.052
Error RMS	0.0129	0.0080	0.0060

The results of the least-squares optimization for Al differ from expected values. While the thickness for both wavelengths is close to what was previously measured, the angular offset is within the expected tolerances, and the error RMS is low, the optical parameters differ from those shown in Figure 2 for Al. Al reacts with oxygen at room temperature forming a thin protective oxide layer. Additionally, in such a thin film, this oxide layer may represent a substantial portion of the aluminum film, thus resulting in different effective optical properties for the Al film. The experimental results for the Mo sample look much closer to the values for Mo in Figure 2. It is known that Mo does not form a surface oxide at room temperature.

It should be noted that the scaling factor had a large effect on the fitting of the optical parameters, providing inaccurate results if allowed to range too far or not far enough. Performing some reflectance measurements below the critical angle at more normal angles of incidence should help to give a better estimate for this scaling factor, as the reflectance profile at angles below the critical angle is relatively constant. This, together with possible stage alignment and precision inaccuracies, could explain the angular offset for the 193nm Al case.

6. CONCLUSIONS

Reflectance calculations were performed using the thin film matrix formalization to examine a large number of materials to verify their support of surface plasmons in the DUV spectrum. ATR experiments were performed to verify the calculated results.

Aluminum is the best choice for plasmonic applications of the materials explored, as it exhibits a strong SP resonance throughout the wavelength range between ~150-700nm. However, aluminum readily forms an oxidized surface in air. Any oxidation on the surface changes the effective optical parameters of the film and the SP resonance properties, ultimately increasing absorption. This thin oxide layer combined with any deposition related thin film structural properties leads to a different optical properties than might be expected when examining published aluminum optical constants.

REFERENCES

[1] V. Veselago, "The electrodynamics or substances with simultaneously negative values of ε and μ", Soviet Physics Uspekhi 10(4), 509-514 (1968).

[2] J.B. Pendry, "Negative refraction makes a perfect lens", Phys. Rev. Lett. 85, 3966-3969 (2000).

[3] R.J. Blaikie, S.J. McNab, "Simulation study of 'perfect lenses' for near-field optical nanolithography", Microelectronic Eng., 61-62, 97-103 (2002).

[4] N. Fang, H. Lee, C. Sun, X. Zhang, "Sub-diffraction-limited optical imaging with a silver superlens", Science, 308, 534-536 (2005).

[5] D.O.S. Melville, R.J. Blaikie, "Super-resolution imaging through a planar silver layer", Optics Express, 13(6), 2127-2134 (2005).

[6] P.H. Berning, A.F. Turner, "Induced transmission in absorbing films applied to band pass filter design", J. Opt. Soc. Am., 47(3), 230-239 (1957.

[7] A. Salwen, L. Stensland, "Spectral filtering possibilities of surface plasma oscillations in thin metal films", Optics Comm., 2(1), 9-13 (1970).

[8] R. Dragila, B. Luther-Davies, S. Vukovic, "High transparency of classically opaque metallic films", Phys. Rev. Lett., 55(10), 1117-1120 (1985).

[9] I.R. Hooper, T.W. Priest, J.R. Sambles, "Making tunnel barriers (including metals) transparent", Phys. Rev. Lett., 97(5) (2006).

[10] H. Fukuda, T. Teresawa, S. Okazaki, "Spatial filtering for depth of focus and resolution enhancement in optical lithography", J. Vac. Sci. Technol. B, 9(6), 3113-3116 (1991).

[11] Y. Ekinci, H.H. Solak, C. David, "Extraordinary optical transmission in the ultraviolet region through aluminum hole arrays", Optics Letters, 32(2), 172-174 (2007).

[12] Y. Ekinci, H.H. Solak, J.F. Loffler, "Plasmon resonances of aluminum nanoparticles and nanorods", J. Appl. Phys., 104(8) (2008).

[13] Q. Gan, L. Zhou, V. Dierolf, F. Bartoli, "Direct mapping of the UV surface plasmons", Optics Letters, 34(9), 1324-1326 (2009).

[14] H. Raether, [Surface Plasmons on Smooth and Rough Surfaces and on Gratings], Springer, Berlin (1988).

[15] S.A. Maier, [Plasmonics: Fundamentals and Applications], Springer (2007).

[16] W.L. Barnes, "Surface plasmon-polariton length scales: a route to sub-wavelength optics", J. of Optics A, 8, 87-93 (2006).

[17] A. Otto, "Excitation of nonradiative surface plasma waves in silver by the method of frustrated total reflection", Zeitschrift fur Physik, 216, 398-410 (1968).

[18] E. Kretschmann, H. Raether, "Radiative decay of non radiative surface plasmons excited by light", Zeitschrift fur Naturforschung, 23a, 2135-2136 (1968).

[19] RIT thin films material database, http://www.rit.edu/kgcoe/microsystems/lithography/utilities.html

[20] J.L. Garcia-Pomar and M. Nieto-Vesperinas, "Transmission study of prisms and slabs of lossy negative index media", Optics Express 12, 2081 (2004).

[21] D. E. Aspnes, "Optical Characterization by Ellipsometry", Journal de Physique Colloques, 44(C10), 3-10 (1983).

[22] S.A. Ramakrishna, J.B. Pendry, "The asymmetric lossy near-perfect lens", J. Mod. Optics, 49(10), 1747-1762 (2002).

[23] H.A. Macleod, [Thin-Film Optical Filters], Macmillan Publishing Company, New York (1986).

[24] E.D. Palik, [Handbook of Optical Constants of Solids], Academic Press, San Diego (1998).

Toward a Consistent and Accurate Approach to Modeling Projection Optics

Danping Peng,* Peter Hu, Vikram Tolani, Thuc Dam

Luminescent Technologies, 2471 East Bayshore Road Suite 600, Palo Alto, CA 94303

Jacek Tyminski, Steve Slonaker

Nikon Precision Inc, 1399 Shoreway Road, Melmont, CA 94002

ABSTRACT

This paper presents a consistent and modularized approach to modeling projection optics. Vector nature of light and polarization effect are considered from the very beginning at source, through mask and projection lens down into film stack. High-NA and immersion effect are also included. Of particular interest is the formulation of a modularized framework for computing optical images that allows various mask models (a thin-mask model, an empirical approximate mask model, or a rigorous mask 3D solver) to be used. We demonstrate that under Kirchoff thin-mask assumption our formulation is the same as Smythe formula. A compact film-stack model is formulated. The formulation is first presented in Abbe's source integration approach and then reformulated in Hopkins' TCC approach which allows for a SVD decomposition, which is computationally more efficient for a fixed optical setting.

Keywords: Abbe, Hopkins, Maxwell equation, vector optics, scalar optics, polarization, partial coherent, lithography simulation

1. INTRODUCTION

Lithography simulation has been indispensable to research and development in the semiconductor industry in its insatiable drive toward printing ever smaller features with increasingly large designs. This trend has be well-captured by Moore's law.

The traditional simulator was built on the assumption of coherent illumination with low NA, and paraxial approximation was used in the modeling of optics and resist. In the past two decades, anticipating the introduction of new exposure and inspection tools, lithography simulator has been extended in several directions: from coherent illumination to off-axis illumination with different source shapes; from low NA to high-NA for dry lithography, to hyper-NA for immersion lithography; from scalar optics to vector optics that include the polarization effect at source, mask, lens and resist film stack; from a simple film-stack model that is only valid for normal incident light to one that supports high-incident angles and the vector nature of light.

Theoretically development and commercial implementation of these features did not happen over night. It took decades of intensive work in academia and industry. In the process, new features were added over the existing features, not always in a consistent formulation. It seems to the authors that there has been no systematic exposition of these developments in a consistent formulation.

The optical community has benefited tremendously from the classics by Born and Wolf,[1] now in its 7-th edition, and Goodman.[2] But they mostly focus on scalar imaging theory and use paraxial approximation, and thus are only valid for low NA optics. In his 1988 paper[3] Yeung proposed a vector imaging and film-stack model that included high-NA effect. In a later paper[4] Yeung simplified his original approach and proposed the Hopkins formulation that included thin-film effect. In his PhD thesis[7] Flagello proposed a thin-film model in matrix format that is easier to compute. Flagello and Rosenbluth[5,6] considerably simplified Yeung's approach for vector optics and film-stack effect. Cole and his co-authors[8,9] formulated the radiometric correction term for high-NA optics. See also the work of Gallatin[10] on scalar high-NA imaging.

Contact: Danping Peng, dpeng@luminescent.com

Optical Microlithography XXIII, edited by Mircea V. Dusa, Will Conley, Proc. of SPIE Vol. 7640,
76402Y · © 2010 SPIE · CCC code: 0277-786X/10/$18 · doi: 10.1117/12.848252

In this paper, we present the above developments in projection optics in a uniform and consistent formulation with some further simplification. The underlying theme of our approach will be using matrices instead of scalars throughout the whole process to track the polarization state as light propagate through the optical components of projection instrument. We take a critical view on the underlying assumptions of various modeling approaches and comment on their validity for the current exposure tools. In section 2, we define the polarization vector and coherence matrix at a single point source. In section 3, we present the vector diffraction model from mask near-field to entrance pupil and from entrance pupil to wafer plane. In section 4, we present the mask model first in a general setting, and then present some simplifying assumptions, and take a critical view on the validity of Hopkins assumption and thin-mask model. Our imaging formulation is compatible with mask models from the simplest to the full rigorous 3D EMF solution. In section 5, we present the projection lens model. We present the ideal lens operator, derive the radiometric correction term, and present how to add aberrations and Jones pupil. We then present the image formation equation, and briefly mentioned how to include a film-stack model. Compared with the previous approach, our formulation of film-stack model use a 3x2 matrix that is more compact. We then put all the pieces together, concluding with Abbe's approach to projection optics. We briefly comment on how to reformulate the Abbe's approach into Hopkins' formulation.

All results in the paper have been presented in published sources in one form or another by various authors. We simply compile the results scattered in many sources by dozens of authors in a consistent formulation.

A few comments on our choice of notation. The variables on the source side will have either a subscript or super script s, as in (α_s, β_s); mask side variables does not use subscript or super script, as in (α, β); image side variables use sup-script $'$, as in (α', β').

We use two coordinate systems, the right-handed (x, y, z) system and the intrinsic local (s, p) system which depends on unit wave-vector $\hat{k}$. Whenever possible, we use the (s, p) system as the formula is simplest in the local intrinsic system.

We will follow the Matlab convention in denoting $[\alpha; \beta; \gamma]$ as column-vector while $[\alpha, \beta, \gamma]$ as row-vector.

2. POLARIZATION VECTOR AND COHERENCE MATRIX

Consider a monochromatic plane wave propagating in the direction $\hat{k}$, where $\hat{k} = [\alpha; \beta; \gamma]$ is the direction cosine with $\gamma = \sqrt{1 - \alpha^2 - \beta^2}$. We define the TE- or s-direction, and TM- or p-direction, as

$$\hat{e}_\perp = \frac{\hat{z} \times \hat{k}}{|\hat{z} \times \hat{k}|} = \begin{bmatrix} \frac{-\beta}{\rho} \\ \frac{\alpha}{\rho} \\ 0 \end{bmatrix}, \hat{e}_\parallel = \hat{k} \times \hat{e}_\perp = \begin{bmatrix} \frac{-\alpha\gamma}{\rho} \\ \frac{-\beta\gamma}{\rho} \\ \rho \end{bmatrix}, \tag{1}$$

where $\rho = \sqrt{\alpha^2 + \beta^2}$, such that $[\hat{e}_\perp; \hat{e}_\parallel; \hat{k}]$ is right-handed. In the special case when the plane wave propagate along the z-axis, that is when $\alpha = \beta = 0$, we agree that $\hat{e}_\perp = [1; 0; 0]$, and $\hat{e}_\parallel = [0; 1; 0]$.

Since electro-magnetic waves are transverse, an arbitrary E-field can be represented as

$$\mathbf{E} = E_\perp \hat{e}_\perp + E_\parallel \hat{e}_\parallel = [\hat{e}_\perp, \hat{e}_\parallel] \begin{bmatrix} E_\perp \\ E_\parallel \end{bmatrix}, \tag{2}$$

where $E_\perp$ is the electric field in s-state, while $E_\parallel$ in p-state. The vector $[E_\perp; E_\parallel]$ describes the polarization state of the plane wave. We will call it the polarization vector. Over the cause of propagation, the polarization vector will be transformed by different components of the optical system; The modeling of the optical system is accomplished by tracking the changes of polarization vector though the optical apparatus.

At any instance of time, the phases of $E_\perp$ and $E_\parallel$ are rapidly changing, and the changes between the two state may be co-related. It is not the field $E_\perp$ and $E_\parallel$ but the average of their intensity and the correlation between them over a microscopically long enough period of time that are detectable by physical devices. So it only makes sense to talk about the average of their intensity and correlation. The coherence matrix is defined as

$$J = \begin{bmatrix} J_{11} & J_{12} \\ J_{21} & J_{22} \end{bmatrix} = \begin{bmatrix} \langle \bar{E}_\perp E_\perp \rangle & \langle \bar{E}_\perp E_\parallel \rangle \\ \langle \bar{E}_\parallel E_\perp \rangle & \langle \bar{E}_\parallel E_\parallel \rangle \end{bmatrix} \tag{3}$$

where $\langle \cdot \rangle$ denotes the time average. Coherence matrix completely describe the polarization states of plane waves.

When tracking the propagation of light through optical systems, we found it easier to use the (s,p) systems; but when we need to compute the actual field or intensity, it is convenient to use the (x,y,z) coordinate system. The mapping between the two systems is:

$$
\begin{bmatrix} E_x \\ E_y \\ E_z \end{bmatrix} = \begin{bmatrix} \hat{e}_\perp, \hat{e}_\| \end{bmatrix} \begin{bmatrix} E_\perp \\ E_\| \end{bmatrix} = \begin{bmatrix} \frac{-\beta}{\rho} & \frac{-\alpha\gamma}{\rho} \\ \frac{\alpha}{\rho} & \frac{-\beta\gamma}{\rho} \\ 0 & \rho \end{bmatrix} \begin{bmatrix} E_\perp \\ E_\| \end{bmatrix} \tag{4}
$$

Note that since electromagnetic waves are transverse, we only need to specify the two components E_x and E_y,

$$
\begin{bmatrix} E_x \\ E_y \end{bmatrix} = T \begin{bmatrix} E_\perp \\ E_\| \end{bmatrix} = \begin{bmatrix} \frac{-\beta}{\rho} & \frac{-\alpha\gamma}{\rho} \\ \frac{\alpha}{\rho} & \frac{-\beta\gamma}{\rho} \end{bmatrix} \begin{bmatrix} E_\perp \\ E_\| \end{bmatrix}. \tag{5}
$$

The third component E_z is determined from

$$
E_z = -\frac{1}{\gamma}(\alpha E_x + \beta E_y). \tag{6}
$$

Conversely, given E_x and E_y, the (s,p) components can be determined by

$$
\begin{bmatrix} E_\perp \\ E_\| \end{bmatrix} = T^{-1} \begin{bmatrix} E_x \\ E_y \\ E_z \end{bmatrix} = \frac{1}{\gamma} \begin{bmatrix} -\frac{\beta\gamma}{\rho} & \frac{\alpha\gamma}{\rho} \\ -\frac{\alpha}{\rho} & -\frac{\beta}{\rho} \end{bmatrix} \begin{bmatrix} E_x \\ E_y \end{bmatrix} \tag{7}
$$

We will use the mapping T and T^{-1} frequently below to change between the (x,y) and (s,p) coordinate systems. The following representation of T and T^{-1} are quite handy

$$
T = \begin{bmatrix} \hat{e}_{\perp,x} & \hat{e}_{\|,x} \\ \hat{e}_{\perp,y} & \hat{e}_{\|,y} \end{bmatrix}, T^{-1} = \begin{bmatrix} \hat{e}_{\|,y} & -\hat{e}_{\|,x} \\ -\hat{e}_{\perp,y} & \hat{e}_{\perp,x} \end{bmatrix} \tag{8}
$$

For scanners and AIMS tools, the angles of incident and diffracted light with optical axis on mask-side are small, hence $\gamma \approx 1$. Some implementations ignore γ in the above mapping and use the following approximation to T and T^{-1}

$$
\tilde{T} = \begin{bmatrix} \frac{-\beta}{\rho} & \frac{-\alpha}{\rho} \\ \frac{\alpha}{\rho} & \frac{-\beta}{\rho} \end{bmatrix}, \tilde{T}^{-1} = \begin{bmatrix} -\frac{\beta}{\rho} & \frac{\alpha}{\rho} \\ -\frac{\alpha}{\rho} & -\frac{\beta}{\rho}. \end{bmatrix} \tag{9}
$$

We will see later that keeping the factor γ will lead to the celebrated Smythe-Kirchoff formula, and is more consistent with Maxwell's equation even under the thin-mask approximation.

3. DIFFRACTION MODEL

In modeling projection optics, there are two kinds of diffraction that we encounter. One is the diffraction from a flat screen to a spherical surface, the other configuration reverse the former, namely the diffraction from a spherical surface to a flat surface near the focus. The former describes the diffraction from mask near field to entrance pupil of projectors in far-field, and the latter describe the focusing effect of projectors from exit pupil to wafer plane near the focus of projectors. In both setups we assume the radius of the spherical surface is much larger than the wavelength so that we are in the Fraunhofer domain. We first present the scalar diffraction formulas and then extend them to vector case.

It is well known that the propagation of light in free-space is governed by Helmholtz equation

$$
(\Delta + k^2)U = 0, \tag{10}
$$

where U is any of the components of the vector of electromagnetic field. The distribution of U on a spherical surface in the far field from a flat screen Σ can be shown to be[2]

$$U_{\text{diffr}}(\hat{k}) = U_{\text{diffr}}(\alpha, \beta) = \frac{\cos\theta}{j\lambda} \frac{e^{jkr}}{r} \iint_{\Sigma} U(x, y) e^{-j2\pi(fx+gy)} \, \mathrm{d}x\mathrm{d}y, \tag{11}$$

where $\hat{k} = [\alpha; \beta; \gamma]$ is the unit vector in the direction of propagation, $\gamma = \cos\theta$ represents the direction cosine of the propagation direction with optical axis, $k = \frac{2\pi n}{\lambda}$ the wave number, r is the radius of the spherical surface, $U(x, y)$ the field distribution over flat screen Σ, and

$$f = \frac{n}{\lambda}\alpha, g = \frac{n}{\lambda}\beta, h = \frac{n}{\lambda}\gamma, \tag{12}$$

represent the spatial frequencies.

The diffraction from a spherical surface S to the plane that is vertical to the optical axis and through its focus is described by

$$U_{\text{diffr}}(x, y) = \frac{1}{j\lambda} \frac{e^{-jkr}}{r} \iint_{S} U(\alpha, \beta) e^{j2\pi(fx+gy)} \, \mathrm{d}\sigma, \tag{13}$$

where $U(\alpha, \beta)$ is the field distribution on the sphere, $\mathrm{d}\sigma$ is the spherical areal element, which can be expressed using direction cosines

$$\mathrm{d}\sigma = r^2 \frac{\mathrm{d}\alpha\mathrm{d}\beta}{\gamma} \tag{14}$$

Since the three components of a vector field $\mathbf{E}$ in a homogeneous and isotropic media must individually satisfy Helmholtz equation, the above formula is also valid for vector optics. But he three components are not independent, $\mathbf{E}$ must satisfy Maxwell's equation, which translates to the divergence-free condition $\nabla \cdot \mathbf{E}(x, y, z) = 0$ in spatial domain, or $\hat{k} \cdot \mathbf{E}(\alpha, \beta) = 0$ in frequency domain. Because of this, it is only necessary to specify the behavior of two components of E under consideration.

Now we apply the above consideration to vector diffraction, with the assumption that only two of the components are governed by the formula, and the third component is implicitly defined by equation (6) if we need it. For diffraction from the mask near-field to the entrance pupil with radius r,

$$\mathbf{E}^{\text{ent}}(\alpha, \beta) = \frac{\cos\theta}{j\lambda} \frac{e^{jkr}}{r} \iint_{\Sigma} \mathbf{E}^{\text{mask}}(x, y) e^{-j2\pi(fx+gy)} \, \mathrm{d}x\mathrm{d}y. \tag{15}$$

For diffraction from exit pupil to wafer plane with radius r',

$$\mathbf{E}^{\text{wafer}}(x', y') = \frac{n'}{j\lambda} r' e^{jk'r'} \iint_{S} \mathbf{E}^{\text{ext}}(\alpha', \beta') e^{j2\pi(f'x'+g'y')} \frac{\mathrm{d}\alpha'\mathrm{d}\beta'}{\gamma'}. \tag{16}$$

Since $\hat{k} \cdot \mathbf{E}^{\text{ent}}(\alpha, \beta) = 0$, we obtain the full-vectorial diffraction equation

$$\mathbf{E}^{\text{ent}}(\alpha, \beta) = \frac{\cos\theta}{j\lambda} \frac{e^{jkr}}{r} \iint_{\Sigma} \left[\mathbf{E}^{\text{mask}} - \frac{(\hat{k} \cdot \mathbf{E}^{\text{mask}})}{(\hat{k} \cdot \hat{z})} \hat{z} \right] e^{-j2\pi(fx+gy)} \, \mathrm{d}x\mathrm{d}y \tag{17}$$

from the observation the the integration of the term $\frac{(\hat{k}\cdot\mathbf{E}^{\text{mask}})}{(\hat{k}\cdot\hat{z})}\hat{z}$ is zero. The introduction of this term into the mask near-field guarantee the far-field is transverse. This equation has an particularly elegant vector representation

$$\mathbf{E}^{\text{ent}}(\alpha, \beta) = \frac{1}{j\lambda} \frac{e^{jkr}}{r} \hat{k} \times \iint_{\Sigma} (\hat{z} \times \mathbf{E}^{\text{mask}}(x, y)) e^{-j2\pi(fx+gy)} \, \mathrm{d}x\mathrm{d}y. \tag{18}$$

This is the Smythe formula.[14] We will derive this equation in a more restricted setting and comment on it later when we discuss mask models.

4. MASK MODEL

Consider a typical point source s that emanates a plane wave in the direction $\hat{k}_s = [\alpha_s; \beta_s; \gamma_s]$ with wave vector $\mathbf{k}_s = \frac{2\pi n}{\lambda}[\alpha_s; \beta_s; \gamma_s]$, and illuminates the mask.

The incident plane wave is modulated by mask pattern and generates a field distribution just below mask called mask near-field. Let the mask near-field from an incident plane wave in s-mode be $[m_{11}(x,y;\alpha_s,\beta_s); m_{21}(x,y;\alpha_s,\beta_s)]$, and that from a p-mode as $[m_{12}(x,y;\alpha_s,\beta_s); m_{22}(x,y;\alpha_s,\beta_s)]$. For an incident plane wave with polarization vector $\left[E_\perp^s; E_\parallel^s\right]$, the mask near-field can be represented as

$$\mathbf{E}^{\text{mask}}(x,y;\alpha_s,\beta_s) = M(x,y;\alpha_s,\beta_s)\begin{bmatrix} E_\perp^s \\ E_\parallel^s \end{bmatrix} \tag{19}$$

where

$$M(x,y;\alpha_s,\beta_s) = \begin{bmatrix} m_{11}(x,y;\alpha_s,\beta_s) & m_{12}(x,y;\alpha_s,\beta_s) \\ m_{21}(x,y;\alpha_s,\beta_s) & m_{22}(x,y;\alpha_s,\beta_s) \end{bmatrix} \tag{20}$$

is called the mask transfer matrix for point source s. It is a 2 by 2 matrix that has the mask induced polarization effect built-in.

Once the mask near-field is known, the mask diffraction vector in the (s,p) coordinates can be represented by

$$\mathbf{E}^{\text{mask}}(\alpha,\beta;\alpha_s,\beta_s) = \begin{bmatrix} E_\perp^{\text{mask}} \\ E_\parallel^{\text{mask}} \end{bmatrix} = T^{-1}(\alpha,\beta)M(\alpha,\beta;\alpha_s,\beta_s)\begin{bmatrix} E_\perp^s \\ E_\parallel^s \end{bmatrix}, \tag{21}$$

where $T(\alpha,\beta)$ is the mapping from (x,y) coordinates to (s,p) system, and $M(\alpha,\beta;\alpha_s,\beta_s)$ is the Fourier transform of the 2 by 2 mask transfer matrix called the mask diffraction matrix.

In general, the mask near-field has to be computed with a rigorous Maxwell-equation solver, using for example RCWA or FDTD approaches. For an extended source distribution, it is impractical to rigorously solve Maxwell equation for the mask near-field for each source point. At most the rigorous solvers are invoked to compute rigorously the mask near-field for a few reference source points judiciously chosen, and approximate the mask near-field for nearby source points from the information of those of reference source points. For ease of presentation, we pick the on-axis point source as reference, and assume the mask transfer matrix for this reference is obtained through either a rigorous solver or some approximation.

For scalar optics, the mask diffraction matrix reduces to scalar mask diffraction spectrum. In this case, we normally assume that the diffraction amplitude of an off-axis point source has the same amplitude but with a shift in frequency space by the off-axis amount. This is the so-called Hopkins assumption for scalar optics.

For vector optics, extension will be performed in two steps. We first relate the polarization vector at an arbitrary point source s with that of reference r by first map $\left[E_\perp^s; E_\parallel^s\right]$ to (x,y) coordinates using the transfer matrix T_s, and then map the (x,y) components to (s,p) components using the transfer matrix T_r^{-1} for the reference point-source. We then assume the diffraction matrix is a shifted version of that of the reference. When the reference is on-axis, the transfer matrix $T_r = I$. We assume

$$M(\alpha,\beta;\alpha_s,\beta_s) = M(\alpha - \alpha_s, \beta - \beta_s)T(\alpha_s,\beta_s) \tag{22}$$

where $M(\alpha - \alpha_s, \beta - \beta_s) = M(\alpha - \alpha_s, \beta - \beta_s; 0, 0)$ is the shifted diffraction matrix of reference. This is the Hopkins assumption for vector optics. It is this crucial assumption that makes the TCC formulation possible. For scanner and AIMS modeling, this is not a serious constraint.[12] For inspection tools, the incident angles on the mask side can be over 70 degrees. Some modification of this assumption has to be made.

For many applications, the speed of a rigorous solver for even a few reference points is not fast enough. Various approximations have to be used.

The simplest mask model is the thin mask model. The mask diffraction matrix for reference is assumed to be

$$M(\alpha,\beta) = \begin{bmatrix} m(\alpha,\beta) & 0 \\ 0 & m(\alpha,\beta) \end{bmatrix} \tag{23}$$

where $\hat{m}(\alpha, \beta)$ is the Fourier transform of $m(x, y)$, a scalar mask transmission function that is dependent on the mask pattern.

Although in wide use, the thin-mask assumption has been demonstrated to be inaccurate for feature widths much smaller than the wavelength. The boundary-layer model (BLM) attempts to correct some of the inaccuracies. It postulates that the transmission function $m(x, y)$ in a thin layer around the mask edge is a constant different from the bulk regions. We derived and implemented a generalization of the basic BLM to make the transmission function dependent on the polarization state and mask edge normal and tangential direction, and get very good match to the far-field from a rigorous solver.

Putting what we obtained so far together, we obtained

$$\begin{bmatrix} E_\perp^{\text{mask}} \\ E_\parallel^{\text{mask}} \end{bmatrix} = T^{-1}(\alpha, \beta) M(\alpha - \alpha_s, \beta - \beta_s) T(\alpha_s, \beta_s) \begin{bmatrix} E_\perp^s \\ E_\parallel^s \end{bmatrix}, \tag{24}$$

Under the thin-mask assumption, the above formulation becomes.

$$\begin{bmatrix} E_\perp^{\text{mask}} \\ E_\parallel^{\text{mask}} \end{bmatrix} = \hat{m}(\alpha - \alpha_s, \beta - \beta_s) T^{-1}(\alpha, \beta) T(\alpha_s, \beta_s) \begin{bmatrix} E_\perp^s \\ E_\parallel^s \end{bmatrix}. \tag{25}$$

Simple algebra shows that this is exactly the celebrated Smythe-Kirchoff equation[14] under the thin-mask assumption:

$$\mathbf{E}(\hat{k}) = \hat{k} \times \iint (\hat{z} \times \mathbf{E}_s) m(x, y) e^{-j\frac{2\pi}{\lambda}(\alpha x + \beta y)} \, \mathrm{d}x \mathrm{d}y \tag{26}$$

We can replace the transformation matrix T with $\tilde{T}$ in the above formula, and obtained a slightly different model that is used by some implementation. We feel that keeping the γ factor, especially on the image-side, is more consistent with vector formulation and is important to correctly characterize high-NA effects. Our formulation here is a generalization to arbitrary patterns to the formula proposed in Chen[11] for line-and-space patterns. Chen and his co-authors showed that results obtained by keeping the γ factor match with experimental data better.

5. PROJECTOR MODEL AND THE FORMATION OF IMAGES

An ideal projector will transform a diverging spherical wave at the entrance pupil into a converging spherical wave at the exit pupil. Let us denote $\hat{k} = [\alpha; \beta; \gamma]$ as the unit wave-vector on the mask side, and $\hat{k}' = [\alpha'; \beta'; \gamma']$ as the unit wave vector of the corresponding wave-vector at the exit pupil. The unit vectors $\hat{e}_\perp$ is unchanged from entrance to exit pupil, but $\hat{e}_\parallel$ undergoes a rotation and becomes $\hat{e}_\parallel'$.

For a projector with demagnification factor M, the sine-condition says

$$Mn\alpha = -n'\alpha' \tag{27}$$
$$Mn\beta = -n'\beta' \tag{28}$$

where n and n' are the refractive index of the media on the mask and image side, respectively.

The spatial frequency on image and mask side are related by

$$f' = \frac{n'\alpha'}{\lambda} = -M\frac{n\alpha}{\lambda} = -Mf \tag{29}$$

$$g' = \frac{n'\beta'}{\lambda} = -M\frac{n\beta}{\lambda} = -Mg, \tag{30}$$

For an ideal projector without aberration, the projector operation is an identity in the (s, p) coordinate system

$$\begin{bmatrix} E_\perp^{ext} \\ E_\parallel^{ext} \end{bmatrix} = \begin{bmatrix} E_\perp^{ent} \\ E_\parallel^{ent} \end{bmatrix} \tag{31}$$

Our goal next is to first express the field in wafer plane in terms of the diffraction of mask near-field, and eventually in the polarization vector of the point source in source space. For the purpose, let $\mathbf{E}^{\text{mask}}(\alpha, \beta)$ be the Fourier transform of $\mathbf{E}^{\text{mask}}(x, y)$, and $\mathbf{E}^{\text{wafer}}(\alpha', \beta')$ be the Fourier transform of $\mathbf{E}^{\text{wafer}}(x', y')$. We can rewrite the diffraction from mask near-field to entrance pupil as

$$\mathbf{E}^{\text{ent}}(\alpha, \beta) = \frac{\gamma}{j\lambda} \frac{e^{jkr}}{r} \mathbf{E}^{\text{mask}}(\alpha, \beta), \tag{32}$$

and represent the field at wafer plane in terms of the field at exit pupil as

$$\mathbf{E}^{\text{wafer}}(\alpha', \beta') = \frac{1}{\gamma'} \frac{1}{j\lambda} r' e^{jk'r'} \mathbf{E}^{\text{ext}}(\alpha', \beta'). \tag{33}$$

We next derive the radiometric correction term from consideration of energy conservation. Represent the field at exit pupil in terms of wafer-plane field as

$$\mathbf{E}^{\text{ext}}(\alpha', \beta') = \gamma' \frac{j\lambda}{n'} \frac{e^{-jk'r'}}{r'} \mathbf{E}^{\text{wafer}}(\alpha', \beta'). \tag{34}$$

Consider the pencil of rays entering entrance pupil at the direction $[\alpha; \beta; \gamma]$ and exit the pupil at the direction $[\alpha'; \beta'; \gamma']$. For an ideal lens, the conservation of energy requires that

$$n|\mathbf{E}^{\text{ent}}(\alpha, \beta)|^2 r^2 \frac{d\alpha d\beta}{\gamma} = n'|\mathbf{E}^{\text{ext}}(\alpha', \beta')|^2 r'^2 \frac{d\alpha' d\beta'}{\gamma'}. \tag{35}$$

This translates to

$$\frac{|\mathbf{E}^{\text{wafer}}(\alpha', \beta')|}{|\mathbf{E}^{\text{mask}}(\alpha, \beta)|} = c\sqrt{\frac{\gamma}{\gamma'}} \tag{36}$$

where c is a constant independent of diffraction order. The factor

$$R(\gamma, \gamma') = \sqrt{\frac{\gamma}{\gamma'}} \tag{37}$$

is called the radio-metric correction term.

In general, the projector has aberrations, and the wafer plane may not be at focus. All these effect can be represented by multiplying the mask spectrum vector with a scalar or 2 by 2 matrix (Jones matrix). In (s, p) components the wafer field can be expressed as

$$\begin{bmatrix} E_{\perp}^{\text{wafer}} \\ E_{\parallel}^{\text{wafer}} \end{bmatrix} = R(\gamma, \gamma') A(\alpha', \beta') J(\alpha', \beta') C(\alpha', \beta') \begin{bmatrix} E_{\perp}^{\text{mask}} \\ E_{\parallel}^{\text{mask}} \end{bmatrix}, \tag{38}$$

where

$$C(\alpha', \beta') = \begin{cases} 1, & \sqrt{\alpha'^2 + \beta'^2} \leq NA \\ 0, & \text{otherwise} \end{cases} \tag{39}$$

and $R(\gamma, \gamma')$ is the radio-metric correction term, $J(\alpha', \beta')$ a 2 by 2 Jones matrix, $A(\alpha', \beta')$ the aberration term.

When we compute the total field from different diffraction orders at the imaging plane, we can not simply add the s- and p-componets for different plane waves directly, since the unit s- and p-vectors are different for different diffraction orders. We must change from (s, p) coordinate system to the fixed (x, y, z) system. We also need to include the z-component that has been suppressed up to now to account for the high-NA effect.

$$\mathbf{E}_s^{\text{wafer}}(\alpha', \beta') = \frac{1}{\sqrt{1 - \gamma'^2}} \begin{bmatrix} -\beta' & -\alpha'\gamma' \\ \alpha' & -\beta'\gamma' \\ 0 & 1 - \gamma'^2 \end{bmatrix} \begin{bmatrix} E_{\perp}^{\text{wafer}} \\ E_{\parallel}^{\text{wafer}} \end{bmatrix} = B(\alpha', \beta'; \alpha_s, \beta_s) \begin{bmatrix} E_{\perp}^s \\ E_{\parallel}^s \end{bmatrix} \tag{40}$$

where

$$B(\alpha', \beta'; \alpha_s, \beta_s) = \begin{bmatrix} \frac{-\beta'}{\rho'} & \frac{-\alpha'\gamma'}{\rho'} \\ \frac{\alpha}{\rho'} & \frac{-\beta'\gamma'}{\rho'} \\ 0 & \rho' \end{bmatrix} W(\alpha', \beta'; \alpha_s, \beta_s) \tag{41}$$

and $W(\alpha', \beta')$ is a 2 by 2 matrix defined by

$$W(\alpha', \beta'; \alpha_s, \beta_s) = A(\alpha', \beta')J(\alpha', \beta')R(\gamma, \gamma')C(\alpha', \beta')T^{-1}(\alpha, \beta)M(\alpha - \alpha_s, \beta - \beta_s)T(\alpha_s, \beta_s). \tag{42}$$

The interference of different plane waves at the image plane is mathematically a Fourier inversion transformation. Hence the total E-field from points source s can be represented as

$$\mathbf{E}_s^{\text{wafer}}(x', y') = \iint \mathbf{E}_s^{\text{wafer}}(\alpha', \beta')e^{j2\pi(f'x'+g'y')} \, d\alpha' d\beta' = B_s(x', y') \begin{bmatrix} E_\perp^s \\ E_\parallel^s \end{bmatrix} \tag{43}$$

where $B_s(x', y')$ is the component-wise Fourier transform of the 3 by 2 matrix $B(\alpha', \beta'; \alpha_s, \beta_s)$. Its column represent the E-field in physical domain for a source point in pure s-state or p-state.

In the presence of film-stack, we need to introduce change the 3 by 2 matrix in equation (40) to

$$\begin{bmatrix} \frac{-\beta'}{\rho'} A_\perp^q(z) & \frac{-\alpha'\gamma'}{\rho'} A_\parallel^q(z) \\ \frac{\alpha}{\rho'} A_\perp^q(z) & \frac{-\beta'\gamma'}{\rho'} A_\parallel^q(z) \\ 0 & \rho' \frac{n'\gamma'}{n_q\gamma_q} B_\parallel^q(z) \end{bmatrix} \tag{44}$$

where $A_\perp^q(z)$ and $A_\parallel^q(z)$ are the horizontal components of field within layer q of the film-stack, for a unit incident wave to the film-stack in s-state and p-state, respectively, $B_\parallel^q(z)$ is the z-component in p-state. The details will be presented in a future work.

The intensity from point source s can be computed as

$$I_s = |E_s^{\text{wafer}}|^2 = \begin{bmatrix} \bar{E}_\perp^s, \bar{E}_\parallel^s \end{bmatrix} B^* B \begin{bmatrix} E_\perp^s \\ E_\parallel^s \end{bmatrix}. \tag{45}$$

The total intensity from an extended source can be computed by

$$I = \iint e(\alpha_s, \beta_s) \begin{bmatrix} \bar{E}_\perp^s, \bar{E}_\parallel^s \end{bmatrix} B^* B \begin{bmatrix} E_\perp^s \\ E_\parallel^s \end{bmatrix} \, d\alpha^s d\beta^s \tag{46}$$

where $e(\alpha_s, \beta_s)$ represent the source strength at point source s.

The above formulation has followed the Abbe's source integration approach. Under the Hopkins' assumption, equation (46) can be reformulated by change the order of integration into a quadratic functions of the 4 elements of mask diffraction matrix, and the standard spectrum decomposition of a compact operator in the Hilbert space of square integrable functions[13] (the equivalent of SVD in discrete finite-dimensional space) can be applied to obtained a set of TCC-kernels. The details for a 2 by 2 mask transfer matrix will be presented elsewhere.

If we further assume the thin-mask model, great simplification is achieved. We can find a set of TCC-kernels ϕ_k and corresponding eigenvalues λ_k dropping to zeros very fast, and the intensity be represented as

$$I = \sum_k \lambda_k |\phi_k \otimes m|^2 \tag{47}$$

where m as we recall, is the scalar transmission function. This formulation can be extended to cases where more than one transmission functions are used, for example in the rigorous solver. The details will be presented in a future work.

Once the optical image is obtained within the film stack, the exposure step transforms the light intensity distribution (called resist latent image) into acid concentration, and during PEB, acid and base are diffused and quenched, and the concentration of dissolution inhibitor can be computed by solving a reaction-diffusion-integration equations numerically. Using one of the well-known development rate model, the developed resist profile can be computed via a moving interface solver. We have implemented these components in our own work using some of the latest research research in computational mathematics, and the results compares favorably with those from other implementations. We will report the work elsewhere.

6. CONCLUSION

We presented a unified approach to modeling projection optics. We take a critical look at the modeling approach to each optical components, and point to the validity and limitations of the underlying assumptions. Future works will provide more details of some of the fine points briefly touched upon in the paper.

REFERENCES

[1] M. Born, E. Wolf, *Principle of Optics*, Cambridge University Press, 7th Ed. 1999

[2] J. Goodman, *Introduction to Fourier Optics*, 2nd ed, 1996.

[3] M. S. Yeung, *Modeling High Numerical Optical Lithography*, SPIE Vol 922 Optical/Laser Microlithography, 1988.

[4] M. S. Yeung, D. Lee, R. Lee and A.R. Neureuther, *Extension of the Hopkins Theory of Partially Coherent Imaging to Include Thin-Film Interference Effects*, SPIE Vol 1927, Optical/Laser Microlithography VI, 1993.

[5] D. G. Flagello, A.E. Rosenbluth, *Lithographic Tolerances Based On Vector Diffraction Theory*, J. Vac. Sci. Technol. B 10(b), 1992.

[6] D. G. Flagello, A.E. Rosenbluth, *Vector Diffraction Analysis of Phase-Mask Imaging in Photoresist Films*, SPIE Vol. 1927, Optical/Laser Microlithography VI, 1993.

[7] D. G. Flagello, *High Numerical Aperture Imaging in In Homogeneous Thin Films*, PhD Thesis, Univ. Of Arizona, 1993.

[8] D. C. Cole, E. Barouch, U. Hollerbach, S. Orszag, *Derivation and Simulation of High Numerical Aperture Scalar Aerial Images*, Jpn. J. Appl. Phys. Vol. 31, 1992

[9] D. C. Cole, E. Barouch, U. Hollerbach, S. Orszag, *Extending Scalar Aerial Image Calculations To Higher Numerical Apertures*, Jpn. J. Appl. Phys. Vol. 31, 1992

[10] G. M. Gallatin, *High-Numerical-Aperture Scalar Imaging*, Applied Optics, Vol. 40, No. 28, Oct. 2001.

[11] C. K. Chen, T.S. Gau, L.H. Shiu, Burn Lin *Mask Polarization Effects in Hyper NA Systems*, SPIE Vol. 5754, 2005.

[12] T. V. Pistor, A. R. Nuereuther, R. J. Socha, *Modeling Oblique Incidence Effects in Photomasks*, SPIE Vol. 4000, 2000

[13] F. Riesz, B. Sz.-Nagy, *Functional Analysis*, Dover, 1990.

[14] D. Jackson, *Classical Electrodynamics*, 3nd ed, 1999.

High Fluence Testing Of Optical Materials For 193-nm Lithography Extensions Applications

V. Liberman[a], S. Palmacci[a], G. P. Geurtsen[a], M. Rothschild[a] and P. A. Zimmerman[b]

[a]Lincoln Laboratory, Massachusetts Institute of Technology
Lexington, MA 02420
[b]Intel assignee to SEMATECH

ABSTRACT

As next generation immersion lithography, combined with double patterning, continues to shrink feature sizes, the industry is contemplating a move to non-chemically amplified resists to reduce line edge roughness. Since these resists inherently have lower sensitivities, the transition would require an increase in laser exposure doses, and thus, an increase in incident laser fluence to keep the high system throughput.

Over the past several months, we have undertaken a study at MIT Lincoln Laboratory to characterize performance of bulk materials (SiO_2 and CaF_2) and thin film coatings from major lithographic material suppliers under continuous 193-nm laser irradiation at elevated fluences. The exposures are performed in a nitrogen-purged chamber where samples are irradiated at 4000 Hz at fluences between 30 and 50 mJ/cm^2/pulse. For both coatings and bulk materials, in-situ laser transmission combined with in-situ laser-induced fluorescence is used to characterize material performance. Potential color center formation is monitored by ex-situ spectrophotometry. For bulk materials, we additionally measure spatial birefringence maps before and after irradiation. For thin film coatings, spectroscopic ellipsometry is used to obtain spatial maps of the irradiated surfaces to elucidate the structural changes in the coating.

Results obtained in this study can be used to identify potential areas of concern in the lens material performance if the incident fluence is raised for the introduction of non-chemically amplified resists. The results can also help to improve illuminator performance where such high fluences already occur.

Keywords: High fluence, 193 nm lithography, optical materials, SiO_2, CaF_2

1. INTRODUCTION

As device features approach the 32-nm node, linewidth roughness of chemically-amplified resists becomes a major contributor to limiting the printed resolution. Thus, a move to non-chemically amplified resists has been contemplated in several studies.[1,2] Since these resists inherently have lower sensitivities than currently-used chemically-amplified resists, the transition would require an increase in exposure dose delivered to the wafer.

One way to increase the exposure dose without raising the energy per pulse is through increasing the laser pulse repetition rate. The pulse repetition rate of the 193-nm lithography-grade excimer lasers has been climbing over the last several years, currently reaching 6000 Hz.[3] While the reported performance data from excimer laser manufacturers does not suggest that 6 kHz is a fundamental limit to the repetition rate, the engineering challenges would make it difficult to extend performance beyond this point.[4,5] Therefore, for the same repetition rate, a higher dose at the wafer plane implies more fluence per pulse on the illuminator and lens components to keep up the wafer throughput of the lithographic tool.

We report the results of a six-month long assessment of bulk optical materials and thin film coatings under 193-nm irradiation at elevated fluences. Up to 700×10^6 (700 Mp) pulses have been accumulated on individual samples with fluences up to 50 mJ/cm^2/pulse. In-situ transmission data and laser-induced fluorescence are used to assess sample performance throughout irradiation. Ex-situ spectroscopic ellipsometry, spectrophotometry and birefringence are used to supplement the in-situ information. We find that while bulk CaF_2 and thin film coatings undergo only subtle changes

Optical Microlithography XXIII, edited by Mircea V. Dusa, Will Conley, Proc. of SPIE Vol. 7640,
76402Z · © 2010 SPIE · CCC code: 0277-786X/10/$18 · doi: 10.1117/12.853094

during laser irradiation, bulk SiO_2 shows rapid transmission degradation and, thus, does not appear to be suitable for use at fluences beyond 35 mJ/cm^2/pulse.

2. EXPERIMENTAL OVERVIEW

2.1 Laser irradiation setup

A diagram of our exposure chamber is shown in Figure 1. The chamber is nitrogen-purged. Irradiation is performed with a 4000 Hz NovaLine A4030 Coherent laser, which does not have a line-narrowed option. The pulsewidth of the laser is 11 nanoseconds full-width half-maximum (FWHM) or 22 ns Time Integral Squared (TIS), where TIS metric is defined in Reference 6. The laser beam is polarized in the horizontal direction. The beam shape at the laser output is flat-top along the vertical direction with a width of 14 mm and Gaussian along the horizontal direction with a 1/e^2 beam width of 2.5 mm. To achieve the required elevated fluence at the sample plane with minimum variation along the sample axis, we focus the beam along the vertical direction with a CaF$_2$ cylindrical lens with a 50-cm focal length at 193 nm. At focus, the laser beam is approximately square with dimensions of ≈2.5x2.5 mm^2, as ascertained by exposing a photosensitive polymer film from SensorPhysics Corporation.[7] Total beam energy is obtained by placing a calibrated thermopile detector from Coherent Corporation into the beam but away from focus. The fluence is then obtained by dividing the beam energy by the spot size measured above. We verified that over the sample length of 2 cm the laser fluence did not change by more than 10%.

Samples are mounted on a computer-controlled vertical translation stage for exact positioning inside the laser beam. For all samples, the beam is centered on the sample in vertical and horizontal direction to ease the overlapping of the exposed spot with ex-situ measurement probes.

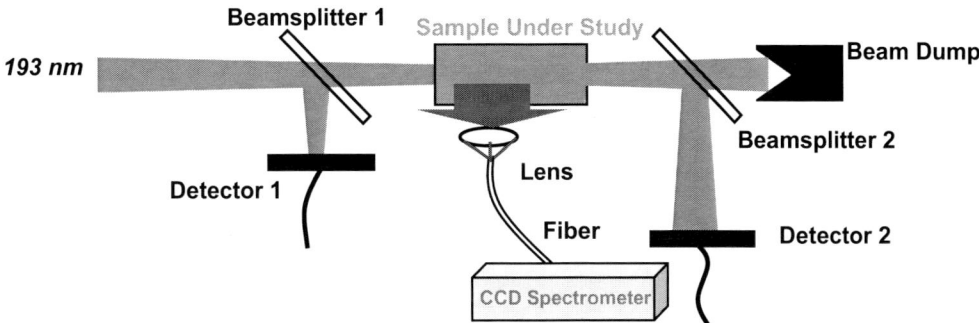

Figure 1. Diagram of the laser exposure setup.

2.2 Samples geometry and in-situ transmission.

Samples were received from the leading suppliers of 193-nm lithographic optical materials. Two suppliers provided bulk SiO_2 and CaF_2 samples; three suppliers provided thin film coatings for testing. For the bulk testing, the samples were rectangular with dimensions of 2x2x1 cm, polished on all sides. Both CaF_2 and SiO_2 materials were tested. All the materials were lithographic grade, optimized for 193-nm exposure. The laser was incident on one of the 2x1 cm faces; thus, the pathlength inside the sample was 2 cm.

For the thin film testing, either one-sided or two-sided coatings were tested. Coatings were deposited on a 2-mm thick CaF_2 substrate, provided by the coating suppliers. Either single-layer films or multiple layers were tested.

Samples were provided in optically clean packaging. No attempt was made to pre-clean any of the samples prior to irradiation.

To measure in-situ transmission, a fraction of the beam (≈17% at s-polarization of the incident beam) was picked off with a blank CaF_2 beamsplitter in front of the sample and with another blank CaF_2 beamsplitter after the sample. Light from the two beamsplitters was directed onto two pyroelectric detectors from Coherent Corporation (Figure 1). The signal from the two pyroelectrics detectors was fed into a dual channel energy meter from Coherent Corporation, EPM 2000. A pulse-by-pulse ratio of the detector signals was obtained for two sample positions: 1) the "baseline" measurement with the sample retracted out of the laser beam and 2) the "sample" measurement with the sample inserted into the laser path. Two thousand pulses were averaged for each of the sample positions. Because of the limited response time of the energy meter and detectors, the laser repetition rate was decreased to 500 Hz for the transmission measurement. Transmission was measured every 10 Mp for long-term irradiation and every 1 Mp for shorter pre-screening runs. The experiment was fully computer-controlled; thus, exposures were performed 24 hours a day.

2.3 Laser induced fluorescence.

Laser-induced fluorescence (LIF) was obtained by positioning a fiber with a focusing lens attachment a fixed distance away from the sample. Different fiber positioning fixtures were used for bulk and coating samples due to differences in respective geometries of the two types of samples; however, for each type of sample, the signals from samples of different grades could be directly compared. Typically, the focusing lens was positioned 1–1.5 cm away from the sample surface. For the bulk samples, fluorescence was measured at right angles to the sample irradiation (Figure 2A). The focusing lens was positioned perpendicular to the 2x2 cm sample face at the same vertical level as the irradiation beam. For the thin film coatings, the samples were tilted so that the angle of incidence of the irradiating beam was 25 degrees with respect to normal (Figure 2B). This tilt prevented 193-nm reflected light from reaching the focusing lens and, thus, avoided spurious 650-nm fluorescence from the SiO_2 lens material itself. The angular tilt caused a slightly higher apparent transmission of the coating due to p-polarization with respect to the laser. The apparent transmission increase as compared to normal-incidence transmission is a result of reduced reflection and depends on the reflectivity of the initial sample. Thus, for an uncoated CaF_2 blank, transmission increases by about 2%, whereas for a one-sided AR coating on a CaF_2 substrate, transmission increases by about 1.2%, as compared to normal incidence.

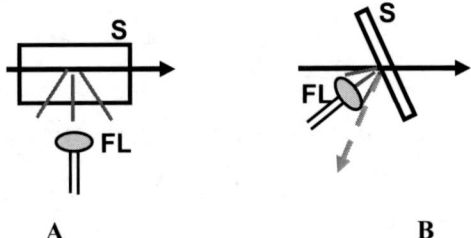

A   **B**

Figure 2. LIF measurement configuration for A) bulk and B) coating samples. Top view is shown. S is the sample and FL is the focusing lens. Solid arrows are incident 193 nm light, and the dashed arrow is reflected 193 nm light; thin lines refer to fluorescence.

Once the fluorescence was focused onto the fiber tip, it was guided through the fiber onto an inline filter holder, which contained a 193 nm normal-incidence dielectric mirror for rejection of Rayleigh scattering. Thereafter, fluorescent light was detected by a charge-coupled device (CCD) spectrometer from Ocean Optics Corporation. The spectrometer has a wavelength response from 190 nm to 800 nm and a bandwidth of ≈2 nm. The spectrometer signal was corrected by the appropriate grating efficiency curve.[8] Typical acquisition times varied between 10 ms and 100 ms, depending on the signal strength. Fluorescence was always measured at the same repetition rate and the same per-pulse fluence as the sample exposures. Fluorescence was measured on the same schedule as the in-situ transmission measurements described in section 2.2.

2.4 Ex-situ diagnostics

In addition to in-situ transmission and fluorescence measurements described above, several ex-situ measurements were performed. For the bulk materials, spectral transmission was measured before and after irradiation with a Hitachi U4001 spectrophotometer. The spectrometer spot size was 2x2 mm and the measured wavelength range was from 190 to 600 nm. Bulk sample birefringence at 633 nm was measured with a Hinds Exicor 150AT tool with a 0.5 mm spot. Birefringence spatial maps were obtained with a 1x1 mm X-Y grid resolution.

For thin film coatings, additional ex-situ measurements included spectroscopic ellipsometry. Ellipsometric data were acquired with a Woollam M2000 instrument over the spectral range of 250 nm to 1000 nm. Spatial ellipsometric maps were obtained over the irradiated area with a 1x1 mm grid resolution.

3. IRRADIATION RESULTS

3.1 Bulk CaF₂

Irradiation of bulk CaF_2 sample for 650 Mp pulses at an incident fluence of 50 mJ/cm^2/pulse (a total dose of 32 MJ/cm^2) showed no change of in-situ transmission within the measurement drift of $\approx 0.3\%$, or, correspondingly, no change in absorption coefficient within 7×10^{-4}/cm, base 10. (Fig. 3) We can compare these results to our earlier comprehensive study of bulk CaF_2 under a lower fluence irradiation at 1 mJ/cm^2/pulse.[9] In that earlier study, lithographic-grade bulk CaF_2 was irradiated for pulse counts up to 4×10^9 (a total dose of 4 MJ/cm^2) with the total induced absorbance varying from $<5 \times 10^{-4}$ to 3×10^{-3}/cm, base 10, depending on the material grade. Our high-fluence test results confirm excellent durability of lithographic-grade CaF_2 over the wide fluence range tested to date.

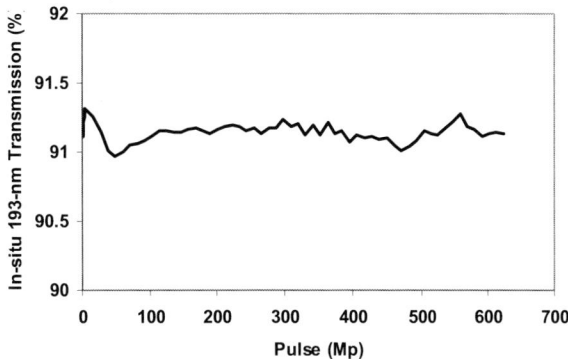

Figure 3. In-situ transmission of a bulk CaF_2 sample.

Spectral transmission before and after irradiation of the sample (Fig. 4) revealed substantial recovery which can be attributed to a combination of surface cleaning and bleaching.[9] Since no corresponding increase in transmission was observed in-situ, we conclude that the transmission recovery is rapid, occurring during the first few thousand pulses as the first in-situ measurement is obtained.

No laser-induced birefringence was detected in irradiated CaF_2 within the sensitivity of the measurement, which is <0.2 nm/cm at 633 nm. This is consistent with past studies in which CaF_2 did not display laser-induced density changes.[9]

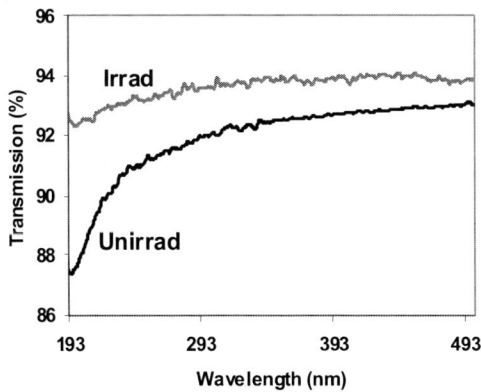

Figure 4. Ex-situ transmission spectrum of the bulk CaF_2 sample from Fig. 3 comparing irradiated and unirradiated areas.

The LIF signal from the sample at the start and at the conclusion of irradiation shows a slight decrease in overall signal intensity, but no major changes in spectral shape (Fig. 5A). The spectrum has a major peak around 320 nm with a repeatable fine structure between 315 and 330 nm, as indicated by arrows in Figure 5A. Additionally, a weak feature is seen around 500 nm. The peak LIF signal intensity dropped within the first 6 million pulses and then remained constant for the rest of the irradiation. (Fig. 5B)

Previous LIF studies of CaF_2 have been reported for 157-nm,[10] 193-nm[10-12] and 248-nm[13] incident irradiation. From those studies, the peak positions of the fine structure marked by arrows in Figure 5A correspond to very low levels of metallic impurities, such as rare earth ions. Impurities at such low levels are not expected to impact CaF_2 transmission. The major emission peak for a high purity CaF_2 corresponds to a self-trapped exciton (STE), arising from two-photon absorption. However, the peak emission wavelength of the STE peak observed here is considerably red-shifted from the expected 278 nm[10-13] to 320 nm in this work. A previous study of CaF_2 STE luminescence when excited with synchrotron irradiation observed a similar red-shift of the emission peak and attributed it to a perturbation resulting from residual impurities or defects in the crystal.[14] At present, we are not able to fully understand the STE shift in our test sample without knowing the details of the sample processing.

The intensity of the STE peak should have a quadratic dependence on fluence, as it is caused by two-photon absorption.[10] The ≈20% emission intensity drop observed in Figure 5B may then correspond to a 10% drop in laser energy per pulse. Such an energy drop is, indeed, observed at the beginning of the exposure run as the laser tube temperature reaches its constant operating value.

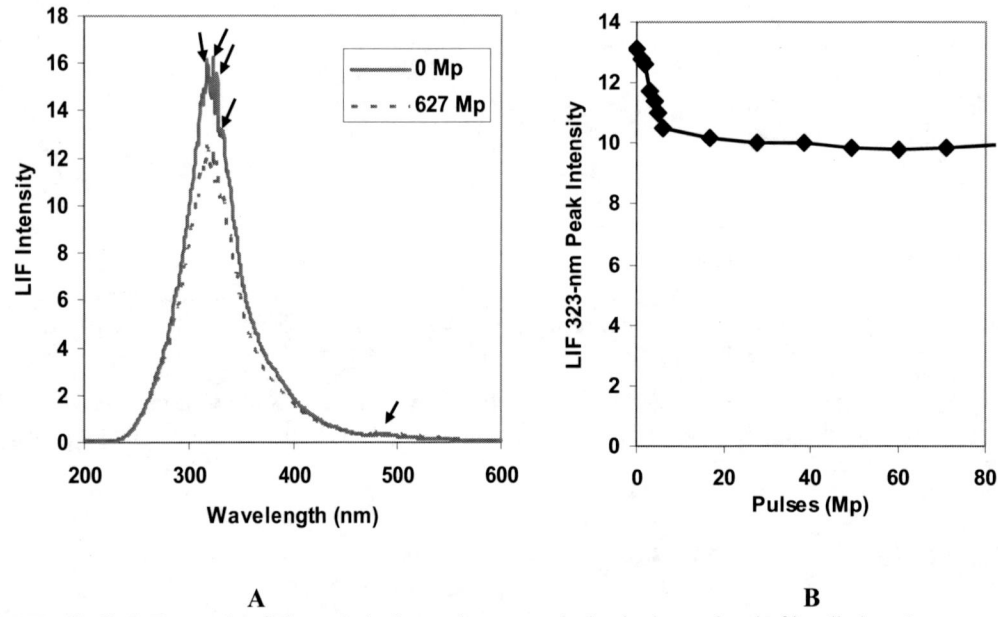

A   **B**

Figure 5. LIF of bulk CaF_2 sample of Figure 3. A. Spectral traces at the beginning and end of irradiation. Arrows point to peaks corresponding to a repeatable fine structure, described in the text. B. Evolution of the major peak intensity with laser irradiation.

3.2   Bulk SiO₂

Bulk SiO_2 was tested at a laser fluence of 35 mJ/cm^2/pulse. Four samples of different material grades, labeled 1 – 4, were irradiated. From the transmission results, all the samples were found to degrade to unacceptable levels within $<5 \times 10^6$ pulses. (Fig. 6A) Two of the samples, labeled 3 and 4, revealed substantially lower transmission with the very first in-situ measurement: less than 80% at 193 nm. By contrast, spectrophotometer pre-measurement of these samples showed 193-nm transmission of ≥87 %, a more reasonable value considering that the surfaces of the samples were not pre-cleaned. (For reference, theoretical 193-nm maximum transmission of SiO_2 material is 90.6%, accounting for surface reflections.) From this discrepancy between the laser-based and lamp-based initial measurements, we postulate that samples 3 and 4 exhibited the rapid damage phenomenon (RDP), which has been observed previously in bulk SiO_2

irradiated at 248 nm[15] and 193 nm.[16,17] To highlight the differences in sample degradation rates, we re-plot the sample data in terms of absorbance (Fig. 6B). Absorbance is calculated from

$$A = \frac{\log_{10}\left(\frac{90.6}{T}\right)}{l} ,\qquad(1)$$

where A is absorbance, base 10 per cm at 193 nm, T is the in-situ 193-nm transmission and l is the sample length. From Figure 6B, we estimate linear slopes for absorbance growth (or damage rate) for the four samples in units of absorbance/cm/Mp: 0.0032 for sample 1, 0.022 for sample 2, 0.0012 for sample 3 and 0.0027 for sample 4. The two samples that exhibit RDP (3 and 4) have the lowest damage rate.

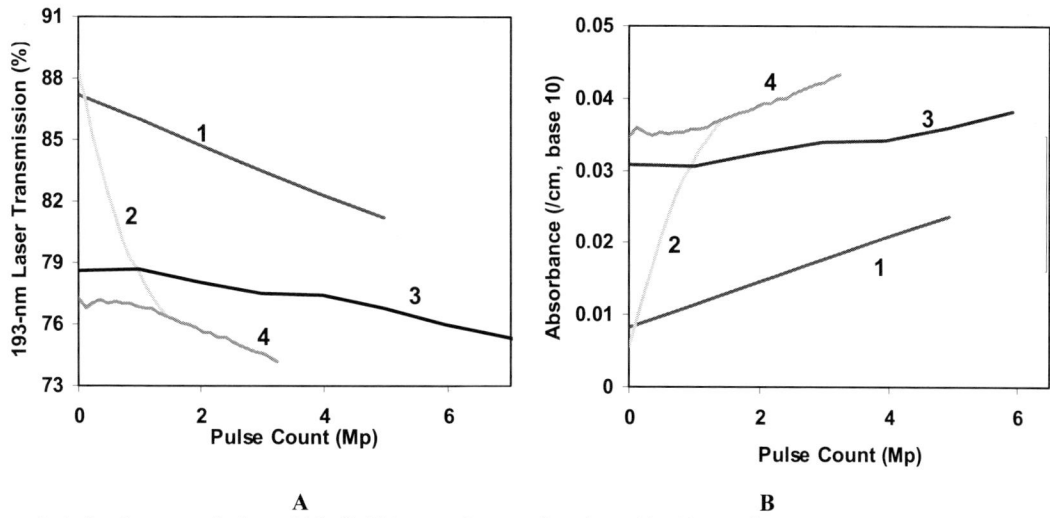

A   **B**

Figure 6. A. In-situ transmission of 4 bulk SiO_2 samples as a function of incident pulse count. B. Corresponding 193-nm absorbance per cm of the 4 bulk SiO_2 samples vs. pulse count. For both figures, samples numbers are shown on the graphs next to appropriate traces

No microchannel formation was observed in any of the SiO_2 samples after laser irradiation. Such filamentary defect growth, which originates at the exit sample face, has been seen by us and others in accelerated 193-nm laser SiO_2 testing.[18, 19]

While none of the samples would be acceptable for use at these elevated fluences, we present detailed damage analysis of samples 2 and 4 in order to illustrate the damage mechanisms. Spectral transmission measurements of the bulk SiO_2 samples 2 and 4 after irradiation revealed the well-known E' color center ($\equiv$Si•) in samples 2 and 4 and NBOH center ($\equiv$Si–O•) in sample 2, located at 220 nm and 270 nm, respectively (Fig. 7).[20] Comparing the spectrometer-based 193-nm transmission of sample 4 at the conclusion of irradiation, we observe a transmission recovery to 85%, compared to the 74% in-situ laser-based measurement. This behavior of sample 4 is consistent with the RDP phenomenon.

Laser-induced birefringence of irradiated fused silica samples was assessed by comparing pre-and post-irradiated spatial birefringence maps. A radially-symmetric birefringence pattern, consistent with unconstrained densification under circular beam irradiation,[9,21,22] was observed in samples 1, 3 and 4. An example of the magnitude map of such a pattern is shown in Figure 8 for sample 3, which received the highest exposure dose. Table 1 summarizes peak birefringence signals obtained in all the samples versus dose. Sample 2 received the lowest dose and showed no measurable laser-induced birefringence. The amount of developed birefringence appears to scale with incident dose across the four samples tested.

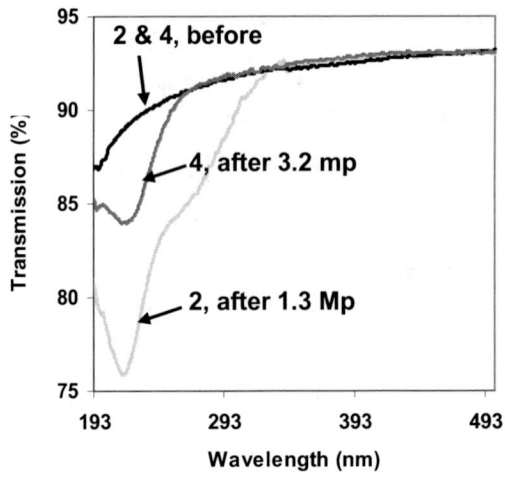

Figure 7. Spectral transmission of bulk SiO_2 samples 2 and 4 before and after laser irradiation.

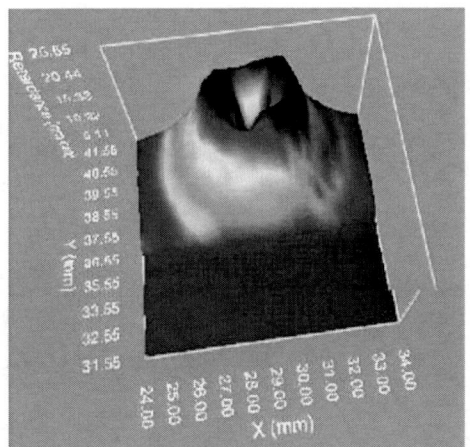

Figure 8. Laser-induced birefringence map of bulk SiO_2 sample 3 acquired at 633 nm.

Table 1. Summary of the peak laser-induced birefringence data for all SiO_2 samples tested

Sample Number	Peak Birefringence (nm/cm)	One-photon dose (NI) in kJ/cm^2	Two-photon dose (NI^2/τ) in $J^2/cm^4/nsec$
1	3.5	180	280
2	<1	60	100
3	13	700	1100
4	2.5	100	200

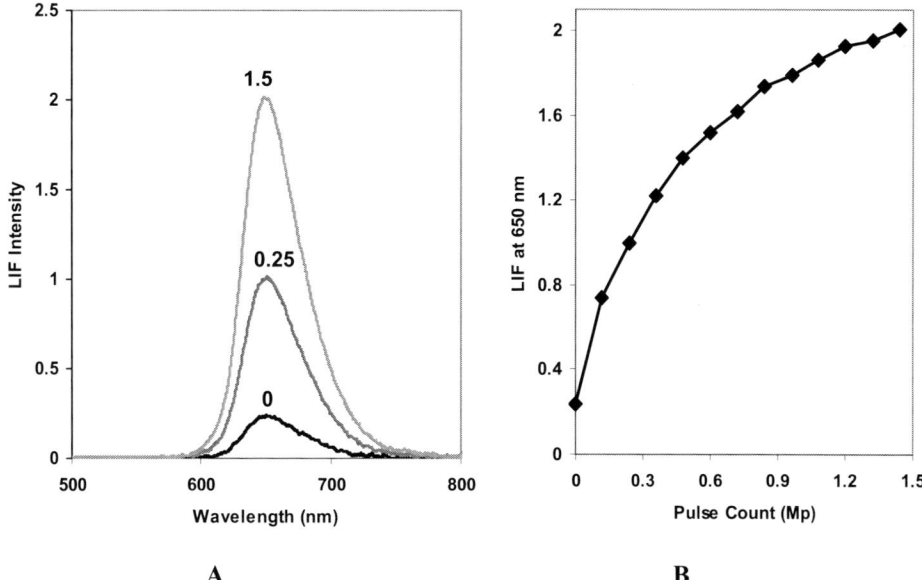

Figure 9. A. LIF spectrum of bulk SiO2 sample 2. Numbers next to the traces indicate pulse count in million pulses. B. Evolution of the 650-nm peak with laser irradiation.

LIF spectrum of sample 2 is shown in Figure 9A. The spectrum consists of a single peak, centered around 650 nm, whose magnitude increases with irradiation. Figure 9B shows the evolution of the peak magnitude with increasing laser pulse count. From previous studies, such a 650 nm emission feature corresponds to a NBOH center in SiO_2.[16] Thus, the peak growth of Figure 9B corresponds to a growth of the absorption feature at 270 nm in Figure 7. This sample also exhibits the highest 193-nm damage rate of the four samples tested.

By contrast, LIF spectrum of sample 4 shows at least three distinct spectral features in the course of laser irradiation: 650 nm, 580 nm and 290 nm (Figure 10A). Since all three features show different behavior with laser irradiation, they arise from different defect centers in the fused silica material. The 650-nm feature for this sample remains constant throughout irradiation. Correspondingly, no growth of the NBOH center is observed for sample 4 in Figure 7. We believe that the lack of growth of this feature is related to a relatively low sample damage rate (Fig. 6B) The 580-nm feature disappears within <0.5 Mp, as shown in Figure 10B. This feature is related to oxygen-deficient centers in the fused silica, such as Si clusters.[23] From the disappearance of this feature, we postulate that oxygen-deficient defects are being destroyed by laser irradiation. The 290-nm emission peak continues to grow throughout laser irradiation. As this peak grows, we also observe a broad shoulder developing between 450 and 600 nm. The defect assignment of these features is less certain, but at least one publication assigns these bands to a specific oxygen vacancy, $O_3 \equiv Si\text{-}Si \equiv O_3$.[24]

Based on the data presented above, we now postulate on the different damage mechanisms in samples 2 and 4. From previous work of Smith et al.,[17] a well-known approach to controlling laser-induced degradation in fused silica involves doping samples with high levels of molecular hydrogen. Such doped samples also tend to exhibit RDP, or transient absorption. The transient absorption phenomenon involves absorbance fading when irradiation ceases and re-darkening in the presence of laser irradiation. The rates for both these processes are defined by kinetics of the dissolved hydrogen recombination at the E' defect centers to form a non-absorbing state. The quasi-steady-state transmission of the sample under laser irradiation depends, among other parameters, on the amount of dissolved hydrogen and the initial concentration of E' centers. Sample 4 , which exhibits transient absorption, appears to have more initial defects than sample 2 that are rapidly converted to E' centers upon irradiation. However, sample 4 also appears to have a higher level of dissolved hydrogen than sample 2, which is responsible for its slower damage rate and the observed transient absorption phenomenon.

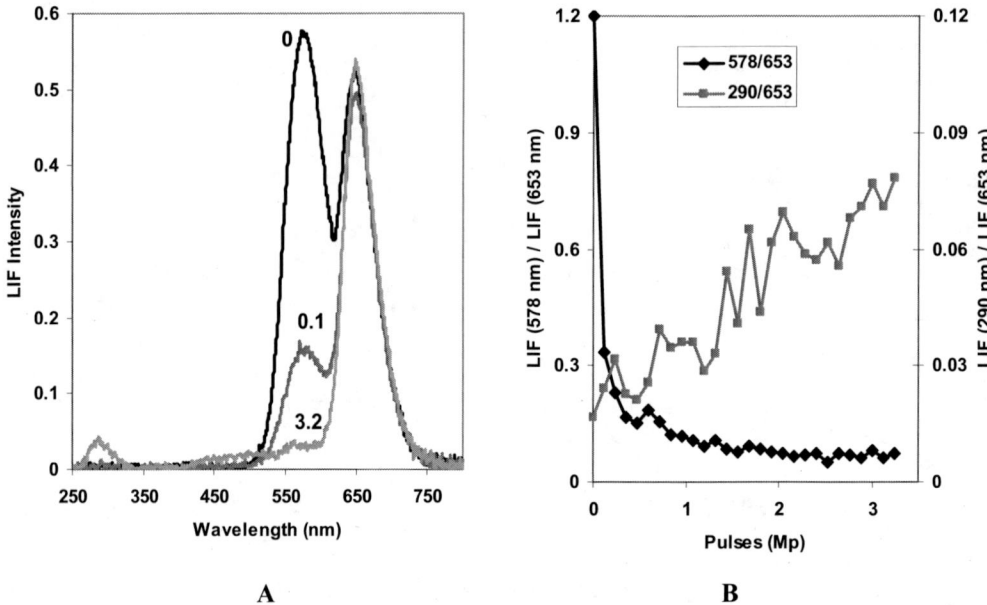

Figure 10. A. LIF spectrum of bulk SiO_2 sample 4. Numbers next to traces indicate pulse count in million pulses. B. Evolution of the 578-nm and 290-nm peaks with laser irradiation.

3.3 Thin film coatings

Thin film coatings from three different suppliers were tested. All coatings were provided on CaF_2 substrates. Coating designs included both single layers and multilayer stacks. In-situ transmission for four representative coatings is shown in Fig. 11. Coatings 1 and 4 were irradiated at an incident fluence of 35 mJ/cm^2/pulse, whereas coatings 2 and 3 were irradiated at an incident fluence of 50 mJ/cm^2/pulse. The transmission of coatings in Figure 11 is obtained at 25 degrees with respect to normal as discussed in section 2.3. Excepting initial changes, no transmission degradation was observed for any of the samples tested. Common to most samples was a transmission recovery upon irradiation, most likely caused by surface cleaning. A spectroscopic trace of such a transmission recovery for sample 1 is shown in Figure 12. No catastrophic failure for any of the tested thin film samples was observed in the course of irradiation.

LIF spectra of coating 1 at the beginning of irradiation and after 100 Mp are shown in Figure 13A. For reference, Figure 13A also shows an LIF signature from a witness CaF_2 substrate that was provided with the coating. Two main spectral features are evident in the emission spectra. The main peak around 320 nm corresponds to the CaF_2 substrate peak, as seen in Figure 5A. The smaller broader feature around 500 nm is present only at the start of irradiation and disappears in less than one million pulses (Fig. 13B). Previous work by Heber and co-workers analyzed LIF in fluoride thin film stacks.[25] By correlating emission signatures to IR spectroscopy, that study concluded that the transient emission peak at 450 –500 nm was caused by evolution of hydrocarbons trapped in relatively porous thin film layers. By analogy, we interpret the disappearance of our 500-nm emission (Figure 13B) as an indication of in-situ hydrocarbon removal from the coating. As expected, the cleaning is quite rapid at the high incident fluences employed here. Hydrocarbon removal is also likely responsible for the rapid transmission recovery of trace 1 in Figure 11 and an upward spectral shift in Figure 12.

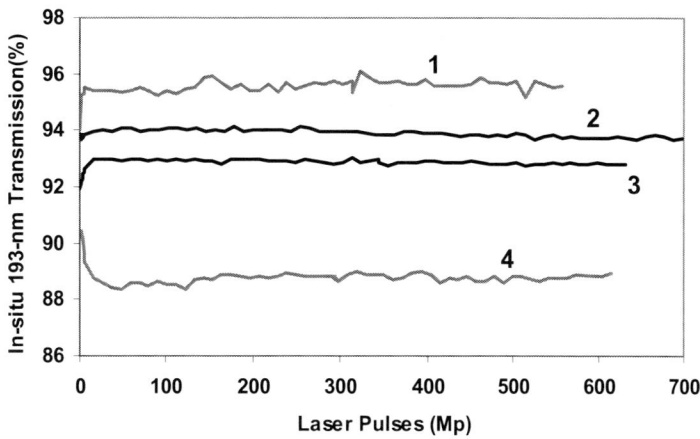

Figure 11. In-situ 193-nm transmission of four thin film coatings vs. pulse count.

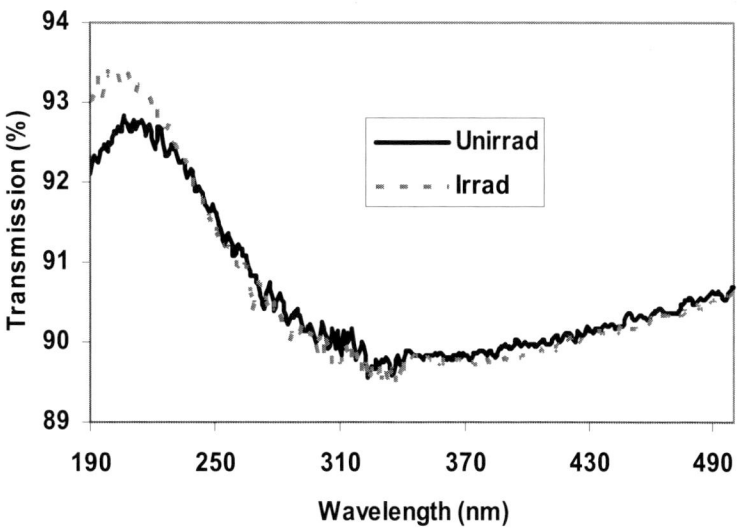

Figure 12. Spectral transmission of coating sample 1 from Figure 11 in the irradiated and unirradiated areas.

To gain more understanding about potential physical changes in the coating layers during long-term irradiation, we used spectroscopic ellipsometry combined in a spatial mapping mode. In the past, we successfully used this technique to study laser-induced changes in thin film coatings irradiated with 157-nm[26] and 193-nm[27] lasers. By comparing spectral ellipsometric data inside and outside the irradiated spot, we were able to detect subtle changes in the structure of the coating layers that anticipated evolution of layers with long-term irradiation.

As an example of ellipsometric data collected in this study, we analyze a one-sided antireflecting coating stack on a CaF_2 substrate that was subjected to 300 Mp at 30 mJ/cm^2/pulse. In-situ 193-nm transmission of this coating remained constant at 96.5% throughout irradiation. However, a spatial ellipsometric map of the Δ parameter at 450 nm reveals a laser-induced change (Fig. 14 A). During ellipsometric data acquisition, such spatial maps are simultaneously collected for wavelengths between 250 nm and 1000 nm. Figure 14 B compares the Δ spectrum in the unirradiated vs. irradiated areas. A spectral shift is observed over most of the measured wavelength range.

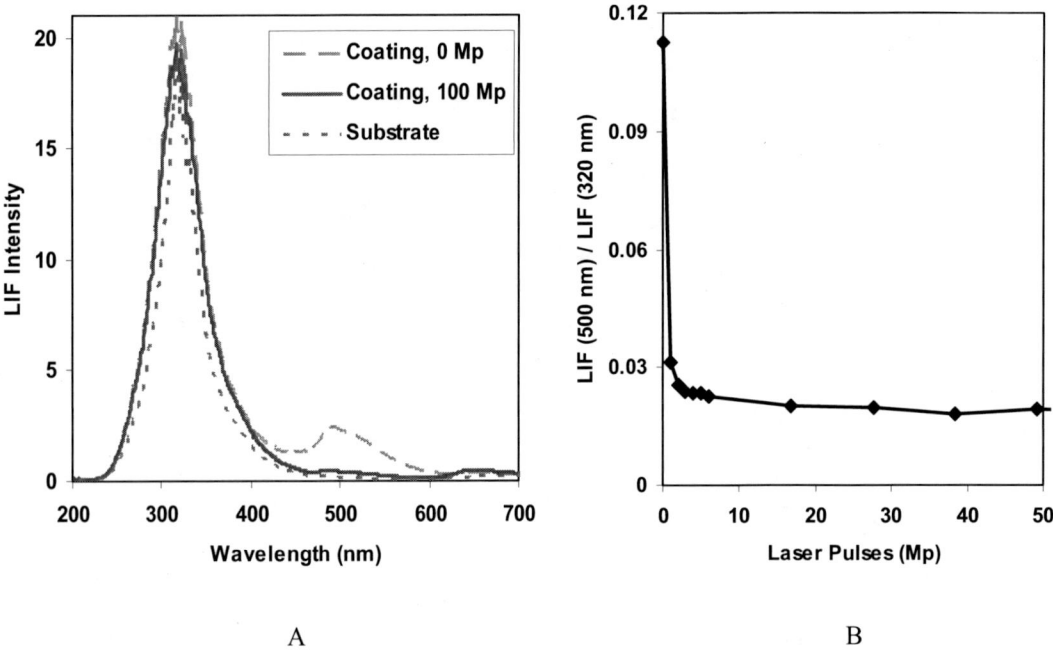

A   B

Figure 13. A. LIF spectra of coating 1 at the beginning of irradiation and after 100 Mp. For reference, an LIF spectrum of substrate witness is also shown. B. Evolution of the 500-nm peak in sample 1 with irradiation.

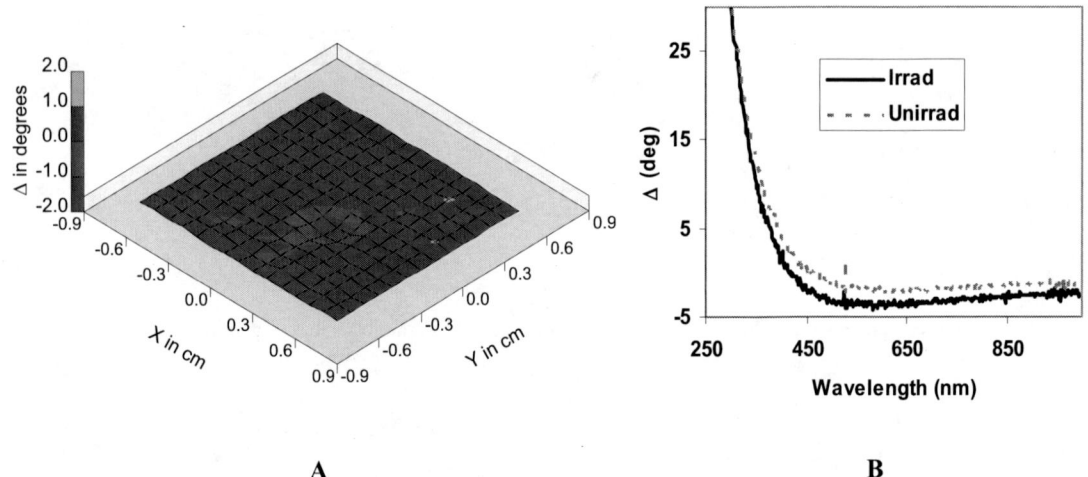

A   B

Figure 14. Spectroscopic ellipsometry from a laser-exposed antireflecting coating. A. spatial map of Δ parameter at 450 nm. B. Δ spectrum inside and outside of an irradiated spot.

To convert ellipsometric changes into thin film properties, such as thickness and refractive index, we need to assume a layer model for the coating. Since the supplier did not provide information about the coating stack, we made reasonable assumptions about its design following guidelines suggested in the literature.[28] We modeled the data of Figure 14 as a three-layer quarter-wave fluoride antireflecting stack, with the starting thin film parameters for high and low index films from previous work.[29-31] By adjusting the starting parameters, we were able to obtain a robust solution for layer thicknesses and indices in the unirradiated area of the coating. We then assumed that the laser-induced changes in Figure 14 were confined to the top layer of the coating stack. With this assumption, the changes of the top layer due to laser irradiation were modeled as an index increase. Expressing the index in the Cauchy formulation as

$$n_{top} = A + B/\lambda^2 \,, \tag{2}$$

where A, B are Cauchy coefficients and λ is the wavelength, we modeled the laser-induced changes as a shift in the A Cauchy coefficient (Fig. 15). No thickness change in the irradiated area was required to explain the data. From Figure 15, the maximum index shift is ≈0.01 due to irradiation; this is a wavelength-independent index shift that may be caused by radiation-induced densification of the thin film. This small index shift would have a minimum impact on the 193-nm transmission: ≈0.1 % transmission change for a one-sided coating may be expected.

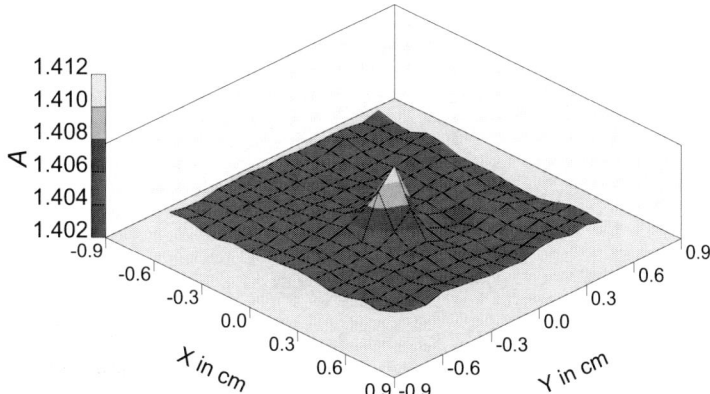

Figure 15. Modeled spatial map of the A Cauchy coefficient of a top-layer refractive index corresponding to the ellipsometry data of Figure 14.

4. SUMMARY AND OUTLOOK

This study of 193-nm laser irradiation of optical materials addressed their suitability for high fluence applications. Optical materials from two suppliers of bulk CaF_2, two suppliers of bulk SiO_2 and three suppliers of thin film coatings were obtained for the study.

We found no degradation in bulk CaF_2 when irradiated up to 650 Mp at a fluence of 50 mJ/cm²/pulse. No laser-induced birefringence was observed. Thus, the material appears to be suitable for high fluence applications.

By contrast, we found unacceptable transmission degradation in bulk SiO_2 for pulse counts < 5 Mp at fluences above 35 mJ/cm²/pulse. Measured degradation rates of the tested materials varied by more than an order of magnitude, depending on the sample grade. Some grades displayed a fluence-dependent rapid damage (or transient absorption) process, which could present additional concerns for dose control. Laser-induced birefringence was observed in three out of four samples and its magnitude followed the incident dose.

Thin film coatings on CaF_2 substrates showed no long-term degradation for pulse counts up to 700 Mp and fluences of up to 50 mJ/cm²/pulse. Some initial transmission changes were observed and attributed primarily to the evolution of trapped hydrocarbons. Detailed post-irradiation ellipsometric studies of irradiated coatings revealed only subtle changes in some samples, attributable to top-layer densification.

Table 2 summarizes the exposure conditions of the samples above and shows the equivalent "tool time" in days, assuming exposure conditions of today's lithographic tools. The tool time in column 3 is calculated from the exposure conditions in column 2 and the assumption of 6 kHz laser operation, 100% up time and a pulse fluence of 0.25 mJ/cm²/pulse. From the table, bulk CaF_2 and coatings show reasonable tool times, even when tested at elevated fluences, whereas bulk SiO_2 clearly shows rapid degradation when tested at elevated fluences. The accelerated degradation rate at these high fluences does not imply a higher-order photon process. From the data presented in this report, one-photon absorption is still likely to be responsible for the fused silica transmission degradation. Instead, since the damage rate is a net effect of several competing reactions, in the higher fluence regime, the incident fluence becomes the damage-rate-determining factor.

Table 2. Summary of exposure conditions for all sample types and an equivalent tool time in days. See text for assumptions in tool time calculations.

Sample Type	Exposure Conditions	Equivalent Tool Time (days)
Bulk SiO$_2$	5 Mp at 35 mJ/cm^2/pulse	1.5
Bulk CaF$_2$	650 Mp at 50 mJ/cm^2/pulse	250
Coatings	700 Mp at 50 mJ/cm^2/pulse	270

5. ACKNOWLEDGEMENT

The Lincoln Laboratory portion of this work was sponsored by a Cooperative Research and Development Agreement between Lincoln Laboratory and SEMATECH, Inc. and the Research Foundation of State University of New York. Opinions, interpretations, conclusions, and recommendations are those of the authors and are not necessarily endorsed by the United States Government. SEMATECH and the SEMATECH logo are registered service marks of SEMATECH, Inc. All other service marks and trademarks are the property of their respective owners.

6. REFERENCES

[1] Blakey, I., L. Chen, Y-K Goh, K. Lawrie, Y-M Chuang, E. Piscani, P. A. Zimmerman and A. Whittaker, "Non-CA resists for 193-nm immersion lithography: Effects of chemical structure on sensitivity," *Proc. SPIE* **7273**, 72733X (2009).

[2] Nishimura, I., W. H. Heath, K. Matsumoto, W-L Jen, S. S. Lee, C. Neikirk, T. Shimokawa, K. Ito, K. Fujiwara and C. G. Wilson "Non-chemically amplified resists for 193nm immersion lithography," *Proc. SPIE* **6923**, 69231C (2008).

[3] Wakana, K., H. Tsushima, S. Matsumoto, M. Yoshino, T. Kumazaki, H. Watanabe, T. Ohta, S. Tanaka, T. Suzuki, H. Nakarai, Y. Kawasuji, A. Kurosu, T. Matsunaga, J. Fujimoto and H. Mizoguchi, "Optical performance of laser light source for ArF immersion double patterning lithography tool," *Proc. SPIE* **7274**, 72743J-1 (2009).

[4] Gillespie, W. D., T. Ishihara, W. N. Partlo, G. X. Ferguson and M. R. Simon, "6 kHz MOPA light source for 193 nm immersion lithography," P*roc. SPIE* **5754**, 1293-1303 (2005).

[5] Hori, T., T. Yabu, T. Ishihara, T. Watanabe, O. Wakabayashi, A. Sumitani, K. Kakizaki and H. Mizoguchi, "Feasibility study of 6 kHz ArF excimer laser for 193 nm immersion lithography," *Proc. SPIE* **5754**, 1285-1292 (2005).

[6] Algots, J. M., R. Sandstrom, W. Partlo, P. Maroevic, E. Eva, M. Gerhard, R. Linder, and F. Stietz, "Compaction and rarefaction of fused silica with 193-nm excimer laser exposure," *Proc. SPIE* **5040**, 1639-50 (2003).

[7] "http://www.sensorphysics.com/377035.html"

[8] Grating H2,"http://www.oceanoptics.com/products/bench_gratingcharts.asp"

[9] Liberman, V., M. Rothschild, J. H. C. Sedlacek, R. Uttaro, A. Grenville, A. K. Bates and C. V. Peski, "Excimer-laser-induced degradation of fused silica and calcium fluoride for 193-nm lithographic applications," *Optics Letters* **24,** 58 (1999).

[10] Muhlig, C., W. Triebel, G. Topfer, J. Bergmann, S. Bruckner, C. Chojetzki and R. Martin, "Laser induced fluorescence of calcium fluoride upon 193 nm and 157 nm excitation," *Proc. SPIE* **5188**, 123-133 (2003).

[11] Burkert, A., C. Muehlig, W. Triebel, D. Keutel, U. Natura, L. Parthier, S. Gliech, S. Schroeder and A. Duparre, "Investigating the ArF laser stability of CaF$_2$ at elevated fluences," *Proc. SPIE* **5878**, 58780E (2005).

[12] Mizuguchi, M., H. Hosono, H. Kawazoe and T. Ogawa, "Time-resolved photoluminescence for diagnosis of resistance to ArF excimer laser damage to CaF_2 single crystals," J. Opt. Soc. Am. **B16**, 1153-9 (1999).

[13] Burkert, A., D. Keutel and U. Natura, "248 nm high fluence irradiation of CaF_2 crystals," *Proc. SPIE* **6720**, 67201K (2007).

[14] Tsujibayashi, T., M. Watanabe, O. Arimoto, I. Minoru, S. Nakanishi, H. Itoh, S. Asaka and M. Kamada, "Two-photon excitation spectra of exciton luminescence in CaF_2 obtained by using synchrotron radiation and laser," *J. Luminesc.* **87-89**, 254-256 (2000).

[15] Thomas, S. and B. Kuhn, "KrF laser induced absorption in synthetic fused silica," *Proc. SPIE* **2966**, 56-64 (1996).

[16] Natura, U., R. Martin, G. von der Goenna, M. Kahlke and G. Fasold, "Kinetics of laser induced changes of characteristic optical properties in Lithosil with 193nm excimer laser exposure," *Proc. SPIE* **5754**, 1312-19 (2004).

[17] Smith, C. M., N. F. Borrelli and R. J. Araujo, "Transient absorption in excimer-exposed silica," *Appl. Opt.* **39**, 5778-5784 (2000).

[18] Wright, E. M., M. Mansuripur, V. Liberman and K. Bates, "Spatial pattern of microchannel formation in fused silica irradiated by nanosecond ultraviolet pulses," *Appl. Opt.* **38**, 5785-8 (1999).

[19] Burkert, A., W. Triebel, U. Natura, R. Martin, "Micro-channel formation in fused silica during ArF excimer laser irradiation," *Phys. Chem. Glasses B* **48**, 107-122 (2007).

[20] Muhlig, C., W. Triebel, S. Kufert and U. Natura, "Accelerated life time testing of fused silica upon ArF laser irradiation," *Proc. SPIE* **7132**, 71321G (2008).

[21] Seward, T. P., III, C. Smith, N. F. Borrelli and D. C. Allan, "Densification of synthetic fused silica under ultraviolet irradiation," *J. Non-Cryst. Sol.* **222**, 407-414 (1997).

[22] Piao, F., W. G. Oldham and E. E. Haller, "The mechanism of radiation-induced compaction in vitreous silica," *J. Non-Cryst. Sol.* **276**, 61-71 (2000).

[23] Sakurai, Y, "Photoluminescence band near 2.2 eV in γ-irradiated oxygen-deficient silica glass," *J. Non-Cryst. Sol.* **342**, 54-8 (2004).

[24] Sakurai, Y. "Correlation between the 2.7eV and 4.3eV photoluminescence bands in silica glass," *J. Non-Cryst. Sol.* **352**, 2917-20 (2006).

[25] Heber, J., C. Muhlig, W. Triebel, N. Danz, R. Thielsch and N. Kaiser, "Deep UV laser induced fluorescence in fluoride thin films," *Appl. Phys. A* **76**, 123-128 (2003).

[26] Liberman, V., T. M. Bloomstein, M. Rothschild, S. T. Palmacci, J. H. C. Sedlacek and A. Grenville, "Photo-induced changes in 157-nm optical coatings," *Proc. SPIE* **5377**, 131-140 (2004).

[27] Liberman, V., M. Switkes, M. Rothschild, S. T. Palmacci, J. H. C. Sedlacek, D. E. Hardy and A. Grenville, "Long-term 193-nm laser irradiation of thin-film-coated CaF_2 in the presence of H_2O," *Proc. SPIE* **5754**, 646-654 (2005).

[28] Heber, J., R. Thielsch, H. Blaschke, N. Kaiscr, K. Mann, E. Eva, U. Leinhos and A. Gortler, "Stability of optical interference coatings exposed to low-fluence 193 nm ArF radiation," *Proc. SPIE* **3334**, 1031-1047 (1998).

[29] Kaiser, N., A. Zuber and U. Kaiser "Evaluation of thin MgF_2 films by spectroscopic ellipsometry," *Thin Solid Films* **232**, 16-17 (1993).

[30] Uhlig, H., R. Thielsch, J. Heber and N. Kaiser, "Lanthanide tri-fluorides: a survey of the optical, mechanical and structural properties of thin films with emphasis of their use in the DUV-VUV-spectral range," *Proc. SPIE* **5963**, 59630N (2005).

[31] Thielsch, R., J. Heber, H. Uhlig and N. Kaiser, "Optical, structural, and mechanical properties of gadolinium tri-fluoride thin films grown on amorphous substrates," *Proc. SPIE* **5963**, 59630O (2005).

Stepwise fitting methodology for Optical Proximity Correction modeling

Artak Isoyan[*], Jianliang Li, Lawrence S. Melvin III

Synopsys, Inc. 2025 NW Cornelius Pass Road, Hillsboro, OR 97124, United States

ABSTRACT

Optical proximity correction (OPC) models consist of a large number of components and parameters that must be optimized during model fitting process for best possible matching with empirical data. There are several optimization methods for OPC models. Most of the published methods, if not all, are based on a global optimization method, where all the model parameters are regressed in their search regions to provide a global minimum of the OPC model. However, there are potential risks of overweighting one OPC model component versus another and as a result loosing the physicality of the final model, which reduces model quality in terms of fit and prediction. In this work a stepwise fitting methodology based on staged optimization of the OPC model components is presented. Components are added into an OPC model in the order of more physical to less physical, starting from mask and optics. In each optimization stage a component is optimized using global regression methods and then the optimized parameters are locked and not regressed during further model optimization. The effectiveness of this approach in terms of accurate correction and comparison with global search regression method is demonstrated through computational experiments.

Keywords: OPC, modeling

1. INTRODUCTION

As integrated circuits continue to shrink device sizes to the 32nm, 22nm technology nodes and beyond with 193nm wafer immersion lithography, there is a need for an improved methodology to increase lithography process simulations robustness for optical proximity correction (OPC) and process window predictability [1,2]. The complexity of OPC models and the number of model components (parameters) are dramatically increasing as the future technology nodes become more complex, and by the need to fully capture the optical lithography and resist chemistry processes in an accurate and predictive OPC model. OPC calculations have become increasingly complex as the size of semiconductor devices and the allowed error tolerances becomes progressively smaller. One of the problems in OPC model optimization is the determination of the optimized values of all the model parameters involved in the OPC model fit. OPC models themselves are complex mathematical models, and can contain hundreds of parameters. A straight forward and simple approach for determination of the OPC model parameters is the global search over the entire OPC model to find the model global minimum. OPC models are usually based on large numbers of components such as mask, optics, resist, development, etc. Due to large amount of model components, some of them can have certain level of similarity in effects, and as a result, in the case of the global model optimization method, there is a potential risk of overweighting one component versus another and which can lead to reduction in the physicality of the final optimized OPC model. A trivial example of similarity of model parameters can be seen in apodization and diffusion parameters, where both parameters can blur the optical signal. The physicality loss of the OPC model reduces model quality in terms of fit, prediction, interpolation and extrapolation.

In this work a stepwise fitting methodology for OPC model optimization is presented. The optimization method is based on staged optimization of model components step by step. The OPC model optimization starts from an optics-only part and then more components are added in following stages in the order of more physical to less physical. After each optimization stage, the optimized parameters are locked and removed from further model regression.

---

[*] e-mail: isoyan@synopsys.com; phone +1-503-547-6128; www.synopsys.com

Optical Microlithography XXIII, edited by Mircea V. Dusa, Will Conley, Proc. of SPIE Vol. 7640,
764030 · © 2010 SPIE · CCC code: 0277-786X/10/$18 · doi: 10.1117/12.846635

2. STEPWISE FITTING FLOW

In lithography process modeling the challenge is to create an OPC model that transforms the input design polygons onto a wafer image. The OPC models are based on complex mathematical calculations of actual physical processes of optical lithography and resist exposure, and as well as on statistical methods. The quality of the OPC models defines the success of the model's usage for proximity corrections. A good OPC model should have the low fitting error and should describe the physics of the resist and optics. In some cases, efforts to lower the fitting error can lead to the degradation of physical meaning. Usually, the OPC models consist of modeling elements from mask, optics, lithography process parameters, resist model, threshold, etc. (figure 1). The components of the OPC models can be categorized into two groups: physical and statistical. Physical components are based on appropriate physical effects and can be well described for some processes. The typical physical components of the models are the mask and optics. Optical physics modeling is well known and the dominant component of the model [3,4]. However, there are also several other physical effects which require complex mathematical methods. Statistical methods are used to find properties that enhance the optical output to better match empirical process data. By considering model components in deference to their performance rate, a more physical prediction occurs when physical model components dominate the model, while a less physical model is developed when lesser physical parameters are the most significant contributors to the output signal. The statistical part of the model should be used only for compensation of effects that are in some cases impossible to represent by physical components. Both physical and statistical parts of the model consist from several components.

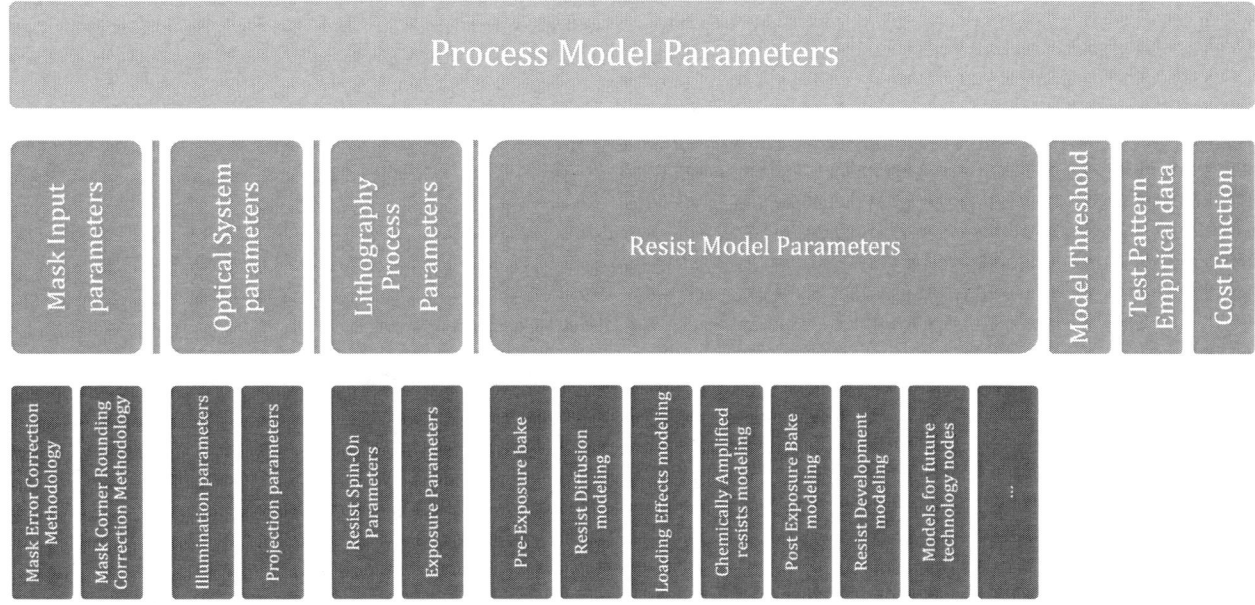

Figure 1. Lithography process OPC model structure.

The conventional approach to the creation of the OPC model is to use a global optimization methodology [5]. The method is well described and has been used in many applications over past decades. The idea of a model global optimization is to add all the components into model at first and then run the regression over entire model in order to find the global minimum. However, OPC model components, both physical and statistical, can have some level of similarity and in some cases the global optimization method can lead to failed model optimization by degradation of model physicality. In order to control the model optimization new techniques are required. One of the solutions to this problem is to build a model which is entirely physical, without any statistical part. But even in this scenario, staged model optimization is more appropriate, since physical components also share effects which could be covered by different components of the model. Staged optimization of OPC models can provide a good solution for controlling the model components and have several benefits versus conventional global optimization method. The stepwise or staged optimization of the model separates the OPC model components optimization, which increases model flexibility for different lithography process parameters changes. In case of stepwise optimized model it is possible to change the optical part of the model by keeping the rest of the model components untouched. This helps to make the models more flexible and adjustable and doesn't require a new optimization for every parameter change. The proposed optimization method is based on stepwise optimization of the model components in a step by step manner, in the order of more physical to less physical, and finally adding statistical parts of the model (figure 2). The optimized parameters are locked and are not regressed in followed stages of the model optimization. A cost function (based on ΔCD) is minimized over the OPC model parameters. At the beginning the model contains only optical component. The optimization of the optical part is done by maximizing the aerial image modulation on specially selected 1D type of gauges (selected location on pattern layout). On this stage a focus centering methodology is also used. The optical part of the model does not contain any modeling of resist effects and hence cannot be used as a final model. Then, the resist diffusion effects are added into the model. Next, in order to compensate line end shortenings introduced by photomask fabrication, mask corner rounding and mask error correction methodologies are added into the model. Photomask fabrication errors can have a huge impact on 2D type of structures. After finishing the main three dominant component optimizations, which are the mask, optical system and the resist diffusion effect, more components are added into the model. After each additional stage, if the last added component did not improve the model quality, the component can be removed from the model. More resist effects are added into model, such as chemically amplified resist models, post-exposure baking effects, resist development, lithography process pattern-dependent effects such as resist micro loading during development, as well as optical flare. The list of effects which could be added into the model strictly depends upon the modeler ability to represent such effects in mathematical forms. As already have been mentioned, some of the effects cannot be represented in physical model forms, so semi-physical approximations or stable statistical components are used. After adding all possible physical components into the model, stable statistical components are added in order to compensate some of the unknown effects in resist chemistry and lithography processes. Thus, the model optimization is finished at this stage.

The stepwise optimization method has been compared to the global optimization method on several data sets. The comparisons showed that in most cases the model statistics, stability, predictability are improved. The process window interpolation and extrapolation of stepwise optimized OPC models also have been improved (Table 1). An example of simulated Bossung plots (Line Width CD across several Depth of Focus (DOF) positions and across several exposure doses) from stepwise and global optimization methods are shown on figure 3. As can seen from plots, a stepwise optimized OPC model shows more physical behavior on selected test location. The comparison of stepwise and global optimization methods have shown similar calculation run time, for the same number of components in each final model.

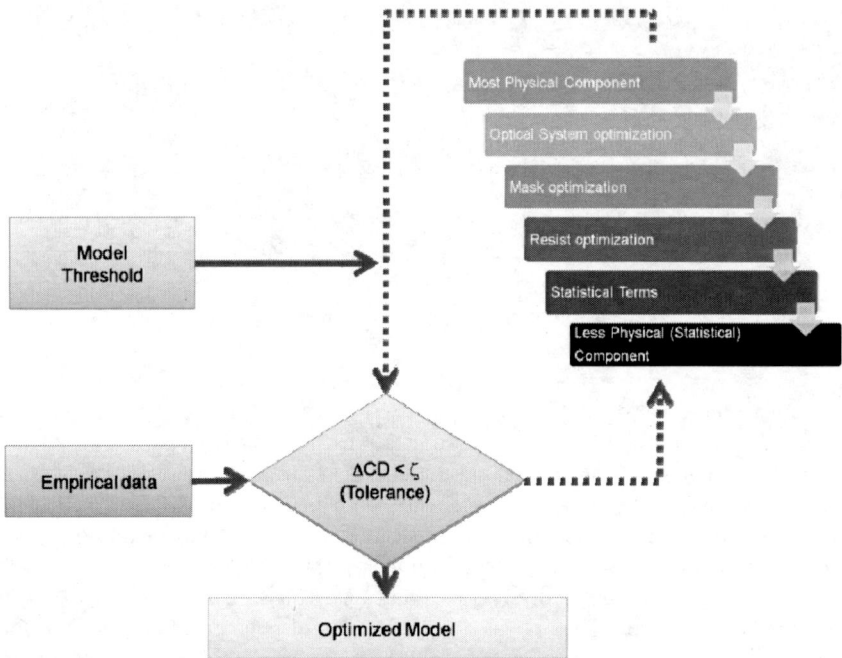

Figure 2. Stepwise optimization flow chart.

Table 1. OPC model RMS (root means square) values for process window (PW) test data set. The calibration of the OPC model is done by using only nominal dose and defocus condition. The rest of the PW conditions are used only for model verification purposes. Stepwise optimization (left column) and Global optimization (right column, red) are used for model calibration. Both methods contain equal number of physical and statistical components.

Stepwise	Global	Negative Defocus 3		Negative Defocus 2		Negative Defocus 1		Nominal Defocus		Positive Defocus 1		Positive Defocus 2		Positive Defocus 3	
Under Dose		6.70	18.54	2.85	6.87	NA	NA	3.27	1.67	NA	NA	2.80	2.62	5.06	4.58
Nom. Dose		5.37	11.37	3.70	5.38	5.10	8.47	2.69	2.29	3.62	2.95	3.62	2.88	4.19	3.93
Over Dose		5.49	16.37	3.42	9.10	NA	NA	4.78	4.77	NA	NA	3.01	3.39	7.31	6.30

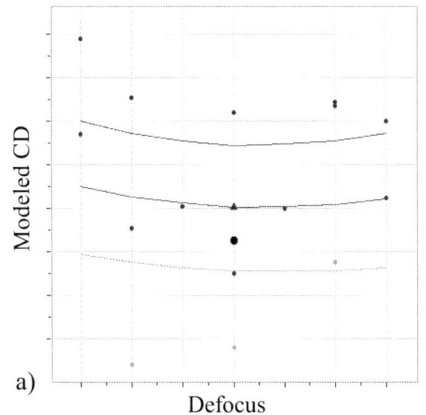
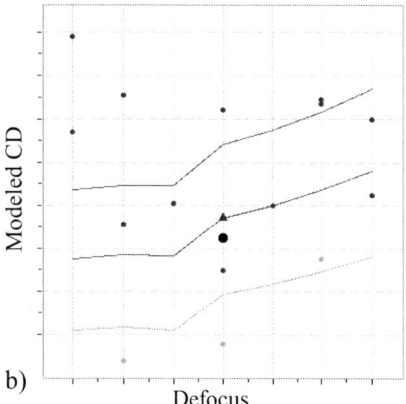

a)　　　Defocus　　　b)　　　Defocus

Figure 3. Modeled CD (solid lines) at several Depth of Focus (DOF) positions and across several exposure doses. Stepwise (a) and Global (b) optimized OPC model Bossung plots on exact same location of test pattern.

3. CONCLUSION

A stepwise fitting methodology based on staged optimization of the OPC model components is presented. Components are added into the OPC model in the order of more physical to less physical. In each optimization stage a component is optimized using global regression methods and then the optimized parameters are locked and not regressed during further model optimization. The stepwise optimization method has been compared to the global optimization method on several data sets. The stepwise optimized OPC models showed improvement in model stability and predictability. One of the major advantages of stepwise optimization is the process window modeling, where only nominal condition data is required for OPC model creation, the rest of the PW conditions can be used for verification purposes. This significantly reduces OPC model optimization runtime. The presented method was successfully implemented into existing tools for full chip mask corrections.

REFERENCES

[1] H. Fukuda, K. Hattori and T. Hagiwara, Proc. SPIE, 4346, 319 (2001).
[2] T. A. Brunner and R. A. Ferguson, Proc. SPIE, 2726, 198 (1996).
[3] H. H. Hopkins, Proceedings of the Royal Society of London. Series A, Mathematical and Physical, Vol. 217, No. 1130, 408-432 (1953).
[4] H. H. Hopkins, Proceedings of the Royal Society of London. Series A, Mathematical and Physical, Vol. 208, No. 1093, 263-277 (1951).
[5] R. Horst, P.M. Pardalos and N.V. Thoai, Introduction to Global Optimization, Second Edition. Kluwer Academic Publishers, 2000.

Automatic numerical determination of lateral influence functions for fast-CAD

Marshal A. Miller*, Kenji Yamazoe*+, and Andrew R. Neureuther*

*UC Berkeley (271 Cory Hall), Berkeley, CA 94704, +Canon Inc.

ABSTRACT

This paper presents kernel convolution with pattern matching (KCPM), which is an updated version of fast-CAD pattern matching for assessing lithography process variations. With KCPM, kernels that capture lateral feature interaction between features due to process variations are convolved with a mask layout to calculate a match factor, which indicates approximate change in intensity at the target location. The algorithm incorporates a custom source, a mask with electromagnetic effects, and an arbitrary pupil function. For further accuracy improvement, we introduce a source splitting technique. Though the evaluation speed is decreased, R^2 correlation of the match factor and change in intensity is increased. Results are shown with R^2 correlation as high as 0.99 for nearly coherent and annular illumination. Additionally, with a numerical aperture of 1.35, unbalanced quadrapole illumination, $10m\lambda$ RMS random aberration in projection optics and complex mask with EMF effects included, R^2 correlation of more than 0.87 is achieved. This process is extremely fast ($40\mu s$ per location) making it valuable for a wide range of applications, most commonly hot spot detection and optimization.

Keywords: pattern matching, boundary layers, edge effects, focus shift, kernel convolution, hotspot detection, mask 3D topology effect, process variation

1. INTRODUCTION AND MOTIVATION

Electromagnetic field (EMF) mask edge effects in photomasks do not scale with device generation and are of increasing importance. They must be accounted for in accurate simulation with techniques such as boundary layer (BL)[1-3] modeling and domain decomposition methods (DDM).[4] The edge effects are polarization dependent and produce both in-phase and out-of-phase components. The out-of-phase component is particularly problematic as it interacts with lateral spillover from defocus to produce feature dependent through-focus tilts of the process window. Thus it is necessary to examine the through focus behavior of the entire chip with EMF effects.

Fortunately, fast-CAD techniques based on kernel convolution with boundary layer models can quantitatively assess hot spots due to mask edge effects. Based on pattern matching software designed by Frank Gennari,[5] kernel convolution with pattern matching (KCPM) can be used to very rapidly scan a layout for susceptibility to a spillover function. BL values found from comparing rigorous FDTD simulation to thin mask models can be added to the layout for including EMF modeling in the framework. Automatic pattern generation based on the pupil function enables the flexibility to assess layout susceptibility to any effect which can be described by the pupil function or drawn on the mask layout. In verifying the accuracy of KCPM, it was found to be somewhat problematic for high off-axis illumination.

This paper characterizes the nature of the sources of inaccuracy in KCPM and then through introducing an empirical method demonstrate highly accurate results. Section 2 summarizes the KCPM algorithm, and presents the current capabilities. Section 3 steps through two controlled examples of using KCPM, along with a key limitation. Section 4, tackles some complicated source pupil configurations and techniques for improving accuracy.

Optical Microlithography XXIII, edited by Mircea V. Dusa, Will Conley, Proc. of SPIE Vol. 7640,
764031 · © 2010 SPIE · CCC code: 0277-786X/10/$18 · doi: 10.1117/12.846630

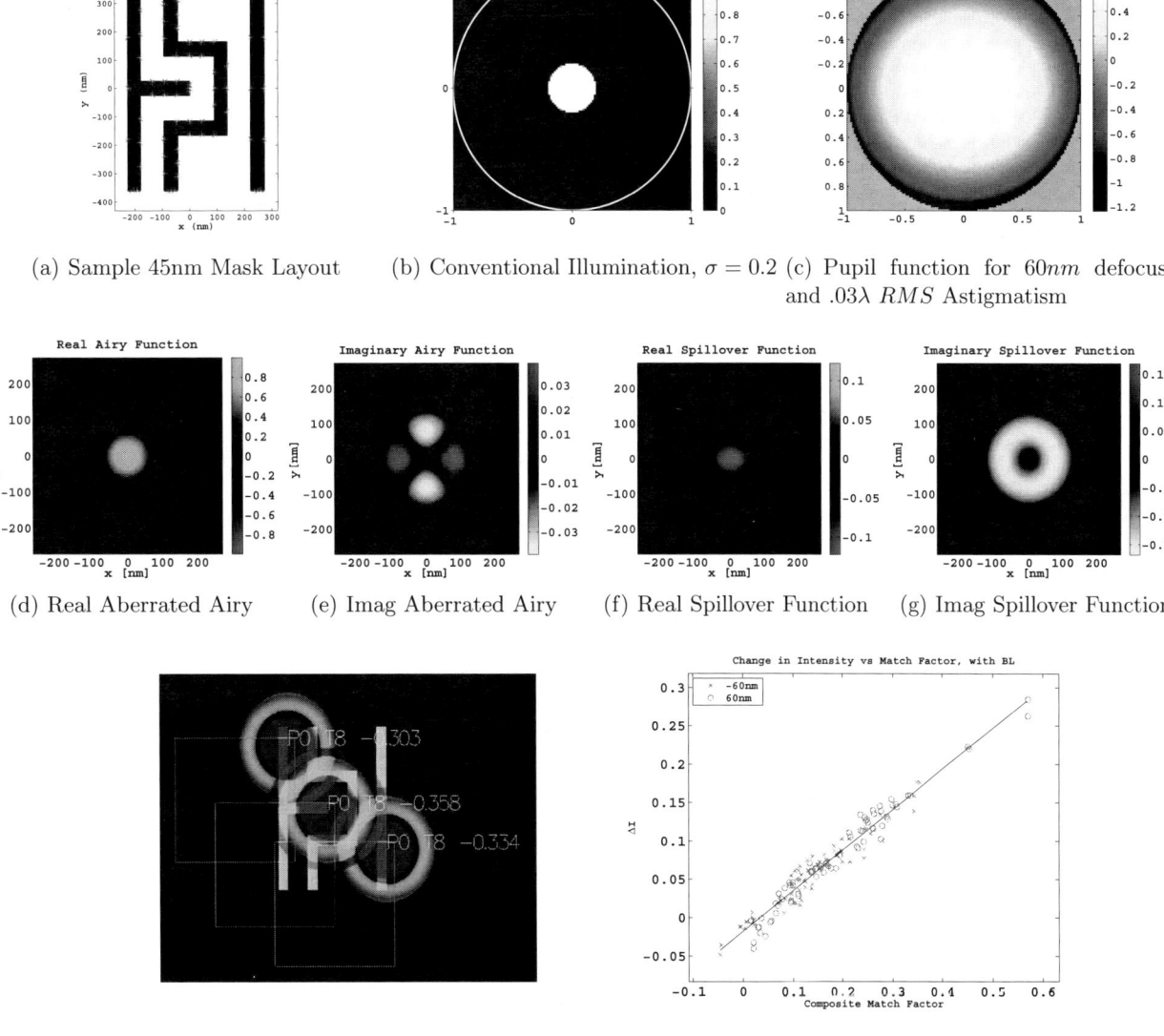

(a) Sample 45nm Mask Layout   (b) Conventional Illumination, $\sigma = 0.2$ (c) Pupil function for $60nm$ defocus and $.03\lambda$ RMS Astigmatism

(d) Real Aberrated Airy   (e) Imag Aberrated Airy   (f) Real Spillover Function   (g) Imag Spillover Function

(h) Pattern Matcher Schematic   (i) KCPM vs. Intensity change for $60nm$ defocus

Figure 1. Full flow for KCPM algorithm.

2. KERNEL CONVOLUTION WITH PATTERN MATCHING

An outline of KCPM was presented at BACUS.[6] Here we summarize and revise this theory. The algorithm is based on using the point response of a single mask pixel to estimate the intensity at a specific location. The procedure for using KCPM to quantitatively assess EMF edge effects is outlined in Fig. 1. For a given source, mask, and pupil, KCPM provides a fast estimate of intensity change for a specified change in process parameters. For this flow, a 45 nm mask layout (Fig. 1(a)) is used with conventional on-axis illumination with $\sigma = 0.2$ (Fig. 1(b)). The pupil has a small level of astigmatism, and we are concerned with looking at the change through focus (Fig. 1(c)).

The point spread function is calculated from the Fourier transform of the pupil ($FT[P]$), which describes the point response for coherent illumination. With no aberrations present, this is simply the Airy function. Throughout this paper, the source distributions describe field, rather than intensity. This introduces an unwanted correlation among source pixels, which will be treated and examined in sections 3 and 4. The benefit of this approximation is that the $FT[S]$ describes the diffraction orders incident on the mask via the pupil. To include partial coherence effects, the $FT[P]$ is multiplied by the $FT[S]$ to produce a source dependent aberrated airy function, shown in 1(d) and 1(e). Convolving this pattern with the layout gives a first order estimate of the field. To get the field at a specific location, the convolution operation can be simplified into a single pixel multiplication calculated at the target. Intensity is then calculated from the square of the absolute value of the field. Computing the match of one pattern with a layout at one location takes $40\mu s$ including complex math on a standard laptop with a 2.4 GHz Intel Core 2 Duo. Runtime is simply $40\mu s*$(number patterns)*(number of target locations). There is a small fixed cost associated with pre- and post-processing which is relatively small when running large layouts.

In general the pupil function describes the optical path difference for a set of aberrations such as focus, coma, astigmatism, and any other higher order Zernike aberrations. KCPM as implemented here uses rigorous defocus, in addition to specified Zernike aberrations. The pupil function based framework is a convenient method for including many of these effects in one simulation run. It is possible to include N effects into the original pupil function, and then make some change of M effects to produce a new pupil function. Now the change of the new effects can be examined in the presence of the previous condition, allowing for a great deal of flexibility. For example, consider a lithography system with background aberrations described by P_N. When moving out of focus, an new pupil function is needed, P_{N+M}, where for this case the subscript M indicates the addition of defocus. The sensitivity of a layout to defocus in the presence of background aberrations is calculated by using the difference:

$$Spillover = \Phi = FT[Source](FT[P_{N+M}] - FT[P_N]) \tag{1}$$

For the general case, the spillover function (Φ) is complex. Figures 1(f) and 1(g) show the spillover function for assessing $60nm$ of defocus in the presence of 0.03λ RMS astigmatism. For a real mask, only the real part is needed to calculate the mask interactions, however EMF effects introduce complex transmissions. Full complex KCPM can be combined with boundary layer mask modeling to monitor both pupil function and mask effects. Potentially polarization and resist effects can also be combined into the pupil function and incorporated into KCPM.

Once the airy estimate kernel and spillover kernels are automatically generated, they are scanned across the layout. The initial field estimate for the original N effects can be expressed as E_{object}, where:

$$E_{object} = E_R + jE_I \tag{2}$$

The spillover due to the additional M effects is defined using the convolution operator $\otimes$ as:

$$\begin{aligned} \Phi &= \Phi_R + j * \Phi_I \\ &= MF_{RR} - MF_{II} + j(MF_{RI} + MF_{IR}) \end{aligned} \tag{3}$$

where: $MF_{RR} = Re[E_{mask}] \otimes Re[\Phi]$, $MF_{II} = Im[E_{mask}] \otimes Im[\Phi]$
$MF_{RI} = Re[E_{mask}] \otimes Im[\Phi]$, $MF_{IR} = Im[E_{mask}] \otimes Re[\Phi]$

E-mail: mamillr@eecs.berkeley.edu

Now, with two complex match patterns, we can estimate the intensity using two pattern convolutions (Fig. 1(h)) to calculate Φ and E_{object}.

$$Intensity \approx (E_{object} + \Phi)(E_{object} + \Phi)^*$$
$$= E_{object}E_{object}^* + E_{object}\Phi^* + E_{object}^*\Phi + \Phi\Phi^* \qquad (4)$$

This gives a KCPM approximation for intensity at the target location. Finally, by separating the component of intensity influenced by the spillover function, we arrive at the expression for the composite match factor to estimate ΔI:

$$\Delta I \propto Composite\ Match\ Factor = E_{object}\Phi^* + E_{object}^*\Phi + \Phi\Phi^* \qquad (5)$$

This value is compared with the aerial image to gauge the accuracy of the KCPM solution (Fig. 1(i)). This process can be used to calculate either intensity or change in intensity depending on the application. For this paper, ΔI is calculated to estimate the change caused by defocus.

3. CALIBRATION EXAMPLES

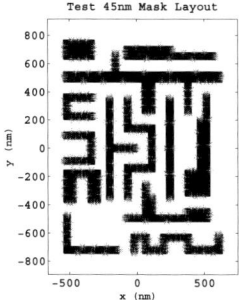

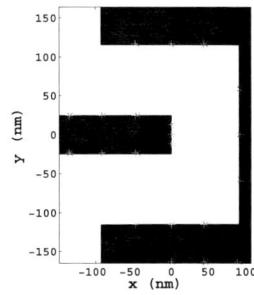

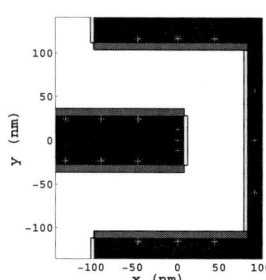

(a) Full layout with match locations marked

(b) Zoom without Boundary Layers

(c) Polarization dependent boundary layers added

Figure 2. 45nm layout used for all KCPM examples in sections 3 and 4.

For the examples throughout this paper, a test pattern (Fig. 2) was used to test typical layout concerns such as dense, semi-isolated, jogs, T-junctions. Three hundred sixty points were examined on all segments of the layout to capture performance across all of these layout features. For all cases, layouts with a binary mask were simulated both without (Fig.2(b)) and with (Fig.2(c)) BL additions to model EMF effects. Representative polarization dependent boundary layers were used to illustrate KCPM flexibility. On vertical segments, TE boundary layers with 8 nm bias and 4nm 90° transmission were added. On the horizontal edges, TM values of 4 nm bias and 8nm 270° transmission were used. All BL values are in wafer dimensions. Various source configurations are explored in the following sections, with simulation conditions of $\lambda = 193nm$, with 1.35 NA immersion lithography. For all tests, KCPM was compared to full aerial image simulation.

For the following two examples in this section, the match patterns used were 512x512 4nm pixels, making the patterns span 2048nm. Two metrics were used to assess the validity of KCPM. Here R^2 correlation indicates the fitness of the KCPM solution. Also looking at the residuals to calculate the absolute error in predicting ΔI is useful for measuring the prediction accuracy in terms of clear field intensity. Three σ deviation is used for calculating errors in the intensity calculation. For the first two cases, there are no background aberrations, so the N initial case is just the unaberrated pupil function. The M effects in this case are $\pm 60nm$ rigorous defocus, each described by its own spillover function.

3.1 Off-axis Illumination

The characterization of how the accuracy depends on illumination begins with the case of a small monopole, $\sigma = 0.25$, that was shifted off axis (Fig. 3). The aggregate R^2 correlation factor and 3σ standard deviation of

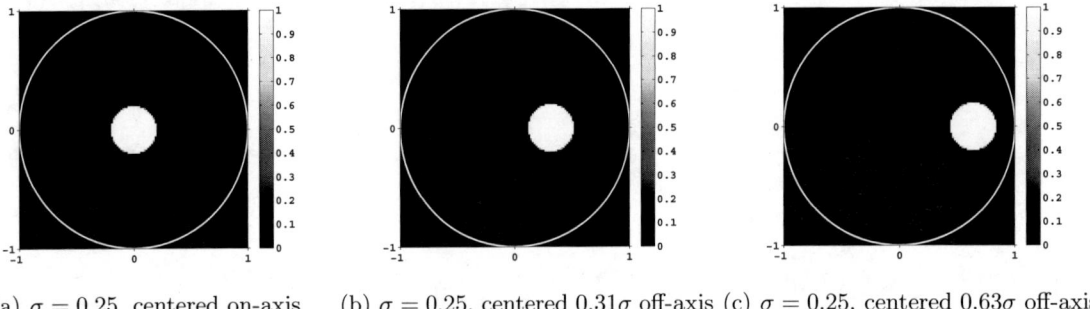

(a) $\sigma = 0.25$, centered on-axis   (b) $\sigma = 0.25$, centered 0.31σ off-axis   (c) $\sigma = 0.25$, centered 0.63σ off-axis

Figure 3. Source configurations for off-axis calibration.

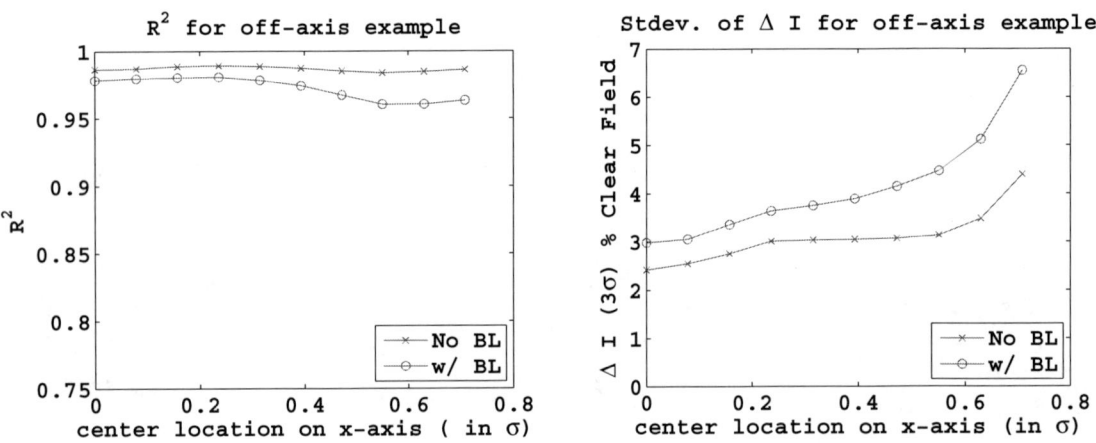

Figure 4. Left, R^2 correlation and right ΔI error plotted vs center location of monopole source.

ΔI are shown in Fig. 4. There was a small amount of movement in the R^2 correlation, but the values remain high, above .97 for all cases, indicating very close agreement between aerial image and KCPM. The results for ΔI consistently got worse as the source moves further off axis. This is explained by the asymmetric source causing an image shift when out of focus. Because the whole image is shifting, the absolute error in predicting ΔI is larger, though correlation remains high. This effect is less significant for symmetric sources, but it is an important property to consider when using KCPM. The absolute error in predicting ΔI increases if the illumination produces a through focus image shift.

3.2 Effect of Partial Coherence

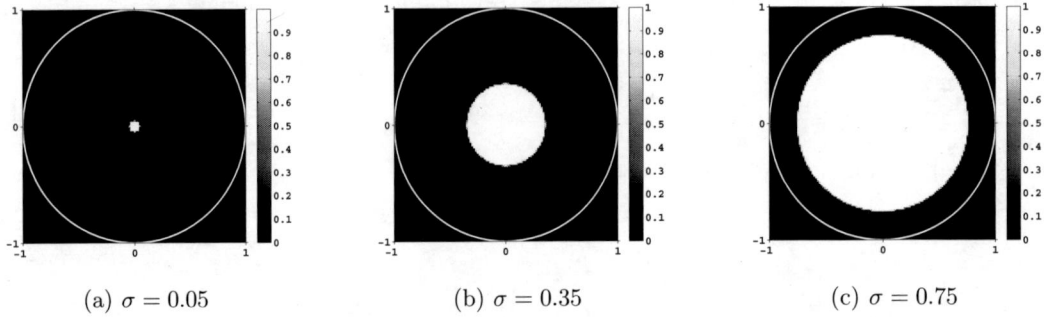

(a) $\sigma = 0.05$ (b) $\sigma = 0.35$ (c) $\sigma = 0.75$

Figure 5. Source configurations for on-axis coherence calibration.

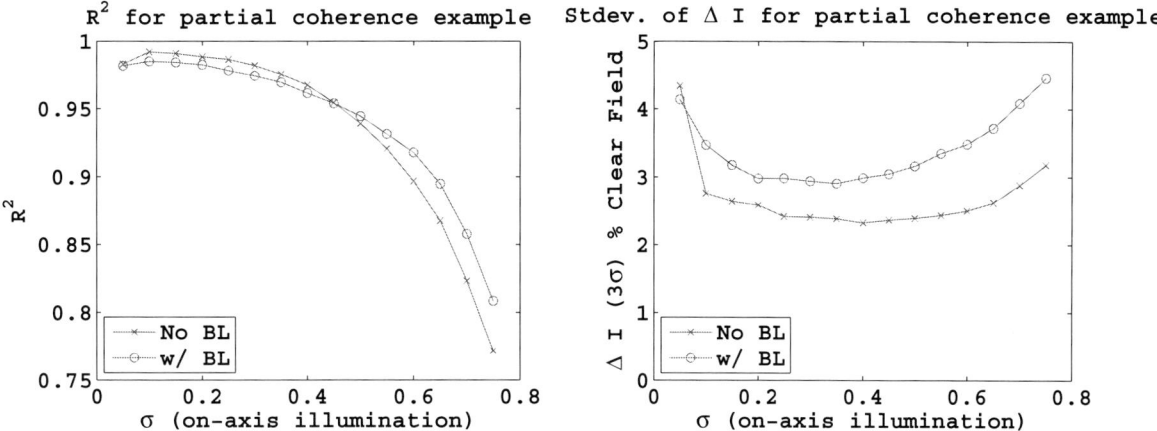

Figure 6. Left, R^2 correlation and right ΔI error plotted vs degree of partial coherence for sources centered on-axis.

Next, KCPM was evaluated for several partially coherent top hat sources (Fig. 5), where the source was varied from very small, nearly coherent σ of 0.05, in steps of .05 up to 0.75. The results for the R^2 correlation factor and 3σ ΔI are shown in Fig. 6. Unlike the previous off-axis example, the results were not well behaved when the source size is increased. At $\sigma = 0.5$, the correlation has dropped to .95, and then falls off sharply when increased further.

The ΔI error also grew as the source size increases. For extremely small σ, there was a drop in correlation, which is related to the kernel size. For a small, nearly coherent source, the spillover function decays more slowly as features interact from a greater distance. For this example, when the kernels were cut to 2048nm, information was lost from the edge of the spillover function leading to reduced correlation and larger ΔI. Once σ was increased slightly, the $FT[S]$ became more compact resulting in the spillover interaction distance becoming smaller and a larger portion was captured by the 2048nm pattern. This illustrates an important quality of KCPM, where the source influences the necessary conditions for kernel size to obtain a given ΔI tolerance.

3.3 Capabilities and Limitations

The previous section showed a clear limitation when dealing with partially coherent illumination. This is caused by unwanted correlation introduced between source pixels. In some cases, this can be ignored, but for sources with a wide range of incident angles, this must be taken into account. To illustrate this problem, consider the example of two source pixels S_1 and S_2 (confined to the xz plane where the z-axis corresponds to the optical axis) that produce the following fields:

$$\vec{E_1} = E_1(\sin\theta_1\hat{x} + \cos\theta_1\hat{z}) * e^{i2\pi k(\cos\theta_1 z + \sin\theta_1 x)} \tag{6}$$
$$\vec{E_2} = E_2(\sin\theta_2\hat{x} + \cos\theta_2\hat{z}) * e^{i2\pi k(\cos\theta_2 z + \sin\theta_2 x)}$$

In a lithography system, each source pixel is uncorrelated relative to one another, therefore we can add intensities to calculate the total on wafer intensity:

$$\begin{aligned} I &= |\vec{E_1}|^2 + |\vec{E_2}|^2 \\ &= |E_1|^2 + |E_2|^2 \end{aligned} \tag{7}$$

But the way KCPM is formulated, the source pixels are treated as correlated, leading to the following intensity calculation:

$$\begin{aligned} I' &= |\vec{E_1} + \vec{E_2}|^2 \\ &= |\vec{E_1}|^2 + |\vec{E_2}|^2 + \langle\vec{E_1}, \vec{E_2}\rangle + \langle\vec{E_2}, \vec{E_1}\rangle \\ &= |E_1|^2 + |E_2|^2 + 2|E_1||E_2|\cos(\theta_1 - \theta_2) \\ &= I + 2|E_1||E_2|\cos(\theta_1 - \theta_2) \end{aligned} \tag{8}$$

For the case where $\vec{E}_1 = \vec{E}_2 = \vec{E}$:

$$\begin{aligned} I &= |E_1|^2 + |E_2|^2 = 2|E|^2 \\ I' &= |E_1 + E_2|^2 = 4|E|^2 \end{aligned} \tag{9}$$

In this situation the difference between I and I', is just a normalization factor, and KCPM still accurately predicts intensity. However, now if we change the problem such that $|E_1| \neq |E_2|$, and $\theta_1 \neq \theta_2$, we can no longer use the same normalization. As shown in Eq. 8 the cross-term is dependent on the magnitudes of $\vec{E}_1$ and $\vec{E}_2$, as well as the angle between them.

When $\vec{E}_1$ is significantly different from $\vec{E}_2$, the normalization correction is no longer sufficient to accurately predict intensity. Furthermore, when we complicate the system with many source pixels, it becomes impossible to correctly offset this cross-term without greatly reducing the runtime. Fortunately, there is a solution where with an increase in match kernels, we can increase the accuracy by reducing the contribution of the cross-term.

4. SOURCE SPLITTING: HYBRID ABBE-KCPM

In order to suppress the errors in the cross-term in Eq. 8 described in the previous section for complicated sources with a large variation in incident angle, we must treat different source regions as separate image calculations. This modification is called Hybrid Abbe-KCPM, as the source is split into regions where the intensity from each section is summed as in Abbe imaging. For source splitting, k-means clustering was used to cluster the source into the regions of a given size with minimum distance from the center of the clusters. The criteria we are concerned with is related to coherence, so the number of clusters is increased until a specified size of σ is reached for each subregion. For example, looking at Fig. 6, if the minimum allowable correlation is 0.95, clusters can span $\sigma = 0.5$ in the source plane. However, if correlation of 0.97 is required, the clusters must have $\sigma \geq 0.25$, and the source is separated into more regions requiring more kernels to match. Unfortunately, this solution adds to the computational cycles for KCPM, but as will be shown through the following examples, depending on the necessary accuracy tolerances, this step may not be needed.

4.1 Annular Illumination: Breakdown vs Normal

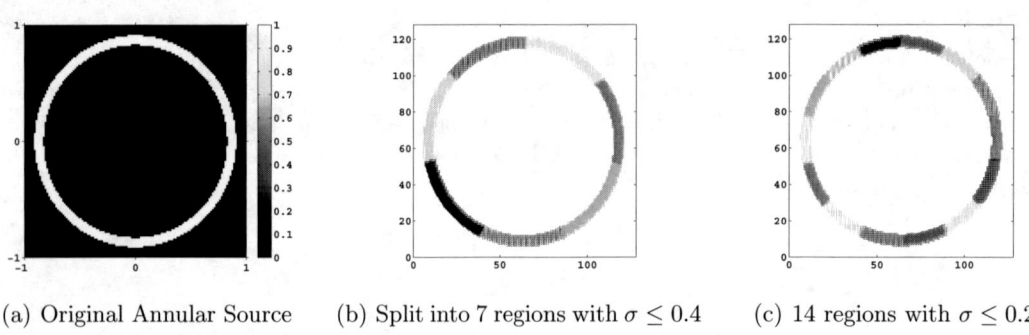

(a) Original Annular Source   (b) Split into 7 regions with $\sigma \leq 0.4$ (c) 14 regions with $\sigma \leq 0.2$

Figure 7. Source and decomposition scheme for annular illumination with $0.82 \leq \sigma \leq 0.9$.

KCPM was used to evaluate the same test pattern for annular illumination with $\sigma_{out} = 0.9$, and $\sigma_{in} = 0.82$. The source was split into 7 and 14 regions such that $\sigma \leq 0.4$ and $\sigma \leq 0.2$ respectively. The source and automatically k-means generated split configurations are shown in Fig. 7. To illustrate the value of Abbe-KCPM, the simulation was run with full annular illumination for comparison. For full annular, two match patterns are needed to monitor one process change: one is for the initial field estimate (E_{object}) and the other is for the spillover Φ. For multiple focal levels and other changing aberrations, an additional Φ kernel is required for each change, but the same field estimate (E_{object}) is used for all calculations. In comparison, splitting the source into regions of $\sigma = 0.4$ or smaller, 7 sources are required, therefore there are 14 patterns for 7 E terms and 7 Φ terms. For regions of $\sigma = 0.2$, 14 source partitions are required for a total of 28 match patterns. Clearly

Table 1. Summary of R^2 correlation and intensity error values for annular illumination test case

	Pattern Pixels	Binary Mask		Boundary Layers	
		R^2	$3\sigma\Delta I$	R^2	$3\sigma\Delta I$
Full Annular	310	0.8821	.0279	0.8615	.0469
7 Regions	302	0.9563	.0173	0.9461	.0303
14 Regions	310	0.9636	.0158	0.9535	.0282

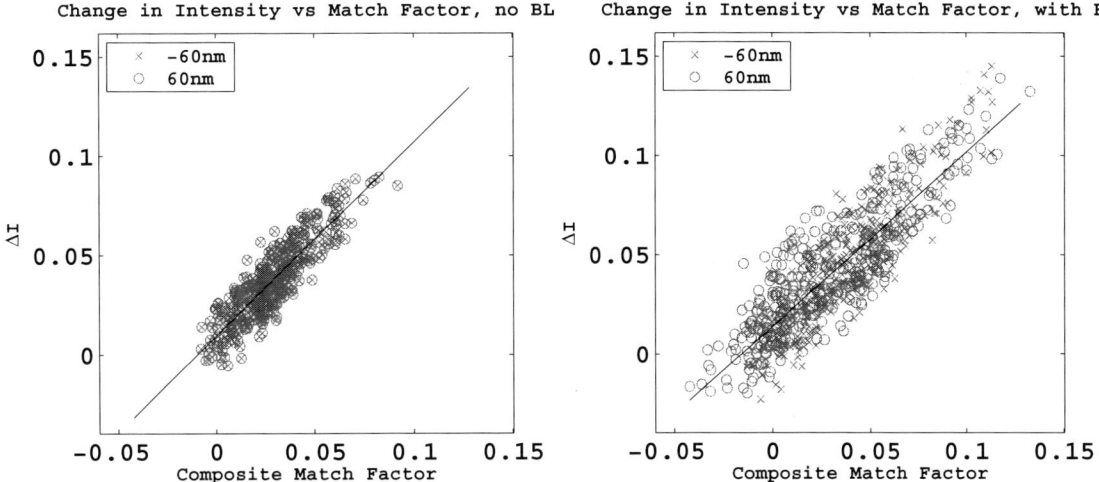

Figure 8. Match factors plotted vs. ΔI for KCPM for entire annular source.

a 7x or 14x increase in computation is undesirable, but if high accuracy is needed, source splitting is a useful tool.

The results are summarized in Table 1. Comparing the full annular case to the split into 7 regions, there is a significant increase in correlation, and the error in ΔI is reduced by more than 35%. Going a step further, increasing to 14 source regions decreases the error band by an additional 7% to 9%. The necessity of source splitting is both source and application dependent. Hot spot detection places a premium on speed, so full source KCPM may be usable. For optimization, accuracy is more important, and the increased accuracy of source splitting can offset the need for an increased number of computations.

4.2 Quadrapole Examples: Full Generality

The final example was designed to explore the full generality of KCPM. Thus far, boundary layers have been used, making the mask complex. Also, splitting the source leads to each region being asymmetric which introduces a complex source dependence. The last generalization is to make the pupil function also complex by adding background aberrations. To highlight the full flexibility, three configurations of quardapole illumination (Fig.11) were tested. The first two cases are simply balanced quadrapole with poles of $\sigma = 0.1$ and $\sigma = 0.2$, with an unaberrated pupil (Figs.11(a) and 11(c)). For the third case, source asymmetry was introduced by shifting the source poles in both x and y, combined with intensity imbalance(Fig.11(d)). Additionally, random values were assigned to fringe Zernikes 5-36,[7] bound by $10m\lambda$ RMS (Fig.11(e)). All cases are compared with and without source splitting (11(b)). Because several different source configurations were used, rather than fixing the pixel size of the match pattern, a threshold was used to determine kernel size. For these three examples, the kernels were cut such that 90% of the field amplitude is contained in the match pattern. This led to a range of kernel sizes from 202 pixels to 424. As shown in the data, there is not much correlation between kernel size and accuracy when total energy is kept constant.

The resulting KCPM data is plotted vs. aerial image in Fig 12. The correlation values and ΔI values are summarized in Table 2. For all cases, splitting the source greatly improved accuracy. For the $\sigma = 0.1$, balanced

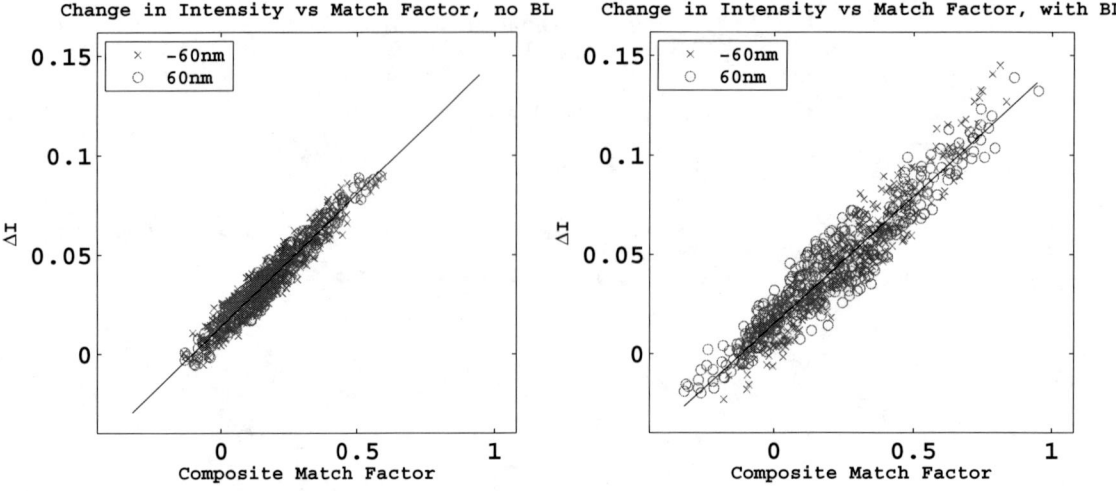

Figure 9. Match factors plotted vs. ΔI for KCPM for annular source split into 7 regions.

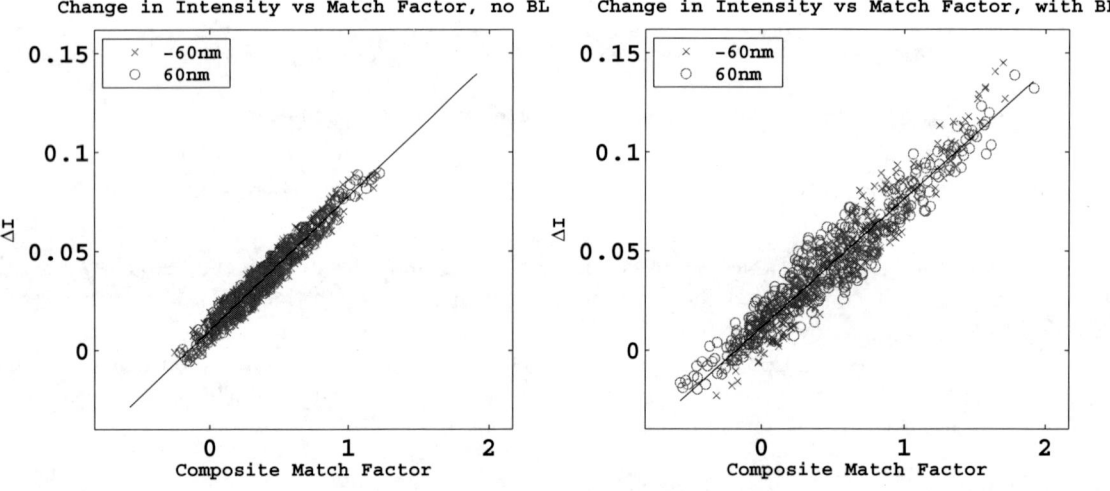

Figure 10. Match factors plotted vs. ΔI for KCPM for annular source split into 14 regions.

Table 2. Summary of R^2 correlation and intensity error values for quardapole illumination test cases

	Pattern Pixels	Binary Mask		Boundary Layers	
		R^2	$3\sigma\Delta I$	R^2	$3\sigma\Delta I$
Quad $\sigma = 0.1$	304	0.858	.0328	0.690	.0490
Split	304	0.989	.0094	0.964	.0180
Quad $\sigma = 0.2$	202	0.788	.0354	0.559	.0532
Split	202	0.936	.0202	0.876	.0310
$\sigma = 0.1$, Zernike	424	0.833	.0408	0.696	.0562
Split	424	0.940	.0253	0.916	.0314

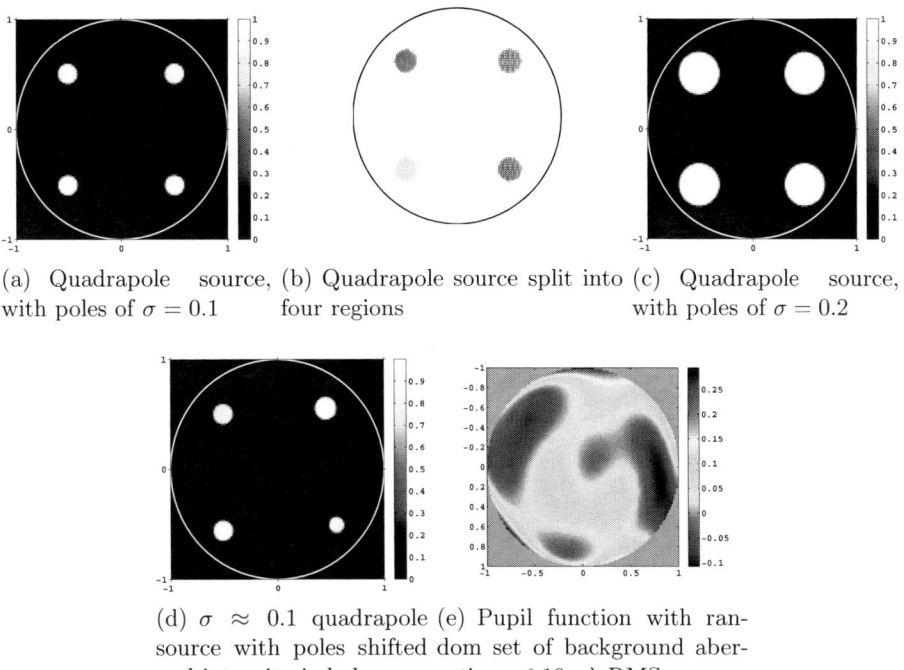

(a) Quadrapole source, with poles of $\sigma = 0.1$

(b) Quadrapole source split into four regions

(c) Quadrapole source, with poles of $\sigma = 0.2$

(d) $\sigma \approx 0.1$ quadrapole source with poles shifted and intensity imbalance

(e) Pupil function with random set of background aberrations $\leq 10m\lambda$ RMS

Figure 11. Source and decomposition schemes for 3 quadrapole examples: Balanced $\sigma = 0.1$, Balanced $\sigma = 0.2$, and Unbalanced with random Zernike profile.

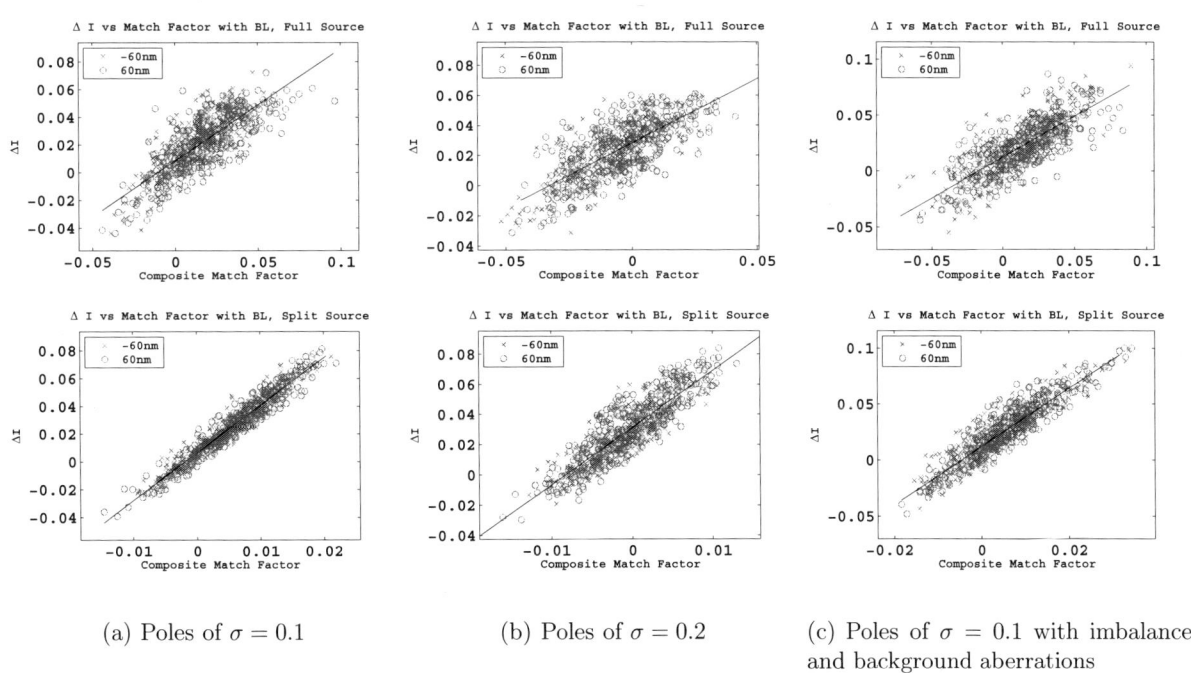

(a) Poles of $\sigma = 0.1$

(b) Poles of $\sigma = 0.2$

(c) Poles of $\sigma = 0.1$ with imbalance and background aberrations

Figure 12. Match factors plotted vs. aerial image ΔI for KCPM for quadrapole examples with boundary layers.

case correlation was 0.99 for no boundary layers and above 0.96 for the case with boundary layers. As expected, this case had the best correlation between KCPM and aerial image. Comparing the other 2 cases, both showed a significant drop in correlation from the small sigma case. Note that the unbalanced, aberrated case performed better than the large σ. This demonstrates that KCPM can handle the full complexity of source, mask, and pupil function, but is still mostly limited by coherence concerns. For several of the configurations, correlation is above 0.8 without source splitting. This indicates that KCPM can still be used for hotspot detection, even for complicated sources. However, for the $\sigma = 0.2$ case correlation as low as 0.56 was observed, indicating that in some cases, source splitting would likely still be necessary.

5. CONCLUSIONS

Kernel convolution with pattern matching (KCPM) has been shown to be a very useful tool for fast image estimation in the presence of process modifications expressed in the pupil. Convolution calculation takes $\approx 40\mu s$ per pattern per location, making KCPM extremely fast and useful for hot spot detection and optimization applications. For simple, relatively coherent sources, R^2 correlation as high as 0.99 was calculated between KCPM and rigorous aerial image simulation. Accuracy limitations due to coherence effects were uncovered and treated in a way to increase accuracy with minimal increase in computation through source splitting. Though source splitting is slower, it offers a convenient knob to turn when accuracy is needed at the sake of speed. Through source splitting, examples with annular, quadrapole and unbalanced quadrapole were tested, showing correlation varying from 0.88 to 0.99. KCPM was demonstrated to be accurate for a fully general case of complex source, mask, and pupil function. Because of the general formulation, any effect that can be captured in the pupil function, mask, or source can potentially be added into the algorithm. Eventually, polarization and resist effects may be treated in this fast approximate framework.

ACKNOWLEDGMENTS

This work was supported by an IBM PhD fellowship and by the IMPACT grant with contributions from AMD, Applied Materials, ASML, Cadence, Canon, Ebara, Hitachi, IBM, Intel, KLA-Tencor, Magma, Marvell, Mentor Graphics, Novellus, Panoramic, SanDisk, Spansion, Synopsys, Tokyo Electron Limited, and Xilinx, with donations from Photronics, Toppan and matching support by the U.C. Discovery Program.

REFERENCES

[1] Miller, M. A. and Neureuther, A. R., "Analysis and modeling of photomask edge effects for 3d geometries and the effect on process window," *Proc. SPIE* **7274** (2009).

[2] Tirapu-Azpiroz, J. and Yablonovitch, E., "Incorporating mask topography edge diffraction in photolithography simulations," *J. Opt. Soc. Am. A* **23**, 821–828 (2006).

[3] Tirapu-Azpiroz, J., Rosenbluth, A. E., Graur, I., Burr, G. W., and Villares, G., "Isotropic treatment of emf effects in advanced photomasks," *Proc. SPIE* **7488** (2009).

[4] Adam, K. and Neureuther, A. R., "Simplified models for edge transitions in rigorous mask modeling," *Proc. SPIE* **4346**, 331–344 (2001).

[5] Gennari, F. and Neureuther, A. R., "A pattern matching system for linking tcad and eda," *Proc. 5th International Symposium on Quality Electronic Design*, 165–170 (2004).

[6] Miller, M. A. and Neureuther, A. R., "Extensions of boundary layer modeling of photomask topography effects to fast-cad using pattern matching," *Proc. SPIE* **7488** (2009).

[7] Shannon, R. R. and Wyant, J. C., [*Applied Optics and Optical Engineering*], Academic Press Inc., San Diego (1992).

Aerial image model and application to aberration measurement

Anatoly Y. Burov*, Liang Li, Zhiyong Yang, Fan Wang, Lifeng Duan
Shanghai Micro Electronics Equipment Co., Ltd. (China)

ABSTRACT

In this paper, we present a streamlined aerial image model that is linear with respect to projection optic's aberrations. The model includes the impact of the NA, partial coherence, as well as the aberrations on the full aerial image as measured on an x-z grid. The model allows for automatic identification of image's primary degrees of freedom, such as bananicity and Y-icity among others. The model is based on physical simulation and statistical analysis. Through several stages of multivariate analysis a reduced dimensionality description of image formation is obtained, using principal components on the image side and lumped factors on the parameter side. The modeling process is applied to the aerial images produced by the alignment sensor in a 0.75NA ArF scanner while the tool is integration mode and aberration levels are high. Approximately 20 principal components are found to have a high signal-to-noise ratio in the image set produced by varying illumination conditions and considering aberrations represented by 33 Zernike polynomials. The combined coefficients are extracted and the measurement repeatability is presented. The analysis portion of the model is then applied to the measured coefficients and a subset of projection lens' aberrations are solved for.

Keywords: Wavefront error, Zernike coefficient, Aerial image, bananicity

1. INTRODUCTION

The aerial image formed in a projection lithographic system is affected by many system parameters. One of the parameter groups is lens aberrations, or wavefront errors. It is estimated that the aberration variation are a major tool-related contributor to across-the-field CD uniformity for modern lithographic processes[1]. It is then critical for the system designers to maintain a careful control of the distribution of the residual wavefront error across the field, and understand the impact of such error on system performance. This understanding has to be maintained throughout the various phases of system lifecycle, such as design, manufacturing, integration and use.

To facilitate and support aberration control during system assembly, integration, and use, in-situ aberration metrology is key. Several aberration metrology methods have been proposed by lithographic tool manufacturers to date. Of these methods, those utilizing direct sensor-based image measurements coupled with mathematical analysis of such images have been most successful[1]. It is thus the goal of this work to develop a mathematical model of aerial image formation that can be applied to the analysis of images captured by the alignment sensor in an ArF scanner manufactured by SMEE and apply this model to the extraction of aberrations present in the lens.

2. AERIAL IMAGE MODELING PROCESS

2.1 Impact of Wavefront Error on Lithographic Aerial Image

The wavefront error of the projection optic has a distinct signature in the aerial image space for several of the specific terms in the Zernike polynomial expansion. The distortion shape can be calculated explicitly using the physical simulation of image formation, as shown in Figure 1.

* Anatoly@smee.com.cn, 86-21-5131-5131x2091

Optical Microlithography XXIII, edited by Mircea V. Dusa, Will Conley, Proc. of SPIE Vol. 7640,
764032 · © 2010 SPIE · CCC code: 0277-786X/10/$18 · doi: 10.1117/12.848421

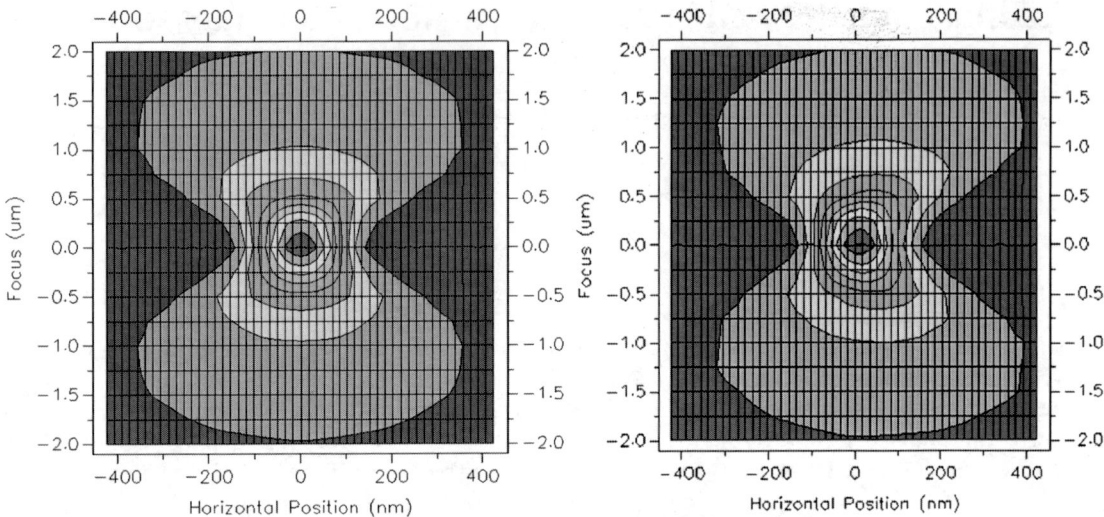

Figure 1 The wavefront error's impact on the lithographic image. On the left is an aerial image of an isolated line produced by a perfect lens with no wavefront error. On the right side is an image produced by a lens with some amount of Z_7, or coma. Because of wavefront error's effect, the right image presents a non-symmetric character.

In the approximation of small wavefront error, the impact can be demonstrated to have a quadratic form with respect to Zernike coefficients c_m [2]. Distortions of the aerial image corresponding to a particular combination of Zernike coefficients $c_m c_n$ can be calculated directly if using the Hopkins imaging formalism. Such individual distortions, while exactly known, may not be orthogonal to each other in aerial image space. Although the Zernike polynomials Z_n are orthogonal to each other in the pupil space, their impact on the aerial image, upon undergoing several integration steps, is no longer orthogonal.

For the measurement purposes, however, identifying degrees of freedom that are orthogonal in the image space is highly desired. This approach is expected to separate the measurement repeatability, or random error, from the fitting accuracy, or the systematic error. It is the goal of the subsequent section to find and utilize such a model.

2.2 Aerial Image Model

In order to find aerial image orthogonal degrees of freedom, a principal components analysis is used.

A set of aerial images is pre-calculated using physical simulation, with varying Zernike coefficients, NA and illumination settings. The values of the settings are generated using Box-Behnken design scheme, resulting in approximately 25,000 combinations. These images are then stacked into a single array, and analyzed for principal components using singular value decomposition. As a result, a set of orthogonal principal components I_i is obtained, along with their weights, V_i.

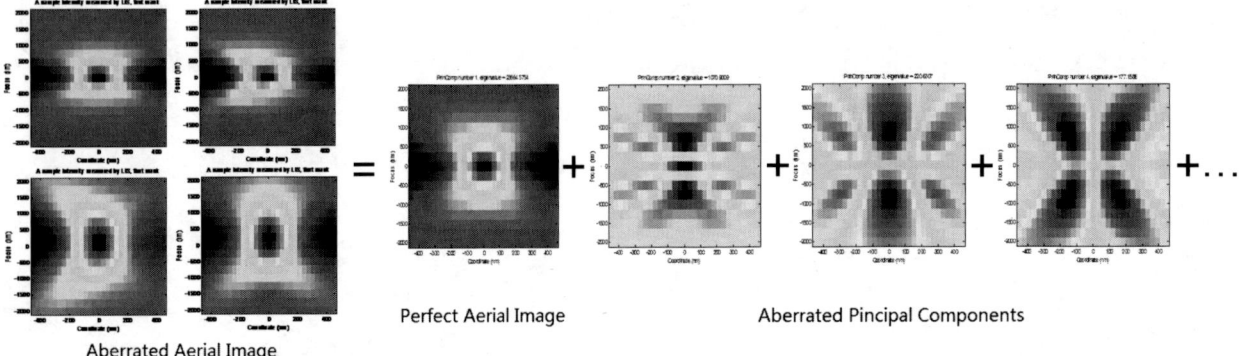

Figure 2 Aerial image principal components decomposition. Typical degrees of freedom, such as "NA pinch", "bananicity", or "Y-icity" are automatically identified via a singular value decomposition of the large set of aberrated aerial images.

A linear regression is then performed on the weights V_i, with respect to NA, σ, and Zernike coefficients. The final form of the obtained model is (in the simple case of single NA and σ)

$$I(x,f) = a_0 \cdot I_0(x,f) + \sum_{i=1}^{n} a_i \cdot I_i(x,f)$$

$$\vec{a_i} = \vec{Z_i} * TM$$

(1)

The final shape of the full model is schematically shown in Figure 3.

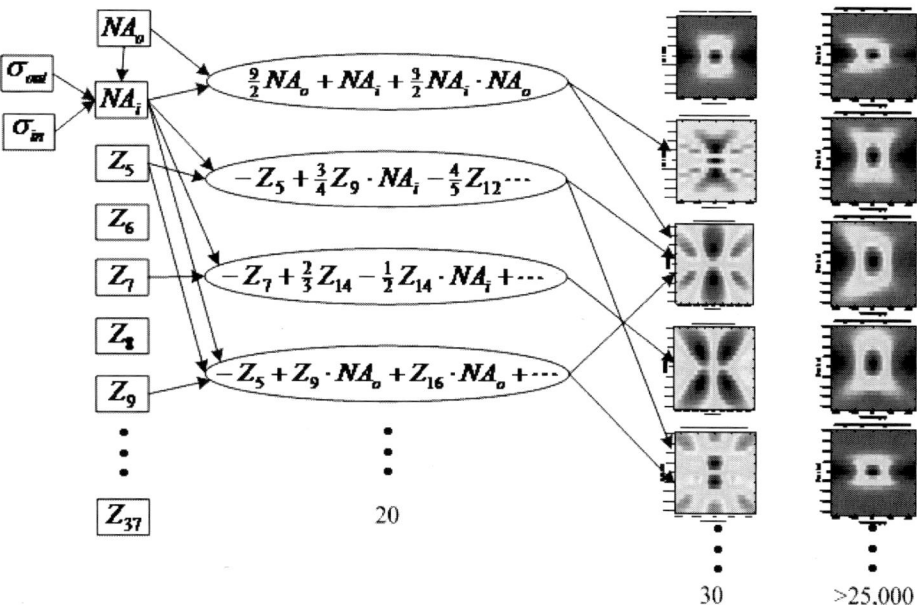

Figure 3 Schematic representation of Aerial Image model. A set of approximately 20 factors is found to predict weights for 30 principal components.

3. ABBERATION MEASUREMENT

3.1 Tool's Configuration and Setting

Aberration measurement for lithography tool's projection lens is a typical in-situ method. This means that most of the lithography tool's sub-systems would be used for aberration measurement. Shown in Figure 4, the measurement system includes the laser source, illuminator, mask and intensity sensor on wafer stage. Projection lens wavefront is the measuring object. Laser source generates the light to form the aerial image; illuminator reshapes the pupil intensity to a special distribution; the projection lens NA should also be set as maximal for whole pupil information; and the sensor captures the intensity of aerial image.

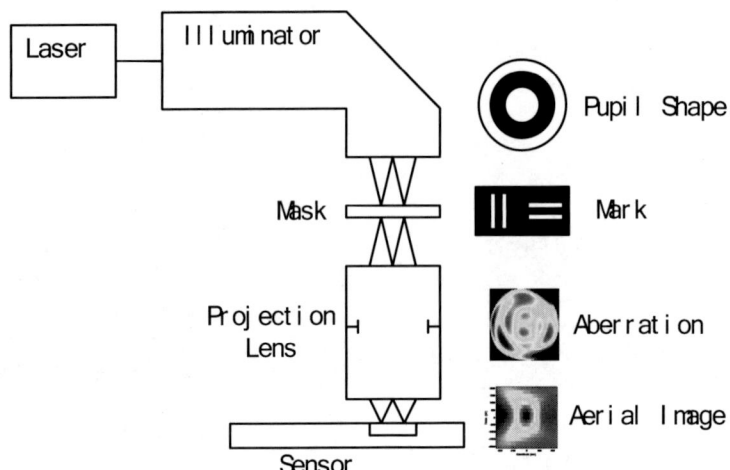

Figure 4 Hardware configuration for an in-situ aerial image measurement. The measurement system includes the laser source, illuminator, mask, projection lens and intensity sensor which is mounted on the wafer stage.

Image formation depends on several factors that include wavelength, NA, mask feature, illuminating shape and wavefront errors. The measurement method is based on imaging process, so the model user needs define all the factors as the tool setting values except the wavefront error. Generally, this method could be used for narrow bandwidth projection lens because this type lens' Zernike coefficients don't change significantly with wavelength. Pupil shape and mask pattern should be optimized for aberration measurement to make the aerial image's information sensitive to Zernike coefficients. To measure the wavefront error, we should get whole pupil's information and sample the pupil uniformly. Pupil shape and mask feature design should follow this principle, 2-D mask and 2-D pupil shape are necessary. Generally, we use two separate 1-D features at different direction for easily capture by sensor and lower X-Y cross-talk.

The intensity sensor which has a small detection window is mounted on the wafer stage and feeds back the intensity value during the image scanning process. At same time, the wafer stage's position data should be recorded. The intensity sensor's detection window has a fixed size. It means that the detected aerial image is the real image's convolution result by sensor window. To account for this, a convolution is performed on all the simulated aerial images.

3.2 Aerial Image Measurement

The data are collected by the scanning hardware and stored as a table after scaling. An example of a single scan is shown below, for both the horizontal and the vertical orientation sensors.

Table 1. A portion of a typical data set collected during an aerial image measurement

X-Z Image Data										
Intensity	0.076	0.143	0.121	0.315	0.702	0.544	0.293	0.125	0.116	...
X-Position (nm)	..92.5	..95.2	..95.1	..92.9	..82.2	..89.8	..92.6	..94.3	..80.8	...
Z-position (nm)	..46.5	..51.3	..59.1	..66.4	..74.0	..81.3	..88.8	..96.6	..03.9	...
Y-Z Image Data										
Intensity	0.082	0.091	0.100	0.161	0.296	0.365	0.070	0.079	0.072	...
Y-position (nm)	..59.9	..50.3	..71.3	..55.5	..77.4	..54.5	..86.6	..48.4	..95.1	...
Z-Position (nm)	..46.4	..12.6	..82.4	..41.5	..56.6	..21.8	..73.4	..13.6	..59.6	...

In these raw data, the X, Y, Z value come from wafer stage's absolute coordinate, and because of stage's servo error, the raw data's X, Y, Z positions are not on the regular grid.

For all of the above reasons, some steps are necessary to processes the raw data. Firstly, the image needs to be centered. Secondly, the data is interpolated onto a regular grid for plotting and visualization needs.

Sensor noise and stage position error are random contributors. Shown in Figure 4 are the re-gridded data from 5 consecutive measurements.

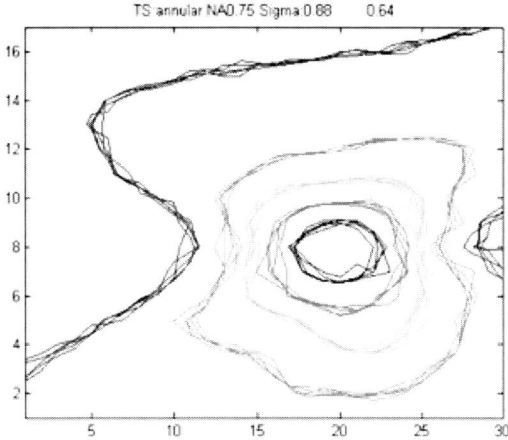

Figure 4 The measured aerial image after five distinct scans at one field point. Typical aerial image for measurement includes two parts: X-Z image and Y-Z image. The intensity sensor should move along the wafer stage's X-axis and step along Z axis after one level's scanning. In these two processes, the intensity data and position data should be synchronized rigorously. Table 1 is the example of raw measurement data.

While random noise does have a noticeable contribution, its impact on the principal components' weights can only be estimated after their fitting.

3.3 Principal Components Extraction

In aerial image model, we have identified some orthogonal principal components to reconstruct the aberrated aerial image. The aberrated aerial image can also be decomposed to the principal components and the weights represent wavefront error's level. This step's purpose is to calculate the weights' vector $\vec{a_i}$ from the detected aerial image. Based on equation (2), Least-Square-Fitting should be used to calculate the $\vec{a_i}$.

$$I(x, f) = a_0 \cdot I_0(x, f) + \sum_{i=1}^{n} a_i \cdot I_i(x, f) \qquad (2)$$

Shown as Figure 5, the regression image is very close to the measurement image, with the residual error coming from two kinds of contributor: one is model's high order residual; the other is noise from stage and sensor.

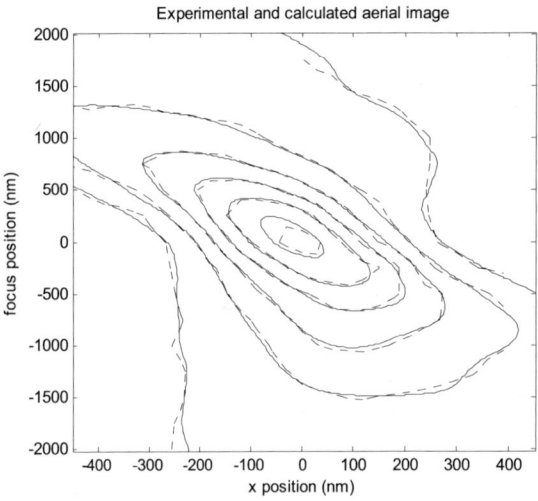

Figure 5 The difference between raw data and fitting data of aerial image. Dashed lines are the sensor measured raw data and solid lines are the fitting data that used 30 principal components. The difference is acceptable for aberration measurement.

The individual principal component images I_i are interpolated onto the actual data x-z grid and the weights are fitted using raw data. This approach was found to yield better repeatability than using re-gridded and interpolated data. An example of repeatability of extracted weights is shown in Figure 6.

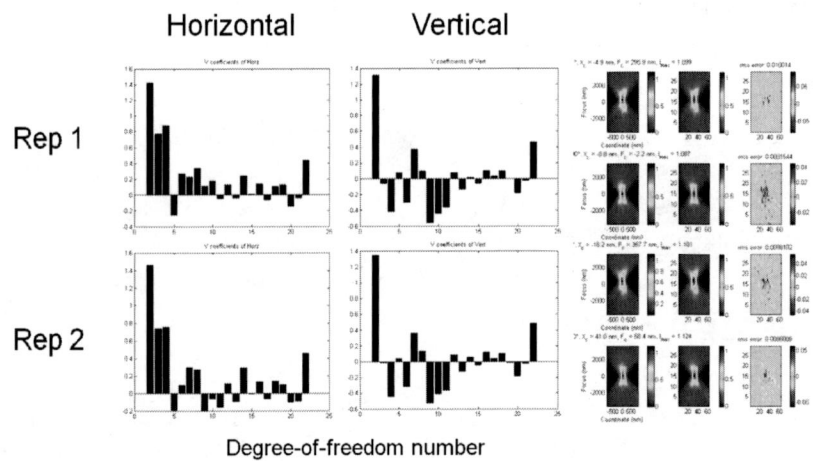

Figure 6. The result of decomposition of two aerial images into their principal components. The two repetitions represent two images, obtained during two scans. The bar plots show the weights of the principal components model, while the raw data, the fitted image, and the difference are shown on the right

3.4 Zernike Coefficient Calculating

For aberration estimation with this method, there are many systematic contributors for error from the imaging process. Main contributors include pupil shape error, mask error, sensor noise and stage position error. Because this method is an in-situ measurement, for good measurement result, the mask feature size and pupil shape should be perfect as the modeling process used. Otherwise, we can use the machine's real data of mask and illuminator to build the model. Machine's mask and illuminator data should be calibrated and recorded as a database for measurement using.

The final step is calculating the aberration values. Since the translation matrix TM is not of full rank and fewer than 33 principal components are available, only combinations of c_n can be estimated. The results of solution for a set of such combinations are presented in the next section.

4. RESULT

A repeatability of measured combined Zernike coefficients is shown in Figure 6 and Figure 7. The number of combinations is lower than the full 33 coefficients of the desired wavefront description.

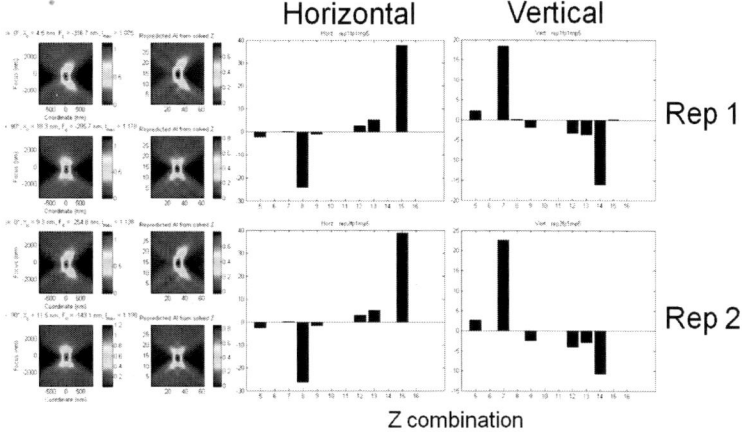

Figure 6 The result of the solution for the combination of Z coefficients, using the observed principal components weights for two repeated measurements.

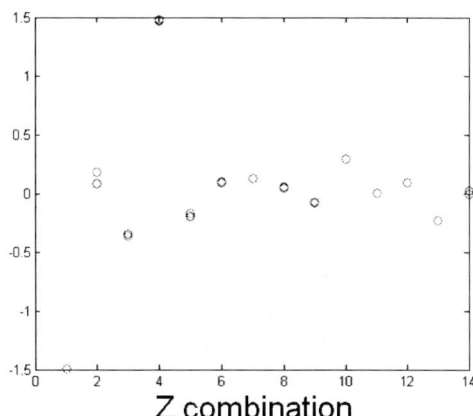

Figure 7 The values of the combined Zernike coefficients, by the combination number. Values from two successive measurements are plotted.

5. CONCLUSIONS

A method to calculate the Zernike coefficients of a lithographic projection tool's lens is presented. An aerial image model linear in Zernike coefficients has been constructed. Several sets of aerial images have been collected and analyzed with this model. The random error of the data collection method has been estimated. In the current implementation, the method lacks resolution to solve for 33 terms of the Zernike wavefront error description, however, the repeatability of the measurements is offering promise that the desired accuracy may be achieved with some modifications.

ACKNOWLEDGEMENTS

The authors wish to thank Mr. Jianrui Cheng and Mr. Lifeng Duan for useful discussions.

REFERENCES

[1] Lai, Kafai, Gallattin, Gregg, Kerkhof, Mark van de, Boeij, Wim de, Kok, Haico, Schriever, Martin, Morillo, Jaime, Fair, Bob, Bennett, Stephanie, Corliss, Daniel, "New paradigm in Lens metrology for lithographic scanner: evaluation and exploration", Proc SPIE 5377, 160 (2004)

[2] Flagello, Donis G., Klerk, Jos de, Davies, Guy, Rogoff, Rich, Geh, Bernd, Arnz, Michael, Wegmann, Uli, Kraemer, Michael, "Towards a comprehensive control of full-field image quality in optical photolithography", Proc SPIE 3051, 372 (1997)

[3] Booth, N. and Smith, A. S., [Infrared Detectors], Goodwin House Publishers, New York & Boston, 241-248 (1997).

Methods for Benchmarking Photolithography Simulators:
Part V

Trey Graves, Mark D. Smith, and Sanjay Kapasi

KLA-Tencor Corp.,

ABSTRACT

As the semiconductor industry moves to double patterning solutions for smaller feature sizes, photolithography simulators will be required to model the effects of non-planar film stacks in the lithography process. This presents new computational challenges for modeling the exposure, post-exposure bake (PEB), and development steps. The algorithms are more complex, sometimes requiring very different formulations than in the all-planar film stack case. It is important that the level of accuracy of the models be assessed.

For these reasons, we have extended our previous papers in which we proposed standard benchmark problems for computations such as rigorous EMF mask diffraction, optical imaging, PEB, and development [1-4]. In this paper, we evaluate the accuracy of the new PROLITH wafer topography models. The benchmarks presented here pertain to the models (and their associated outputs) most affected by the switch to non-planar film stacks: imaging at the wafer (image intensity in-media) and PEB (blocked polymer concentration). Closed-form solutions are formulated with the same assumptions used in the model implementation. These solutions can be used as an absolute standard and compared against a simulator. The benchmark can then be used to judge the simulator, in particular as it applies to speed vs. accuracy tradeoffs.

Keywords: Lithography simulation, numerical accuracy, image intensity, post-exposure bake, PROLITH

1. INTRODUCTION

Single patterning lithography has pushed k1 values near the theoretical limit (0.25). Double patterning will be the method used to print features at the 32nm and 22nm nodes and possibly beyond. Double patterning presents some new challenges for lithography simulation. In the past, it was a good assumption that the wafer stack was made of planar homogeneous films. The first patterning step in a double patterning process is modeled well under this assumption. However, in the second step this is no longer the case. Imaging, PEB, and develop computations must handle the non-planar topography introduced by processes such as etch, spin coating, deposition, etc. The algorithms that do these computations are more complex, sometimes requiring very different formulations than in the all-planar film stack case

The accuracy of simulators is important, especially if lithographers are to use these simulators to make quantitative assessments that lead to critical decision making. It is also important to realize that there are many ways to model various steps in the photolithographic process. FDTD, RCWA, and FEM are just a few of the methods that can be used to determine the image-in-media in a non-planar film stack. Regardless of the implementation of the model, the algorithm should have good characteristics in terms of accuracy and convergence. It is also necessary that any numerical implementation of a model in a simulator be reasonably fast, as well as free of bugs or algorithm problems. For these reasons, we have extended our previous papers [1,2,3,4] where we proposed standard benchmark problems for aerial image calculations, image in resist calculations, EMF mask topography effects, PEB, and development. The benchmarks presented here are for image-in-media and post-exposure bake (PEB) under wafer topography conditions. We will use closed-form solutions as an absolute standard to judge the accuracy of a simulator. The benchmark can then be used to judge the simulator as it applies to speed vs. accuracy tradeoffs.

Optical Microlithography XXIII, edited by Mircea V. Dusa, Will Conley, Proc. of SPIE Vol. 7640,
764033 · © 2010 SPIE · CCC code: 0277-786X/10/$18 · doi: 10.1117/12.846376

2. IMAGE-IN-MEDIA BENCHMARK

We are interested in determining the light intensity in materials such as resist, hard masks, anti-reflective coatings, etc. under wafer topography conditions. We call this the "image-in-media" to distinguish it from the image-in-resist which is of primary importance in a lithography simulator.

This benchmark problem follows a method presented by Botten et al. [5], except that here we have adapted the method so that the incident beam passes through the immersion fluid (water) instead of air. The algebraic details of the solution will not be presented here, but a brief outline of the method is as follows. First, the eigenfunctions for the electric field (S polarization) or for the magnetic field (P polarization) are found inside the grating material. These eigenfunctions can be found analytically but the eigenvalues must be found numerically by solving a transcendental equation. The eigenfunctions represent plane waves propagating in the oxide and resist which are periodic and continuous across the material interfaces. Because the grating is lossless, the eigenfunctions have the properties that they are self-adjoint, and they form a complete, orthogonal basis for continuous functions [5]. This means that the electric field (or magnetic field) inside the grating can be represented by an infinite series of eigenfunctions.

The electric (or magnetic) fields in the regions above and below the grating are represented by Rayleigh expansions (sets of plane waves). We find the coefficients in the eigenfunction expansion and in the Rayleigh expansions by using the method of moments to enforce the boundary conditions. This leads to a large set of algebraic equations that can be solved using any standard linear algebra package. Of course, the infinite series that is used to represent the fields in the grating and in the regions above and below the grating must be truncated in order to find a numerical solution to the problem. We chose expansions with 31 eigenfunctions and 83 Raleigh modes for the transmitted and reflected fields. Finally, we used MATLAB to find the eigenvalues and to solve the set of linear algebraic equations.

Reference 5 assumes that the imaginary part of the index of refraction is zero. In the following comparison, we use only the real part of the refractive index. The immersion fluid is water (n = 1.44), the grating is made of "resist" (n = 1.7) and "oxide" (n = 1.563). The substrate is "silicon" (n = 0.88). The periodic grating of the wafer pattern uses a 1:1 duty with 150nm of resist and 150 nm of oxide (giving a 300nm pitch). The depth of this grating is 120 nm. The wavelength is 193nm. Normal incidence is used. The setup is shown in Figure 1.

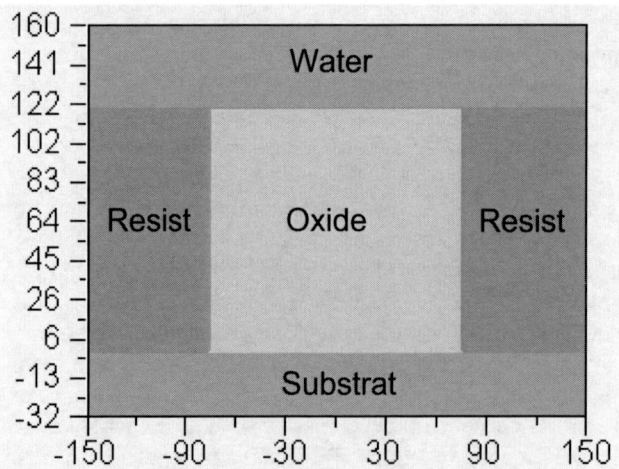

Figure 1. Dimensions and geometry of image-in-media benchmark.

By using the methods described above, we calculated the intensity of the electric field (electric field squared) in the resist and oxide. We can also calculate the results using the PROLITH Wafer Topography simulator. Results are shown in Figures 2 and 3. The results indicate that even at low truncation order, PROLITH has good accuracy. It also converges very well with increasing truncation order.

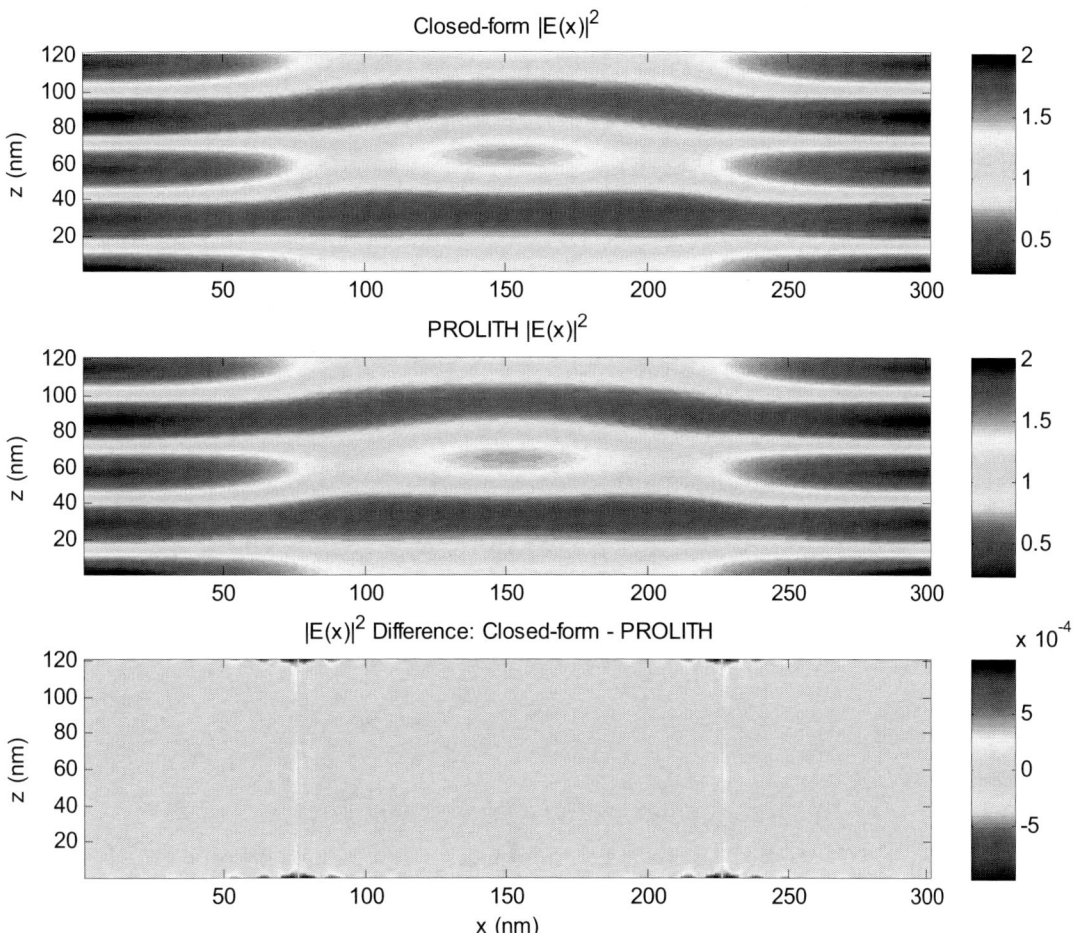

Figure 2. The upper plot shows the image intensity in resist and oxide for the closed-form method. The middle picture is the result from PROLITH with a maximum truncation order of 40. In the lower plot the difference is plotted and is seen to be very small. The regions with slightly larger error at 75 nm and 225 nm in x are the interfaces of the resist and oxide.

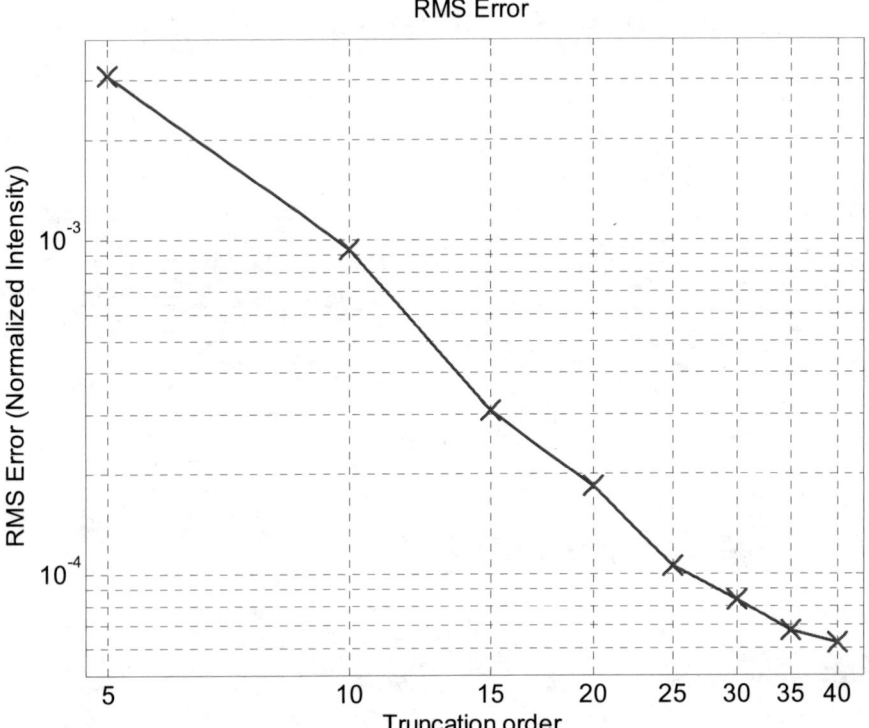

Figure 3. RMS error of the electric field intensity vs. truncation order.

3. POST-EXPOSURE BAKE BENCHMARK

During PEB, the photoacid (H), created during the exposure process, diffuses through the resist and catalytically reacts with blocked polymer sites, M. At the same time, the base quencher, Q, also diffuses through the resist and neutralizes the acid [6]. Mathematically, we model this as:

$$\frac{dM}{dt} = -\mathrm{k}_a \cdot H \cdot M \tag{1}$$

$$\frac{dH}{dt} = -\mathrm{k}_{loss} \cdot H - \mathrm{k}_Q \cdot H \cdot Q + D_\mathrm{H} \nabla^2 H \tag{2}$$

$$\frac{dQ}{dt} = -\mathrm{k}_Q \cdot H \cdot Q + D_Q \nabla^2 Q. \tag{3}$$

Here D_H is the diffusivity of the acid, k_{loss} is the acid loss reaction rate constant, k_Q is the acid-base quench rate constant, D_Q is the diffusivity of the base quencher, and k_a is the deblocking reaction rate constant.

3.1. POST-EXOSURE BAKE SOLUTION

The neutralizing of acid and base makes an analytical solution of equations (*1*)-(3) difficult to impossible. However, a simplified solution is possible with no quencher. Setting $Q=0$, equations (*1*)-(3) reduce to two equations:

$$\frac{dM}{dt} = -\mathrm{k}_a \cdot H \cdot M \qquad (4)$$

$$\frac{dH}{dt} = -\mathrm{k}_{loss} \cdot H + D\nabla^2 H \qquad (5)$$

D now represents the diffusivity of the acid. Assuming periodic boundary conditions, the following solutions for H and M can be shown to satisfy (4) and (5):

$$H = \sum_n \hat{H}_n Cos\left(\frac{2\pi nx}{p}\right) \exp\left(-\left[\frac{2\pi n}{p}\right]^2 Dt\right) \qquad (6)$$

$$M = \exp\left[-k_a \sum_n \hat{H}_n Cos\left(\frac{2\pi nx}{p}\right) \frac{1 - \exp\left(-\left[\frac{2\pi n}{p}\right]^2 Dt\right)}{\left[\frac{2\pi n}{p}\right]^2 D}\right]. \qquad (7)$$

x is the spatial coordinate in this 1D test problem. The pitch is labeled by p. $\hat{H}_n$ is the Fourier coefficient of the initial acid concentration. In the following section a solution will be derived for $\hat{H}_n$ for a specific test problem.

Another solution to (4) and (5) can be found for an infinite domain instead of periodic boundaries. The solution is obtained with a Green's function approach. This is the "resist blur" function from the IBM group [7].

3.2. POST-EXPOSURE BAKE SOLUTION VERIFICATION

In a previous benchmarking paper, we chose a set of conditions that gave a closed-form solution for $\hat{H}_n$ [4]. The illuminator and mask settings were set up to provide 2-beam interference at the wafer. This choice allowed $\hat{H}_n$ to be written in terms of Bessel functions. This can still be done; however, in this study we have determined $\hat{H}_n$ by Fourier transforming the initial acid concentration. The mask is alternating PSM with 65nm lines and spaces. For the imaging system, we use a 193nm exposure wavelength, NA =1.2, coherent illumination ($\sigma = 0$), unpolarized light and a reduction ratio = 4.0. The film stack has all optical properties set to the properties of a non-absorbing "resist" (n = 1.72, k = 0). The exposure dose is 2 mJ/cm^2 with the Dill C parameter set to 0.06 cm^2/mJ. Post-exposure bake time is 60 seconds. The diffusion coefficient for acid is 4.482 nm^2/s. An amplification rate of 0.3678 sec^{-1} is used.

The region of interest is T-shaped, as shown in Figure 4. The image-in-media is calculated with the same refractive indeces inside and outside the T. The PEB solution is then computed only in the T-shaped domain. The above solution is valid as a benchmark in this case because the no-flux boundary conditions are implemented in PROLITH at all edges of the domain. As can be seen in Figure 5, the PROLITH model has very good convergence.

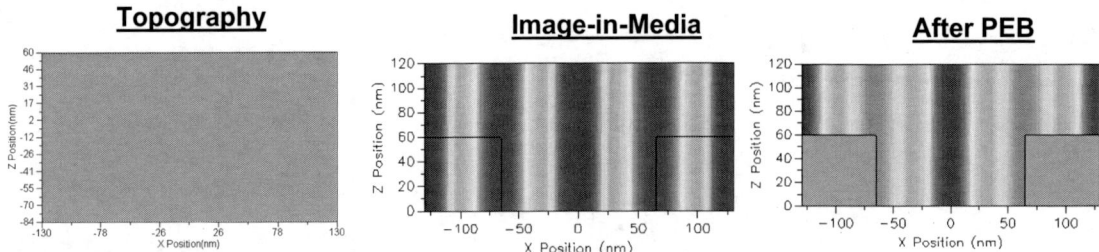

Figure 4. The left figure shows the topography of the problem. The middle and right figures show the image-in-media used as input for the problem. The figure on the right shows the acid concentration after PEB.

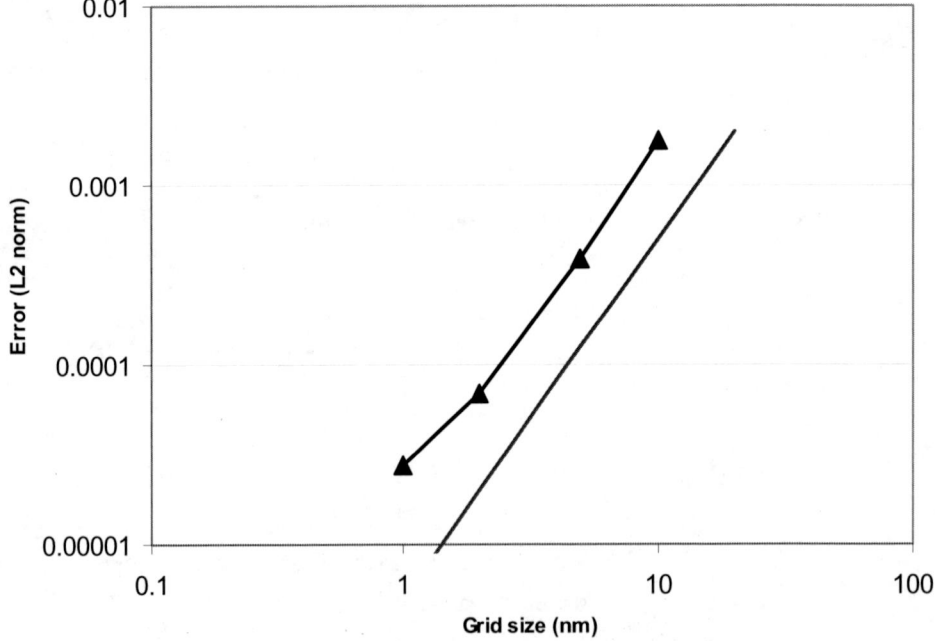

Figure 5. The L_2-norm is shown as a function of grid size. The convergence is very good. The tapering off of the error at small grid size is due to 6 digits of output in PROLITH results. The straight line is shown as a guide to the eye to indicate that the convergence is quadratic.

4. CONCLUSION

We have reviewed closed form solutions for image-in-media and PEB when the wafer film stack is no longer made of all-planar films. These solutions can be used to quantitatively determine the numerical accuracy of a wafer topography simulator, which is currently highly relevant to optical lithography. PROLITH's image-in-media and PEB models converge rapidly to the exact solutions.

5. REFERENCES

1. M.D. Smith, C.A. Mack, "Methods for Benchmarking Photolithography Simulators", *Proc. SPIE*, Vol. 5040 (2003) pp. 57-68.
2. M.D. Smith, J. D. Byers, C.A. Mack, "Methods for Benchmarking Photolithography Simulators: Part II", *Proc. SPIE,* Vol. 5377 (2004) pp. 1475-1486.
3. Mark D. Smith, Trey Graves, Jeffrey D. Byers, Chris A. Mack, "Methods for Benchmarking Photolithography Simulators: Part III," *Optical Microlithography XVIII, Proc.*, SPIE Vol. 5754-99 (2005).
4. Trey Graves, Mark D. Smith, C. A. Mack, "Methods for Benchmarking Photolithography Simulators: Part IV," *Optical Microlithography XIX, Proc.*, SPIE Vol. 6154 (2006).
5. I.C. Botten, M.S. Craig, R.C. McPhedran, J.L. Adams, J.R. Andrewartha, "The dielectric lamellar diffraction grating", *Optica Acta*, Vol. 28 (1981) pp. 413-428.
6. M.D. Smith, J. D. Byers, C.A. Mack, "The lithographic impact of resist model parameters", *Proc. SPIE,* Vol. 5376 (2004) pp. 322-332.
7. Hinsberg, et al. "Extendibility of chemically amplified resists: another brick wall?", *Proc. SPIE*, Vol. 5039 (2003) pp. 1-14.

Selective Inverse Lithography Methodology.

ChinTeong Lim, Vlad Temchenko
Logic Lithography Development, Infineon Technologies Dresden GmbH and Co. OHG
PO Box 10 09 40, D-01079, Dresden, Germany

Martin Niehoff
Mentor Graphics Corporation, Arnulfstr. 201, 80634 München

ABSTRACT

Selective Inverse Lithography (ILT) approach recently introduced by authors [1] has proven to be advantageous for extending life-span of lower-NA 193nm exposure tools to achieve satisfactory 65nm contact layer patterning. We intend to find an alternative solution without the need for higher NA tools and advanced light source optimization. In this paper we explore possible region selection criteria for ILT application based on pitch for a full chip optical proximity correction (OPC). Through studying the impact of a given selection criteria on runtime, resolution, and the process window we recommend an optimal combination. With a justified choice of an ILT selection criteria, we construct a hybrid OPC flow comprising a recursive sequence of direct assist features generation, selective ILT application, layout repair, model OPC and hot spots screening.

1. INTRODUCTION

Among a number of available OPC solutions ILT provides the best process window, its application to a full chip is however problematic [4]. Typical mask pattern outcome of ILT is represented by smooth shapes and has to be further simplified or pixelized in order to improve mask writing time. As a mask is becoming another optical element and a pattern doesn't look like original design, arial image mask inspection has to be adapted in order to cope with mask defects characterization. A degree of ILT pattern simplification depends on a trade-off between reticle manufacturability and overall process window combined with careful control of SRAF printability and possible hot spots.

Lithographers are continuously searching for a full chip solution that provides manufacturable mask shapes and acceptable runtimes. With a goal to establish the best ILT practice, our idea of applying ILT selectively to the areas of a chip where it's needed the most has been verified on a 65nm gate contact layout using low NA 90nm node tools. [1] Selective or partial ILT was found to be a bridge for the transition towards eventually full-chip ILT due to higher mask writing time, inspection and resultant mask CD non-uniformity.

We suggested combining different simplification schemes to address process window for different pitch ranges. We have scanned a typical C65 contact chip logic layout excluding dense SRAM and considered only projecting CH (projection length exceeding 45nm) to find out the counts of various pitch' occurrences. Dense pitch range had the highest count and if dense pitch range can be resolved with direct OPC and special illumination choice then approximately 30% of remaining pitches would require inverse OPC treatment as they belong to forbidden pitch range.

Complete ILT treatment of a chip would lead to unreasonably high mask writing time. Typical pattern simplification such as minimal polygon area restriction would lead to slightly shorter mask writing time. However by selectively applying ILT treatment, only a fraction of a chip would require longer mask writing time. We will further explore pitch-based selection criteria, and investigate this hybrid OPC flow in more detail.

Optical Microlithography XXIII, edited by Mircea V. Dusa, Will Conley, Proc. of SPIE Vol. 7640,
764034 · © 2010 SPIE · CCC code: 0277-786X/10/$18 · doi: 10.1117/12.845464

2. HYBRID SELECTIVE ILT & OPC MASK DATA PREPARATION FLOW & DISCUSSION

The mask data preparation flow chart in **Figure 1** describes our approach of composing the hybrid combination of rule generated assist feature, OPC and ILT operations on a given layout. The aim is to provide a solution to further improve process window on layouts that fall within certain weakness filtering mechanism.

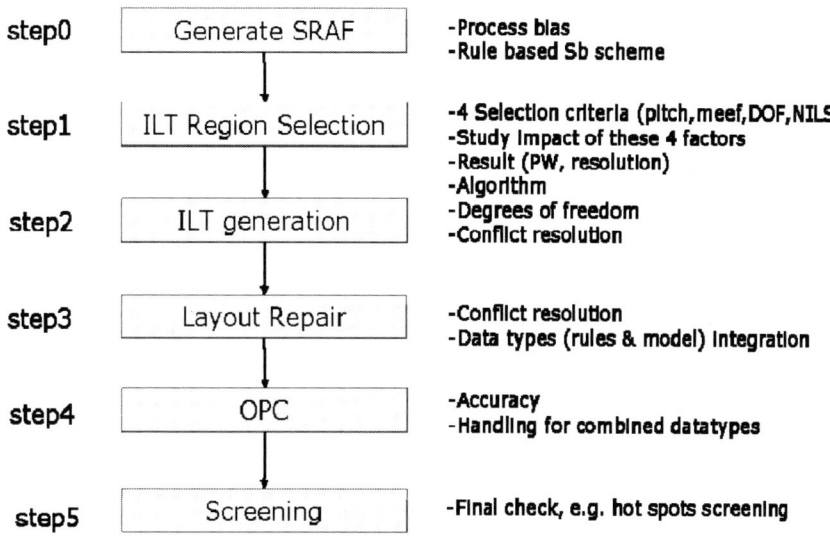

Figure 1: Hybrid flow chart.

2.1 Rule Generated Assisted Feature (RGAF) Operation

It has become a standard practice in sub 90nm lithographic process that assisted features are used in combination with off-axis illumination. We used standard rule generated assisted features operation in this step. The aim is to reduce as much as possible the amount of ILT during full chip operation, which subsequently results in shorter run time. The limited capability of generating assisted features within and near forbidden pitch range makes ILT a good candidate to provide process window enhancement solution on this region.

2.2 ILT Assisted Region Selection

There are two main purposes of using ILT, namely process window improvement and layout weakness elimination. In our case on 65nm node contact, PW is the main concern, particularly on pitch range which RGAF can't reach. Important consideration is whether pre-simulation on post standard OPC layout is required, which will significantly increase the overall runtime. The process window within forbidden pitch range is our main concern. From **Figure 2**, the forbidden pitch range is between where the first diffraction order light weighting is heading zero while the second diffraction has yet no influence, based on the illumination chosen. The criteria can be defined with rule-based approach which is the result learned from a number of pre-simulation cycles, together with empirical result analysis from wafer. This approach helps to move much closer toward realizing an acceptable hybrid OPC and ILT operation runtime and achieving 0.2um DOF across full proximity range.

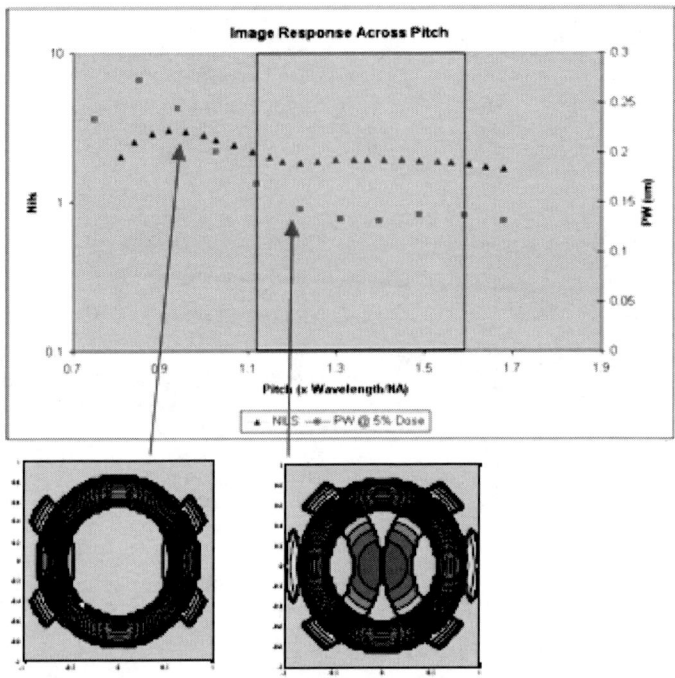

Figure 2: Image response across pitch simulation based on NAPC 0.75/0.85/0.55, highlighted is ILT assisted pitch range which was defined beginning on region where the influence of the first order is decreasing while the second order is yet to be captured.

2.3 ILT Assisted Operation

ILT methodology is paid a lot of interest in recent years. One of the arguments for ILT is that more structural dimensions (such as transmission, MoSi and quartz thickness) and optical parameters (such as contrast, NILS, DOF) are assessed to achieve better quality in terms of process window, whereas standard OPC operations concentrate on dimension as the main target. In our case, full ILT was applied within selected region together with certain surrounding region. During the initial development phase, the basic algorithm used is to co-optimize contrast and critical dimension. There are approximately 11% of polygons from the targeted 65nm product that fall into the rule-defined criteria. We concentrate on assisted region border generation and integration with neighboring areas as shown in **Figure 3** without deteriorating the beyond the standard OPC PW, while at the same time without creating unnecessary burden on runtime and layout complexity.

2.4 Layout Repair & Simplification

Simplification was done next to fulfill mask-house manufacturing requirements. More details of this simplification can be found in **[1].** The main work is finding a working simplification scheme for mask manufacturability while not sacrificing much on imaging quality. Realistically it is feasible based on current mask manufacturing requirements, but it doesn't quite address the mask inspection concern, thus mask AIMS inspection remains a future topic which needs to be solved in collaboration with mask house for a workable solution development (**Figure 3**). With the consideration of more work required to improve CD accuracy due to integration of the two quite different layout manipulation mechanism - OPC and ILT within a chip, all main ILT assisted targets were removed at this step, the rest of other ILT-generated assisted features (ILTG-AF) together with original design will be the input to the next OPC step. Main consideration on this step is to optimize the clean-up of overlapping assist features surrounding borders of the two different data types as shown in **Figure 4**. Insertion of ILT-assisted main features will be considered next to further improve common process window.

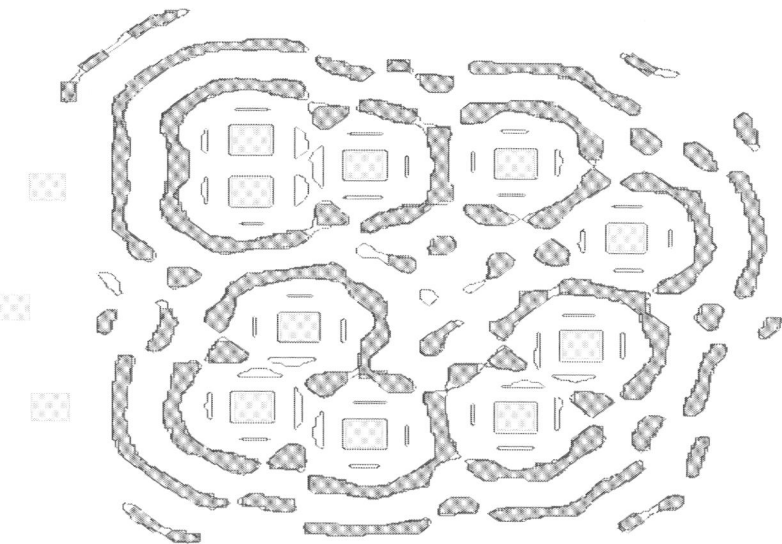

Figure 3: Blank polygons –full ILT layout. Dark polygons – simplified ILT layout.

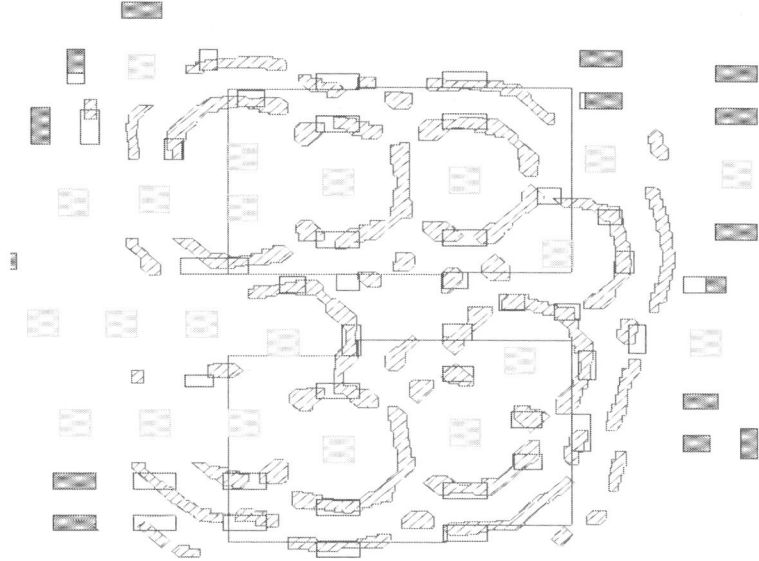

Figure 4: Blank polygons –RGAF, striped – simplified ILT AF, dark – RGAF beyond ILT selection region, outline – border generation.

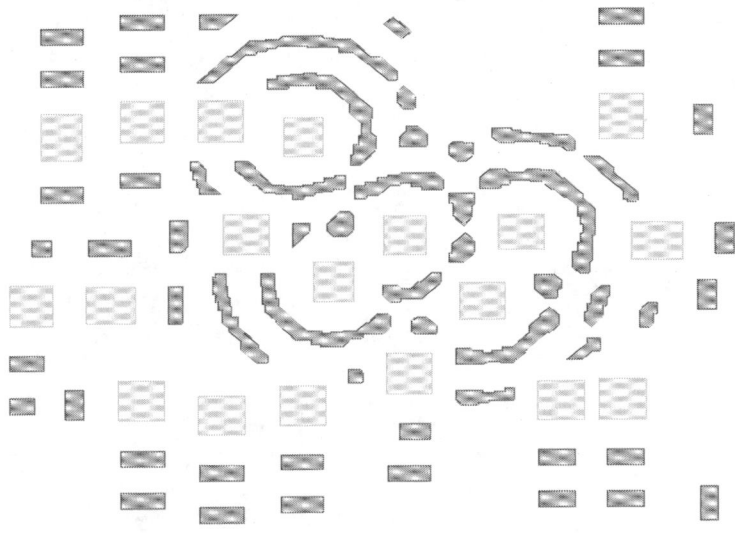

Figure 5: Example of final hybrid flow layout

2.5 OPC Operation

Normal OPC operation is carried out in this step on original design, taking into consideration both rule-based and ILT PW enhanced assisted features to ensuring CD accuracy across chip as shown in **Figure 5**. Future work will require applying OPC only for non-ILT generated region in order to realize full ILT benefit.

3. FULL CHIP SIMULATION RESULTS DISCUSSION

We have scanned a typical C65 contact chip logic layout considering only projecting CH (projection length exceeding 45nm) to find out the counts of various pitches. From **Figure 6**, it can be seen that dense pitch range has the highest count. As we have discussed previously, the dense pitch range can be resolved with direct OPC and optimized illumination. Approximately 11% of remaining pitches require inverse OPC treatment as they belong to forbidden pitch range. We define such range as the range starting from a pitch where process window at 5% exposure dose falls below 200nm and to the semi-dense pitch where rule-based assist features insertion is allowed and can improve process window. In our case, forbidden pitch is the range of CH contacts is between 200 and 300nm. Complete ILT treatment of a chip filled with our random CH layout would lead to unreasonably high mask writing time of around 130 hrs. Typical pattern simplification such as minimal polygon area restriction would lead to slightly shorter mask writing time. However by selectively applying ILT treatment, only a fraction of a chip would require longer mask writing time.

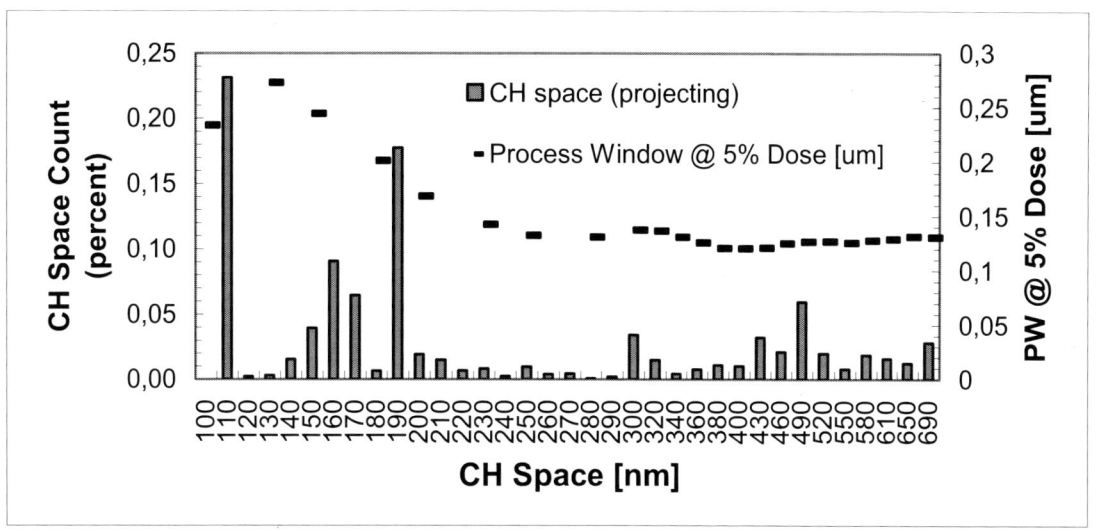

Figure 6. Various pitch occurrence in a typical C65 contact chip layout – projecting CH only. Process window calculated at 5% exposure dose for unassisted CH is below 200nm for CH spaces larger than 200nm.

We can compare ILT to conventional OPC performance with higher NA plus advance illumination mode, based on process window enhancement parameter such as process variation (PV) bands. PV bands in particular enable quantitative comparison of process windows for various schemes. For each run, the full chip results were sorted based on the number of occurrences in the design for each PV band interval, ranging from 1 to 5nm. The objective of a comparison was to find which of the setups produces fewer occurrences of large PV band widths. The process window conditions for the test included contours at nominal focus, +/-75nm defocus and +/-5% exposure dose.

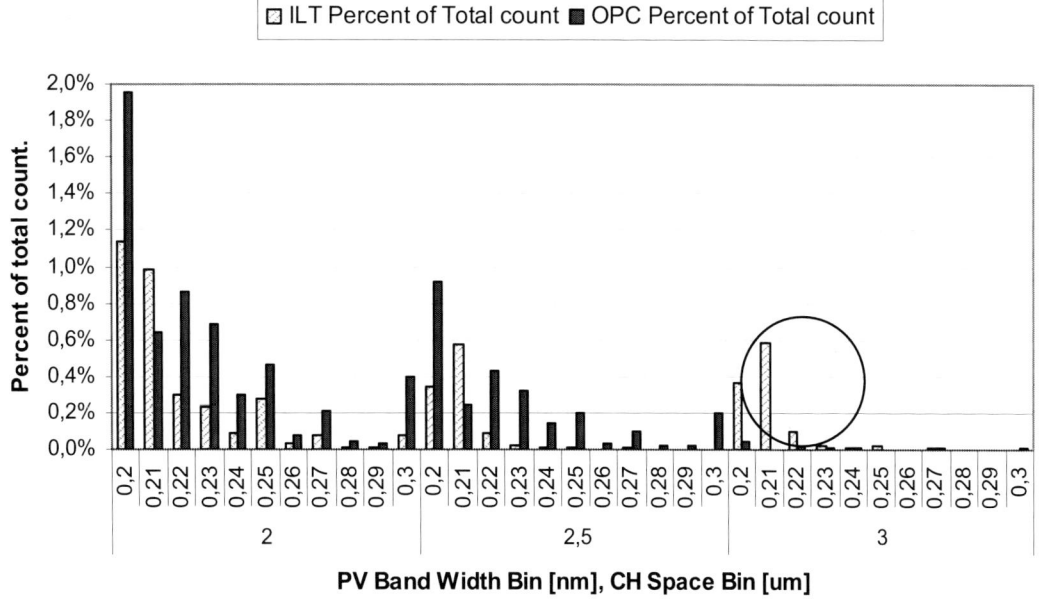

Figure 7. Through-pitch, full-chip PV band scan for direct OPC (standard 65nm process) and ILT cases in forbidden pitch range. Larger than 3nm PV Band bins had zero counts. Standard 65nm OPC process is based on higher NA and advanced illumination: 0.85NA with Windmill illumination. ILT process is based on 0.75NA annular illumination.

From **Figure 7** it can be seen that in forbidden pitch area, the low NA ILT solution approaches the performance of standard direct 65nm OPC. It is worth mentioning that some of the thicker PV band counts represent contacts with not optimized rule-based assist-features occurring due to non-projecting pitch being outside forbidden pitch region. This issue can be addressed through more thorough SRAF placement rules.

4. SELECTIVE ILT LAYOUT PROCESS WINDOW PERFOMANCE

Process window extension was verified by applying inverse layout on forbidden pitch patterns and comparing in-line results to direct OPC on same NA and illumination mode. Although Direct OPC can achieve comparable process windows to ILT on dense and Iso pitches through sizing and rule-based SRAF application, its performance is poor for forbidden pitch range. The main weakness of rule-based assist feature insertion applied to semi dense pitch range within forbidden region has been poor MEEF even though a good depth of focus (DOF) could be achieved. Verification patterns of 100nm contacts across proximity have shown a range of forbidden pitch to lie between 310nm to 410nm – as shown in **Figures 8-10**. Process window gain in the range of forbidden pitch allowed us to meet our DOF target of around 200nm at 5% EL.

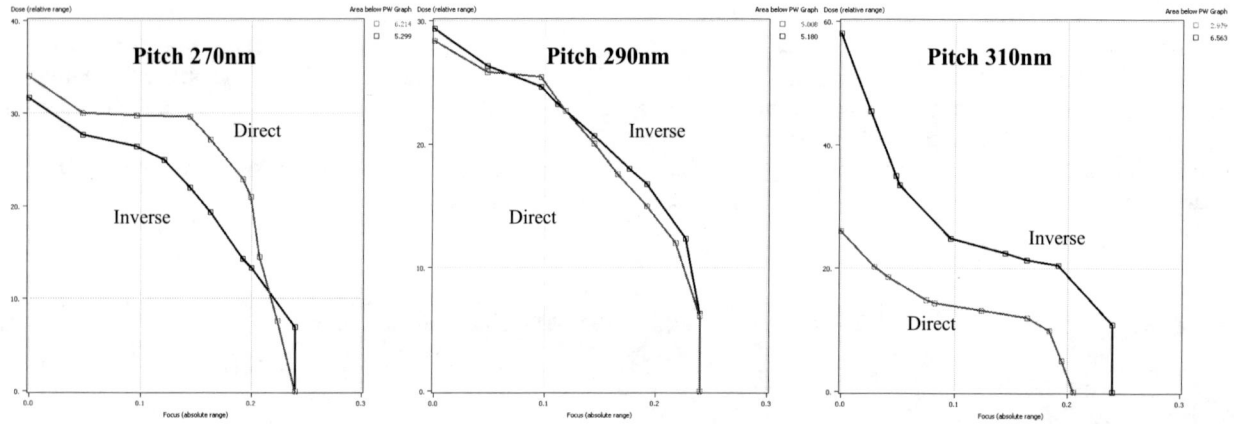

Figure 8. Comparison of direct and inverse OPC performance for the beginning of forbidden pitch range. Optimized resist system and annular illumination with Numerical Aperture of 0.75 and Sigma In/Out of 0.85/0.55 were used.

With the improvement of forbidden pitch region DOF, common process window across proximity has increased substantially as can be seen from **Figure 9**, where a range of pitch between 220-2000nm for 100nm CH is represented.

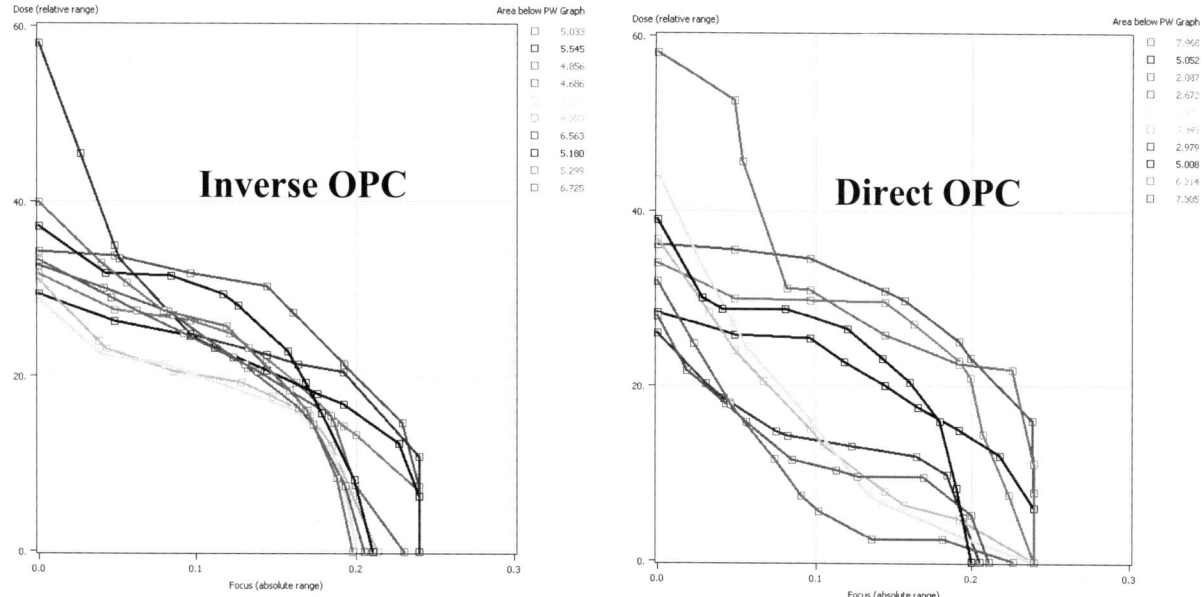

Figure 9. Process windows comparison of direct and inverse OPC patterns across proximity. Left: direct OPC with rule-based assist feature generation method. Right: inverse OPC approach.

In addition to the gain in process window, ILT patterns are closer to the design rule even at pattern edges. As can be seen from top-down SEM measurement images on **Figure 9**, inverse OPC performs better that Direct OPC at pattern edges.

Pitch(nm)	310	330	350	370	400	410
Direct OPC						
Inverse OPC						

Figure 10. OPC test patterns' top-down view comparing direct and inverse OPC in the forbidden pitch region for 100nm CH. Contact holes deteriorate towards pattern edges for the direct OP, while showing better imaging for inverse layouts.

5. CONCLUSION

We have verified the feasibility of hybrid OPC approach for full chip 65nm node contact layer without the need for higher NA tools advanced illumination source. Selective ILT applied to forbidden pitch range provided a run-time of 5 days on 64CPUs for a typical 65nm logic contact layer. Our aim is to reduce the runtime further by another 20%. Future work will concentrate on realizing our flow in productive environment. This includes reticle inspection strategy, quantification of mask error, better RGAF and ILT integration scheme to eliminate image quality issue around hybrid border region. Possible advanced extension of hybrid approach by adding more degrees of freedom such as reticle parameters (chrome, MoSi thicknesses, profile undercut and rounding) for contact and 2D line-space patterns will be explored once the supporting infrastructure is ready.

REFERENCES:

[1] Lim C., Temchenko V., et al., "Manufacturability of ILT patterns in low-NA 193nm environment," Proc. SPIE 7274, (2009).

[2] Lim C., Temchenko V., et al., "Investigation of DFM-lite ORC Approach During OPC Simulation," Proc. SPIE 6520, (2007).

[3] Temchenko V., Lim C., et al., "Coupled-dipole modeling for 3D mask simulation," Proc. SPIE 6924, (2008).

[4] Hendrickx E., Tritchkov A., et al., "Hyper-NA imaging of 45nm node random CH layouts using inverse lithography," Proc. SPIE 6924, (2008).

[5] Granik Y., Sakajiri K., et al., "On objectives and algorithms of inverse methods in microlithography," Proc. SPIE 6349, (2007).

CDU linear model based on aerial image principal components

Zhiyong Yang*, Anatoly Y. Burov, Liang Li, Fan Wang, Zhaoxiang Chu
Shanghai Micro Electronics Equipment Co., Ltd. (China)

ABSTRACT

In this paper, we present an image quality model and a process window model that is linear or quadratic with respect to common pupil space errors. Similar to other CDU models in its simplicity, our model expands linear representation to comprehensive image quality specs in a large focus-dose grid. With this model we identify corrections to the full Bossung curve or process window shapes that are proportional to aberration levels.

Keywords: CDU linear model, Zernike coefficient, Bossung curve, process window

1. INTRODUCTION

For lithography machine, wafer side image quality control is the most important item. There are many different kinds of contributors that could impact to image quality at wafer side. Typical contributors include DOSE error, focus error, source error, wavefront error, flare, motion blur, mask error and process error. In the conditions of k_1 less than 0.5, wavefront error has great impact to image quality. Projection lens or lithography tool designers must pay much more attentions to wavefront error control and should understand the impact behavior from wavefront error to image quality.

Some previous papers [1,2] present several CD variation models to describe the wavefront error's impact behavior. In this paper, we introduce a comprehensive model and using a different method to build it. Rather than other general methods, our calculating process is based on aerial image principal components model.

Additionally, our CDU (Critical Dimension Uniformity) linear model is not only focused on the CD variation at a single point in the process window. This paper has introduced a comprehensive model to describe all the main specs of image quality, such as H-V bias, FPD (Focus Position Deviation) and so on. Bossung curve and Process Window's behavior are also analyzed in our model. The purpose of this work is to provide a method to translate the customer's requirement to machine's specs succinctly but accurately.

2. AERIAL IMAGE MODEL

Aberrated lithographic aerial image can be described as the sum of some weighted principal components. In this aerial image model, we decompose the aberrated aerial image into some principal components; the first component is the un-aberrated aerial image whose weight is fixed. Other principal components express the aberration's impact and the weights can be calculated from Zernike coefficients' values. The linear aerial image model is shown below.

*yangzy@smee.com.cn, +86-21-51315131-2090

Optical Microlithography XXIII, edited by Mircea V. Dusa, Will Conley, Proc. of SPIE Vol. 7640,
764035 · © 2010 SPIE · CCC code: 0277-786X/10/$18 · doi: 10.1117/12.848429

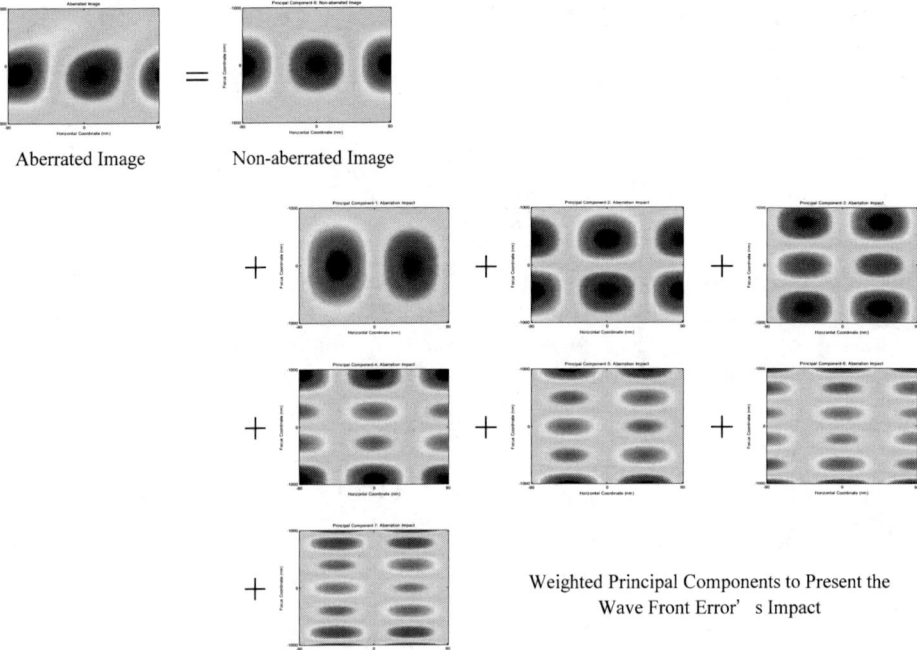

Aberrated Image   Non-aberrated Image

Weighted Principal Components to Present the
Wave Front Error's Impact

Figure 1. Aberrated aerial image can be represented as the sum of weighted principal components; the first principal component is an un-aberrated aerial image. Other principal components express the wavefront error's impact and their weights are a linear transformation of Zernike coefficients input.

$$I(x,f) = a_0 \cdot I_0(x,f) + \sum_{i=1}^{n} a_i \cdot I_i(x,f)$$

$$\vec{a_i} = \vec{Z_i} * TM$$

(1)

Here, I_i is the principal components and TM is the translation matrix from Zernike coefficients to principal components' weights; and a_i is the weight. Inputting a Zernike vector, we can calculate the aberrated aerial image. For example, if we build the aerial image which could describe the 33 orders (Z5~Z37) Zernike coefficients' impact to image quality, the Z_i is a 33 elements vector and a_i is an n-elements vector which includes every principal components' weight. TM is a 33 by n matrix and n is the number of principal components.

Modeling process is based on lithographic image's physical calculation and statistical methods. The first step of the modeling process is the definition of the mask feature and illuminator settings. Then, the un-aberrated aerial image is obtained. The second step is calculation of a set of aberrated aerial images by inputting a set of structured Zernike coefficients into a physical simulation engine[3]. The third step is identifying the principal components from all the aberrated and no-aberrated aerial images using Karhunen-Loeve expansion. Finally, using linear regression, the translation matrix TM is obtained. After all the steps, the linear aerial image model is built and this model is available for a defined mask feature and illuminator setting.

Principal components have some special symmetry or period characters. For symmetric features, such as dense line or contact, the principal components have the symmetry characters listed as below. These symmetrical properties could simplify the follow modeling process.

Table 1. Aerial Image Principal Components' Symmetric Type and Impact Character

Symmetric Shape			Symmetric Character	Impact to Aerial Image Quality
	+	−	Up and down mirror symmetry; Left and right mirror inversion. $I(x,f) = -I(-x,f)$ $I(x,f) = I(x,-f)$	This kind of principal component causes feature position shift and bananicity; Typical aberration is spherical, astigmatism and trefoil.
	+	−		

		Up and down mirror inversion; Left and right mirror symmetry. $I(x,f) = I(-x,f)$ $I(x,f) = -I(x,-f)$	This kind of principal component causes Iso-Focal-Tilt and focus position shift; Typical aberration is spherical, astigmatism and trefoil.
+	+		
-	-		
+	+	Up and down mirror symmetry; Left and right mirror symmetry. $I(x,f) = I(-x,f)$ $I(x,f) = I(x,-f)$	This kind of principal component causes CD changing at best focus position and process window size degreasing; All the Zernike coefficients could cause this kind of impact.
+	+		

3. MODELING FOR IMAGE QUALITY EVALUATION

In order to evaluate the lens performance in a lithographic process, many kinds of specifications are used to inspect the image quality. Typical specification is CDU (CD variance through exposure field or whole wafer) while others include H-V bias, EPE (Edge Placement Error – feature position shift) and FPD (Focal Position Deviation). For some special features, such as Two-Bar and Brick Wall, the feature's symmetric character is also inspected. In this section, we present a comprehensive CDU model to calculate all the specifications mentioned above. Because the lithographic imaging process can be described as a linear model, all the calculations can be expressed as a linear sequence.

3.1 CD Variance and Edge Placement Error Linear Model

CD variation and edge placement error calculation are based on the intensity distribution character of aerial image and that of the principal components. As is common in simplified analysis, we find the aerial image's edge by a defined threshold. Firstly, we define the threshold and find the un-aberrated aerial image edge position at a specified focus position. Secondly, based on the un-aberrated aerial image's intensity curve, we find out the curve's normal vector at this edge position. Thirdly, we calculate the edge position shift based on the normal vector and aberrated principal components' intensity value at the same edge position. In this model, every aberrated principal component's weights are calculated by inputting Zernike coefficients and the transition matrix.

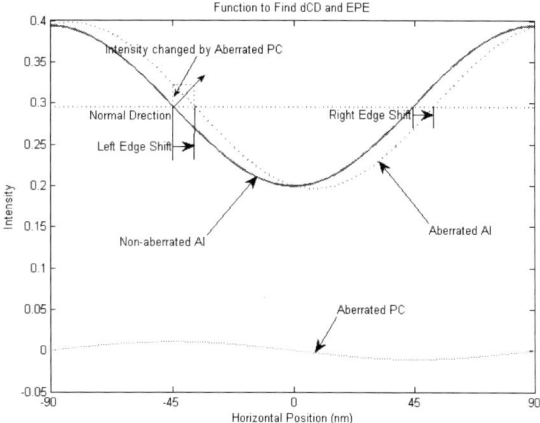

Figure 2. In this figure, BLUE line is the un-aberrated aerial image's intensity distribution at a defined focus position, the GREEN line is the sum of aberrated principal components' intensity distribution. Edge position is found at defined threshold of the aerial image. Based on the un-aberrated aerial image intensity's normal vector and aberrated principal components' intensity, we can calculate every edge's position shift. Both CD change and EPE are obtained.

After the calculating sequence above, we can obtain the two edge's position shift. CD variation is the difference of these two edge shifts and EPE is the average shift. Because all the principal components have some special symmetric character, the process of CD variance and EPE's calculations could be simplified. Finally, we can obtain the CD variation and EPE's model in the format below.

$$dCD\left(I, f, \overrightarrow{Z_i}\right) = \overrightarrow{Z_i} \cdot \overrightarrow{V_{dCD}\left(I, f\right)}$$
$$EPE\left(I, f, \overrightarrow{Z_i}\right) = \overrightarrow{Z_i} \cdot \overrightarrow{V_{EPE}\left(I, f\right)}$$

(2)

We have tested the model's accuracy by comparing with Prolith[TM] 10.2. Inputting intensity, focus position and Zernike coefficients randomly, we can compare our model and Prolith's output CD variation and EPE data. The comparison test result is shown in Figure 3.

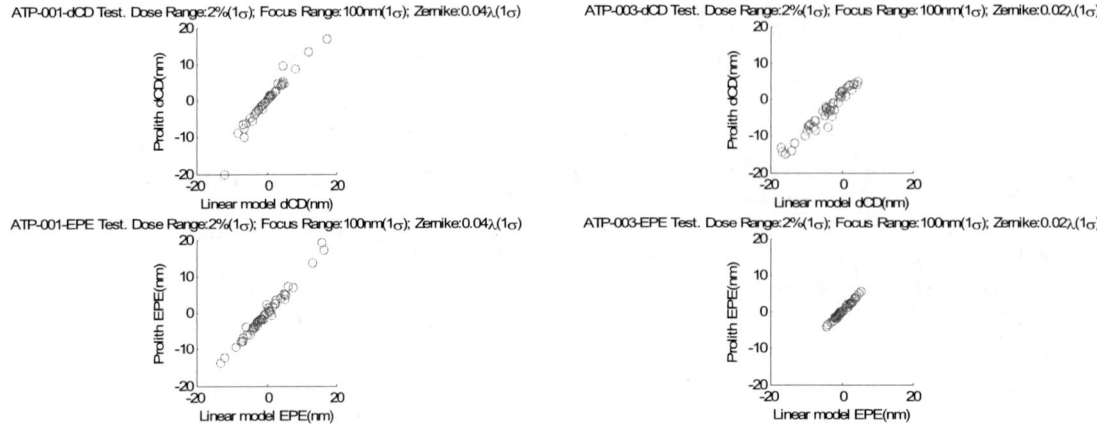

Figure 3. The comparison result between Prolith[TM] 10.2.and our linear model. The first column is 90nm dense line's dCD and EPE; the column on the right is the 120nm dense contact dCD and EPE. We input different focus/threshold setting and wavefront error randomly in large range, comparison result is acceptable.

3.2 H-V Bias Linear Model

H-V bias is a key spec to inspect the CD difference between horizontal feature and vertical feature. Based on CD variance model, we can get the horizontal feature's CD variance; rotating the pupil map 90 degree and input the rotated Zernike coefficients to the same model, we can obtain the vertical feature's CD variance. H-V bias is obtained. As shown below, the rotated Zernike coefficients can be expressed as the original Zernike coefficients multiplying the 90° rotation matrix M_{rot}. The H-V bias model can be simplified while maintaining its linear format.

$$H - V(I, f, Z_i) = \overrightarrow{Z_i} * (M_{rot} - I) * \overrightarrow{V_{dCD}(I, f)}$$
$$= \overrightarrow{Z_i} \cdot \overrightarrow{V_{H-V}(I, f)}$$

(3)

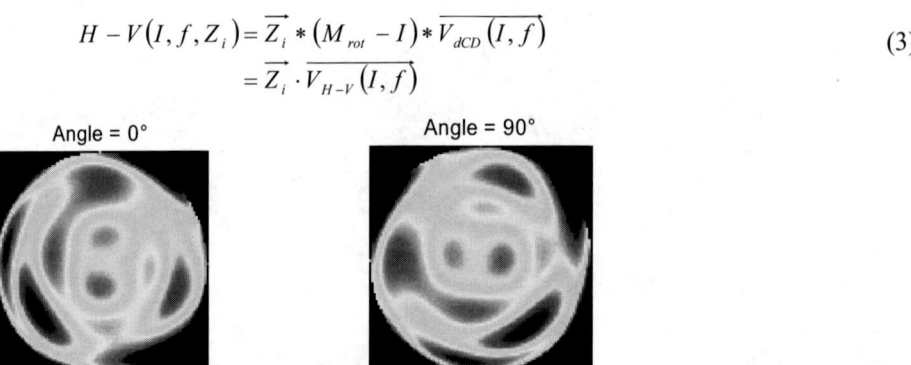

Figure 4. In order to calculate H-V bias quickly, we rotate the wavefront 90° and input to our linear model. H-V bias is the difference between rotated result and the original result. Because the dCD model is a linear model, we can calculate the difference map between original wavefront map and rotated wavefront map. Input the difference map to dCD linear model, we can obtain the H-V bias directly.

3.3 FPD and AST Linear Model

In our model, the best focus position is defined as the extremum point of horizontal gradient in aerial image, because at this vertical position, the aerial image has the largest contrast or NILS. We consider that the Zernike coefficients (higher than Z4) could cause an additional focus point deviation and our model needs to represent this kind of impact. To find the FPD, we first extract the horizontal gradient curve at feature's edge and calculate the un-aberrated aerial image's focus position. Then, extraction of the aberrated principal components' horizontal gradient curve is performed. We can then calculate the FPD from these two curves' 1st-order and 2nd-order derivative.

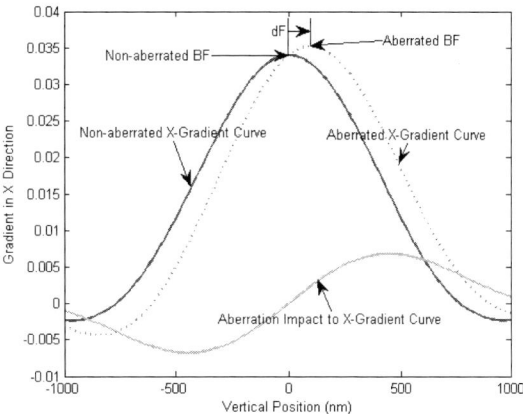

Figure 5. In this figure, we present the method to calculate FPD. BLUE line is the horizontal gradient curve which is extracted from un-aberrated aerial image and the extremum point is the focus position. GREEN line is the horizontal gradient curve which is extracted form aberrated principal component. The aberrated aerial image's gradient curve is the PINK line and the extremum point is the aberrated focus position. Calculating the BLUE curve's 2nd derivative and GREEN curve's 1st derivative, we can obtain the FPD caused by high-order Zernike coefficients.

Wavefront error causes separation of horizontal and vertical feature's focus position, and the distance is ASTigmatism. The calculation method is similar to the H-V bias' method, rotating the pupil map (Zernike coefficients multiplying the 90 degree rotated matrix) and calculating the difference related to the original FPD.

Finally, the FPD and AST model can be expressed as below format.

$$FPD\left(I, f, \overrightarrow{Z_i}\right) = \overrightarrow{Z_i} \cdot \overrightarrow{V_{FPD}\left(I, f\right)}$$
$$AST\left(I, f, \overrightarrow{Z_i}\right) = \overrightarrow{Z_i} \cdot \overrightarrow{V_{AST}\left(I, f\right)}$$

(4)

3.4 Two Bar and Brick Wall Feature Balance

Two Bar and Brick Wall are the typical features in lithographic process and this kind of features have symmetric character (shown as Figure 6). In this case, we consider the feature's symmetric specs. For Two Bar, we inspect the left and right Bar's CD difference; for Brick Wall, we inspect one cell's upper and lower CD difference. Generally, these specifications are very important to evaluation of the lens performance.

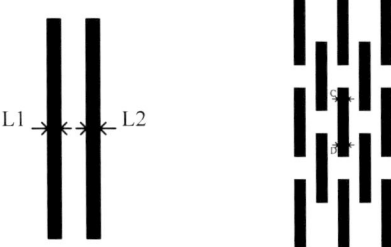

Figure 6. Left figure is the typical Two Bar feature and in lithographic imaging process, we concern the left and right difference (L1-L2); right figure is the typical Brick Wall feature and we concern every cell's upper and lower difference. All of this kind of features has symmetrical character.

Similar to the H-V bias's modeling method; we calculate one position's CD variance and rotate the pupil map 180 degree to obtain the symmetric position's CD variance. Using the Zernike coefficient rotation matrix, the model can be simplified as below format.

$$L1 - L2\left(I, f, \overrightarrow{Z_i}\right) = \overrightarrow{Z_i} \cdot \overrightarrow{V_{L1-L2}\left(I, f\right)}$$
$$C - D\left(I, f, \overrightarrow{Z_i}\right) = \overrightarrow{Z_i} \cdot \overrightarrow{V_{C-D}\left(I, f\right)}$$

(5)

Some special order Zernike coefficients would impact to the feature's balancing performance. In the modeling process, we rotate the pupil map by 180°, meaning that the odd order Zernike coefficients, such as coma and trefoil aberration has big impact to feature's balancing character.

4. BOSSUNG AND PROCESS WINDOW LINEAR MODEL

In lithographic process, Bossung curve and Process Window are often used to calculate CDU (DOF and EL) to evaluate the process ability. In sections above, we discussed the image specification calculations at any defined intensity and focus position. In this section, we present a method to calculate aberrated Bossung curve and Process Window. Wavefront errors change the Bossung curve and Process Window's shapes. It means that the lithographic process capability has also been changed by wavefront error. In this section, the Bossung curve and Process Window shape's change vector is calculated.

Bossung curve and Process Window's calculation is also based on the aerial image model. First step is folding the un-aberrated aerial image and combining the left and the right side of the aerial image. Second step is projection of the combined aerial image's contour to CD/Focus space to obtain the Bossung curve, or to Intensity/Focus space to obtain the Process window. Calculating the aberrated aerial image and using the above method, we can obtain the aberrated Bossung curve and Process Window. Projecting the un-aberrated aerial image's surface normal to CD/Focus or Intensity/Focus space and scaling by aberrated principal components' intensity distribution, we can obtain the Bossung or Process Window changing vectors caused by wavefront error.

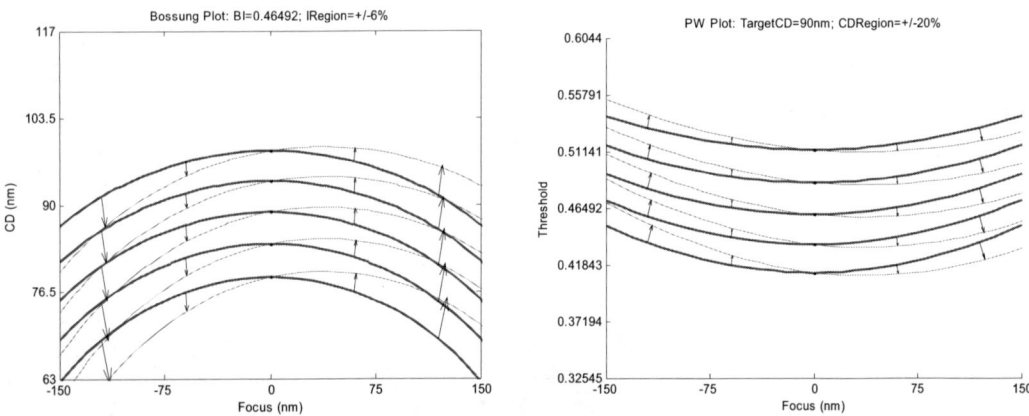

Figure 8. These two figures present the 90nm Iso line's Bossung curve and Process Window Aberrated Bossung curve and Process Window can be calculated from the actual aerial image (RED line).The arrows represent the changes calculated using the linear model and the projecting the distortion vectors into the respective space. Note the agreement between the linear model and the full Bossung curve and process window across the whole Focus / CD / Threshold space.

The wavefront error varies through the lithography tool's exposure field. It means that every field point's Bossung curve and Process Window are also different. In this Bossung curve and Process Window model, the distortion vector calculation is the key point. The distortion vector can represent the CD variation in whole Intensity/Focus region. All of the feature's information is contained in one simple model. It is critical for CDU calculation and the model user can create overlapping Process Windows through the field to analyze the whole chip performance. It also helps us to visually understand the wavefront error's impact in Bossung or Process window space.

5. DISCUSSION

In modeling process, we found that all the linear model's principal components have the left and right or upper and lower symmetric character. On the symmetrical axis, intensity is zero. It means that the model can't predict the CD variance at the best focus position.

Looking back to the aerial image modeling process, at the principal component identify step, we use the principal component which only includes linear term's Zernike coefficient impact. It's a linear transition from Zernike coefficients to principal component's weight and cross-term's impact is ignored. The purpose is to make the model simple and can be used inversely. But if we want to obtain a higher accuracy model, we can prepare the aerial image model which includes cross-term's impact.

$$I(x,f) = a_0 \cdot I_0(x,f) + \sum_{i=1}^{m} a_i \cdot I1_i(x,f) + \sum_{i=1}^{n} b_i \cdot I2_i(x,f)$$

$$\vec{a_i} = \vec{Z_i} \cdot TM_1$$

$$\vec{b_i} = \vec{Z_i} \cdot TM_2 \cdot \vec{Z_i}^T \qquad (6)$$

Quadratic aerial image model includes three parts: first one is un-aberrated aerial image, second one is linear principal components part and third one is quadratic principal components part. In the quadratic model, we have to obtain two transition matrices using regression method, TM_1 for the linear model and TM_2 for the quadratic model. There, TM_2 is a $(m \times n) \times (m \times n)$ size matrix, m is the cross-term principal components' number and n is the Zernike coefficients' number.

Based on the quadratic aerial image model, we can develop the CDU quadratic model using same method as previously. CD variation, EPE, H-V bias, FPD and AST, all the specifications can be modeled. Because quadratic model's principal components symmetric axis intensity is not zero, this model can represent the best focus position's CD variance. The comparison result is shown as Figure 9.

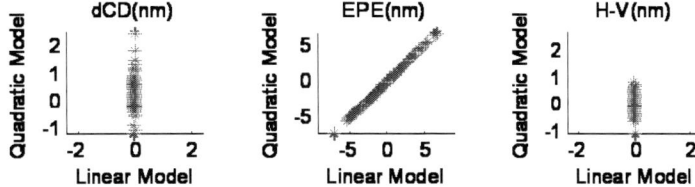

Figure 9. Difference between linear model and quadratic model. The intensity and wavefront error values are random. At best focus position, EPE result is same, but the dCD and H-V linear model's output is zero. Quadratic model can present CD change caused by wavefront error.

Based on quadratic model, we can also plot the Bossung curve, Process Window and changing vector using same geometric projection method. Comparing the linear model's plots and quadratic model's plots, we find that, except for the Iso-Focal-Tilt, quadratic model's plot can represent the Process Window's shift and scaling. It means that the wavefront error could change the best DOSE and best Focus in process. DOF and EL could also be decreased by wavefront error's impact, at defined DOSE control and focus control repeatability in lithography machine, the CDU should be worse.

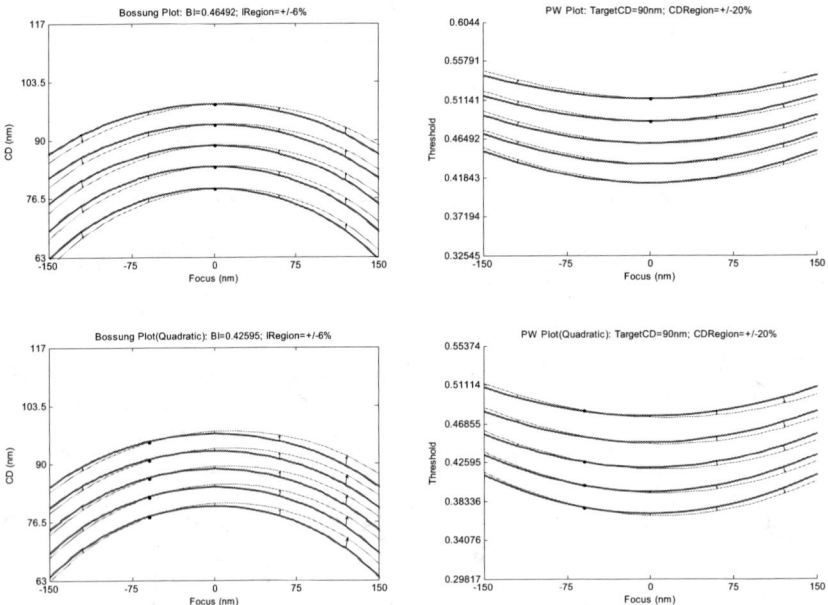

Figure 10. The comparison results of linear model's and quadratic model's Bossung curve or Process Window. Linear model can present the Iso-Focal-Tilt, but except for Iso-Focal-Tilt, quadratic model can present the Bossung curve or Process Window's center shift and scale change.

Linear model and quadratic have different advantages, if the model users want fast calculation time and inversion ability, they can choose the linear model; if the model users want to get higher accuracy, they can use quadratic model.

6. CONCLUSION

Whole modeling process includes three parts:

1. Aerial image modeling process uses physical simulation and statistic method to obtain the whole aerial image model which includes wave front error's impact.

2. We analyze all the lithographic image specs based on the aerial image and principal components' intensity distribution character. So, the image quality specs model can be present as below format and it's easy to translate to system budget function or project lens design merit function.

$$spec\left(I, f, \overrightarrow{Z_i}\right) = \overrightarrow{Z_i} \cdot \overrightarrow{V(I, f)} \tag{7}$$

When lithography or project lens designers want to use this method, they should define the parameters I and f based on DOF and EL requirement.

3. Bossung curve and Process Window geometrically projection model is given. In this model, we can calculate the Bossung curve and Process Window's reshaping character to present wavefront error's impact

In this paper, we present a comprehensive model for the machine design persons to evaluate the project lens' imaging performance. Some simple but ingenious mathematic and statistic methods are used in modeling process. It shows our perspective of wavefront error's impact to lithographic image. We hope this model can help engineers to solve complex system problems in lithography machine design and application process.

ACKNOWLEDGEMENTS

The authors wish to thank Mr. Jianrui Cheng and Mr. Lifeng Duan for useful discussions.

REFERENCES

[1] Paul Grauper *et al.*, "Impact of wavefront errors on low k1 processes at extreme high NA", Proc. SPIE Vol. 5040, 2003

[2] T. Nakashima *et al.*, "Evaluation of Zernike sensitivity method for CD distribution", Proc. SPIE Vol. 5040, 2003

[3] Prolith v 10.2, software package, KLA-TENCOR, Austin TX

Impact of Illumination on Model –Based SRAF Placement for Contact Patterning

John L. Sturtevant, Srividya Jayaram, Omar El-Sewefy, Aasutosh Dave, Pat LaCour
Mentor Graphics Corp, Wilsonville, OR 97070

ABSTRACT

Sub-Resolution Assist Features (SRAFs) have been used extensively to improve the process latitude for isolated and semi-isolated features in conjunction with off-axis illumination. These SRAFs have typically been inserted based upon rules which assign a global SRAF size and proximity to target shapes. Additional rules govern the relationship of assist features to one another, and for random logic contact layers, the overall ruleset can become rather complex. It has been shown that model-based placement of SRAFs for contact layers can result in better worst-case process window than that obtained with rules, and various approaches have been applied to affect such placement. The model comprehends the specific illumination being used, and places assist features according to that model in the optimum location for each contact hole. This paper examines the impact of various illumination schemes on model-based SRAF placement, and compares the resulting process windows. Both standard illumination schemes and more elaborate pixel-based illumination pupil fills are considered.

Keywords: Illumination, simulation, PV Band, SRAF

INTRODUCTION

For several technology generations, assist features have been used in conjunction with off-axis illumination to increase the depth of focus and exposure latitude for semi-isolated and isolates features. It has been known that for a specific illumination profile and pattern pitch, there is a precise placement of assist feature which will result in the maximum increase in process latitude. In general, the larger the dimension of the assist feature, the larger the process window improvement, but the restriction of non-printing places an upper limit on this size. Rules-driven polygon processing engines have been used in to place the assist features across full-chip layouts. With complex two-dimensional layouts, placement conflicts become plentiful, and often result in regions with sub-optimal coverage. As a result, model-based techniques have been developed to optimally guide the placement for each pattern configuration[1-2]. Such techniques have been found to result in much more complex mask shapes, and to deliver significant improvement over rules for contact hole layers[3].

Post-OPC simulation tools can be extremely valuable for evaluating process and mask options, providing quantitative process window information through a wide range of variables[4]. The concept of a process variable (PV) band was introduced by Torres et al., and has become an industry-standard metric for expressing process capability[5].

In this paper, we present the use of model-based assist feature placement with subsequent PV band analysis to compare different illumination mode options.

EXPERIMENTAL

The simulated 193 nm process had a numerical aperture of 1.35. A simple constant threshold resist model was utilized for each case. The test layout used was a 3X3 array of contact holes (CH). CH target size was 60 nm, with a continuum of hole/line duty cycles from 1:1, 1:2, 1:N, … 1:9, and fully isolated. For each illumination condition, assist features were placed with PIXbar, then the target contacts were corrected with a simple Calibre nmOPC recipe. Thus at nominal condition, the printing was on target. Then OPCVerify was used to measure various process window metrics including maximum PVband width, average PV band width and area ratios of the contacts.

Optical Microlithography XXIII, edited by Mircea V. Dusa, Will Conley, Proc. of SPIE Vol. 7640,
764036 · © 2010 SPIE · CCC code: 0277-786X/10/$18 · doi: 10.1117/12.846620

Several different source shapes were used including annular, quasar, cquad, and dipole. Additionally, SMO was used to generate a pixelized source. For each source shape the appropriate degrees of coherence freedom were varied over a wide range, as summarized in Table 1 and Figure 1.

Process window variability was determined for the center most contact in each array. The process window conditions for simulations include exposure dose variation of +/- 3.5%, focus variation of +/- 60nm.
The process variability (PV) band predicts yield failure in several ways, including the four most common failure configurations: pinching, bridging, area overlap and critical-dimension variability. The PV band was created by calculating the silicon-printed image at various process conditions and combining the resulting images into a composite one. PV bandwidth shows how printing responds to process variations by representing the simulated contour extrema as a function of process variation (focus and dose). The "worst" PV band is obtained by including all the contours obtained from the various defocus (+/-60nm) and exposure (+/-3.5%). The different metric used for quantifying the PV bands are Maximum PV bandwidth, Average PV bandwidth and Area Ratio of the outer to inner pv band. Process variability bands were generated on the center contact hole only.

In the case of the SMO source, the source was determined by using only the 1:1, 1:2, and 1:3 pitches.
The maximum width of this PV band is calculated by comparing the difference between two contours of the PV band (defined as the Euclidean distance between them) and returns the maximum distance between them. The value of this maximum PV bandwidth should be as small as possible, ideally close to zero, which means the process variability is negligible across the given process window conditions. . This is automatically calculated by the verification tool by measuring the distance between the two contours of the pv band.

The average width of this PV band is calculated by obtaining the ratio of the area to the perimeter of the "worst" PV band. The value of this average PV band width should be minimum, ideally close to zero, which means the process variability is negligible across the given process window conditions This gives an understanding of how stable the PV band width is through the pw conditions throughout the layout

	Outer sigma	Inner Sigma	Illum Angle	Rotation
Annular	1.0 0.8 0.6	0.5 0.3 0.1		
Quasar	1.0 0.8 0.6	0.5 0.3 0.1	60 30 10	
CQUAD	1.0 0.8 0.6	0.5 0.3 0.1	60 30 10	
Dipole	1.0 0.8 0.6	0.5 0.3 0.1	60 30 10	90 45 0

Table 1. Range of parameters associated with each Illumination shape which were explored in this work.

RESULTS

1.1. Annular

For each illumination condition, the average PV bandwidth was determined for the central contact of each pitch array. The smaller the PV band, the better the process latitude, and a given process would ultimately be limited by its highest PV band value pitch. In turn, the best illumination settings for a given primitive shape can be determined from the lowest worst-pitch value for PV bandwidth. So for instance in the case of annular illumination, the sigma 0.6/0.3 condition at 1:2 pitch had a PV bandwidth of 7.13 nm, and all other pitch values were lower. As such, it was the best annular process condition among those sampled. (See Figure 2). It is noted that for the nine different outer / inner sigma settings swept for annular illumination, there is not a huge difference in PV band values. This is reasonable given the nature of an annulus which exposes the reticle at all azimuthal angles regardless of the inner or outer settings. Figure 3 shows examples of the assist feature placement for an isolated contact as well as the best and worst iso contact PV band cases.

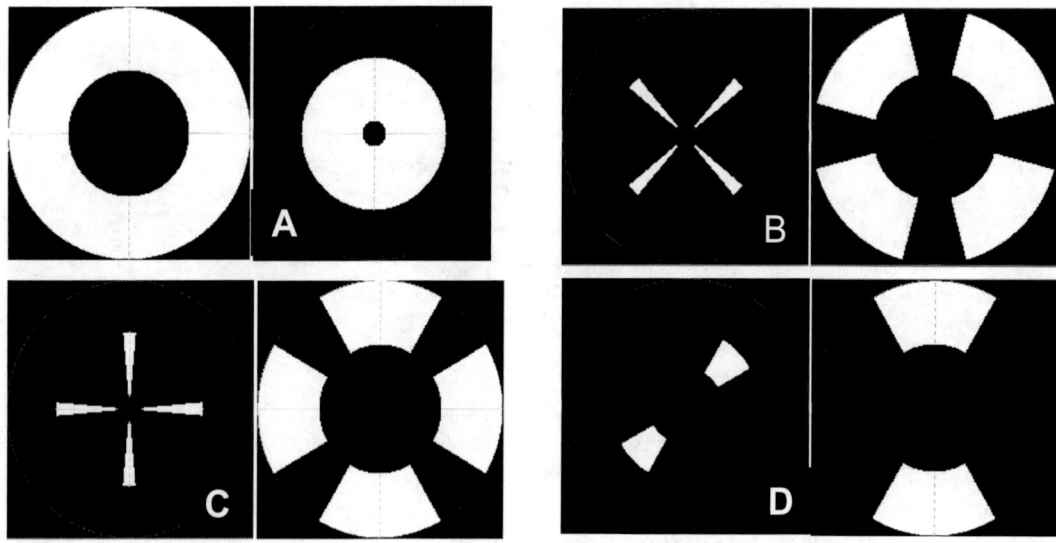

Figure 1. Examples of the four different illumination types, and different manifestations of inner, outer sigma, angle, and rotation. A: Annular, B: Quasar, C: CQUAD, D: Dipole.

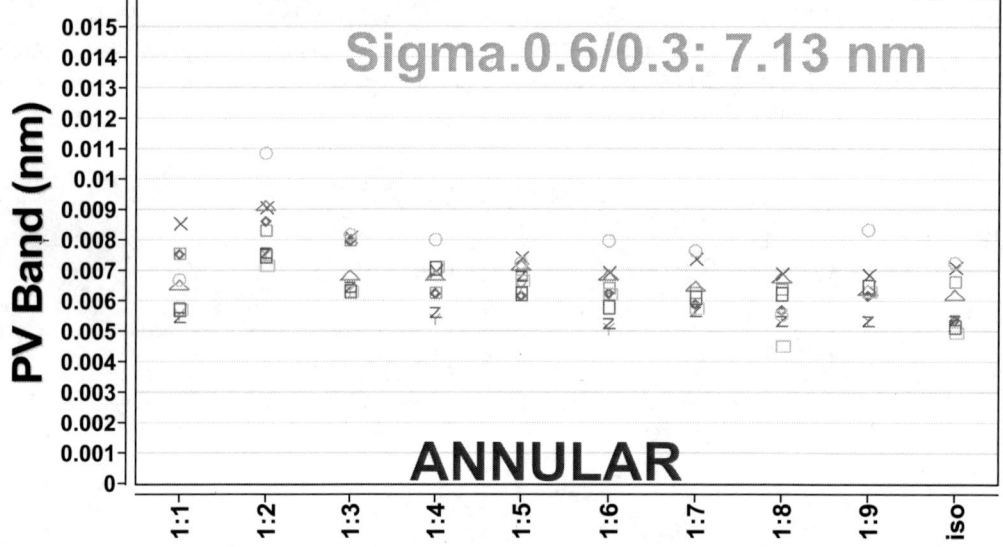

Figure 2. Through pitch PV bandwidth for the nine different outer/inner sigma settings. The sigma 0.6/0.3 case featured the lowest maximum through pitch PV band value.

1.2. Quasar

The quasar illuminator sweep featured an additional degree of freedom beyond annular with pole angle. As such, there was a significant range in PV band through pitch as a function of the 27 different Quasar conditions. Figure 4 shows the results and the best case is seen to be sigma 1.0 / 0.5 and illumination angle 30 degrees which had a worst case through pitch PV bandwidth of 9.7 nm. This was the highest value of the four illumination modes studied.

1.3. Dipole

With the addition of rotation angle, the dipole mode featured 81 unique settings. Not surprisingly, with dipole illumination, the post-OPC contact shapes were decidedly rectangular. Like the quasar case the PV band values for pitches beyond 1:2 were in general greater than 7 nm.

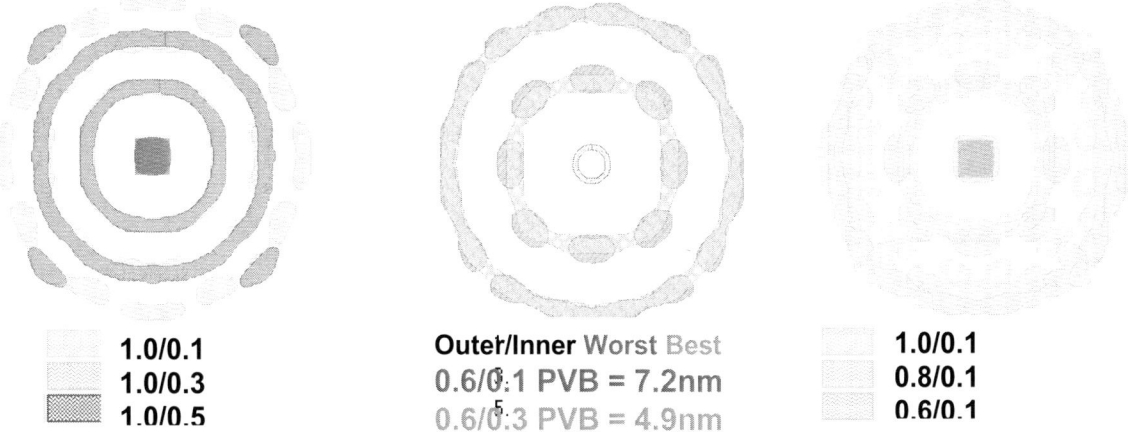

1.0/0.1	
1.0/0.3	
1.0/0.5	

Outer/Inner Worst Best
0.6/0.1 PVB = 7.2nm
0.6/0.3 PVB = 4.9nm

1.0/0.1	
0.8/0.1	
0.6/0.1	

Figure 3. Isolated contact assist feature placement for annular illumination. Left: placement as a function of increasing inner sigma. Right: placement as a function of decreasing outer sigma. Middle: Worst isolated PV band performance (0.6/0.1) and best isolated PV band performance (0.6/0.3)

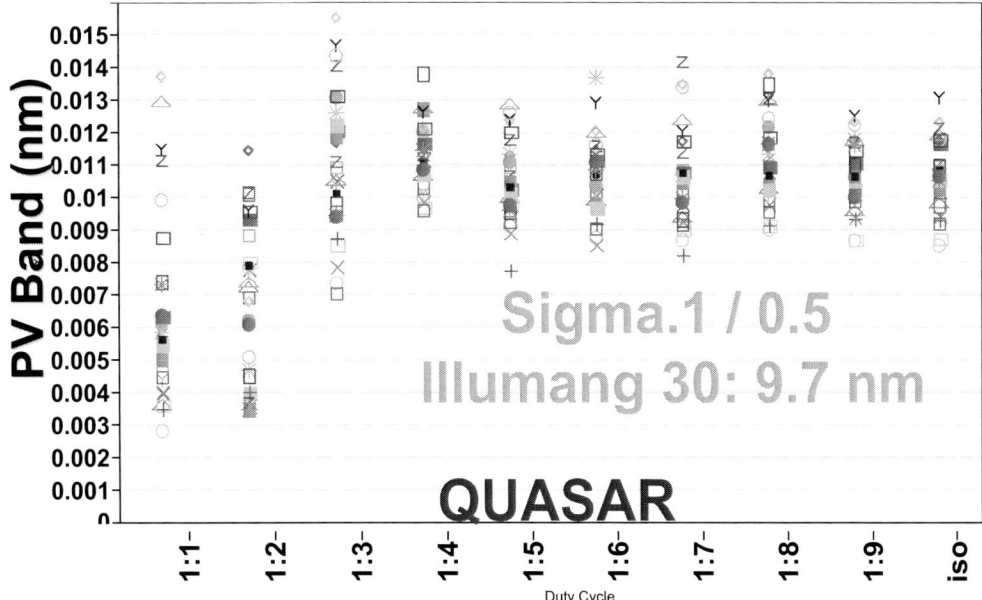

Figure 4. Through pitch PV bandwidth for the 27 different QUASAR outer/inner/illumangle sigma settings. The sigma 1.0/0.5 30 degree case featured the lowest maximum through pitch PV band value.

1.4. CQUAD

Because of the interaction of the vertical and horizontal contact hole edges with the illuminating poles, it is not a surprise that CQUAD illumination was shown to give the best process windows and featured the lowest worst pitch case, as shown in Figure 6. A worst case pitch PV band of 4.3 nm was obtained for the case of sigma 0.8/0.3 at 30 degree angle. It can be seen that many pitch/source combinations result in PV bandwidths as low as 2 nm.

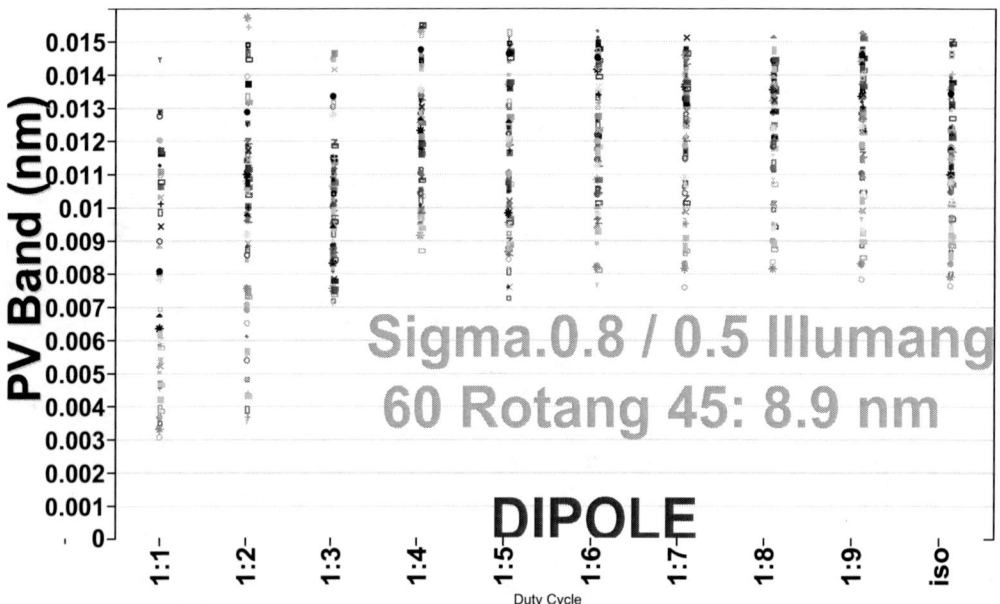

Figure 5. Through pitch PV bandwidth for the 81 different dipole outer/inner/illumangle sigma settings. The sigma 0.8/0.5 with 60 degree illumination angle and 45 degree rotation featured the lowest maximum through pitch PV band value.

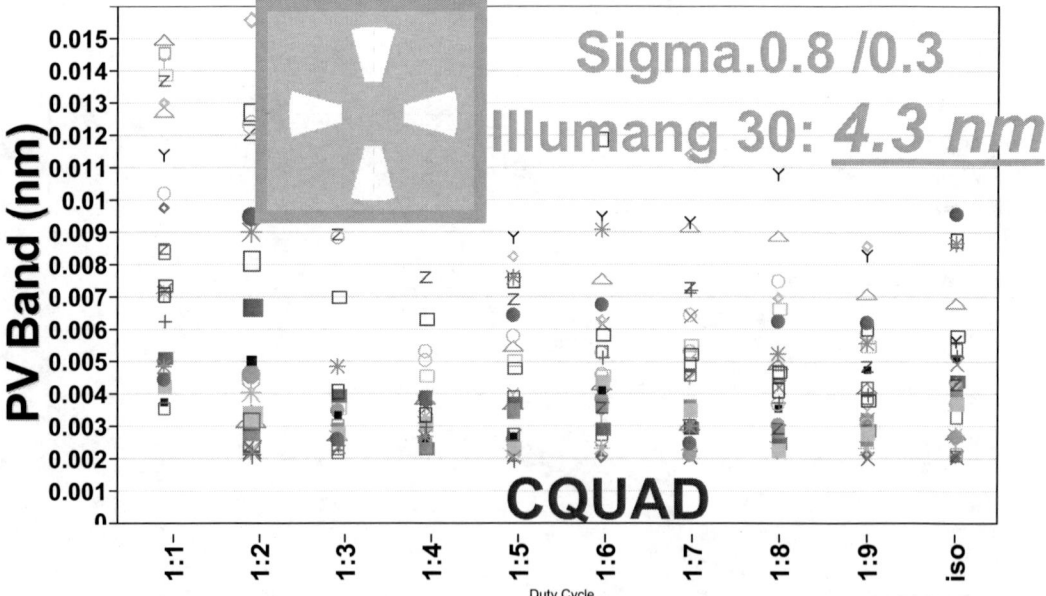

Figure 6. Through pitch PV bandwidth for the 27 different CQUAD outer/inner/illumangle sigma settings. The sigma 0.8/0.5 with 60 degree illumination angle and 45 degree rotation featured the lowest maximum through pitch PV band value.

1.5. Pixelized Source

Calibre SMO was used to determine a pixelized source based upon the 1:1, 1:2, and 1:3 clips. The pixelized source is shown in Figure XX. Next, the assist features were placed referencing that source, and OPC performed. The PV bandwidth through pitch is shown in Figure 7, along with snapshots of the assist feature placement. It is noted that the mask correction used in this case was different from the other sources, and was not optimized, thus the process latitudes may not represent the best achievable.

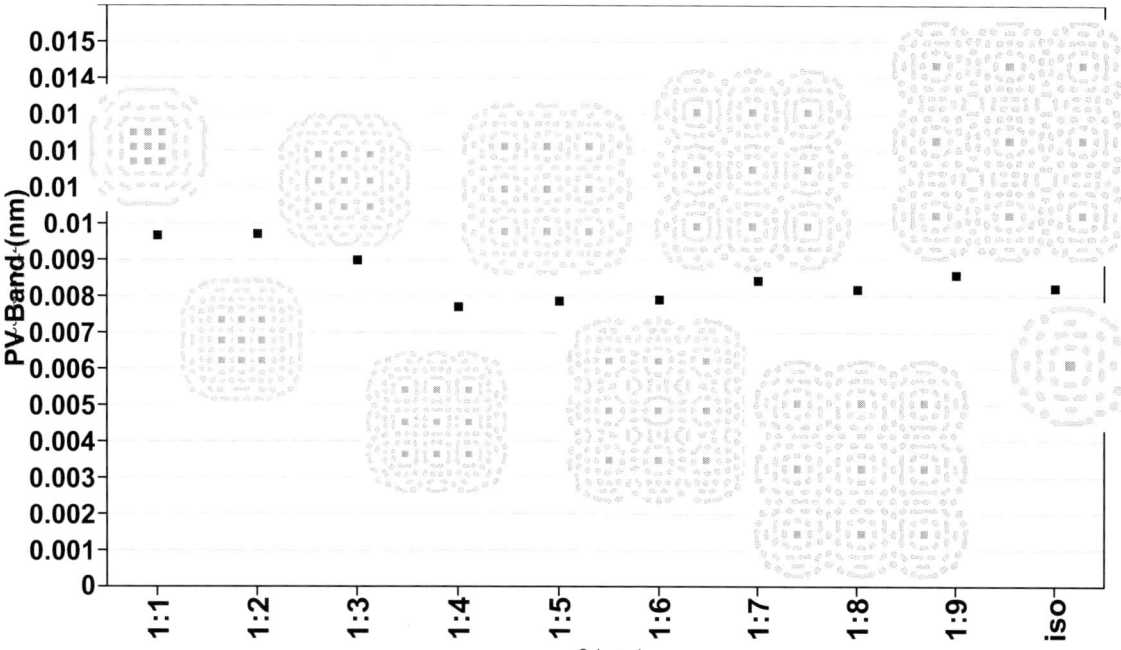

Figure 7. Through pitch PV bandwidth for the pixelized illumination source.

CONCLUSION

The ideal placement of assist feature shapes for a given illumination source is accomplished with model-based tooling. After subsequent OPC on the target contact, the PV bandwidth for each pitch is determined, and the process window limiting worst case pitch is identified. Hundreds of different illumination sources can quickly be compared and the global optimum source and assist feature configuration is determined. The analysis in this study featured regular arrays, but in principal any range of target layout configurations could be used. Rules can be developed which match the ideal placement, and in conjunction with further optimization during OPC, can deliver the optimum process window solution.

REFERENCES

1. Chen, K., Liao, C., Chen, S., Wey, T., Cheng, P., Chou, R., Schacht, J., Chou, D., Jayaram, S., "The PIXBAR OPC for Contact-Hole Pattern in sub-70-nm Generation", Proc. SPIE 7275, 72750Z (2009).
2. Shang, S., Granik, Y., Swallow, L., Zhang, L., Brist, T., Torres, A., Hung, C., Liu, Q., "Model-based Insertion and Optimization of Assist Features with application to Contact Layers", Proc. SPIE 5992, 59921Y (2005).
3. Farys, V., Chaoui, F., Entradas, J., Robert, F., Toublan, O., Troullier, Y., "SRAF enhancement using Inverse Lithography for 32 nm hole patterning and beyond », Proc. SPIE 7488, 74883M (2009).
4. Sturtevant, J., Jayaram, S., Hong, L., "Process Variability Band Analysis for Quantitative Optimization of Exposure Conditions", Proc. SPIE,7275, 72751Q (2009).
5. Torres, J.A., Berglund, C.N., "Integrated circuit DFM framework for deep sub-wavelength processes", Proc. SPIE 5756, 39 (2005).

A Novel Decomposition of Source Kernel for OPC Modeling

C.T. Hsuan, T.S. Wu, Fred Lo, Elvis Yang, T. H. Yang, K. C. Chen and Chih-Yuan Lu
Macronix International Co. Ltd, No. 16, Li-Hsin Rd., Science Park, Hsinchu 300, Taiwan

ABSTRACT

The accuracy and efficiency of OPC (Optical Proximity Correction) modeling have become paramount important at the low k1 lithography. However the accuracy of OPC model has to compromise with the efficiency of model calibration and pattern correction, since the model accuracy is usually improved by using more kernels to represent the model but the runtime of model setup and pattern correction also increase as kernel count increasing.

A novel decomposition of source kernel for OPC model calibration was presented in this study to maintain the model accuracy and preserve the OPC runtime at acceptable level. Firstly, the source kernel was decomposed into multiple sub-source kernels and then the magnitude of electric field for each decomposed sub-source was modulated in frequency domain. Finally, the resultant source can be the combination of many different sub-sources to represent the tool-specific characteristics. The model accuracy, model stability and modeling runtime were compared among decomposed source, ideal source and measured source models. The results showed modeling residual RMS error, predictive capability of decomposed source can be reduced to be comparable to measured source and superior to the ideal source. As for the modeling efficiency, the decomposed source is up to 5 times faster than the measured source but just few percentages slower than the ideal source approach.

Keyword: OPC, ideal source, KIF, measured source, decomposed source, model accuracy, runtime

1. INTRODUCTION

The challenges of ever-smaller CD and overlay budgets for leading-edge products have imposed stringent demand on the OPC model accuracy. In OPC-ed mask, the mask layout was corrected for compensating optical proximity effect to meet CD (Critical Dimension) target based on the OPC model. Thus, it is important to develop accurate OPC models to predict the behavior of the proximity effect for lithographic processes. The OPC model is constructed from a fitting of empirical data, and it consists of optical and resist simulation components. In the optical simulation, ideal source kernel is too primitive to sufficiently describe the actual illumination pupil fill. Since the ideal source is always assumed as a binary-like pattern that have a near constant and uniform intensity for all bridge area of illumination pupil fill and zero intensity for the remainder. The difference of pupil fill will impact the pattern transfer function from the mask to aerial image in lithography projection system [1]. While the impact becomes even more significant with the increasingly sophisticated and dense layout in low k1 process that the specific illumination optimization and extreme illumination setting are indispensable. The unrealistic assumption of binary-like ideal source instead of actual illumination pupil fill leads to a loss in accuracy of optical part in process simulation. Therefore, the changes or deviations of illumination system in OPC model are non-negligible. The KIF (Kernel Import Facility) was thus implemented to OPC kernel for enhancing model accuracy [2-3] but at the cost of longer runtime and lower efficiency.

In this study, a decomposed source modified from measured source was presented for optical part in OPC modeling. The comparison on model accuracy, predictive capabilities and modeling runtime were carried out among decomposed source, ideal source and measured source using 65nm memory contact layer as a vehicle.

Optical Microlithography XXIII, edited by Mircea V. Dusa, Will Conley, Proc. of SPIE Vol. 7640,
764037 · © 2010 SPIE · CCC code: 0277-786X/10/$18 · doi: 10.1117/12.845763

2. OPTICAL SIMULATION MODEL

The Hopkins equation is normally used for calculating aerial image intensities on a photo-mask for fast OPC modeling. In Hopkins equation, the aerial image intensity, I, at the image plane can be expressed as a 2D convolution:

$$I(x,y) = \int\int\int\int M(x_1,y_1)T(x-x_1,y-y_1;x-x_2,y-y_2)M^*(x_2,y_2)dx_1dx_2dy_1dy_2, \qquad (2.1)$$

where

$M(x, y)$ is the two-dimensional mask transmission function

M^* is the complex conjugate of M

$T(x-x_1, y-y_1; x-x_2, y-y_2)$ is the TCC (Transmission Cross Coefficient) matrix

the TCC matrix describes the physical properties of the optical system and is given by:

$$T(x-x_1,y-y_1;x-x_2,y-y_2) = J_0(x_1-x_2,y_1-y_2)K(x-x_1,y-y_1)K^*(x-x_2,y-y_2), \quad (2.2)$$

in which $J_0(x, y)$ is the mutual intensity of the source that describes the coherent properties of the illumination system, $K(x, y)$ is the pupil function of the projection lens that represents the properties of the projection system and K^* is the conjugate of the K. When TCC is represented as a matrix, $\overline{\overline{T}}$, that it also can be described as the pupil function $\overline{\overline{P}}$ and the source definition $\overline{\overline{S}}$ [4].

$$\overline{\overline{T}} = \overline{\overline{P}} * \overline{\overline{S}} * \overline{\overline{P}}^* \qquad (2.3)$$

$$\overline{\overline{P}} = f(\frac{NA}{\lambda}, Mag) \qquad (2.4)$$

$$\overline{\overline{S}} = f(\sigma, \lambda) \qquad (2.5)$$

The optical model is composed of the matrices $\overline{\overline{P}}$ and $\overline{\overline{S}}$ which use the parameters such as NA(Numerical Aperture), σ (partial coherence), λ (wavelength), and Mag (magnification). To improve the primitive shape of source, the new matrices $\overline{\overline{S}}'$ is given by

$$\overline{\overline{S}}' = f(\sum_{i=1}^{n}W_i\sigma_i, \lambda) \qquad (2.6)$$

In Equation 2.6, W_i is the transmission for each decomposed σ and $\overline{\overline{S}}'$ describes the resultant source in frequency domain. From the equation above, the source could be combination of many sub pupils with different transmissions and that a suitable integration of sub pupils can make the source more close to the measured one from scanner.

3. SOURCE DECOMPOSITION

The source definition in traditional OPC model has been expressed as equation 2.5, it is usually a binary-like pattern before varying the slope of the source in frequency domain, as shown in Figure 1(a). As also displayed in Figure 1 that the ideal source is usually much different from the measured source (shown in Figure 1(b)), the difference of pupil fill will impact the pattern transfer function from the mask to aerial image in lithography projection system. The unrealistic assumption of ideal source instead of actual illumination pupil fill results in a loss in accuracy of optical part in process simulation. Hence the source decomposition method in this study decomposed the source to many segmented pupils and recombined them to simulate the measured source hence retain the tool-specific characteristics, as schematically illustrated in Figure 2.

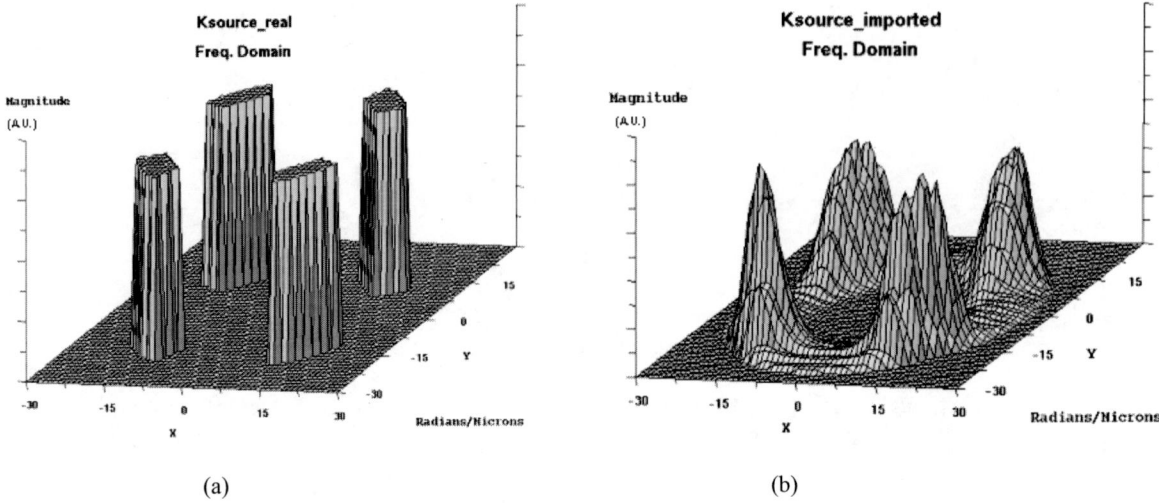

Figure 1. The Quasar illumination pupil fill measurement: (a) binary-type ideal source, and (b) measured source derived from scanner. The ideal source is much different from the measured source.

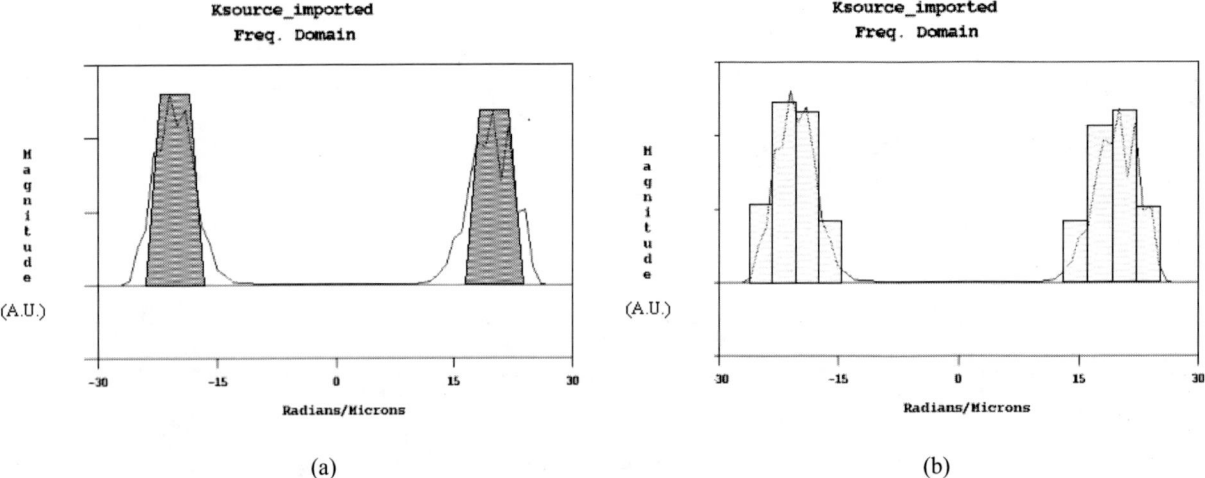

Figure 2. (a) The difference between ideal source and measured source in the cross-sectional view along the 45d diagonal of figures in Figure 1. (b) The difference between resultant decomposed source and measured source in the cross-sectional view. The combination of decomposed sub-sources can be more similar to the measured source.

4. EXPERIMENTAL

This paper takes contact layer of 65nm node memory product as a vehicle to compare three different source implementations in OPC modeling. Those three methods are idea source in traditional OPC model, measured source by KIF approach and decomposed source. A wide range of 1D and 2D contact OPC calibration test patterns was created and measured for OPC modeling. Figure 3 schematically displays the linearity calibration patterns with scattering bar configuration, in which the mask dimension ranges from 160nm to 210nm. While the pitch of the calibration patterns covers from 420nm to 2000nm. A 150nm-pitch 1D pattern was used as dose-to-target anchor for the data collection. The illumination conditions for model calculation is ArF Quasar illumination of NA= 0.85, sigma-out/sigma-in= 0.88/0.64.

OPC model fitting was performed with constant threshold and pure optical kernel was carried out first, followed by adding the resist kernel after fixing the parameters in optical kernel. The fitting accuracy, predictability and modeling runtime were compared among three type models.

	1D_contact	2D_contact
Dense		
Semi_dense		
ISO		

Figure 3. The examples of through-pitch linearity calibration test patterns for OPC modeling.

5. RESULTS AND DISCUSSION

The OPC model can be roughly divided into optical and resist simulation components. The optical part is the most important portion in the simulation, since the use of a more accurate pupil measurement fill to create more physically valid lithographic model and retain the tool specific characteristics that yields a better decoupling of optical and resist effects. And therefore a larger range where the OPC model is with better predictive capability. Hence, the impact of various illumination pupils on the aerial image simulations was first compared by their modeled CD from optics-only simulations, as shown in Figure 5. The three optics-only models behave the similar trend on modeled CD for 1D and 2D through-pitch contact patterns. It can be seen that the measured and the decomposed sources are better fitting to the proximity curve of wafer measurements than the ideal one. The overall result of 1D and 2D through-pitch patterns reveals that the measured source is better fit to the wafer measurements and the ideal source is the worst. The possible cause is that the ideal source is too primitive to present the actual illumination pupil fill and yields a loss in accuracy in process simulation.

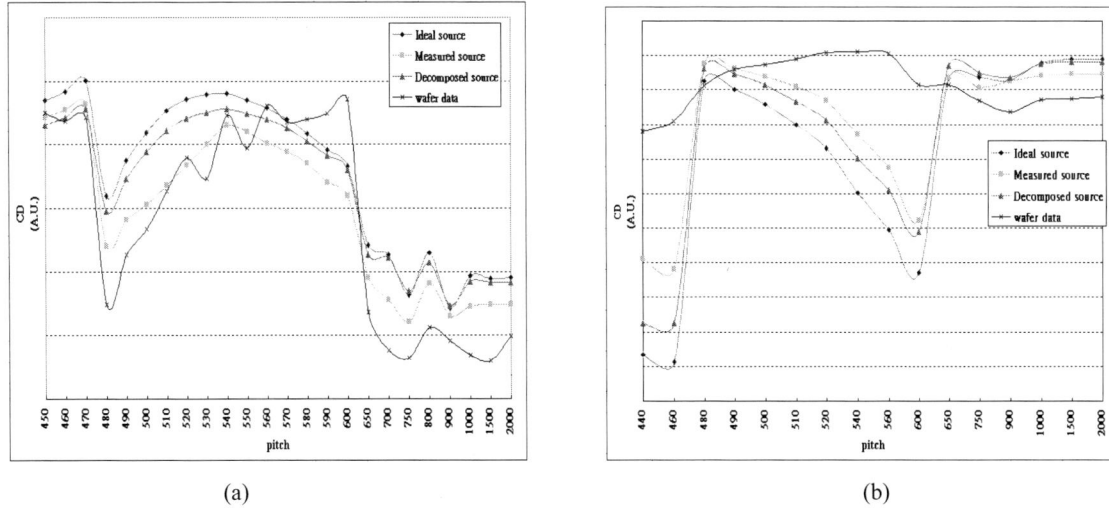

(a)   (b)

Figure 5. The comparison of proximity curve of wafer measurements and modeled CD of three optics-only models, including ideal source, measured source and decomposed source. (a) 1D 0.2um through-pitch contact patterns, and (b) 2D 0.2um through-pitch contact patterns.

When we combine the optical model with the resist model for model calibration, the fitting examples were displayed Figures 6(a) and 6(b). These figures compare the model simulation versus calibration data measurements for the model building. In the fitting of 1D calibration dataset, the decomposed source model is slightly better fitting to the measurements in <470nm and 550~2000nm pitches. And there is insignificantly different in 480~540nm pitches for three models. For the 2D calibration dataset fitting, the decomposed source and measured source models are near identical fitting to the measurements. While the ideal source model has the largest fitting errors for both 1D and 2D dataset. The resist model is mainly an empirical fit, any optical effect is not accounted for in the optical simulation will be coupled into the OPC model simulation, and the resist part in the OPC model will to some degree correct the weakness of the optical model. It is clear that the resist part of OPC models do compensate some bad optical simulations but not all in the model fitting hence the trends get more similar in the fitting than optics-only simulation for all the ideal source, decomposed source and measured source models, in addition the fitting errors to the measurements also decrease significantly. However the measured source model is still fitting the model calibration dataset slightly better than the decomposed and ideal source models, implying the important role of optical model can't be totally ignored.

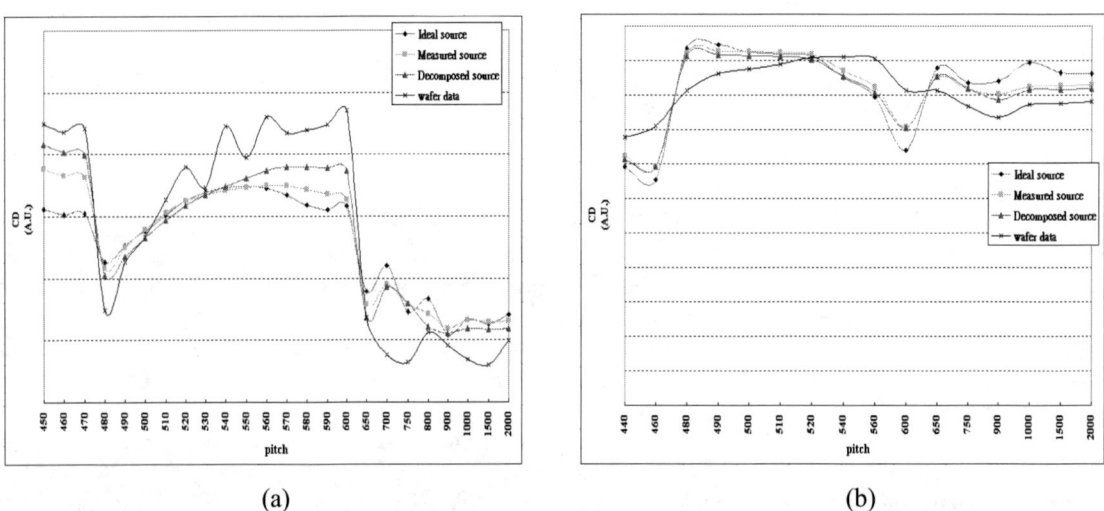

(a)   (b)

Figure 6. The fitting accuracy comparison for ideal source, decomposed source and measured source models for through-pitch calibration patterns. (a) 1D 0.2um through-pitch contact patterns, and (b) 2D 0.2um through-pitch contact patterns.

The ideal source, decomposed source and measured source models were also tested for the linearity calibration patterns. Figure 7 shows the model statics for the three models for linearity and through-pitch calibration patterns. The EPE (Edge Placement Error) is the difference between the layout and the simulated of model. As can be seen from Figures 7(a) and (b) that the residual RMS (Root Mean Square) error of the decomposed source is comparable to that of the measured source model, and it is much improved from the ideal source model especially in the 2D pattern fitting. In the Figure 7, the maximum to minimum EPE was also expressed as error bar with the residual RMS error. The longer bar of EPE represents a larger range of deviations to the targets and can be explained as the model accuracy is not good enough at some calibration data range. Obviously, the decomposed source model is also comparable to the measured source model on EPE range for both 1D and 2D patterns, but the ideal source model shows the longer EPE bar for 1D and 2D patterns.

The quality of an OPC model cannot only be determined by the residual RMS error for the dataset which was adopted for the model fitting, but rather by the range of validity outside of the calibration dataset. In another word, good predictive capability of an OPC model outside the original calibration space is another one important indicator for model quality. The model verification consists of two tests in this study, the first test was conducted via the CD sizes located at the edge of calibration test patterns, and the second test was carried out for 1D, 2D and isolated test patterns with different CD targets to the calibration dataset. All the aforementioned verification test patterns were not included in the OPC model calibration. Figure 8 depicts EPE of 1D contact x and y dimensions as well as 2D contact of the first verification pattern from three models. It is apparent that the ideal source model predicts the verification patterns with larger error than the decomposed and measured source models.

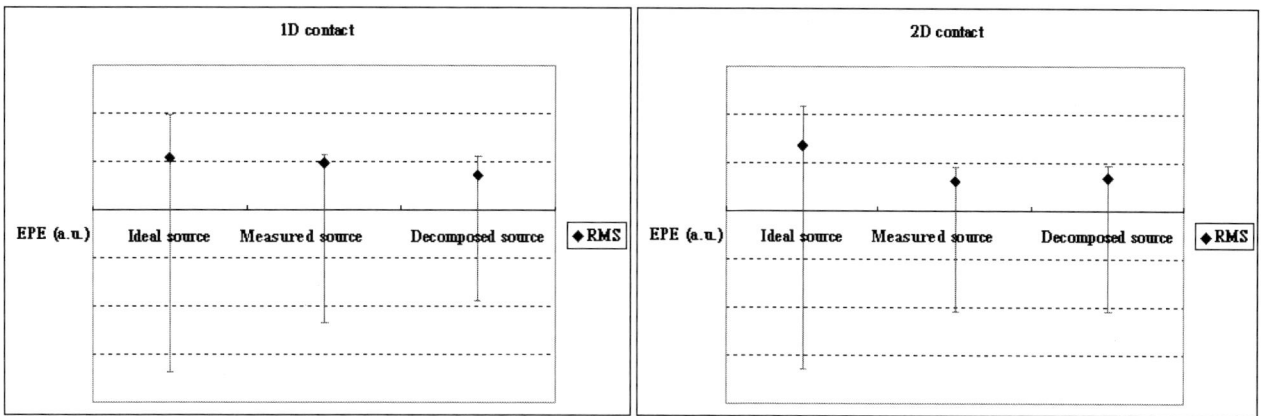

Figure 8. The comparison of residual RMS errors and EPE range of ideal source, decomposed source and measured source models.

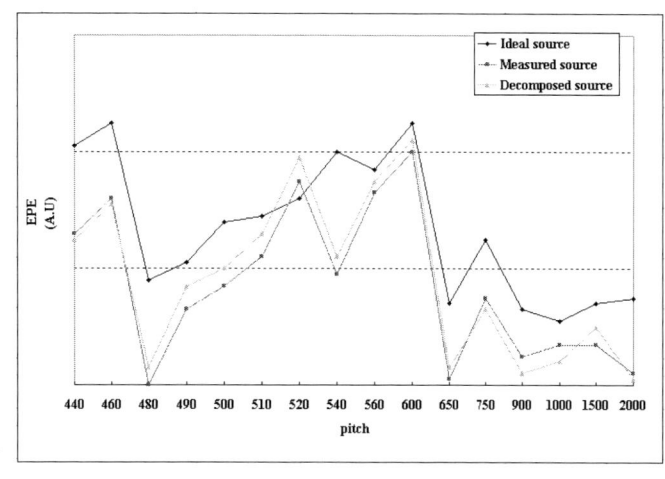

(a)

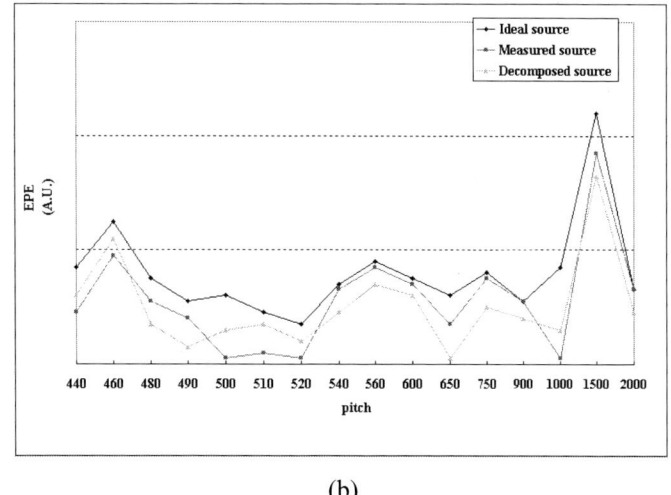

(b)

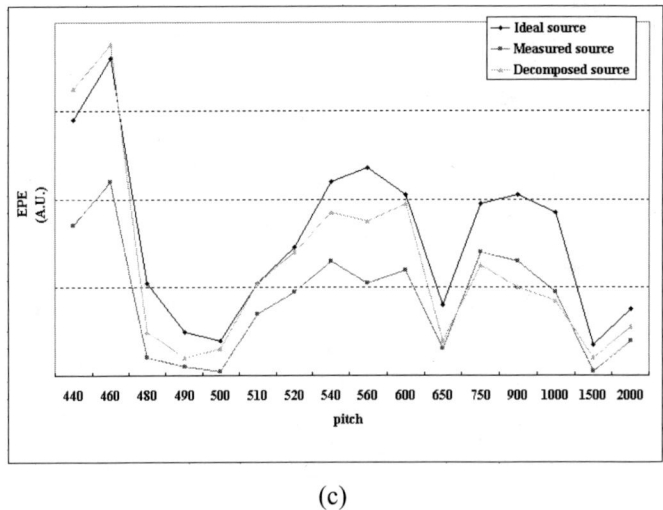

(c)

Figure 7. The comparison of EPE on through-pitch verification patterns among three models: (a) x-dimension of 1D contact at edge, (b) y-dimension of 1D contact at edge, and (c) x-dimension of 2D contact at edge.

In the second verification, the modeled errors were expressed in Figure 9. For all the verified patterns, the ideal source model performs the largest error and decomposed source model is comparable to the measured source model with smaller error. The verifications summarized that the model stability of the decomposed source model is good enough and comparable to the measured source model.

	Anchor point (EOP)	1D_contact	2D_contact	ISO contact
Pattern schema				
MCD(nm)	143*124	196*211	200*200	202*202

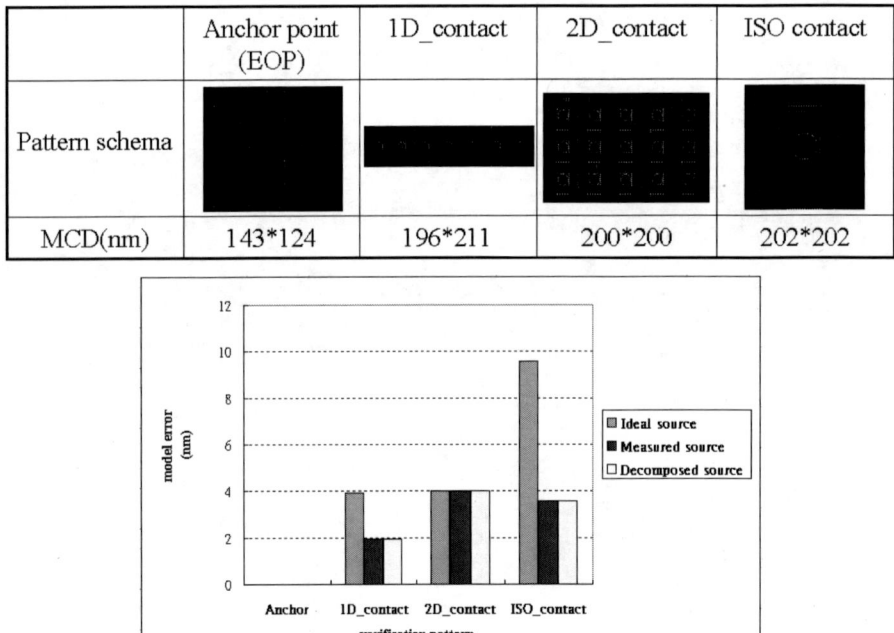

Figure 9. The verification results of three models by 1D, 2D and isolated test patterns with different CD targets to the calibration dataset.

Finally, the efficiency on modeling was compared among the three source models. Figure 10 shows the modeling runtime versus the regression items for the same step. The runtime of measured source model is the longest as expected, while the decomposed source model is at 5 times faster than the measured source model and comparable to the ideal source model on runtime. Besides, unlike the quick increasing rate of measured source model on runtime with the regression item increasing, the runtimes of decomposed source and ideal source models also increase but as a much weaker function of regression item. The conclusion is hence quite obvious that the model accuracy and model stability of decomposed source model is improved from the ideal one and comparable to the measured source model, furthermore the decomposed source model enhances the modeling efficiency from the measured source model.

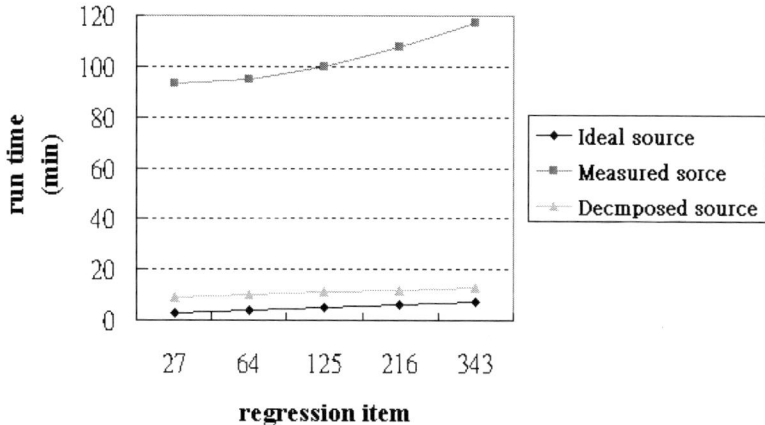

Figure 10. The modeling runtime comparison for three models.

6. CONCLUSIONS

In this study, a so-called "decomposed source", was implemented for OPC modeling and compared with measured source and ideal source on fitting accuracy, model predictability and modeling efficiency. The decomposed source was generated by decomposing the source to many segmented pupils and recombined them to simulate the measured source. In terms of edge placement error for model fitting, the decomposed source model is superior to ideal source model and is comparable to the measured source model. The ability of interpolation and extrapolation for decomposed source model is also comparable to measured source model and superior to the ideal source model. Furthermore, the decomposed source model is at 5 times faster than the measured source model and comparable to the ideal source model in modeling runtime.

REFERENCES

1. T. C. Barrett, "Impact of illumination pupil-fill spatial variation on simulated image performance," *Proceedings of SPIE 4000*, pp. 804-817, 2000.
2. Christof Bodendorfl, R. E. Schlief, Ralf Zieboldc, "Impact of Measured Pupil Illumination Fill Distribution on Lithography Simulation and OPC Models," *Proceedings of SPIE 5377*, pp1130-1145, 2004.
3. Te Hung Wu, C. L. Lin, Ming Jui Chen, Zen Hsiang Tsai, Chen Yu Ao, H. C. Thuang, Jian Shin Liou, Chuen Huei Yang, Ling Chieh Lin, "Improvement of OPC Accuracy for 65 nm Node Contact Using KIF ," *Proceedings of SPIE 6155*, 61550M, 2006.
4. Lawrence S. Melvin III, Kevin D. Lucas, "A Discussion of the Regression of Physical Parameters for Photolithographic Process Models," *Proceedings of SPIE 6520*, 65201V, 2007.

Methods for assessing empirical model parameters and calibration pattern measurements

Xin Zhou[*], Eldar Khaliullin, Lan Luan

Luminescent Technologies, Palo Alto, CA, USA 94303

ABSTRACT

Assessing an empirical model for ILT or OPC on a full-chip scale is a non-trivial task because the model's fit to calibration input data must be balanced against its robust prediction on wafer prints. When a model does not fit the calibration measurements well, we face the difficult choice between readjusting model parameters and re-measuring wafer CDs of calibration patterns. On the other hand, when a model does fit very well, we will still likely have the nagging suspicion that an overfitting might have occurred. Here we define a few objective and quantitative methods for model assessment. Both theoretical foundation and practical use are presented.

Keywords: model calibration, model effectiveness, Fisher Information Matrix, Cramér-Rao Lower Bound

1. INTRODUCTION

Assessing an empirical model for ILT or OPC on a full-chip scale is a non-trivial task because the model's fit to calibration input data must be balanced against its robust prediction on wafer prints. When a model does not fit the calibration measurements well, we face the difficult choice between readjusting model parameters and re-measuring wafer CDs of calibration patterns. On the other hand, when a model does fit very well, we will still likely have the nagging suspicion that an overfitting might have occurred. Apparently we are in need of objective and quantitative methods for model assessment.

In practice, when we are afforded a cross-validation data set, a positive confirmation will make us feel comfortable with our model in hand. However, when a significantly larger error shows up in the cross-checking procedure than in the original calibration, we will be at a loss whether those data points in the verification set are suspect or the model itself ought to be made substantially different. The coupling between the model and the data is so strong that the cross-validation alone cannot answer all the questions when we are in doubt.

Some existing tools and concepts in the estimation theory of statistics provide a suitable theoretical framework for our purpose in model calibration and assessment. The next section summarizes the theoretical foundation detailed in a previously published paper [1].

2. THEORETICAL BACKGROUND

If a discrete data set represented by a vector $x = \{x[0], x[1], ..., x[N-1]\}$ depends on a series of parameters $\boldsymbol{\theta} = \{\theta_1, \theta_2, ..., \theta_p,\}$, we may design an estimator that finds the true value of $\boldsymbol{\theta}$ based on the data set

$$\hat{\boldsymbol{\theta}} = f(x[0], x[1], ..., x[N-1])$$

Here f is the parameter estimation function, which could be as simple as an algebraic function, or as complex as a lithographic process model calibration process.

If we use $P(x; \boldsymbol{\theta})$ to denote the probability density function (PDF) of obtaining measurements at x with the parameter at $\boldsymbol{\theta}$, the $p \times p$ Fisher information matrix $\boldsymbol{I}(\boldsymbol{\theta})$ is defined as

---

[*] xzhou@luminescent.com

Optical Microlithography XXIII, edited by Mircea V. Dusa, Will Conley, Proc. of SPIE Vol. 7640,
764038 · © 2010 SPIE · CCC code: 0277-786X/10/$18 · doi: 10.1117/12.847734

$$[\boldsymbol{I}(\boldsymbol{\theta})]_{ij} = -E\left[\frac{\partial^2 \ln P(\boldsymbol{x};\boldsymbol{\theta})}{\partial \theta_i \partial \theta_j}\right] = -\int \frac{\partial^2 \ln P(\boldsymbol{x};\boldsymbol{\theta})}{\partial \theta_i \partial \theta_j} P(\boldsymbol{x};\boldsymbol{\theta})d\boldsymbol{x}$$

The Cramér-Rao Lower Bound (CRLB) of parameter variance is

$$\text{var}(\hat{\theta}_i) \geq \left[\boldsymbol{I}^{-1}(\boldsymbol{\theta})\right]_{ii} \tag{1}$$

where the true value of $\boldsymbol{\theta}$ is used when the derivatives are taken.

In an example of a signal sequence depending on a vector parameter $\boldsymbol{\theta}$

$$x[n] = s[n;\boldsymbol{\theta}] + w[n], \quad n = 0,1,...,N-1 \tag{2}$$

The Fisher information matrix elements are

$$[\boldsymbol{I}(\boldsymbol{\theta})]_{ij} = \frac{1}{\sigma^2}\sum_{n=0}^{N-1}\frac{\partial s[n;\boldsymbol{\theta}]}{\partial \theta_i}\frac{\partial s[n;\boldsymbol{\theta}]}{\partial \theta_j} \tag{3}$$

In the above formula, σ^2 is uniform variance of the white Gaussian noise w. The form of the deterministic signal $s(\boldsymbol{\theta})$ is not yet restricted [2].

2.1 Uncertainty of parameters

We consider the entire calibration process of finding the true model parameters from measurements as a problem of computing a p-dimensional estimator $\boldsymbol{\theta}$. Furthermore, we assume that whatever systematic errors of metrology have been adequately absorbed in simulated CDs $s[n; \boldsymbol{\theta}]$, and that all aspects of lithographic process have been adequately modeled, so that σ^2 represents the true variance of noise $w[n]$ from measurement. Now we are ready to apply Eq.1 and Eq.3 to our calibrated empirical model to obtain parameter variance

$$\text{var}(\theta_i) \geq \left|\sigma^2\left(\boldsymbol{J}^T \cdot \boldsymbol{J}\right)^{-1}\right|_{ii} \tag{4}$$

where $\boldsymbol{J}$ is the $N \times p$ Jacobian matrix of simulated CDs w.r.t. parameters

$$\boldsymbol{J} = \begin{bmatrix} \dfrac{\partial s[0;\boldsymbol{\theta}]}{\partial \theta_1} & \dfrac{\partial s[0;\boldsymbol{\theta}]}{\partial \theta_2} & \cdots & \dfrac{\partial s[0;\boldsymbol{\theta}]}{\partial \theta_p} \\ \dfrac{\partial s[1;\boldsymbol{\theta}]}{\partial \theta_1} & \dfrac{\partial s[1;\boldsymbol{\theta}]}{\partial \theta_2} & \cdots & \dfrac{\partial s[1;\boldsymbol{\theta}]}{\partial \theta_p} \\ \vdots & \vdots & \ddots & \vdots \\ \dfrac{\partial s[N-1;\boldsymbol{\theta}]}{\partial \theta_1} & \dfrac{\partial s[N-1;\boldsymbol{\theta}]}{\partial \theta_2} & \cdots & \dfrac{\partial s[N-1;\boldsymbol{\theta}]}{\partial \theta_p} \end{bmatrix} \tag{5}$$

The inequality Eq.4 specifies the minimal uncertainty of any calibrated model parameter.

2.2 Uncertainty of predictions

Now we present the result of the CRLB of $N \times N$ covariance matrix $\boldsymbol{C}_s$ whose diagonal elements correspond to the simulation variance at test patterns respectively. Assuming that we have performed Singular Value Decomposition (SVD) of $\boldsymbol{J}=\boldsymbol{U}\boldsymbol{\Sigma}\boldsymbol{V}$

$$C_s \geq = \sigma^2 U \begin{bmatrix} 1 & 0 & 0 & 0 & \cdots & 0 \\ 0 & \ddots & 0 & \vdots & & \vdots \\ 0 & 0 & 1 & \vdots & & \vdots \\ 0 & \cdots & \cdots & 0 & & \vdots \\ \vdots & & & & \ddots & \vdots \\ 0 & \cdots & \cdots & \cdots & \cdots & 0 \end{bmatrix} U^T \tag{6}$$

where the $N \times N$ diagonal matrix in the middle of two left-singular-vector unitary matrix U has p 1's and $(N-p)$ 0's; in other words, its upper left corner is a $p \times p$ identity matrix, and the rest of the elements are zeros.

The variance σ^2 is related to fitting error by

$$\sigma^2 \approx \frac{SSE}{N-p} = \frac{N}{N-p}(rms)^2 \tag{7}$$

where SSE is the sum-squared-error between simulation and measurement, N is the number of data points, p the number of parameters, and rms is the fitting error i.e. the usual metric for model accuracy.

The square root of CRLB of the right hand side of Eq.6 serves as a metric of prediction error of each individual test pattern. This will also be called the simulation uncertainty.

2.3 Effectiveness of a model

We define the trace of the right hand side of Eq.6 to be $\varXi$ Model Effectiveness Index (*MEI*), which is the lower bound of total variability of a model:

$$\varXi \equiv \sigma^2 Tr \begin{bmatrix} 1 & 0 & 0 & 0 & \cdots & 0 \\ 0 & \ddots & 0 & \vdots & & \vdots \\ 0 & 0 & 1 & \vdots & & \vdots \\ 0 & \cdots & \cdots & 0 & & \vdots \\ \vdots & & & & \ddots & \vdots \\ 0 & \cdots & \cdots & \cdots & \cdots & 0 \end{bmatrix} = \sigma^2 p = \frac{Np}{N-p}(rms)^2 \tag{8}$$

This figure of merit $\varXi$ characterizes modeling uncertainty by balancing the fitting error (*rms*) with the number of parameters used (*p*), and the number of measurements taken (*N*). We can also use the square root of $\varXi$ as another figure of merit, which has the same dimension as *rms*, Model Effectiveness Standard Index (*MESI*):

$$MESI \equiv \sqrt{\varXi} = \sqrt{\frac{Np}{N-p}} \cdot rms \approx \sqrt{p} \cdot rms \tag{9}$$

The right hand side of this definition holds when $N \gg p$. The smaller the value of $\varXi$ is, the more effective the model is.

2.4 Identification of measurement outliers

We can use Eq.6 to compute the lower bound of expected simulation variance for each individual calibration test pattern. If we find that the actual measurement differs from the simulation by a certain factor, e.g. 6, of the square root of variance, we can be quite confident that that particular measurement point is an outlier and thus better to be re-measured or temporarily ignored.

3. EXPERIMENTAL

Using two sets of data points of wafer CD measurements from two different layers, we calibrated two models to demonstrate our method. For each model we have computed the model parameter variance and six times the square root of simulation variance at each data point, which we call "Uncertainty" in the plots.

3.1 Uncertainty of the model parameters

Fig.1 shows the variance of each model parameter, computed with Eq.4. Low variance of a parameter indicates that it is a stiffer parameter which is most sensitive to calibration measurement data.

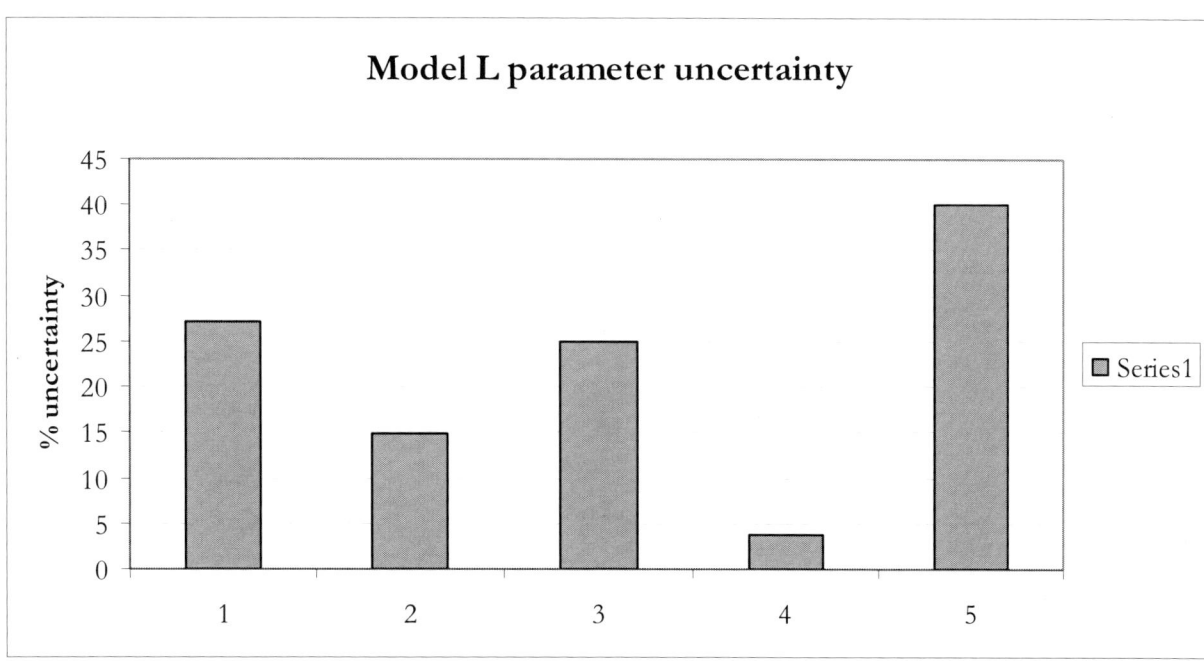

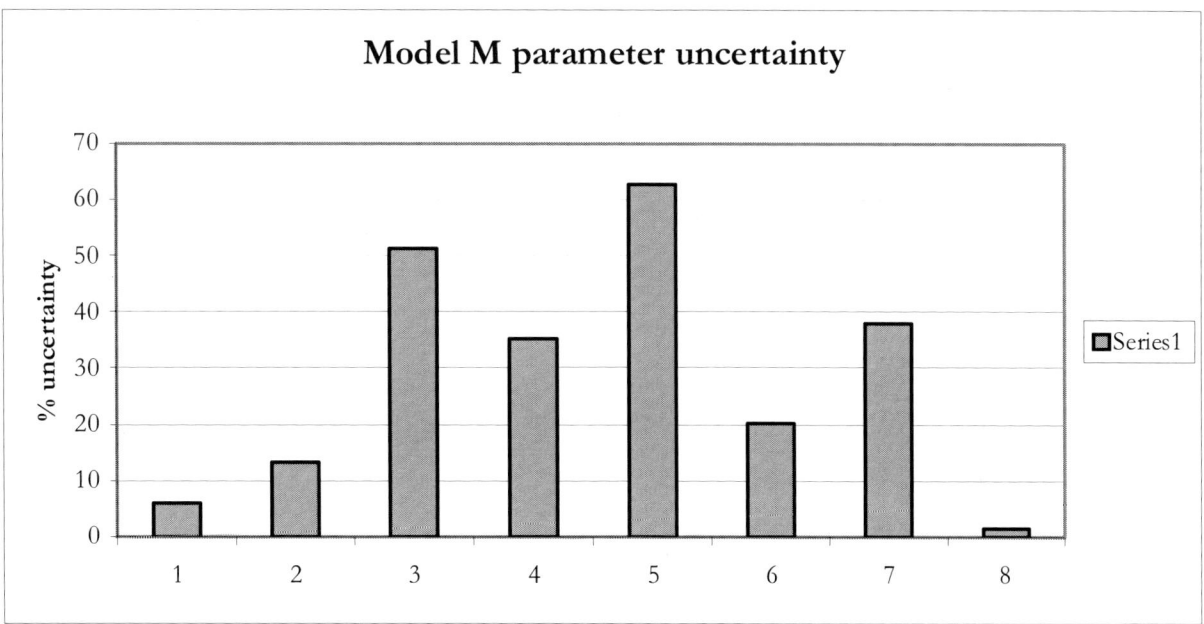

Figure 1. Expected uncertainty of calibrated model parameters

3.2 Comparing simulations with measurements

Judging by the model effective metric **MESI**, Model L matches its data better than Model M : MESI of model L is 6.1, while that of Model M is 22.5.Using Eq.6 we compute model prediction error for each test pattern. Fig.2 Model L shows smaller (fitting error/simulation uncertainty) ratio than Fig.3 Model M.

Model *L*

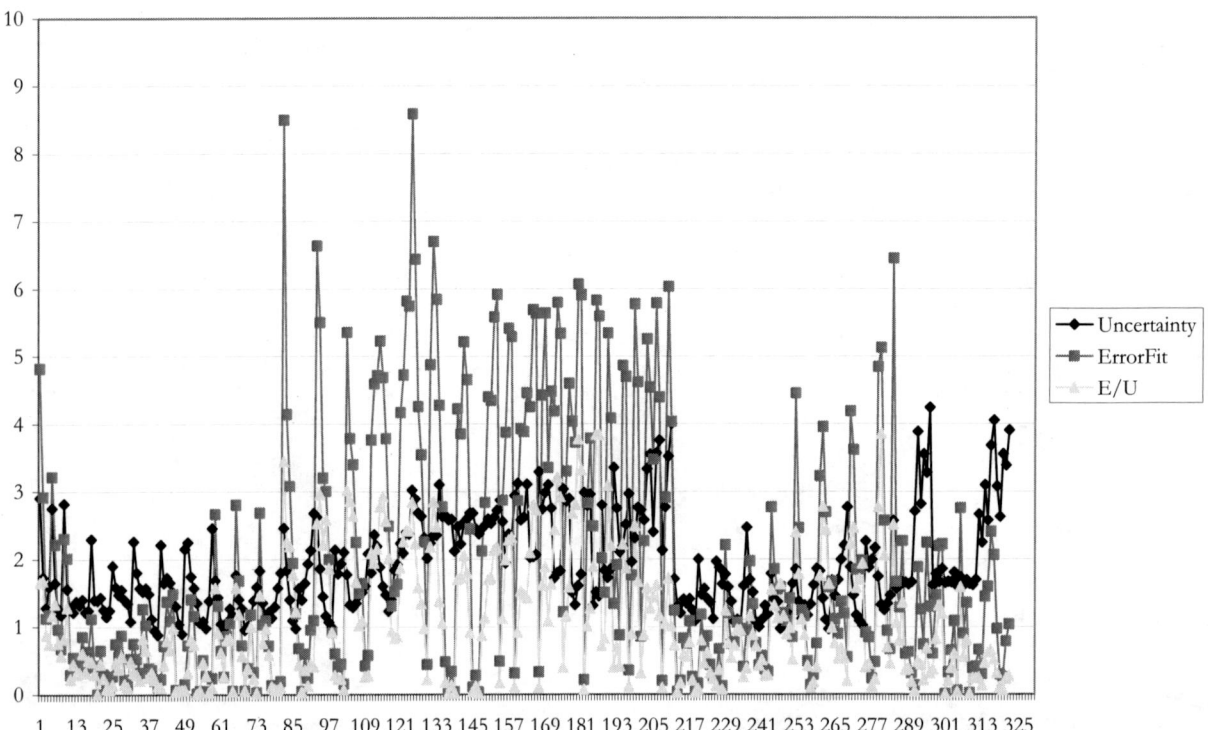

Figure 2.Model L has MESI = 6.1

The two examples show that when a model does not seem effective, as in the case of Model M, we should not yet jump to the easy conclusion that there is something wrong with the calibration process or the model formulation. The ErrorFit/Uncertainty ratio plot can tell us if examining raw data is the right direction to pursue. If there appear many outliers, we should take a step back and review the data collection process and see if printing or metrology plays the spoiler here.

Discussion with our customers revealed that Model M input data were not taken with repeated measurements of same patterns across wafer dies, but that they were measured just once per test pattern. The statistical variance of each data point in this case would be intrinsically infinite. However, the model calibration process takes into account all the input data simultaneously, so that an averaging process is taking place and a statistical significance can be reached even though any given individual data point is suspect. Because we don't have an independent and reliable measure of the metrology uncertainty in the single shot, single measurement data collection scheme, the resulting model fit of Model M is much less informative than it could be had its data been collected multiple times.

Model *M*

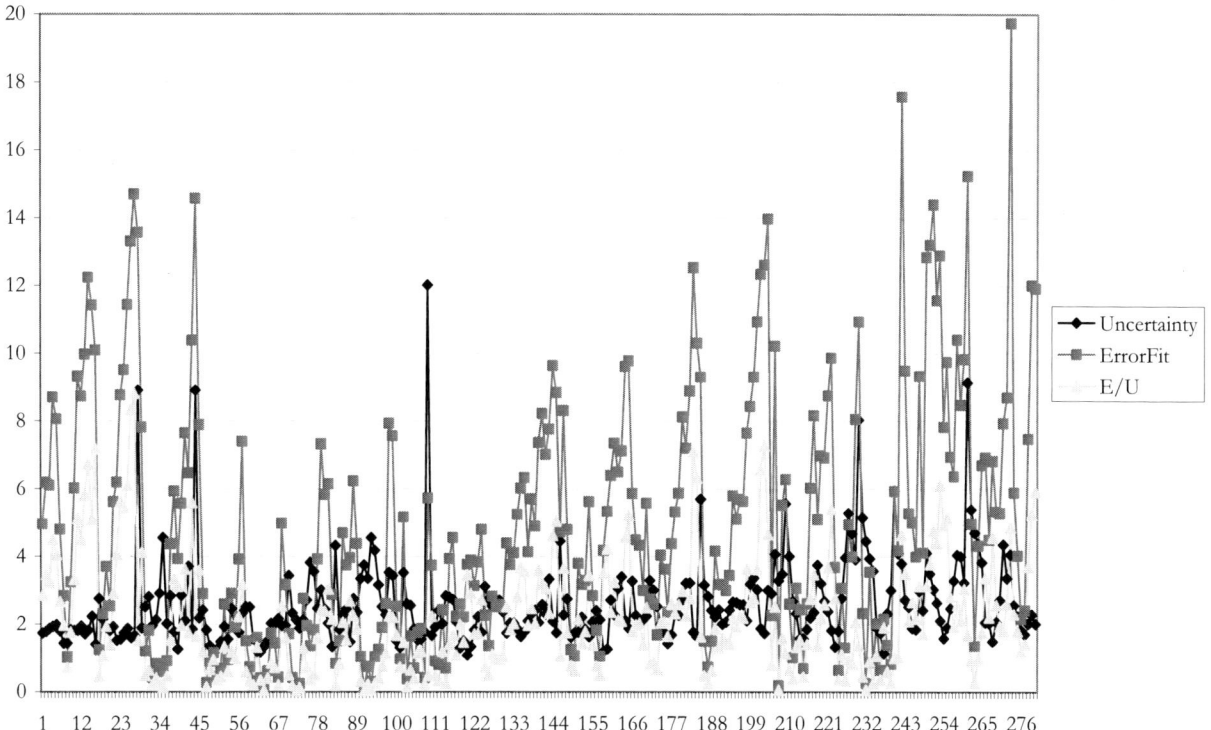

Figure 3. Model M shows larger measurement error, and MESI=22.5

4. CONCLUSIONS

We have set out to define some quantitative and objective methods to help evaluate empirical model parameters and the input measurement data. Model Effectiveness Standard Index (***MESI***) is defined as the square root of the trace of the CRLB of the simulation covariance matrix. The lower the ***MESI***, the better the model. We have also analyzed simulation uncertainty relative to measurement data, with the hope that the calibrated model itself can shed light on the question of whether some data points are true outliers.

Future research into this fascinating topic of model calibration will look at the coupling between model and data [3], and attempt to address the sufficiency question, i.e. whether our model formulation has adequately captured all underlying physical and chemical process.

REFERENCES

[1]   Eldar Khaliullin, Yaogang Lian, Mark Davey, Xin Zhou, "What is a good empirical model?", Proc. SPIE 7488, 74883P-1 (2009).
[2]   Samit Basu, Yoram Bresler, "The Stability of Nonlinear Least Squares Problems and the Cramér-Rao Bound", IEEE Transactions on Signal Processing, Vol.48, No.12, 3426-3436 (2000).
[3]   Ralph E. Schlief, "Effect of data selection and noise on goodness of OPC model fit", Proc. SPIE 5754, 1147-1158 (2005).

A Simplified Reaction-diffusion System of Chemically-Amplified Resist Process Modeling for OPC

Yongfa Fan[a], Moon-Gyu Jeong[b], Junghoon Ser[b], Sung-Woo Lee[b], Chunsuk Suh[b], Kyo-Il Koo[c], Sooryong Lee[c], Irene Su[d], Lena Zavyalova[e], Brad Falch[e], Jason Huang[a], Thomas Schmoeller[f]

[a]Synopsys Inc., 700 East Middlefield Road, Mountain View, CA 94043, USA
[b]Samsung Electronics, San #16, Banwol-Dong, Hwasung-City, Gyeonggi-Do, 445-701, Korea
[c]Synopsys Inc., 12th FL., Star Tower 737, Yeoksam-dong, Kangnam-gu, Seoul 135-984, Korea
[d]Synopsys Inc., 4F-1, #28, Tai-Yuan street, Chupei City Hsinchu Hsien 302, Taiwan
[e]Synopsys Inc., 1301 S MO Pac Expy, Austin, TX 78746, USA
[f]Synopsys Inc.,Karl-Hammerschmidt-Strasse 34 Europa-Forum-II Building,Aschheim/Dornach,D-85609,Germany

ABSTRACT

As semiconductor manufacturing moves to 32nm and 22nm technology nodes with 193nm water immersion lithography, the demand for more accurate OPC modeling is unprecedented to accommodate the diminishing process margin. Among all the challenges, modeling the process of Chemically Amplified Resist (CAR) is a difficult and critical one to overcome. The difficulty lies in the fact that it is an extremely complex physical and chemical process. Although there are well-studied CAR process models, those are usually developed for TCAD rigorous lithography simulators, making them unsuitable for OPC simulation tasks in view of their full-chip capability at an acceptable turn-around time. In our recent endeavors, a simplified reaction-diffusion model capable of full-chip simulation was investigated for simulating the Post-Exposure-Bake (PEB) step in a CAR process. This model uses aerial image intensity and background base concentration as inputs along with a small number of parameters to account for the diffusion and quenching of acid and base in the resist film. It is appropriate for OPC models with regards to speed, accuracy and experimental tuning. Based on wafer measurement data, the parameters can be regressed to optimize model prediction accuracy. This method has been tested to model numerous CAR processes with wafer measurement data sets. Model residual of 1nm RMS and superior resist edge contour predictions have been observed. Analysis has shown that the so-obtained resist models are separable from the effects of optical system, i.e., the calibrated resist model with one illumination condition can be carried to a process with different illumination conditions. It is shown that the simplified CAR system has great potential of being applicable to full-chip OPC simulation.

Key Words: OPC, modeling, Chemically Amplified Resist.

1. INTRODUCTION

Rapid advances in computing technology have made it possible to perform trillions of computational operations each second on data sets that are sometimes as large as trillions of bytes. These advances can be attributed to the dramatic improvements in semiconductor manufacturing technologies which have made it possible to integrate hundreds of millions of devices onto a single chip. The advent of immersion technique has enabled 193-nm wavelength scanners to image 40 nm half pitch features at a numerical aperture (NA) of 1.35 with a k_1 factor of

Optical Microlithography XXIII, edited by Mircea V. Dusa, Will Conley, Proc. of SPIE Vol. 7640,
764039 · © 2010 SPIE · CCC code: 0277-786X/10/$18 · doi: 10.1117/12.846737

0.28. [1-4] At the same time, double exposure, double patterning etc. are being actively pursued to extend 193-nm immersion lithography further. The main goal of these techniques is to counter the pronounced image fidelity loss since the k_1 factor is being pushed towards 0.25. Therefore Optical Proximity Correction (OPC) has become more and more important and must accompany the RET techniques. [1] In a lithographic process, the final pattern image such as resist image or etched image on a silicon wafer result from a series of sub- processes, which may include effects from mask-writing errors, resist film coating, post-apply-bake (PAB), exposure, post-exposure-bake (PEB), development, etching, etc. Among these sub-processes, PEB is a critical one. [5-6] In this paper, after a brief overview is given on modeling PEB process, a simplified reaction-diffusion system is probed.

2. PHYSCAL DESCRIPTION OF POST-EXPOSURE-BAKE EFFECTS

2.1 Rigorous description

Post-exposure-bake (PEB) is a critical step for a chemically amplified resist process. [5-6] After exposure with 248 nm or 193 nm wavelength of deep ultra-violate (DUV) laser, a latent image of photo-generated acid is preserved in the resist film. During the PEB step that follows, the photo-generated acid modules catalyze the polymer chain de-protections, which in turn make the material soluble in a base developer solution. Diffusion and reaction are the two key components in the PEB process. The system is a classical diffusion-reaction problem in physical chemistry.[4] The photo-generation of initial acid concentration, as described in Equation (1), follows an exponential law, where C is the Dill's C parameter (cm^2/mJ), *dose* is the exposure dose (mJ/cm^2), and *I(x,y)* is the normalized optical intensity. The initial base concentration B_0 is uniform across the wafer (Equation 2). Equation (3) describes the acid catalyzed deprotecting reaction, where P is the protected site concentration, k_r is the reaction rate constant (1/s), H is the acid concentration. Equation (4) describes the consumption of acid by neutralization reaction (quenching) and acid diffusion, where k_l is the quenching rate constant (1/s), B is the base concentration, D_H is the acid diffusivity (m^2/s). Equation (5) describes the consumption of quencher by the neutralization reaction and base diffusion, where D_B is the base diffusivity (m^2/s). The diffusion in the photoresist is usually a non-Fickian process. Higher concentration of protecting group typically enhances the diffusivity. Equation (6) and (7) define the typical exponential relation between the acid/base diffusivity and the protecting group concentration in the photoresist.

$$H \mid_{t=0} = H_0(x,y) = 1 - \exp[-C \times dose \times I(x,y)] \tag{1}$$

$$B \mid_{t=0} = B_0 \tag{2}$$

$$-\frac{\partial P}{\partial t} = k_r P H \tag{3}$$

$$\frac{\partial H}{\partial t} = -k_l H B + \nabla \cdot (D_H \nabla H) \tag{4}$$

$$\frac{\partial B}{\partial t} = -k_l H B + \nabla \cdot (D_B \nabla B) \tag{5}$$

$$D_H = D_{H0} \exp(w_H \times P) \tag{6}$$

$$D_B = D_{B0} \exp(w_B \times P) \tag{7}$$

A numerical method is often employed to solve the above partial differential equations since there is not a close form solution. Usually, PEB time is divided into multiple steps and reaction and diffusion is calculated in a step-by-step fashion to approximate the physical process. The calculation using the above method is slow, hence not suitable for OPC simulation. For practical OPC simulation, some levels of simplification are conducted. The acid-quencher mutual diffusion model described in the next paragraph is one of the simplified models.

2.2 Acid-quencher mutual diffusion model

Fukuda et al proposed a simple FEB model called acid-quencher mutual diffusion model. [7-8] The Fukuda model considers both acid and quencher diffusion. The optical signal profile is used as the initial acid concentration profile and the initial quench concentration assumes a constant background. Upon heating, both acid and quencher molecules diffuse in accordance with Fick's diffusion law. The acid quencher neutralization reactions occur during the diffusion process. The model assumes that neutralization reactions are fast enough in comparison to the diffusion process. The assumption implies that acid and quencher molecules do not exist simultaneously at one location. If acid concentration is higher than quencher concentration, the neutralization result is that all quencher molecules are neutralized and the concentration of the left over acid is the concentration difference. If quencher concentration is higher than acid concentration, the neutralization result is that all acid molecules are neutralized and the concentration of the left over quencher is the concentration difference. The model also assumes that the diffusion-reaction process can be approximated with multiple time steps. The model is illustrated in detail in Figure 1. for convenience of discussion in the following sections.

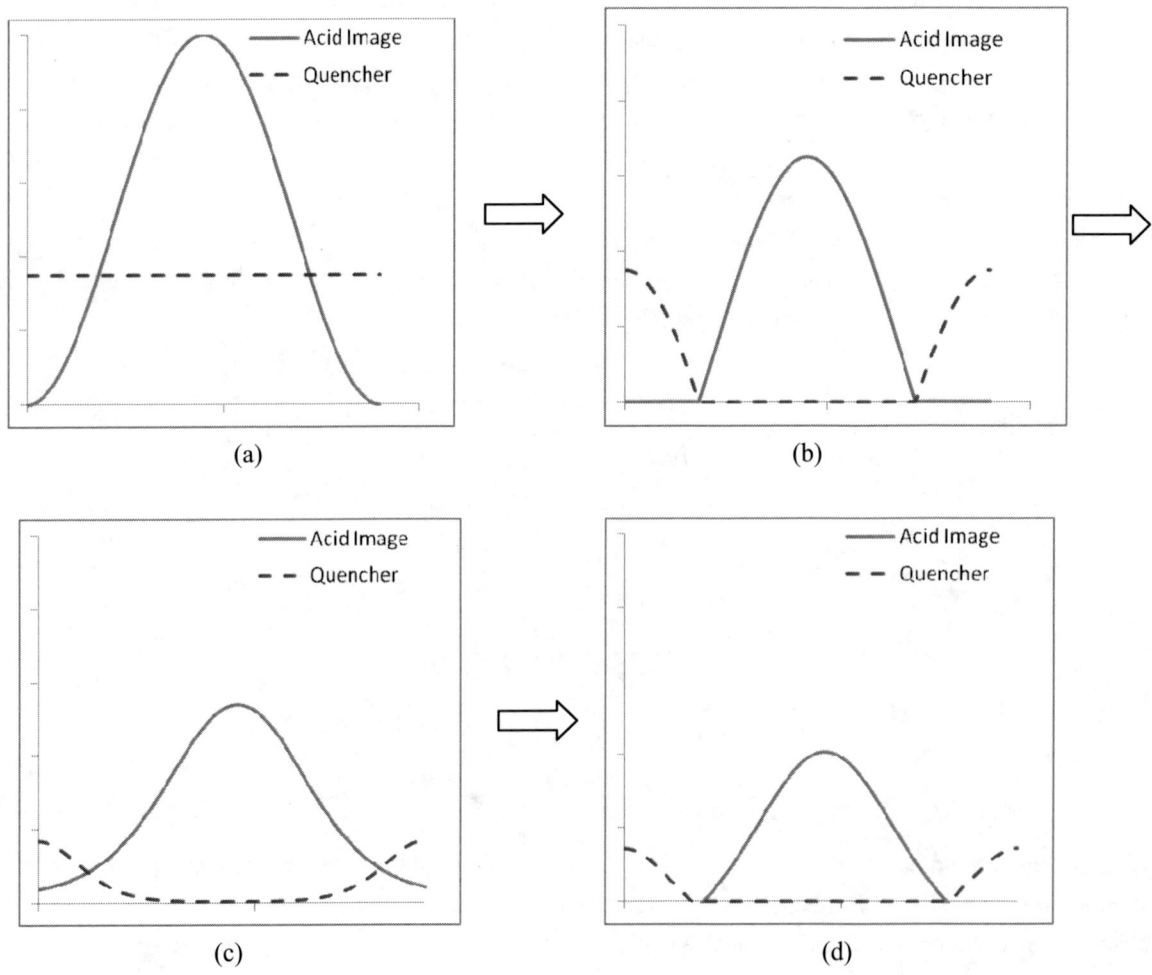

Figure 1. Acid-quencher mutual diffusion model. (a) initial acid and quencher concentration before PEB; (b) neutralization; (c) mutual diffusion; (d) neutralization after mutual diffusion.

3. A NEW SIMPLIFIED REACTION-DIFFUSION MODEL

3.1 Description of the model

The acid-quencher mutual diffusion model described in the above section assumes that neutralization reactions are fast enough in comparison to the diffusion process, implying that acid and quencher molecules do not exist simultaneously at one location. The assumption makes the model form very simple so that it is computationally efficient. However, it is not an accurate description of a real PEB process. We know that resist films are in the solid state. Even with an elevated PEB temperature of 90-120 °C which is around the glass transition temperature of the resist material, the molecules are in a high viscosity environment, making them move very slowly. Hence, it is very unlikely that there are "complete neutralization" zones during the PEB process since it is known that molecular collision is a necessary condition of chemical reaction. A second factor worth consideration is the thermal equilibrium of acid-base reactions. That is, acid and quencher can co-exist in one local zone although they collide with each other millions of times. Therefore, it is logical to argue that the assumption of complete neutralization in the acid-quencher neutral diffusion model may not account for PEB processes accurately.

To address the factors raised in the above paragraph, a new model is proposed in this work to improve modeling accuracy. The new model adopts the idea from the acid-quencher mutual diffusion model that the diffusion-reaction system can be approximated by multiple neutralization-diffusion iterations. Instead of assuming a complete neutralization, the model uses a partial neutralization concept as in a chemical equilibrium. That is, acid and quencher molecules can co-exist in one local area. The actual ratio between acid and quencher molecules after neutralization depends on their initial concentrations. The concept is illustrated in Figure 2 as an example. The initial acid and quencher profiles are plotted in Figure 2 (a). Upon neutralization, the profiles are changed to the plots in Figure 2 (b). In areas where acid concentration is much higher than quencher concentration, quencher molecules are completely neutralized. In areas where quencher concentration is much higher than acid concentration, acid molecules are completely neutralized. In areas where none of them are predominant, both acid and quencher molecules exist. It is noticed that the acid and quencher profiles become smoother and broader by encroaching into each other's territories. Followed by an acid/quencher mutual diffusion, the acid and quencher profiles in Figure 2(b) become plots in Figure 2(c). The acid profile in Figure 2 (c) can be used to model a chemically amplified resist process for OPC. If not accurate enough, a second neutralization process can follow as shown in Figure 2 (d), which in turn can be operated with an acid/quencher mutual diffusion process. In theory, multiple iterations will make the approximation close to the actual chemical/physical process. However, with real wafer data it is found that after one or at most two iterations it will reach a saturated state. That is, no more accuracy is observed with increasing the neutralization/diffusion iterations. The results will be discussed in section 4.0.

The aforementioned partial neutralization approach is distinguished from the complete neutralization approach in the acid-quencher mutual diffusion model. One approach to estimate partial neutralization is described as follows. The kinetic equation for acid concentration obeys the following equation:

$$\frac{\partial H}{\partial t} = -k_l H B + \nabla \cdot (D_H \nabla H) \tag{4}$$

Assuming that neutralization reaction is independent of diffusion, the above equation becomes:

$$\frac{dH}{dt} = -k_l H B \tag{8}$$

If the acid concentration is much higher than quencher, it can be assumed that quencher molecules are consumed completely, then the acid concentration becomes:

$$H = H_0 - B_0 \tag{9}$$

Equations (8) and (9) can be used to construct acid concentration profiles with partial neutralization approximation.

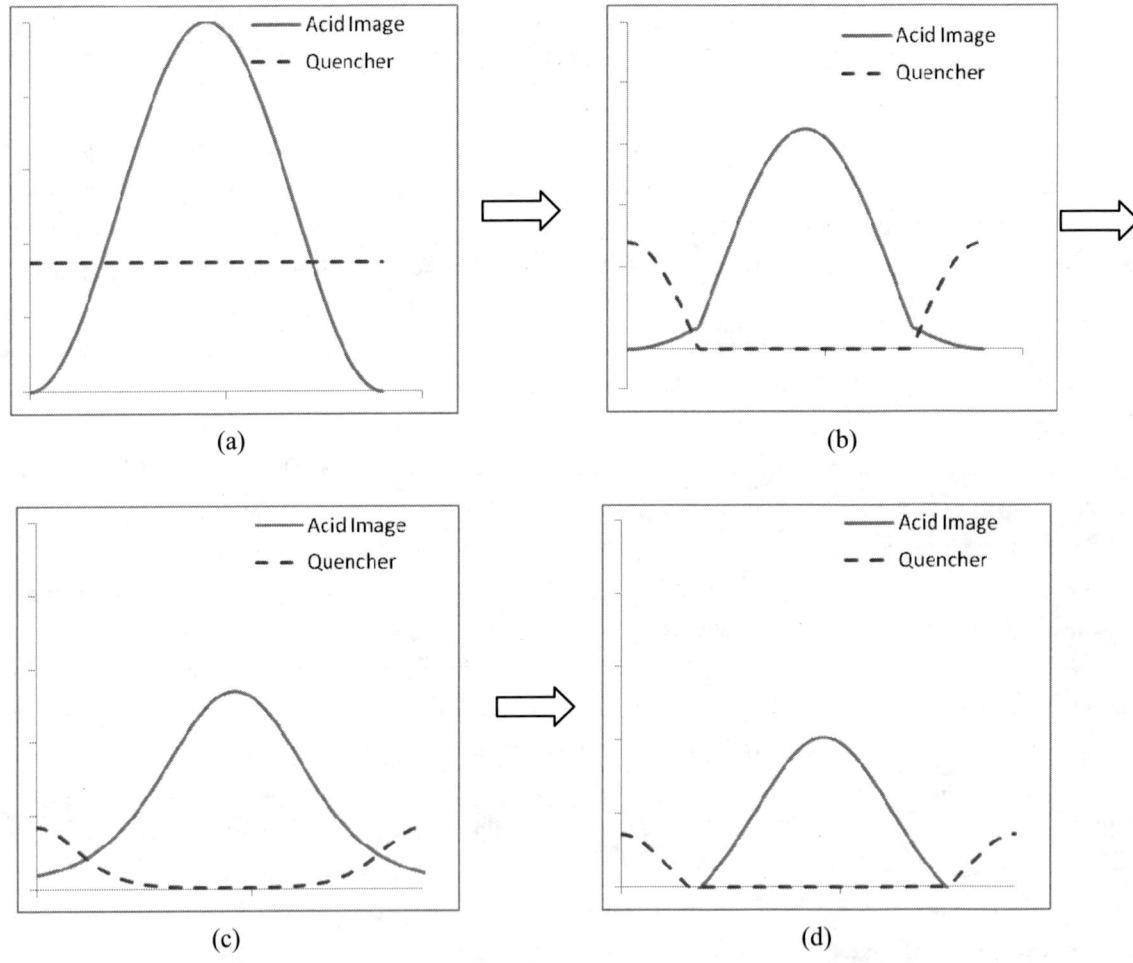

Figure 2. Partial neutralization - diffusion model. (a) initial acid and quencher concentration before PEB; (b) partial neutralization; (c) mutual diffusion; (d) partial neutralization after mutual diffusion.

3.2 Characteristics of the model.

To examine how the model responds to the neutralization mechanisms proposed in the preceding section, it is helpful to look at the trend of model CD with through pitch structures. An example is given in Figure 3. In the plot, model CD is plotted versus pitch as a function of the neutralization factor. The mask CD is 50nm with pitch varying from 100nm to 700nm. It is a 6% transmission phase shifting clear tone mask. With an NA of 1.35, a cross quadrupole illuminator is used in the test. The CD trend of the pure optical image model is plotted in the curve with a diamond legend. A fixed diffusion length of 15.0 nm is used for all the curves and a fixed 0.2 background of quencher is used. Pitch of 250nm is used as an anchor structure for targeting a 50nm print CD. As the neutralization factor m (a factor of relative neutralization) varies, the CD trend deviates from optical model trend. At $m=0$, the model is reduced to the Fukuda Acid-Quencher Mutual Diffusion model. At $m=1$, the model is essentially a diffused optical model without any neutralization involved. For m being between 0.0 and 1.0, the concept of partial neutralization is used in the model, with m being correlated to the degree of partial neutralization.

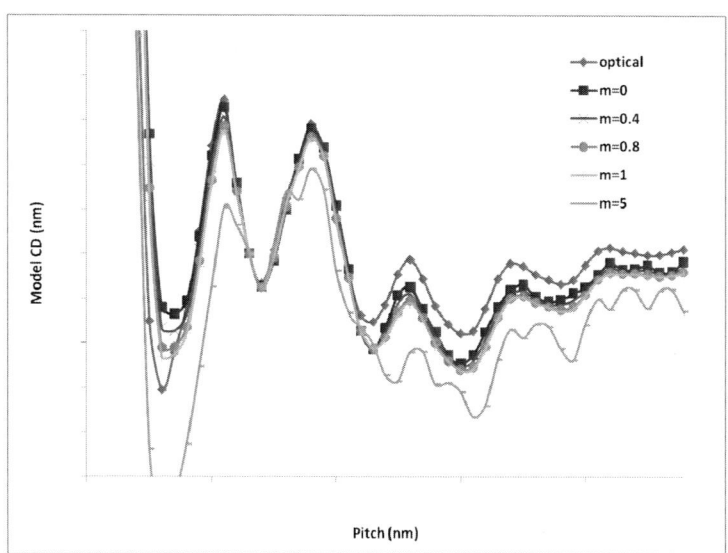

Figure 3. Model CD is plotted versus pitch as the neutralization factor varies. The mask CD is 50nm with pitch varying from 100nm to 700nm. The mask is a 6% transmission phase shifting clear tone mask. NA of system is 1.35. A cross quadrupole illuminator is used .

As the neutralization factor varies, the CD trend is very interesting. At $m=0$, the model CD is lower than optical image model CD for pitches larger than anchor pitch and higher than optical image CD for pitches lower than anchor pitch. As the factor m increases, the model CD decreases for pitches larger than anchor pitch and pitches lower than anchor pitch. In other words, partial neutralization drives the model to behave more like optical models at lower pitches but deviate more from optical models for larger pitches. The observation is consistent with the analysis that the complete neutralization mechanism overestimate the neutralization effects since the dense features are more affected by the over neutralization. Another interesting case is with factor m of 5.0, which makes the model CD lower than optical model CD for pitches other than the anchor pitch.

The example given above indicates that this new model is suitable for simulating quenching/diffusion processes for chemically amplified resist. The partial neutralization mechanism gives model the ability to handle more complex situations than the Acid-Quencher Mutual Diffusion model.

4. EXPERIMENTAL STUDY OF THE MODEL

In this section, actual wafer data are used to validate the model proposed in Section 3. The model is first validated on a nominal data set using fit residual RMS and CDSEM image overlay as model error criteria. The second test is model portability validation (resist model calibrated under one illumination condition should be able to be extrapolated to other illumination conditions).

4.1 Model fit with a nominal wafer data set

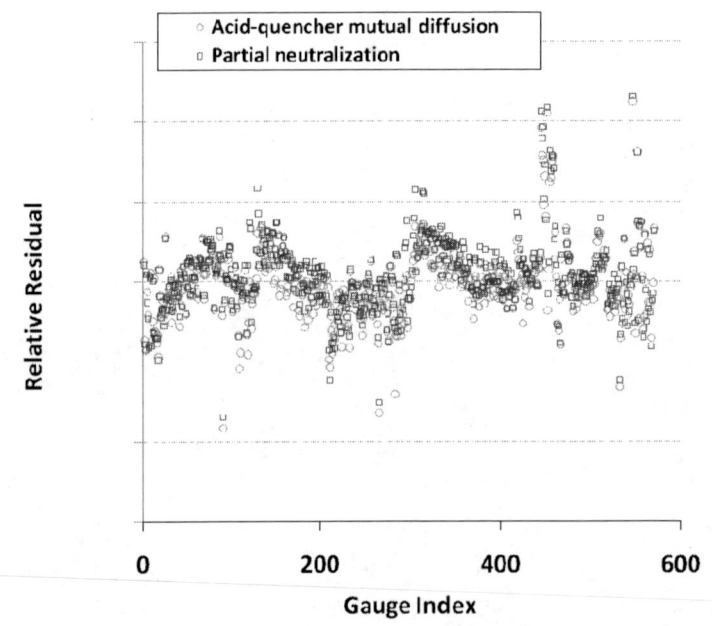

Figure 4. Relative Residual versus gauge index. Circle: Fukuda model. Square: model by this work.

An actual 193 nm wafer process with an NA of 1.35, an azimuthally polarized cross quadrupole illuminator is used for calibration. It has a 6% transmission phase shifting clear tone mask. The diffusion length is regressed to 13.1nm, quencher background regressed to 0.152 and the m factor regressed to 0.0706. Model relative residual (model residual normalized to measurement CD) is plotted versus gauge index in Figure 4. The results for Acid-quencher-mutual-diffusion model are plotted for comparison. It is shown that the relative model residuals are confined between +/-0.01 except a few possible outliers and partial neutralization method achieved better results. Root mean square (RMS) of relative residuals are summarized in Table 1 for various gauge classifications for the two models.

In addition to error statistics, the model is also validated with CDSEM image overlays, as shown in Figure 5. The model contours are in red. In Figure 5 (a), there are two bridging sites and one pinching site. The model contour precisely predicts those critical problematic sites. In Figure 5 (b), there is one hard pinching site and a few potential pinching sites, which are successfully caught by the model as well. The results show that the chemically amplified resist process is precisely modeled using the proposed quencher/diffusion model.

Table 1. Relative error RMS for various gauge classes.

Gauge Class	GaugeType	Fukuda	Partial Neutralization
All	1D/2D	0.0233	0.0217
Class 1	1D	0.0199	0.0186
Class 2	1D	0.0162	0.0107
Class 3	1D	0.0606	0.0553
Class 4	1D	0.0096	0.0112
Class 5	1D	0.0185	0.0161
Class 6	2D	0.0106	0.0138
Class 7	2D	0.0330	0.0322

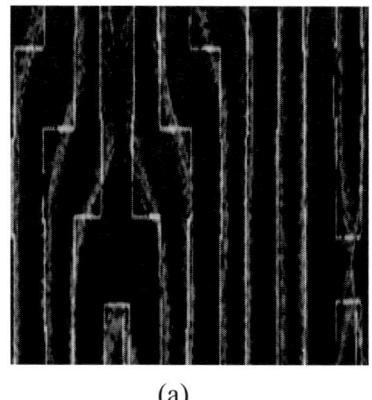 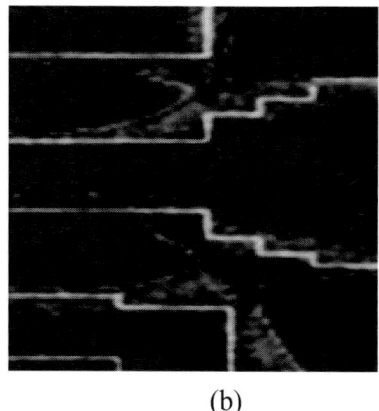

(a)   (b)

Figure 5. Model contour overlays on CDSEM images.

4.2 Portability test of the model

Assuming that resist effects can be decoupled from optical effects, the resist model calibrated under one illumination condition should be able to be extrapolated to other illumination conditions with the same resist process. A resist model possessing such property is called to be portable or separable. The following two schemes are used to conduct the portability test. In Scheme 1, a model is calibrated with wafer data printed using an azimuthally polarized cross-pole illuminator at an NA of 1.35. The model is assessed with wafer data for fit residual analysis. The optics and resist parameters are noted down. Then the cross-pole illuminator is replaced with an annular illuminator to extrapolate a new model. The new model is assessed with wafer data printed with the annular illuminator for fit residual analysis. Scheme 2 is a same test but the other way around. In Scheme 2, the model with an annular illuminator is first calibrated then it is extrapolated to a cross-pole illuminator.

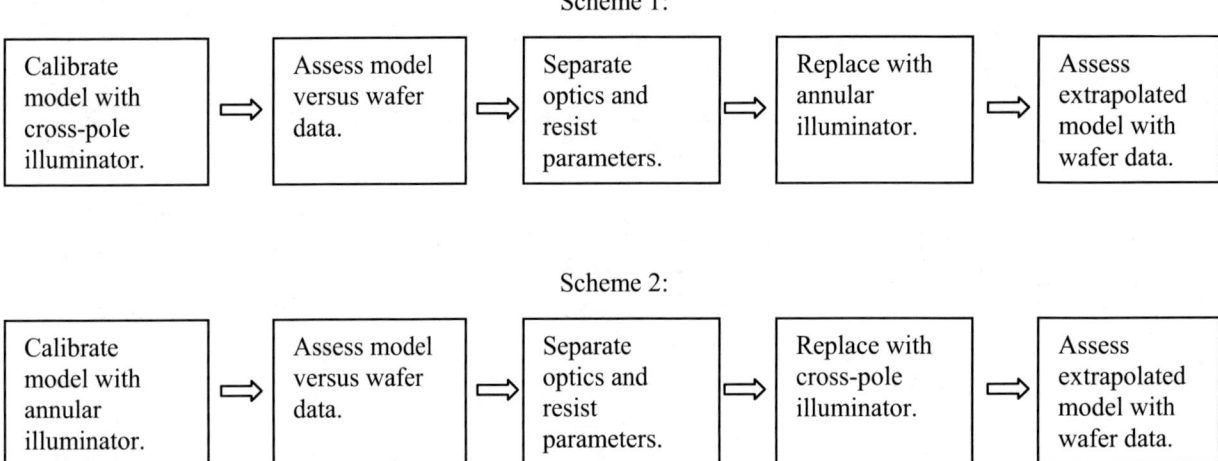

The test results with Scheme 1 are summarized in Figure 6 where relative model residual is plotted versus test gauge index. The diamond legend is for calibration results with annular illumination and the square legend is for cross-pole illumination with resist model extrapolated from annular illumination. It shows that the extrapolated resist model from annular illumination still works well with cross-pole illumination. The model residuals are comparable to each other. The test results with Scheme 2 are summarized in Figure 7 where relative model residual is plotted versus test gauge index. The diamond legend is for calibration results with cross-pole illumination and the square legend is for annular illumination with resist model extrapolated from cross-pole illumination. It shows that the extrapolated resist model from annular illumination still works well with cross-pole illumination. The model residuals are comparable to each other. The two tests demonstrate that the proposed diffusion/quenching model is capable of being extrapolated to different optical settings. In other words, the model is separable or portable.

5. CONCLUSIONS

It is challenging to model the process of Chemically Amplified Resist (CAR) with a compact model form suitable for full-chip simulation. The difficulty lies in the fact that it is a complex physical and chemical process. The acid-quencher mutual diffusion model developed by Fukuda et al is a compact form CAR model and has shown promising results. In this work, a partial neutralization concept is introduced to improve the acid-quencher mutual diffusion model. Analysis has shown that partial neutralization can help the model to simulate complex reaction/diffusion system. This method has been tested to model numerous CAR process with wafer measurement data sets. Model residual of 1nm RMS and superior resist edge contour predictions have been observed. Analysis has shown that the so-obtained resist models are separable, that is, the calibrated resist model with one illumination condition can be carried to a process with different illumination conditions. It is shown that the simplified CAR system has great potentials of being applicable to full-chip OPC simulation.

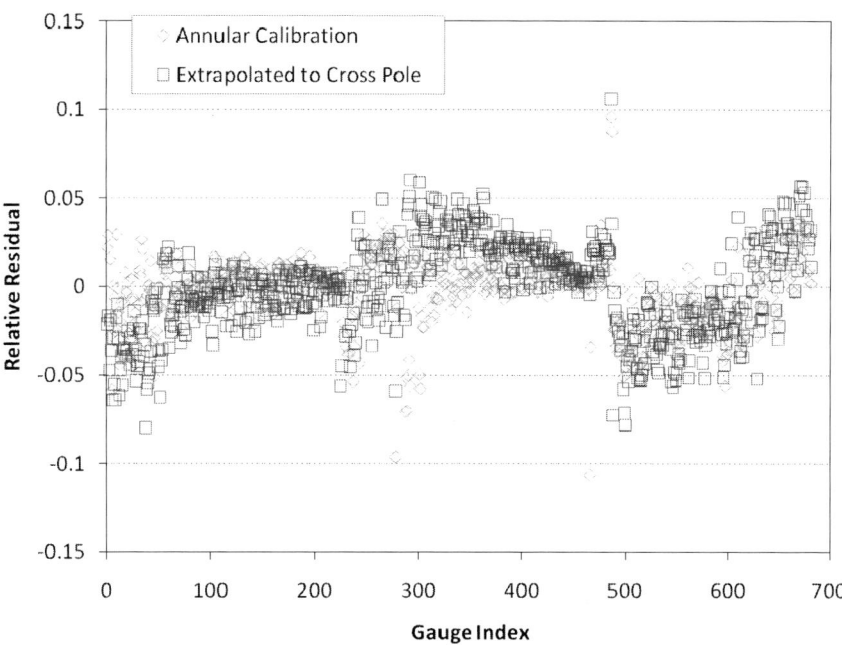

Figure 6. Relative model residual is plotted versus test gauge index. Diamond legend: calibration results with annular illumination. Square legend: cross-pole illumination with resist model extrapolated from annular illumination.

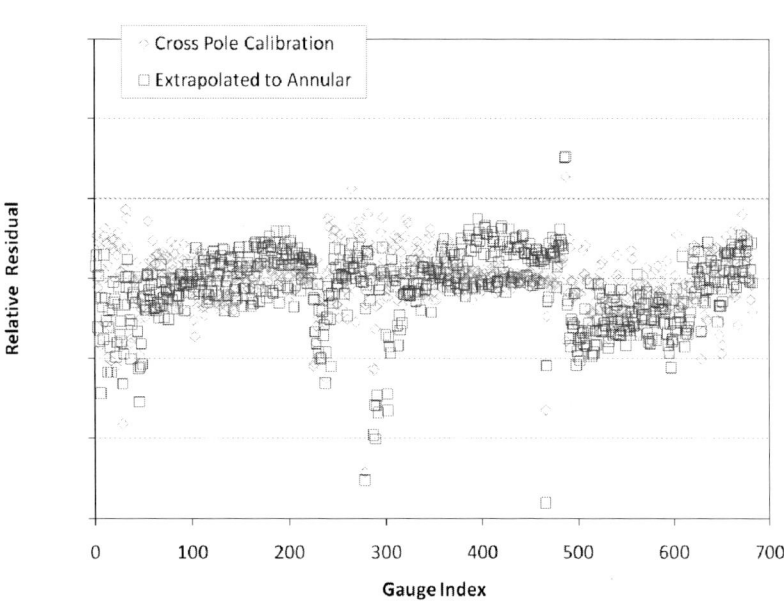

Figure 7. Relative model residual is plotted versus test gauge index. Diamond legend: calibration results with cross-pole illumination. Square legend: annular illumination with resist model extrapolated from cross illumination.

REFERENCES

[1] F. M. Schellenberg, Luigi Capodieci, Bob Socha , "Adoption of OPC and the impact on design and layout", IEEE Design Automation Conference, 38th, 89-92 (2001)

[2] B. Fay, "Advanced optical lithography development, from UV to EUV," Microelectronic Engineering 61-62, 11-24 (2002).

[3] B. W. Smith, H. Kang, A. Bourov, F. Cropanese, Y. Fan, "Water immersion optical lithography for the 45 nm node," in Optical Microlithography XVI, A. Yen eds. Proc., SPIE 5040, 679-689 (2003).

[4] B. W. Smith, Y. Fan, M. Slocum, L. Zavyalova, "25 nm immersion lithography at 193 nm wavelength," B. W. Smith ed., in Optical Microlithography XVII, B. W. Smith ed., Proc. SPIE 5754, 141-147 (2005).

[5] C. Mack, "Modeling solvent effects in optical lithography", Dissertation, 3-32 (1998)

[6] E. Barouch, B. Bradie, V. Babu, "Simulation of three-dimensional positive photoresist images", Jpn J. of App. Phys., Vol. 28, No. 12, 2624-2628 (1989)

[7] H. Fukuda, K. Hattori, T. Hagiwara, "Impact of acid/quencher behavior on lithography performance", in Optical Microlithography XIV, C. J. Progler ed., Proc. SPIE 4346, 319-330 (2001)

[8] K. Hattori, J. Abe, T. Hagiwara, "The accuracy of simulation based on the acid-quencher mutual diffusion model in KrF processes", in Optical Microlithography XV, A. Yen ed., Proc. SPIE 4691, 1243-1253 (2002)

Improved Process Window Modeling Techniques

Christian Zuniga and Tamer Tawfik

Mentor Graphics Corp., 1001 Ridder Park Dr. San Jose Ca USA

ABSTRACT

The continuous reduction of device dimensions and densities of integrated circuits increases the demand for accurate process window models used in optical proximity correction. Beamfocus and dose are process parameters that have significant contribution to the overall critical feature dimension error budget. The increased number of process conditions adds to the model calibration time since a new optical model needs to be generated for each focus condition. This study shows how several techniques can reduce the calibration time by appropriate selection of process conditions and features while maintaining good accuracy. Experimental data is used to calibrate models using a reduced set of data. The resulting model is compared with the model calibrated using the full set of data. The results show that using a reduced set of process conditions and using process sensitive features can yield a model as accurate as the model calibrated using the full set but in a shorter amount of time.

Keywords: Lithography, OPC model, process window, calibration, Bossung, focus

1. INTRODUCTION

In a modern photolithographic system, the overall critical dimension (CD) error budget, across process window (PW), is due to the contribution of multiple sources including lens aberrations, source non-uniformity, or mask errors[1]. The process window is the region of focus and dose values that keeps the CD within the required tolerance. Modern processes continue to push the CD to smaller values, causing the process window to shrink as well.

Optical proximity correction (OPC) models need to accurately predict CD variations across the process window. Traditional OPC models have been calibrated with CD data at nominal conditions (at best dose/best focus). At large technology nodes (90 nm and above), this nominal model is good enough. For smaller technology nodes, however, the faster variation of CD with focus and dose causes the nominal model to cover a smaller process window than is required. Process window models become essential for the 45 nm node and below[2, 3, 4]. Figure 1 shows the focus exposure matrix (FEM) of an experimental data set (small circles) and two models. Figure 1 shows that a model calibrated with nominal data only (dashed lines) may predict CD values well at other process conditions only if the focus and dose range is small. Eventually the predicted values diverge from the actual measured value. Figure 1 also shows that a model calibrated with process window data (solid lines) tracks the CD variations over a greater range than the nominal model.

The nominal model does not do as well because of two main reasons. One reason is that beamfocus, the position within the resist of the Gaussian image plane, is not known beforehand and must be calibrated. Calibration with only best focus data may not adequately determine beamfocus. The second reason is due to the semi-empirical nature of the resist model. The model can fit CD data well at given process conditions, but will not be as good at extrapolating to process conditions further away from the one used for calibration. These limitations introduce the need to calibrate OPC models using process window data.

Optical Microlithography XXIII, edited by Mircea V. Dusa, Will Conley, Proc. of SPIE Vol. 7640,
76403A · © 2010 SPIE · CCC code: 0277-786X/10/$18 · doi: 10.1117/12.846792

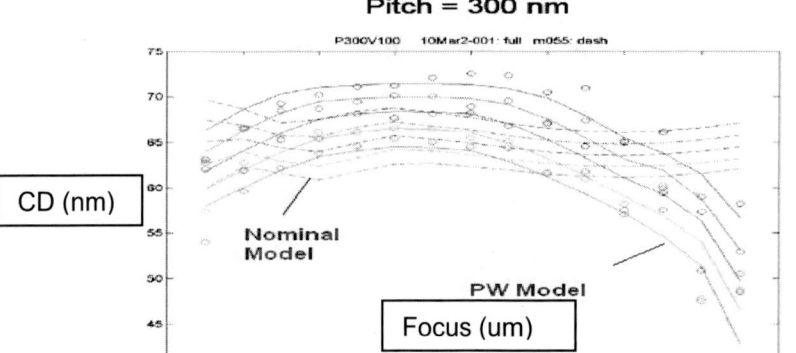

Pitch = 300 nm

Figure 1: Comparison of a nominal model and a process window model.

Process window data consists of a range of CD values for a given feature under different dose and focus values. This data forms the focus exposure matrix (FEM data) of a feature. Although using all of the data to calibrate a model can give the required accuracy, it will also result in a significant increase in calibration time. For every focus condition, a new optical model has to be created. Fortunately, it is not necessary to use all of the data. It is sufficient to use FEM data from a few focus sensitive features like isolated lines to calibrate beamfocus and defocus start. Figure 2 illustrates schematically the reduction in calibration time when using fewer process conditions. Other optical parameters can be calibrated with nominal data only. Once these are calibrated, the optical model can be fixed for the resist calibration. The resist model can be calibrated only with a set consisting of nominal data and process corner data, either at the principal axes or the diagonal dose and focus axes. Nominal data will usually have most of the measurements so it is necessary to include it in the calibration.

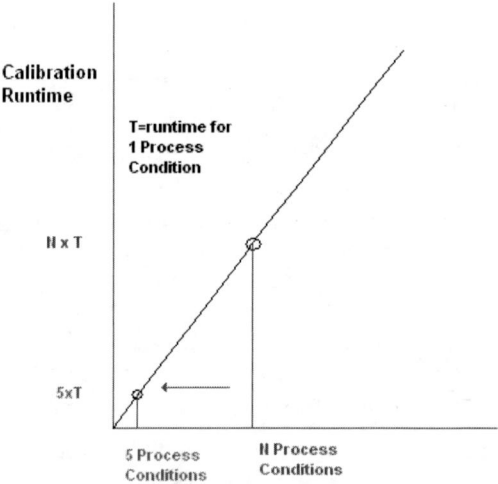

Figure 2: Calibration time vs. number of process conditions.

The next sections describe how to calibrate a process window model using a smaller subset of all the data. The authors used Mentor Graphics proprietary Calibre software to calibrate all models. Section 2 reviews the traditional calibration methods and types of models used. Section 3.1 introduces the process window calibration flow. Section 3.2 describes the actual selection process of the focus sensitive data for optical calibration. Section 3.2 demonstrates using the nominal and process corner data gives a resist model with comparable accuracy as a model calibrated with all the data. Finally Section 4 summarizes the results.

2. OPC MODEL CALIBRATION

OPC models are made of two components, an optical model and a resist model. The optical model takes the mask description and produces its corresponding intensity inside the resist. It is based on physical parameters like NA, wavelength, and beamfocus. The resist model takes the intensity from the optical model and produces a CD. Although the resist model is motivated by the physical process, it is more of an empirical model and requires more extensive calibration than the optical model. Generally an OPC model is calibrated by taking multiple SEM measurements of a representative set of test patterns and calibrating the model parameters to make the simulated CDs match the measurements as best as possible. Figure 3 shows a representative calibration flow. Usually the measurements are at best dose/ best focus. However, the smaller process window size of the latest nodes requires measurements at several dose/focus conditions to ensure the models are predicting these process variations well.

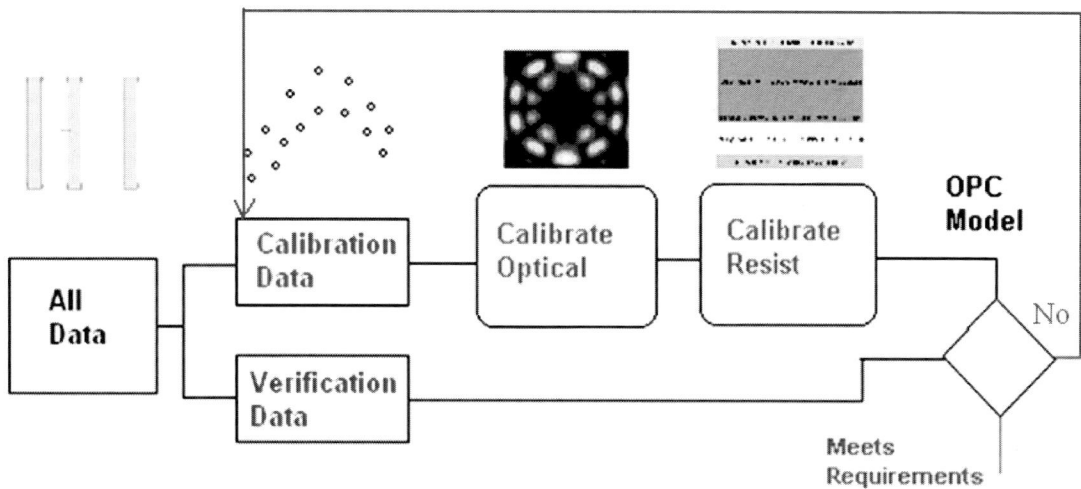

Figure 3 General OPC model calibration flow

The optical model is based on Hopkins' imaging formulation and describes the optical system through a four-dimensional filter called the transmission cross coefficient or TCC. The TCC is decomposed into its eigenfunctions and produces the intensity through a sum of convolutions of the eigenfunctions or kernels with the mask [5, 6]. The optical model accounts for vector, thin-film, and pupil effects. Every optical model has two parameters, beamfocus and defocus start. They specify the location of the Gaussian image plane in the resist and the location in the resist at which the CD is extracted, respectively. Beamfocus is an actual physical parameter but is not generally known at the OPC calibration stage. Beamfocus correlates with the focus parameter in the scanner and equal changes in focus correspond to equal changes in beamfocus. The value of zero beamfocus usually corresponds with the best focus in the scanner. The optical model tracks changes in focus by changing beamfocus. Defocus start is more of a fitting parameter but necessary to extract the simulated CDs. Beamfocus and defocus start are usually correlated[7]. Every different pair of these parameters requires building a new optical model through a new TCC and eigenfunction decomposition. Dose changes are modeled by a simple rescaling of the intensity so they do not require a new TCC. These parameters could be calibrated using only nominal data. However, the resulting values do not necessarily produce a good model. The problem lies in the fact that the calibration blindly proceeds to vary the model parameters to minimize the modeling error. In this context, beamfocus and defocus start become another set of fitting parameters that can be used to reduce the calibration error. Accurately calibrating beamfocus and defocus start requires using process window data where the changes in CD are known to come from changes in focus and dose.

The resist model falls under the category of compact models that are based on a series of operations on the intensity [8, 9]. These operations usually consist of convolutions and truncations of the intensity. The final resist intensity or latent image is formed by the sum modified intensities. The final CD is extracted by the measuring the width of the modified intensity at a particular level or threshold. Many different types of features including lines, spaces and line ends are needed to calibrate a resist model. Compact models have been very successful for full chip correction and verification. Although the resist model could be calibrated at nominal conditions only, a more accurate model is obtained when using process window data as seen in Figure 1.

3. PROCESS WINDOW MODEL CALIBRATION

3.1 Calibration flow

Figure 4 shows the suggested calibration flow for creating OPC models. Focus sensitive features are first used to calibrate beamfocus and defocus start. Once these are found, other optical parameters can then be calibrated using nominal data. Finally, the resist model is calibrated using data at nominal and process corner conditions.

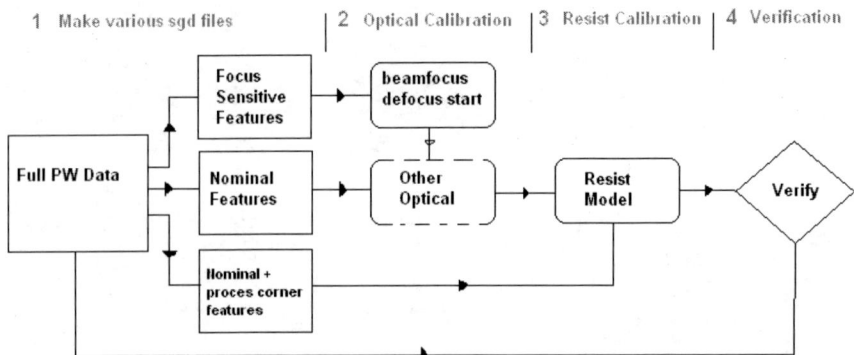

Figure 4 Process window calibration flow

3.2 Optical model calibration

The optical calibration essentially consists of calibrating beamfocus and defocus start. It is not necessary to use all of the process window measurements to determine beamfocus and defocus start. Using the most focus sensitive features can adequately determine them. The features can be ranked in order of focus sensitivity by fitting a quadratic function of focus to the CD measurements of the form:

$$CD_{meras} \approx a_0 + b_0 f + Qf^2$$

The coefficients are found through a least-squares fit of the measured CDs. Although higher order polynomials can also be used, a quadratic usually gives a low enough fitting error[1, 10]. The quadratic coefficient Q gives a measure of the focus sensitivity of each feature. Features with a higher absolute value of Q are more focus sensitive. Figure 5 shows an example of the features that are classified as most focus sensitive and least focus sensitive by this procedure. It is worthwhile to include enough features to eliminate the effect of measurement errors in the data and variations in best focus among the features. Selecting 10% of the features with the highest |Q| values is generally good enough. There is no exact number since the final verification RMS error varies slightly with different number and type of features. Figure 6 shows how the final verification RMS error decreases as more focus sensitive features are included in the beamfocus/defocus start calibration for a 45 nm poly process. The figure shows that using the full set of 73 features gives little improvement in accuracy. Using only 10% of the features for optical calibration cuts down on the calibration time but retains the accuracy.

In reality no feature is completely insensitive to focus. Every feature has a finite Q value. It would also be conceivable to select 10% of the features in a random fashion and avoid the overhead of fitting the quadratic to every feature. However, it makes little sense to leave the selection to chance since many times features with low sensitivity may be included and produce optical models with less than optimal beamfocus and defocus start values. The resist model will then have to compensate for the errors in focus and give a final model with less predictive power. Fitting the features with a polynomial has the added advantage of establishing a method of eliminating outliers and of doing process window analysis. It is much better to select the most focus sensitive features. Once beamfocus and defocus start have been calibrated, additional optical parameters like pupil apodization and image diffusion can be calibrated.

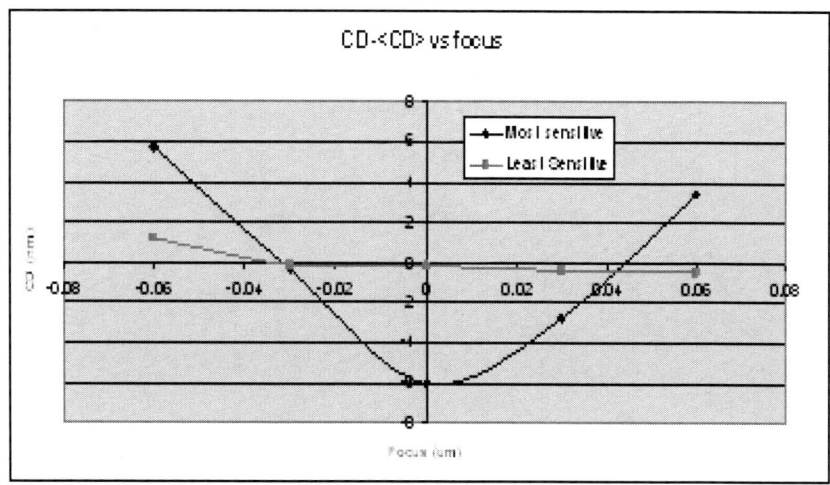

Figure 5: Most focus sensitive and least focus sensitive features at best dose.

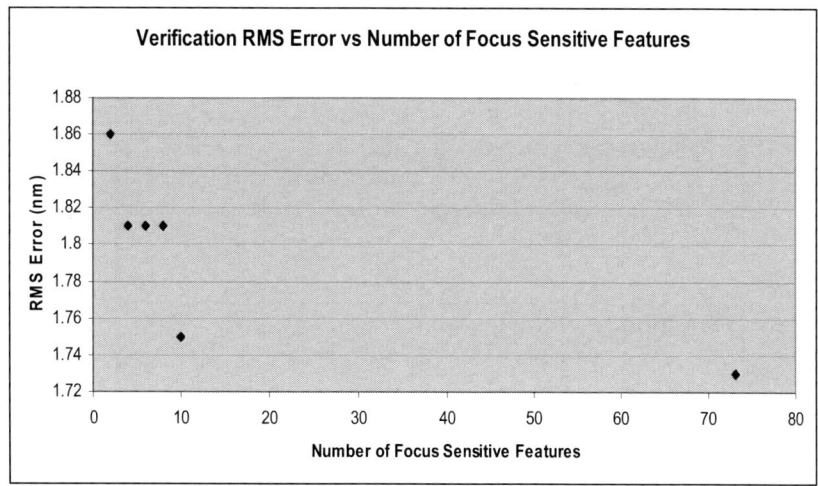

Figure 6: Verification RMS error vs. the number of focus sensitive features

3.3 Resist model calibration

Once the optical model is calibrated, it is fixed for resist calibration. Although the resist model could be calibrated using only nominal data, using process corner data in addition to the nominal data gives a more accurate model. As mentioned, the semi-empirical nature of the resist model increases the need for more calibration points. The process corner data lie at the extremes of the expected process window. Figure 7 shows two possible sets of process corner data, those along the diagonal axes (red) and those along the principal axes (blue). It is not desirable to use all of the process conditions since each one would require building a new TCC and also add to the final number of gauges. Furthermore, many more SEM measurements would be needed. Fortunately, using only the nominal and process corner data combined with correctly establishing beamfocus produces a continuous model of focus and dose that can be used to predict the measurement at any of the process conditions. Nominal data is necessary since it usually has most of the measurements. Table 1 shows a comparison of the verification error on the 55 process conditions shown in Figure 7 of 30 line space patterns. Table 2 shows the results for a second 45 nm poly process confirming that using the nominal and process corner data give a model with good accuracy. In addition, the table shows that the optical parameters do not differ significantly.

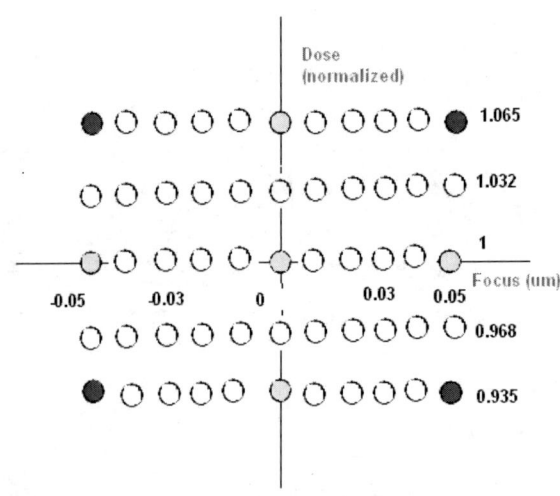

Figure 7 Nominal and process corner data locations

Table 1 Verification errors using process corner conditions

Verification on all gauges	Full (9 conditions)	Diagonal (RED 5 conditions)	Principal (BLUE 5 conditions)
RMS (nm)	2.42	2.42	2.22
MAX (nm)	12.82	12.61	10.68

Table 2 Verification error on 15 process conditions

Verification	FULL From 15 conditions	RED From 5 conditions	BLUE From 5 conditions
RMS	1.80	1.82	1.73
MAX	12.96	12.38	10.62
Beamfocus (nm)	113.6	114.7	115.2
Defocus_start (nm)	146	146	142

4. SUMMARY

Current processes require OPC models that can accurately predict dose and focus changes. This paper showed how process window models can be calibrated faster by properly selecting subsets of the process window data. First, the optical model is calibrated using FEM data from a few focus sensitive features. Then, the resist model is calibrated using nominal and process corner data. The final OPC model using this process gives the same level of accuracy as using the entire set of data.

REFERENCES

[1] Mack, *Fundamental Principles of Optical Lithography* John Wily & Sons Copyright 2007
[2] Borjon, Belledent, Trouiller, Patterson, Lucas, Couderc, Sundermann, Urbani, Baron, Rody, Gardin, Foussadier, and Schiavone " Through Process Window Resist Modeling Strategies for the 65 nm Node" Proc. SPIE Vol. 5992 599219 2005
[3] Torres, Roessler, and Granik "Process Window Modeling Using Compact Models" Proc. SPIE Vol. 5567 2004 pp. 638-548
[4] Zhang, Croffie, Fan, Li, Lucas, Falch, and Melvin "Process Variation Aware OPC Modeling for Leading Edge Technology Nodes" Proc. SPIE Vol. 7275 72751J 2009
[5] Cobb, Zakhor, and Miloslavsky " Mathematical and CAD Framework for Proximity Correction" Proc. SPIE Vol. 2726 pp. 208-222 1996
[6] Adam, Granik, Torres, and Cobb "Improved Modeling Performance with an Adapted Vectorial Formulation of the Hopkins Imaging Equation" Proc. SPIE Vol. 5040 pp. 78-91 2003
[7] Zuniga and Tawfik "New Approach to Determine Best Beamfocus" Proc. SPIE 7274 72742S 2009
[8] Granik, Cobb, and Medvedev "Application of CM0 resist model to OPC and verification" Proc. of SPIE 6154 (2006) 61543E-1
[9] Granik, Medvedev, and Cobb "Toward standard process models for OPC" Proc. of SPIE 6520 (2007) 652043

[10]   De Bisschop, Lalovic and Trintchouk "Impact of finite laser bandwidth on the critical dimension of L/S structures" J. Micro/Nanolith MEMS MOEMS 7(3) 033001 (Jul-Sep 2008)

Lithography Cycle Time Improvements Using Short-Interval Scheduling

David Norman, Scott Watson, Michael Anderson, Steve Marteney, Ben Mehr

Applied Materials Incorporated, 5245 Yeager Road, Salt Lake City, UT USA 84116

1 INTRODUCTION

Partially and fully automated semiconductor manufacturing facilities around the world have employed automated real-time dispatchers (RTD) as a critical element of their factory management solutions. These RTD has allowed manufacturers to consistently implement complex dispatching decisions, react in real time to frequent unplanned events, and to continually improve their lot selection policies.

The success of RTD is attributable to a detailed and extremely accurate data base that reflects the current state of the factory, consistently applied dispatching policies and continuous improvement of these dispatching policies.

The initial results of RTD solutions were extremely compelling: 10 percent reductions in product cycle time, increased factory throughput, and increased predictability of both manufacturing capacity and order delivery.

However, many manufactures are now reaching the benefit limits of pure dispatching-based or other "heuristic-only" solutions. A new solution is needed that combines locally optimized short-interval schedules with RTD policies to target further reductions in product cycle time.

This paper describes an integrated solution that employs four key components:

1. real-time data generation,
2. simulation-based prediction,
3. locally optimized short-interval scheduling, and
4. schedule-aware real-time dispatching.

The authors describe how this solution was deployed in lithography and wet / diffusion areas, and report the resulting improvements measured.

This integrated approach is the result of lessons learned from installations of Applied Materials Real-Time Dispatcher™ in over 90% of all worldwide 300mm semiconductor wafer fabs, and the collaboration with customer partners.

One key finding is that any solution must integrate with existing manufacturing systems and adapt quickly to the wide variety of manufacturing practices and business metrics that are unique characteristics of semiconductor manufacturing.

The authors assert that achieving measurable manufacturing benefits is not merely a matter of a better technical solution (that is, an optimized schedule). It is also vital that the fab operations understand, trust, and follow the schedule.

In early implementations of RTD and trial implementations of short-interval scheduling, it became clear that fostering trust in the schedule rapidly was a fundamental requirement for a better scheduler/dispatch solution. Trust could be achieved by providing more detailed insight in to the scheduler's decision making process, which offers more finite control over the data model, objective functions, and scheduling constraints. The authors describe their experience, data, and effort required to systematically create the trust needed to implement the generated schedules.

Optical Microlithography XXIII, edited by Mircea V. Dusa, Will Conley, Proc. of SPIE Vol. 7640,
76403B · © 2010 SPIE · CCC code: 0277-786X/10/$18 · doi: 10.1117/12.848442

We also examine the potential for other "schedule-aware" components in the manufacturing system. When all of the systems that orchestrate fully automated fabs are aware of the detailed manufacturing schedule, they can make local optimizations that further improve factory throughput and reduce product cycle time. System Architecture

The integrated short-interval scheduler incorporates four primary processes that are essential, both from a technical perspective and to fulfill the requirement to integrate with existing manufacturing systems, without requiring major changes to those systems. These include efficient data preparation, prediction, scheduling, and dispatching.

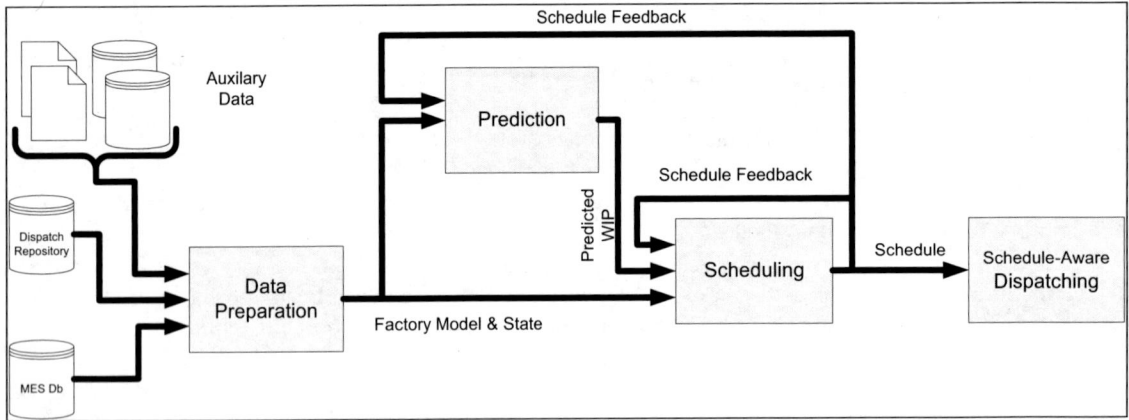

Figure 1 Applied Short-Interval Scheduler Functional Pipeline

Figure 1, shows a pipeline of the key functional components and feedback loops used for each generation of a scheduling solution.

1.1 Data Preparation

A prerequisite for any scheduling solution is complete and accurate information. The data pipeline begins with a data generation process that imports detailed information from an **existing** manufacturing execution system (MES), RTD systems, equipment automation and APC systems, automated material handling systems (AMHS), and other auxiliary data systems that are part of the overall factory automation system. This process transforms the information into a common and extremely detailed model used by both the prediction and scheduler components. The section titled, *Data Preparation,* describes the details of this data generation process and the major functional and architectural roles it plays in the solution.

1.2 Prediction

The prediction process shown in figure 1 represents a detailed discrete-event simulation of the manufacturing facility. As its name implies, this process forecasts the work-in-process (WIP) and it's expected when it will arrival in the targeted manufacturing area. The scheduler component uses the prediction results to ensure that lots are not scheduled to process prior to their predicted arrival times. A later section titled, *"Prediction"*, describes details of this process, and contrasts the quality of the schedules produced using a prediction model to other solutions that use simple cycle time calculations.

1.3 Scheduler

The scheduling process produces and optimizes the short-interval schedule for the target area. The scheduler component is responsible for the actual calculations that produce an optimized short-interval schedule for the target area. The scheduler requires three major inputs as shown in the diagram: a detailed model of the current state of the fab, the predicted lot arrival times, and the previously generated schedule.

A later section titled, *"Scheduling"*, describes details of this component, including the constraints and the objective function used in the scheduling and optimization processes.

1.4 Real-Time Dispatching

RTD is an essential part of the scheduling solution. An RTD system such as the Applied Real-Time Dispatcher™ continues to play a crucial role as the component responsible for all "what next?" and "where next?" decisions. When running as part of a scheduling solution the dispatcher is schedule-aware and is responsible for executing according to the schedule generated by the scheduling engine. The dispatcher retains its responsibility to respond to inevitable, and unpredicted, factory events that occur between each successive schedule generation. The section titled, *Schedule-Aware Dispatching*, below describes the modifications in dispatching logic necessary to for it to become "schedule-aware".

1.5 Schedule-Aware Applications

Many of the manufacturing systems used in the fab can take advantage of the short-interval schedules created to optimize targeted areas of the fab. Applied Materials terms these applications as, "schedule-aware", and has begun to develop concepts for these applications that can improve utilization of other fab assets and increase their capacity simply because they are schedule-aware.

2   DATA PREPARATION

The foundation of any good schedule is accurate data. Both dynamic and static data is required to precisely understand the current and future states of systems that will use the schedule. A successful scheduling solution solves the problem of providing the scheduler with accurate, detailed, and real-time data in a way that is extremely efficient and adaptable.

In order to adapt to a variety of complex semiconductor and other high-tech manufacturing environments, the scheduler requires a normalized and very detailed data model. Thus the developers started with a proven data model based on Applied's discrete event simulation capabilities. This data model includes the detail necessary for the prediction and scheduler components and has a history of easily adapting to many different semiconductor manufacturing environments.

The data required for accurate short-interval scheduling can come from many different sources including the site's manufacturing execution system (MES), real-time data feeds channeled to a cached data store used for real-time dispatching, and other auxiliary manufacturing systems. An effective real-time data integration technology is needed to support the scheduling system, including enhancements that expose internal caches; and that have the capability to read and transform data into tables needed for prediction and scheduling.

These enhancements leverage RTD rules and reports, to create and cache the data tables used in the short-interval scheduling solution. This enables RTD to dispatch at a higher level of performance at critical manufacturing times.

Using real-time data from the fab has a number of benefits, starting with the reduced cost of ownership achieved by using existing infrastructure and tools. During early scheduling trials, the project team chose APF real-time data tools to provide real-time data because they were readily available, familiar, and provided access to the data required to develop early prototypes. The team quickly learned, however, that these tools performed even greater critical functionality. The APF tools delivered nearly all of the data needed to develop prediction and scheduling projects, allowing the team to focus on developing scheduling solutions, rather than on data integration, custom development, or data extracts. Consequently, these tools were added as a core component of the scheduling solution.

3   PREDICTION

The new short-interval scheduling solution produces a schedule that spans as much as one factory shift. Therefore, it needs to know about not just the available WIP at the start of the schedule, but also WIP that will arrive at various steps being plotted during the scheduling interval. For example, if a high-priority lot will be arriving at a schedule step two hours after the start of the scheduling interval, the best schedule will arrange for the lot to be processed shortly after those two hours.

Current fab computer integrated manufacturing (CIM) systems occasionally generate rough predictions, For example, it is sometimes useful to check if a high-priority lot will be arriving shortly when trying to decide if a batch should be formed. However, such predictions are typically implemented as ad hoc parts of a dispatching system, which limits their sophistication.

The most common methods currently used are formulated with simple calculations based on average or target cycle time, for example:

Predicted arrival time = Arrival time at previous step + Target cycle time for previous step

Here, the target cycle time might depend on the lot priority or other lot attributes. Because these calculations depend on a single target cycle time, they are inherently flawed. All steps in a fab have a variable cycle time, and most lots do not have a cycle time that exactly matches the target or average cycle time. These simple prediction calculations transform cycle time variability into prediction error. If a step has a highly-variable cycle time, then these simple calculations will inherently have a high error rate. As an example, consider a batching step. Some lots need to wait for other lots to arrive so that a batch can be formed, while other lots have no waiting time for the batch. Yet the calculation above uses the same cycle time for both types of lots, so some lots will arrive much earlier than the predicted time, and some much later.

To avoid these sorts of errors the solution uses a more sophisticated system based on Applied's simulation capabilities. Normally, fab simulation models are run over time periods of weeks or months and the results of thousands of lots and equipment state changes are averaged to generate reports that describe average fab behavior. However, this new solution uses existing simulation technology in a fundamentally new way: The prediction model is initialized with detailed information about the existing state of the factory and then run for a short interval with the goal of producing a run that closely matches the events and decisions that actually occur in the fab over an interval of time. In particular, the model can keep track of when lots arrive for processing at scheduled steps.

It is important to realize that using simulation-based predictions do not have the inherent errors contained in target cycle time-based calculations. Because the simulation can model the factory in as much detail as necessary, it can produce results that respond to the current loading of the fab: the amount of WIP, the current state of equipment, scheduled equipment PM's, product mix, etc. Static calculations do not consider the dynamics of the fab and therefore deviations from nominal result in poor input data to the scheduler. For instance, in the batching step above, the simulation contains detailed information about the current WIP waiting for a batch, the batching parameters, and the time future lots will arrive. This information is used in the simulation to predict when the batch will actually be formed, thus estimating the variable wait time for forming a batch and producing an accurate predicted arrival time.

A simulation model that generates accurate predictions requires an accurate model of that the upstream steps. This requires that the model include things like the dispatching rules for the upstream tools, processing times for the upstream tools, sampling strategies, etc. The Applied modeling environment provides a well-understood framework for doing such modeling.

4   SCHEDULING

The core of the solution is the scheduling process. In order to produce and optimized schedule, the scheduling process considers the equipment being scheduled, the current state of the fab, the predicted lot arrival times, and the objective function which is a collection of components that describes the relative cost of schedule metrics that are optimized when generating a schedule. This schedule describes the operations to perform at each tool in the module during a specified time horizon. The resulting schedule includes lot-equipment assignments, lot processing orders, lot and reticle transports and a reticle inspection schedule.

4.1 Lithography Scheduling – Constraints

One of the keys to producing accurate schedules is comprehensive modeling of the various parts of semiconductor manufacturing. These manufacturing activities need to be converted into mathematical constraints that can be provided to the core solver. The complexity of semiconductor fab operations creates a large number of complicated, non-linear constraints. The new short-interval scheduling solution includes constraints that model the following parts of the lithography area:

Send-ahead wafers: The user can specify send-ahead lots and reticles, which process first at a tool that requires initial calibration. If multiple lots process with that reticle on that station, one lot is selected to be the send-ahead lot and the other lots must wait for a metrology delay before processing.

Reticle use limits, time limits, and reticle inspections: Each reticle in the data set can have a use limit and or a time limit before it must be inspected. The capacity of the inspection stations is also modeled.

Carrier and reticle transport delays: The user can specify transport times for carriers and reticles at various levels of detail from just bay-to-bay times to specific tool-to-tool and tool-to-stocker times.

Qualification lots: The user can specify that certain tools require a qualification lot to be processed, along with a due date for the qualification.

Equipment recipe qualifications: The user can specify which tools are qualified for which recipes and steps

Lot-tool dedications: The user can specify lot-specific exceptions to the recipe qualifications. This includes dedicating a lot to a given tool and indicating that a lot cannot process on a tool that is otherwise certified for the lot's recipe.

Time-bound operations that start in photolithography: Some manufacturing routes include pairs of operations that must complete within a certain time of one another. If the second step in the pair isn't completed in time, then the lot must be reworked or possibly scrapped. To prevent this, most dispatching implementations will check downstream WIP and tool availability before starting a lot that begins a time-bound operation. The new short-interval scheduling solution provides a user-configurable constraint that performs much the same sort of check: the data generation provides how much downstream capacity is available over various time intervals and also describes how much capacity is used by each lot. The constraint then appropriately limits the number of lots that start time-bound operations over each interval.

Tool preventative maintenance constraints: The data generation can specify a time window for preventive maintenance (PM) on a tool. The engine then decides the best time to schedule the PM within that window.

Carriers containing multiple lots: The user can specify that a carrier contains several lots. If the lots require the same recipe, then the scheduler will prefer scheduling them together. If they require different recipes, then the scheduler will ensure that if they are scheduled on different tools, there is sufficient time to transport the carrier between the tools.

Train length constraints: To prevent overloading of downstream stations, the user can specify a maximum train length for each recipe on each tool. The user can also specify hot lots that can exceed the maximum train length.

Tool setups: The scheduler models temperature setups for both the coat and develop portions of the track, along with reticle setup times.

The realities of data collection in a fab mean that occasionally the data provided to the scheduler will be incomplete or incorrect. This can mean that the scheduler may be given an input data set that is not internally consistent and may appear to violate one or more of the modeled constraints. Because the scheduler must produce a schedule even in the presence of bad data, it needs to check the data input for such inconsistencies as part of the process of converting the data input into the mathematical optimization problem. If the scheduler finds inconsistencies in its data, it fixes these inconsistencies, logs errors, and then proceeds to produce a schedule with its corrected data.

4.2 Scheduling – Objective Function

The objective function is the mechanism through which the user can prioritize manufacturing objectives driving business benefit. These objectives are often called key performance indicators (KPI). In particular, by specifying different weights for different KPIs, the user can prioritize performance metrics. The objective function defines a cost for each schedule, so that schedules with a lower cost are preferred over those with a higher cost.

The short-interval scheduling solution for photolithography uses the calculations below within the objective function to optimize the schedule. This allows different fabs to choose, and weight the appropriate objective function components to target specific KPI's.

Cycle time: Describes the relative importance of minimizing the cycle time for lots with a given set of characteristics. For example, high priority lots could be singled out for processing before lower priority lots.

Lot-step due dates: This component allows the user to assign each of a list of lots a date by which it must complete its lithography step. If the lot completes the step past the due date, it increases the schedule cost. If the lot is early, the cost is not changed.

Move targets: Specifies a category of lots (for example, a technology or product), and then defines the target number of wafers for that category that should finish over a specified time interval. The schedule cost is increased if the number of wafers that complete during the interval either exceed or fall short of the target.

Reticle moves: Defines the cost of various reticle moves in the schedule. Different moves can have different costs, so that a tool-to-tool move within a bay might have a lower cost than a move between two different bays.

Stability: When the scheduler is generating a new schedule, it attempts to keep it similar to the previously generated schedule. This gives fab operations a consistent objective to work towards. This is particularly important in a fab with manual reticle movement so that the operators moving reticles are not given a continually changing set of moves to implement. The stability objective function allows the user to quantify the tradeoffs between minimizing changes in the schedule and the other objective functions.

4.3 Scheduling – Technology

The scheduling engine uses constraint programming to generate optimized schedules. This allows scheduling engine to guarantee that it will create schedules for all data sets, particularly in the presence of complicated equipment qualifications and dedications and the non-linear constraints involved in lithography manufacturing. It also allows the scheduling engine to make the lot-tool assignment a first-class part of the optimization rather than doing it outside the mathematical optimization proper.

5   SCHEDULE-AWARE DISPATCHING

The final process in the short interval solution is dispatching the production and transport activities. Initially, the short-interval scheduling development teams made minimal changes to existing dispatching rules.

They assumed that only minor changes would be required because

- The schedule acts as a basic prioritized and time-sequenced work list for each tool.
- Processing logic contained in existing dispatch rules must be maintained by each fab that uses the schedule.
- Complete dispatching logic is necessary when factory events that occur between schedule generations invalidate specific schedule tasks.

However, as the trials progressed, the concept of "schedule-aware" dispatching emerged and was refined because the dispatcher needs to understand the intent of the schedule, it cannot treat the schedule as it would a simple prioritized list of lots.

Schedule-Aware dispatching acknowledges the fact that one of the primary inputs to the dispatching rule is, in fact, a schedule. In contrast to a simple prioritized list of available lots. The schedule is a list of lots for a tool, but they are organized not simply by priority but also by temporally designating an earliest and expected start time for each task.

When the dispatcher simply treats the schedule as a prioritized list of tasks, effectiveness of the schedule drops slightly and the benefit of using optimized short-interval schedules begins to fade. When the dispatcher is "schedule-aware" it can effectively conform to the schedule and realize the predicted benefits.

Consider the following simple but illustrative example. A lithography tool is running a train of lots that use a single reticle and recipe. The tool completes one of the lots in the train, releases the carrier, and triggers the dispatcher's "what-next?" dispatching rule. The short-interval schedule for the tool has planned the last lot in the train; it is a high priority lot that is within a few minutes of finishing an upstream inspection and moving into the lithography tool. We will say that the high priority lot will arrive in lithography within 25 percent of the nominal transport time. There is also a lower priority lot that is in storage and ready to process. The lithography tool can only process one of these lots before the reticle requires re-inspection. Which of the two lots should the dispatcher choose?

The schedule-aware dispatcher is capable of looking at the schedule and deciding to "do nothing" until the high priority lot is available. The scheduler has optimized the schedule to reduce the cycle time of the higher priority lot and the dispatcher is able to wait and choose the higher priority lot as soon as it is ready to make a tool-to-tool move.

Some fabs may want to have a liberal dispatching policy that allows the dispatcher to wait for the high priority lot for some time before dispatching a lower priority lot, resulting in higher schedule conformance and lower cycle time for the high priority lot.

Other fabs may want to have a very conservative dispatching policy that overrides the schedule when a lot is not currently available for dispatch. In this case the "schedule-aware" dispatcher overrides the schedule and chooses a lower priority lot, validates processing rules, and sends the lot to the tool for processing.

A short-interval scheduling system with "schedule-aware" dispatching incorporates the policies of both conservative and liberal conformance to the schedule. This is the essence of schedule-aware dispatching, which is a key component in the overall short-interval scheduling solution and was critical to the success of both trial projects conducted by Applied Materials and its customer partners.

6   OTHER SCHEDULE-AWARE FACTORY SYSTEMS

The short-interval scheduler has the potential to integrate with other automated systems in the fab to create opportunities to improve overall efficiency and cost management. Consider a "schedule-aware" material control system (MCS) or AMHS. The schedule-aware MCS can anticipate carrier moves and position carriers in overhead storage, combining moves and reducing empty vehicles. It is a simple concept but one that yields significant results. Future applications of such schedule-aware capabilities are being explored.

7   RESULTS AND ECONOMIC BENEFITS

The Applied Materials short-interval scheduler is currently undergoing production trial tests to confirm efficacy and economic impact. Prior to the production trials the engineering team expected to see an improvement in cycle-time of high priority lots and negligible cycle-time change in lower priority lots. Engineering studies were used to validate this theory and showed that using optimized short-interval scheduling, cycle time for high priority lots could be reduced by nearly 25 percent. The engineering study made one significant assumption; perfect conformance to the schedule.

The results of the engineering study were compared with a baseline model that did not use the scheduler but did use highly refined heuristics (dispatching rules).

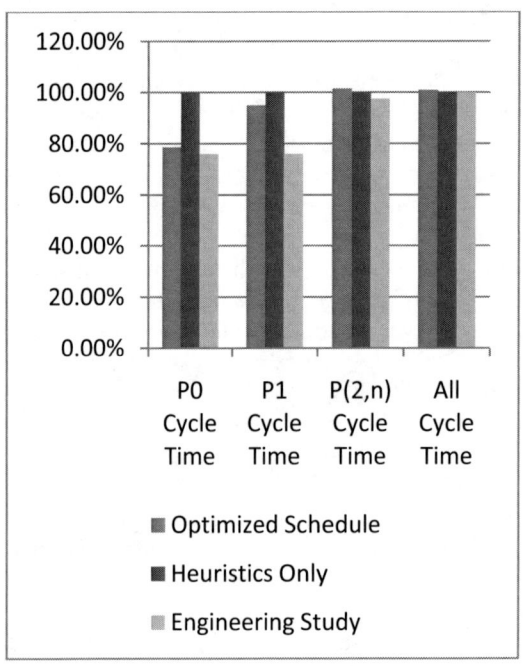

Figure 2 Cycle Time Improvements

The study was repeated using production data from a trial implementation of the short-interval scheduler. In this study, 214 samples of actual production data were taken used to generate a production schedule. The same production data was processed a second time to determine the production schedule resulting from pure heuristics-only real time dispatching. The results showed that the short-interval scheduler was able to significantly improve cycle time for the highest priority lots (P0) by approximately 20 percent but only improved the second highest priority lots (P1) by less than 5 percent. The cycle time for lower priority lots remained unchanged.

To complete the study, the economic impact of the improvement in cycle time for high priority lots was computed using the Leachman Model (Robert C. Leachman, Leachman & Associates LLC;, 2007).

Table 1 Economic Analysis of Cycle Time Improvement

Cycle Time Improvement	High Priority Lots	Normal Priority Lots
Lithography	20.83 %	- 1.67%
Overall	3.04 %	- 0.04 %
Capacity Increase	3.13%	- 0.04 %
First Year Value	$ 6,194,887	($324,239)

The analysis showed that the cycle time reduction for high priority lots while holding lower priority lots at relatively unchanged, results in a significant economic benefit both in increased total capacity of the fab and increased product value.

First year value, shown in Table 1, includes both the value of additional throughput and price erosion avoidance calculated using the Leachman Model. The market data used to perform the Leachman calculations and determine the value of addition capacity are based on data collected by Applied Materials for this purpose.

The conservative results shown lead to a return on investment that is realized within five to seven months.

These results have shown our development team that in a mature fab, short-interval scheduling can reduce cycle-times in photolithography and that these reductions indeed have a significant value.

8 SUMMARY

The complexities of determining an optimal scheduling solution in the dynamic environment of wafer manufacturing remains challenging. The constant change of product mix, large volumes, processing dependencies, and equipment and tool availability contribute to the difficulties in optimizing system performance.

The solution combines real-time data management and integration, relying on innovative applications for prediction, short-interval schedule optimization, and schedule-aware dispatching. Through the use of existing technologies, a commercially deployable and integrated advanced scheduling approach has been successfully developed and is providing near optimal scheduling solutions first in critical photolithography areas, and in the near future, in other critical fab production areas.

9 ACKNOWLEDGEMENTS

We would like to thank our customer partner and dedicated development team for their effort creating and deploying the new short interval scheduling solution.

We also wish to thank, Jeremy Webb, Shannon McGinley, and Sheryl Root for their editorial review and advise.

10 REFERENCES

[1] Govind, N., Bullock, E. W., He, L., Iyer, B., Krishna, M., & Lockwood, C. S. (2008). Operations Management in Automated Semiconductor Manufacturing With Integrated Targeting, Near Real-Time Scheduling, and Dispatching. *IEEE Transactions on Semiconductor Manufacturing , 21* (3), 363-370.

[2] Leachman, R. C., Kang, J., & Lin, V. (2002). SLIM: Short Cycle Time and Low Inventory in Manufacturing at Samsung Electronics. *Interfaces , 32* (1), 61-77.

[3] Robert C. Leachman, Leachman & Associates LLC;. (2007). The Economics of Speed. *Fab Engineering and Operations* (1), 98-102.

Topography-aware BARC optimization for double patterning

Shijie Liu,[1] Tim Fühner,[1] Feng Shao,[1] Aliaksandr Barenbaum,[1,2] Johannes Jahn[2] and
Andreas Erdmann[1]

[1]Fraunhofer Institute for Integrated Systems and Device Technology (IISB), Schottkystrasse 10,
91058 Erlangen, Germany

[2]Applied Mathematics II, University of Erlangen-Nuremberg, Martensstrasse 3,
91058 Erlangen, Germany

ABSTRACT

This paper aims at identifying appropriate bottom anti-reflective coatings (BARCs) for double patterning techniques such as Litho-Freeze-Litho-Etch (LFLE). A short introduction into the employed optimization methodology, including variables, figures of merit, models and optimization algorithms is given. A study on the impact of a refractive index modulation caused by the first lithographic step is presented. Several optimization surveys taking the index modulation into account are set forth, and the results are discussed. In addition to optimization procedures aiming at optimizing one litho step at a time, a co-optimization study for both litho steps is proposed. Finally, two multi-objective optimization procedures that allow for a post-optimization exploration and selection of optimum solutions are presented. Numerous solutions are discussed in terms of their anti-reflectance behavior and their manufacturing feasibility.

Keywords: lithography simulation, bottom anti-reflective coating (BARC), waferstack optimization, double patterning

1. INTRODUCTION

To achieve the required feature and pitch sizes for the 32-nm technology node and beyond, double patterning with 193 nm immersion lithography is one of the most promising candidates [1]. For example, Litho-Etch-Litho-Etch (LELE) has been introduced as one of the standard processes for line/space and contact hole patterning. To further improve the throughput, alternative processes such as Litho-Freeze-Litho-Etch (LFLE) have been widely investigated [2]. Whatever patterning process is selected, efficient anti-reflective coatings are indispensible in order to reduce the CD sensitivity to resist thickness variations, standing waves, pattern irregularity and pattern collapse.

For double patterning, organic bottom anti-reflective coating (BARC) materials are commonly used, because they can be processed in the scanner track [3]. In general, organic BARC materials are composed of polymer, cross-linker, cross-linking catalyst and solvent in order to meet different requirements such as chemical freezing, contamination blocking and reflection control at the same time. During the double pattering process, the first lithographic exposure and subsequent process steps modify the properties of the BARC and impact its performance in the second lithography stage. For example, an exposure with a lines and spaces pattern can result in a laterally and sinusoidally modulated refractive index of the BARC. Experimental results showed that the difference of the index of refraction (n) before and after the first exposure and develop can be as large as -0.08 [4]. The change of the n value and its modulation in the BARC layer alters its reflectivity behavior in the second litho step. Moreover, in immersion lithography systems, the increased NA leads to a significant polarization dependent behavior of the wafer stack. Large angles of incidence are even close to Brewster's angle [5]. So, it becomes increasingly difficult to determine BARCs with an adequate performance, even more in the case of double patterning processes. A thorough optimization study on different BARC layouts that incorporates the refractive index modulation resulted from the first lithographic step is therefore required. However, to render such an optimization approach feasible, efficient models for both the planar case (litho 1) and the electro-magnetic field computations in the second litho step are essential.

2. OPTIMIZATION AND SIMULATION METHODOLOGY

In this section, the formulation of the optimization problem, including variables, targets and underlying optimization algorithms are briefly introduced.

Optical Microlithography XXIII, edited by Mircea V. Dusa, Will Conley, Proc. of SPIE Vol. 7640,
76403C · © 2010 SPIE · CCC code: 0277-786X/10/$18 · doi: 10.1117/12.846441

2.1 Optimization variables and figure of merit

The anti-reflective performance of a BARC is determined by the optical and topographical properties of its individual layers. Therefore, the refractive index (n), the extinction coefficient (k) and the thickness (d) of each BARC layer serve as optimization variables for the approach proposed here (see Figure 1). The optimization goal is given by the reflectance returned at the interface between the top BARC layer and the photoresist. In order, however, to optimize the BARC performance for a large range of pitches of features on a photomask, the reflectivity (R) is evaluated for a number of different incidence angles (θ_i). The objective function for one lithography step (l) can hence be stated as:

$$\mu_l : \mathbf{x} \mapsto \frac{1}{n} \sum_{i=1}^{n} R_l^{\theta_i}(\mathbf{x})^2 \, , \tag{1}$$

where $\mathbf{x}$ is the optimization variables vector (i.e., the optical properties and thickness values of the layers) and θ_i ($i = 1, \ldots, n$) is the range of incidence angles for which the reflectance is evaluated.

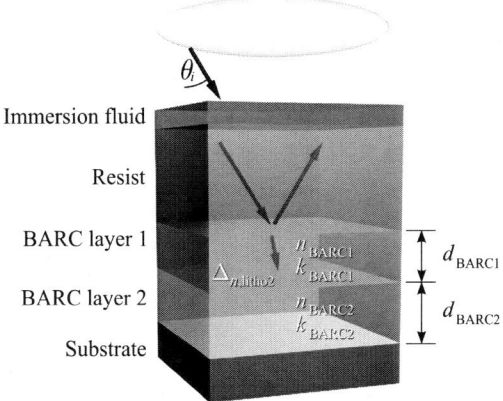

Figure 1: Bi-layer BARC with optimization variables.

As mentioned in the introduction, this work aims at indentifying BARC configurations that are suited for double patterning approaches and that are for this reason required to show a persistent reflectivity behavior after the first lithography step in which BARC properties are altered. Here, this is modeled by a laterally sinusoidal modulation of the refractive index in the top layer of the BARC, as could be resulted from the exposure with a line/space pattern. As shown below, the amount of index modulation ($\Delta_{n,\text{litho2}}$) is studied in the scope of this work—it is of course fixed per optimization run. In the double patterning case, the evaluation of the BARC has thus to be performed in two stages: In the first lithography step, the layers can be assumed to be planar, allowing for the application of the time-efficient transfer-matrix method. The second litho step, however, exhibits an inhomogeneous BARC top layer so that a rigorous electro-magnetic field computation has to be carried out, for which the Waveguide Method is used [6]. Applying, for example, a weighted sum approach, the total objective can be stated as:

$$\mu_\Sigma : \mathbf{x} \mapsto \sum_{l=1,2} w_l \sqrt{\mu_l(\mathbf{x})} \, , \tag{2}$$

where w_l is the weight assigned to each litho step ($w_1 + w_2 := 1$).

The computation time for 35 incidence angles (9, ..., 43 degrees) is less than 0.1 seconds for the planar case (litho 1) and less than 90 seconds for the second lithography step (using the Waveguide method) on our computing cluster nodes (AMD Opteron 2352 @ 1 GHz and Intel Xeon E5405 @ 2 GHz, for which computations are slightly faster).

2.2 Single- and multi-objective genetic algorithms

The optimizer used in this work is a genetic algorithm (GA), which belongs to the class of meta-heuristic search and optimization approaches. In GAs, a set of solution candidates is iteratively improved by means of combinatorial operations. The corresponding operators resemble natural selection, reproduction and mutation processes and are hence called genetic operators. Genetic algorithms can be considered global optimizers in three respects: (1) They are largely independent of any local information such as gradients; (2) the solution candidates explore different regions of the search space in parallel but globally share and exploit the gathered information, and (3) GAs converge to a global optimum—provided certain pre-conditions are met. Because of these characteristics, GAs are well suited for combinatorial problems and problems that evince discontinuities in the search space. They have therefore been successfully applied to similar optimization tasks to the one presented here (e.g., see Ref. [7]).

Many real-world optimization problems employ various often conflicting objectives. For example, in order to optimize a product, its performance or quality has to be maximized while at the same time the production effort should be minimized. Formally, a (merely box-constrained) multi-objective optimization problem can be stated as follows:

$$\arg\min_{\mathbf{x}} \quad \boldsymbol{\mu}(\mathbf{x}) = [\mu_1(\mathbf{x}),...,\mu_n(\mathbf{x})]^T$$
$$\text{subject to} \quad x_j^{\min} \le x_j \le x_j^{\max} \quad (i=1,...,m) \tag{3}$$
$$\text{where} \quad \boldsymbol{\mu}: R^m \to R^n$$

An ideal solution would be a minimizer for all objectives. With contradicting criteria, this is hardly achievable: Instead, an adequate solution in such a context has to present a compromise between the figures of merit. A common approach is to construct a scalar replacement problem. In this work, a weighted sum method is applied where the total figure of merit is the dot product of the objective vector ($\boldsymbol{\mu}$) and a weight parameter vector. This technique poses two main deficiencies: (1) In most cases the weight parameters, that have a strong impact on the final solution and on the convergence behavior of the optimization routine, have to be manually selected. (2) Not all solutions that would intuitively be considered to perform equally well are guaranteed to be discovered.

On that score, a genuine multi-objective optimization strategy has additionally been applied to the BARC optimization task. This approach utilizes a genetic algorithm and a strict partial order relation, defined as follows:

$$x \text{ dominates } y \ (\mathbf{x} \prec \mathbf{y}) :\Leftrightarrow$$
$$((\forall x_i)(x_i \le y_i)) \wedge ((\exists x_i)(x_i < y_i)) \tag{4}$$

That is, a solution is said to dominate another solution if and only if its performance is at least equal in all objective components and superior in at least one component. An optimality set is then given by all solutions that are not dominated by any other solution candidate. Such a set is often termed Pareto front (or set) after the Italian economist Vilfredo Pareto (1848–1923), who introduced this concept.

The genetic algorithm used here is called the Strength Pareto Evolutionary Algorithm 2 (SPEA2). It has been proved to meet one of the key challenges in multi-objective GAs, the preservation of diversity of solutions [8].

In order to exploit high performance computing facilities, both the single-objective and the multi-objective GAs incorporate distributed computing techniques.

3. RESULTS

In this section, results obtained from different optimization studies are presented. First, optimization results for both a single and a bi-layer BARC are displayed and compared. Then, the impact of the refractive index modulation in the top BARC layer is shown, and several optimization studies for the second litho step are presented. Finally, three different optimization studies co-optimizing the BARC for both litho steps are demonstrated.

3.1 Optimization of single and bi-layer BARCs for the first litho step

In a first approach, the BARC performance for the first lithography step with a mono-layer BARC is optimized. The objective is to obtain a low reflectance within an incident angle range from 9 to 43 degrees, which corresponds to a feature pitch of 360–83 nm. The figure of merit is given by the mean of the squared reflectivities over the angle range (1). Let the incident medium be a photoresist with a refractive index of 1.7 at a 193 nm wavelength and the outgoing medium be silicon with a complex refractive index of about $0.91 - 2.8j$. Since all layers can be assumed to be planar, the transfer-matrix method is employed. Constraining the thickness of the BARC layer to 0–200 nm and its refractive index to 1–2.2, an optimum reflectance behavior is achieved at a BARC layer thickness of 23.4 nm and a refractive index of $2.05 - 0.57j$ (cf. Figure 2). The corresponding reflectivity curve shows a strong pitch dependence: The minimum reflectivity is almost zero (at 29 degrees), and the maximum reflectivity is about 1.03 percent (at 43 degrees).

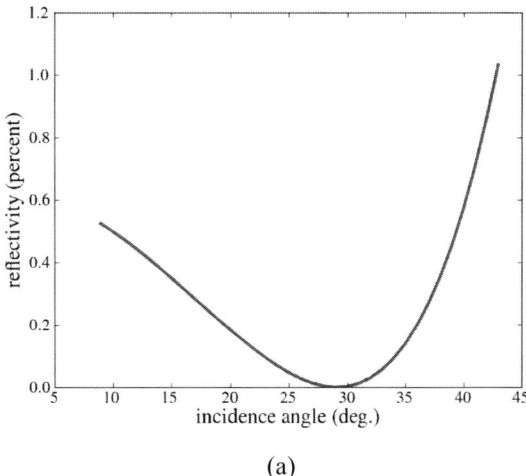

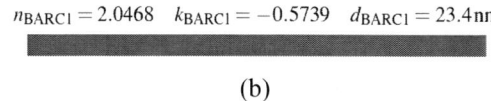

$n_{BARC1} = 2.0468$ $k_{BARC1} = -0.5739$ $d_{BARC1} = 23.4\,nm$

(a)   (b)

Figure 2: Optimized single layer BARC: (a) reflectance performance, (b) optimum solution properties.

One of the main motives to apply BARCs is to achieve a high degree of pattern stability. For example, the critical dimension (CD) should exhibit only a low sensitivity to resist thickness variations. One way to assess the impact of the performance of the optimized BARC on the pattern stability is given by a swing curve analysis, in which the line width variation is plotted as a function of the thickness of the resist. Figure 3 shows this interrelation for the optimized monolayer BARC. The analysis is performed for the same pitches as used in the optimization. The target CD on the wafer is about 41.5 nm. Because of higher reflectivities at larger incident angles, the wafer CD shows a non-linear relation to the resist thickness (Figure 3(a)). At the same time, due to the resist absorption, the wafer CD linearly changes with the resist thickness at smaller incident angles (Figure 3(b)).

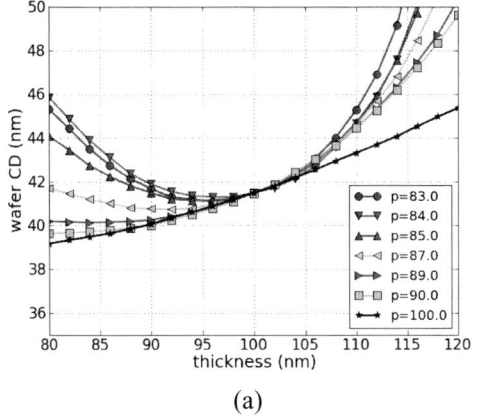

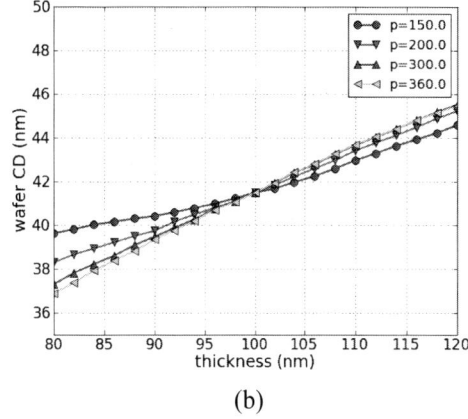

(a)   (b)

Figure 3: Pattern stability achieved with a mono-layer BARC optimized for the first lithography step. The target CD is 41.5 nm: (a) For small pitches, a non-linear CD dependence on the resist thickness is observed, caused by the decreased antireflection performance of the BARC at higher incidence angles; (b) larger pitches show a resist absorption-induced linear behavior.

In a second optimization step, a bi-layer BARC is optimized. Again, the goal is to minimize the reflectivity within the incident angle range from 9 to 43 degrees. The same thickness and optical property constraints as in the former optimization task are applied—this time, however, to each of the two layers. Moreover, the same merit function as before is used.

The genetic algorithm is run for 800 iterations (generations). However, after about 300 generations a local minimum that closely resembles the final solution is obtained. As shown in Figure 4(b), the optimum bi-layer stack consist of a top layer with a thickness of 13.2 nm and a refractive index of $2.08 - 0.13\,j$. The second layer has a thickness of 46.7 nm and a refractive index of $1.675 - 0.633j$. The reflectivity performance is drastically improved compared to the best single layer BARC (see Figure 4(a)). The maximum reflectance, which occurs at an angle of 43 degrees, is less than 0.04 percent. No significant pitch dependence occurs.

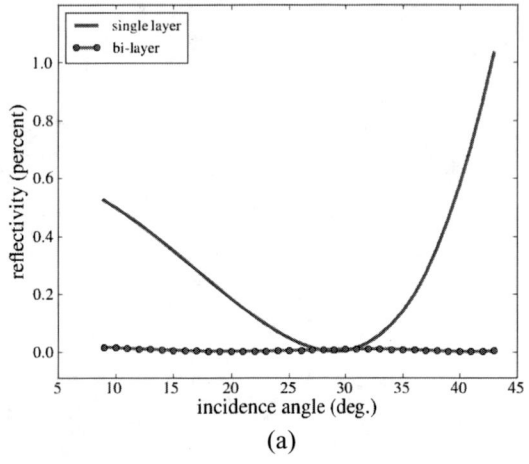 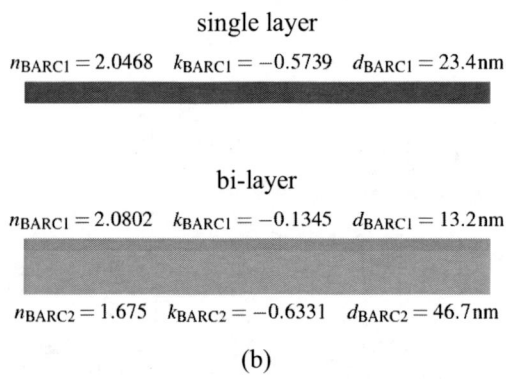

(a)

(b)

Figure 4: Comparison of the optimized single and bi-layer BARCs. (a) Anti-reflective performance within the incidence angle range of [9,43] degrees; (b) thickness and optical properties of the optimized single and bi-layer BARC stacks.

The CD swing curves obtained with the optimized bi-layer BARC is demonstrated in Figure 5. The target CD on the wafer is about 41.5 nm. The swing curve shows an almost perfectly linear behavior in incident angle range of concern, allowing for an adequate control of the CD values on the wafer.

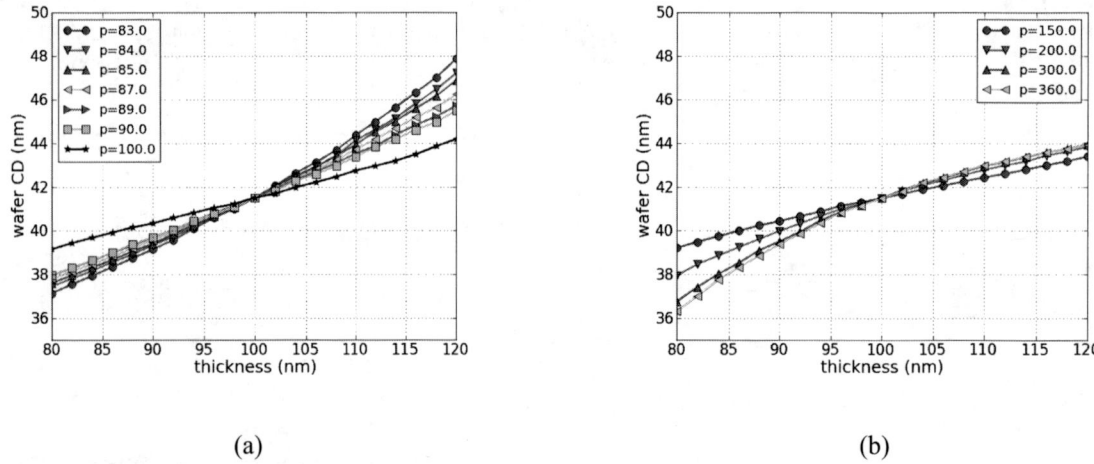

(a)   (b)

Figure 5: CD swing curves obtained with the optimized bi-layer BARC: for both (a) small and (b) large pitches an almost perfectly linear behavior is achieved.

3.2 Bi-layer BARC with index modulation

After the first lithography step, the optical properties of the top layer of the BARC are modified due to both the exposure itself and subsequent process steps including baking. The topography of this index modulation can be assumed to be related to the patterns imaged in the first litho step. For example, line/space (L/S) patterns will result in a sinusoidal modulation of the top layer index, whose period of this modulation is determined by the pitch of the L/S features. Figure 6 schematically demonstrates the resulted topography.

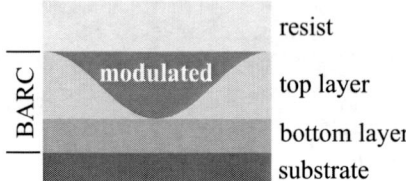

Figure 6: Structure of bi-layer BARC with index modulation in one period

In order to assess the impact of the refractive index modulation, a simulation study is conducted in which the influence of an assumed index modulation factor in dependence of the pitch is investigated. In the modulated region, the refractive index is given by the sum of the original refractive index and the modulation factor (Δn). Both the BARC and the resist properties are taken from the former optimized bi-layer BARC (cf. Figure 4): A refractive index of $2.08 - 0.135j$ and a thickness of 13.2 nm are assumed for the top layer, and for the second layer, a refractive index of $1.68 - 0.633j$ and a thickness of 46.7 nm are applied. Index modulation values in the range of -0.5 to 0.5 are investigated, giving a refractive index variation of the first layer from 1.58 to 2.58. As before, the range of incidence angles is 9–43 degrees.

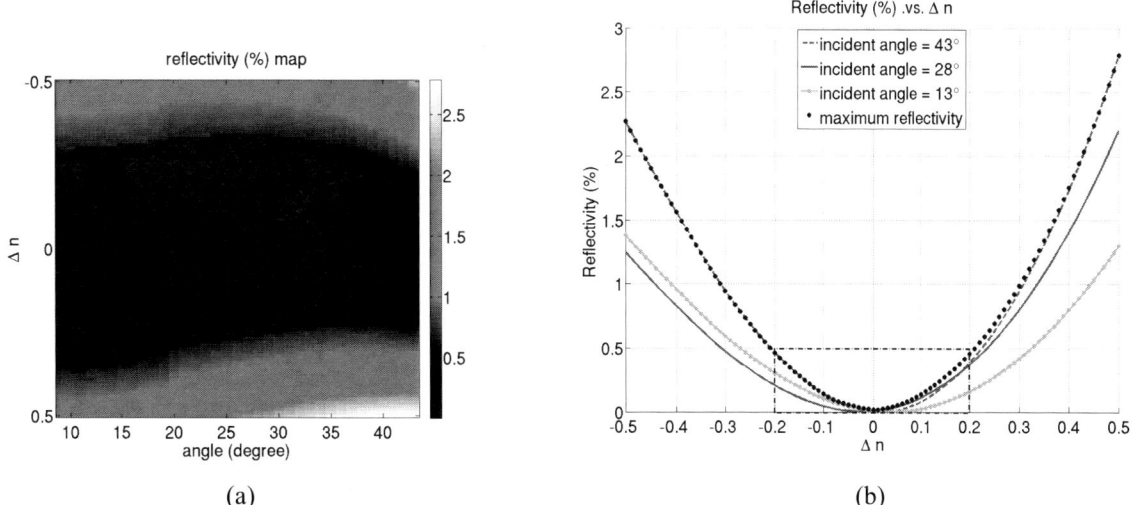

(a)   (b)

Figure 7: Reflectivity of BARC in litho step two (a) as a function of index modulation of top layer and incidence angle.
(b) Cross-section for selected incidence angles.

The resulting bi-layer BARC performance in the second litho step as a function of the incidence angle and the index modulation value is shown in Figure 7(a). The period of the sinusoidally modulated region in the top layer is determined by the angle of incidence, i.e., the pitch of a given L/S feature (see Figure 6); only one period is evaluated. For small pitches (that is, large angles of incidence), the modulation impact is significantly larger than for large pitches. Figure 7(b) depicts corresponding cross-sections for selected angles.

In order to account for the effects of index modulation resulted from litho 1, the bi-layer BARC is next optimized for the second lithography step. Simulations for typical settings have shown that most of the energy of the reflected light is contained in the zeroth and minus first diffraction order. We have therefore modified the objective function (1) by replacing the reflectance by the average reflectance of both orders. In addition to the same constraints as applied in the previous optimization study, an index modulation (Δn) of 0.2 was assumed. As with the modulation impact study, the period of the modulated region is determined by the pitch (incidence angle) under consideration. The same incidence angle range of [9, 43] degrees is applied.

After 500 generations, an optimum solution was obtained—a very similar solution is already achievable after about 250 generations. The corresponding stack showing all optical properties and the thickness of the two BARC layers is illustrated in Figure 8(a). A significantly thinner top layer than in the BARC optimized for litho 1 (Figure 4) is obtained. The optical properties of the layer are almost identical to the litho 1 optimization step. The second BARC layer shows a considerably larger thickness but a smaller extinction coefficient. In Figure 8(b), the reflectivity performance in litho 2 for both the stack optimized for litho 1 and the stack optimized for litho 2 is compared. For both stacks, an index modulation value of 0.2 is assumed. An index modulation of 0.2 in the BARC top layer has significantly increased the total reflectivity of the wafer stack, as illustrated by the dashed line in Figure 8. The maximum reflectivity is about 0.54 percent at 38 degrees. If the index modulation is considered in the optimization process, the total reflectivity can be greatly decreased, as illustrated by the solid line in Figure 8(b): The maximum reflectivity is reduced to about 0.066 percent.

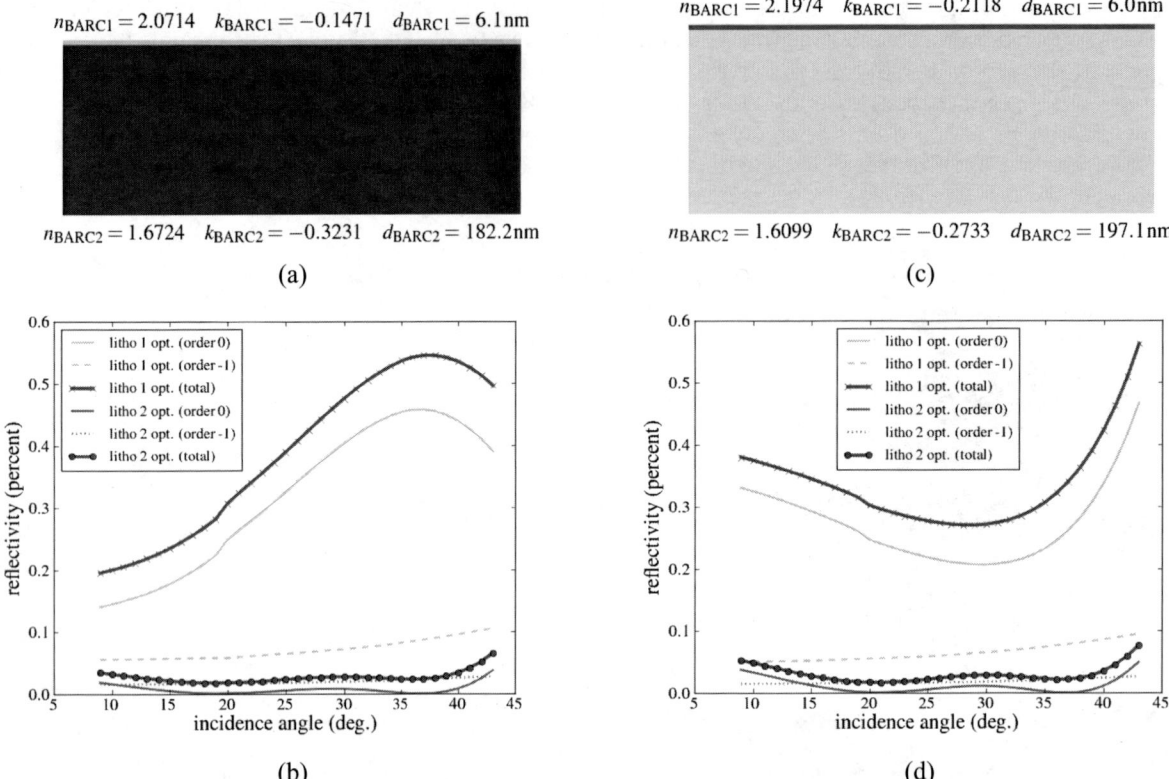

$n_{BARC1} = 2.0714 \quad k_{BARC1} = -0.1471 \quad d_{BARC1} = 6.1\,nm$

$n_{BARC2} = 1.6724 \quad k_{BARC2} = -0.3231 \quad d_{BARC2} = 182.2\,nm$

(a)

$n_{BARC1} = 2.1974 \quad k_{BARC1} = -0.2118 \quad d_{BARC1} = 6.0\,nm$

$n_{BARC2} = 1.6099 \quad k_{BARC2} = -0.2733 \quad d_{BARC2} = 197.1\,nm$

(c)

(b)

(d)

Figure 8: Bi-layer BARC optimized for the second litho step, assuming a sinusoidal index modulation of 0.2. (a) Resulted stack, (b) comparison of litho 2 performance (total reflectivity of both contributing orders) of stack optimized for litho 2 (line marked with dots) and stack optimized for litho 1 (line marked with crosses). The other lines show the reflectance of the individual minus first and second orders; (c) stack optimized for negative index modulation, (d) evaluation of litho 1 and litho 2 optimized stacks in the negative modulation regime

Assuming a negative modulation value (in this example: -0.2) yields a slightly different optimum BARC (see Figure 8(c)). Comparing the litho 2 performance—under the assumption of a negative modulation value—of both the litho 1 and the litho 2 optimal BARC shows a qualitatively comparable difference as in the positive modulation value case (Figure 8(d)).

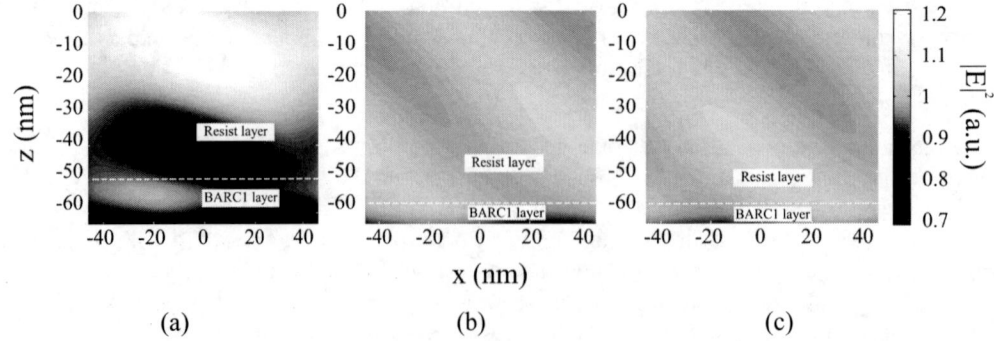

Figure 9: Squared electric field ($|\mathbf{E}|^2$) in the bi-layer BARCs in litho 2, under oblique incidence (38 deg.): (a) litho 1 optimal BARC, (b) litho 2 optimal BARC; (c) litho 2 optimal BARC, assuming a negative index modulation.

The performance of the BARC stacks can also be analyzed by an examination of the internal electric field ($|\mathbf{E}|^2$) distribution in the resist layer, as shown in Figure 9. The asymmetry of the intensity distributions results from the oblique incidence (38 degrees). If the index modulation in the BARC is not considered in the optimization process, pronounced standing waves occur in the resist layer in the second litho-process. In contrast, the optimized bi-layer BARCs exhibits little standing waves in electric field distribution. All BARCs optimized for the second litho step possess an extremely thin top BARC layer, so exhibiting a low diffraction efficiency at higher orders.

3.3 Co-optimization of BARCs for both litho steps

As shown above, BARCs optimized for only the first lithographic process step, cannot be guaranteed to show an adequate performance also in the second litho step. Thus, in order to identify an appropriate bi-layer BARC for double patterning applications, a co-optimization taking both litho steps into account has to be performed. For that purpose, we have combined both evaluation routines—for litho 1 using TMM and for litho 2 using the Waveguide method—in a weighted sum approach (see Equation (2)). As in the litho 2 only case, the average of the two contributing orders (-1 and 0) is assessed. A refractive index modulation of 0.2 is assumed.

After 800 generations using the single-objective genetic algorithm, a bi-layer BARC with an optimum anti-reflection behavior in both litho steps is obtained; a solution that closely resembles the final one is already achieved after 400 generations. The optimized stack and its performance are portrayed in Figure 10. The maximum reflectivity in litho 1 is about 0.1 percent at an incidence angle of 43 degrees and 0.13 percent at 9 degrees in litho 2. Again, the resulted top layer is extremely thin, jeopardizing the manufacturability of the BARC and suggesting the introduction of a feasibility criterion.

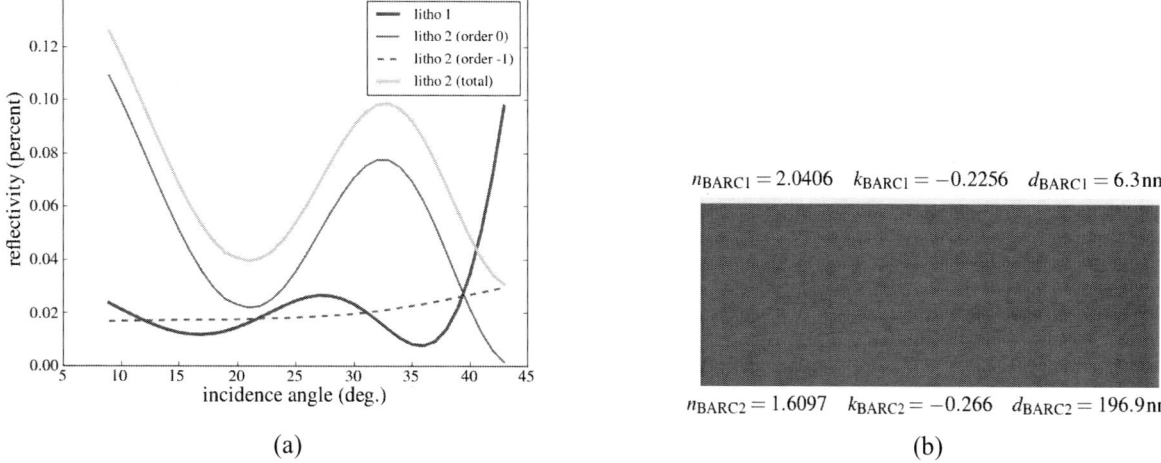

$n_{BARC1} = 2.0406$ $k_{BARC1} = -0.2256$ $d_{BARC1} = 6.3 nm$

$n_{BARC2} = 1.6097$ $k_{BARC2} = -0.266$ $d_{BARC2} = 196.9 nm$

(a)   (b)

Figure 10: Bi-layer BARC co-optimized for litho 1 and litho 2: (a) Anti-reflective performance, (b) geometry and optical properties of the stack.

3.4 Bi-objective bottom layer optimization

In the previous simulation studies, high index modulations of 0.2 or -0.2 were assumed. Related experimental studies by Sakamoto *et al.* suggest smaller maximum modulation values [4]. In a further optimization study, we have therefore applied BARC top layer properties reported by Sakamoto *et al.*: A refractive index of $1.8 - 0.3j$ and an index modulation factor of -0.08. In addition, the thickness of the top layer is fixed to 30 nm, also resolving the manufacturability issue resulting from the extremely thin top layers obtained in the previous study. Both the thickness and the optical properties of the second BARC layer are subject to optimization. The thickness range is set to [10, 200] nm, the refractive index was optimized in a range of [1, 2.2] nm, and the extinction coefficient between 0.1 and 0.5. A bi-objective optimization approach using a multi-objective genetic algorithm was applied for the optimization. All other settings, including the number of incidence angles was unaltered compared to the previous studies.

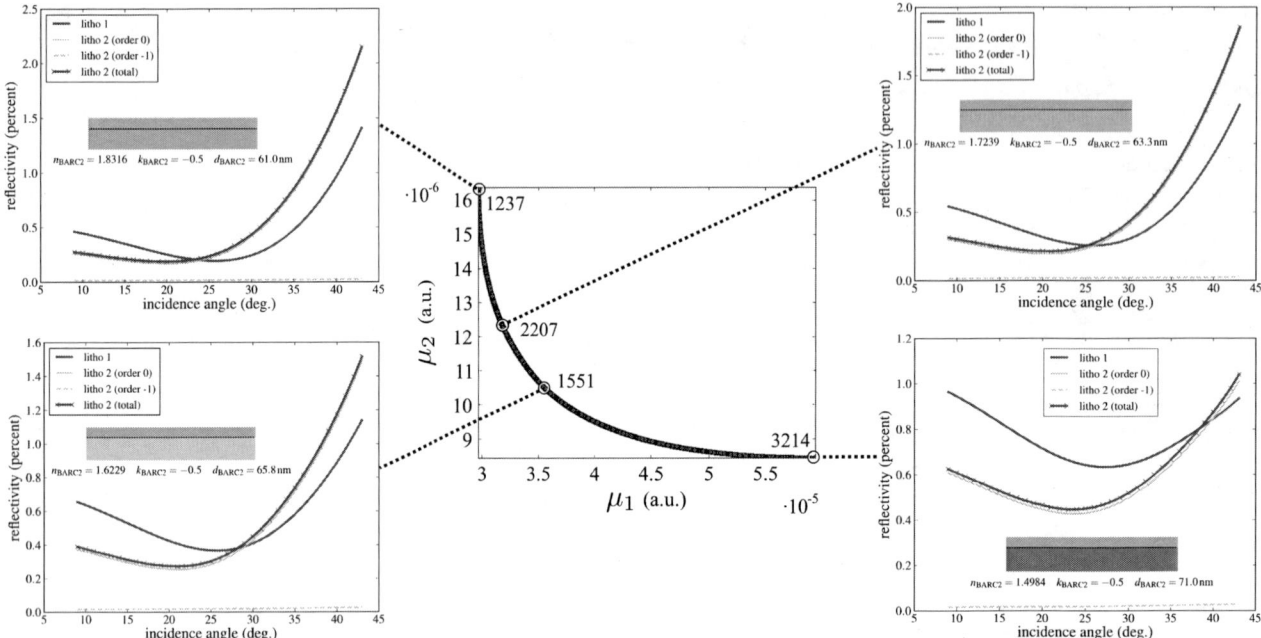

Figure 11: Pareto front (center) and example solutions obtained with a bi-objective optimization study: The figures of merit are constituted by the two litho steps (litho 1: μ_1, litho 2: μ_2). For the second litho step, the reflectivity of the minus first, the zeroth and the total reflectivity are displayed. Solutions performing well in the first litho step (e.g., 1237) show a high refractive index of the bottom layer of the BARC. Solutions with a low reflectivity in step two (e.g., 3214) exhibit a higher index of refraction in the bottom layer. Compromise solutions can be found where the Pareto front is close to the bisectrix of the coordinate system, for example 1551 and 2207.

Figure 11 illustrates the solutions obtained with the bi-objective optimization approach. The Pareto front, depicted in the center of the figure, forms the optimality set. That is, solutions on the Pareto front outperform other candidates either with respect to litho 1 (μ_1) or litho 2 (μ_2), while still showing an acceptable performance in the respective other litho step. Solutions with a perfect compromise between litho 1 and litho 2 can be found close to where the Pareto front intersects the bisectrix of the coordinate system. The circumjacent graphs demonstrate the anti-reflective performance of example solutions taken from the Pareto front. Despite the reduced index modulation, the performance for both litho 1 and litho 2 is significantly degraded compared with the previous study in which also the top layer of the BARC was subject to optimization. Solutions optimal in litho 1 (e.g., solution 1237) have a bottom BARC layer with a extensively higher index of refraction than solutions optimal in litho 2 (e.g., solution 3214). The two example compromise solutions (1551 and 2207) corroborate this tendency: The closer solutions are situated to the minimum of μ_2, the lower is the refractive index of the bottom layer. This behavior can be attributed to the fact that higher refractive indices in the second layer of the BARC may increase back-reflections from the bottom layer into the top layer and hence increase the sensitivity of the solution to the index modulation of the top layer, which plays a role only in litho 2. For the same reason, the extinction coefficient in all solutions is set to the maximum value of 0.5. No significant variation of the thickness of the second BARC layer in the Pareto set can be observed.

3.5 Tri-objective bottom layer optimization with variable top layer thickness

In order to further improve the performance of the bi-layer BARC, we have carried out another optimization study, allowing for an additional degree of freedom, by subjecting the thickness of the top layer to optimization. In contrast to the single-objective approach demonstrated in Section 3.3, the thickness of the first layer is constrained to [10, 200] nm. To strictly account for the feasibility of solution candidates but also to investigate the impact of the top layer thickness on the performance of BARCs, we have also incorporated an additional figure of merit, which can be considered a thickness penalty:

$$\mu_d : (x_1, ..., x_k = d_{\text{BARC1}}, ..., x_m) \mapsto d_{\text{BARC1}}^{\max} - d_{\text{BARC1}}, \tag{5}$$

where $d_{\text{BARC1}}^{\max}$ is the upper bound of the top layer thickness (200 nm) and d_{BARC1} is the top layer thickness of the solution candidate to be evaluated. Hence, by minimizing Equation (5), BARCs with a thicker top layer are preferred.

Figure 12 shows the Pareto set and example solutions obtained after 1000 iterations. As before, the X and the Y axes denote the figures of merit in litho 1 and litho 2, respectively. The thickness penalty criterion is constituted by the Z axis and also by the gray levels of the solution points. Depending on the preferred figure of merit, both the layer thicknesses and the optical properties of the second BARC layer of the optimum solution candidates show significantly different values. Solutions with a thick top BARC layer show a relatively thin second layer. All solutions optimal in litho 2 (μ_2) exhibit thinner BARC top layers.

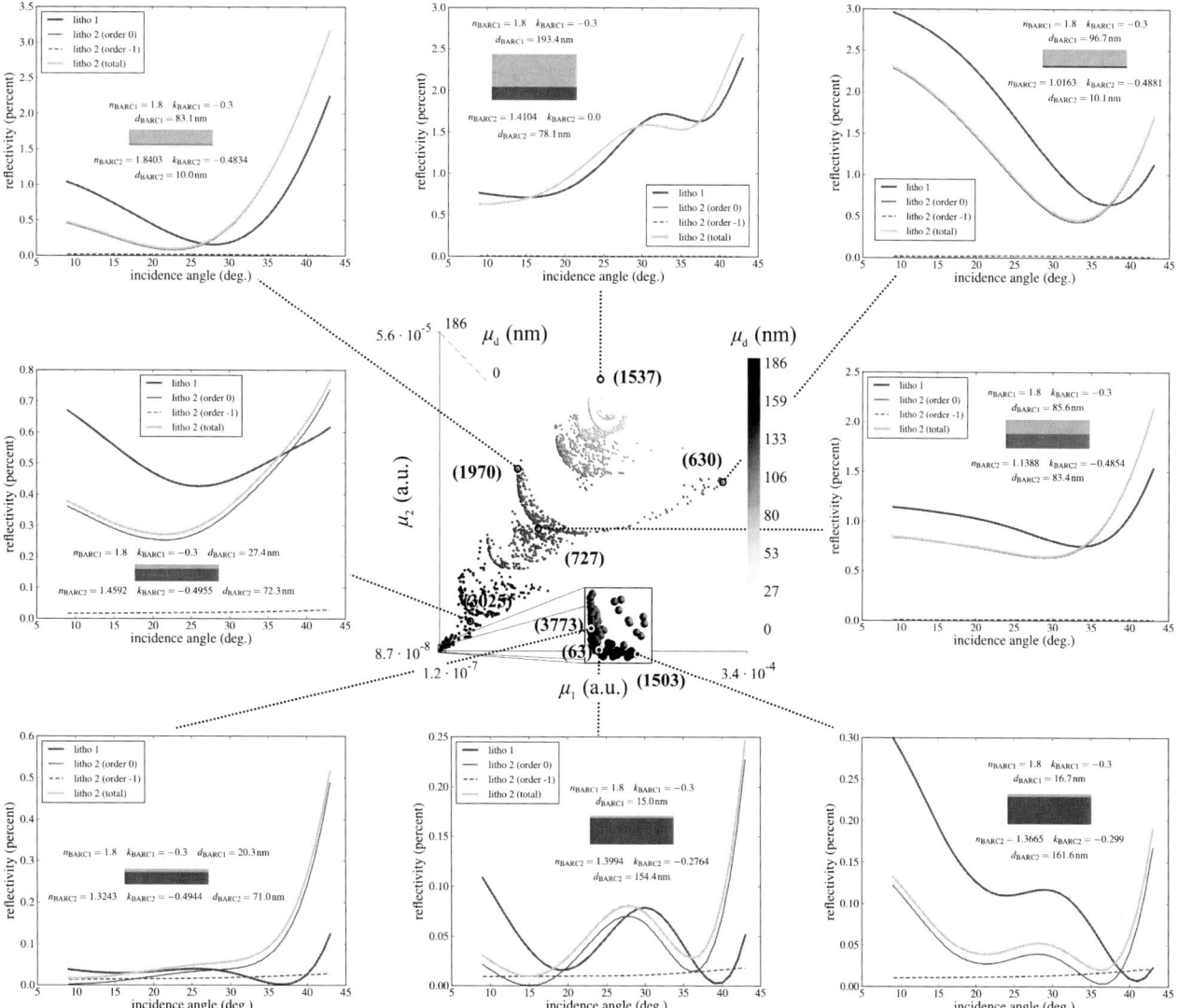

Figure 12: Different bi-layer BARC solutions obtained with the multi-objective optimization approach. In the center, the three dimensional Pareto front is shown. The X and Y axes denote the objective of litho 1 and litho 2, respectively. The Z direction and the gray levels of the solution points indicate the thickness penalty of the first BARC layer. The anti-reflection performance of the example solution candidates are displayed by their stacks, the total reflectivity of the first litho step, and the reflectance of the zeroth and the minus first order, and the total reflectance of litho 2. The depicted solutions are taken from different regions of the Pareto front. For example, solution 1537 exhibits a large top BARC layer. However, its reflectivity performance is worse compared with solutions having thinner top layers (e.g., 3773, 63 and 1503). Compromise solutions can be found in the center of the Pareto set (e.g., 727).

As observed in the previous studies, thin top layers lead to a decrease of reflectance in the second litho step due to a lower sensitivity to the refractive index modulation. In addition, an inverse effect can also be observed: Thin bottom layers lead to lower diffraction efficiencies in the modulated top layer, which in turn also decreases the impact of modulation and thus results in smaller reflectance differences between litho 1 and litho 2—however, leading to higher absolute reflectance values. Especially, BARCs with a top layer thickness close to the maximum of 200 nm (e.g., candidate 1537), where the impact of the bottom BARC layer is negligible, exhibit a poor reflectance performance at large angles. Solution candidates with a medium top layer thickness (e.g., 630, 727 and 1970) show an increased impact of the second BARC layer on the overall anti-reflective performance.

Candidate 3025 constitutes a feasible solution both in terms of manufacturability (27.4 nm thickness of top layer and 72.3 nm thickness of the bottom layer) and reflectivity performance (below one percent). In addition, the difference between litho 1 and litho 2 is small both absolutely but also relatively, since both total reflectance curves show a similar behavior. Furthermore, the angle-dependent difference is small compared to all other results.

For top layers with a thickness of 20 nm and below (e.g., 63, 1503, 3773), the reflectivity of both litho steps drops to below 0.5 percent. The difference between litho 1 and litho 2 is extremely small on the entire angle range in terms of absolute values. There is, however, a stronger reflectance dependence on the incidence angle than in the previous solution (3025). The manufacturability of these solutions may not be guaranteed.

4. CONCLUSIONS AND FUTURE WORK

We have shown that bi-layer BARCs show a significantly improved anti-reflectivity performance compared to mono-layer BARCs, which despite the added manufacturing effort might render it a viable option for double patterning. The performance of both the mono-layer and the bi-layer BARC can be drastically improved by the application of rigorous optimization studies. The first litho step in a double patterning regime such as litho-freeze-litho-etch (LFLE) leads to a non-homogeneous refractive index modulation of at least the top layer of the BARC. Depending on the amount and the topography of this index modulation, the same BARC shows different characteristics in the first and in the second litho steps. Because of the non-planar nature of the index modulation by litho 1, the performance evaluation in litho 2 has to be performed using rigorous electro-magnetic field solvers such as the Waveguide method. We have shown that even then extensive simulation and optimization studies, which are required to identify ideal BARC configurations for double patterning, can be conducted. Furthermore, we have demonstrated and discussed several results obtained with a single-objective optimization approach assuming a relatively high value for the index modulation and allowing a high degree of freedom in terms of constraints both for the optical properties and the thickness of the BARC layers. Optimum solutions showed a very small impact of the index modulation on the BARC performance, which can, however, be attributed to a very low thickness of the first BARC layer of the very optimum solutions. Although such a BARC would indeed show a very low sensitivity to index modulations, it is hardly manufacturable. We have therefore conducted another optimization study with a top layer with fixed properties taken from published experimental results. We have shown that the proposed approach is suited to indentify appropriate BARC set-ups that show both an adequate performance in litho 1 and litho 2, and are manufacturable. In addition, by applying a multi-objective genetic algorithm, compromise solutions for litho 1 and litho 2 are found, and the set of optimum solutions can be explored following the actual optimization routine. The same optimization technique was applied to a set-up in which the thickness of the top layer of the BARC and the optical properties and the thickness of the bottom layer were co-optimized. There, in addition to the reflectance behavior in litho 1 and litho 2, a thickness penalty for the top layer was introduced in order to account for manufacturability. By introducing this additional degree of freedom, the performance of the BARC could be further improved, yet not compromising the manufacturability of the BARC.

The materials presented in this work are purely hypothetical. In order to further improve the feasibility of the proposed solutions, a look-up mechanism for existing materials would have to be employed. In practice, the performance of BARCs also depends on their chemical properties. For example, BARC materials are often modified with certain pH levels to control footing and undercut. In some cases a small constant reflectivity value is used to control these effects. Zhu *et al.* proposed to replace the reflectivity objective by a foot exposure criterion to avoid footing effects or undercuts [9]. An according extension of the merit function could be used to cover these and other contributing effects. Moreover, the algorithms could be extended to cover exposure and process induced variations of the BARC thickness and specific pitches and mask layouts of the first lithography step.

Although the studies presented in this work suggest and aim at bi-layer BARCs, this approach can certainly be reduced to an investigation of single layer set-ups.

REFERENCES

1. Maenhoudt, M., Versluijs, J., Struyf, H., Van Olmen, J. and Van Hove, M., "Double patterning scheme for sub-0.25 k1 single damascene structures at NA=0.75, lambda=193nm," Proc. SPIE 5754 (2005)
2. Bae, Y. C., Liu, Y., Cardolaccia, T., McDermott, J. C., Trefonas, P., Spizuoco, K., Reilly, M., Pikon, A., Joesten, L., Zhang, G. G., Barclay, G. G., Simon, J. and Gaurigan, S., "Materials for single-etch double patterning process: surface curing agent and thermal cure resist," Proc. SPIE 7273 (2009)
3. Guerrero, D. J., Gibbons, S., Lowes, J. and Mercado, R., "Anti-reflective coating for multipatterning lithography," Proc. SPIE 6923 (2008)
4. Sakamoto, R., Endo, T., Ho, B.-C., Kimura, S., Ishida, T., Kato, M., Fujitani, N., Onishi, R., Hiroi, Y. and Maruyama, D., "Bottom-anti-reflective coatings (BARC) for LFLE double patterning process," Proc. SPIE 7520 (2009)
5. Brunner, T. A., Seong, N., Hinsberg, W. D., Hoffnagle, J. A., Houle, F. A. and Sanchez, M. I., "High-NA lithographic imagery at Brewster's angle," Proc. SPIE 4691 (2002)
6. Shao, F., Evanschitzky, P., Fühner, T. and Erdmann, A., "Efficient simulation and optimization of wafer topographies in double patterning," Proc. SPIE 7274 (2009)
7. Shokooh-Saremi, M., Nourian, M., Mirsalehi, M. M. and Keshmiri, S. H., "Design of multilayer polarizing beam splitters using genetic algorithm," Optics Communications 233(1-3) (2004)
8. Zitzler, E., Laumanns, M. and Thiele, L., "SPEA2: Improving the Strength Pareto Evolutionary Algorithm for Multiobjective Optimization," Proc. EUROGEN (2001)
9. Zhu, Z., Piscani, E., Edwards, K. and Smith, B., "Reflection control in hyper-NA immersion lithography," Proc. SPIE 6924 (2008)

A Novel Method to Reduce Wafer Topography Effect for Implant Lithography Process

Lei Yuan[*], Sanggil Bae, Yong Feng Fu, Ao Chen, Hui Peng Koh, Qun Ying Lin
GLOBALFOUNDRIES Singapore, 60 Woodlands Industrial Park D, Singapore 738406
[*]Tel: 01-845-894-4413, Email: yuan@us.ibm.com

ABSTRACT

Wafer topography structures in the implant lithography process, which include the shallow trench isolation and the poly gate, can result into a severe degradation of the resist profile and significant critical dimension variation. While bottom anti-reflective coating (BARC) is not suitable for the implant lithography because of the plasma induced substrate damage, developable bottom anti-reflective coating (DBARC) is now the most promising solution to eliminate wafer topography effects for the implant layer lithography. Currently, some challenges still remain to be solved and DBARC is not ready for mass production yet. In this study, a novel method is proposed to improve wafer topography effects by use of sub-resolution features. Compared with DBARC, this new approach is much more cost effective. Numerical study by use of Sentaurus-Litho simulation tool shows that the new method is promising and deserves more comprehensive investigation.

Keywords: wafer topography, implant lithography, sub-resolution feature, lithography simulation

1. INTRODUCTION

The optical reflection and diffraction arising from the wafer substrate is a major source causing the process variation for the advanced photolithography and therefore must be harnessed. The most common and successful solution to eliminate the substrate reflection and diffraction is to utilize bottom anti-reflective coating (BARC). Unfortunately, BARC is not favorable in the implant lithography process due to the potential substrate damage resulted from the dry plasma etch, as well as the process complexity and high cost of ownership. With BARC being out of scope, the wafer topography, including the shallow trench isolation (STI) and the poly gate, presents a serious challenge to process control of the implant layer patterning[1,2]. In the implant lithography process, the STI structure will cause the substrate optical reflection and the poly gate will generate the substrate optical diffraction, both of which may lead to the unwanted irradiation in the targeted resist pattern.

Currently, the most promising solution to eliminate the substrate reflection and diffraction in the implant lithography process is to utilize developable bottom anti-reflective coating (DBARC), which is soluble in aqueous base development and the dry plasma etch can be avoided[3~5]. Two types of DBARC materials are present, which are photo-sensitive and non-photo-sensitive. The challenges in the DBARC application include preventing intermix occurring between DBARC and resist, profile control and defect control. Currently, DBARC is not yet ready for mass production. In this study, a new method is proposed to reduce the wafer topography induced process variation in the implant lithography process by utilizing sub-resolution features. Compared with the DBARC approach, this new method is more cost efficient and avoids process complexity.

This paper consists of seven sections. In Section 2, the experimental and modeling conditions are introduced. In Section 3, the wafer topography modeling in Sentaurus-Litho is verified by comparing the simulation with the experimental observation of a resist line formed on a STI substrate. In Section 4, a new method to reduce the wafer topography effects is proposed and illustrated. The effectiveness of the new method is then verified by numerical study in Section 5. In Section 6, an example is given to show the concept of optimizing sub-resolution features as used in the new method by numerical simulation. At last, a summary is given in Section 7.

Optical Microlithography XXIII, edited by Mircea V. Dusa, Will Conley, Proc. of SPIE Vol. 7640,
76403E · © 2010 SPIE · CCC code: 0277-786X/10/$18 · doi: 10.1117/12.846002

2. EXPERIMENTAL AND MODELING CONDITION

In this study, all wafers are exposed on an ASML KrF scanner, wherein, a conventional illumination with a numerical aperture of 0.8 and sigma of 0.7 is used. A commercially-available KrF resist and top anti-reflective coating (TARC) are coated and processed on a TEL LITHIUS resist track.

Lithography simulations in this paper are done by use of Synopsys lithography software — Sentauraus-Litho version D-2009.12, where a vector imaging model is applied to calculate the optical image intensity in the resist film stack and a reaction-diffusion system is assumed to model the post-exposed bake (PEB) process. A wafer topography model is provided in Sentauraus-Litho to simulate the resist pattern formation on a three dimension wafer topography, wherein, a finite difference time domain (FDTD) algorithm is used to solve electro-magnetic field in the resist film stack as well as the wafer substrate.

Two types of wafer topography structures common in the implant lithography process are the shallow trench isolation (STI) silicon oxide and the poly gate, a simplistic view of which are shown in Fig. 1. For the shallow trench isolation, a trench depth of 300nm and a sidewall angle of 85 are assumed in this study. For model simplicity, it is assumed that there is zero step height passing from silicon to silicon oxide. In the poly gate simulation, the gate spacer structure is neglected also for model simplicity.

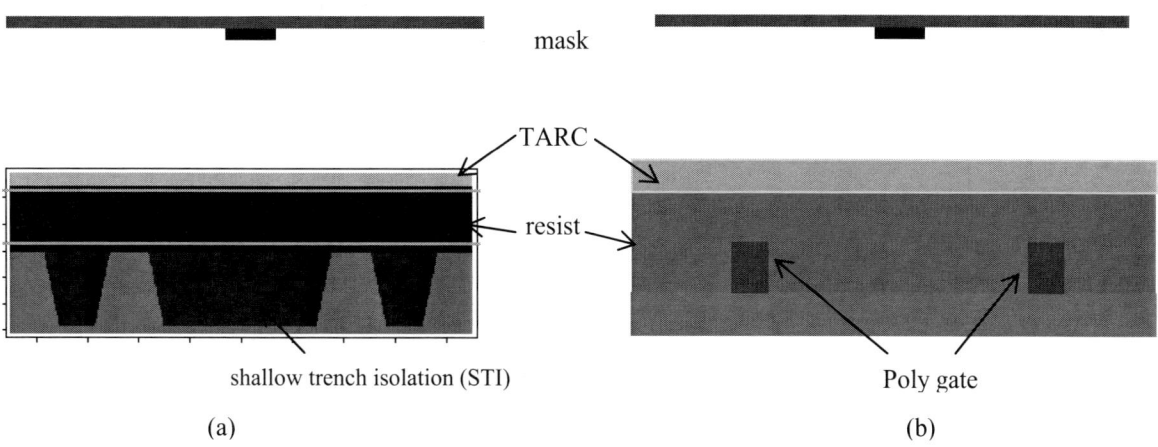

Figure 1 (a) shallow trench isolation (STI) substrate (b) poly gate substrate

3. EXPERIMENTAL VERIFICATION OF WAFER TOPOGRAPHY MODELING

The accuracy of the wafer topography modeling in Sentauraus-Litho is verified by comparing the model simulation of a resist line formation on a STI substrate with experimental measurements.

The first experiment is to print 150nm isolate resist line on both blank substrate and STI substrate. As shown in Fig. 2a~b, a straight resist line is produced on the blank substrate while severe resist undercut (reflection notching) and CD shrinkage are observed on the STI substrate. Both substrates are simulated by use of the wafer topography model and the resulting resist profiles are presented in Fig. 2c~d. It shows that the wafer topography modeling can successfully predict the formation of the resist undercut as observed on the STI substrate.

The second experiment is to produce 180nm resist line on a STI substrate having a trench isolation of 740nm wide. A set of overlay shifts ranging from -260nm to 260nm is applied intentionally to observe the resulting CD variation. Experimental CD results are shown in Fig. 3a. It shows that, while the resist line moves from the STI center to active silicon, its CD firstly increases to reach a peak value at about 200nm overlay shift and then starts to decrease slightly.

The simulation results as shown in Fig. 3b correctly predicts the experimental measurement within the experimental and simulation error.

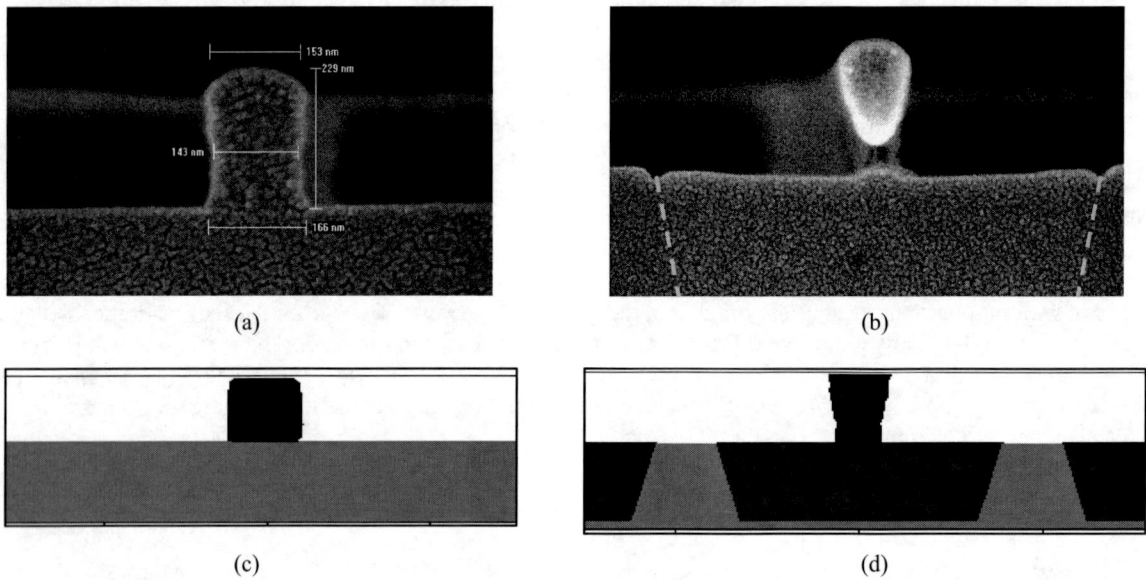

(a)

(b)

(c)

(d)

Figure 2 (a) Experiment of isolate resist line on the blank wafer (b) experiment of isolate resist line on the STI substrate (c) simulation of isolate resist line on the blank wafer (d) simulation of isolate resist line on the STI substrate

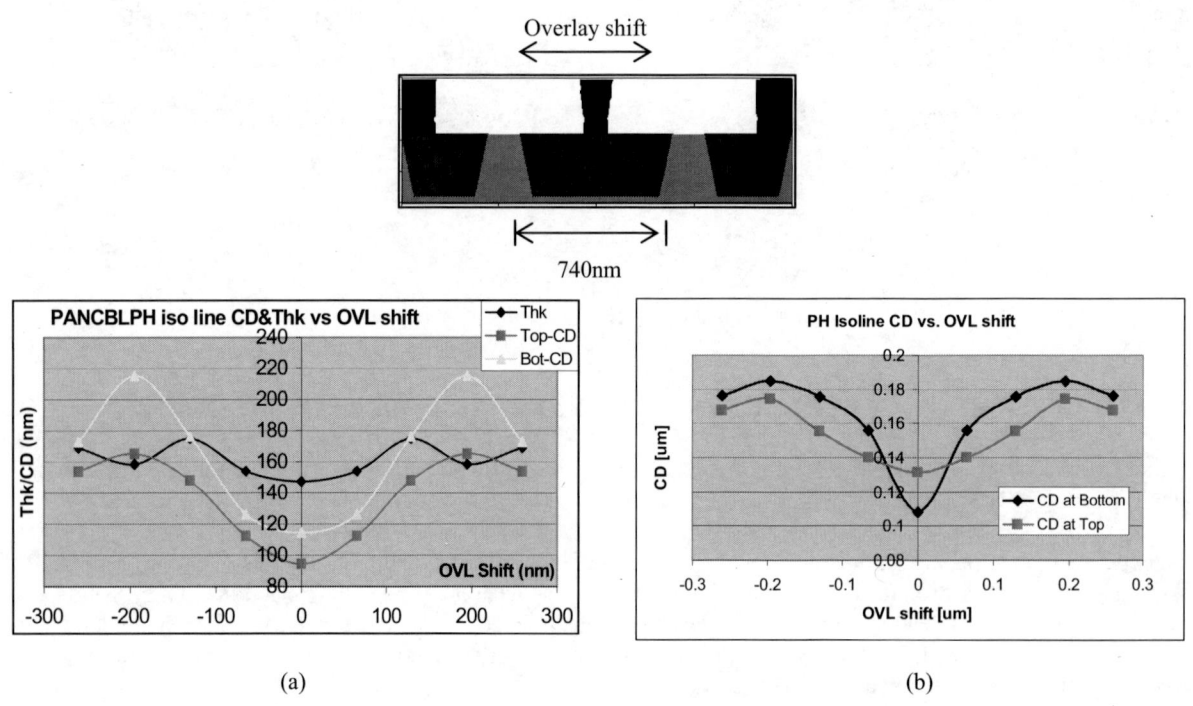

(a)

(b)

Figure 3 Top and bottom CD vs. overlay shift (a) experiment measurements (b) simulation results

4. METHOD OF UTILIZING SUB-RESOLUTION FEATURE TO REDUCE WAFER TOPOGRAPHY EFFECT

This section firstly illustrates the mechanism of wafer topography effects on the implant lithography performance for both STI substrate and poly gate substrate. According to the insight, a method of reducing the wafer topography effects is proposed and illustrated.

Sec. 4.1 Mechanism of wafer topography induced patterning degradation

The performance degradation of implant lithography on the STI substrate as revealed in section 3 is caused by a destructive bottom reflection as shown in Fig. 4a, wherein, optical light of certain incident angle is reflected back to the resist pattern through a series of reflections on the silicon-SiO2 interface. The unwanted reflection light that propagates into the main feature will result into a significant change of the main feature CD and a resist undercut, which is also called destructive bottom reflection in this study.

Different from the STI substrate, wafer topography effects on a poly gate substrate are resulted from the optical diffraction on the poly gate as shown in Fig. 4b. A fraction of the diffracted light may irradiate the main feature and, consequently, changes its critical dimension as well as the resist profile.

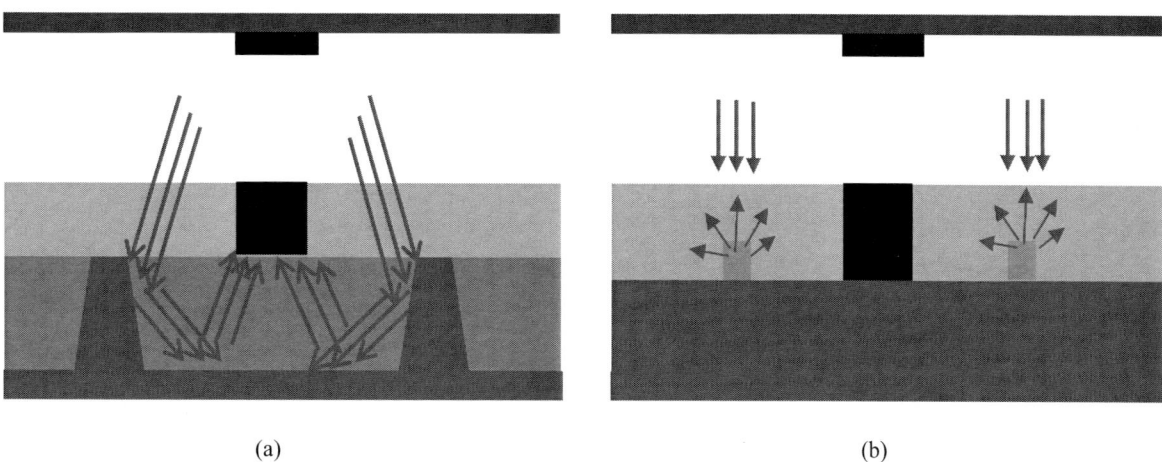

(a)   (b)

Figure 4 (a) Destructive bottom reflection on a STI substrate (b) destructive optical diffraction on a poly gate substrate

Sec. 4.2 Method of applying sub-resolution features to improve patterning on a topography substrate

A new method is proposed to reduce wafer topography effects by utilizing sub-resolution features to produce "dark" regions in the resist film that will mitigate the incoming optical light attributed to the destructive bottom reflection or diffraction. The new method is illustrated by two schematic plots in Fig. 5, wherein, the "dark" region projected by the sub-resolution features perform like a light blockade that greatly alleviates the unwanted light into the main feature. Therefore, the sub-resolution feature that is used in this approach is also called sub-resolution light blockade (SRLB) in this paper. The "dark" region is quoted because it is not completely dark and will not print on wafer.

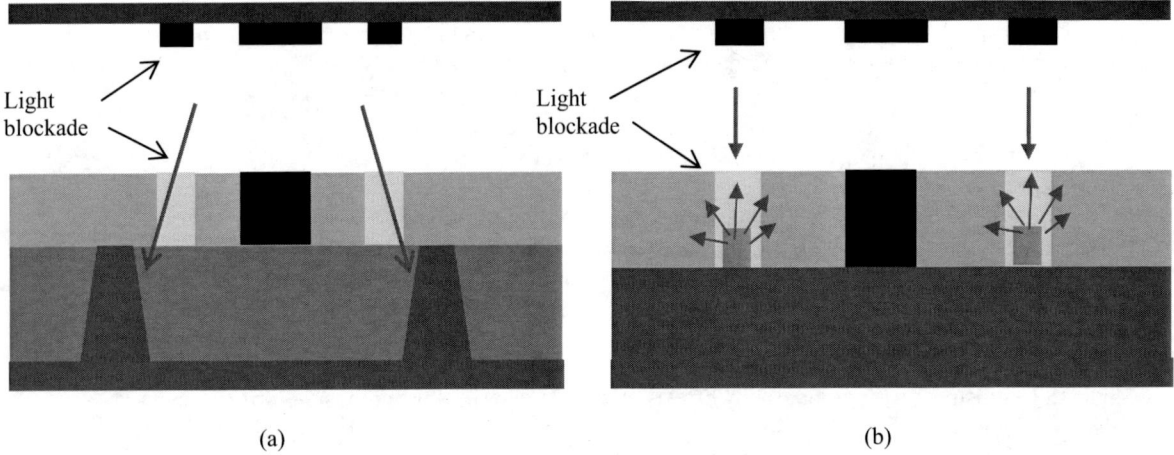

(a)   (b)

Figure 5 Concept of applying sub-resolution light blockades for (a) STI substrate (b) poly gate substrate

The principle of the sub-resolution feature used in this study is different from that of the conventional sub-resolution assistant feature (SRAF). SRAF is applied to enhance the process window of isolate and semi-isolate features by taking advantage of the optical interference between the main features and the assistant features. In contrast, the sub-resolution feature used here performs as a light blockade. The applications of SRAF and SRLB are also very different. SRAF need be inserted to mimic the optimum pitch under a determined illumination condition and has no relevance with the underlying wafer topography at all. Conversely, SRLB need be placed to produce a light obstruction that most effectively alleviates the destructive light arising from the underlying wafer topography.

5. COMPUTATIONAL VERIFICATION

In this section, the effectiveness of the new method will be verified by a numerical study of isolate resist line patterning on both STI substrate and poly gate substrate. Simulations are performed for masks with and without SRLB and the results are compared with that on blank wafer.

Sec. 5.1 On STI substrate

Fig. 6a and 6b show the simulated resist profile of 200nm resist line formed on a blank wafer and a STI substrate respectively by use of conventional mask with no SRLB. On the STI substrate, the trench isolation is 600nm wide. It is found that STI substrate results into both a severe resist undercut and about 50% CD shrinkage. A new mask using 55nm SRLB is also simulated and the results are shown in Fig. 6c. It shows that, by adding SRLB, a straight resist line can be obtained the same as on blank wafer and CD on target can be achieved. It is worthy to point out that some minor line edge roughness is observed for the mask using SRLB, which is a consequence of the image contrast degradation.

Sec. 5.2 On poly gate substrate

Firstly, the image profile generated by taking average of image intensity along the vertical direction is examined. Fig. 7a shows an obvious discrepancy between the image profiles resulted from blank wafer and the poly gate substrate. As a comparison, the image profiles as generated by using mask with SRLB are shown in Fig. 7b. It is found that, by applying SRLB, the imaging on blank substrate and poly substrate becomes very close.

Fig. 8a and 8b show the cross section image intensity of 250nm isolate resist line printed on blank wafer and a poly substrate respectively by use of conventional mask with no SRLB. On the poly substrate, two poly lines that are 80nm wide and 100nm high are located 200nm away from the main feature. Fig. 8c shows the simulation result for the same resist pattern on the same poly substrate by use of mask having two 50nm SRLB. Table 2 shows the comparison of the resulting critical dimension for all three cases. It is found that the poly gate substrate results into about 25% CD shrinkage while the new method produces the similar CD as on the blank wafer.

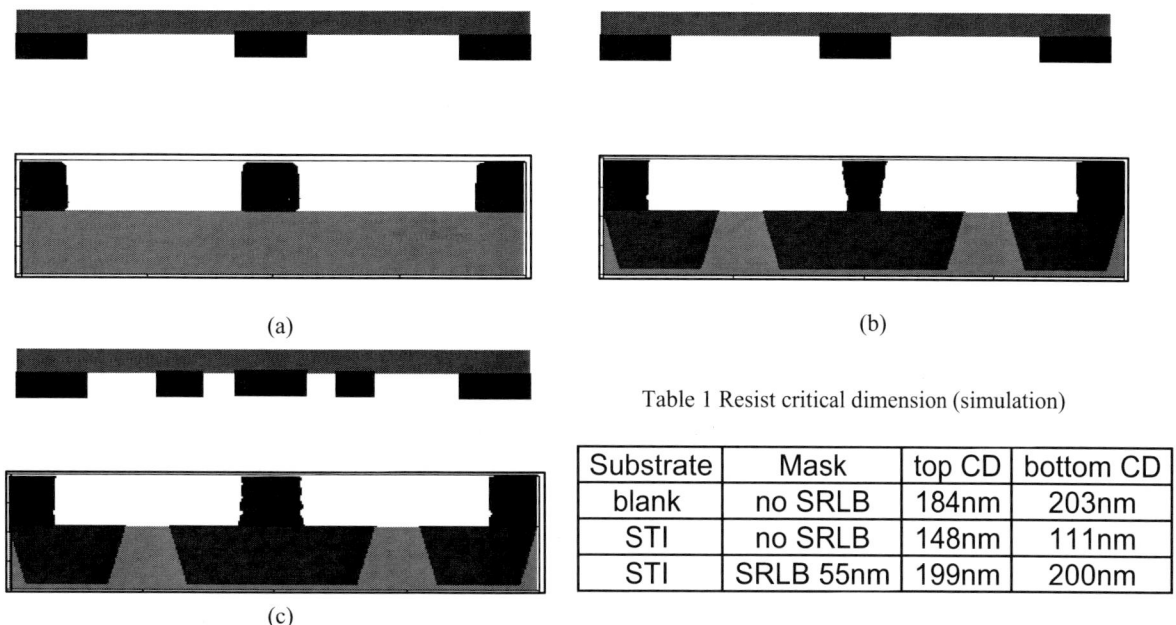

(a)

(b)

(c)

Table 1 Resist critical dimension (simulation)

Substrate	Mask	top CD	bottom CD
blank	no SRLB	184nm	203nm
STI	no SRLB	148nm	111nm
STI	SRLB 55nm	199nm	200nm

Figure 6 Resist profile simulation of (a) resist line printing on a blank wafer (b) resist line printing on a STI substrate (c) resist line printing on a STI substrate by the utilization of SRLB

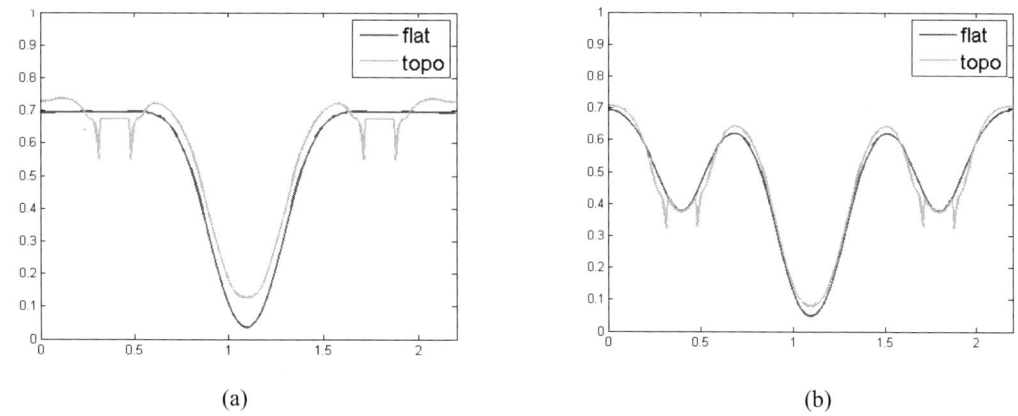

(a)

(b)

Figure 7 Image profile simulation results (a) with no SRLB (b) with SRLB

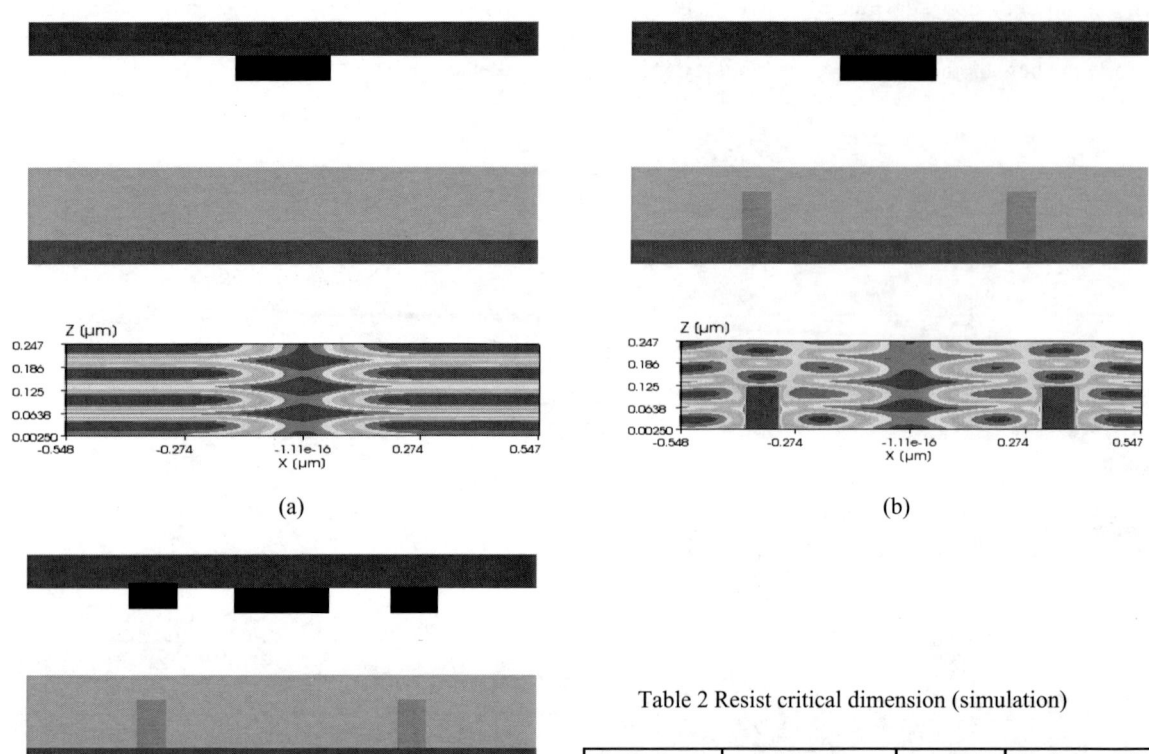

Table 2 Resist critical dimension (simulation)

Substrate	Mask	top CD	bottom CD
blank	no SRLB	247nm	253nm
poly gate	no SRLB	199nm	206nm
poly gate	SRLB 50nm	247nm	253nm

(a)

(b)

(c)

Figure 8 Image intensity (a) on a blank substrate (b) on a poly gate substrate without using SRLB (c) on a poly gate substrate by use of SRLB

6. OPTIMIZATION OF SUB-RESOLUTION FEATURE

The effectiveness of SRLB application is dependent on its size and position. In general, a large size is preferred for SRLB application before it starts to print on wafer. The optimum position for SRLB placement is dependent on the underlying topography and can be determined with the assistance of numerical simulation.

Fig.9a shows an example of 200nm resist line patterned on a substrate with a trench isolation of 600nm wide, where a 55nm SRLB is in use. To determine the optimal SRLB placement, resist patterning under various spaces is simulated and the resulting CDs and sidewall angles are shown in Fig. 9b. In this specific example, the 150nm space generates CD on target as well as 90 degree sidewall angle, and, therefore, is the optimal.

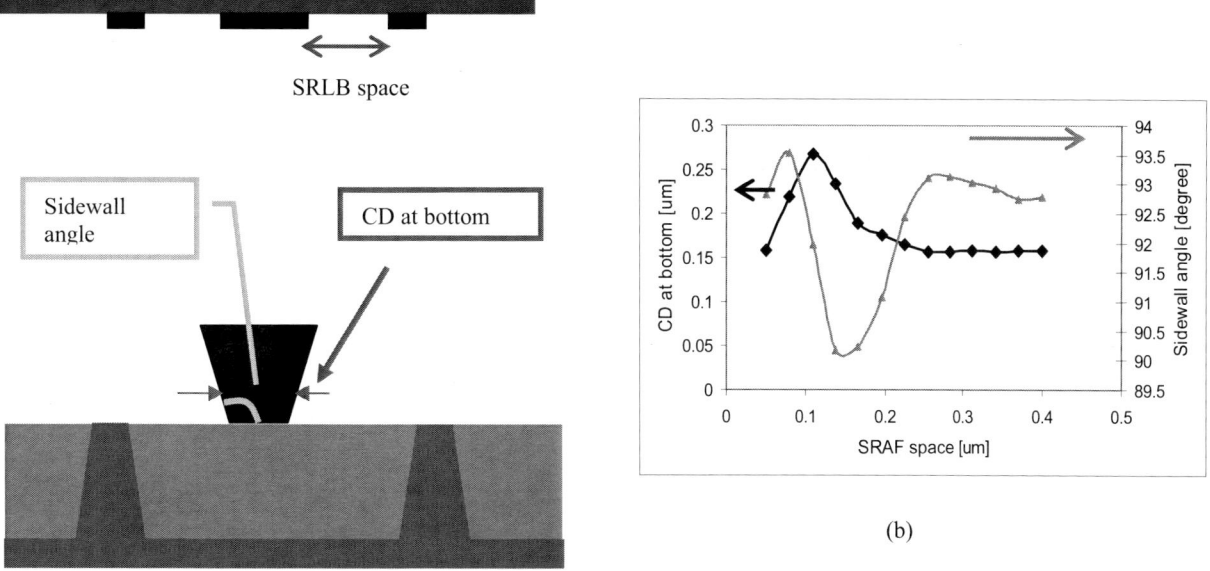

Figure 9 (a) 200nm resist line on a substrate with a trench isolation of 600nm span (b) CD and sidewall angle vs. SRLB space

7. SUMMARY

This study first analyzed the mechanism of the destructive substrate reflection and diffraction induced by the wafer topography, including the shallow trench isolation (STI) and the poly gate. According to this insight, a new method is proposed to improve the wafer topography effects by use of sub-resolution features, which can form sub-resolution light blockade (SRLB) in the resist layer to alleviate the destructive substrate reflection and diffraction. Preliminary numerical verification has been done and the new method is found to be very promising. In the future, more comprehensive simulation and experimental study are needed to prove its effectiveness on more complex substrates.

ACKNOWLEDGEMENTS

Authors would like to thank Puay Cheng of GLOBALFOUNDRIES Singapore for providing X-SEM pictures; thank Weiming Gao and Swee Guan Ker of Synopsys for the help on Sentaurus-Litho application.

REFREENCES

1. Word, J., Chou, D., Gu, Y. and Sturtevant, J., "Reduction of reflection notching through illumination optimization", Proc. Of SPIE, vol. 4691, p.990, 2002.
2. Bailey, T.C., McIntyre, G., Zhang, B. *et al*, "Reflectivity induced variation in implant layer lithography", Proc. Of SPIE, vol. 6924, 2008
3. Cameron, J., Amara, J., Prokopowicz, G. *et al*, "Progress toward production worthy developable BARCs (DBARCs)", Proc. Of SPIE, vol. 7273, 2009
4. Kudo, T., Chakrapani, S., Dioses, A. *et al*, "Latest development in photo-sensitive developable bottom anti-reflective coating (DBARC)", Proc. Of SPIE, vol.7250, 2009
5. Houlihan, F.M., Dioses, A., Kudo, T. *et al*, "Radiation sensitive developable bottom anti-reflective coating (DBARC): recent results", Proc. Of SPIE, vol.7250, 2009

Immersion BARC for Hyper NA Applications ll

Yu-Chin Huang, Kai-Lin Chuang, Tsung-Ju Yeh, Steven Wu, Bill Lin, Wen-Liang
Huang, Bo-Jou Lu, E.T. Liu, and Chun Chi Yu
Advanced Lithography Department, ATD Advanced Modules.
United Microelectronics Corporation, Tainan County 744, Taiwan, ROC

Chaoyang Lin, Jeong Yun Yu, Greg Prokopowicz, Sue Ryeon Kim, Sabrina Wong, and
George Barclay
Dow Electronic Materials, The Dow Chemical Company
455 Forest Street, Marlborough, Massachusetts 01752, USA

ABSTRACT

Reflectivity control through angle is challenging at hyper NA, especially for Logic
devices which have various pitches in the same layer. A multilayer antireflectant system
is required to control complex reflectivity resulting from various incident angles. In our
previous works, we showed the successful optimization of multilayer antireflectant
systems at hyper NA for BEOL layers. In this paper, we show the optimization of new
multilayer bottom anti-reflectant systems to meet new process requirements at 28nm
node Logic device. During the manufacturing process, rework process is necessary when
critical dimension or overlay doesn't meet the specifications. Some substrates are
sensitive to the rework process. As a result, litho performance including the line width
roughness (LWR) could change. The optimizations have been done on various stack
options to improve LWR. An immersion tool at 1.35NA was used to perform lithography
tests. Simulation was performed using Prolith[TM] software.

Keywords: Immersion lithography, Reflectivity control, Substrate effect

1. INTRODUCTION

As the semiconductor industry moves to immersion lithography, reflectivity control has
become more difficult to achieve. There are many types of substrates being used
depending on the application, making reflectivity control even more complex. The
multilayer stacks at hyper NA for critical layers are needed to improve the aerial image[1-3].
Some substrates are sensitive to the rework process. Litho performance including the line
width roughness (LWR), process window and CDU (critical dimension uniformity) could
be changed. Adding the layer A into the stack is desired to improve the litho performance.
In our previous work[4-5], we had shown the successful optimization of a multilayer
antireflectant system at 1.35NA for metal and via layers at 32nm node logic devices. We
demonstrated the optimization with CDU, through pitch performance, LWR and profiles.
We also showed the multilayer bottom antireflectant system with thinner films in a
simplified multilayer stack at 1.35NA for the critical metal trench layer of the 28nm node
Logic device. Four different organic antireflectants in terms of n/k, and thickness have
been evaluated.

Optical Microlithography XXIII, edited by Mircea V. Dusa, Will Conley, Proc. of SPIE Vol. 7640,
76403F · © 2010 SPIE · CCC code: 0277-786X/10/$18 · doi: 10.1117/12.848454

In this paper, the layer A was added into the stacks on the silicon to improve litho performance at hyper NA with immersion lithography. Three organic antireflectants were compared at various thicknesses as dictated by their n&k of 1st and 2nd minimum applications, respectively.

2. EXPERIMENTAL

ProlithTM version 10.2 was used for reflectivity simulations. Exposure was performed by using a 1.35NA immersion tool to print 90nm, 280nm and 1500nm pitches. Previously, a SiON film was used in the stack of the metal trench layers of the 32nm nodes. On the 28nm node, the Layer A was added into the stacks to prevent rework effect. Three organic antireflectants with different n/k were evaluated. Sample 1 was selected for a 50nm thickness target as the 1st minimum antireflectant, and Sample 2 and Sample 3 were selected for 84nm and 85nm thickness, respectively, as 2nd minimum antireflectants.

These three Dow Chemical spin on organic antireflectants, covering a thickness range of 50nm to 88nm, n range of 1.62 to 1.91, k range of 0.21 to 0.25, and a range of etch selectivity were used, as indicated in Table 1 of results. Percent solids were targeted to obtain the optimal thickness at 1,500rpm. All the samples were cured at 205°C/60sec. VUV Vase ellipsometer and WVASE32 software from J.A. Woollam were used for obtaining n/k of the antireflectants.

Layer	METAL
Stack	PR
	Antireflectant
	Layer A
	Layer B
	Substrate

3. RESULTS

Several steps are required for the optimization. First, one must find the optimal n/k target for a given thickness by n/k contour plots. Second, the substrate reflectivity curves must be generated for each BARC being fine tuned by looking at angle dependency

3.1 Table 1. Sample Summary

Samples	n/k	Lowest %R	Optimum thickness, nm (t_{op})	Etch Rate, Relative to Sample 1	Etch Time to Open BARC t_{op}, Relative to sample 1
Sample 1	1.62/0.22	~0.41	50	1.00	1.00
Sample 2	1.91/0.25	~0.58	84	1.24	1.35
Sample 3	1.87/0.21	~0.43	85	1.15	1.48

3.2 Reflectivity Simulation

Figure 1 shows n/k contour plots of organic antireflectants at thicknesses of 50nm, 80nm, and 90nm for the new metal layer stack of the 28nm node device.

a) 50nm antireflectant

b) 80nm antireflectant

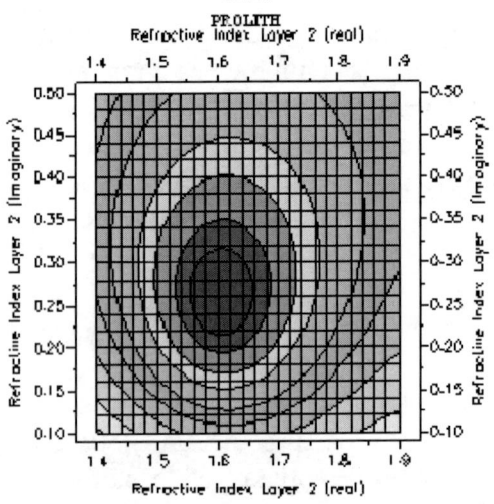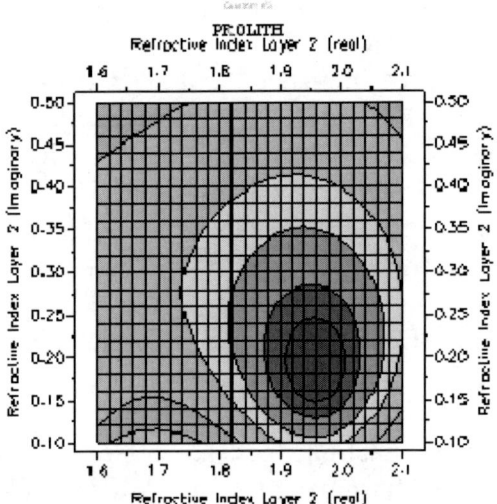

c) 90nm antireflectant

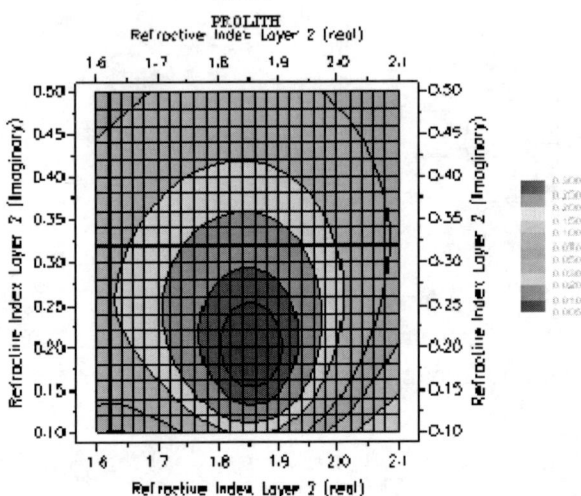

Figure 1: n/k contour plots for a) 50nm antireflectant, b) 80nm antireflectant, and c) 90nm antireflectant over the metal layer for the 28nm node.

According to the contour plots, 50nm antireflectants having optical constants of 1.52-1.68n and 0.2-0.35k would give less than 1.0% substrate reflectivity. The second and third plots show that 1.88-2.03n and 0.13-0.28k would be suitable for 80nm antireflectants and 1.76-1.94n and 0.14-0.25k would be appropriate for 90nm antireflectants to achieve less than 1.0%.substrate reflectivity. Three organic antireflectants around the optimum n/k at the three thickness targets were selected for the evaluation.

Figure 2 shows the reflectivity curves of the evaluated antireflectants on the new stack of the metal layer for the 28nm node device. Sample 1 has 1st minimum at the target thickness, and Samples 2 and 3 have 2nd minimum at the target thickness. The minimum reflectivity of all the samples is less than <0.6%, regardless of 1st or 2nd minimum antireflectants thickness.

a) Sample 1(1st minimum)

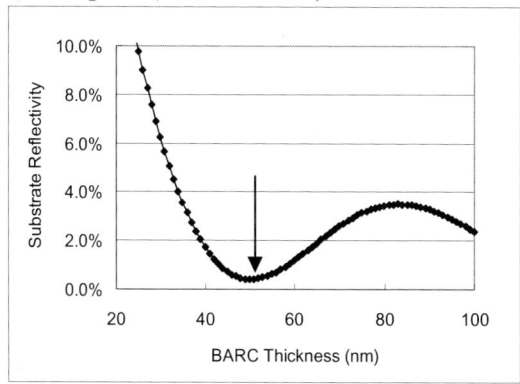

b) Sample 2(2nd minimum)

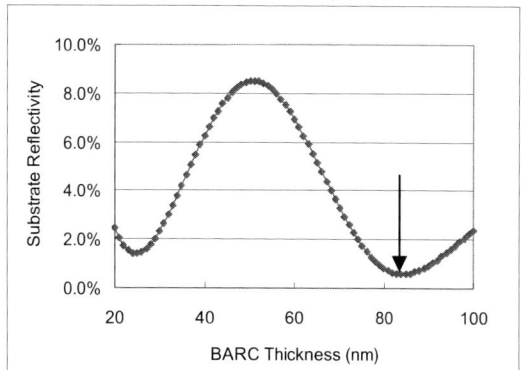

c) Sample 3(2nd minimum)

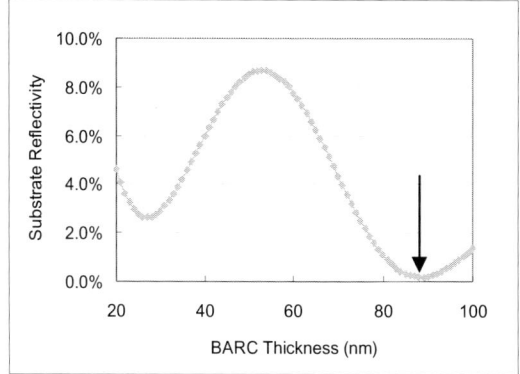

Figure 2 Reflectivity curves of the evaluated antireflectants on the new stack of the metal layer for the 28nm node device.

3.3 Lithography

The resist process windows for these three evaluated samples are shown in Figure 3. Samples 2 and 3 showed similar process windows, but smaller than Sample 1. Usually, thick organic antireflectant will better compensate for high topography but need longer etch time. However, in our study, the thin antireflectant (Sample 1) showed the widest process window and shorter etch time making it effective in terms of the pattern transfer.

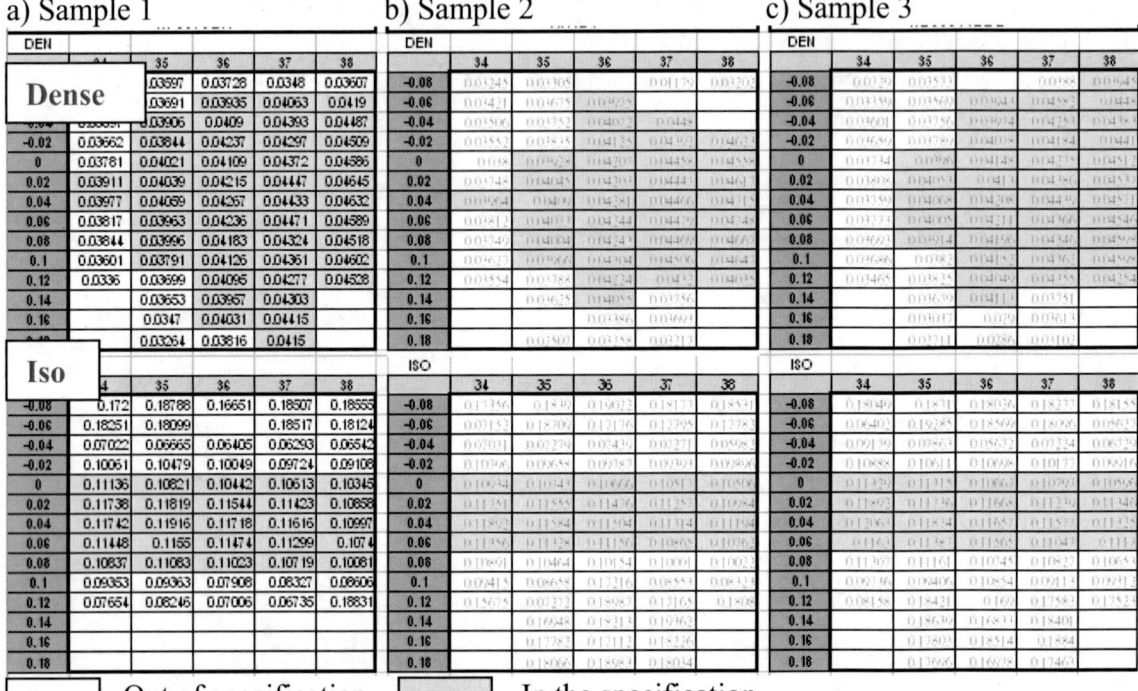

a) Sample 1   b) Sample 2   c) Sample 3

Out of specification   In the specification

Figure 3 Process windows of the evaluated antireflectants on the new stack of the metal layer for the 28nm node device.

a) Sample 1 (Dense)   (Iso)

b) Sample 2 (Dense)   (Iso)

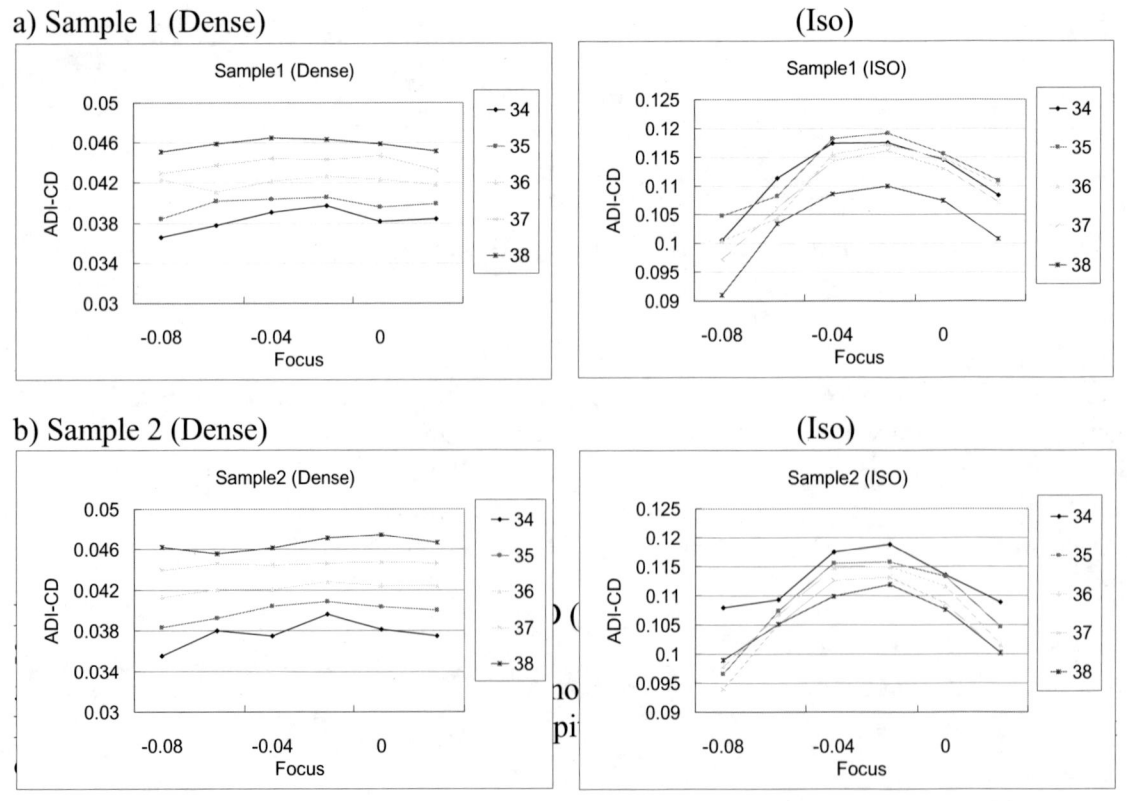

c) Sample 3 (Dense)   (Iso)

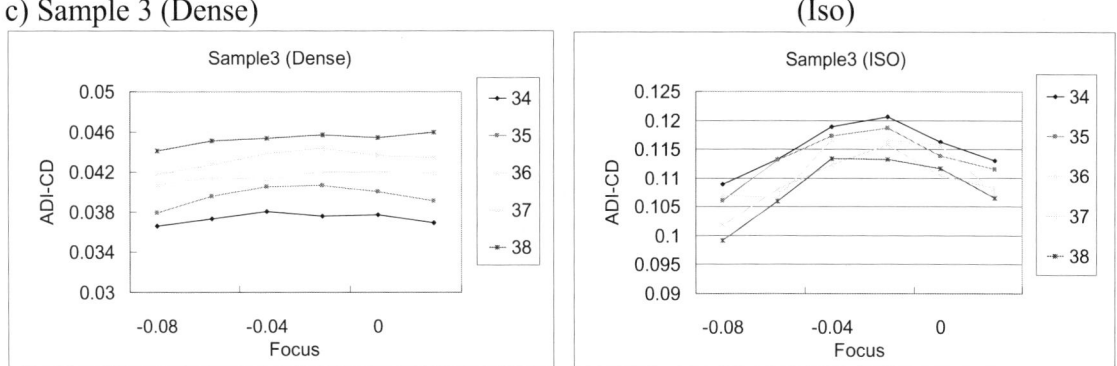

Figure 4 Iso and Dense Process windows of Metal ADI-CD on antireflectants a) 50nm Sample 1 b) 84nm Sample 2 c) 85nm Sample 3 over the stack of the 28nm node device.

a) Sample 1 (Dense)   (Iso)

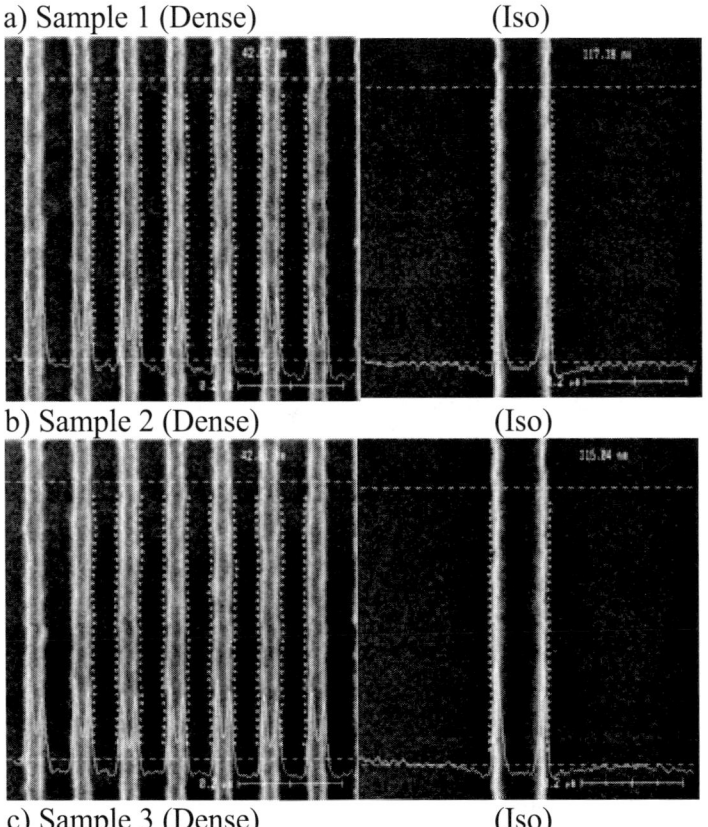

b) Sample 2 (Dense)   (Iso)

c) Sample 3 (Dense)   (Iso)

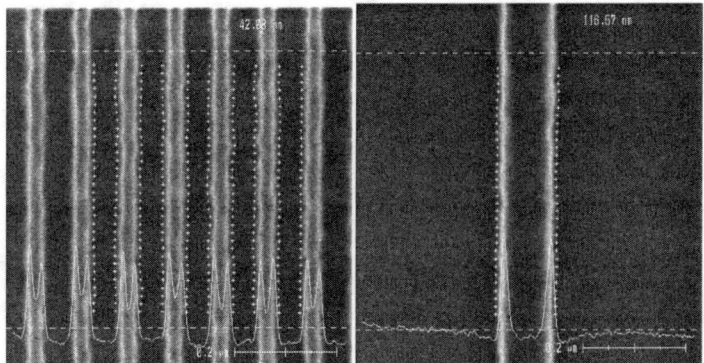

Figure 5 Iso and Dense Top-down profiles of Metal ADI-CD on antireflectants a) 50nm Sample 1 b) 84nm Sample 2 c) 85nm Sample 3 over the stack of the 28nm node device.

Figure 4 show the ADI-CD data of these three antireflectants on the stack of 28nm node device measured by Iso and Dense metal trench, respectively.

From the top-down Iso and Dense profiles Figure 5, the line width roughness for sample 1 is better than sample 3. Sample 2 is the worst. LWR and CD were measured by Hitachi CD-SEM CG4000 and the data is shown in Table 2.

Table 2 LWR and Process window Summary

	Sample 1	Sample 2	Sample 3
LWR	3.85	5.31	4.59
Process Window	widest	small	small

Over 120 points were measured to obtain the CDU as shown in Figure 9.
CDU for 42nm dense pitch has been well controlled on all samples over the stack with the layer A, and the 3 sigma within a wafer is an acceptable level at 1.6 to 1.7nm.

a) 50nm Sample 1   b) 84nm Sample 2   c) 85nm Sample 3

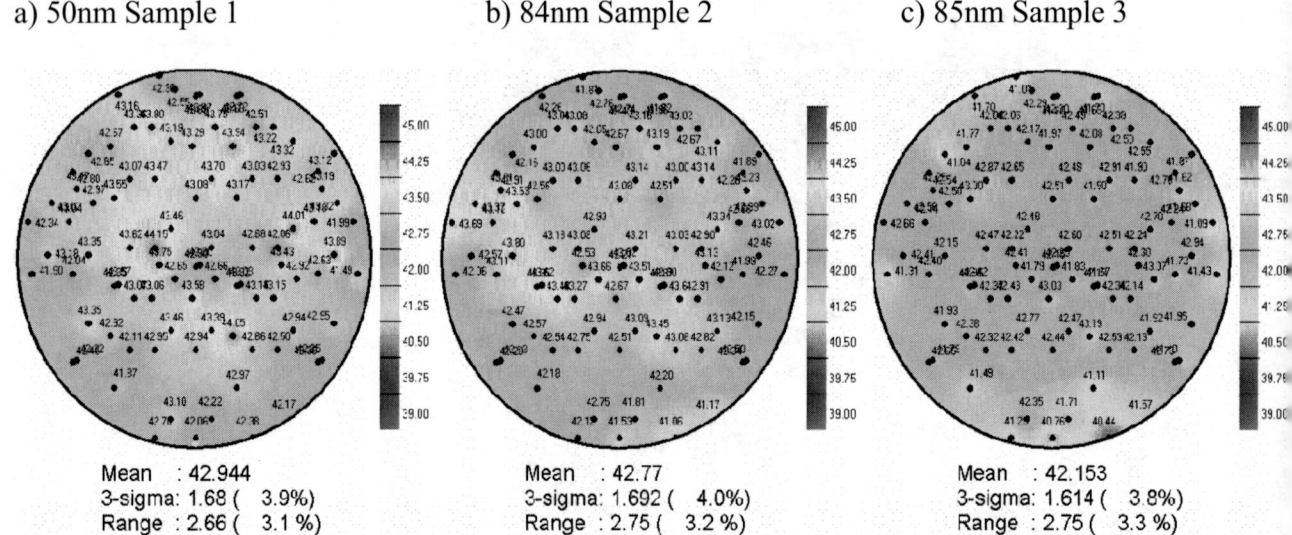

Mean : 42.944   Mean : 42.77   Mean : 42.153
3-sigma: 1.68 (3.9%) 3-sigma: 1.692 (4.0%) 3-sigma: 1.614 (3.8%)
Range : 2.66 (3.1 %) Range : 2.75 (3.2%) Range : 2.75 (3.3 %)

Figure 6 CDU of Metal ADI-CD on antireflectants a) 50nm Sample 1 b) 84nm Sample 2
c) 85nm Sample 3 over the stack of the 28nm node device.

4. SUMMARY

A number of antireflectants options are available for metal layer stack with the layer A on
the 28nm node application. Through 1st or 2nd minimum thickness, range of n& k, process
window, LWR and CDU, good litho performance can be obtained. The stack including
layer A has been well optimized based on simulations and lithographic experiments using
various organic antireflectants.

REFERENCES

[1] Andreas Erdmann, et al., "Mask and Wafer Topography Effects in Immersion
lithography", Proc. SPIE, 5754, 383-394 (2005).

[2] Vincent Farys, et al., "ARC Stack Development for Hyper NA Imaging", Proc. SPIE,
6520, 65204O (2007).

[3] Jeong Yun Yu, et al., "Analysis of Resist Pattern Collapse and Optimization of DUV

Process for Patterning Sub-0.20µm Gate Line", Proc. SPIE, 3333, 880-889 (1998).

[4] Wan-Ju Tseng, et al., "Reflection Control for Immersion Lithography at 45/32nm
Nodes", Proc. SPIE, 6923,69232Z (2008)

[5] Sue Ryeon Kim, et al., "Immersion BARC for Hyper NA Applications", Proc. SPIE,
(2009).

[6] Sabrina wong, et al., "Reflection Control for Immersion Lithography: A single
organic antireflectant over high reflective substrates for double patterning", Proc. SPIE,
(2009).

Methods and Challenges to extend existing dry 193nm medium NA lithography beyond 90nm

Dr. Jens Schneider, Andreas Greiner, ChinTeong Lim, Vlad Temchenko, Dr. Felix Braun, Dieter Kaiser, Tarja Hauck, Dr. Ingo Meusel, Dietrich Burmeister, Dr. Stephan Loehr, Susanne Volkland, Astrid Bauch, Hendrik Kirbach and Daniel Sarlette
Infineon Technologies Dresden GmbH
PO Box 10 09 40, D-01079, Dresden, Germany

Katrin Thiede
Canon Deutschland GmbH
Buchenstr. 16, D-01097 Dresden, Germany

ABSTRACT

In order to fulfill the demands of further shrinkage of our mature 90nm logic litho technologies under the constraints of costs and available toolsets in a 200mm fab environment, a project called "Push to the Limits" was started. The aim is to extend the lifetime and capabilities of existing dry 193nm litho toolsets with medium to low numerical aperture, coupled with the availability of materials and processes which were known to help up CD miniaturization and to shrink the 90nm logic litho process as far as possible. To achieve this, various options were explored and evaluated, e.g. optimization of illumination conditions, evaluation of new materials, usage of advanced RET techniques (OPC, LfD, DfM and ILT) and resolution enhancement by chemical shrink (RELACS®). In this project we demonstrate how we were able to extend our existing 90nm technology capability, down close to 65nm node litho requirements on most critical layers. We present overall result in most critical layer generally and specifically on most difficult layer of contact. Typical contact litho target at 100nm region was enabled, while realization of 90nm ADI target is possible with addition of new process materials.

Keywords: sub 90nm, chemical shrink, dry 193nm, medium NA

1. INTRODUCTION

As the demands for further shrinkages of existing technology nodes are increasing, the development of a suitable lithography process with acceptable lithographic margin becomes more and more challenging, especially under the constraints of costs and available tool sets. Due to escalating costs of high-NA exposure tool systems, one of the most important tasks in a semiconductor device fabrication is the extension of existing tool capability to fulfill the increased requirements with respect to CD/pitch minimization and overlay accuracy consequently.

Table 1. Design rule comparison

	L65 Min Width/Space	L90FLR Min Width/Space	L90N Min Width/Space
RX	100 / 100nm	100 / 120nm	120 / 140nm
PC	60 / 130nm	80 / 200nm	80 / 160nm
CA	90 / 100nm	90/ 130nm	120 / 160nm
M1	90 / 90nm	100 / 100nm	120 / 120nm
V1	100 / 100nm	120 / 120nm	140 / 180nm
M2	100 / 100nm	120 / 120nm	140 / 140nm

Optical Microlithography XXIII, edited by Mircea V. Dusa, Will Conley, Proc. of SPIE Vol. 7640, 76403G · © 2010 SPIE · CCC code: 0277-786X/10/$18 · doi: 10.1117/12.848207

This paper will describe the lithography process results for a sub 90nm node using dry 193nm lithography within a project called "Push to the Limits" (PTL). The process development was done on Canon FPA-6000 AS4 Scanner. The Canon AS4 Scanner is a 193nm λ scanner targeted for volume chip production at 90nm node. The AS4 system has a variable numerical aperture (NA) from 0.55 to 0.85. Target of this sub 90nm project was the achievement of sufficient process margins through pitch, such as DOF, EL, CD uniformity and overlay performance for gate, contact, via and metal levels; key parameters for good device performance, especially within a fabrication of random logic devices. The sub 90nm process was installed by optimization of our Canon AS4 scanner and advanced tool maintenance and calibration with dynamic distortion and focus control. Resolution Enhancement Techniques (RET) like Optical Proximity Correction (OPC), Litho friendly Design (LfD) and Design for Manufacturing (DfM) are important parts of our process development, as well as common resolution enhancement techniques such as off-axis illumination, edge movement, or applying sub-resolution assist features. Inverse Lithography Technology (ILT) [4,5,6] is under development to extend the tool resolution further and to enlarge the process window for random as well as periodic mask patterns. In addition to the above mentioned RET evaluations, the Resolution Enhancement of Lithography by Assist of Chemical Shrinkage (RELACS®) [7,8] was installed and evaluated. The results of the RELACS® process evaluation for hole and trench levels will be shown.

2. EXPOSURE TOOL OPTIMIZATION AND RESULTS

In order to fulfill the requirements of the PTL project, the Canon AS4 scanners were optimized. The illumination modi were adapted with respect to the different layer requirements (Table 1). Main tool modifications were done together with Canon with respect to CD uniformity and lens aberrations optimization. CD uniformity within the exposure shot was improved by optimization of IUC/slit settings and the use of Active Dose Control (ADC) machine parameter. The front glass, which influences the light intensity and uniformity, was replaced by a newly polished and coated one. Figure 1 shows the exposure results without any optimizations for a dense contact hole pattern with CD = 120nm. Figure 2 displays the CD uniformity status for the same contact hole pattern after the above described optimization procedure.

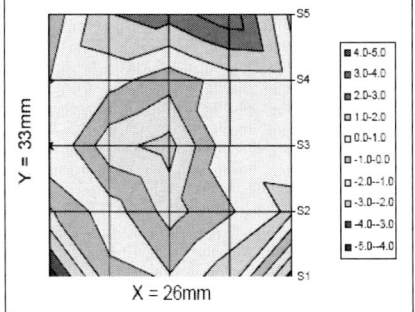

Figure 1.. deviation to average CD
without any optimization
CD range = 15,6nm
CD 3sig = 8.6nm

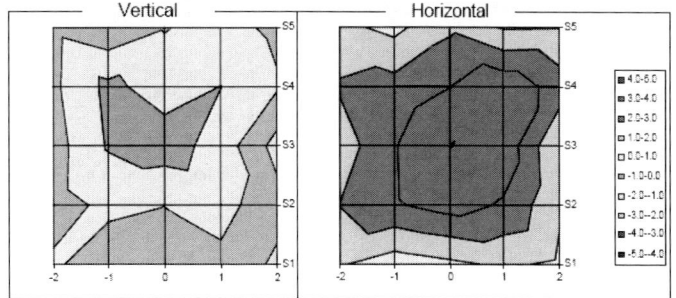

Figure 2. deviation to average CD
after optimization
CD range: V = 8.8nm, H = 8.4nm
CD 3sig: V = 3.6nm, H = 4.4nm

The lens aberrations optimization was done by implementation of a new front glass, by lens adjustment and sheet glass adjustment. The optimization was done together with Canon, whereas the lens adjustment was carried out in field. Table 2 provides an overview about the aberration topics and the corresponding adjustments. The resulting improvements on the scanner performance is shown in table 3.

Table 2. Optimization topics

topic	adjustable by
spherical aberration	reticle stage position
horizontal coma	sheet glass
45 deg astigmatism tilt	lens adjustment
45 deg astigmatism	sheet glass
vertcial coma tilt	lens z drive
distortion	front glass
	lens adjustment

Table 3. Lens aberrations improvement results

			before optimization	after optimization
C6 on-axis	45deg As	average across slit	25.7mλ	1.7mλ
C6 off-axis	45 deg As	range across slit	70.8mλ	11.2mλ
C7 off-axis	V coma	range across slit	72.3mλ	10.3mλ
C8 on-axis	H coma	average across slit	46.9mλ	1.4mλ
C9 on-axis	spherical	average across slit	11.3mλ	2.6mλ

The above described optimization procedures, done on our Canon AS4 scanners, result in tool capabilities, which are suitable for the PTL project. The resulting distortion improvement was 5nm for the new front glass and 2nm in x and y for the lens adjustment, respectively. The lens aberrations optimization results in a value 3.3mλ for spherical aberration and 5.6mλ for Coma. The maximum achievable CD uniformity is 4nm and the maximum achievable overlay (tool to itself) is ≤15nm. In order to maintain this tool status an advanced maintenance and calibration procedure was introduced with respect to dynamic distortion and dynamic focus control, as well as monitoring of illumination conditions and lens aberrations (Zernike polynomials). In addition feedback and feedforward loops for CD and OVL are available.

3. RESOLUTION ENHANCEMENT TECHNIQUES

3.1 Optical Proximity Correction and Litho Simulations

OPC modeling is used to enlarge DOF and CD uniformity. It is also used for the adaption of litho processes regarding device requirements. The model-generation is done in-house with Synopsis Mentor Calibre® software, as well as litho simulations to identify process-window weaknesses, design- and process issues. The Design for manufacturing methodology is used to verify new designs with respect to fab requirements. Detailed descriptions can be found in the reference links [1,2,3]

3.2 Inverse Lithography Technology (ILT)

In addition to common RET procedures, ILT [4,5,6] was used to evaluate the process window performance of random 90nm contact hole patterns with minimum pitch of 190nm up to isolated contacts and complex irregular 2D patterns. The exposures were done using the Canon AS4 exposure tool with NA = 0,85 and below on an standard attenuated phase-shift mask (Att-PSM) with 6% transmission. ILT, which is also known as a computational lithography or pixilated mask approach, inverses the model-based OPC techniques. It creates the optimized OPC pattern by inverting the direct physical problem from wafer to mask. Imaging investigations using both ILT approach and a combination with conventional model-based OPC were done. The goal was to achieve extended resolution capability of our existing exposure tool and to enlarge process window of a current technology (Figure 3). Detailed information on the ILT project can be found here [7].

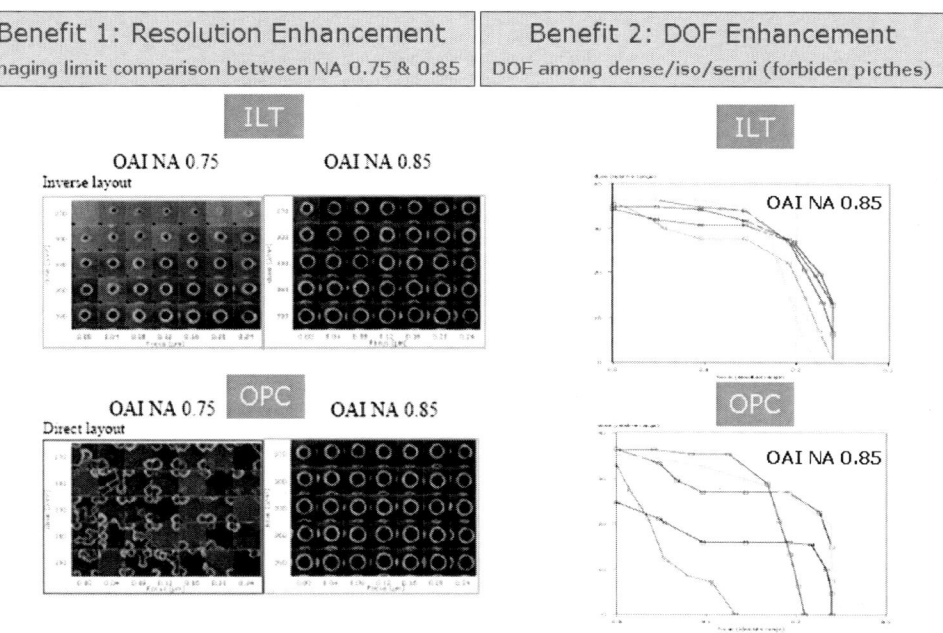

Figure 3. Extension of resolution capability and DOF enhancement by Inverse Lithography Technique (ILT)

3.3 Resolution Enhancement Lithography by Assist of Chemical Shrinkage

A further technique to extend the capability of the current lithography tool sets and to enlarge the process window is the introduction of chemical shrink methodologies, like SAFIER® (TOK) and RELACS® (AZ) or plasma-assisted shrink methods like MOTIF® (Applied Materials, LAM). These methodologies are developed by a number of companies in order to decrease the contact hole diameter without increasing the NA. [8,9,10,11]. We used the RELACS® [12,13] process with water-based SH-114 material (AZ® Electronic Materials) on contact hole and on trench levels. The shrink material was processed on a TEL Act8 track within a TARC unit. For contact hole applications the wafers are coated with 300nm photoresist and 75nm BARC. For trench applications the wafers are coated with 210nm photoresist and 45nm BARC. Table 4 displays the RELACS® process data.

Table 4. RELACS® process data

Supplier	AZ Electronic Materials
Material	water based SH-114
Unit	TARC unit
Dispense volume	2,0 ml
Mixing bake temperature	140°C/90s → 20nm shrink
Film thickness	135nm
Developement	DI rinse, 60s puddle

The evaluation of the shrink bias with respect to the applied mixing bake temperature is done on bare Si wafers. Figure 4 shows the achievable CD shrink bias evaluated on contact hole patterns. The maximum achievable CD shrink was 30nm for reasonable mixing-bake temperatures. CD measurements on product wafers, for contact holes, as well as for trench levels, after the shrink process showed that a shrink bias of 20nm causes no problems with respect to CD uniformity and range, whereas increased shrink biases and accordingly mixing bake temperatures lead to decreased CD performance data. Therefore we decided to setup our shrink process to a mixing bake temperature of 140°C for 90s resulting in a 20nm CD shrinkage.

CD shrinkage versus mixing bake temperature on bare Si

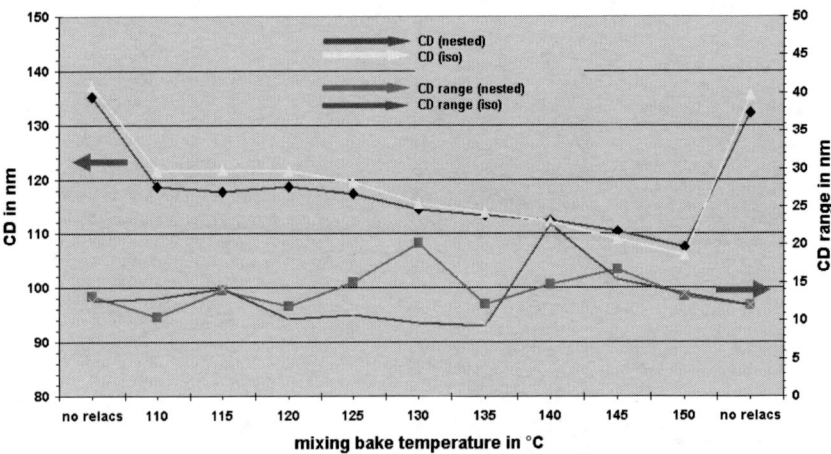

Figure 4. Evaluation of shrink bias with respect to mixing-bake temperature

The sensitivity to mixing bake temperature variations is about 2nm/1°C taking into account, that the shrinkage starts at a mixing-bake temperature of 125°C. Figure 5 shows the CD uniformity/range data for dense, semi and isolated contact hole patterns after litho for a CD targeted on 127nm. Figure 6 shows the CD uniformity/range data for the same contact hole patterns after the shrink process.

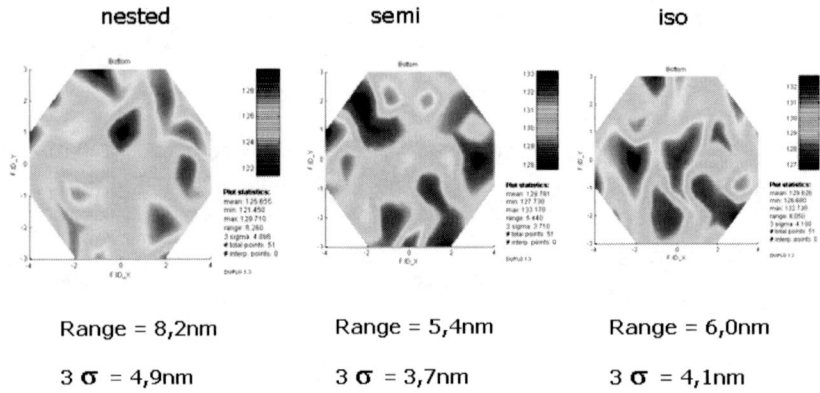

nested   semi   iso

Range = 8,2nm   Range = 5,4nm   Range = 6,0nm

3 σ = 4,9nm   3 σ = 3,7nm   3 σ = 4,1nm

Figure 5. CD uniformity data for nested, semi and isolated contact hole patterns after litho

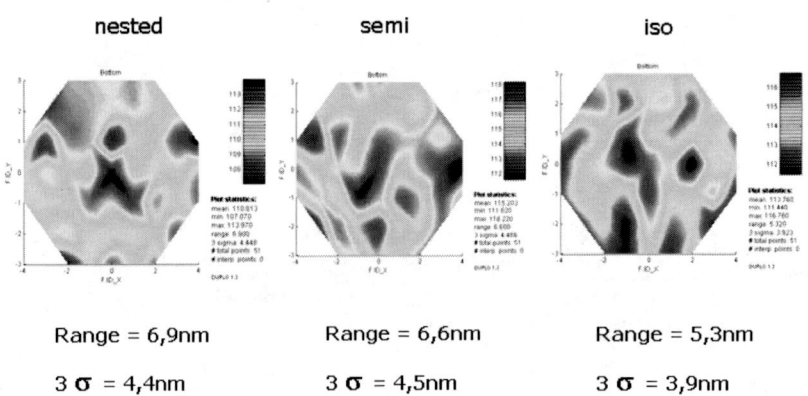

nested   semi   iso

Range = 6,9nm   Range = 6,6nm   Range = 5,3nm

3 σ = 4,4nm   3 σ = 4,5nm   3 σ = 3,9nm

Figure 6. CD uniformity data for nested, semi and isolated contact hole patterns after shrink process

Up to a shrink bias of 20nm no CD uniformity or CD range degradation could be observed. Figure 7 displays the same measurements for a shrink of 40nm. A strong increase of the CD range was observed.

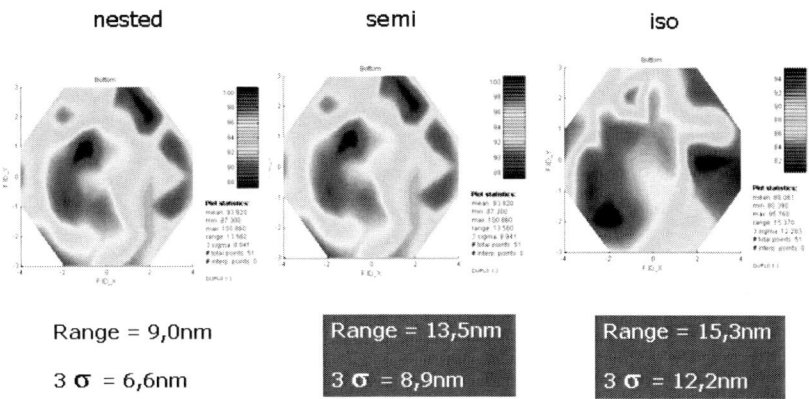

Range = 9,0nm   Range = 13,5nm   Range = 15,3nm

3 σ = 6,6nm   3 σ = 8,9nm   3 σ = 12,2nm

Figure 7. CD uniformity data for nested, semi and isolated contact hole patterns after shrink process

Figure 8 shows top-down SEM pictures of nested, semi and isolated contact holes after litho and after the chemical shrink process as well as cross-sections of the dense structure.

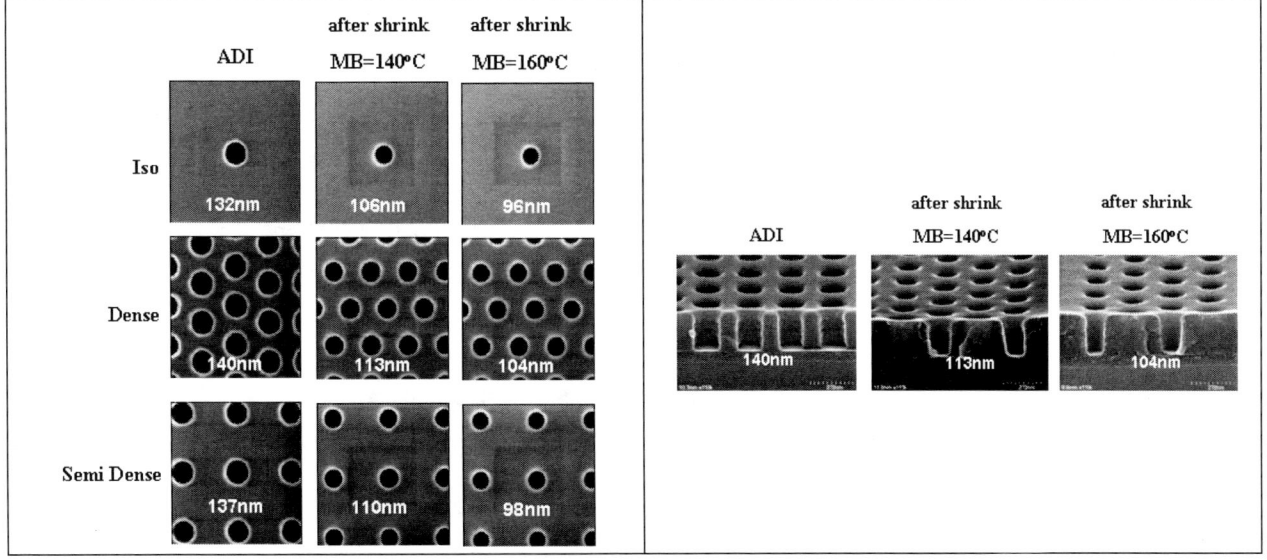

Figure 8. Top-down SEM images and cross-sections of contact hole patterns after litho and after chemical shrink process

The comparison of process window data with and without the resist shrink showed no degradation in DOF, but rather an improvement, which is expected to be one of the advantages of the chemical shrink process in order to enlarge the life span of current litho equipments. Figure 9 displays the DOF data for a dense contact hole pattern, centered on 127nm and 105nm after litho and the DOF data for same contact hole pattern after the shrink process. In order to achieve a after shrink CD target of 127nm and 105nm, the CD after litho was centered on 147nm and 127nm, respectively, resulting in a shrink of 20nm.

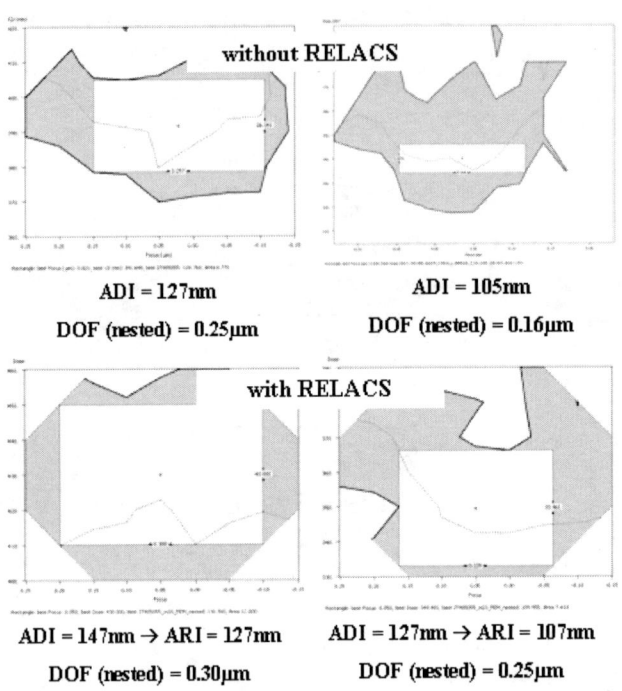

without RELACS

ADI = 127nm

DOF (nested) = 0.25μm

ADI = 105nm

DOF (nested) = 0.16μm

with RELACS

ADI = 147nm → ARI = 127nm

DOF (nested) = 0.30μm

ADI = 127nm → ARI = 107nm

DOF (nested) = 0.25μm

Figure 9. Process-window comparison for a dense contact hole pattern with and without shrink process

The chemical shrink process was not only used for hole level applications, it was also used for the shrinkage of trench levels. The overall behavior with respect to CD uniformity, CD range, DOF was identical to the hole level evaluations. Figure 10 shows SEM pictures and CD measurements of a trench level exposed on productive wafers after litho and after shrink. The target CD for this application is 100nm for nested trench patterns with a pitch of 250nm.

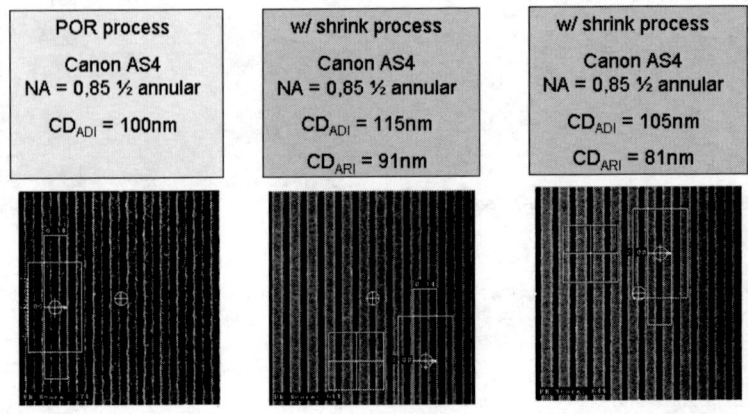

POR process	w/ shrink process	w/ shrink process
Canon AS4 NA = 0,85 ½ annular	Canon AS4 NA = 0,85 ½ annular	Canon AS4 NA = 0,85 ½ annular
CD_{ADI} = 100nm	CD_{ADI} = 115nm	CD_{ADI} = 105nm
	CD_{ARI} = 91nm	CD_{ARI} = 81nm

Figure 10. Top-down SEM images of a trench level targeted on 100nm (POR). For the RELACS® process the after litho CD was targeted to 115nm and 105nm, respectively

Another advantage of the chemical resist shrink process for trench levels seems to be the improvement in line edge roughness, visible in the SEM pictures in Figure 10. A reason for this behavior might be the additional heat treatment of the resist by the mixing bake, but this topic needs further evaluations on an increased statistical background.

An important topic for the reliable usage of such chemical shrink materials will be their influence on the chemical properties of the treated resist material. This may result in etch bias and taper profile changes, which will results in the need of etch process adoptions. We observed a decrease in the etch resistivity of the RELACS® processed resist materials. This effect was also reported by the material vendor and is in the range of 1.2 times of the etch resistivity without RELACS®. This effect was more obvious on the trench levels, than on the hole levels. Therefore we started further etch evaluations on shrinked structures, which are ongoing. The comparison of cross-sections after etch of wafers with and without shrink process did not show any dramatic differences with respect to the taper profile. Figure 11 displays cross-sections for a trench level, etched with and without chemical shrink process.

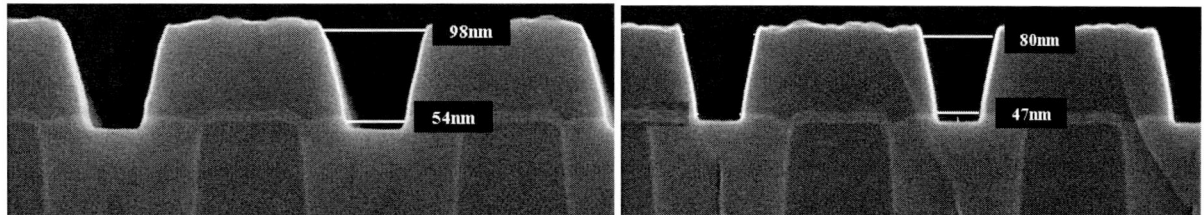

Figure 11. Cross-sections of trench level after etch with and without RELACS®

The chemical shrink process was evaluated regarding the defectivity behavior on blank Si, as well as on structured wafers. No defectivity problems could be observed on structured wafers, there is no dependency of defect count and defect type on the mixing-bake temperature observed.

4. RESULTS

Table 5 displays a comprehensive list of the PTL process settings and the results, achieved on the different levels. The PTL processes are used for the litho setup of our technology L90FLR (see designrules in Table 1), which is in development. The PTL CA process was implemented into the L90FL technology which achieved the T7 milestone (process freeze).

Table 5. PTL process settings and results

Level	Vx	Mx	M1	CA	RX
BARC/resist thickness (nm)	75 / 300	44 / 210	44 / 210	75 / 300	44 / 210
Exposure tool	Canon FPA-6000 AS4	Canon FPA-6000 AS4	Canon FPA-6000 AS4	Canon FPA-6000 AS4	Canon FPA-6000 AS4
NA / Sout / Sin	0.85 / 0.8	0.75 / 0.93 / 0.537	0.75 / 0.93 / 0.537	0.85 / 0.8	0.85 / 0.7 / 0.375
illumination mode	conventional	annular	annular	conventional	annular
evaluated DR (line/space)	100nm/100nm	100nm/100nm	90nm/90nm	90nm/100nm	100nm/100nm
Common DOF @ 5% EL (1 site, critical pitches=N,S/N,Iso)	0.2μm	0.2μm	0.2μm	0.2μm	0.2μm
CDU across wafer (min pitch/iso)	6nm/11nm	5nm/10nm	4nm/11nm	6nm/11nm	7nm/-
recommended OVL performance (nm)	<30	<30	<30	<30	<30

The process results in excellent profiles, as shown in the cross-section image of a line/space level after litho (Figure 12). The level was exposed with the PTL RX settings on L90FLR wafers.

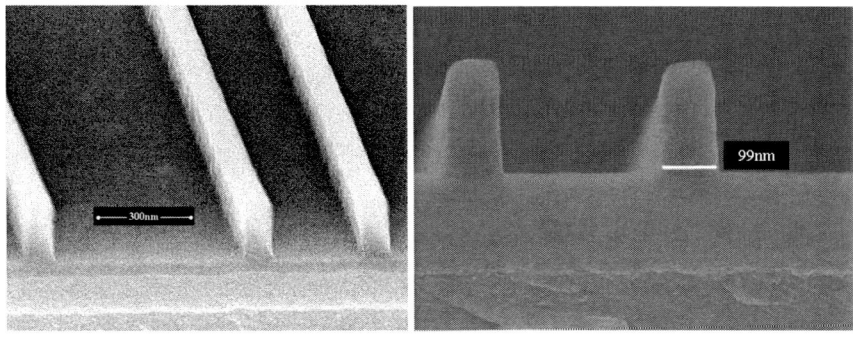

Figure 12. Cross-sections of line level after litho with a CD of 105nm.

Figure 13 displays the cross-section of different levels, exposed with the PTL processes on a L90FLR stack after final processing. With the introduction of the PTL processes, supported by tool optimizations and the usage of resolution enhancement techniques, we were able to fulfill the CD and overlay requirements of sub 90nm design rules.

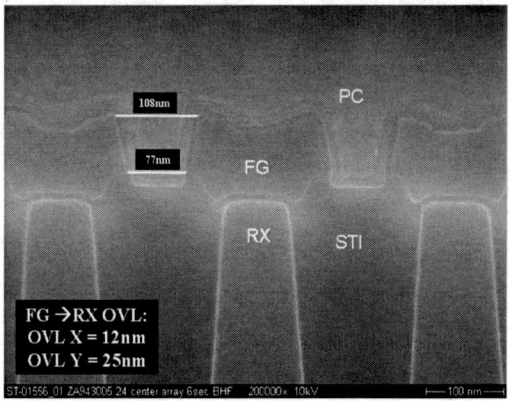

Figure 13. Cross-section of L90FLR stack, exposed with PTL processes.

5. SUMMARY

We have demonstrated how we were able to extend our existing 90nm technology capability, down close to 65nm node litho requirements on most critical layers with our standard tool park. The processes developed within a project called "Push to The Limits" where implemented into actual 90nm and sub 90nm technologies, called L90FL and L90FLR. In order to achieve the required CD and overlay performance of a sub 90nm node various options were explored to enable the processes and to secure the development of further 90nm shrinkage technologies, e.g. optimization of illumination conditions, evaluation of new materials, usage of advanced RET techniques (OPC, LfD, DfM and ILT) and resolution enhancement by chemical shrink (RELACS®). Typical contact litho, as well as line/space target at 100nm region was enabled. Figure 14 displays a schematic picture of the necessary ingredients, which are in place, in preparation and which are possible for Infineon Dresden to extend existing dry 193nm medium NA lithography beyond 90nm.

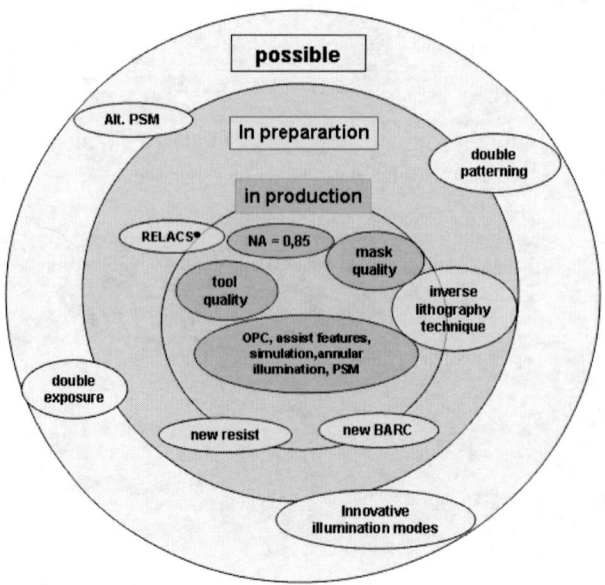

Figure 14. Ingredients which are in production, in preparation and which are possible to go beyond 90nm

REFERENCES

[1] Temchenko V., Lim C., Wallis D., Schneider J., Niehoff M., "Coupled-dipole modeling for 3D mask simulation," Proc. SPIE 6924, (2008).

[2] Lim C., Peter K., Temchenko V., Wallis D., Kaiser D., Meusel I., Schmidt S., Niehoff M.,"Investigation of DFM-lite ORC Approach During OPC Simulation," Proc. SPIE 6520, (2007).

[3] Sayegh S., Saleh B., and Nashold K., "Image design: Generation of a prescribed image through a diffraction limited system with high-contrast recording," IEEE Transaction on Acoustics Speech and Signal Processing 33, 460-465 (1985)

[4] Granik Y., Sakajiri K., and Shang S., "On objectives and algorithms of inverse methods in microlithography," Proc. SPIE 6349, (2007).

[5] Hendrickx E., Tritchkov A., Sakajiri K., Granik Y., Kempsell A., Vandenberghe G., "Hyper-NA imaging of 45nm node random CH layouts using inverse lithography," Proc. SPIE 6924, (2008).

[6] Lim C., Temchenko V., Niehoff M., "Selective Inverse Lithography Methodology," Proc. SPIE 7640, (2010).

[7] Maaike Op de Beeck et al, "A novel plasma-assisted shrink process to enlarge process windows of narrowtrenches and contacts for 45nm node applications and beyond" ," Proc. SPIE Vol. 6519, 65190U, (2007)

[8] Jun Hwan Hah et al, "Most Efficient Alternative Manner of Patterning sub-80 nm Contact Holes and Trenches with 193 nm Lithography," Japanese Journal of Applied Physics, Vol. 43, No. 6B, 2004, pp. 3663-3667

[9] Richard Peters et al, "A novel contact hole shrink process for the 65-nm-node and beyond," Proc. SPIE, Volume 5753, pp. 195-205

[10] Hsuen-Li Chen, et al, "Shrink Techniques for Sub-100 nm Contact Hole Fabrication in Electron Beam Lithography," Japanese Journal of Applied Physics Vol. 45, No. 8A, 2006, pp. 6539-6539

[11] Mamoru Terai, et al., "Newly Developed Resolution Enhancement Lithography Assisted by Chemical ShrinkProcess and Materials for Next-Generation Devices," JJAP Vol. 45, No. 6B, pp. 5354-5358, 2006

[12] Wallace C., Schacht J., Huang H I, "Optimization of resist shrink techniques for contact hole and metal trench ArF lithography at the 90nm technology node," Proc. SPIE 5376, p238-244 (2004).

Examining Reflectivity Criterion for Various ArF Lithography

Meng-Feng Tsai, Chia-Chi Lin, Wei-Chun Chao, Chan-Tsun Wu, Jun-Cheng Lai

Powerchip Semiconductor Corp., No. 12, Li-Hsin Rd. 1, SBIP, Hsinchu, Taiwan, R. O. C.

ABSTRACT

When the feature size keep shrinking to 4Xnm, ArF lithography has already proceed to immersion process and became mature enough. There is an important factor that will obviously influence photo process window in the initial phase development is the optical reflection from imperfect substrate design. From previous experience, reflection would be optimized to fine level by adjusting TARC (Top Anti-Reflection Coating) or BARC (Bottom Anti-Reflection Coating) thickness through index of reflectivity. However, actual criteria of reflectivity for various ArF lithography process are unlikely the same, e.g. different system type (wet/dry), node (feature size), illumination type, or even substrate effect, and also need to be examined to retain a decent process window. In this paper, experimental result of various above-mentioned ArF process have been compared with reflectivity index from prolith simulation engine, and distinctly clarified criteria of reflectivity for each case. Furthermore, effects of reflection to several optics caused patterning-related results, e.g. IDB (Iso-Dense Bias), OPC (Optical Proximate Correction) accuracy, will also be discussed. The result also shows severe criterion of reflection is requested as feature size getting smaller to 4Xnm node, and RET-applied (Resolution Enhancement Technology) process has opposite result on it. From experimental results, IDB has been obviously affected by reflection and become one important factor that influences reflection criterion examination.

Keywords: reflection, ArF, BARC (bottom anti-reflection coating), RET (Resolution Enhancement Technology)

INTRODUCTION

Common ArF lithography process would consider reflection control by optimizing film scheme including materials deposited before and in photo process, e.g. etch multi-layers hard mask and bottom anti-reflection coating (BARC). For the purpose of minimizing reflection from substrate, specific thickness of substrate film scheme and barc would be chosen to apply for meeting criteria referred to reflectivity index results from conventional simulation. Actually, for current ArF process, reflection caused standing wave effect is not so influential due to current ArF resist photo-acid normalization. In practice, it is unreasonable to set a reflectivity criterion for all kind of ArF process; it should depend on different optical setup, pattern type and even substrate types to avoid severe limitation of substrate selection.

In this paper, various lithography process parameters have been collected under different ArF processes and reflectivity index from simulation results to examine which parameter would be affected by reflection and to figure out how tendency existing for various ArF processes. Furthermore, because of application of double patterning for next generation, topography beneath patterning layer is another factor that would possibly influence photo process due to reflectivity behavior change. The effect has also been studied by simulation method to understand the situation on topography case.

Optical Microlithography XXIII, edited by Mircea V. Dusa, Will Conley, Proc. of SPIE Vol. 7640,
76403H · © 2010 SPIE · CCC code: 0277-786X/10/$18 · doi: 10.1117/12.848324

Experiments

Based on figure (1a), prolith simulation tool results, ranges of reflectivity index from simulation results have been split by adjusting one of hard mask layer thickness, silicon dioxide, to form distinguishing bias in reflectivity for ArF dry and wet process. Each case of ArF system could further separate to two different illumination case, dipole and annular, under equal Lens NA and RET ratio due to same out-put reflectivity curve from simulation. All experiments completed in 12" wafer production fab of Powerchip Semiconductor by using Nikon 193nm ArF S308 and ArF immersion S610C scanner with corresponding resist system.

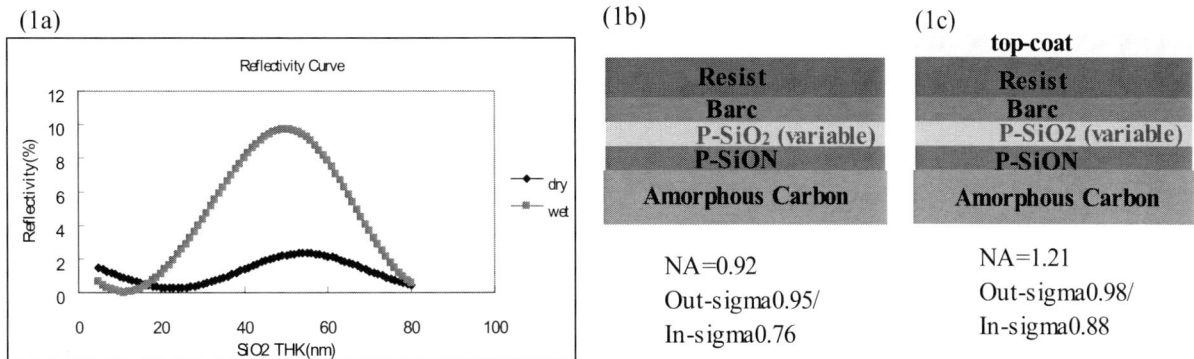

Figure 1: (1a) reflectivity results of dry and wet from simulation tool, (1b) film scheme and optic setting of dry system, (1c) film scheme and optic setting of dry system.

Results and discussions

The experimental result shows that dipole illumination is more sensitive to variation of reflectivity (table 1). The dense window and line width roughness performance would decay on dipole while reflectivity became higher, but it is not found on annular illustrated as figure 2, it may attribute to higher standing wave amplitude which is calculated from UV intensity distribution of accumulated light interference in resist as figure 3 and also combine cross-effect of resist photo-acid diffusion. However, expected dose window of resolution decrease caused by standing wave effect have not been observed from all evaluating set, it seems that current ArF resist system can minimize standing wave effect effectively.

Table 1: dense process results of dry system under reflectivity bias

Oxide THK(nm)	Reflectivity(%)	Dipole dense results				
		R-DOF(um)	CD-DOF(um)	Eo/Ec	Ed/Eo	LWR(nm)
25	0.2946	0.48	0.45	1.33	>1.16	4.78
17	0.4421	0.48	0.33	1.33	>1.16	5.24
72	1.0894	0.45	0.30	1.26	>1.16	5.34
40	1.4011	0.48	0.39	1.33	>1.16	5.45
55	2.3383	0.51	0.33	1.26	>1.16	5.54

Oxide THK(nm)	Reflectivity(%)	Annular dense results				
		R-DOF(um)	CD-DOF(um)	Eo/Ec	Ed/Eo	LWR(nm)
25	0.2946	0.45	0.42	1.19	1.29	8.19
17	0.4421	0.45	0.39	1.15	>1.35	7.69
72	1.0894	0.48	0.39	1.15	1.29	8.29
40	1.4011	0.42	0.39	1.15	1.29	8.02
55	2.3383	0.48	0.45	1.19	1.29	8.47

*Eo: energy of target, Ec: under-energy margin, Ed: over-energy margin

(2a)　　　　　　　　　　　　　　　(2b)

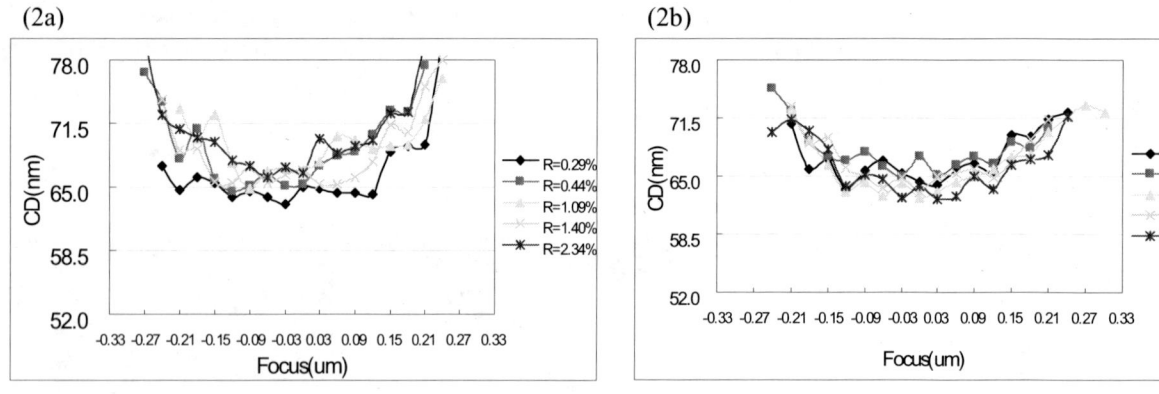

Figure 2: (2a) dense CD results under bias reflectivity for dipole illumination of dry system, (2b) dense CD results under bias reflectivity for annular illumination of dry system.

(3a) **Standing wave amplitude=0.101**　　(3b) **Standing wave amplitude=0.097**

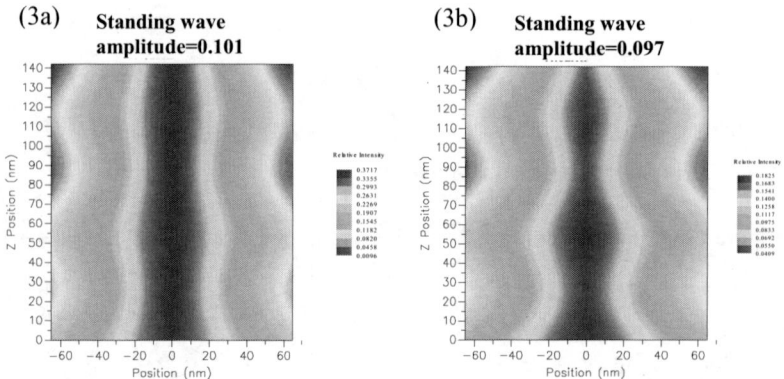

Figure 3: UV intensity distribution in resist and standing wave amplitude, (3a) dense of dipole/dry case, (3b) dense of annular/dry case.

Furthermore, OPC relative item, proximity, has also been checked as figure 4 and 5, similar result as dense was found, feature of proximity shifted as reflectivity index changed on dipole case and even bias > 5% while reflectivity index > 1.4%, result also indicated tendency of more sensitive to iso-pattern. Therefore, more severe RET applied in same scanner system indicated more severe reflectivity criteria needed for maintaining qualified litho performance which be verified on this work.

(4a)   (4b)

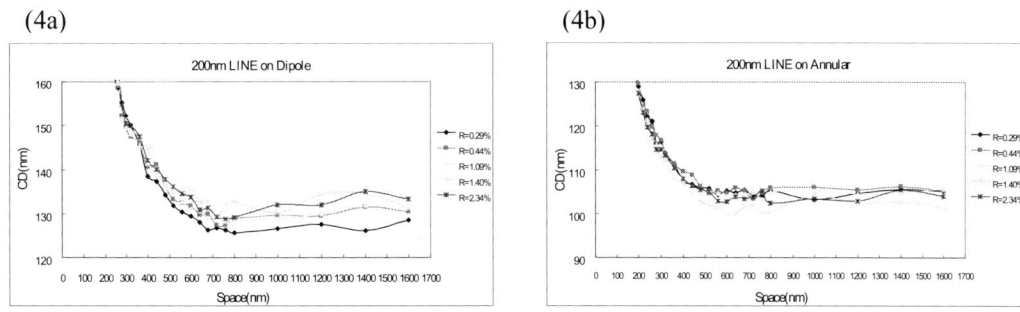

Figure 4: Proximity feature of 200nm line of (4a) dipole/dry case, (4b) annular/dry case.

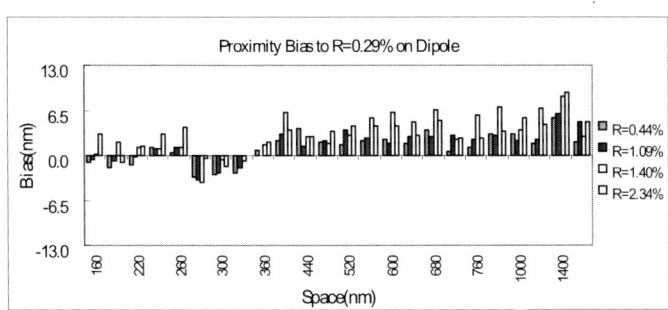

Figure 5: Proximity bias to lowest reflectivity of 200nm line on dipole/dry case.

In wet system, same feature also appeared on ArF immersion process, dipole is still more sensitive to reflectivity than annular on OPC relative items as figure 7, and tendency of influence on iso-pattern still be observed, however, it became mitigated compared to dry dipole case, and dense result revealed same mitigated effect from reflectivity index on dipole. It is far away from expected that extreme RET would bring more impact. According to standing wave amplitude of immersion system from simulation result indicated lower index than dry system even extreme NA=1.21 was used to be studying baseline, substance NA should be taken into consideration to compare with dry system, therefore, lower substance NA than dry system that was checked should be the factor resulted in minor effect to process parameter from reflectivity as table2 and figure 6.

Table 2: dense process results of wet system under reflectivity bias

Oxide THK(nm)	Reflectivity(%)	Dipole dense results					Annular dense results				
		R-DOF(um)	CD-DOF(um)	Eo/Ec	Ed/Eo	LWR(nm)	R-DOF(um)	CD-DOF(um)	Eo/Ec	Ed/Eo	LWR(nm)
15	0.2491	0.42	0.33	1.20	1.29	3.69	0.24	0.24	1.10	1.12	6.88
17	0.5273	0.48	0.33	1.20	1.29	3.95	0.30	0.30	1.10	1.18	7.32
20	1.1350	0.48	0.39	1.20	1.29	3.60	0.27	0.27	1.10	1.15	6.96
25	2.5685	0.45	0.36	1.20	1.29	4.27	0.24	0.24	1.10	1.15	5.69
32	5.1231	0.42	0.33	1.20	1.25	4.19	0.24	0.21	1.10	1.15	7.22
55	9.2473	0.42	0.33	1.20	1.25	4.03	0.24	0.24	1.10	1.12	7.04

*Eo: energy of target, Ec: under-energy margin, Ed: over-energy margin

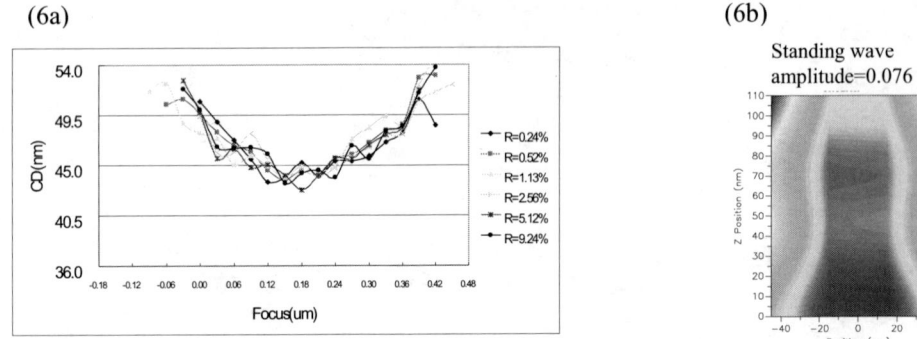

(6a)

(6b)

Standing wave
amplitude=0.076

Figure 6: (6a) dense CD results under bias reflectivity for dipole illumination of wet system, (6b) UV intensity
distribution in resist and standing wave amplitude for dipole illumination of wet system.

(7a)

(7b)

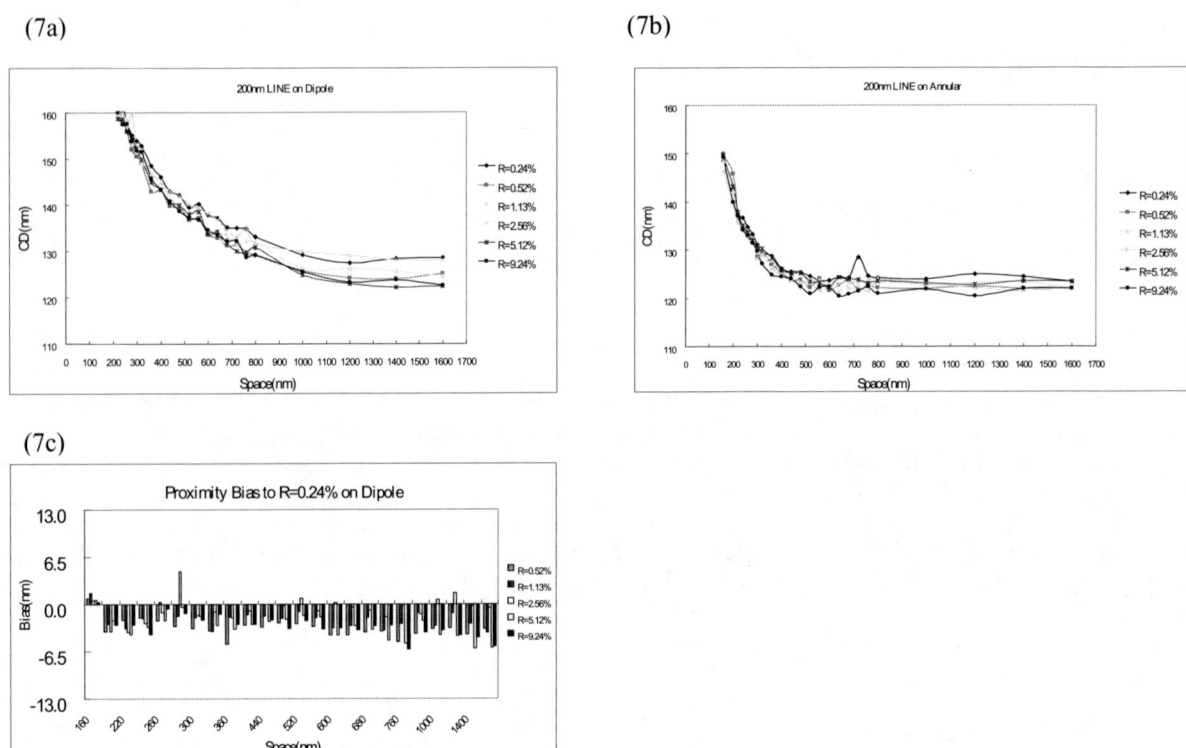

(7c)

Figure 7: Proximity feature of 200nm line of (7a) dipole/wet case, (7b) annular/wet case; (7c) bias to lowest
reflectivity of 200nm line on dipole/wet case

Another feature that was found is iso-pattern window. From images of printed different iso-line size as figure 8, obvious
resolution difference appeared as reflectivity changed, and better window on smallest reflectivity; this feature could
attribute to un-friendly contribution from extreme RET on immersion and enhanced effect of reflectivity to more
sensitive on iso-pattern.

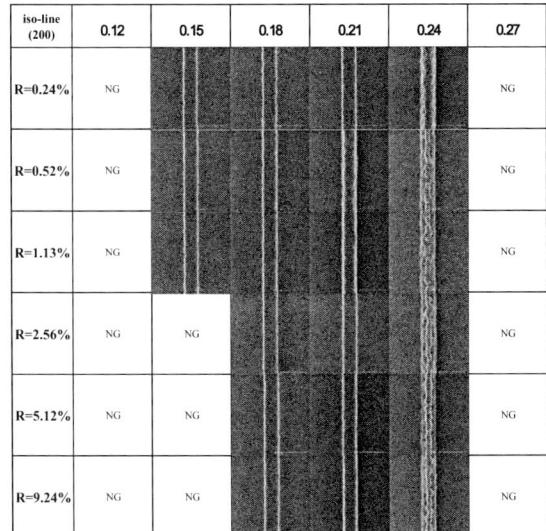 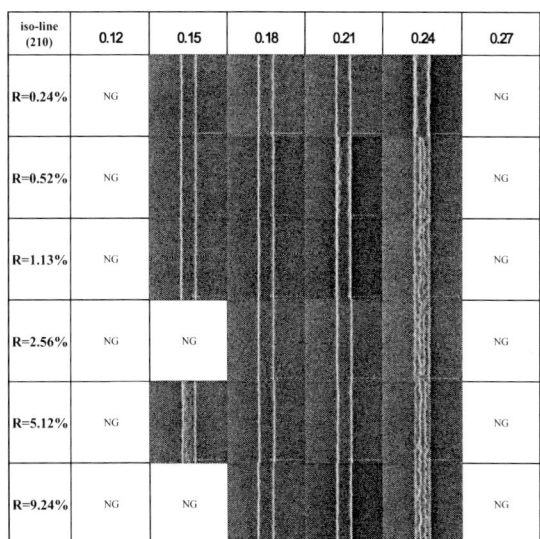

Figure 8: iso-line resolution images on dipole/wet case

Substrate effect

Substrate effect has also been studied by using rigorous 3D wafer topography simulator to understand whether same result and tendency will appear under combined topography beneath substrate that is common structure type of gate, as figure 9. From simulation results of figure 10, as patterning above topography, we found that there isn't apparent difference to topography in standing wave amplitude under the presence of common used hard-mask. For the other set, severe effect from modifying topography shape was observed due to local reflectivity variation and also revealed tendency of stronger effect to dipole than annular illumination. This result reminds that reflection would have more impact to next generation due to the situation of topography substrate will become more often on various double patterning technology for 3x node application.

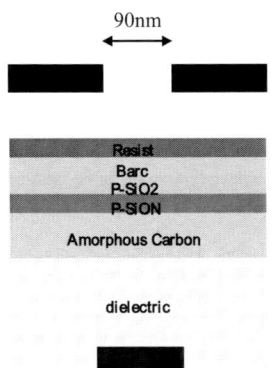

Figure 9: substrate with topography design for simulation study

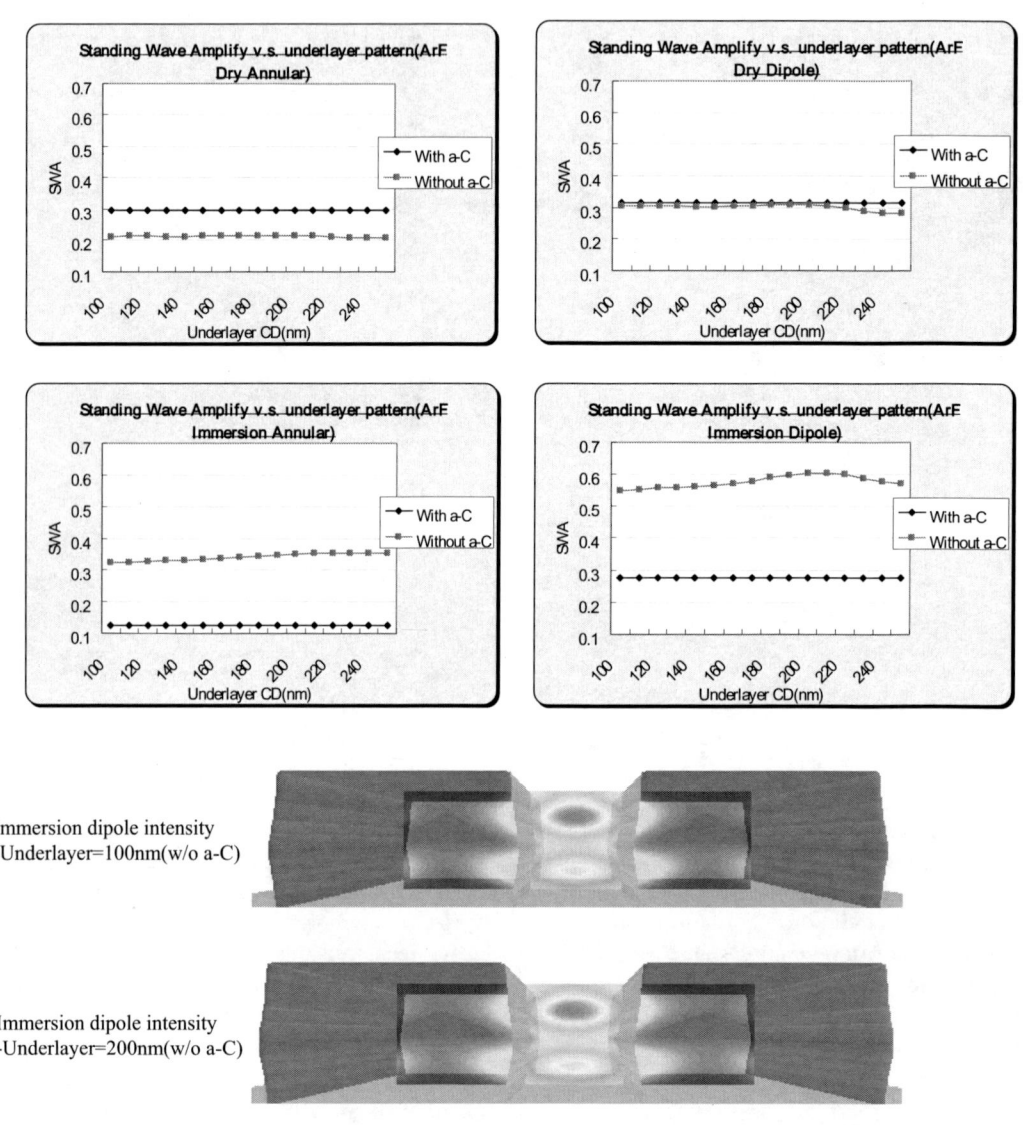

Figure10: Simulation results of various simulation set

Conclusion

In this paper, various ArF lithography processes were evaluated under reflectivity bias referred from simulation tool. The results printed by annular illumination indicated almost no impact found on dense line, iso line and OPC behavior, even reflectivity up to ~10% in immersion system. The opposite result of dipole case revealed obvious influence on same parameter as annular and tendency of high NA and RET enhancement. In the case of topography effect, simulation presented same tendency of stronger effect on dipole case while lack of effective anti-reflection layers, e.g. amorphous carbon. These results would provide important information to the user that applied RET normally, e.g. flash and Dram manufacturer. For the purpose of maintaining process requirement, suitable reflectivity criteria should be controlled for high NA and extreme RET application. However, effect of reflectivity should depend on optic setup, pattern types and even substrate to balance reflectivity criteria.

Reference

1. Zhimin Zhu, "Reflection control in hyper-NA immersion lithography" Proc. SPIE, 6924 69244A-1, 2008
2. Todd C. Bailey, "Reflectivity-induced variation in Implant Layer Lithography" Proc. SPIE, 6924 69244F-1, 2008
3. Itaru Kamohara, "Split, overlap/stitching and process design for double patterning considering local reflectivity by using rigorous 3D wafer-topography/lithography simulation" Proc. SPIE, 7274 72740H, 2009

CD-Uniformity for 45nm NV Memory on product-stack

Umberto Iessi [a], Brian Colombo [a], Johannes Plauth [b], Benedetta Triulzi [a], Elio De Chiara [a], Paolo Canestrari [a]

[a] Numonyx, Via C. Olivetti 2, Agrate Brianza (MI) 20041, Italy
[b] ASML Italy srl, Via Grandi 1, 20060 Pessano-con-Bornago (MI), Italy

ABSTRACT

CD uniformity budget for a 45-nm NV memory device requires the analysis and compensation of each single contributor factor. A dedicated simulation tool "CDU Predictor" helps to quantify the impact of main scanner and process factors for a comprehensive study of the CD Uniformity for an ideal flat wafer. However this analysis could under estimate the real CD distribution on a real production wafer if artefacts induced by thin-film effects and underling device topography significantly increase the contribution of the optical leveling-device to the total focus-error and hence spread the CD distribution for processes with low DOF. Such artefacts can be eliminated by application of an offset-map obtained by probing the mechanical top-surface of the resist-stack with an AirGauge (AirGaugeImprovedLEvelling, AGILE). The systematic variation of CD across the wafer, no matter whether due to fingerprints of the reticle, the device-topography, the track-process or the exposure-tool, can be mapped into dose-corrections for compensation (DoseMapper). We discuss an experimental case with a combination of both tools for an effective CD Uniformity optimization.

Keywords: CD Uniformity, Dose Mapper, Agile, LithoCruiser, CDU Predictor.

1. INTRODUCTION

A lithographic process for vertically oriented dense lines and spaces with pitch-interruption as depicted in fig. 1 was implemented using immersion-lithography at NA=1.2 and ArF dipole-illumination (193nm wavelength). The process-window was found rather tight, in particular as to the usable Depth-of-Focus (DOF) of approx. 90nm: this made this process suitable to examine margins of improvement by application of AGILE and DoseMapper, two advanced options available on the exposure-tool of type ASML XT:1700Fi.

After providing in the remaining of this introduction details on the experimental setup, the pattern to be printed and its process-window, section 2 shows the results obtained by AGILE (AirGaugeImprovedLEvelling) on the finger-print of best focus across the shot. Section 3 reports simulated and experimentally obtained CD-distributions from fine-tuning of the exposure-dose as function of pattern-location in the shot by means of DoseMapper.

1.1 Experimental setup

All exposures were done on double-side polished Si-wafers (200mm diam) with full product-stack beneath the resist-stack composed of BARC, ArF-resist and top-coat, the latter to ensure a suitable contact-angle during the immersion-lithography. The scanner of type XT:1700Fi (manuf: ASML) was interfaced to a track of type RF3i (manuf: SOKUDO). CD-inspection was done on CD-SEM of type S-9380 (manuf: HITACHI). Defect-inspection was done on PUMA-9000 (manuf: KLA-T).

The AirGaugeImprovedLEvelling [1] generates correction-maps for both gain and offset for each spot of the LevelSensor which provides the full-wafer height-map to the exposure-process. Due to its optical principle the LevelSensor may suffer from artefacts that depend on the thickness and composition of the thin films deposited on the bulk Si-wafer. These artefacts may lead to wrong information on the height-position of the top-surface of the resist-stack. The 'AirGauge' employed by AGILE is insensitive to the underlying stack, it only 'sees' the top-surface of the stack. The correction-maps for gain and offset are generated once from one single AGILE-measurement and then applied on demand in the form of subrecipes attached to the exposure-job.

The DoseMapper [2] comprises a set of actuators that allow accurate control of the exposure-dose as function of the pattern-location on the wafer and in the exposure-shot, along with the software required for feedback of measured CD,

Optical Microlithography XXIII, edited by Mircea V. Dusa, Will Conley, Proc. of SPIE Vol. 7640,
76403I · © 2010 SPIE · CCC code: 0277-786X/10/$18 · doi: 10.1117/12.846538

its linking to relevant info generated during the exposure-process, and its mapping into a dose-recipe. Legendre-polynomials of up to the 6th order are available for correction in slit-direction, scan-direction and across the wafer. Up to 6 different sets of shot-maps can be derived and applied across the wafer. The 'CD-Analyzer' of DoseMapper allows simultaneous analysis and dose-correction for more than one feature. In our case we measure and optimize 3 features, the 'LINE A', 'LINE B ', and 'SPACE' as defined in fig. 1. In our case we discarded CD-data that was gathered at positions within 3mm from the wafer-edge, as well as such found outside a +/- 3-Sigma-interval from average ('flyer removal').

1.2 Dense L/S with pitch-interruption

Fig. 1 depicts the pattern under investigation. It was printed from a 6%-attenuated PhaseShiftMask at NA=1.20 using dipole illumination at 193nm wavelength. The dipole was horizontally oriented and had 60-deg opening angle. It was radially delimited by SigmaInner=0.65, SigmaOuter=0.85. The exposure-light was polarized in y-direction.

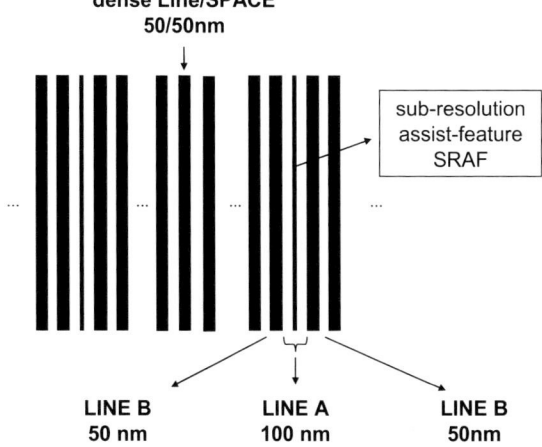

fig. 1: Pattern to be printed: Dense lines and spaces with pitch-interruption. Investigated are Critical Dimensions CD of 'LINE A', 'LINE B' and 'SPACE'. Dimensions indicated in the figure are target-values.

1.3 Process-window

Fig.2 shows BOSSUNG curves for LINE A,LINE B and SPACE. At best focus, the sensitivities of CD vs dose were found as -0.864 [nm/%], -0.5764 [nm/%] and +0.577 [nm%] for LINE A, LINE B and SPACE, respectively. The overlapping process-window for LINE A, LINE B and SPACE was found limited in focus by line-collapse towards negative defocus and thinning of the LINE A at positive defocus (fig. 3). This left an overlapping Depth of Focus of approx. 90nm. Fig. 4 shows the process-window as simulated by LithoCruiser using a calibrated resist-model. For +/- 10% CD-tolerance depth-of-focus was found 90nm and exposure-latitude 10%.

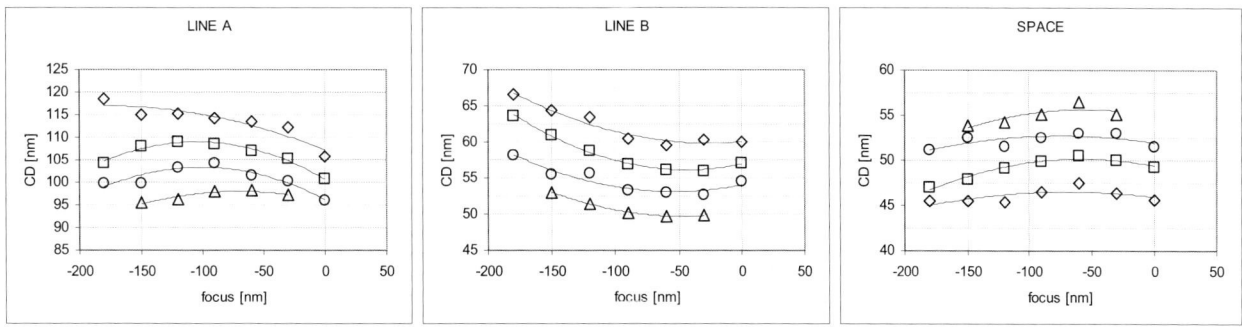

fig.2: BOSSUNG-curves of LINE A, LINE B and SPACE of pattern under investigation. Different curves belong to exposure-doses ranging from 18 to 21 mJ/cm2

fig.3:   left: line-collapse at negative defocus;   right: thinning of the LINE A at positive defocus

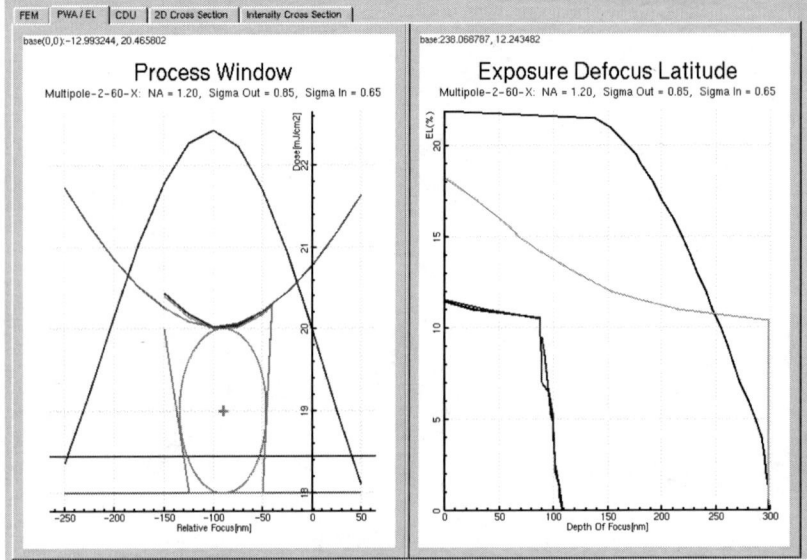

fig. 4: overlapping process-window for pattern of fig. 1. A calibrated resist-model was used for this simulation
with LithoCruiser

1.4 Feature-dependent best focus

A previous investigation [3] had experimentally determined a maximum difference of 1.5[nm] across the slit between CDs of first lines to the left and to the right of the pitch-interruption (LINE A). This was then found consistent with simulations based on wave-front-error measured with the built-in interferometer ILIAS.

That investigation [3] had revealed best focus as determined from BOSSUNG-curves to differ 50nm between LINE A, LINE B and SPACE (see also fig. 2). This was a key limitation to the overlapping DOF. The investigation had identified the composition of the resist-stack and exposure-dose (bias) as significant to the amount of feature-dependant focus-difference.

At present state of the implementation neither bias nor composition of the resist-stack were free variables. In this work here we limited our attention to improvements that were available without modification of the imaging-process itself (reticle, illumination-mode, resist-stack etc.).

2. BEST FOCUS FLATTENED ACROSS THE SHOT

Determination of best focus from BOSSUNG-curves taken for feature LINE A across the shot revealed a rather large variation. Analysis allowed distinction of two populations that were found correlated to locations in the shot: best focus determined for LINE A in parts of the shot used for test-patterns ('TP'), with relatively large unpatterned areas in the vicinity, was approx. 40nm more negative than best focus determined for the same feature in parts of the shot used for a memory-device ('MATRIX'), hence in an area of dense patterning.

Fig. 5 shows on the left the schematic distribution of memory-device (sampling points denominated 'MATRIX') and test-pattern ('TP') across the shot. In the middle the shot-map of non-correctable leveling-error before correction with

AGILE is shown as grey-scale-map (full range 90nm). On the right this map is repeated after correction with AGILE, at unchanged scale. 3-Sigma of the distribution of non-correctable leveling-error reduces from 30nm to 16nm and its range (Max-Min) shrinks from 56nm to 25nm.

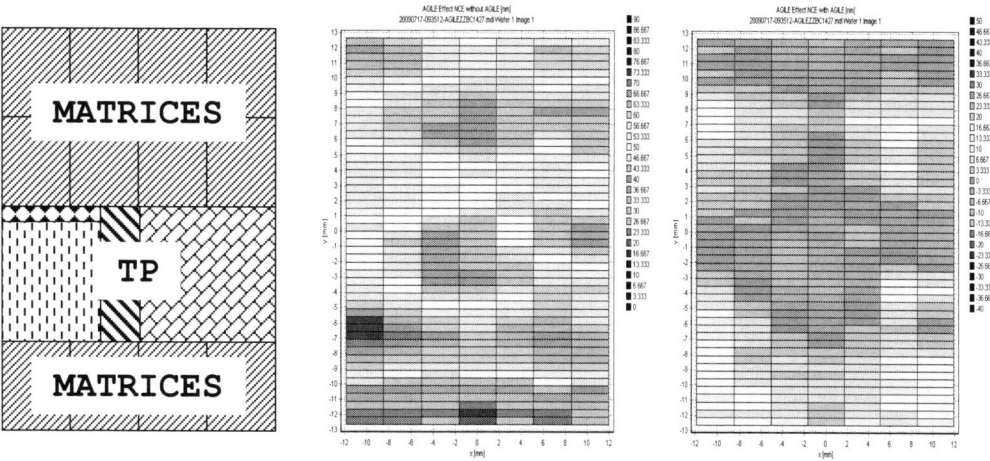

fig.5: left: locations of memory-device 'MATRIX' and test-pattern 'TP' in the shot;
        middle: shot-map of non-correctable leveling-error before correction with AGILE
        right: shot-map of non-correctable leveling-error after correction with AGILE
                  (in both cases the full range of the grey-scale is 90nm)

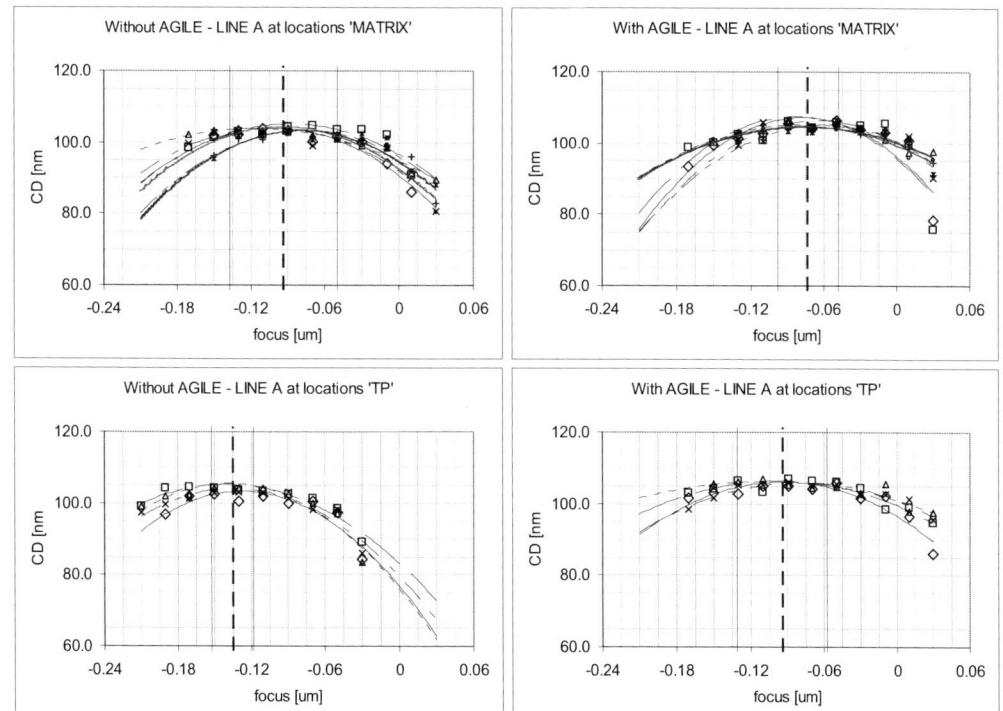

fig.6: Bossung curves (CD vs defocus) for feature 'LINE A' at different shot-locations (vertical lines: avg(BF)+/-3sig)
        - left: BEFORE correction with AGILE;   top: locations 'MATRIX'; bottom: locations 'TP'
        - right: AFTER correction with AGILE;   top: locations 'MATRIX'; bottom: locations 'TP'

Determination of best focus for the feature 'LINE A' from exposures with a leveling-map corrected by AGILE yields a clear improvement as shown in fig. 6: The difference of average best focus determined at locations 'MATRIX' to that at locations 'TP' is reduced to approx. 20nm.

3. CD-DISTRIBUTION AFTER DOSE-MAPPING

3.1 Dose Recipe derived and CD-distribution simulated from measurements at best focus

CD was measured by SEM for LINE A, LINE B and SPACE in full wafer layout and with very dense sampling (ca 1200 samples per feature per wafer) from wafers processed at best focus with AGILE (AGL) and without AGILE (NOA). This input was mapped into dose-corrections by means of the 'cd_analyzer' as described in section 1.2, applying the sensitivities as given in section 1.3. Coefficients of a 4th-order Legendre polynomial were calculated for dose-correction along the slit-axis and of a 5th order Legendre-polynomial for that along the scan-axis of the exposure-shot [3]. We chose not to apply corrections across the wafer since that impact had been found very small. During the calculation equal weight was assigned to CD of each of the 3 features LINE A, LINE B and SPACE. The estimated CD after correction was exported and analyzed. Fig. 7 compares for each feature the 3-sigma of measured CD to that of CD expected after correction. 'Ful / NOA' and 'Ful / AGL' refer to evaluation of full-shot data obtained from exposures without and with AGILE, respectively. As bench-mark of ultimately achievable performance a 3rd pair of bars is added, obtained from evaluation of CD measured in the device ('MATRIX') and optimized exclusively for CD-distribution of LINE A, hence excluding limitations discussed in the previous section.

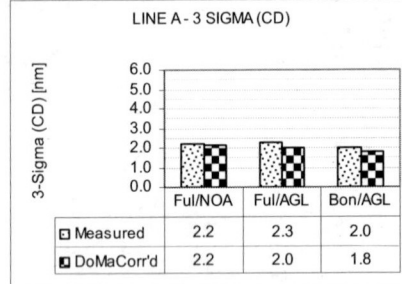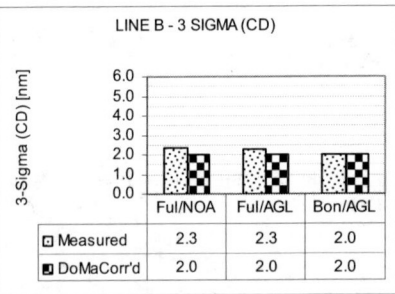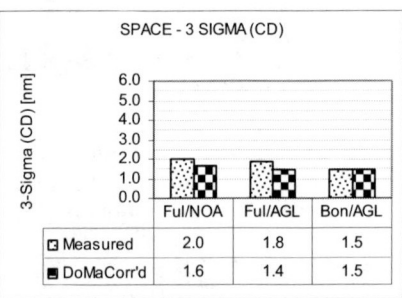

fig. 7: 3-Sigma of measured and simulated CD from wafers exposed at best focus without AGILE (NOA) and with AGILE (AGL). 'Ful' refers to inclusion of the full set of measured CD, 'MTR' refers to CD taken at locations in the memory-device (as opposed to CD taken at locations in the test-pattern 'TP').

Fig. 7 shows only a negligible improvement can be expected both from AGILE and/or fine-tuning of a dose-recipe for 3-Sigma of CD for exposure at best focus. This is somewhat not surprising, since at best focus the BOSSUNG-curves are rather flat, a local focus-error has little impact on the CD.

This is expected to change for exposures done closer to the edge of the process-window as investigated in the next paragraph.

3.2 Dose Recipes derived and CD-distribution simulated from measurements across focus-range of +/-45nm

CD-measurement and analysis reported in the previous paragraph was repeated over a set of wafers exposed at intentionally defocused conditions. Focus was varied +/-30nm and +/-45nm with respect to best focus. The software of DoseMapper is not designed for processing of CD-data from multiple wafers and in particular not for such of wafers processed at different exposure-conditions. To circumvent this issue, CD-measured on features LINE A, LINE B and SPACE from the wafer exposed at defocus -30nm was treated as such of new features 'LINE A_M30NM', 'LINE B_M30NM' and 'SPACE_M30NM'. To ensure conservation of the average CD characteristic to this defocus, the target-value for the optimization was set to the average measured CD (after exclusion of data obtained within 3mm from the wafer-edge and flyers outside a +/- 3-sigma interval from the average). The same was done with CD-data obtained from wafers at the other defocus-conditions. Eventually for both conditions 'NOA' and 'AGL' the set of features had increased from original 3 (LINE A, LINE B, SPACE) to 15 (LINE A, LINE A_M30NM, LINE A_P30NM, LINE A_M45NM, LINE A_P45NM, LINE B, ...). For a first simulation ('weight-set 0') weights were left at 1.0 for CD's measured at best-focus condition, weights for all other features were set to 0.0. This corresponds to monitoring the impact of a dose-recipe identical to the one derived in the previous paragraph on CD measured at defocus-conditions. Sensitivities of CD vs dose were left at values found for best focus. In reality the sensitivities are higher at out-of-focus condition but this is neglected at this point.

Fig. 8 depicts 3-sigma of CD for each of the 3 features LINE A, LINE B and SPACE as measured and simulated over a set of 5 wafers, processed at BF, BF+/-30nm and BF+/-45nm. The simulation uses a dose-recipe obtained using 'weight-set 0', hence set to minimize CD-distribution only from measurements found at best focus.

Whereas the CD-distribution for SPACE is rather small (3-sig = 2.5nm) across these 5 wafers even without AGILE and without applying a dose-recipe, it has approximately doubled for LINE A and LINE B, if compared to values of respective feature obtained at best focus (compare fig. 7).

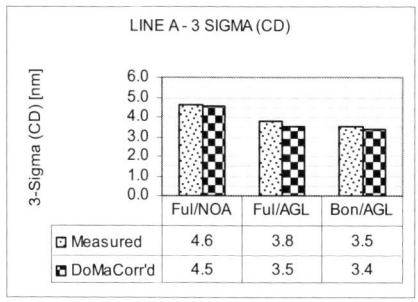 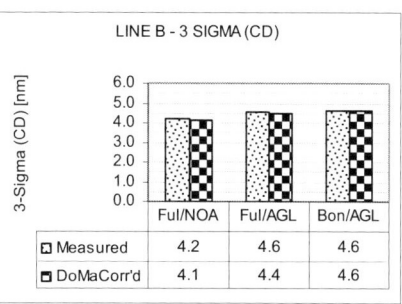 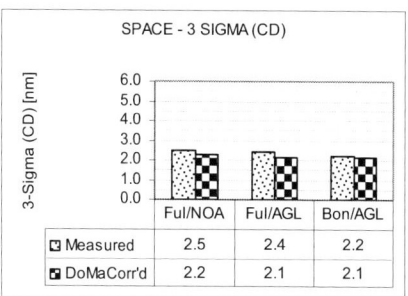

fig. 8: 3-Sigma of measured and simulated CD from 5 wafers, one each exposed at best focus, +/-30nm and +/-45nm without AGILE (NOA) and with AGILE (AGL). 'Ful' refers to inclusion of the full set of measured CD, 'MTR' refers to CD taken at locations in the memory-device (as opposed to CD taken at locations in the test-pattern 'TP').

In the case of LINE A, the correction of AGILE yields a significant reduction of 3-Sigma for measured CD, from 4.6nm to 3.8nm. After application of the dose-recipe, this is expected to reduce further to 3.5nm. This is rather close to 3.4nm, the value simulated for 3-Sigma of LINE A only from locations 'MATRIX' and optimized exclusively for LINE A. Hence for this feature, AGILE and DoseMapper together yield a significant reduction in 3-Sigma of CD of approx. 1 nm over a set of 5 wafers processed over a focus-range of +/-45nm.

Unfortunately the feature LINE B shows an opposite trend. While starting off without AGILE, without any dose recipe at a value of 3-Sigma similar to that of the LINE A, it does not benefit from the correction of AGILE nor from the dose recipe. On the contrary, 3-Sigma even slightly increases, from 4.2 nm to 4.4 nm. We suspect this behaviour is related to the significant difference in best focus between LINE A and LINE B (see previous paragraph) and to the determination of best focus based mainly on characteristics of LINE A. A slight recentering of focus may improve the results.

Eventually, two other sets of weights were examined (see tab. 1,2) to explore potential improvements if distribution of CD also out-of-focus is considered. However, no improvements were found.

	weight-set 0			weight-set 1			weight-set 2		
	BF	+/-30nm	+/-45nm	BF	+/-30nm	+/-45nm	BF	+/-30nm	+/-45nm
LINE A	1.0	0	0	1.0	0.78	0.57	0.2	0.6	1.0
LINE B	1.0	0	0	1.0	0.78	0.57	0.2	0.6	1.0
SPACE	1.0	0	0	1.0	0.78	0.57	0.2	0.6	1.0

tab. 1: weight-sets used for DoseMapper simulation of Ful/NOA and Ful/AGL

	weight-set 0			weight-set 1			weight-set 2		
	BF	+/-30nm	+/-45nm	BF	+/-30nm	+/-45nm	BF	+/-30nm	+/-45nm
LINE A	1.0	0	0	1.0	0.78	0.57	0.2	0.6	1.0
LINE B	0	0	0	0	0	0	0	0	0
SPACE	0	0	0	0	0	0	0	0	0

tab. 2: weight-sets used for DoseMapper simulation of MATRIX/AGL

3.3 CD measured from exposures with AGILE- and DoseMapper-corrections applied

A set of 5 production-wafers, different from the lot used in the previous paragraphs, was exposed at identical nominal dose and focus (hence 1 wafer each at BF, +/-30nm and +/-45nm), but now with both AGILE- and DoseMapper corrections applied. These corrections were identical to the ones applied in the preceding paragraphs. CDs were measured and analyzed as in the preceding paragraphs. Fig. 9 shows 3-Sigma as function of defocus for each of the 3 features: Plots on the left the show the state before, those on the right the status after application of AGILE- and DoseMapper-correction. The top pair of graphs reports 3Sigma from CD measured across the full shot, the center pair that of CD measured in the 'MATRIX' and the bottom pair that measured in 'TP'. The improvement in CD-uniformity, in particular at the edges of the focus-window, is evident.

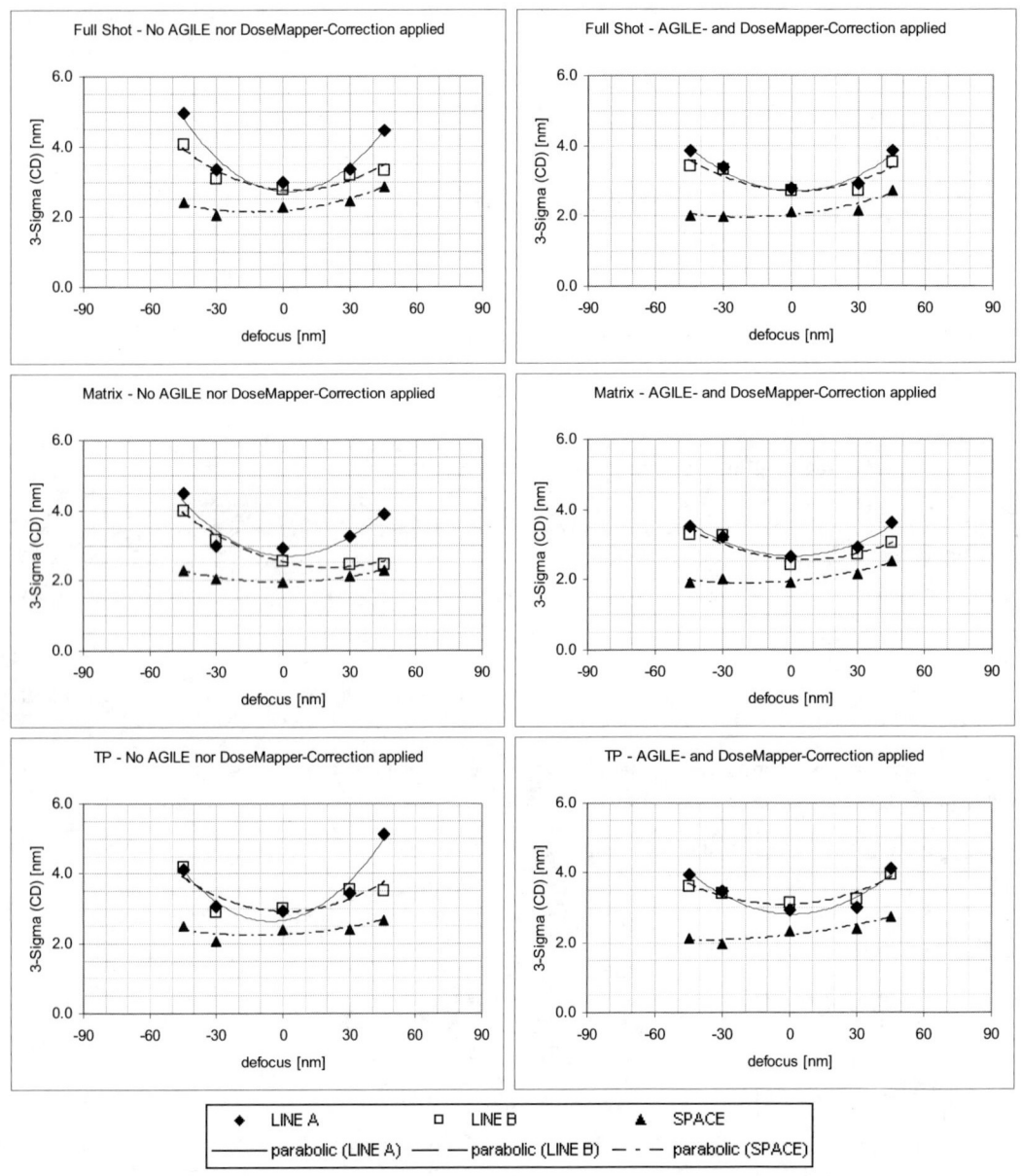

Fig. 9: 3-Sigma obtained per defocus-condition, before and after AGILE- and DoseMapper corrections
      - top: CD as sampled from the full shot (= both types of location 'Matrix' and 'TP')
      - middle: CD as sampled from locations 'Matrix'
      - bottom: CD as sampled from locations 'TP'

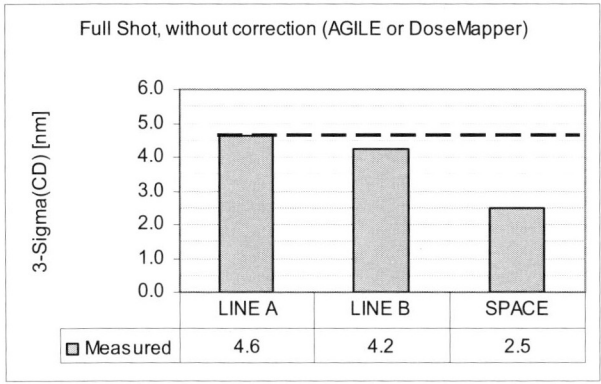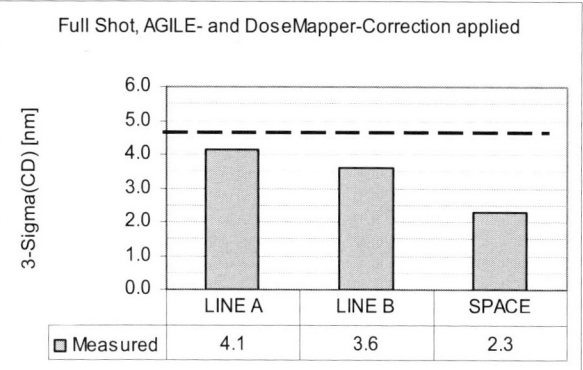

fig. 10: 3-Sigma obtained across 5 wafers (exposed at best focus, +/-30nm and +/-45nm from best focus)
- left: exposed without AGILE- and DoseMapper corrections
- right: exposed with AGILE- and DoseMapper corrections

The improvement also shows in 3-Sigma calculated from total population across 5 wafers that cover the full focus-range of 90nm (see fig. 10): 3-Sigma for LINE A is reduced by 0.5 nm compared to the distribution obtained without AGILE- and DoseMapper-correction. This improvement is inferior to the simulation which predicted 3-Sigma to diminish by 1nm. At the same time 3-Sigma for LINE B is found 1nm better than simulated.

We suspect these differences between measured and simulated CD-distributions may be due to tiny shifts of best focus effective during processing one and the other set of wafers: With 2 out of the 5 wafers exposed intentionally at the very edge of the usable focus-range, very small changes in best focus (whether induced by machine-drift or by slight variations in the composition of the product-stack) immediately affect the distribution of CD. Possibly an update of the AGILE-corrections by repetition of the appropriate measurement might reduce such effects.

4. SUMMARY

In a lithographic process of dense Lines and Spaces with pitch-interruption, implemented in immersion-lithography at NA=1.20 and dipole-illumination at 193nm wavelength and characterized by a small overlapping DOF OF 90nm, correction of the shot-leveling map by means of AGILE and subsequent fine-tuning of the exposure-dose as function of the location in the shot by means of DoseMapper are simulated to reduce 3-sigma of CD-distribution of the pitch-interruption (LINE A) across the focus-range of 90nm from 4.6nm to 3.5nm, while 3-sigma of CD of the LINE B is expected to remain essentially unvaried at 4.6 nm. Experimental verification yields 3-Sigma of LINE A to reduce to 4.1nm (-11%) and 3-Sigma of LINE B to drop to 3.6nm (-22%). Given the tight process-window even these tiny improvements are found worthwhile implementation.

REFERENCES

1. ASML product-documentation: "AirGauge Improved process LEveling (AGILE)", doc-ID 50236, issue2, 13-feb-2008
2. ASML product-documentation: "DoseMapper for PAS5500 and Twinscan", doc-ID 95235, issue 4, 27-nov-2006
3. Bernardini, S.: "Studio delle aberrazioni per il nodo tecnologico 45nm nella fabbricazione di memorie non volatili", tesi di laurea alla Facoltà di Scienze MM.FF.NN, Università degli Studi di Roma Tor Vergata, Anno Accademico 2007-2008

Analysis of Photoresist Edge Bead Removal
Using Laser Light and Gas

V. Chaplick, E. Degenkolb, D. Elliott, K. Harte, R. Millman Jr., M. Tardif

UVTech Systems Inc., 490 Boston Post Road, Sudbury, MA 01776

ABSTRACT

Wafer edge defects are currently considered a major problem as they negatively impact device yields in integrated circuit manufacturing, especially in immersion lithography. A primary source of edge defects is from particles of photoresist originating from the edge bead of resist caused by spin coating. In this paper, photoresist edge bead removal (EBR) is studied in a series of experiments using a laser and gas cleaning system. One goal of the experiments was to reduce the edge exclusion by gradually reducing the area cleaned by the laser and gas system. Reduction of EBR width will increase die yield. A number of varying exposure algorithms were tested, and are described along with microscope and SEM photos of the resulting edge geometry and surface condition. Another goal of these experiments is time-efficient removal of thick edge beads, a problem for conventional expose/develop methods. A matrix of varying laser parameters and gas types was run to produce a best-known-method (BKM) to meet these goals.

Keywords: defectively, edge bead, edge exclusion, apex, fluence, laser

1. INTRODUCTION

Semiconductor manufacturing technology, driven by Moore's Law, has provided significant performance in computing and communications fields at continually lower costs since the invention of the transistor. The deceleration of Moore's Law, coupled with increased costs of energy and global economic market changes have caused all semiconductor manufacturers to look hard at new ways to reduce costs. The most highly leveraged area for cost reduction is yield improvement, and recent industry attention is being directed to the edge area of silicon wafers as a place to improve device yields. Immersion lithography has resulted in edge defects caused by the re-deposition of particles and film delamination from the immersion fluid. More specifically, when wafers are spin coated with photoresist, excess resist forms on the top, bottom and sidewalls or apex of the wafer, and in subsequent processing, breaks up and re-deposits flakes of resist on the inner critical portions of the wafer containing devices. Flakes of resist also contaminate the processing equipment, causing tool downtime. Reducing edge exclusion will potentially increase the number of good die per wafer, or die yield. The combination of improved EBR coupled with reduced edge exclusion can, according to recently published studies, potentially increase device yields by up to 50% at the wafers edge. In summary, photoresist edge bead removal and edge exclusion have become a major topic and focus for yield improvement in integrated circuit manufacturing.

Edge bead removal methods now in use involve backside solvent rinse (BSR), wafer edge exposure and develop (WEED), and plasma EBR. Solvent droplets can cause defects on good die, and are difficult to control. Expose and develop methods may cause unacceptable throughput reduction, especially on thick resist edge beads. All of the liquid-based EBR cleaning methods have been shown to create delamination and peeling problems with BARCs, porous low K films, and tensile films such as amorphous carbon. The industry is seeking a wafer edge cleaning method that does not require liquids, has low cost of ownership, eliminates or greatly reduces defects, and can provide the reduced edge exclusion for increased die yield per wafer. This paper will present the results from a number of experiments using a new technology, based on laser light and gas, for cleaning the top, bottom, and apex portions of silicon wafer edges. This paper will also present data on how a laser-based EBR method can provide controllable reduced edge exclusion (REE).

Optical Microlithography XXIII, edited by Mircea V. Dusa, Will Conley, Proc. of SPIE Vol. 7640,
76403J · © 2010 SPIE · CCC code: 0277-786X/10/$18 · doi: 10.1117/12.853444

2. METHODOLOGY

2.1 Photoreactive cleaning

Photoreactive cleaning is the method used in this paper to perform edge bead removal from the top edge, apex, and bottom edges of silicon wafers. A laser beam is programmed to deliver a specific fluence and dose of radiation to the wafer surfaces while a gas injection nozzle delivers reactive gas. There are both issued and pending patents on this process. Figure 1 shows a schematic of the reaction mechanism (left) and a photo of the 'live' gas reaction zone (GRZ) on the right. A beam of pulsed laser radiation is directed to the wafer edge in the presence of a stream of gas. The resist is ablated and combusted into by-products and removed by a vacuum nozzle.

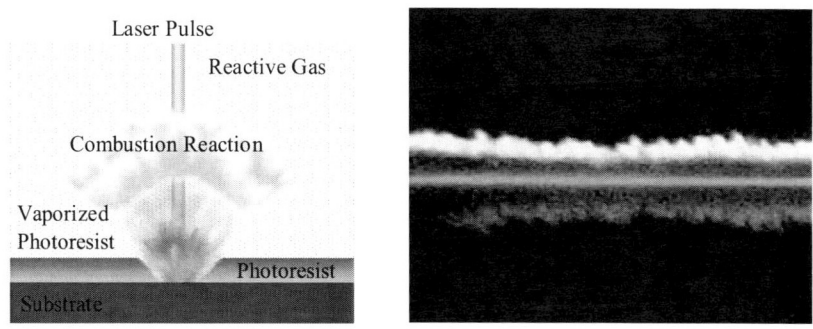

Figure 1: Schematic of photoreactive cleaning mechanism (left) and a photograph of reaction (right)

Figure 2 shows the absorption coefficients of photoresist and some selected reactive gasses at several laser wavelengths. This data is useful in deriving process recipes using combinations of photoresist, gasses, and laser wavelength.

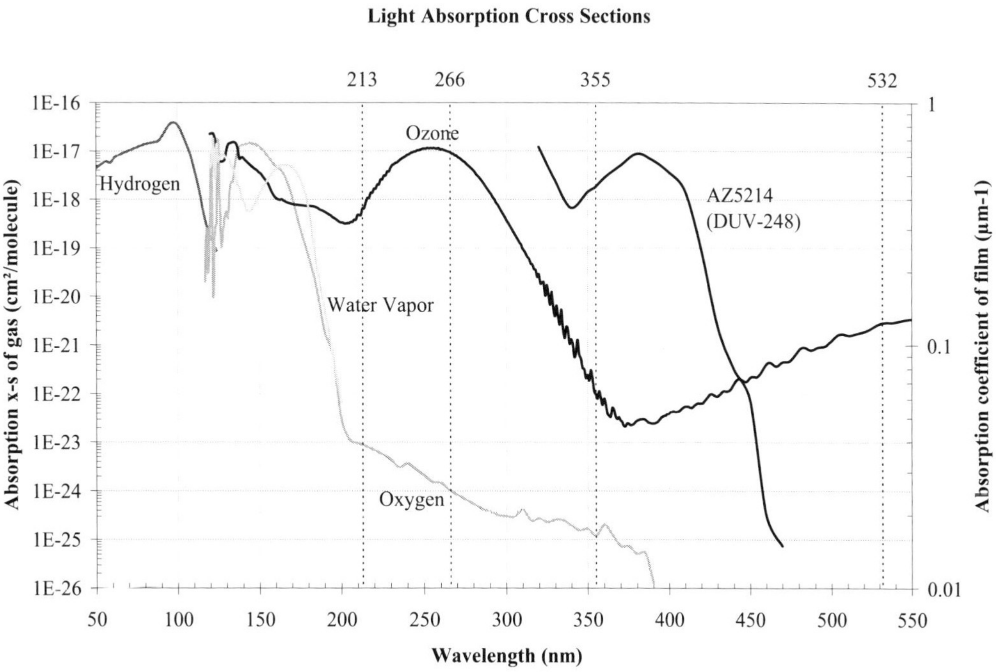

Figure 2: Reaction components

2.2 Method of cleaning

The method used in this study to clean the photoresist edge bead involves vacuum clamping the wafer, then directing the laser beam to the area of the excess resist on wafer edges. The beam is programmed to deliver laser pulses at a specific repetition rate, and each repetition rate will have a characteristic pulse energy. The dose used to remove the edge bead will then be proportional to the pulse energy multiplied by the number of scans. In this study, various laser doses were evaluated on samples with varying edge bead profiles. Two different types of injector nozzle heads were evaluated, while the vacuum exhaust nozzle remained fixed.

On the EBR-Evaluator tool used in these tests, the laser beam rapidly scans a portion of the edge area being cleaned, while the wafer remains stationery. In the next generation ALPHA-EBR system, the wafer will rotate while the optical system will be fixed. As shown in Figure 3 below, the injector sends a stream of gas to the site of the edge bead. The direction of the injection nozzle is set to react and direct reaction by-products away from the active device area of the wafer, and into the vacuum nozzle, as shown below.

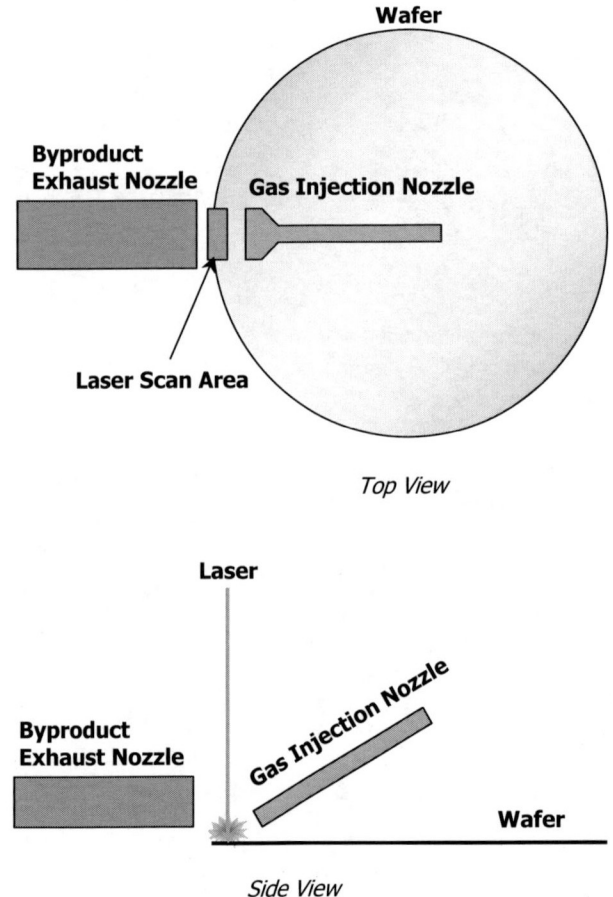

Figure 3: Diagram of experiment set up for edge bead removal with laser and gas

2.3 Wafer bevel cleaning

Figure 4 shows a side view of a silicon wafer edge to illustrate the surfaces to be addressed by the laser and gas injection and exhaust systems. Spin coating leaves unwanted photoresist on the top, apex, and bottom edges of the wafer bevel in varying thicknesses. The method used to remove this excess resist must be able to address all of the wafer edge regions, and remove the resist without leaving defects or residues. In our experiments, we tested the photoreactive cleaning method on the top edge and bottom edge. Since the laser beam strikes the beveled apex, this resist is also removed.

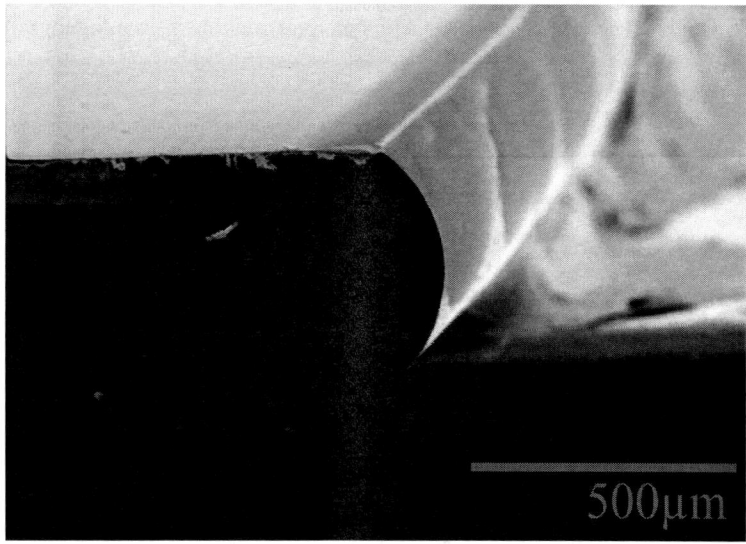

Figure 4: Side view of silicon wafer edge showing areas to be cleaned of excess photoresist

3. DATA AND RESULTS

3.1 Edge bead thickness

This set of tests was made to determine the shape and thickness of several edge beads on different wafers using Rohm and Haas 3012 photoresist. Figure 5 shows data taken with a DEKTAK profilometer of the top edge bead. Note the wide variation in thickness of the resist profiles.

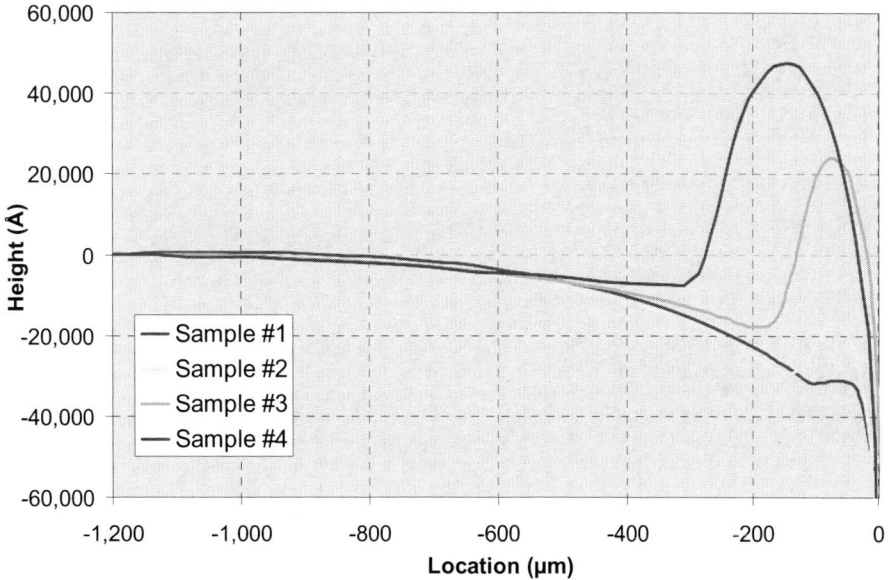

Figure 5: DEKTAK data of photoresist edge beads on the top edge of a wafer

3.2 Results after Cleaning

Figure 6 below shows a wafer edge cleaned with the laser with gas injection, but without an optimized exhaust nozzle.

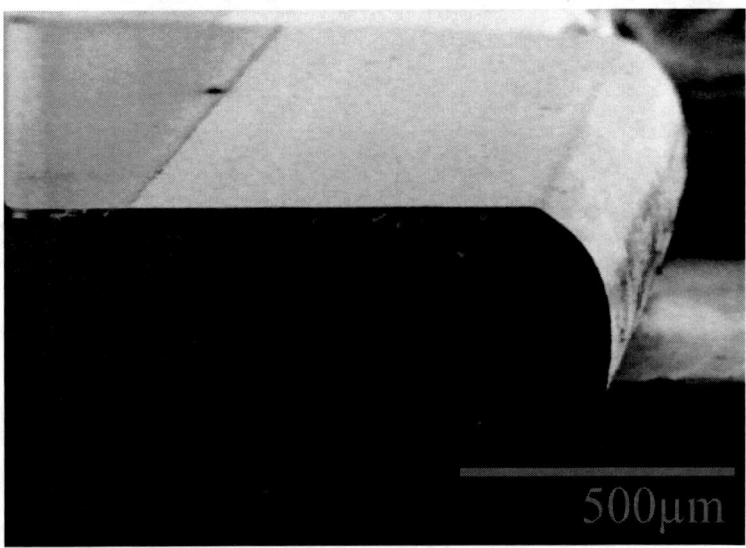

Figure 6: SEM of wafer bevel after topside edge bead removal

Figure 7 below shows a DEKTAK profile of a large edge bead before removal (red) and after removal (blue). The photoresist creating the bead is Rohm and Haas 812.

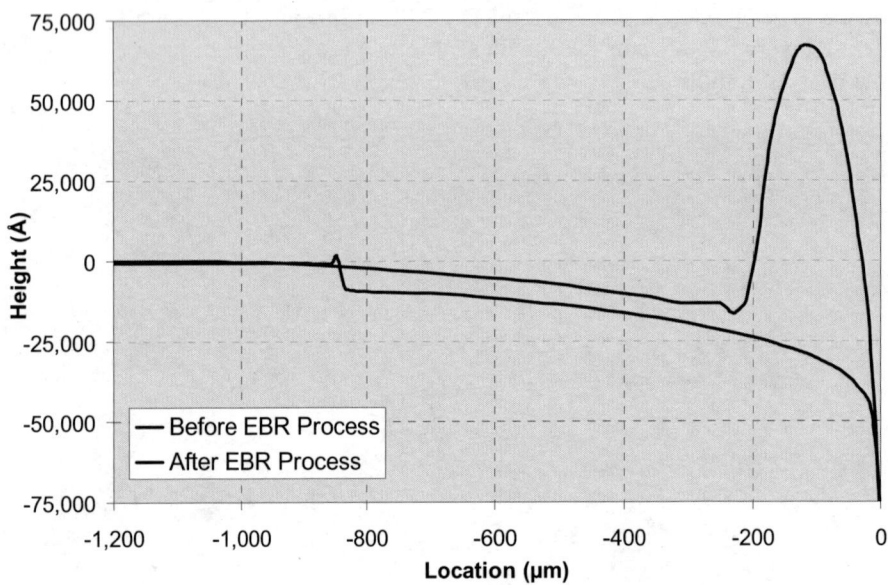

Figure 7: Profile of wafer edge before and after edge bead removal

3.3 Gas Type

Three different gases were evaluated in this paper. Oxygen, nitrogen, and air were each tested for photoresist edge bead removal using the laser and gas on the EBR-Evaluator. The removal time and overall results were similar, but in each case, a reactive gas is slightly better than the inert gas, measured by the number of particles remaining. The particle sizes used to optimize the gas types were very large (>5um) to permit rapid assessment of these preliminary results.

3.4 Dose vs. particle level

Another parameter evaluated was the dose of laser energy delivered in the edge bead removal process. In this test, three widely varying doses were chosen. As with the gas-type evaluation in 3.3 above, the most effective dose was determined by microscope evaluation at 40x magnification, of very large (>5um) particles. The relative doses used were 1x, 4x and 16x, translated from the scan numbers in the figure below. No surface damage was observed or measurable.

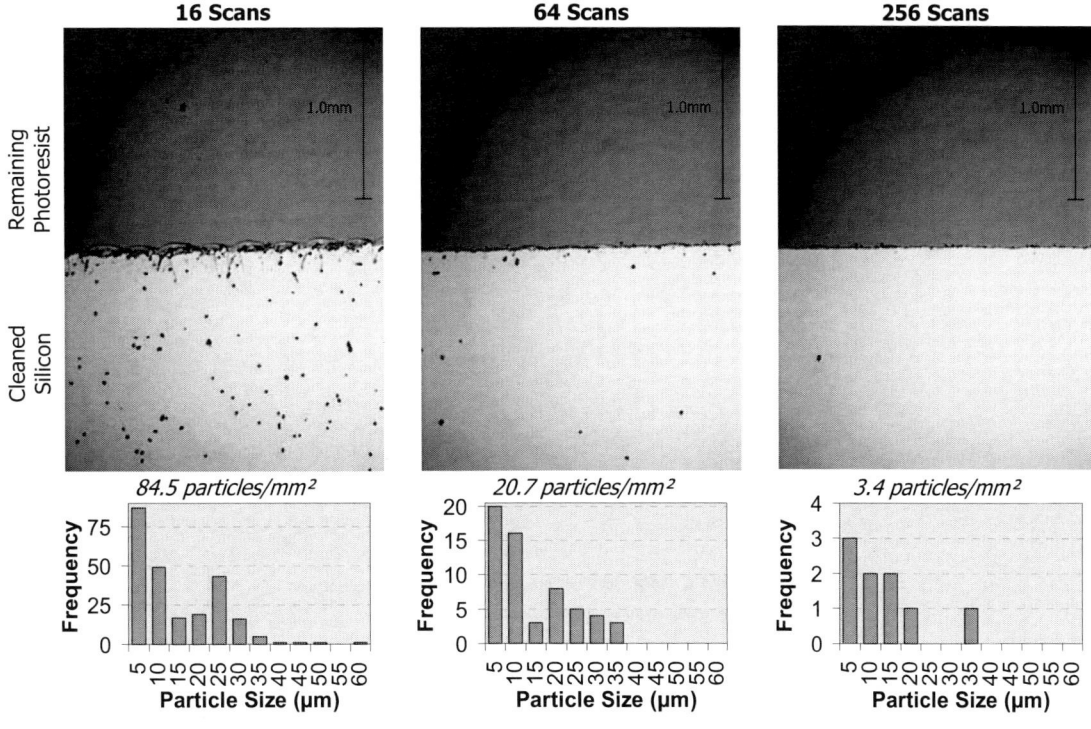

Figure 8: Laser dose vs. particle level

3.5 Bottom Edge Removal

A test was conducted to remove the excess resist on the bottom edge of a wafer. Figure 9 below shows the 'before' and 'after' photos of this test. The same approximate dose of laser energy was used for this test as used on the top edge EBR process.

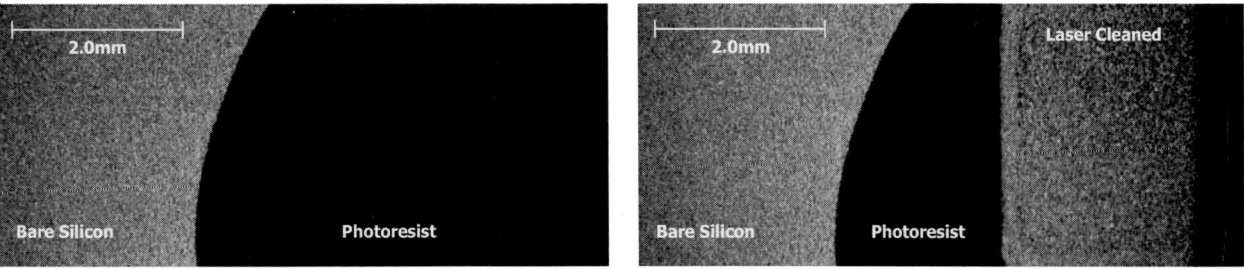

Figure 9: Bottom edge removal of photoresist before (left) and after (right) processing

3.6 Edge exclusion

The ability to use a laser tool permits controllable edge exclusion geometries down to 0.5mm beam diameter. The experiment result photos in Figure 10 show three different edge exclusions: 2.0mm, 1.5mm, and 1.0mm.

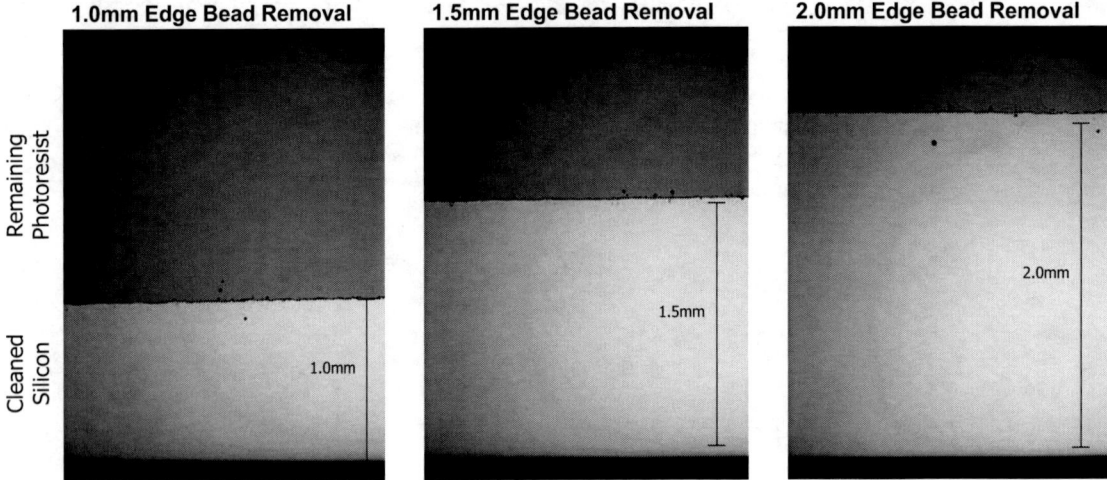

Figure 10: Three cases in the edge exclusion test

4. CONCLUSION

4.1 Resist edge bead removal

In this paper we evaluated a photoresist edge bead removal method using laser light and gas. The thickness of the photoresist edge bead on the top surface varies considerably depending on the coating condition and the type of resist. This variation did not have a strong impact on the removal efficiency time or gas type. Reactive gas provided the overall cleanest removal, but there were no significant variations between reactive gas injection and inert gas injection. The injector used was not optimized for pressure or nozzle configuration. Additional defect analysis will be needed to identify the contribution of reactive gas vs. inert gas. The dose of laser radiation was varied significantly. In general, the higher the dose, the lower the defect level. A closed chamber with low-level vacuum was used initially, but the open chamber with the process run at atmosphere produced the best result.

4.2 Edge exclusion and die yield

Edge exclusion was varied from 2.0mm to 1.0mm, to establish the degree of resolution and control of the laser beam with this method. The experiments showed that controllable edge exclusions down to better than 0.5mm can be achieved without special beam focusing optics. To take advantage of the ability of a laser-based EBR system to reduce edge exclusion to 1.0mm and below, improved coating methods and/or modified resist formulations will be required. The laser method may permit controllable edge exclusion to <0.5mm by focusing the beam with simple optics, an experiment planned for future work.

The benefit of reducing the edge bead was calculated for one case where four more die were yielded by reducing the edge exclusion from 2.0mm to 1.5mm using the laser and gas process described. The wafer was a 200mm wafer with a 97 die per wafer, increased to 101 good die per wafer by reducing the edge exclusion from 2.0mm to 1.5mm. This is shown in figure 11; die yield with 300mm wafers should be considerably higher based on previously published studies.

4.3 Summary

In conclusion, the basic goals of this first set of experiments using a prototype EBR-Evaluator in the lab have been met. The ability to use laser light and gas to remove a variety of varying thickness photoresist edge beads has been established. The degree of controllable edge exclusion with the laser beam has been tested and measured, and will comply with ITRS guidelines. Further, the laser and gas EBR method described in this paper can be used to remove resist from both top and bottom edges of silicon wafers with similar results in terms of dose, process time, and defectivity.

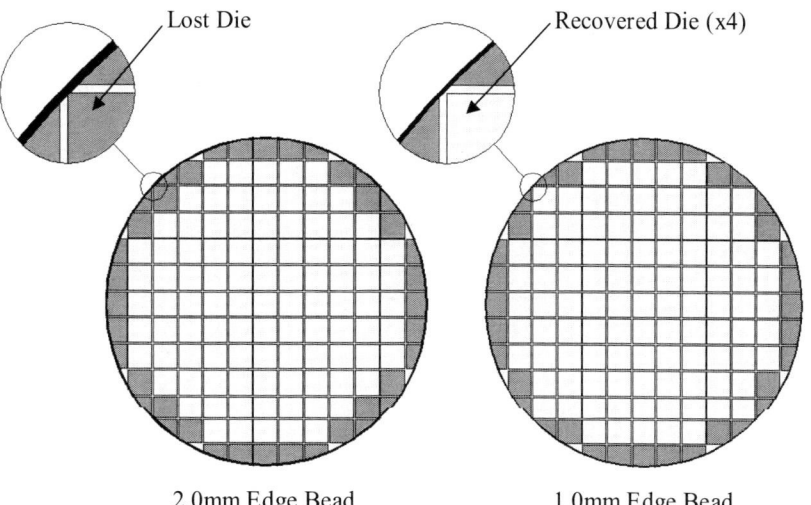

Figure 11: Example of die recovery when edge bead is reduced from 2.0mm to 1.0mm

5. FUTURE WORK

5.1 Experiments
Future work on this project will focus on optimizing cleaning recipes for a variety of film types, including organic and inorganic films. Defectivity analysis on a wafer inspection system will be used to measure particles before and after the EBR laser process on the top, apex, and bottom edges of the wafer. Multiple gas recipes will also be evaluated.

5.2 Next Generation Tool
The next set of tests will be with the ALPHA-EBR tool using a modular-integrated gas injector/exhaust nozzle positioned at the wafer edge, with a rotating wafer and a fixed optical system. Defectivity will be the primary goal of these tests, using the KLA-Tencor Vis-Edge wafer edge inspection system or equivalent. This system will be configured to establish the final configuration of a track-integrated EBR module, with focus on low cost of ownership (CoO).

6. ACKNOWLEDGEMENTS

We would like to gratefully acknowledge the following for technical insights, sample processing, and valuable discussions:

JSR Corporation
Rohm and Haas
Lou Steen, Tokyo Electron Limited, private communication
Harry Levinson, Global Foundries, private Communication

7. REFERENCES

[1] Kalyan Jami, Srini Vedula, Gerry Blumenstock, KLA-Tencor Corp., Milpitas, CA, USA; Jack chen, Keechan Kim, Yunsang Kim, Yung Kim, Lam Research Corp., Fremont, CA USA, "Optimization of edge die yield through defectivity reduction", Solid State Technology, 28 October, 2009.

[2] I. Pollentier, IMEC, A. Somanchi, F. Burkeen, S. Vedula, KLA Tencor, "Influence of Immersion Lithography on Wafer Edge Defectivity", Yield Management Solutions, Issue 1, 2008.

3. F. Burkeen et al., "Visualizing the Wafer's Edge", Yield Management Solutions, Winter, 2007

[3] Mai Randall, Michael Linnane, Chris Longstaff, etc. "A universal process development methodology for complete removal of residues from 300mm wafer edge bevel", Proceedings of SPIE, Vol. 6153-61533C-2, (2006).

[4] Quang Tran, "Efficient Resist Edgebead Removal for Thick I-Line Resist Coating Application on TELMARK 7 Tra.ck System", Proceedings of SPIE, Vol. 4181 (2000).

Author Index

Numbers in the index correspond to the last two digits of the six-digit citation identifier (CID) article numbering system used in Proceedings of SPIE. The first four digits reflect the volume number. Base 36 numbering is employed for the last two digits and indicates the order of articles within the volume. Numbers start with 00, 01, 02, 03, 04, 05, 06, 07, 08, 09, 0A, 0B ... 0Z, followed by 10-1Z, 20-2Z, etc.